THE LIGHT ELEMENTS AND THEIR EVOLUTION

IAU SYMPOSIUM NO. 198

COVER ILLUSTRATION:

A painting by Licius Bossolan showing the "Forte dos Reis Magos" (the Fort of the Three Magi), the main historical building of the city of Natal. It was constructed in the 16th century by the Portuguese to defend the town from the pirates and invaders. According to R. de la Reza and L. da Silva, the fort can represent a Li rich giant star, losing mass.

(The Editors thank the artist for permitting us to reproduce this painting).

Information on other IAU Symposium proceedings
is given at the back of this volume

INTERNATIONAL ASTRONOMICAL UNION
UNION ASTRONOMIQUE INTERNATIONALE

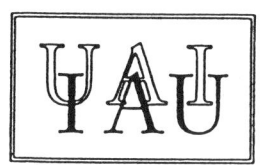

THE LIGHT ELEMENTS AND THEIR EVOLUTION

Proceedings of the 198th Symposium of the
International Astronomical Union
held in Natal, RN, Brazil
21-27 November, 1999

Edited by

LICIO DA SILVA
Observatório Nacional, Rio de Janeiro, Brazil

MONIQUE SPITE
Observatoire de Paris-Meudon, Meudon, France

and

JOSÉ RENAN DE MEDEIROS
Universidade Federal do Rio Grande do Norte, Natal, Brazil

Publisher

THE LIGHT ELEMENTS AND THEIR EVOLUTION

All Rights Reserved
Copyright © 2000

INTERNATIONAL ASTRONOMICAL UNION
98bis, bd Arago – 75014 Paris – France

Tel: +33 1 4325 8358;   Fax: +33 1 4325 2616;
E-mail: iau@iap.fr;   Web Site: www.iau.org

No part of the material protected by this copyright notice may be reproduced or utilized in any form or by any means, electronic or mechanical including photocopying, recording or by any information storage and retrieval system, without written permission from the IAU.

Published on behalf of the
INTERNATIONAL ASTRONOMICAL UNION

by
Astronomical Society of the Pacific
First published 2000

Managing Editor, D. H. McNamara
Associate Managing Editor, J. W. Moody
LaTeX Computer Consultant, T. J. Mahoney
Production Manager, Enid Livingston

EDITORIAL/PUBLISHING OFFICE:			CATALOG/BOOK ORDERS:	
Managing Editor			IAU Publications	
PO Box 24463			390 Ashton Avenue	
211 KMB Brigham Young University			San Francisco CA 94112-1722	
Provo UT 84602-4463			USA	
USA				
(801) 378-2298	Phone		(415) 337-1100	Phone
(801) 378-2265	Fax		(415) 337-5205	Fax
pasp@astro.byu.edu	E-mail		catalog@aspsky.org	E-mail
			www.aspsky.org	Web Site

Printed by Sheridan Books, Inc., Chelsea, Michigan

Library of Congress Catalog Card Number: 00-108013
ISBN: 1-58381-048-X

TABLE OF CONTENTS

Preface xi
List of Participants xiii
Conference Photo xvii

1. INTRODUCTION

The light elements : what is known, what is controversial (*) 3
    F. Spite

2. PRODUCTION AND DESTRUCTION OF THE ELEMENTS

Primordial Nucleosynthesis For The New Millennium (*) 13
    G. Steigman
Alternative Solutions to Big Bang Nucleosynthesis (*) 25
    H. Kurki-Suonio
The Superbubble Model for LiBeB Production and Galactic Evolution 35
    E. Parizot and L. Drury
LiBeB Production and Associated Astrophysical Sites (*) 41
    E. Vangioni-Flam and M. Cassé
LiBeB Evolution: Three Models (*) 51
    R. Ramaty, R. E. Lingenfelter and B. Kozlovsky
Sinks of Light Elements in Stars – Part I (*) 61
    C. P. Deliyannis, M. H. Pinsonneault and C. Charbonnel
Sinks of Light Elements in Stars - Part II (*) 74
    M. H. Pinsonneault, C. Charbonnel and C.P. Deliyannis
Sinks of Light Elements in Stars – Part III (*) 87
    C. Charbonnel, C. P. Deliyannis and M. Pinsonneault
Creation and Destruction of ^{7}Li and ^{3}He in RGB and AGB Stars (*) 98
    I.-J. Sackmann and A. I. Boothroyd
Cosmological Gravitons Back Reaction and the Primordial Nucleosynthesis 108
    M. R. G. Maia, J. C. Carvalho, J. S. Alcaniz and J. M. F. Maia
Constraints from Big Bang Nucleosynthesis on a Time-Varying 111
    Cosmological Constant
    J. A. S. Lima, J. M. F. Maia and N. Pires
Photon Creation in the Universe and Primordial Nucleosynthesis 113
    J. A. S. Lima, J. S. Alcaniz, J. Santos and R. Silva Jr.
Change in Primordial Abundances Due to a Change in the Primordial
    Plasma Energy Density 116
    M. Opher and R. Opher
The Recombination in a FRW Universe with a Variable Cosmological
    Term 118
    N. Pires and J. A. S. Lima
Improved Use of Inputs to Primordial Nucleosynthesis 120
    K. M. Nollett and S. Burles

(*) Invited Paper

3. ABUNDANCES OF D, ^{3}He AND ^{4}He

Measurements of The Primordial D/H Abundance Towards Quasars (*) ... 125
    D. Tytler, J. M. O'Meara, N. Suzuki, D. Lubin, S. Burles
    and D. Kirkman
The Deuterium Abundance in QSO Absorption Systems:
    A Mesoturbulent Approach 135
    S. A. Levshakov
Deuterium Observations in our Galaxy - View A) (*) 141
    J. L. Linsky and B. E. Wood
Deuterium observation in our Galaxy - View B (*) 151
    A. Vidal-Madjar
The D/H Ratio in Interstellar Gas toward the Hot, White Dwarf
    G191-B2B 161
    M. S. Sahu
The Deuterium Abundance In The Galactic Center 50 km/s Molecular
    Cloud: Evidence For A Cosmological Origin Of D 167
    D. A. Lubowich, J. M. Pasachoff, R. P. Galloway,
    T. J. Balonek, C. Tremonti, T. Millar and H. Roberts
Blue Compact Galaxies and the Primordial ^{4}Helium Abundance (*) 176
    T. X. Thuan and Y. I. Izotov
Inhomogeneous H II Regions and the Helium Abundance 188
    S. M. Viegas and R. Gruenwald
The Magellanic Clouds and the Primordial Helium Abundance (*) 194
    M. Peimbert and A. Peimbert
Some aspects of the chemical evolution of ^{4}He in the Galaxy:
    the He/H radial gradient and the $\Delta Y/\Delta Z$ enrichment ratio (*) 204
    W. J. Maciel
The Primordial 3-Helium Abundance At Last? (*) 214
    T. M. Bania, R. T. Rood and D. S. Balser
Deuterium and Helium-3 in the Protosolar Cloud (*) 224
    G. Gloeckler and J. Geiss
Helium and Oxygen Abundances in SMC Planetary Nebulae 234
    R. D. D. Costa, J. A. de Freitas Pacheco and T. P. Idiart
Interstellar D/H on the Sightline of Sirius 236
    G. Hébrard, A. Vidal-Madjar, R. Ferlet, C. Mallouris,
    D. York and M. Lemoine
Deuterium Balmer Emission from Nebulae 238
    G. Hébrard, D. Péquignot, A. Vidal-Madjar, J. R. Walsh
    and R. Ferlet
FUSE Spectra of Sk 80, an O7 Supergiant in the Small Magellanic Cloud . 240
    R. C. Iping, G. Sonneborn, D. Massa, A. W. Fullerton,
    J. B. Hutchings and the FUSE Science Team
Spatial Variations in the Atomic D/H Ratio in the ISM 242
    G. Sonneborn, E. B. Jenkins, T. Tripp, P. Wozniak,
    R. Ferlet, A. Vidal-Madjar and U. J. Sofia

(*) Invited Paper

In-Orbit Performance of the Far Ultraviolet Spectroscopic Explorer 244
    G. Sonneborn, H. W. Moos, K. R. Sembach and
    the FUSE Science Team

4. LITHIUM ABUNDANCES

^{7}Li in Metal-Poor Stars: The Spread of the Li Plateau (*) 249
    S. G. Ryan
Observations of ^{6}Li in Metal Poor Stars (*)..............................259
    P. E. Nissen
Li Abundance in Pop I Stars (*) 269
    Luca Pasquini
Evolution of Lithium Abundance in Pop I Giants..........................279
    S. V. Mallik
Lithium in the Open Cluster NGC 6475...................................287
    S. Randich, R. Pallavicini and J.-C. Mermilliod
Lithium in the Old Open Cluster NGC 2243 293
    V. Hill and L. Pasquini
Lithium in Brown Dwarfs (*)..299
    R. Rebolo
Lithium in Giant Stars (*) 310
    R. de la Reza
The Properties of the PDS Li-rich Giant Stars...........................320
    C. A. O. Torres, G.R. Quast, R. de la Reza and L. da Silva
Search for Lithium-Rich Stars Among G–K Giants with IR–excess.........325
    G. Jasniewicz, M. Parthasarathy, P. de Laverny,
    F. Thévenin, N. Mauron and M. Chadid
Be vs. Li Abundance in Li-Rich Giants: an Evidence of Li Production in
Red Giants 331
    B.V. Castilho
The Interstellar Lithium Isotope Ratio Toward Per OB2...................338
    D. C. Knauth, S. R. Federman, D. L. Lambert and P. Crane
New Determination Method of Primordial Li Abundance 344
    T. Kajino, T.-K. Suzuki, S. Kawanomoto and H. Ando
Lithium in Young Open Clusters 350
    R. Pallavicini, S. Randich, J. R. Stauffer and
    S. C. Balachandran
Li Abundance in Evolved Stars of NGC 6397.............................354
    D.M. Allen, B.V. Castilho, L. Pasquini, B. Barbuy and P. Molaro
Lithium Depletion in a [Fe/H]= -3.4 star? 356
    M. Spite, F. Spite, R. Cayrel, V. Hill, E. Depagne,
    B. Nordström and T.C. Beers
Lithium in Metal Deficient K Giant Stars: The Absence of Dust Signature.358
    R. de la Reza, L. da Silva, N. A. Drake and M. A. Terra

(*) Invited Paper

Lithium in Binary Systems with Evolved Components 360
    J. M. Costa, L. da Silva and J. R. De Medeiros
Lithium and Rotation on the Subgiant Branch. A Theoretical Analysis
    of Observations 362
    J. D. do Nascimento Jr, C. Charbonnel, A. Lèbre, P. De Laverny and J.R. de Medeiros
Lithium Abundances in Bright Giant Stars 364
    A. Lèbre, P. de Laverny and J. R. de Medeiros
Lithium in Cool Stars Detected in EUV Surveys 366
    G. Tagliaferri, L. Pastori, G. Cutispoto and R. Pallavicini
Lithium Abundance in Late-Type Stars 368
    L. Pompéia, B. Barbuy and M. Grenon
Understanding the Li Production in AGB stars: the J-type Stars 370
    C. Abia and J. Isern
The Origin of the Lithium Rich Giants 373
    C. Charbonnel and S. Balachandran
Detailed Analysis of Li-rich Giants 375
    J. Gregorio-Hetem, B.V. Castilho, B. Barbuy, F. Spite and M. Spite
The Lithium Abundance and Mass Loss Rate in Galactic Super-Li-Rich
    Carbon and S Stars 377
    D. A. Lubowich, V. V. Smith, B. E. Turner and R. Sahai
Measurements of Li Abundance in a Sample of T Tauri Stars 379
    M. J. Sartori, J. Gregorio-Hetem, B. V. Castilho and J. R. D. Lépine

5. ABUNDANCE OF BERYLLIUM AND BORON

Beryllium in the Sun: Re-Measurement and Implications (*) 383
    S. C. Balachandran
The Galactic Evolution of Beryllium 389
    A. M. Boesgaard
Galactic Evolution of Beryllium and Oxygen 397
    G. Israelian, R. J. G. López and R. Rebolo
The Galactic Evolution of Boron (*) 405
    F. Primas
The Abundance of Boron in Disk-Metallicity Stars (*) 415
    K. Cunha
The Light Elements Be and B as Stellar Chronometers in the Early 425
    Galaxy
    T. C. Beers, T. K. Suzuki and Y. Yoshii
A Very Reduced Upper Limit on the Interstellar Abundance of Beryllium .. 432
    G. Hébrard, M. Lemoine, R. Ferlet and A. Vidal-Madjar

(*) Invited Paper

6. STELLAR KNOWLEDGE TO AND FROM LIGHT ELEMENTS

Effects of Photospheric Temperature Inhomogeneities on Lithium
abundance Determinations (2D) (*) 437
   R. Cayrel and M. Steffen
The Light Elements in the Light of 3D Hydrodynamical Model
Atmospheres 448
   M. Asplund
Formation of the Optical Spectra of the Coolest M- and L-dwarfs and
Lithium Abundances in their Atmospheres 454
   Y. V. Pavlenko
Constraints on Stellar Hydrodynamics from Abundance Anomalies of
LiBeB and Metals (*) 460
   G. Michaud, J. Richer and O. Richard
Transport Phenomena and Light Element Abundances in the Sun and 470
Solar Type Stars
   S. Vauclair
AGB Stars Interferometric Signatures: Effects of Possible Li-rich Spots. ... 476
   P. de Laverny and B. Lopez
Lithium Abundances in Main-Sequence F Stars and Sub-Giants 478
   J. D. do Nascimento Jr, S. Théado and S. Vauclair
He Abundance in Planetary Nebulae 480
   R. Gruenwald and S. M. Viegas
Non-LTE Effects in Berylium Abundances 483
   T. P. Idiart and F. Thévenin
White Dwarf Probes of Interstellar Deuterium 485
   W. Landsman
IR Boron Lines in Stellar Spectra 487
   J. Meléndez, B. V. Castilho and B. Barbuy
Lithium in Cool Magnetic CP Stars: Some New Results of Observations,
Using CAT (ESO), 2.6m (CrAO) and (NOT) La Palma Telescopes .. 489
   N. Polosukhina, D.Kurtz, M. Hack, P. North, I. Ilyin and J. Zverko
Lithium Abundances in Solar-Type Stars 495
   L. da Silva and G. F. Porto de Mello
On meridional Circulation in Stars 498
   S. Talon, G. Michaud and A. Vincent
On the Link Between Rotation and Lithium in Giant stars 500
   J. R. De Medeiros, J. D. Nascimento Jr, S. Sankarankutty,
   J. M. Costa, J. R. P. Da Silva and M. R. G. Maia
Lithium-Rich K Giants with Infrared Excesses: 502
Fundamental Parameters and CNO Abundances
   N. A. Drake, R. de la Reza and L. da Silva
Lithium as Probe of the Scenarios of the Chemical Enrichment of the
Galaxy 504
   P. François, V. Hill, M. Spite and F. Spite
Peculiar J-type Carbon Stars and Li 506
   S. Lorenz-Martins and N.A. Drake

(*) Invited Paper

The Behavior of the Rotational Velocity in Lithium-Rich Evolved Stars ... 508
    C. H. F. Melo, B. B. Soares, A. C. Miranda, J. R. P. Da Silva
    and J. R. De Medeiros
Lithium in Post T Tauri Stars 510
    G.R. Quast, C. A. O. Torres, R. de la Reza and L. da Silva
Li in Chromospherically Active Stars with Large Velocity Components 512
    H. J. Rocha-Pinto, B. V. Castilho and W. J. Maciel
The Ideal Stars for Exploration of Early-Epoch ^{7}Li Abundances 514
    S. Rossi and T. C. Beers
Meridional Circulation, Turbulence and Lithium in Sub-Giants
    Originating from the Hot Side of the Dip 516
    S. Talon and C. Charbonnel
On the Formation of Lithium Emission Lines in Nova Shells 518
    M. Diaz
Self-regulated Hydrodynamical Process in Halo Stars:
    a Possible Explanation of the Lithium Plateau 520
    S. Théado and S. Vauclair

7. EVOLUTION OF THE LIGHT ELEMENTS

Evolution of D and ^{3}He in the Galaxy (*) 525
    M. Tosi
Implications of Early Cooling Flows and Galactic Winds for the 535
    Evolution of Deuterium
    A. C. S. Friaça
The Evolution of ^{3}He, ^{4}He and D in the Galaxy 540
    C. Chiappini and F. Matteucci
The Evolution of ls ^{4}He and LiBeB (*) 547
    K. A. Olive
Stellar and GCR Production of Lithium in the Milky Way 558
    F. Matteucci and D. Romano
Light Element Evolution at the Solar Neighborhood 563
    A. Alibés, J. Labay and R. Canal
Evolution of ^{6}LiBeB in Inhomogeneous Early Galaxy 565
    T. Ken Suzuki, Y. Yoshii and T. Kajino
One Zone Numerical Model for the Galactic Evolution of Lithium 567
    M. Terra and L. I. Arany-Prado

8. CONCLUSIONS

Conclusions I (*) 571
    B.E.J. Pagel
Conclusions II (*) 578
    H. Reeves

(*) Invited Paper

PREFACE

The 198th I.A.U. Symposium "The Light Elements and Their Evolution" was held between Sunday 21st November and Friday 27th November, 1999, in Natal, a city in the North-East of Brazil.

The light elements deserve special attention because of their relationship to several important astrophysical problems, ranging from stellar structure to cosmology. The previous large meeting on this topic was held in May 1994, so five years have elapsed between the two meetings, which, for this rapid progressing field, is a long time. During this symposium, many new results and many new problems were presented and discussed. It was attended by 102 astronomers from all over the world and 24 Brazilian PhD students. The participants were: 25 astronomers from North America, 36 from Latin America, 34 from Europe, 3 from the Middle-East, 3 from Asia and 1 from North Africa, which is impressive for a astronomical symposium taking place outside the USA-Europe axis.

This book contains all the invited papers and contributed talks presented at the meeting, organized by topics. Likewise, due to the large number of young astronomers participating in the symposium, we have also decided to include also the posters.

This symposium was sponsored by Division IV (Stars) and Commission 29 (Stellar Spectra), and was supported by the Divisions VI(Interstellar Matter) and VII(Galactic System) and the Commissions 34 (Interstellar Matter), 35 (Stellar Constitution), 36 (Theory of Stellar Atmospheres) and 47 (Cosmology) of the IAU. This broad support shows the scope of the subject. The organizers would like to thank all their members, specially their respective presidents, and the IAU Executive Committee, specially Dr. Johannes Andersen, the IAU General Secretary.

The Scientific Organizing Committee consisted of:

M. Spite (co-chairperson), France L. da Silva (co-chairperson), Brazil
G. Michaud, Canada B. E.J. Pagel, UK-Denmark
L. Pasquini, ESO, Italy Y. V. Pavlenko, Ukraine
M. Peimbert, Mexico M. Rugers, USA
S. G. Ryan, Australia I.-J. Sackmann, USA
K. Sato, Japan V.V. Smith, USA
G. Steigman, USA E. Terlevich, Argentina-UK
T. L. Wilson, USA-Germany

The location of the symposium, Natal, is a beautiful city of 600,000 inhabitants and is located on the estuary of the River Potengi, on the coast of the Atlantic Ocean. The symposium was the most important scientific event in the celebration of Natal's 4th centenary. The importance of this symposium in these celebrations was demonstrated by the participation of the Lord Mayor of the city, Mrs. Wilma de Faria, at the opening of the meeting. This session counted also with the presence of Dr. Johannes Andersen, the IAU General Secretary, who received from Madam the Mayor, in name of the organizers, a celebrative plate of the event.

The very local organization of the symposium was carried out by the Astronomical Group of the Physics Department of the Federal University of Rio Grande do Norte (UFRGN). They had the assistance from colleagues of the University of São Paulo (USP) and the Observatório Nacional, Rio de Janeiro (ON).

The Local Organizing Committee consisted of:

J.R. de Medeiros (Chairperson), UFRGN	B. Barbuy, USP
E. Bonelli, UFRGN	J. C. Carvalho, UFRGN
K. Cunha, ON	S. R. Gomes Jr., UFRGN
M. R. G. Maia, UFRGN	R. de la Reza, ON
J. A. S. de Lima, UFRGN	L. C. Jafelice, UFRGN
N. Pires, UFRGN	W. Maciel, USP
J. Santos, UFRGN	

The "The Light Elements and Their Evolution" symposium was possible thanks to generous support of:
- the International Astronomical Union,
- the Conselho Nacional de Desenvolvimento Científico e Tecnológico (Brazil),
- the Coordenação de Aperfeiçoamento de Nível Superior (Brazil),
- the Observatoire de Paris,
- the Fundação de Amparo a Pesquisa do Estado de Sao Paulo ,
- the Fundação de Amparo a Pesquisa do Estado de Rio de Janeiro,
- the Universidade Federal do Rio Grande do Norte and
- the Prefeitura do Municipio de Natal.

The editors would like to thanks the participants for providing their papers, many of them promptly. The organizers give their warmest thanks to all the institutions and individuals who have contributed to the success of the symposium. They thank Mrs. Marina Freitas, for having been in charge of the secretary of the event, and Miss Maria Jose F. Ferreira, for her help during its realization. Licio da Silva knowledges the ESO for the computational infrastructure used to prepare this volume.

L. da Silva, M. Spite and J. R. de Medeiros
July 2000

LIST OF PARTICIPANTS

PARTICIPANT	INSTITUTION
Abia, Carlos	Universidad de Granada, Spain
Alcaniz, Jailson	Universidade Federal do Rio Grande do Norte, Brazil
Alibes, Andreu	Universidad de Barcelona, Spain
Allen, Dinah Moreira	Universidade de São Paulo, Brazil
Andersen, Johannes	University of Copenhagen, Denmark
Arany-Prado, Lilia	Universidade Federal do Rio de Janeiro, Brazil
Asplund, Martin	Uppsala Astronomiska Observatorium, Sweden
Balachandran, Suchitra	University of Maryland, USA
Bania, Thomas M.	Boston University, USA
Barbuy, Beatriz	Universidade de São Paulo, Brazil
Beers, Timothy	Michigan State University, USA
Marassi de Almeida, Lucio	Universidade Federal do Rio Grande do Norte, Brazil
Bezerra, Wellington	Universidade Federal do Rio Grande do Norte, Brazil
Bonelli, Enivaldo	Universidade Federal do Rio Grande do Norte, Brazil
Braga Bezerra, Lupercio	Universidade Federal do Pernambuco, Brazil
Casse, Michel	CEA - Service d'Astrophysique , France
Castilho, Bruno	Universidade de São Paulo, Brazil
Cayrel, Roger	Observatoire de Paris, France
Charbonnel, Corinne	Lab. d'Astrophysique de Toulouse, France
Chiappini, Cristina	Observatorio Nacional, Brazil
Costa, João da Mata	Universidade Federal do Rio Grande do Norte, Brazil
Costa, Roberto D.D.	Universidade de São Paulo, Brazil
Cunha, Katia	Observatorio Nacional, Brazil/ Univ. of Texas, USA
Cutispoto, Giuseppe	Catania Astrophysical Observatory, Italy
Da Silva, Jose Ronaldo P.	Universidade Federal do Rio Grande do Norte, Brazil
da Silva, Licio	Observatorio Nacional, Brazil
de Garcia Maia, Mrcio R.	Universidade Federal do Rio Grande do Norte, Brazil
de la Reza, Ramiro	Observatorio Nacional, Brazi
l de Laverny, Patrick	Observatoire de la Côte d'Azur, France
De Medeiros, Jose Renan	Universidade Federal do Rio Grande do Norte, Brazil
Deliyannis, Constantine	Indiana University, USA
Diaz, Marcos	Universidade de São Paulo, Brazil
do Nascimento Jr., Jose D.	Observatoire Midi-Pyrennes, France
Drake, Natalia	Observatorio Nacional, Brazil
Espichan Carrillo, Jorge	Universidade Federal do Rio Grande do Norte, Brazil
Ferlet, Roger	Institut d'Astrophysique de Paris, France
Friaa, Amâncio	Universidade de São Paulo, Brazil
Gloeckler, George	University of Maryland, USA
Gomes Jr., Samuel	Universidade Federal do Rio Grande do Norte, Brazil
Gregorio-Hetem, Jane	Universidade de São Paulo, Brazil
Gruenwald, Ruth	Universidade de São Paulo, Brazil
Hbrard, Guillaume	Institut d'Astrophysique de Paris, France

PARTICIPANT	INSTITUTION
Hill, Vanessa	European Southern Observatory, Germany
Idiart, Thais	Universidade de São Paulo, Brazil
Iping, Rosina	NASA/Goddard Space Flight Center, USA
Israelian, Garik	Instituto de Astrofisica de Canarias, Spain
Jafelice, Luiz Carlos	Universidade Federal do Rio Grande do Norte, Brazil
Jasniewicz, Gerard	Universite Montpellier II, France
Kajino, Taka	University of Tokyo, Japan
Knauth, David	University of Toledo, USA
Kurki-Suonio, Hannu	University of Helsinki, Finland
Landsman, Wayne	Raytheon ITSS, NASA/GSFC, USA
Lbre, Agnes	Universite Montpellier II, France
Levshakov, Sergei	Ioffe Physico-Technical Institute, Russia
Lima, Jose Ademir Sales	Universidade Federal do Rio Grande do Norte, Brazil
Linsky, Jeffrey	University of Colorado, USA
Lorenz-Martins, Silvia	Universidade Federal do Rio de Janeiro, Brazil
Lubowich, Donald	Hofstra University and American Ins. of Physics, USA
Maciel, Walter	Universidade de São Paulo, Brazil
Mallik, Sushma V.	Indian Institute of Astrophysics, India
Matteucci, Francesca	Depart of Astronomy, University of Trieste, Italy
Mehdi, Si-Khaled	Universite' des Sc. et de la Technologie HB, Algeria
Melendez-Moreno, Jorge	Universidade de São Paulo, Brazil
Michaud, Georges	Universite de Montreal/CERCA, Canada
Miranda, Antonio Carlos	Universidade Federal do Rio Grande do Norte, Brazil
Nissen, Poul E.	University of Aarhus, Denmark
Nollett, Kenneth	University of Chicago, USA
Nordstrom, Birgitta	Niels Bohr Ins. for Astro, Phy. Geophy., Denmark
Olive, Keith	University of Minnesota, USA
Opher, Reuven	Universidade de São Paulo, Brazil
Pagel, Bernard	Sussex University, United Kingdom
Pallavicini, Roberto	Osservatorio Astronomico di Palermo, Italy
Parizot, Etienne	Dublin Institute for Advanced Studies, Ireland
Pasquini, Luca	European Southern Observatory , Germany
Pavlenko, Yakiv	Main Astron. Obs. of Ukrainian Acad. Sc., Ukraine
Peimbert, Manuel	Universidad Nacional Autonoma de Mexico, Mexico
Pinsonneault, Marc	The Ohio State University, USA
Pires, Nilza	Universidade Federal do Rio Grande do Norte, Brazil
Polosukhina, Nina	Crimean Astrophysical Observatory, Ukraine
Pompia, Luciana	Universidade de São Paulo, Brazil
Primas, Francesca	Europen Southern Observatory, Germany
Quast, Germano	Laboratorio Nacional de Astrofisica, Brazil
Ramaty, Reuven	NASA/Goddard Space Flight Center, USA
Rauscher, Thomas	University of Basel, Switzerland
Rebolo, Rafael	Instituto de Astrofisica de Canarias, Spain
Reeves, Hubert	CEA - Service d'Astrophysique, France
Rocha-Pinto, Helio	Universidade de São Paulo, Brazil
Rood, Robert T.	University of Virginia, USA

PARTICIPANT	INSTITUTION
Rossi, Silvia	Universidade de São Paulo, Brazil
Ryan, Sean	The Open University, United Kingdom
Sackmann, I.-Juliana	California Institute of Technology, USA
Sahu, Meena	NASA/Goddard Space Flight Center, USA
Sankarankutty, Shobha	Universidade Federal do Rio Grande do Norte, Brazil
Santos, Janilo	Universidade Federal do Rio Grande do Norte, Brazil
Sartori, Marilia	Universidade de São Paulo, Brazil
Smith, Verne	University of Texas at El Paso/McDonald Obs., USA
Soares, Braulio Batista	Universidade Federal do Rio Grande do Norte, Brazil
Sonneborn, George	NASA/Goddard Space Flight Center, USA
Spite, François	Observatoire de Paris-Meudon, France
Spite, Monique	Observatoire de Paris-Meudon, France
Steigman, Gary	The Ohio State University, USA
Suzuki, Takeru	University of Tokyo, Japan
Talon, Suzanne	CERCA/Universite de Montreal, Canada
Terra, Marco	Universidade Federal do Rio de Janeiro, Brazil
Thvenin, Frederic	Observatoire de Nice, France
Thuan, Trinh Xuan	University of Virginia, USA
Torres, Carlos Alberto O.	Laboratorio Nacional de Astrofisica, Brazil
Tosi, Monica	Osservatorio Astronomico di Bologna, Italy
Tytler, David	University of California at San Diego, USA
Vauclair, Sylvie	Laboratoire d'Astrophysyque, Tolouse, France
Vidal-Madjar, Alfred	Institut d'Astrophysique de Paris, France
Viegas, Sueli M.	Universidade de São Paulo, Brazil
Wichoski, Ubi	Centra, Instituto Superior Tecnico, Portugal

Mrs. Wilma de Faria, Lord Mayor of the Natal City,
presenting a celebratory plate of the event
to Dr. Johannes Andersen, the IAU General Secretary.

We wish to give credit to Ana Aguiar Maia
for all the conference photographs reproduced in this volume.

CONFERENCE PHOTO

IAU Symposium 198 "The Light Elements and their Evolution" Natal, Brazil 1999 November 22-26

INTRODUCTION

The opening session -- from right to left:
Dr. Beatriz Barbuy, President of the IAU Commission 29,
Dr. Johannes Andersen, the IAU Secretary,
Mrs. Wilma de Faria, Mayor of Natal,
Dr. Renan de Medeiros, President of the LOC,
and
Dr. Paulo Fulco, Heard of the Department of Physics-UFRN
representing the University

The Light Elements and Their Evolution
IAU Symposium, Vol. 198, 2000
L. da Silva, M. Spite, J. R. de Medeiros, eds.

The light elements : what is known, what is controversial

F. Spite

Observatoire de Paris, DASGAL UMR 8633 du CNRS, F-92195 Meudon CEDEX, France

Abstract. The light elements are essential, because their primordial abundances are linked to the general parameters of the universe (at least in the Big Bang theory). Some of the light elements are fragile, and the interpretation of their abundances in stars requires a good knowledge of the stellar structure. The stellar abundances have to be known with a considerable accuracy, challenging the current level of representation of the stellar atmospheres. It is also difficult to reach accurate abundances in the gas : interstellar, H II regions, PN, intergalactic...

On the other hand, the fragile light elements are important probes of the stellar interiors, and their observation helps to the determination of reliable models, which in turn will improve the accuracy of stellar abundances. A part of this symposium is devoted to this probing aspect.

The talks (and many high quality posters) which build this symposium, cover all these aspects. I will try here to point out the controversial points and the reasonably well known facts. We expect from discussions, both formal and informal, a number of (eagerly awaited for) clarifications.

1. Introduction

Among the theories of the early evolution of the Universe, the Big Bang is currently the one which seems the less open to criticisms (this point will be discussed), and its acceptance leads naturally to the scenario of a progressive enrichment by heavy elements of the matter in the galaxies. I will use here the concepts and the wording of these classical theories. I also have to stress that, owing to the enormous amount of literature about the subject, I will be unable in this limited space, to quote all the authors who have made significant contributions to the field, and I apologize for this uncompleteness.

In the standard Big Bang nucleosynthesis (BBN), the light elements D, ^{3}He, ^{4}He, Li are formed in the first minutes of the expansion of the Universe, and the abundances reached at this time are linked with the ratio of photons to baryons, and therefore to the baryonic density of the Universe. Later on, these elements (except probably D) are also formed by various processes in stars, supernovae, cosmic rays etc. Some elements are fragile. The problem of the light elements is then to derive the primordial abundances from the observed abundances in more or less old objects currently observable, where the initial abundance has been more or less altered by subsequent production and/or depletion.

2. Pop III stars

A solution would be the analysis of the so-called Population III stars, i. e. the old stars formed from the primordial matter. However, in spite of considerable efforts, these "first stars" have not yet been found (Beers et al. 1998, Cayrel 1996), and several hypotheses have been advanced for this scarcity : for example SN triggered star formation (e. g. Cayrel, 1986, Tsujimoto et al. 1999, etc.).

3. Methodology

We have therefore to collect the abundances in more or less young objects, and use the theories of the chemical evolution of the galaxies (essentially the evolution of our Galaxy) in order to restore the abundances in the old Pop III stars. The evolution theories are in reasonable agreement (far from perfect) with the observations (Chiappini et al. 1997).

4. Deuterium

D is observed :

-in the (young) interstellar gas, with a good uniformity in the local interstellar cloud (Linsky, this symposium) ; however some scatter is noted by Jenkins et al. (1999), see also Mullan & Linsky (1999), Sahu, Vidal-Madjar (this symposium) : the problem needs some clarification.

-in the solar system, the observations (solar wind, planets, meteorites) provide a coherent value within error bars

-in intergalactic gas, a few authors find, on one hand, high values (Songaila et al. 1994, Carswell et al. 1994), on the other hand Burles & Tytler (1998) find consistently rather low values. Molaro et al. (1999) find an object with a surprisingly low one (but see Levshakov, this symposium).

A better understanding of the exact nature, history and evolution of the intergalactic gas would be essential.

Some global information may be extracted from the currently available data, and a general trend extracted (Tosi 1998, Lemoine et al. 1999). Some uncertainty remains : is the scatter real (see Vidal-Madjar, this symposium) ? what is the source of the scatter ? is D only destroyed as generally assumed ? or significantly produced in stellar flares (Mullan & Linsky 1999) ? The data, to be obtained by FUSE, will help.

5. ^{3}He

In the young interstellar gas, the obtained abundance is about $2.5 \ 10^{-5}$. The H II regions provide a plateau (Balser et al. 1999), in contradiction with the standard model, which predicts some production in low-mass stars. Refined models (e. g. Charbonnel et al. 1998, Charbonnel & Do Nascimento 1998) explain the contradiction, as well as the high abundances found in some Planetary Nebulae.

Formally, the plateau (the best configuration for a safe extrapolation), as well as the model of Tosi, provide a reliable value of the primordial abundance, but a better understanding of the evolution of ^{3}He in the Galaxy is desirable.

6. ^{4}He

Notation : the abundance is computed :
 -by number of atoms : y = N(He)/N(H)
 -by mass : Y, with $X + Y + Z = 1$ and $Y \sim 4y/(1+4y)$.

The primordial values are y_p and Y_p. Helium is essentially produced by the Big Bang. Later, all stars produce He, the bulk of the production is by massive stars. A general Galactic trend (progressive enrichment in He) is obvious, with however a large scatter and large error bars.

In the galaxies :

Blue compact galaxies (BCG) provide a better defined trend. Izotov & Thuan (1999) derive a high primordial He abundance, higher than in previous determinations : to be discussed.

The Y_p value depends rather heavily on the observations of the most metal-poor galaxies, such as I Zw 18. The history and structure of such objects are not yet fully understood.

7. Lithium

7.1. The ^{7}Li isotope

This isotope is produced in the Big Bang, with additional subsequent production
 -by the Cosmic Rays
 -possibly by the ν-process (Woosley et al. 1990) in SN II (controversial)
 -by novae (controversial)
 -by moderate-mass stars (AGB, Red Giants) on a long time scale
 -in possibly significant amount by compact objects, cataclysmic variables, flares etc.

Notations : $A(Li) = \log(\varepsilon) Li = \log(N(Li)/N(H)) + 12.00$

The observations show a trend in the Galaxy :

1) Young objects :

The field stars and clusters provide $A(Li) = 3.0 - 3.3$ dex, whereas the interstellar gas suggests a slightly larger value (3.5 dex).

2) Solar system :

$A(Li) = 3.31$ (meteoritic), but only 1.16 (photospheric), indicating a depletion of Li in the Sun, by a processus (or several ones) extending the convection. But the problem is not well understood : see for example the puzzling lack of correlation beween Li and age in the solar-type stars of the solar neighborhood (Spite and Spite 1982, Pasquini et al. 1994).

3) Old stars :

The Li abundance is uniform in Pop II dwarfs and subgiants
 - for "warm" stars : $T_{\text{eff}} > 5700$ K
 - for stars of low metallicity : [Fe/H] < -1.4 dex

defining a two-dimensional plateau versus temperature and/or metallicity. Let us recall the notation $[X] = \log X_* - \log X_\odot$. The slopes of A(Li) versus $T_{\rm eff}$ and versus [Fe/H] are small or negligible. The abundance is around A(Li) = 2.2 dex (Bonifacio & Molaro 1997) when the temperature is in the IRFM scale (Alonso et al. 1996).

Finally, there is a definite trend of Li abundance in the Galaxy, but any interpretation has to take into account that Li is a fragile element. A number of contributions to this symposium discuss (section 9) the behavior of Li in stars in relation with various parameters (e. g. rotation, internal structure etc.).

4) Other galaxies :

In the Magellanic Clouds, the less processed interstellar matter has a low value of Li abundance : an argument in favor of a low primordial abundance. In the supergiants of the field, the behavior is similar to that found in the Galactic Pop I (Hill 1997).

DISCUSSION OF THE PLATEAU

Two interpretations have been proposed :

1 - A depletion of a high (primordial therefore uniform) abundance.

2 - A negligible depletion of a low (primordial therefore uniform) abundance.

In both cases, the *two slopes* of the plateau and the amplitude of the real *scatter* of the Li abundance around the plateau are both crucial AND controversial.

In the first interpretation, it is proposed that the same model of depletion, which works rather well for the Pop I stars, should apply to Pop II stars. The stellar structure is however clearly different in these two populations, in particular the depth of the convective zone (Cayrel 1998).

A detailed, documented and argumented review of the problem, by Cayrel (1999) from the observational point of view (and on the theoretical aspects by Vangioni-Flam et al. 1999), has been presented in the LiBeB workshop (Ramaty et al., ed. 1999).

A remark : The modest progressive enrichment of Li by cosmic rays (moderated by a modest depletion by stellar processing of Li in the massive stars) changes slightly the slope (Li versus [Fe/H]) of the plateau. The other slope of the plateau (Li versus $T_{\rm eff}$) needs a good determination of the effective temperature : a fundamental problem (not yet solved) of stellar atmospheres.

Recently, Ryan et al. (1999) discussed a well delimited sub-sample of high quality observations, and in this way they obtained an observed scatter of Li abundances comparable to the errors, suggesting that the(intrinsic scatter is *very* small. The high accuracy, needed for reaching cosmological conclusions, is pushing the theory of stellar model atmospheres to its current limit.

We need

-best temperature determinations : interferometry, accurate multi-color photometry, accurate computed colors and threfore better stellar models

-2D or 3D convective atmosphere NLTE analysis (Asplund et al. 1999) : a considerable task (Cayrel and Asplund, this meeting).

Let us note that the model of Pinsonneault (1998) adjusted to the small observed scatter of the plateau, requires a limited depletion of lithium in Pop II dwarfs, implying a rather low Big Bang abundance, and consequently a large production of Li in the Galaxy.

How is made this large production ?
- Novae : controversial
- Carbon stars, AGB (a significant source) and probably red giants.

A small proportion of red giants are found Li-rich, but this is important, since the red giants are very numerous and also since such an enrichment may remain local, perhaps explaining an inhomogeneous Li abundance in the Galaxy (e. g. the famous star BD 23 3912).

Our Brazilian colleagues have made a lot of work and observations about the Li-rich giants, as testified by their numerous contributions in this symposium (and see e. g. de la Reza et al. 1998, da Silva et al. 1995, Castilho et al. 1996). An important theoretical work has been done (Sackmann & Boothroyd 1999). The observations of Castilho et al. (1999) have shown that Li is not preserved, but produced in such giants. The discovery of a Li-rich giant in the cluster B 21 (Hill & Pasquini 1999) provides for the first time, in a moderately metal-poor open cluster, the chemical composition and the age of the star and its precise evolutionary phase.

7.2. ^{6}Li and the isotopic ratio

This isotope is produced in the Big Bang at a very low level. It is produced :
   - in Cosmic Rays : see for example the LiBeB workshop (Ramaty et al., ed., 1999)
   - in situ (not likely : Lambert, 1995)
   - in flares (controversial)

The observed abundances :

Pop II and disk :

^{6}Li is measured only in two (three ?) Pop II stars (e. g. Smith et al. 1998, Cayrel et al. 1999) and two disk stars (Nissen et al. 1999).

Solar system : from meteorites, the ^{7}Li/^{6}Li isotopic ratio is around 12.

Interstellar matter : The isotopic ratio (and the abundance of ^{6}Li) is variable. The ratio : ^{7}Li/^{6}Li is around 1.1 towards ρ Oph, 6.8 and 5.5 towards ζ Oph and ζ Per respectively, and the revised value for ζ Oph (2 clouds) is a combination of two individual ratios : 8.6 et 2 (Lemoine et al. 1995). This variation suggests some local inhomogeneous production, the high value of the ratio in some clouds suggests a local production of the ^{7}Li isotope.

The ^{6}Li isotope is much more fragile than ^{7}Li, the presence of ^{6}Li in Pop II dwarfs, is an argument for a small (perhaps negligible ?) depletion of ^{7}Li on the plateau (controversial !).

8. Beryllium and Boron

Those elements are not supposed to be produced by the Big Bang, and do not provide any *direct* information about it (see however Suzuki, Yoshii & Kajino 1999). But their similarities with the other light elements (produced simultaneously with the Li isotopes, depleted in stars but less than D and Li) make that they are quite important in the general study of primordial abundances. These elements are not produced in stars, however B could be produced in SN II by the ν-process (Woosley et al. 1990). They are produced by the Cosmic Rays (see e. g. the LiBeB workshop, Ramaty et al., eds. 1999).

Recent observations in old "warm" metal-poor stars (Molaro et al. 1997, Duncan et al. 1997, Boesgaard et al. 1999, for Be, Duncan et al. (1998), Primas et al. (1999) for B), show that the abundances of Be and B vary in locksteps with the iron abundance, and the slope versus [Fe/H] is about one, suggesting a primary origin. However, the slope versus [O/H] is between 1 and 2 (García López et al. 1999), suggesting a mixed mode production, in agreement with the new models of Cosmic Rays.

The slopes (versus either Fe or O) are similar for Be and B, and the ratio B/Be remains constant (around 20).

There is a marked scatter, which suggests local inhomogeneities in the production of Be and B by the Cosmic Rays.

Interstellar matter : The abundance of B is there smaller than the meteoritic one (grain condensation). More interestingly, the ratio $^{11}B/^{10}B$ is 3.4 ± 0.7, in 3 clouds : it is similar to solar (4.05), different from the spallation ratio (2.5) suggesting a non classical production by the Cosmic Rays and/or by the ν-process (Woosley et al. 1990).

B and Be are fragile. The meteoritic abundance of Be is 1.42 dex, the photospheric abundance is 1.15 dex, classically interpreted by a depletion of the fragile Be in the Sun, but Balachandran & Bell (1998) suggest rather an underestimation of the photospheric Be due to a bad estimation of the photospheric opacity in the UV : controversial.

The meteoritic abundance of B is 2.88 ± 0.04, the photospheric one is 2.6 ± 0.3dex

B is less fragile than Be : small (or no) depletion is expected in Pop II "warm" dwarfs (controversial ?), but some dilution of both Be and B is expected (and observed) in cool subgiants and giants.

9. Light elements as probes of stellar structure

As seen all along the previous sections, the fragile elements are probes of the stellar structure. This second aim of the observation of the light elements is essential for stellar studies, and by helping to gain a better knowledge of stellar structure (including convection), enables to gain a better accuracy and reliability of stellar abundances : an accuracy needed for reaching cosmological conclusions. However, the large variety of the sub-topics makes that it is extremely difficult to summarize briefly the numerous important and/or controversial points. Most of them are covered in the talks and posters of this symposium, in particular the complex relations between stellar rotation and lithium depletion.

10. Conclusion

The numerous problems and controversial points indicated here are discussed in this meeting. The accuracy needed for the abundances of the light elements, in order to reach cosmological conclusions, is at the limit of the possibilities of our current knowledge of stellar atmospheres and stellar structure. And the observed behavior of the fragile elements contribute to this knowledge. The interpretation of the abundances in gas is also difficult. The interaction of these different points of views is of course complex, and the aim of this symposium is to favor and develop fruitful exchanges. The excellent organisation of this colloquium, in an enchanting place, is definitely favoring discussions (both formal and informal), and also collaborations and progresses in this particularly exciting field : we all are grateful to the Local Organizing Committee.

References

Alonso, A., Arribas, S., Martínez-Roger, C. 1996, A&AS, 117, 227

Asplund, M., Nordlund, A., Trampedach, R., Stein, R. 1999, A&A, 346, L17

Beers, T. C., Rossi, S., Norris, J. E., Ryan, S. G., Molaro, P., Rebolo R. 1998, Space Sci.Rev. 84, 139

Balachandran, S., Bell, R. A. 1998, Nature, 392, 791

Balser, D. S., Rood, R. T., Bania, T. M. 1999, ApJ, 522, L73

Boesgaard, A. M., Deliyannis, C. P., King, J. R., Ryan, S. G., Vogt, S. S., Beers, T. C. 1999, AJ, 117, 1549

Bonifacio, P., Molaro, P. 1997, MNRAS, 285, 847

Burles, S., Kirkman D., Tytler, D. 1999, ApJ, 519, 18

Carswell, R. F., Rauch, M., Weymann, R. J., Cooke, A. J., Webb, J. K. 1994, MNRAS, 268, 1

Castilho, B. V., Spite, F., Barbuy, B., Spite, M., de Medeiros, J. R., Gregorio-Hetem, J. 1999, A&A, 345, 249

Castilho, B. V., Barbuy, B., Gregorio-Hetem, J. 1996, Rev. Mex. A&A Conf. Ser. 4, 94

Cayrel, R., 1986, A&A, 168, 81

Cayrel R., 1996, A&A Rev 7, 217

Cayrel, R., 1998, Space Sci.Rev. 84, 145

Cayrel, R., 1999 in ASP Conf. Ser. Vol. 171, LiBeB, Cosmic rays and related X- and Gamma-Rays, ed. R. Ramaty, E. Vangioni-Flam, M. Cassé and K. Olive (San Francisco: ASP), 261

Charbonnel, C., Brown, J. A., Wallerstein, G. 1998, A&A 332, 204

Charbonnel, C., Do Nascimento, J. D. Jr. 1998, A&A, 336, 915

Chiappini, C., Matteucci, F., Gratton, R. 1997, ApJ, 477, 765

da Silva, L., de la Reza, R., Barbuy, B. 1995, ApJ, 448, L41

de la Reza, R., Drake, N. A., da Silva, L. 1998, Ap&SS 255, 291

Duncan, D. K., Rebull, L. M., Primas, F., Boesgaard, A. M., Deliyannis C. P., Hobbs, L. M., King, J. R., Ryan, S. G. 1997, ApJ, 488, 338

Duncan, D. K., Rebull, L. M., Primas, F., Boesgaard, A. M., Deliyannis, C. P., Hobbs, L. M., King, J. R., Ryan, S. G. 1998, A&A, 332, 1017

García López, R. J. 1999 in ASP Conf. Vol. 171, LiBeB Cosmic rays and related X- and Gamma-Rays, ed. R. Ramaty, E. Vangioni-Flam, M. Cassé and K. Olive, (San Francisco: ASP), 77

Hill, V. 1997, A&A, 324, 435

Hill, V., Barbuy, B., Spite, M. 1997, A&A, 323, 461

Hill, V., Pasquini, L. 1999 A&A, 348, 21

Izotov, Y. I., Thuan, T. X. 1999, ApJ, 511, 639

Jenkins, E., Tripp, T. M., Wozniak, P. A., Sofia U. J., Sonneborn G. 1999, ApJ 520, 182

Lambert, D. L. 1995, A&A, 301, 478

Lemoine, M., Ferlet, R., Vidal-Madjar, A. 1995, A&A, 298, 879

Lemoine, M., Audouze, J., Ben Jaffel, L. 1999, NewA, 4, 231

Molaro, P., Bonifacio, P. 1997, MNRAS, 285, 847

Molaro, P., Bonifacio, P., Centurion, M., Vladilo, G. 1999, A&A, 349, L13

Molaro, P., Bonifacio, P., Castelli, F., Pasquini, L. 1997 A&A 319, 593

Mullan, D. J., Linsky, J. L. 1999, ApJ, 511, 502

Nissen, P. E., Lambert, D. L., Primas, F., Smith, V. V. 1999, A&A, 348, 211

Pasquini, L., Liu, Q., Pallavicini, R. 1994, A&A, 287, 191

Pinsonneault, M. H., Walker, T. P., Steigman, G., Narayanan, Vijay K. 1999, ApJ, 527, 180

Primas, F., Duncan, D. K., Peterson, R. C., Thorburn, J. A. 1999, A&A, 343, 545

Ramaty, R., Vangioni-Flam, E., Cassé M. and Olive K., ed.,1999, ASP Conf. Vol.171, LiBeB, Cosmic rays and related X- and Gamma-Rays, (San Francisco: ASP)

Ryan, S. G., Norris, J. E., Beers, T. C. 1999, ApJ, 523, 654

Sackmann, I.-J., Boothroyd, A. I. 1999, ApJ, 510, 217

Smith, V. V., Lambert, D. L., Nissen, P. E. 1998, ApJ, 506, 405

Songaila, A, Cowie, L. L., Hogan, C. J., Rugers, M., 1994, Nature, 368, 599

Spite, F., Spite, M. 1982, A&A, 115, 357

Suzuki, T. K., Yoshii, Y., Kajino, T. 1999, ApJ, 522, L125

Tosi, M. 1998, Space Sci.Rev. 84, 207

Tsujimoto, T., Shigeyama, T. Yoshii, Y. 1999, ApJ, 519, L 63

Vangioni-Flam, E., Ramaty, R., Cassé, M., Olive, K. A. 1999 in ASP Conf. Vol.171, LiBeB, Cosmic rays and related X- and Gamma-Rays, ed. R. Ramaty, E. Vangioni-Flam, M. Cassé and K. Olive (San Francisco: ASP), 268

Woosley S. E., Hartmann D. H., Hoffman R. D., Haxton W. C. 1990, ApJ 356, 272

PRODUCTION AND DESTRUCTION OF THE ELEMENTS

Francois Spite presenting his Introduction to the Symposium

Primordial Nucleosynthesis For The New Millennium

G. Steigman

Departments of Physics and Astronomy; The Ohio State University; 174 West 18th Avenue; Columbus, OH 43210 USA

Abstract. The physics of the standard hot big bang cosmology ensures that the early Universe was a primordial nuclear reactor, synthesizing the light nuclides (D, ^{3}He, ^{4}He, and ^{7}Li) in the first 20 minutes of its evolution. After an overview of nucleosynthesis in the standard model (SBBN), the primordial abundance yields will be presented, followed by a status report (intended to stimulate further discussion during this symposium) on the progress along the road from observational data to inferred primordial abundances. Theory will be confronted with observations to assess the consistency of SBBN and to constrain cosmology and particle physics. Some of the issues/problems key to SBBN in the new millenium will be highlighted, along with a wish list to challenge theorists and observers alike.

1. Introduction

Among the quantitative, "hard" sciences, astronomy has traditionally been scorned, with particular disdain reserved for cosmology. No more. In the decade of the nineties the combination of an avalanche of high quality observational data and theoretical advances driven by enhaced computer (and brain) power, have succeeded in transforming cosmology to a precise science. In this introductory lecture to IAU Symposium 198 on The Light Elements and Their Evolution it is my intent to describe primordial nucleosynthesis in this precision era of cosmology and to highlight the challenges, along with some goals, for the new millennium. After a brief review of the important physics during the era of primordial nucleosynthesis in the standard, hot big bang cosmological model (SBBN), I will present an overview of the predicted primordial abundances, emphasizing the generally very small theoretical uncertainties. These will then be compared to the present best estimates (including their uncertainties) of the primordial abundances inferred from current observational data. After assessing the consistency of SBBN, I will explore what SBBN has to offer to Cosmology and to Particle Physics and, what Cosmology may teach us about SBBN. I will conclude with a summary of the key issues/problems confronting SBBN and with a wish list of topics I hope will be addressed during this meeting – and beyond.

2. An Early Universe Chronology

Our story begins when the Universe is a few tenths of a second old and the temperature of the cosmic background radiation has dropped to a few MeV as the Universe expanded and cooled from its denser, hotter infancy. At this time (and earlier) the density and average energy of colliding particles is so high that even the weak interactions occur sufficiently rapidly to establish equilibrium. In particular, at this stage all flavors of neutrinos (e, μ, τ) are in thermal equilibrium with the cosmic background radiation (CBR) photons and with the copius electron-position pairs present $(\nu_i + \bar{\nu}_i \leftrightarrow e^+ + e^- \leftrightarrow \gamma + \gamma)$. However, as the Universe ages beyond a few tenths of a second and the temperature drops below a few MeV, these weak interactions become too slow to keep pace with the rapid expansion of the Universe and the neutrinos decouple from the CBR. The electron-type neutrinos continue to play a role in transforming neutrons into protons and, vice-versa $(p + e^- \leftrightarrow n + \nu_e, n + e^+ \leftrightarrow p + \bar{\nu}_e, n \leftrightarrow p + e^- + \bar{\nu}_e)$. As the temperature continues to drop, less massive protons are favored over the more massive neutrons and the n/p ratio falls (roughly as $e^{-\Delta m/kT}$, where Δm is the neutron – proton mass difference ~ 1.3 MeV). After the temperature drops below 800 keV or so, when the Universe is a few seconds old, even these weak interactions become too slow to keep pace with the expansion and the neutron-to-proton ratio "freezes out" (in fact, the ratio continues to decrease, albeit very slowly). All the while, neutrons and protons have been colliding, occasionally forming deuterons $(p + n \rightarrow D + \gamma)$. However, the deuterons find themselves bathed in a high density background of energetic CBR photons which quickly photodissociate them $(D + \gamma \rightarrow p + n)$ before they can find a proton or neutron and form the more tightly bound, less fragile, ^{3}H or ^{3}He nuclei. Since, as we shall see, there are roughly nine to ten orders of magnitude more CBR photons than nucleons in the Universe, the deuteron "stepping-stone" to further nucleosynthesis is absent until the temperature drops sufficiently low so that even in the high-energy tail of the black-body spectrum there are too few photons to prevent the deuteron from acting as a catalyst for primordial nucleosynthesis. This critical temperature, which is weakly (logarithmically) dependent on the nucleon abundance (the nucleon-to-photon ratio η), is roughly 80 keV. Now, at last, when the Universe is a few minutes old, Big Bang Nucleosynthesis finally commences. However, the Universe was a fatally flawed nuclear reactor, cooling and diluting rapidly as it aged. When the Universe is some 10 – 20 minutes old (~ 1000 sec) and the temperature has dropped below 30 keV or so, the coulomb barriers preventing nuclear reactions between charged nuclei and protons and among charged nuclei become insurmountable (in the short amount of time available) and primordial nucleosynthesis comes to an abrupt end. In this all too brief but shining era there has been time to synthesize (in abundances comparable to those observed or observable) only the lightest nuclides: D, ^{3}He, ^{4}He, and ^{7}Li. In "standard" (a homogeneous Universe, expanding isotropically with the particle content of the standard model of particle physics in which there are three flavors of light ($m \ll$ MeV) or massless neutrinos) big bang nucleosynthesis (SBBN) the abundances (relative to protons $\equiv$ hydrogen) of these four nuclides are determined by only one free parameter, the present epoch nucleon-to-photon ratio η ($\eta \equiv (n_N/n_\gamma)_0$, $\eta_{10} \equiv 10^{10}\eta$).

3. SBBN-Predicted Primordial Abundances

Once the deuterium photodissociation bottleneck is breached primordial nucleosynthesis begins in earnest, quickly burning D to ^{3}H, ^{3}He and ^{4}He. The higher the nucleon density, the faster D is destroyed. The same is true of ^{3}H (which, if it survives will decay to ^{3}He) and ^{3}He. Thus, the primordial abundances of D and ^{3}He are determined by the competition between the nuclear reaction rates and the universal expansion rate. The former rate depends on the overall density of the reactants – the nucleon density. Since all densities decrease as the Universe expands, it is convenient to quantify the nucleon density by specifying the *ratio* of the nucleon density to the photon density (measured after e^+e^- annihilation which enhances the Universe's photon budget) η. Since observations of the cosmic background radiation (CBR) temperature (T = 2.73 K) determine the present density of CBR photons, a knowledge of η is equivalent to a determination of the present mass density in nucleons ("baryons" $\equiv$ B). In terms of the density parameter Ω_B (the ratio of the mass density to the critical mass density) and the present value of the Hubble parameter ($H_0 \equiv 100h$ km/s/Mpc), $\eta_{10} = 273\Omega_B h^2$. As η increases the surviving abundances of D and ^{3}He decrease; since the ^{3}He nucleus is more tightly bound than the deuteron, the decrease of the ^{3}He/H ratio with η is less rapid than that of D/H.

In contrast to D and ^{3}He, the primordial abundance of ^{4}He is not reaction rate limited since the nuclear reactions building helium-4 are so rapid that virtually all neutrons available when BBN commences are incorporated into ^{4}He. As a result the ^{4}He abundance, conventionally presented as the mass fraction of all nucleons which are in ^{4}He, Y_P, is *neutron limited*. Since the neutron-to-proton ratio is determined by the competition between the (charged-current) weak interactions which mediate the transformation of neutrons into protons (and, vice-versa) and the universal expansion rate, Y_P is sensitive to the universal expansion rate at the time the n/p ratio "freezes" and when the deuterium photodissociation barrier disappears. Since the universal expansion rate is controlled by the total energy density, Y_P provides an important test of cosmology and of particle physics in the early Universe (Steigman, Schramm & Gunn 1977). It should be noted that Y_P is not entirely insensitive to the nucleon density since the higher η, the earlier the photodissociation barrier is overcome. At earlier times when the temperature is higher, fewer neutrons have been transformed into protons and are available for incorporation into ^{4}He. As a result, Y_P increases logarithmically with η.

There is no stable nucleus at mass-5 and this presents a gap in the road to the synthesis of nuclei heavier than ^{4}He. In order to bridge the gap nuclear reactions must occur among nuclei with two or more nucleons. But, the abundances of D, ^{3}H, and ^{3}He are small and the coulomb barriers (especially between ^{3}He and ^{4}He and between ^{4}He and ^{4}He) suppress these reactions as the Universe expands and cools. As a result, there is very little "leakage" to nuclei beyond mass-4; as a corollary, virtually all the ^{4}He formed, survives. The only heavier nucleus produced primordially in an abundance comparable to that observed (or, even, observable with current technology) is ^{7}Li, whose BBN abundance is some 4 – 5 orders of magnitude smaller than that of D and ^{3}He. The absence of a stable nucleus at mass-8 provides another gap preventing the production of astrophysically interesting abundances of any heavier nuclei.

As will become clear in our subsequent discussion, the "interesting" range of η is $\eta_{10} = 1 - 10$ ($\Omega_B h^2 = 0.004 - 0.037$), so we focus our discussion here on values of η in this range. In the current precision era of BBN *most* of the nuclear reactions relevant to the synthesis of the light elements have been measured to reasonable accuracy at energies directly comparable to the thermal energies at the time of primordial nucleosynthesis (*e.g.*, see Nollett, this volume). As a result, the theoretical uncertainties in the BBN-predicted abundances are generally quite small. For η in the above range, the 1σ uncertainties in D/H and ^{3}He/H vary from 8 – 10%. Since ^{4}He is most sensitive to the very well measured weak interaction rates, the error in SBBN-predicted Y_P is very small (0.2 – 0.5% or, $\sigma_Y = 0.0005 - 0.0011$). In contrast, larger uncertainties, of order 12 – 21%, afflict the predicted primordial abundance of ^{7}Li.

Since this Symposium devoted much discussion to ^{7}Li, and space-limitations here prevent me from discussing all the light elements in detail, I will concentrate in the following on the two key light elements, deuterium and helium-4. In Figure 1 is shown the relation between the BBN-predicted abundances of D and ^{4}He. The band going from upper left to lower right represents the $\pm 2\sigma$ range of uncertainties in the primordial abundances ((D/H)$_P$ and Y_P). Low D/H (high η) corresponds to high Y_P and high D/H (low η) corresponds to low Y_P. This anticorrelation will be very important when we confront the predictions of SBBN with the observational data.

4. Precise (Accurate?) Primordial Abundances

To test SBBN and fully exploit the opportunities it offers to constrain cosmology (e.g., the baryon density) and particle physics (e.g., new particles with weak or weaker-than-weak interactions) requires that observational data be used to pin down the primordial abundances of the light elements to precisions as good as (or, better than) those of the SBBN predictions. As we approach the new millennium there is good news along with some bad news. The good news is that new detectors on ever larger telescopes which cover the spectrum from radio to x-ray energies and beyond are providing very high quality data, leading to inferred abundances of high statistical accuracy. Furthermore, the abundances of the light elements are determined from observations which differ from element to element in the telescopes and techniques employed as well as in the astrophysical sites explored. As a result, insidious correlated errors between and among the various element abundances are unlikely to be a problem. The good news is also responsible for the bad news. Since the statistical errors have become so small, systematic errors now tend to dominate the uncertainties in the derived primordial abundances. As Bob Rood has said during this Symposium, estimating systematic errors is an oxymoron. When a potential source of systematic error is identified, observations can (and should) be designed to eliminate or bound its contribution to the error budget. It is a pointless and potentially misleading exercise to "estimate" the magnitude of unidentified systematic errors. In part to remind us that our precise abundance determinations may not be accurate, and in part to challenge our observational colleagues who have done such a magnificent job of reducing the statistical errors, I will try to focus on the potential

4.1. Deuterium

As J. Linsky (this volume) has reminded us, the deuterium abundance in the local interstellar medium (the local interstellar cloud: LIC) is known very accurately: $(D/H)_{LIC} = 1.5 \pm 0.1 \times 10^{-5}$ (Linsky 1998). Since deuterium is only destroyed during the evolution of the Galaxy (Epstein, Lattimer & Schramm 1976), the LIC abundance provides a lower bound to its primordial (pre-Galactic) value. This bound is strong enough to bound the nucleon density from above ($\eta_{10} \lesssim 10$; $\Omega_B h^2 \lesssim 0.04$), ensuring that baryons cannot "close" the Universe ($\Omega_B \ll 1$), nor even dominate its present mass density ($\Omega_B \ll \Omega_M \approx 0.3 - 0.4$). Thus, local observations of deuterium, combined with the *assumption* of the correctness of SBBN (which we must test), already reaps great rewards: the mass-energy density of the Universe must be dominated by unseen ("dark") non-baryonic matter. To go beyond (in the quest for the primordial deuterium abundance) we must look for observing targets which are less evolved than the LIC. The presolar nebula is one such site. From solar system observations of ^{3}He reported by G. Gloeckler (this volume), it is possible to infer the presolar deuterium abundance (Geiss & Reeves 1972; Geiss & Gloecker 1998): $(D/H)_\odot = 1.9 \pm 0.5 \times 10^{-5}$. Although marginally higher than the LIC abundance, the larger errors prevent us from using this determination to improve on our previous bounds from the LIC. What this result does indicate is that there has been very little (if any) evolution in the D-abundance in the solar vicinity of the Galaxy in the last 4.5 Gyr. This is consistent with a large class of Galactic chemical evolution models discussed by M. Tosi (this volume) which point to only a modest overall destruction of primordial deuterium by a factor of 2 – 3 (Tosi et al. 1998). If this theoretical estimate is combined with the LIC abundance, we may estimate the primordial abundance: $(D/H)_P \approx 2.6 - 5.1 \times 10^{-5}$ ($\sim 2\sigma$). Although possibly model dependent, this estimate is in remarkable agreement with the 2 – 3 determinations of D/H in high-redshift, low-metallicity (hence very nearly primordial) Ly-α absorbers illuminated by background QSOs described by D. Tytler and S. Levshakov (this volume). The data and analysis of Burles & Tytler (1998a,b: BT) suggests that $(D/H)_P = 2.9 - 4.0 \times 10^{-5}$ ($\sim 2\sigma$). Notice that the 1σ uncertainty in the observationally determined primordial abundance, $\sim 8\%$, is impedance-matched to the $\sim 8\%$ SBBN theoretical uncertainty cited earlier. However, lest we risk dislocating a shoulder while patting ourselves on the back at the triumph of such wonderful data, we should not ignore the claim (Webb et al. 1997; Tytler et al. 1999) that the deuterium abundance in at least one Ly-α absorption system may be much higher. This is a reminder that while any determination of the deuterium abundance anywhere in the Universe (LIC, solar system, Ly-α absorbers, etc.) provides a *lower* bound to primordial deuterium, finding an upper bound is more problematic. Indeed, in some absorbing systems it may be impossible to distinguish D-absorption from that due to hydrogen in an interloping, low column density, "wrong-velocity" system. Thus, the deuterium abundance inferred from absorption-line data may only provide an *upper* bound to the true deuterium abundance. Since the low-Z, high-z QSO absorbing systems hold the greatest promise of revealing for us nearly unevolved, nearly primordial material, we look forward to the time when we can use the *distribution* of D/H values from more than a handful of such systems to eliminate – statistically – the uninvited contribution to the inferred primordial deuterium abundance from such interlopers. Keeping this in mind, in the following I will, nevertheless, use the BT determination when confronting theory with data.

4.2. Helium-4

In contrast to deuterium whose primordial abundance only decreases as pristine gas is incorporated into stars, stars burn hydrogen to helium. As a result, the ^{4}He observed anywhere in the Universe is an unknown mixture of primordial and stellar-produced helium. It has long been appreciated that to minimize the uncertain correction due to the debris of stellar evolution, it is best to concentrate on ^{4}He abundance determinations in the lowest-metallicity regions available. These are the low-Z, extragalactic H II regions which have been discussed by K. Olive, T. Thuan, and S. M. Viegas at this Symposium (this volume). The reader is urged to consult their papers for details; here I will merely summarize my view of the current status of the determination of the primordial ^{4}He mass fraction Y_P. Several years ago Olive & Steigman (1995: OS) gathered together the data from the literature (dominated by the data assembled by Pagel et al. 1992). More recently Olive, Skillman & Steigman (1997: OSS) supplemented this with newer data (some of it, unfortunately, still unpublished). Using a variety of approaches such as the regression of Y on the oxygen and/or nitrogen abundances and the weighted means of Y in the lowest metal-abundance H II regions, OSS concluded that $Y_P = 0.234 \pm 0.003$ (note that, in contrast to the published (OSS) result, this value is obtained when the NW region of IZw18, suspected of being contaminated by underlying stellar absorption, is excluded from the fit, and the newer data of Izotov, Thuan and collaborators is not included). Izotov, Thuan and their collaborators (Izotov, Thuan, & Lipovetsky 1994, 1997; Izotov & Thuan 1998(IT); Thuan, this volume) have been systematically observing a mostly independent set of H II regions. Although, as with the data employed in the OS and OSS studies, they ignore the ionization correction ($icf \equiv 1$), they take special care with the correction for collisional excitation. IT (also Thuan, this volume) find $Y_P(IT) = 0.244 \pm 0.002$. Comparing the IT and OSS estimates of Y_P we find that difference between the two Y_P estimates far exceeds the statistical errors, suggesting systematic differences in the acquisition and/or analysis of the data samples. In a recent discussion which attempted to account for these unidentified systematic differences, Olive, Steigman & Walker (1999: OSW) combined the 2σ ranges for each determination to conclude: $Y_P = 0.238 \pm 0.005$; at the 2σ level, $Y_P \leq 0.248$. Note, that this is also the 2σ upper bound to the IT data alone. Since, as we shall see shortly, it is the upper bound which is crucial to testing the consistency of SBBN, in the following we shall adopt the IT value (and error estimate) for the primordial abundance of ^{4}He.

Recently, Viegas, Gruenwald & Steigman (1999: VGS; see Viegas & Gruenwald, this volume) have emphasized the importance of the ionization correction which has heretofore been ignored. VGS suggest that the IT helium abundance (Y_P) should be reduced by 0.003 to account for unseen neutral hydrogen in regions where the helium is still ionized in H II regions ionized by young, hot, metal-poor stars. In subsequent comparisons I shall explore the implications of adopting $Y_P(\text{VGS}) = 0.241 \pm 0.002$.

4.3. Helium-3 and Lithium-7

The cosmic history of the two other light nuclides produced in astrophysically interesting abundances during SBBN, ^{3}He and ^{7}Li, is considerably more complex than that of D or ^{4}He, which limits their utility as probes of the consistency of SBBN. ^{3}He is destroyed in the hotter interiors of all stars, but some ^{3}He does survive in the cooler, outer layers. For lower mass stars this ^{3}He survival layer increases and, indeed, newly synthesized ^{3}He is produced by incomplete hydrogen burning. The competition between destruction, survival, and synthesis complicates the Galactic history of the ^{3}He abundance. Nonetheless, since any deuterium incorporated into stars is first burned to ^{3}He, the apparent lack of enhanced ^{3}He (see Bania & Rood, this volume) argues against a very large pre-Galactic abundance of deuterium (Steigman & Tosi 1995). For further discussion of the evolution of ^{3}He see Bania & Rood (this volume).

As with ^{3}He, any ^{7}Li incorporated into stars is quickly burned away. However, fusion and spallation reactions between cosmic ray nuclei and those in the interstellar medium are a potent source of ^{7}Li (as well as of ^{6}Li, ^{7}Be, ^{10}B, and ^{11}B). It is also likely that there are stellar sources of ^{7}Li as indicated by the sample of lithium-rich red giants (V. Smith, this volume). Since the abundance of lithium in the solar system and in the interstellar medium ("here and now") greatly exceeds that in the very metal-poor halo stars (T. Beers & S. Ryan, this volume), the latter likely provide the closest approach to a nearly primordial sample. Since a significant fraction of this Symposium is devoted to lithium, I will defer here to those other discussions except to comment that, within the theoretical and observational uncertainties, the primordial abundances inferred from the observational data are consistent with SBBN constrained by the confrontation with D and ^{4}He.

5. Confrontation Of SBBN With Data

Although SBBN does lead to the prediction of the abundances of D, ^{3}He, ^{4}He, and ^{7}Li, the currently best-constrained primordial abundances are those of deuterium and helium-4 which we are concentrating on in this status report. For each value of η, SBBN predicts a pair of $(D/H)_P$ and Y_P values. Therefore, in SBBN there is a unique connection between $(D/H)_P$ and Y_P which, allowing for the theoretical uncertainties discussed above, is shown as the band (solid lines) in Figure 1 going from the upper left to the lower right (2σ uncertainties). Note that high-helium correlates with low-deuterium and, vice-versa. Also shown as the dotted ellipse in Figure 1 is the contour of the (independent) 2σ uncertainties in the BT deuterium abundance and the IT helium-4 mass fraction.

Although the overlap between theory and data is not complete, Figure 1 shows that, at the $\sim 2\sigma$ level, the predictions of SBBN are consistent with current observational data. This is a dramatic success for the standard hot, big bang cosmological model. Of course it is not at all surprising that some value of η may be found to provide consistency with the inferred primordial deuterium abundance. But there was no guarantee at all that the helium-4 abundance

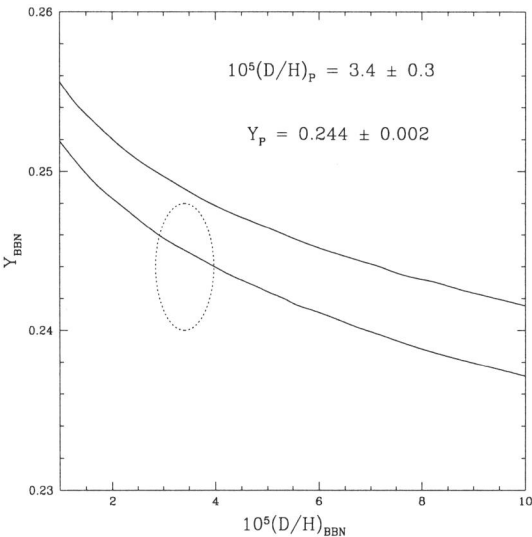

Figure 1. The SBBN-predicted ^{4}He mass fraction Y_P as a function of the SBBN-predicted primordial deuterium-to-hydrogen ratio D/H is shown (at the $\pm 2\sigma$ level) by the solid lines. The dotted ellipse is the 95% contour of the BT deuterium abundance and the IT ^{4}He mass fraction (see the text).

corresponding to this choice would bear any relation to its inferred primordial value. Consistency with the BT D-abundance limits the nucleon abundance to the range (2σ) $\eta_{10} = 4.4 - 5.9$ or, $\Omega_B h^2 = 0.016 - 0.022$. For η in this range there is consistency, within the theoretical and observational uncertainties, between the SBBN-predicted and observationally inferred primordial abundances of ^{3}He and ^{7}Li as well. Four for the price of one! There is, of course, one more test – and opportunity – offered by this result. This SBBN-inferred nucleon abundance must also be consistent with present epoch estimates of the baryon density. Indeed, the SBBN-determined value of Ω_B is *larger* than estimates (Persic & Salucci 1992) of the "luminous" matter in the Universe suggesting that the majority of baryons are "dark". This is good ($\Omega_B > \Omega_{\rm LUM}$); the opposite would have been a disaster. This early-Universe estimate of the baryon density is in good agreement with that inferred from the X-ray cluster baryon fraction (Steigman, Hata & Felten 1999) and with the independent estimate from the Ly-α forest (Weinberg et al. 1997) discussed below.

5.1. What BBN May Do For Cosmology

X-ray clusters likely provide a "fair" sample of the universal baryon *fraction* f_B (White et al. 1993; Steigman & Felten 1995; Evrard, Metzler, & Navarro 1996) which, when combined with the SBBN-inferred baryon density Ω_B, leads to a "clean" prediction, independent of detailed cosmological models, of the overall matter density Ω_M. If the results presented here are combined with the determination of f_B from Evrard (1997), and with a Hubble parameter $h = 0.70 \pm 0.07$ (Mould et al. 1999), we predict $\Omega_M = 0.35 \pm 0.08$, in excellent agreement with several other recent, independent determinations. For example, a lower bound to the cosmic baryon density follows from the requirement that the high-redshift intergalactic medium contain enough neutral hydrogen to produce the Ly-α absorption observed in quasar spectra. According to Weinberg et al. (1997), depending on estimates of the quasar UV background intensity, this lower bound corresponds to $\eta_{10} \gtrsim 3.4 - 4.9$, in excellent agreement with the SBBN prediction based on the BT deuterium determination. Note that this lower bound from the Ly-α absorption forbids (in the context of SBBN) the primordial deuterium abundance to be any larger than $\sim 8 \times 10^{-5}$, largely excluding the one surviving claim of high D (Webb et al. 1997).

Indeed, if the SBBN results are combined with the magnitude-redshift data from surveys of high-redshift Type Ia supernovae (Garnavich et al. 1998; Perlmutter et al. 1999) which bound a linear combination of Ω_M and the cosmological constant $\Omega_\Lambda \equiv \Lambda/3H_0^2$, we may also constrain the cosmological constant ($\Omega_\Lambda = 0.80 \pm 0.20$), the curvature ($\Omega_k \equiv 1 - (\Omega_M + \Omega_\Lambda) = -0.15 \pm 0.25$), and the deceleration parameter ($q_0 = \Omega_M/2 - \Omega_\Lambda = -0.62 \pm 0.18$).

5.2. What Cosmology May Do For BBN

As we have just seen, the SBBN-determined baryon density is consistent with that determined or constrained by observations of the Universe during its present or recent evolution. We may turn the argument around and ask what baryon density is suggested by non-BBN contraints, and then compare the light element abundances which correspond to that density with those inferred from the observational data. As an exercise of this sort, suppose (for reasons of "naturalness" or inflation) that the Universe is "flat": $\Omega_M + \Omega_\Lambda = 1$. When combined with the SN Ia magnitude-redshift data (Perlmutter et al. 1999), this suggests that $\Omega_M = 0.29 \pm 0.07$ (and $\Omega_\Lambda = 0.71 \pm 0.07$). Now, if this mass density estimate (Ω_M) is combined with with the X-ray determined cluster baryon fraction f_B (Evrard 1997; Steigman, Hata & Fcltcn 1999), the resulting nucleon abundance is $\eta_{10} = 4.5 \pm 1.5$. Although the uncertainty is large, it is reassuring that this non-BBN estimate has significant overlap with our SBBN estimate. Indeed, for the baryon density in this range SBBN predicts: $(D/H)_P = 4.3 \pm 2.3 \times 10^{-5}$ and $Y_P = 0.245 \pm 0.004$.

5.3. What SBBN May Do For Particle Physics

The expansion rate of the early Universe is controlled by the density of the relativistic particles present. In the standard model at the time of BBN these are: photons, electron-positron pairs (when $T \gtrsim m_e$) and three "flavors" of neutrinos (ν_e, ν_μ, ν_τ) which, if "light" ($m_\nu \ll 1$ MeV), are relativistic at BBN even if one or more of them may contribute to the present density of non-relativistic ("hot") dark matter. If "new" particles were to contribute to the energy density at BBN, the increase in the density would result in an increase in the universal expansion rate, leaving less time for neutrons to transform into protons. The higher n/p ratio at BBN would result in the production of more primordial ^{4}He (Steigman, Schramm & Gunn 1977). It is convenient (and conventional) to characterize such additional contributions to the energy density by comparing their effects

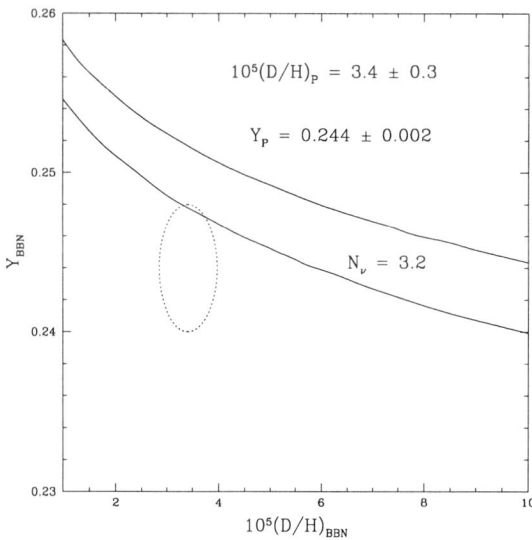

Figure 2. As Figure 1, but for $N_\nu = 3.2$

to that of an additional "flavor" of (light) neutrino: $\Delta\rho \equiv \Delta N_\nu \rho_\nu$. For ΔN_ν small, $\Delta Y \approx 0.01 \Delta N_\nu$. Notice in Figure 1 that the predicted ^{4}He abundance is a little high for perfect overlap with the observations. If ΔN_ν were < 0, ($N_\nu \approx 2.8$) the overlap would improve (e.g., Hata et al. 1995), while if $\Delta N_\nu > 0$, the overlap would be reduced until it disappeared. This is illustrated in Figure 2 which shows the Y versus D/H BBN band that would result if $\Delta N_\nu = 0.2$ (i.e., $N_\nu = 3.2$, in contrast to the SBBN value of 3.0). Notice that due to the faster expansion, more deuterium survives being burnt away so that, for fixed η, the D-abundance also increases; however since D/H is a much more sensitive function of η, this has a much smaller effect on the Y versus D/H relation than does the increase in Y.

6. Conclusions And Outlook

The study of the early evolution of the Universe and, in particular primordial nucleosynthesis, has truly entered the precision era of cosmology. Precise abundances of the light nuclides are predicted and inferred from observations and the two are – apparently – in excellent agreement. As pleased as we may be at this success, it behooves us to avoid the temptation to rest on our laurels and to test this consistency ever more carefully. To this end, it doesn't take much contemplation to identify several clouds looming on the horizon. What follows is my personal list of some problems/issues I would like to see addressed at this Symposium and beyond.

6.1. Problems/Issues

First consider deuterium. On the one hand, any determination of the D/H ratio anywhere, anytime provides a *lower* bound to the primordial abundance. On the other hand, since "wrong" velocity hydrogen can masquerade as deuterium, any observation of "deuterium" is really an *upper* bound to its true abundance. More data tracking the velocity structure of the absorbing features used to identify D and H and exploring variations in D/H in material with similar histories will be very valuable. More data at high-redshift and low-metallicity will be very valuable. After all, at present we are drawing profound conclusions on the basis of only two such systems.

Much remains to be done concerning the primordial abundance of ^{4}He. For the most part, the H II regions from which the helium abundance is inferred have been modelled as homogeneous spheres or plane-parallel slabs. A glance at the beautiful HST images of real H II regions reveals that they are anything but such idealizations. What are the effects of temperature and/or density inhomogeneities, and how large may they be? What of underlying stellar absorption which, if present but neglected, would lead to an *under*estimate of the helium abundance. And, what of the usually neglected ionization correction for neutral hydrogen and helium (Viegas, Gruenwald & Steigman 1999; see Viegas & Gruenwald, this volume)? Considering this latter work, where models of H II regions ionized by realistic spectra of young star clusters were used in a reanalysis of the IT data, a *reduction* in Y_P of order 0.003 was suggested. A comparison with Figure 1 shows that if Y_P were reduced by this amount, the overlap between theory and data would, in fact, disappear.

6.2. Wish List

Given the setting of this Symposium (Natal) and the proximity to the Christmas season, I'd like to conclude with my personal wish list. To avoid being greedy, I'll only ask for two gifts.

A half-dozen or so observations of deuterium in high-z, low-Z systems along the lines-of-sight to distant quasars, with D/H determined in each (on average) to 10% or better. With such a gift, I could determine η to better than 4%, predict Y_P to $\lesssim 0.0007$, and constrain ΔN_ν to an uncertainty less than $\pm\, 0.05$. I'd be a very happy cosmologist indeed.

My second wish is for ^{4}He measured to 3% accuracy (or better) in each of about a dozen low-metallicity, extragalactic H II regions, with care taken to address the several problems outlined above. With such data, Y_P could be fixed to better than the current level of ± 0.002, permitting ^{4}He to be used as a baryometer ($\Delta\eta/\eta \lesssim 20\%$).

Acknowledgments. Much of what I know about this subject I have learned from my collaborators and I would be remiss if I failed to thank them for their contributions. In particular, I wish to acknowledge R. Gruenwald, K. Olive, E. Skillman, M. Tosi, and S. M. Viegas and, of course, my late friend Dave Schramm. L. da Silva, M. Spite and the Scientific and Local Organizing committees deserve great credit for their efficient organization of a very enjoyable and successful meeting. In part, this work is supported at The Ohio State University by DOE grant DE–AC02–76ER–01545.

References

Burles, S., & Tytler, D. 1998a, ApJ, 499, 699 (BT)
Burles, S., & Tytler, D. 1998b, ApJ, 507, 732 (BT)
Epstein, R. Lattimer, J., & Schramm, D. N. 1976, Nature, 263, 198
Evrard, A. E. 1997, MNRAS, 292, 289
Evrard, A. E., Metzler, C. A., & Navarro, J. F. 1996, ApJ, 469, 494
Garnavich, P. M. et al. 1998, ApJ, 509, 74
Geiss, J., & Reeves, H. 1972, A&A, 18, 126
Geiss, J., & Gloeckler, G. 1998, Space Sci.Rev., 84, 239
Hata, N., Scherrer, R. J., Steigman, G., Thomas, D., Walker, T. P., Bludman, S., & Langacker P. 1995, Phys.Rev.Lett, 75, 3977
Linsky, J. L. 1998, Space Sci.Rev., 84, 285
Izotov, Y. I., Thuan, T. X., & Lipovetsky, V. A. 1994, ApJ, 435, 647
Izotov, Y. I., Thuan, T. X., & Lipovetsky, V. A. 1997, ApJS, 108, 1
Izotov, Y. I., & Thuan, T. X. 1998, ApJ, 500, 188 (IT)
Mould, J. R. et al. 1999, ApJ, submitted (astro-ph/9909260)
Olive, K. A., Skillman, E., & Steigman, G. 1997, ApJ, 483, 788 (OSS)
Olive, K. A., &, Steigman, G. 1995, ApJS, 97, 49 (OS)
Olive, K. A., &, Steigman, G., & Walker, T. P. 1999, Physics Reports, in press (astro-ph/9905320) (OSW)
Pagel, B. E. J., Simonson, E. A., Terlevich, R. J. & Edmunds, M. 1992, MNRAS, 255, 325
Perlmutter, S. et al. 1999, ApJ, 517, 565
Persic, M., & Salucci, P. 1992, MNRAS, 258, 14P
Steigman, G., Schramm, D. N., & Gunn, J. E. 1977, Phys. Lett., B66, 202
Steigman, G., & Felten, J. E. 1995, Space Sci.Rev., 74, 245
Steigman, G., Hata, N., & Felten, J. E. 1999, ApJ, 510, 564
Steigman, G., & Tosi, M. 1995, ApJ, 453, 173
Tosi, M., Steigman, G., Matteucci, F., & Chiappini, C. 1998, ApJ, 498, 226
Tytler, D., Burles, S., Lu, L., Fan, X. M., Wolfe, A., & Savage, B. D. 1999, AJ, 117, 63
Viegas, S.M., Gruenwald, R., & Steigman, G. 1999, ApJ, 532 (in press, March 20, 2000; astro-ph/9909213)
Webb, J. K., Carswell, R. F., Lanzetta, K. M., Ferlet, R., Lemoine, M., Vidal-Madjar, A., & Bowen, D. V. 1997, Nature, 388, 250
Weinberg, D. H., Miralda-Escudé, J., Hernquist, L., & Katz, N. 1997, ApJ490, 564
White, S. D. M., Navarro, J. F., Evrard, A. E., & Frenk, C. S. 1993, Nature, 366, 429

The Light Elements and Their Evolution
IAU Symposium, Vol. 198, 2000
L. da Silva, M. Spite, J. R. de Medeiros, eds.

Alternative Solutions to Big Bang Nucleosynthesis

Hannu Kurki-Suonio

Helsinki Institute of Physics, P.O. Box 9, FIN-00014 University of Helsinki, Finland

Abstract. Standard big bang nucleosynthesis (SBBN) has been remarkably successful, and it may well be the correct and sufficient account of what happened. However, interest in variations from the standard picture come from two sources: First, big bang nucleosynthesis can be used to constrain physics of the early universe. Second, there may be some discrepancy between predictions of SBBN and observations of abundances. Various alternatives to SBBN include inhomogeneous nucleosynthesis, nucleosynthesis with antimatter, and nonstandard neutrino physics.

1. Introduction

The success of standard big bang nucleosynthesis (SBBN) in predicting the observed abundances of the light elements has led to the widespread view that SBBN must be correct. According to this view, any remaining disagreements must be due to systematic errors in observations or incorrect, or too crude, chemical evolution models. While this view may well be the right one, we should not be blind to other possibilities.

I will stay within the context of the Hot Big Bang (for an alternative, see Burbidge & Hoyle 1998), and discuss some models of nonstandard big bang nucleosynthesis (NSBBN). NSBBN scenarios range from small modifications to SBBN to a complete change in the decisive physical phenomena, like in the late-decaying massive particle scheme of Dimopoulos et al. (1988).

Motivations for studying NSBBN go in two directions. First, the remarkable success of SBBN allows one to severely constrain the physics of the early universe. If one tries to change the conditions from the standard assumptions the resulting abundances of the light elements differ from the observed ones. For many modifications, BBN provides the strongest constraints. BBN gives also the strongest constraint on the single parameter of SBBN, the baryon density, usually given as the baryon-to-photon ratio,

$$\eta \equiv \frac{n_b}{n_\gamma}, \qquad \eta_{10} \equiv 10^{10}\eta. \qquad (1)$$

Second, one may try to improve on SBBN. From time to time it has seemed that there might be some discrepancy between observations and SBBN, which could then be explained by NSBBN. In particular, there has been tension between D/H and Y_p (see, e.g., Hata et al. 1995). To relieve this tension, either a lower D or a lower ^{4}He yield has been looked for. Also one may want to relax the

SBBN bounds to η. Other astronomical considerations have given motivation for trying to raise the upper limit to η. If one believes that the energy density of the universe is dominated by vacuum energy (the cosmological constant) and accepts the newer observations on D/H and Y_p favoring somewhat larger η within SBBN, this motivation largely disappears.

There is a very large body of work on NSBBN. Extensive reviews are given by Malaney & Mathews (1993) and Sarkar (1996), which contain, respectively, over 500 and over 700 references. Here I will be able to mention only a random few.

Most of the work on NSBBN can be divided into four broad classes:

1. Inhomogeneous BBN. Usually this means inhomogeneity in the baryon-to-photon ratio, η, but there are also other possibilities, like inhomogeneity in the neutrino chemical potentials.

2. Nonstandard neutrino physics, e.g., additional ("sterile") neutrino flavors, neutrino degeneracy (asymmetry), massive ν_τ, or neutrino oscillations.

3. Late-decaying ($\tau = 1\text{--}10^8$ s) massive particles, black holes, cosmic strings, etc.

4. Time-varying fundamental constants.

In the interest of time and space, I will discuss the first two classes only.

2. Inhomogeneous Big Bang Nucleosynthesis

The single parameter of SBBN is the baryon-to-photon ratio η, or the density of baryonic matter. In inhomogeneous big-bang nucleosynthesis (IBBN) one assumes that η is inhomogeneous. To get a significant effect on BBN this inhomogeneity has to be large, $\delta\eta/\eta \gtrsim 1$. Since the baryons make an insignificant contribution to the energy density at nucleosynthesis time, the total energy density may still be essentially homogeneous. The inhomogeneity could be caused by, e.g., first-order phase transitions. The distance scale of this inhomogeneity is of crucial importance for IBBN. Without inflation, causal physics can only produce significant inhomogeneity at subhorizon scales (see Table 1).

Table 1. The approximate temperature and horizon scale (in comoving units) for various events in the early universe.

event	T	horizon
EW phase transition	100 GeV	10^{-3} pc
QCD phase transition	150 MeV	1 pc
^{4}He synthesis	70 keV	1 kpc

Mechanisms connected with inflation can produce inhomogeneity at any scale. The isotropy of the cosmic microwave background (CMB) rules out significant inhomogeneity at $\gtrsim 10$ Mpc scales, and it is difficult to construct an acceptable IBBN scenario which would explain inhomogeneity in observations.

In the usual IBBN models one considers a significantly smaller distance scale, so that while η is inhomogeneous during BBN, resulting in inhomogeneous abundances at first, everything gets mixed and becomes chemically homogeneous before or during galaxy formation. Thus the observable primordial abundances are homogeneous, while different from the SBBN predictions.

The simplest version of IBBN is one where SBBN occurs with different η in different parts of the universe, and the yields get mixed afterwards, so that one obtains the IBBN results by averaging SBBN results over the η distribution, whose average we denote by $\bar{\eta}$. This kind of IBBN has a long history. Typically Y_p goes up, ^{7}Li goes up (down for small $\bar{\eta}$), and D goes up for large $\bar{\eta}$, and down for small $\bar{\eta}$, compared to SBBN with $\eta = \bar{\eta}$. Leonard & Scherrer (1996) concluded that this way one can reduce the lower bound to η from observations (in fact remove it, if arbitrary η distributions are allowed), but the upper bound is essentially unchanged from SBBN, as ^{7}Li and ^{4}He are overproduced for larger $\bar{\eta}$. The tension between D an ^{4}He is worsened at the large end of the SBBN acceptable range. Thus this kind of modification to BBN appears undesirable.

2.1. Small Scale Inhomogeneity and Neutron Diffusion

The above applies to inhomogeneity with distance scales significantly larger than the neutron diffusion scale (~ 0.1 pc). If there is inhomogeneity at smaller scales, neutrons will diffuse out of the high density regions resulting in an inhomogeneous n/p ratio. Especially if this results in $n/p > 1$ in some regions, the consequences for BBN may be dramatic. This scenario (Applegate, Hogan, & Scherrer 1987) looked very exciting about ten years ago when it was noted that the QCD (quark-hadron) transition seemed likely to produce strong inhomogeneity at just the right distance scale, and early IBBN calculations indicated a large reduction in Y_p and increase in D/H allowing very large η, even a critical density in baryons only. More detailed calculations showed that the effects were less dramatic, and the upper limit to η given by D/H and Y_p is raised at most by a factor of 2 or 3 as compared to SBBN, and this only if the inhomogeneity was at near the optimal distance scale ($10^{-3} \ldots 10^{-2}$ pc), and most of the baryon number was in the high density regions. The most severe problem for this kind of IBBN is ^{7}Li overproduction. Some ^{7}Li depletion (by a factor of 2 or 3) in Pop II stars is needed to allow for larger η than in SBBN. Figure 1 is from a recent review of this scenario by Kainulainen, Kurki-Suonio, & Sihvola (1999).

Recent lattice QCD calculations favor a much smaller distance scale, although uncertainties are big enough so that the optimal distance scale cannot be ruled out. The distance scale from the electroweak (EW) phase transition must be so small that the effects on BBN cannot be large; in the best case they could be comparable to other small effects that have recently been included in accurate BBN codes.

2.2. Regions of Antimatter

A less-studied variant of IBBN is one where η is allowed to have negative values, i.e., there are antimatter regions. This is possible in some baryogenesis scenarios (Dolgov 1996). Antimatter in cosmology has been reviewed by Steigman (1976). If the distance scale of antimatter regions is small, antimatter and matter will mix and annihilate in the early universe, and the presence of matter today

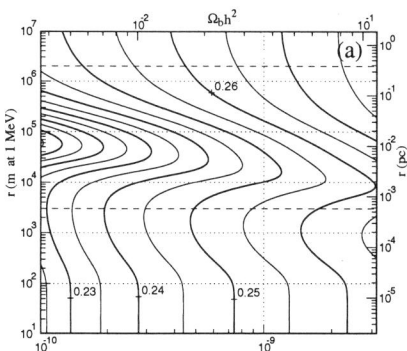

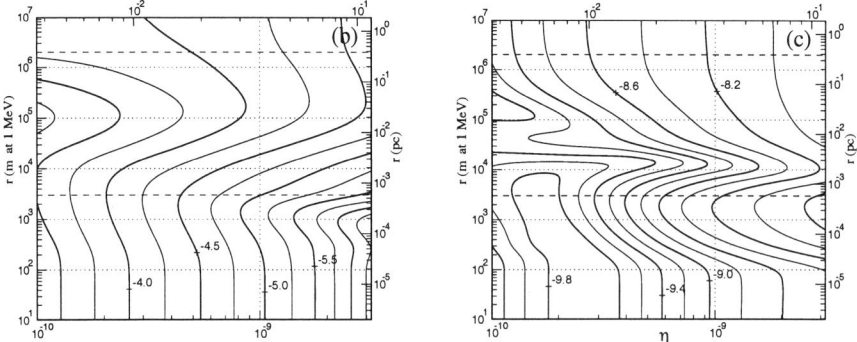

Figure 1. The ^{4}He, D, and ^{7}Li yields from small-scale inhomogeneous nucleosynthesis runs with a centrally condensed geometry, with density contrast $R = 800$ and high-density volume fraction $f_v = 0.125$. The contours of (a) Y_p, (b) $\log_{10}$D/H, and (c) $\log_{10}$ ^{7}Li are plotted as a function of the average baryon-to-photon ratio η and the distance scale r of the inhomogeneity. The two horizontal dashed lines denote the horizon scale ℓ_H at the QCD (upper) and EW (lower) phase transitions. From Kainulainen et al. (1999).

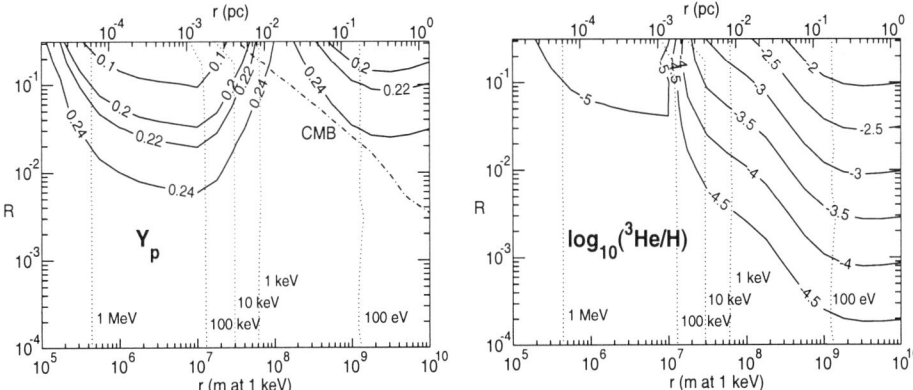

Figure 2. The (a) ^{4}He and (b) ^{3}He yields as a function of the antimatter/matter ratio R and the antimatter domain radius r. The distance scales are given both at $T = 1$ keV (in meters) and today (in parsecs). We plot contours of Y_p and (the logarithm of) the number ratio ^{3}He/H. The dotted lines show contours of the "median annihilation temperature", i.e., the temperature of the universe when 50% of the antimatter has annihilated. Typically the annihilation is complete at a temperature lower than this by about a factor of 3. The dot-dashed line gives the upper limit to R from CMB spectrum distortion. This plot is for $\eta_{10} = 6$. From Kurki-Suonio & Sihvola (1999).

implies that there was initially more matter than antimatter. If the distance scale is large, so that antimatter regions will survive till present, observational constraints require either the amount of antimatter to be very small, or the distance scale to be very large, comparable to the present horizon or larger (Cohen, De Rújula, & Glashow 1998), so that the case of large regions is not of interest for BBN.

The smaller the antimatter regions are, the earlier they annihilate. Rehm & Jedamzik (1998) considered annihilation immediately before nucleosynthesis. Kurki-Suonio & Sihvola (1999) extended these results to larger distance scales where annihilation occurs during or after nucleosynthesis (see Figure 2). So far the focus has been on obtaining upper limits to the amount of antimatter at various scales in the early universe, but clearly there is also potential for obtaining acceptable abundances with nonstandard values of η, although probably only with fine-tuned model parameters.

3. Neutrinos and Big Bang Nucleosynthesis

Neutrinos affect BBN in two ways, through the energy density effect and the ν_e effect. The most significant effect is on Y_p in both cases.

The energy density in neutrinos affects the expansion rate of the universe. The simplest way to increase the energy density of the early universe from the standard model is to have additional particle species (sterile neutrinos or other hypothetical particles). The custom is to parametrize this by an "effective number of neutrino species". The standard case is $N_\nu = 3$. We now know that there are only three "active" neutrino species, so any additional species must be "sterile" neutrinos or other very weakly interacting particles. A higher energy density means faster expansion. This leads to n/p freezeout at a higher temperature, leaving more neutrons, and resulting in a higher ^{4}He yield. The D yield is also increased, so an increased energy density is disfavored by BBN, and one gets an upper limit, e.g., $N_\nu < 3.2$ (Burles et al. 1999) or $N_\nu \lesssim 4$ (Lisi, Sarkar, & Villante 1999), depending on what observational constraints one uses.

Electron neutrinos affect the weak $n \leftrightarrow p$ reactions directly. More ν_e leads to fewer neutrons and thus to less ^{4}He (and everything else), whereas more $\bar{\nu}_e$ leads to more neutrons and more ^{4}He.

3.1. Neutrino Degeneracy

In SBBN one assumes that the neutrino asymmetry (difference between the number of neutrinos and antineutrinos),

$$L_\nu \equiv \frac{n_\nu - n_{\bar{\nu}}}{n_\gamma} = 0.069 \left(\frac{T_\nu}{T}\right)^3 (\pi^2 \xi + \xi^3), \qquad (2)$$

which is related to the neutrino chemical potential μ_ν, or the degeneracy parameter $\xi \equiv \mu_\nu/T$, is small, $\ll 1$. This seems natural, since the comparable baryon asymmetry η is small. However, the neutrino background is unobservable, so we cannot rule out a large neutrino asymmetry. A larger asymmetry always means a larger neutrino energy density, raising N_ν. To have a significant effect on BBN, we must have $|\xi|, |L_\nu| \gtrsim 0.1$. There is a separate contribution from each neutrino flavor. Thus there are three indepedent degeneracy parameters, $\xi_e, \xi_\mu,$ and ξ_τ. The energy density effect is the same for all three flavors, and depends only on $|\xi|$. The electron neutrino effect depends only on ξ_e, but is much stronger, and the direction of the effect depends on the sign.

There are two possible scenarios for affecting BBN. If ξ_e is comparable in magnitude to ξ_μ and ξ_τ, or larger, one can forget the other two in first approximation. One can then adjust ξ_e to dial in the desired value of Y_p. The other elements are hardly affected. A less natural scenario is one where the asymmetries in the other two neutrino flavors are much larger, and the energy density and ν_e effects are balanced against each other to keep Y_p in the acceptable range. This way one can have a significant effect on the other abundances and raise the acceptable range for η. This second scenario is constrained by structure formation, since the large neutrino energy density means that the matter/radiation equality and thus the beginning of structure formation occurs later. Kang & Steigman (1992) used a generous lower limit for matter/radiation equality, $z_{eq} > 10^3$ to widen the SBBN acceptable range from $\eta_{10} = 2.8$–4.7 to $\eta_{10} = 2.8$–19.

3.2. Inhomogeneous Neutrino Degeneracy

The different results from high-z D/H measurements (Tytler, Fan, & Burles 1996; Webb et al. 1997) raised the question whether there might be a large-scale inhomogeneity in primordial abundances. This is very difficult to achieve, since the extreme isotropy of the CMB rules out any significant large-scale inhomogeneity in η or the energy density. Dolgov & Pagel (1999) have come up with a way of getting around this constraint. In their model the asymmetries of the different neutrino flavors are inhomogeneous but balanced with each other so that they add up to a homogeneous total energy density. The inhomogeneous ξ_e is then responsible for the inhomogeneous primordial abundances through the ν_e effect. They suggest that an Affleck-Dine type scenario of generation of leptonic charge asymmetry, respecting the symmetry between different lepton families, could be responsible for creating a domain structure, where the neutrino asymmetries would have the same three values but interchanged with respect to e, μ and τ. To achieve a significant D/H inhomogeneity, a huge Y_p inhomogeneity has to be allowed. But since there are no high-z Y_p determinations, this cannot be used to rule out their model. Table 2 shows an example of what kind of abundances we could have in such a domain structure. The first line would correspond to our local domain; from the other domains we would have only D/H observations.

Table 2. Abundances of light elements for $\eta_{10} = 5$ and nonzero values of all three chemical potentials. One example from Dolgov & Pagel (1999).

ξ_e	ξ_μ	ξ_τ	D/H	Y_p	^{7}Li/H
0.1	-1	1	3.8×10^{-5}	0.23	2.5×10^{-10}
-1	0.1	1	9.2×10^{-5}	0.55	4.5×10^{-10}
1	-1	0.1	2.8×10^{-5}	0.08	1.1×10^{-10}

3.3. Decay of a Massive Tau Neutrino

If the rest mass of a neutrino species is much larger than 100 MeV, then it is becoming nonrelativistic before nucleosynthesis and its contribution to the energy density is different from the standard zero-mass case. The laboratory limits for the neutrino masses leave this as a possibility for ν_τ. Above the neutrino decoupling temperature, $T \sim 3$ MeV, a massive neutrino species contributes less energy density, because of neutrino-antineutrino annihilation, but after neutrino decoupling the annihilation ceases and the rest mass then contributes extra energy density. Neutrinos this heavy must decay to avoid contributing too much to the present energy density. The decay time and mode are of crucial importance to BBN. If the decay time is very short, then the contribution to N_ν will be less than one. The most interesting case is the one where ν_τ decays into ν_e (and a scalar particle), since then the ν_e effect could cause a significant reduction in Y_p.

These calculations are difficult since the decisive effects occur near the neutrino decoupling temperature, so thermal equilibrium is not maintained and the neutrino spectra are distorted. The recent results by Hannestad (1998) and Dolgov et al. (1999) are in disagreement with each other. Hannestad gets the maximum reduction of Y_p, from the SBBN result $Y_p = 0.239$ to $Y_p < 0.20$, for ν_τ mass $m_\nu = 0.2$–0.5 MeV and lifetime $\tau < 100$ s. According to Dolgov et al., the maximum reduction is less, to $Y_p \sim 0.21$, and occurs for larger masses, $m_\nu = 2$–3 MeV, and requires a shorter lifetime $\tau < 1$ s.

The most natural explanation of the SuperKamiokande (1998) result on atmospheric neutrinos is $\nu_\mu \to \nu_\tau$ oscillation. Then ν_τ cannot be heavy and its mass will not affect BBN significantly. To allow the above scenario, the atmospheric neutrino oscillations would have to be into a sterile neutrino species, $\nu_\mu \to \nu_s$, instead (Kainulainen et al. 1999).

3.4. Neutrino Oscillations

Observations of solar neutrinos and atmospheric neutrinos (SuperKamiokande 1998) as well as the LSND (1998) accelerator experiment see different amounts of the different neutrino flavors than predicted by the Standard Model. This can be explained by neutrino oscillations. This is a quantum-mechanical phenomenon where the flavor $(\nu_e, \nu_\mu, \nu_\tau)$ content of the neutrino varies periodically. This requires nonzero neutrino masses and the effect is determined by the difference in mass-squared, Δm^2, and the "mixing angle".

All three (solar, atmospheric, and LSND) "neutrino problems" cannot be simultaneously explained by oscillations among three flavors, but require at least a fourth flavor, ν_s, which must be "sterile", i.e., much more weakly interacting than the three known "active" flavors, in order not to violate the limit $N_\nu \sim 3$ from Z^0 decay width (Particle Data Group 1998). A sterile neutrino would also be useful for supernova nucleosynthesis (Peltoniemi 1996; Caldwell, Fuller, & Qian 1999).

The LSND results are controversial, so the other viewpoint is to ignore them until they are confirmed by independent experiments, in which case the solar and atmospheric neutrino problems can be explained just with the three active neutrinos.

Oscillations among (light, non-degenerate, i.e., $\xi = 0$) active neutrinos do not affect BBN, since they all have equal abundances. If the sterile neutrino exists, it would have thermally decoupled from the other neutrinos very early, much before BBN, so that its contribution to N_ν would be $\ll 1$. Active-sterile neutrino oscillations before BBN would then lead to production of ν_s, increasing N_ν (Enqvist, Kainulainen, & Thomson 1992), which from the BBN point of view is undesirable. The situation is more complicated, however. The oscillation depends on the background temperature, and at a certain temperature there is a resonance. This resonance temperature depends on the neutrino energy, so as the temperature falls, the resonance sweeps through the neutrino spectrum. If there is a small pre-existing asymmetry (this will be the case, since thermal fluctuations suffice), the rates of neutrino and antineutrino oscillation will be different. Resonant active–sterile neutrino oscillations will then lead to a growth of the neutrino asymmetry by a large factor (Barbieri & Dolgov 1991; Foot & Volkas 1995; Shi 1996; Enqvist, Kainulainen, & Sorri 1999; Di Bari & Foot 2000). This may generate a large enough electron neutrino asymmetry to affect BBN (Bell, Foot, & Volkas 1998; Kirilova & Chizhov 1998; Shi, Fuller, & Abazajian 1999).

Depending on the oscillation parameters, the asymmetry may either just grow or oscillate between positive and negative values, so that the final sign of the asymmetry becomes unpredictable. To calculate the effect on BBN is complicated, since the resulting distortion of the ν_e spectrum is also important for BBN, and the process happens near the neutrino decoupling temperature.

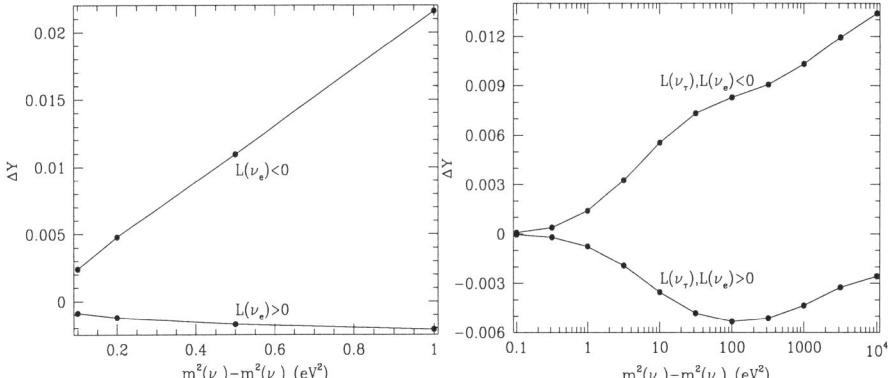

Figure 3. The impact on the primordial ^{4}He abundance Y if an asymmetry in $\nu_e \bar{\nu}_e$ is generated by a resonant $\nu_e \leftrightarrow \nu_s$ mixing (left) or by the indirect neutrino mixing scheme (right), as a function of Δm^2. The baryon-to-photon ratio is set to $\eta = 5.1 \times 10^{-10}$. From Shi et al. (1999).

There are two schemes to generate a large ν_e asymmetry, either directly via $\nu_e \leftrightarrow \nu_s$ oscillations or indirectly via $\nu_{\mu(\tau)} \leftrightarrow \nu_s$ and $\nu_{\mu(\tau)} \leftrightarrow \nu_e$ oscillations.

This scenario is under active study and there is much controversy among the different research groups. In Fig. 3 we show results obtained by Shi et al. (1999). The maximal effect on Y_p seems to be at the ± 0.01 level.

4. Conclusions

At present, no NSBBN scenario appears as convincing as SBBN, which is the simplest of all. Often the real world has turned out to be more complicated in the end than first assumed, but for the early universe a simple picture has been very successful. However, it is healthy to keep in mind the possibility that SBBN might not be the full story, and that any discrepancies between observations and SBBN might actually be telling us something important about the early universe or particle physics.

Acknowledgments. I thank K. Kainulainen, A. Kalliomäki, J. Peltoniemi, and A. Sorri for advice on neutrino physics.

References

Applegate, J. H., Hogan, C. J., & Scherrer, R. J. 1987, Phys.Rev.D 35, 1151
Barbieri, R. & Dolgov, A. 1991, Nucl.Phys.B 237, 742
Bell, N. F., Foot, R., & Volkas, R. R. 1998, Phys.Rev.D 58, 105010
Burbidge, G., & Hoyle, F. 1998, ApJ 509, L1

Burles, S., Nollett, K. M., Truran, J. W., & Turner, M. S. 1999, Phys.Rev.Lett 82, 4176
Caldwell, D. O., Fuller, G. M., & Qian, Y.-Z. 1999, astro-ph/9910175
Cohen, A. G., De Rújula, A., & Glashow, S. L. 1998, ApJ 495, 539
Di Bari, P. & Foot, R. 2000, Phys.Rev.D, to be published, hep-ph/9912215
Dimopoulos, S., Esmailzadeh, R., Hall, L. J., & Starkman, G. D. 1988, ApJ 330, 545
Dolgov, A. D. 1996, in Proceedings of the International Workshop on Baryon Instability, Oak Ridge, March 1996, hep-ph/9605280
Dolgov, A. D., Hansen, S. H., Pastor, S., & Semikoz, D. V. 1999, Nucl.Phys.B 548, 385
Dolgov, A. D., & Pagel, B. E. J. 1999, New Astron 4, 223
Enqvist, K., Kainulainen, K., & Thomson, M. 1992, Nucl.Phys.B 373, 498
Enqvist, K., Kainulainen, K., & Sorri. A. 1999, Phys.Lett.B 464, 199
Foot, R. & Volkas, R. R. 1995, Phys.Rev.Lett 75, 4350
Hannestad, S. 1998, Phys.Rev.D 57, 2213
Hata, N., Scherrer, R. J., Steigman, G., Thomas, D., Walker, T. P., Bludman, S., & Langacker, P. 1995, Phys.Rev.Lett 75, 3977
Kainulainen, K., Kurki-Suonio, H., & Sihvola, E. 1999, Phys.Rev.D 59, 083505
Kang, H.-S., & Steigman, G. 1992, Nucl.Phys.B 372, 494
Kirilova, D. P. & Chizhov, M. V. 1998, Phys.Rev.D 58, 073004
Kurki-Suonio, H., & Sihvola, E. 1999, astro-ph/9912473
Leonard, R. E., & Scherrer, R. J. 1996, ApJ 463, 420
Lisi, E., Sarkar, S., & Villante, F. L 1999, Phys.Rev.D 59, 123520
LSND Collaboration (Athanassopoulos, C., et al.) 1998, Phys.Rev.Lett 81, 1774; Phys.Rev.C 58, 2489
Malaney, R. A., & Mathews, G. J. 1993, Phys.Rep 229, 145
Particle Data Group (Caso, C., et al.) 1998, Eur.J.Phys.C 3, 1
Peltoniemi, J. 1996, in Proceedings of the 3rd Tallinn Symposium on Neutrino Physics, ed. I. Ots, J. Lohmus, P. Helde & L. Palgi, 103
Rehm, J. B., & Jedamzik, K. 1998, Phys.Rev.Lett 81, 3307
Sarkar, S. 1996, Rep.Prog.Phys 59, 1493
Shi, X. 1996, Phys.Rev.D 54, 2753
Shi, X., Fuller, G. M., & Abazajian, K. 1999, Phys.Rev.D 60, 063002
Steigman, G. 1976, ARA&A 14, 339
Super-Kamiokande Collaboration (Fukuda, Y., et al.) 1998, Phys.Rev.Lett 81, 1562
Tytler, D., Fan, X. M., & Burles, S. 1996, Nature 381, 207
Webb, J. K., Carswell, R. F., Lanzetta, K. M., Ferlet, R., Lemoine, M., Vidal-Madjar, A., & Bowen, D. V. 1997, Nature 388, 250

The superbubble model for LiBeB production and Galactic evolution

Etienne Parizot and Luke Drury

Dublin Institute for Advanced Studies, 5 Merrion Square, Dublin 2, Ireland

Abstract. We show that the available constraints relating to ^{6}LiBeB Galactic evolution can be accounted for by the so-called superbubble model, according to which particles are efficiently accelerated inside superbubbles out of a mixture of supernova ejecta and ambient interstellar medium. The corresponding energy spectrum is required to be flat at low energy (in E^{-1} below 500 MeV/n, say), as expected from Bykov's acceleration mechanism. The only free parameter is also found to have the value expected from standard SB dynamical evolution models. Our model predicts a slope 1 (primary) and a slope 2 (secondary) behaviour at respectively low and high metallicity, with all intermediate slopes achieved in the transition region, between 10^{-2} and $10^{-1} Z_\odot$.

1. Introduction

Galactic nucleosynthesis and chemical evolution are about how a given element is produced in the universe and how its abundance evolved from the primordial universe on. In the case of the light elements, it is widely agreed that the nucleosynthesis occurs through spallation reactions induced by energetic particles (EPs) interacting with the interstellar medium (ISM). In these reactions, an heavier nucleus (most significantly C, N or O) is 'broken into pieces' and transmuted into one of the lighter ^{6}Li, ^{7}Li, ^{9}Be, ^{10}B or ^{11}B nuclei. Except for ^{7}Li, this spallative nucleosynthesis is thought to be the main (if not the only) light element production mechanism. The case of ^{11}B is slightly more complicated, as neutrino-induced spallation in supernovae (the so-called ν-process) is sometimes invoked to increase the B/Be and ^{11}B/^{10}B ratios which one would expect should the light elements be produced by nucleo-spallation alone.

Concerning the Galactic evolution of light element abundances, Fields et al. (2000) have recently re-analyzed the available data as a function of O/H, discussing the uncertainties associated with the methods used to derive the O abundance, the stellar parameters and the incompleteness of the samples. According to their results, Be and B evolution can be described in terms of two distinct production processes: i) a primary process dominating at low metallicity and leading to a linear increase of the Be and B abundances with respect to O – 'slope 1' – followed by ii) a secondary process compatible with the standard expectations of the Galactic cosmic ray nucleosynthesis scenario (GCRN) – 'slope 2'. This behaviour is characterized by a transition metallicity, $Z_t \equiv (O/H)_t$, below which the Be/O and B/O ratios are constant and above which they are proportional to O/H. Although the value of Z_t is rather uncertain because very few data points have yet been reported at $Z < Z_t$, energetics arguments show that a primary process *is* indeed required below, say, $10^{-2} Z_\odot$ (Parizot & Drury 1999a,b,2000b, Ramaty et al. 2000), and therefore the very existence of a transition metallicity separating a primary from a secondary evolution scheme seems reasonably well established. In spite of the current large uncertainties on the exact value of Z_t, Fields et al. find a range of possible values between $10^{-1.9}$ and $10^{-1.4} (O/H)_\odot$ (see also Olive, this conference).

This two-slope picture seems to reconcile the two competing theories for light element nucleosynthesis, namely GCRN which predicts a secondary behaviour for Be and B evolution (e.g. Vangioni-Flam et al. 1990, Fields & Olive 1999), and the superbubble model which predicts a primary behaviour at low metallicity (Parizot & Drury 1999b,2000b, Ramaty et al. 2000). However, we show here that when applied to the whole lifetime of the Galaxy (not only to the early Galaxy), the superbubble model *alone* actually predicts the entire two-slope behaviour inferred from observations, and accounts for all the qualitative and quantitative constraints currently available. Implications for particle acceleration inside a superbubble (SB) are also analyzed.

2. Description of the SB model

The superbubble model is based on the observation that most massive stars are born in associations, and evolve quickly enough to explode as SNe in the vicinity of their parent molecular cloud. The dynamical effect of repeated SN explosions in a small region of the Galaxy is to blow large bubbles – superbubbles – of hot, rarefied material, surrounded by shells of swept-up and compressed ISM. The interior of superbubbles consists of the ejecta and stellar winds of evolved massive stars *plus* a given amount of ambient ISM evaporated off the shell and dense clumps passing through the bubble. The exact fraction of the ejecta material inside the SB is not well known, and can be expected to vary with time and from one SB to another. However, this fraction, which we note x, is all we need to know in order to fully determine the mean composition of the matter inside SBs. Noting $\alpha_{ej}(X)$ and $\alpha_{ISM}(X)$ the abundances of element X among the SN ejecta and in the ISM, respectively, we can indeed write the abundance of X inside the SB as:

$$\alpha_{SB}(X) = x\alpha_{ej}(X) + (1-x)\alpha_{ISM}(X). \tag{1}$$

The second assumption of the SB model is that the material inside superbubbles is efficiently accelerated by a combination of shocks produced by SN explosions and supersonic stellar winds, secondary shocks reflected by other shocks or clumps of denser material, and a strong magnetic turbulence created by the global activity of all the massive stars. Two different SB models have been proposed so far, assuming different EP compositions and energy spectra. In our model (Parizot & Drury 1999b,2000b), we follow Bykov & Fleishman (1992) and Bykov (1995,1999) and argue that the SB acceleration process produces a rather flat spectrum at low energy, namely in E^{-1}, as expected from multiple shock acceleration theory (Markowith & Kirk 1999), up to a few hundreds of MeV/n, say. Above this value, the spectrum of the superbubble EPs (SBEPs) is either cut off through a steep power-law or turned into the standard cosmic ray source spectrum (CRS), in E^{-2}. The exact behaviour of this so-called 'SB spectrum' at high energy is important in itself and should be derived from a detailed calculation of the particle acceleration, but we do no consider it here, as it is not relevant to our problem (most of the LiBeB production arises from the most numerous low-energy particles anyway). The other SB model proposed so far (Ramaty & Lingenfelter 1999, Ramaty et al. 2000) assumes that the SBEPs *are* actually the cosmic rays and thus their energy spectrum is the standard CRS spectrum ($Q(p) \propto p^{-2}$).

To summarize, the essence of the SB model is that repeated SN explosions occurring in OB associations lead to the acceleration of EPs having either the CRS spectrum or the SB spectrum, and a composition given by Eq. (1), where the only free parameter is the proportion of the ejecta inside the SB: x. In principle, x can be derived from the study of SB evolution dynamics, coupled with a gas evaporation model. But we shall first study LiBeB evolution for itself with no external prejudice about the value of x, and therefore consider it as a free parameter which we vary from 0 (i.e. SBEPs have the ambient ISM composition) to 1 (i.e. SBEPs are made of pure SN ejecta). Later, we compare the value derived from the LiBeB constraints with the value expected from standard SB dynamical models.

3. Be and B Galactic evolution

3.1. Qualitative features

Having parameterized our problem as above, we can easily calculate the Be/O production ratio in the Galaxy as a function of $Z_{\text{ISM}} \equiv (\text{O/H})_{\text{ISM}}$. We consider SBs blown by 100 SNe exploding continuously over a lifetime of 30 Myr. We then integrate the Be production rates induced by the SBEPs and divide the result by the total O yield (added up assuming a Salpeter IMF and SN yields from Woosley & Weaver 1995). The result is plotted in Fig. 1 for various values of x and the two investigated spectra. The main difference between the latter is the Be production efficiency, i.e. the number of Be produced per erg of SBEP. Apart from the SB spectrum being more efficient, both figures show distinctively the sought-for two-slope behaviour, with a transition metallicity Z_t depending on the actual value of x. This behaviour derives directly from Eq. (1). Replacing X by O there, we see that the abundance of O among the SBEPs is essentially $x\alpha_{\text{ej}}(\text{O})$ at low metallicity, and $(1-x)\alpha_{\text{ISM}}(\text{O})$ above a transition metallicity $Z_t \sim \frac{x}{1-x} Z_{\text{ej}}$ (where $Z_{\text{ej}} \sim 10 Z_\odot$). Therefore, remembering that O is the main progenitor of Be, we find that the SB model predicts a primary behaviour below Z_t (production efficiency independent of Z_{ISM}), and a secondary behaviour above Z_t, since the SB model is then essentially identical to the GCRN model (except maybe for the assumed energy spectrum).

Incidentally, it is interesting to note that the SB model can be considered as a correction of the GCRN scenario, taking into account the chemical inhomogeneity of the early Galaxy. Indeed, since particle acceleration occurs precisely in those places where metals are released (i.e. superbubbles), the SBEP composition is considerably richer in O than the average ISM, as long as the SN ejecta dominate the O content of the SBs. Afterwards, it makes little difference, as far

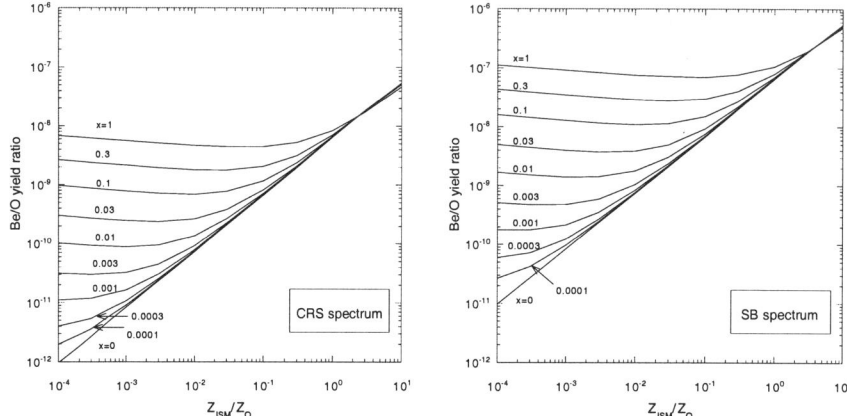

Figure 1. Be/O yield ratios obtained with the CRS spectrum (left) and the SB spectrum (right), as a function of the ambient metallicity, for various values of the mixing parameter, x. The Be yield is calculated for a SBEP total energy of 10^{50} erg per SN.

as composition is concerned, whether the EPs producing LiBeB are accelerated inside SBs or in the regular ISM.

Finally, we see from Fig. 1 that the predicted slope 1 and slope 2 correlations between Be and O are limit behaviours for very low and very high metallicity respectively. Depending on the value of x, any intermediate value for the Be-O slope is reached over a given range of stellar metallicity. This is in contrast with what would arise if the two-slope behaviour were to be explained in terms of two different mechanisms (e.g. the SB model at low Z and GCRN at high Z). In that case, indeed, one would have a sharp change of slope at the precise metallicity where the secondary process becomes dominant, with no intermediate values. Of course, expected physical fluctuations of the parameters would weaken this effect, and current observational error bars prevent us from distinguishing conclusively between the two pictures. But we argue that the observed 'slope 1.45' behaviour reported by Boesgaard & Ryan (this conference) can be explained (in principle) only if there is a continuous *transition* from slope 1 to slope 2 within a *unique* model (as in the SB model above), rather than two unrelated models with a slope 2 eventually superseding a slope 1.

3.2. Quantitative features

Quantitatively, the Be/O ratio at low metallicity derived from the observations is about $4\,10^{-9}$ (Parizot & Drury 2000b). This can be achieve either by the CRS spectrum model, provided $x \sim 50\%$, or by the SB spectrum model, provided $x \sim 2\,or\,3\%$ (see Fig. 1). So far, both models are equally acceptable since we chose not to accept any prejudice about the value of x from outside the restricted field of LiBeB evolution. But when considering the transition metallicity associated with the two possible models, we see that the CRS spectrum implies $Z_t \gtrsim 10^{-1} Z_\odot$, well outside the range derived by Fields et al. On the other hand, the value of Z_t predicted by SB spectrum model falls exactly in the required range.

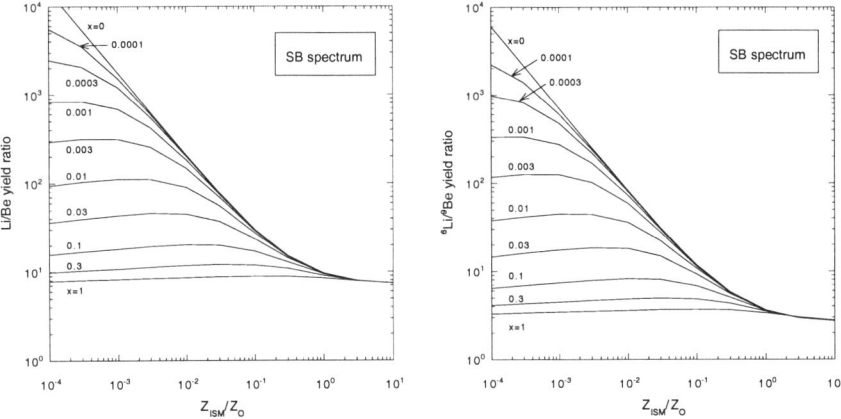

Figure 2. Li/Be (left) and ^{6}Li/^{9}Be (right) production ratios obtained with the SB spectrum model, as a function of the ambient metallicity, for various values of the mixing parameter, x.

As a conclusion, the SB model is fully consistent with the observations provided that i) the SBEP spectrum is flattened at low energy (in E^{-1}), and ii) the SN ejecta amount to a few percent of all the matter present inside SBs.

Now let us extend the scope of our study. Quite remarkably, the first condition above is exactly what is expected from the SB acceleration model developed by Bykov et al. As for the second condition, it is in perfect agreement with the dynamical model for SB evolution worked out by Mac Low & Mc Cray (1988). In other words, had we looked beforehand for a theoretically preferred value of x, we would have chosen just the particular value which turns out to account for the various constraints of Be Galactic evolution. Therefore, our results actually bring support not only to the SB model as the natural framework for Be and B evolution studies, but also to the SB acceleration model and standard SB dynamics.

Concerning B, unfortunately, only qualitative constraints can be checked against the SB model (successfully in this instance), since either a significant ν-process or a LECR component is required anyway to account for the observed B/Be and ^{11}B/^{10}B ratios. However, Li does provide additional quantitative constraints. First, in order not to break the Spite plateau, the Li/Be production ratio must be lower than ~ 100. This is shown to be satisfied for any value of x greater than about 1% in Fig. (2a). Second, the measurement of the ^{6}Li abundance in two halo stars of metallicity $Z \simeq 10^{-2.3} Z_\odot$ indicates that the ^{6}Li/^{9}Be ratio in these stars should be in the range 20–80 (see Vangioni-Flam, Cassé, & Audouze 2000 and references therein), in contrast with the solar value of ~ 6. This could not be explained if the proportion of SN ejecta inside SBs were of the order of 50% (CRS spectrum model). However, it is quite remarkable again that the value of a few percent derived from the SB spectrum model is totally consistent with the observed value of the ^{6}Li/^{9}Be ratio, both a low metallicity and at solar metallicity.

4. Conclusion

The SB model described above has been shown to be fully consistent with the qualitative and quantitative constraints of LiBeB Galactic evolution: 1) it explains the inferred two-slope behaviour in the framework of one sole model; 2) it provides the correct value of Be/O at low metallicity; 3) it predicts the correct value of the transition metallicity; 4) it does not break the Spite plateau; 5) it is consistent with the ^{6}Li/^{9}Be ratio at any metallicity. Most importantly, these successes rely on the value of only one free parameter, namely the proportion of SN ejecta inside a SB. The value which we find is of the order of a few percent, i.e. exactly in the range derived from standard SB dynamical evolution. Likewise, the SB model is found to be successful only if the SBEPs have the SB spectrum, i.e a flattened shape at low energy (in E^{-1}). But this is exactly what is predicted by the SB acceleration model of Bykov et al. As a conclusion, the SB model appears to account for all the available constraints about LiBeB evolution by making only the most standard assumptions about the involved models relating to other fields of astrophysics. This may be considered as lending support to these models as well.

Acknowledgments. This work were supported by the TMR programme of the European Union under contract FMRX-CT98-0168.

References

Bykov A. M. 1995, Space Sci. Rev., 74, 397

Bykov A. M. 1999, in "LiBeB, cosmic rays and gamma-ray line astronomy", ASP Conference Series, eds. R. Ramaty, E. Vangioni-Flam, M. Casse, K. Olive

Bykov A. M., & Fleishman G. D. 1992, MNRAS, 255, 269

Fields B. D., & Olive K. A., 1999, ApJ, 516, 797

Fields B. D., Olive K. A., Vangioni-Flam E., & Cassé M. 2000, submitted to ApJ (astro-ph/9911320)

Mac Low M. M., & McCray R., 1988, ApJ, 324, 776

Markowith A., & Kirk J. G. 1999, A&A, 347, 391

Parizot E., & Drury L. 1999a, A&A, 346, 686

Parizot E., & Drury L. 1999b, A&A, 349, 673

Parizot E., & Drury L. 2000a, A&A, submitted

Parizot E., & Drury L. 2000b, A&A, submitted

Ramaty R., & Lingenfelter R. E. 1999, in "LiBeB, cosmic rays and gamma-ray line astronomy", ASP Conference Series, eds. R. Ramaty, E. Vangioni-Flam, M. Casse, K. Olive

Ramaty R., Scully S. T., Lingenfelter R. E., & Kozlovsky B. 2000, ApJ, in press (astro-ph/9909021)

Vangioni-Flam, E., Cassé, M., Audouze, J., & Oberto, Y. 1990, ApJ 364, 586

Vangioni-Flam, E., Cassé, M., & Audouze, J. 2000, submitted to Physics Reports

Woosley S. E., & Weaver T. A. 1995, ApJSS, 101, 181

The Light Elements and Their Evolution
IAU Symposium, Vol. 198, 2000
L. da Silva, M. Spite, J. R. de Medeiros, eds.

LiBeB Production and Associated Astrophysical Sites

Elisabeth Vangioni-Flam

Institut d'Astrophysique de Paris, 98 bis Bd Arago, 75014 Paris, France

Michel Cassé

*Service d'Astrophysique, Orme des Merisiers, CEA,
91191 Gif sur Yvette, France*

Abstract. The various modes of spallative LiBeB production are summarized, and classified according to their dependence or independence on the abundance of medium heavy elements (CNO) illustrated by that of oxygen in the interstellar medium. The predictions of the models are confronted to the available observational correlations (Be, B vs O). Clearly, a primary mechanism should lead to a slope one in the lg(Be/H) vs [O/H] plot and a secondary mechanism to a slope two. Due to the ambiguity of the O data, another criterion, based on energetics, can help us to select an adequate model. A purely secondary origin in the very early Galaxy is much more energy demanding than a primary one. Indeed, magnesium seems to be a possible surrogate of oxygen and iron since i) it is spectroscopically more easy to cope with and ii) its nucleosynthetic yield is independent of the mass cut and does not depend on metallicity.

1. Introduction

A turning point in the theory of the origin and evolution of light elements has been the observation of a linear relationship between both Be and B and Fe in metal poor halo stars (Rebolo, Molaro & Beckman 1988; Duncan et al 1992, 1997; Gilmore et al 1992, Boesgaard & King 1993; Ryan, Norris, Bessel & Deliyannis 1994, Primas et al 1999). But recently, the debate has taken a complex turn, due to a modification of the O-Fe correlation indicated by the data of Israelian et al (1998) and Boesgaard et al (1999). This revision is however not universally admitted, and the debate is still open (e.g. Fulbright & Kraft 1999; Gustafsson 1999; Reetz 2000). Since the last International Cosmic Ray Conference (Barring 2000) and the most recent LiBeB meeting (Ramaty et al 1999), the situation has not ceased to evolve.

Prior to the proposal of change of the O-Fe relation, the situation was the following: in order to explain the observed proportionality of the Be/H ratio to Fe/H (itself taken proportional to O/H in halo stars as implied at that time by the data and by the SNII calculated yields), it was necessary to invoke a primary production mechanism (i.e. a production rate of LiBeB independent of the interstellar metallicity), driven by the break up in flight of C and O colliding with H and He in the ISM (for reviews see Vangioni-Flam et al 1999b, 2000a,b).

Now the reality of this primary component is questioned on the basis of the "new" O-Fe correlation. Concerning oxygen, however, the situation is not settled. The abundances derived from the forbidden (OI) line, which is certainly the most accurate source when it is not too weak, suggest a plateau but measurements of near IR OH band in dwarfs and subgiants lead to a rising trend with decreasing Fe/H. In contrast, Fulbright & Kraft (1999) have analysed in great details the (OI) spectral region in the two metal poorest stars of the Israelian et al sample and have found a lower O/Fe ratio. As stressed for instance by Pagel (1999), both methods, indeed, have their drawbacks and technical difficulties: the OH bands are subject to uncertainties in UV continuum absorption (Balachandran & Bell 1998) and effective temperature, while the forbidden lines are so weak at low metallicity that the determination of the continuum becomes problematic.

So the situation is wide open and one is inclined to propose a different metallicity index, less ambiguous than O. The closer element to O whose abundance is widely measured is Mg. It has been chosen, due to its various advantages, both observational and theoretical, as the reference element to follow galactic evolution by Thomas et al (1998), Fuhrmann (1998) and Shigeyama & Tsujimoto (1998). Indeed i) Mg seems easier to measure than O and ii) Mg and O are coproduced in SNII explosions, which are the main sources of these elements (Woosley & Weaver 1995, Thielemann, Nomoto & Hashimoto, 1996). Moreover the Mg yield is independent of the mass cut, and does not significantly depend on metallicity (Umeda et al 2000). In the light of existing data, the Mg/Fe vs Fe/H correlation is rather flat up to Fe/H = -1, like that of other alpha elements (Mc Williams 1997, Pagel & Tautvaisiene, 1995). Taken at face value, (assuming Mg proportional to O, on nucleosynthetic grounds, neglecting the peculiar behaviour of the most massive stars which are marginal in the chemical evolution budget), these data indicate the need of a primary component, then we are back to the previous situation (Cassé et al 1995; Vangioni-Flam et al 1996, 1998; Ramaty et al 1996). Thus, on this sole basis, a purely secondary origin of Be in the halo (Fields & Olive 1999a) driven by the standard GCR seems inadequate. However, the situation is not completely settled.

The outline of this paper is the following: in section 2 we recall the basic production mechanisms of LiBeB, in section 3 we decline the astrophysical agents and sites, in section 4, we compare the various models designed, in section 5 we propose key observations to remove the present ambiguities.

2. Nucleosynthesis of LiBeB

2.1. Thermal production and destruction

Thermal nucleosynthesis in the Big Bang produces negligible amounts of Be and B. Only ^{7}Li is synthesized in significant amounts. Moreover, LiBeB do not survive stellar temperatures except in a thin surface layer where they are observed, reflecting the interstellar composition inherited by the star at birth. ^{7}Li is however thought to be produced by AGB stars (Abia et al 1993) and novae (Hernanz et al 1996) and also by SNII through neutrino spallation of carbon. This neutrino spallation in carbon shells of type II supernovae is expected to produce also ^{11}B, but the yields are sensitive to the assumed temperature (energy) of the neutrinos, which is uncertain (see Hartmann et al 1999 for a review).

2.2. Non Thermal production

Nuclear spallation, i.e. the break up of medium heavy elements by collisions with protons and alphas remains the leading production process of light elements in the cosmos (Meneguzzi, Audouze & Reeves 1971). In principle, all isotopes of interest are generated either by the interaction of fast p and alphas on CNO at rest in the ISM, or conversely by the interaction of fast C and O (principally) with ambient H and He, supplemented by the alpha +alpha reaction giving rise exclusively to Li isotopes. (e.g. Reeves 1994).

The cross sections are well measured (Read & Viola 1985; Webber et al 1990 a, b), and have been updated recently by Ramaty et al (1997). The hierarchy of the cross sections reflects that of the abundances of the light nuclei in nature. In a collision between a proton and an oxygen nucleus, the probability of production of ^{11}B is higher than that of ^{10}B which is itself higher than that of ^{9}Be. Thus, it is not surprising that the abundances of these three isotopes go in declining order. This is a stricking example of a direct application of nuclear physics in the understanding of natural abundances. Note that the peaks of the cross sections lie at low energy especially that of alpha + alpha (Read & Viola 1985). Thus low energy particles (about 10 MeV/n) have to be inserted carefully in the treatment of the problem. Note also that at low energy, where the alpha + alpha reaction is operating at full strength, ^{7}Li and ^{6}Li are produced in comparable amounts (1.5), which is at variance with the ^{7}Li/^{6}Li ratio observed in meteorites (12.5). A stellar source of pure ^{7}Li has to be invoked to explain this high value (see above).

The production rate, in the most general case, is function i) of the number density of the target nuclei (N_T) and their composition, ii) of the flux of the projectiles (Φ) and iii) of the cross section averaged over the energy spectrum in the interaction region :

dN(L)/dt = $N_T <\sigma> \Phi$.

If p and alpha are the projectiles and CNO the targets, one deals with a "secondary" production, thus:

dN(L)/dt = $N_{CNO}<\sigma>\Phi_{p,\alpha}$.

It is assumed (quite reasonably) that the flux of energetic particles is proportional to the supernova rate since they are thought to be the main agents of acceleration through the shock waves they produce. Due to the fact that SNII are also the main O producers, one expects that:

$\Phi(t)$ prop. d(SN(t))/dt prop. d(O/H)/dt.

On the other hand, the CNO abundance cumulated in the ISM up to time t, is proportional to the total number of SN having exploded from 0 to t. Summarizing:

d(Be/H)/dt prop. (O/H)d(O/H)/dt or after integration Be/H prop. $(O/H)^2$.

In contrast, if C, O are the projectiles and H, He the targets, a primary production arises governed by the equation:

dN(L)/dt = $N_{HHe}<\sigma>\Phi_{CO}$

Now with the same hypothesis than above, and considering that the target abundances (H, He) do not evolve significantly:

d(Be/H)/dt prop. d(O/H)/dt and Be/H prop. O/H

3. Astrophysical agents and sites

3.1. Galactic Cosmic Rays (GCR)

The energy spectrum of GCR is directly observed above, say, 1GeV/n. Below, it is deduced from various indirect observations (Strong & Moskalenko 1999). It is reasonably well explained by the diffuse shock wave acceleration mechanism (Blandford & Ostriker 1978; Jones & Ellison 1991; Ellison et al 1997). The observed composition is extrapolated back to the sources thanks to a classical propagation model.

It is edifying to compare the elemental and isotopic source composition to that of other materials of known abundances (the solar system for instance, or that of the supernova ejecta computed with stellar models). Indeed SNIa alone do not fit O/Fe, Ne/Fe, Mg/Fe neither the s process elements, whereas, SNII alone do not fit Fe/Co/Ni, neither the s process elements. (see e.g. Meyer 1996, 2000). Indeed, the nucleosynthetic origin of the groups (Mg, Si, Ca, Fe, Ni), (Sr, Zr, Ba, Ce), (Pt peak, actinides) are all different (explosive burning, s-process and r process, respectively) and their production sites are also different, thus the solar mix is a complex mixture of all that, historitically built up, and individual sources are unlikely to lead to the GCR source abundances (since they are similar to that of the solar system). Note that in this example we have chosen only refractory elements that are not supposed to be affected by selective effects (see below).

A stricking fact, of great importance for our purpose, is the similarity between the isotopic composition of the GCRs and of the solar system (Connell & Simpson 1997; Stone et al 1998; Wiedenbeck et al 1999), which is a mixture of the products of generations and generations of stars of different masses indicating at face value that GCR are nuclei accelerated out of a normal reservoir and not strange, exotic, stars or objects.

In recent years, grains have been considered to play a central role in explaining the peculiarities of the GCR source composition (Meyer et al 1997; Ramaty et al 1997). It is assumed that grain debris are more efficiently accelerated than elements in the gas phase and grain models have progressively replaced the traditional two step acceleration mechanism (injection by flare stars and acceleration by supernova shocks) in which grains were undesirable (Cassé & Goret 1978, Meyer 1985; Silberberg & Tsao 1990).

However the situation is not absolutely settled (Shapiro 1999, Silberberg et al 2000; Cassé & Vangioni-Flam 2000, in preparation). Anyway, among the grain supporters themselves, there are divergent views concerning the origin and nature of the grains of interest. In one camp, they are supposed interstellar (Meyer and coworkers) and on the other they are supposed to contain fresh products of nucleosynthesis (Ramaty and colleagues). This has a strong bearing on the primary or secondary character of the LiBeB production mechanism. If the grains that are impacted by the accelerating shock waves have the ISM composition, then, the resulting CR composition should reflect that of the ISM which is the true reservoir of CR particles and, as it is H, He dominated, the process is secondary.

Ramaty and coworkers (see Ramaty et al 2000 and references therein) assume that GCR originate from grains loaded with freshly synthesized nuclei, C and O that are released by SNII and accelerated by shock waves in galactic superbubbles (SB). Subsequently, they interact with the surrounding interstellar medium to give LiBeB in a primary way. However, this proposal has to face the following objections: in the superbubble (SB), high temperature context, according to the observed trend (Cardelli 1996; Savage & Sembach, 1996; Jenkins et al 1998; Howk, Savage & Fabian 1999), grains have little chances to remain intact. Only their refractory cores would survive in the hot phase. Silicon, for instance would be significantly evaporated (see fig 5 in Savage & Sambach 1996) with respect to Fe, and the Si/Fe ratio in the accelerated particles should be different than solar contrary to what is deduced from the GCR observations. Thus grains in SB are unlikely to account, in a detailed manner, for the present GCRS composition. Moreover, the $^{34}S/^{32}S$ ratio, derived at the cosmic ray sources poses also a problem (Cassé & Vangioni-Flam 2000 in preparation). Finally, the grain mechanism does not go without gas (H, He) acceleration and thus particle accelerated in superbubbles cannot induce a purely primary process all time long and as the Galaxy evolves, the secondary process becomes more and more important. In this case, the transition primary-secondary will take place at about [Fe/H] about -1 (taking an average superbubble metallicity of about 5 times solar, Higdon et al 1998). Thus a purely primary component seems unlikely.

On this basis we are tempted to conclude that standard GCR, as traditionally thought, act as a secondary source of LiBeB, and if a primary component is made necessary by the data, it should be different from the standard GCR one. It could come from SB's under the condition that this SB component is confined to low energy not to spoil the observed CGR composition.

3.2. Superbubble Accelerated Particles (SAP)

Superbubbles gather a great number of massive stars which explode as type II supernovae, enriching the surrounding medium in fresh products of nucleosynthesis, and among them, oxygen. The shock wave and turbulence sutained by a given supernova, accelerate the material enriched by the previous ones (Bykov & Fleisshman 1992; Bykov 1995; Bykov et al 2000). The energy spectrum obtained could depart significantly from the GCR one, depending on the detail of the mechamism and more precisely on the escape time of the fast particles, which is poorly known (Klepach et al 1999). As said previously, an energy cut off should be imposed to avoid contamination of the (observed) GCR abundances. This energy cut off is still a free parameter and we can choose it in order to avoid energetic problems (Ramaty et al 1996), i.e. in the range 30-300 MeV/n. Then, all in all, superbubbles appear to be the best agents of a primary production process of LiBeB (Parizot & Drury 1999; Bykov et al 2000). Admittedly, this theoretical proposal remains to be firmly (experimentally) substantiated by X ray and gamma-ray line observations (see the proceedings edited by Ramaty et al 1999 for a general discussion). The search for a non thermal low energy component as the one once claimed to be discovered by the COMPTEL experiment but discarded afterwards (Bloemen et al 1999) is, in our minds, one of the major objectives of the European INTEGRAL satellite to be lauched in 2001. For the time being, we can only speculate on the composition and the spectrum of this hypothetical (but somewhat necessary) component. In the halo phase, the composition is taken as representative of the ejecta of massive low metallicity stars. It is highly enriched in O w.r.t present CRs. (Woosley & Weaver 1995). This composition however is expected to vary in time due to metallicity dependent mass loss rate (Vangioni-Flam et al 1997).

4. Production and evolution of LiBeB

4.1. Observed correlations

Turning to observations, we are confronted to a certain ambiguity (as mentionned in the introduction). According to the choice of the conversion between Fe and O abundances we get different conclusions. Relying on the Israelian et al (1998) and Boesgaard et al (1999) relation we get Be proportional to $O^{1.7}$, i.e. nearly secondary, but taking the conventional O-Fe relation, strengthened by the Mg data, we get Be proportional to O, i.e. purely primary, at least in the halo phase. Thus the O/Fe behaviour at low metallicity is a central issue, once again. Furthermore non LTE effects on B (Kiselman 1999) make the situation even more complex (see also Primas 2000). Thus we must rely on a independent argument to answer the question: is a primary component really necessary? This argument of energetic nature has been essentially developed by Ramaty et al (1996, 1997, 2000). To briefly summarize: it seems that a primary component is required by both the Mg-Fe relation (more secure than O-Fe) and by energetic arguments.

4.2. The three evolutionary models

Concerning the galactic evolutionary models, there exists three types of them i) a pure secondary standard GCR based on the variable O/Fe ratio (Fields & Olive 1999a) ii) a pure primary GCR from SB, which, suprisingly is still valid with flat or variable O/Fe (Ramaty et al 2000) iii) an hybrid model (Standard GCR + SAP based on a flat O/Fe ratio (Vangioni-Flam et al 1998) or based on a variable O/Fe ratio (Fields et al 2000).

4.3. Energetic requirements

The number of atoms of Be produced per erg injected per supernova is promoted to the role of a selection criterion for the theoretical models (both primary and secondary). Stated differently, are there enough SNII and are they sufficiently efficient to produce all the Be observed in halo stars in the primary and/or secondary cases? According to Ramaty et al (2000) the pure secondary standard GCR produce BeB at high energy cost in the early Galaxy. SAP or primary GCR are much more economical. Thus a very plausible solution is that a primary component appears first, and a secondary component takes over afterwards, at a metallicity (O/H ratio) which remains to be determined precisely. The position of the break depends on the Fe-O correlation used. Relying on the new one, and using the analysis based on the IRFM data, (Fields et al 2000) one finds a break point at about [O/H] = -1.6. On the other hand, if the ancient correlation is chosen, motivated by the magnesium data, the break point (if any) is at higher [O/H] (about -1). Anyway, the existence of a primary component in the early stages of the evolution of the Galaxy seems mandatory, irrespective of the abundance data used, and this is a strong conclusion. Definitively, a primary component is required to fulfill the energetic constraint. What is the nature of this primary component? In our opinion, once again, it is distinct from GCR (assumed by Ramaty et al 2000 to originate from SBs), since, as said previously, we do not think that it is possible to identify SAP and GCR, as Higdon et al (1998) did, due to their different inferred composition.

Thus, we support the view that a low energy component is at work, complementing GCR to produce its lot of LiBeB, specifically in the halo phase. Is this component still active? Nuclear gamma ray line astronomy will say. Thus our hopes are related to the INTEGRAL satellite. Our conclusion is that an hybrid model combining both a primary and a secondary components reconcile the abundance observations and energy requirement. The two components dominate sequentially. Of course the primary component, related to SAP would be overwhelming in the halo phase and is afterwards dominated by the secondary one, which constitutes the standard GCR. The SAP component plays a major role in the LiBeB production when the galactic gas is almost devoid of medium elements i.e. in the early Galaxy. Now, what is the ultimate reservoir of GCR? Grains in the ISM or stellar surfaces? This point is left to a future discusssion.

5. Conclusion

Ambiguities on abundance data preclude definitive conclusion: is O/Fe flat or not at low Z? For the time being, the answer to this question depends on the observer to whom it is posed. Hopefully the debate will clarify in the next years. In the mean time we propose to rely on magnesium which is used as a secure metallicity index by a growing number of people. Special care should be taken to make NLTE corrections on Fe and B at low Z. A primary process probably is made necessary by energetic requirements. We endorse the view that two different components are responsible for the synthesis and evolution of LiBeB in the Galaxy. This hybrid model invoking the operation of both GCR (extracted by flares from stellar surfaces and/or grain debris in the ISM) subsequently accelerated by shock waves and a primary component of lower energy coming probably from superbubbles fulfil (or at least do not violate) all the composition constraints on i) the present cosmic radiation ii) the light element abundances in stars of all metallicities, including ^{6}Li (Vangioni-Flam et al 1999a; Fields & Olive 1999b).

What to do next? Obviously measure. One would like idealy to get simultaneously the abundances of ^{6,7}Li, Be, B, O, Mg and Fe in the same star, this for many halo members of various metallicities, which does not seems out of reach of a dedicated VLT program (Cayrel, private communication). The solution of the LiBeB riddle is definitly in the hands of observers. The best way to reveal the presence of a low energy flux of C and O in the Galaxy related supperbubbles is the observation of broad gamma ray lines arising from their excitation and deexcitation in flight. The best hope to detect them in Vela, Orion and other star forming regions is offered by the european INTEGRAL satellite to be launched around 2001.

Aknowledgements

We thank R. Cayrel and C. Furhmann for having drawn pure attention on the reliability of using magnesium as a metallicity index. We thank also Andrei Bykov and Vladimir Ptuskin for illuminating discussions on superbubbles.

References

Abia, C., Isern, J. & Canal, R. 1993, A&A, 275, 96

Balachandran, S. & Bell, R.A. 1998, Nature, 392, 791

Barring, M. 2000, Summary-Rapporteur Volume of The 26th International Cosmic ray Conference, ed. B.L. Dingus (AIP, New-York). astro-ph/9912058

Blandford, R.D. & Ostriker, J.P. 1978, ApJ, 221, L29

Bloemen, H. et al 1999, in "The Extreme Universe", 3rd INTEGRAL Workshop, Astrophysical Letters and Communications, vol. 38, 349

Boesgaard, A.M. & King, J.R. 1993, AJ, 106, 2309

Boesgaard, A.M., King, J.R., Deliyannis, C.P. & Vogt, S.S. 1999, AJ, 117, 492

Bykov, A.M. 1995, Space Sci. Rev., 74, 397

Bykov, A.M. & Fleishman, G.D. 1992, MNRAS, 15, 269

Bykov, A.M., Gustov, M.Y. & Petrenko, M.V. 2000, preprint

Cardelli, J.A. 1996, Science, 265, 209

Cassé, M. Lehoucq, R. & Vangioni-Flam, E. 1995, Nature, 373, 38

Cassé, M. & Goret, Ph. 1978, ApJ, 221, 703

Connel, J.J. & Simpson, J.A. 1997, ApJ, 475, L61

Duncan, D., Lambert, D.L. & Lemke, M. 1992, ApJ, 401, 584

Duncan, D. et al 1997, ApJ, 488, 338

Ellison, D., Drury, L.O'C. & Meyer, J.P. 1997, ApJ, 487, 197

Fields, B.D. & Olive, K.A. 1999a, ApJ, 516, 797

Fields, B.D. & Olive, K.A. 1999b, New Astron., 4, 255

Fields, B.D., Olive, K.A., Vangioni-Flam, E. & Cassé, M. 2000, astro-ph/9911320

Fuhrmann, K. 1998, A&A, 338, 161

Fulbright , J.P. & Kraft, R.P. 1999, AJ, 118, 527

Gilmore, J. et al 1992, Nature, 357, 379

Gustafsson, B. 1999, in "Chemical evolution from zero to high redshift", eds: J.R. Walsh & M.R. Rosa, Springer, 1

Hartmann, D., Myers, J. , Woosley, S., Hoffman, R. & Haxton, H. 1999, in "LiBeB, Cosmic rays, and related X-and gamma-rays", edts: Ramaty et al. ASP Conf. Ser. vol 171, 235

Hernanz, M., José, J., Coc, A. & Isern, J. 1996, ApJ, 465, L27

Higdon, B., Lingenfelter, R.E. & Ramaty, R. 1998, ApJ, 309, L33

Howk, J.C., Savage, B.D. & Fabian, D. 1999, ApJ, 525, 253

Israelian, G., Garcia-Lopez, R.J. & Rebolo, R. 1998, ApJ, 507, 805

Jenkins, E.J. et al 1998, ApJ, 492, L147

Jones, F.C. & Ellison, D.C. 1991, Sp. Sci. Rev., 58, 259
Klepach, E.G., Psuthskin, V.S. & Zirakashvili, V.N. 1999, Astroparticle Physics, in press
Kiselman, D. 1999, in"LiBeB Cosmic Rays, and Related X-Ray and Gamma-rays", Edts Ramaty et al, ASP Conf. Ser. vol 171, 85
McWilliam, A. 1997, ARA&A, 35, 503
Meneguzzi, M., Audouze, J. & Reeves, H. 1971, A&A, 15, 337
Meyer, J.P. 1985, ApJS, 57, 173
Meyer, J.P. 1996, in "The Sun and beyond", 2nd Rencontre du Vietnam, Eds: Tran Than Van et al, Edts frontières, 27
Meyer, J.P., Drury, L.O'C. & Ellison, D.C. 1997, ApJ, 488, 730
Meyer, J.P. 2000, ISSI Workshop, Kluwer Academics Publishers, in press
Pagel, B.E.J & Tautvaisiene, G. 1995, MNRAS, 276, 505
Pagel, B.E.J. 1999, astro-ph/9911204
Parizot, E. & Drury, L.O'C. 1999, A&A, 346, 339
Primas, F., Duncan, D.K., Peterson, R.C. & Thorburn, J.A. 1999, A&A, 313, 545
Primas, F. 2000, this conference
Ramaty, R., Kozlovsky, B. & Lingenfelter, R.E. 1996, ApJ, 456, 525
Ramaty, R., Kovlowsky, B., Lingenfelter, R.E. & Reeves, H. 1997, ApJ, 488, 730
Ramaty, R., Vangioni-Flam, E., Cassé, M. & Olive, K.A. 1999, Proceedings "LiBeB, Cosmic Rays, and Related X-and Gamma-Rays, ASP Conf. Ser., vol. 171
Ramaty, R., Scully, S., Lingenfelter, R.E. & Kozlowsky, B. 2000, astro-ph/9909021, ApJ, in press
Read, S. & Viola, R. 1985, Atomic Data Nucl. Data tables, 31, 359
Rebolo, R., Molaro, P. & Beckman, J.E. 1988, A&A 192, 192
Rcctz, J. 2000, in "Galaxy Evolution: Connecting the distant Universe with Local Fossil Record", Eds M. Spite and N. Crifo, Astroph. Spc. Sci., in press
Reeves, H. 1994, Rev. Mod. Phys., 66, 193
Ryan, S.G., Norris, I., Bessel, M. & Deliyannis, C. 1994, ApJ, 388, 184
Savage, B.D. & Sembach, K.R. 1996, ARA&A, 34, 279
Shapiro, M. 1999, in "LiBeB, Cosmic rays, and related X-and Gamma-Rays, Edts Ramaty el al, ASP Conf. Ser. vol. 171, 138
Shigeyama, T. & Tsujimoto, T. 1998, ApJ, 507, L139
Silberberg, R. & Tsao, C.H. 1990, ApJ, 352, L49
Silberberg, R. Tsao, C.H. & Barghouty, A.F. 2000, in the 26th International Cosmic Ray Conference eds B.L. Dingus AIP (New York), OG.3.1.06, in press
Stone, E. et al 1998, Spc. Sci. Rev., 96, 285
Strong, A. & Moskalenko, I. 1999, in "LiBeB, Cosmic rays, and related X-and Gamma-Rays, Edts Ramaty el al, ASP Conf. Ser. vol. 171, 162
Thomas, D., Greggio, L. & Bender, R. 1998, MNRAS, 296, 119
Thielemann, F.K., Nomoto, K. & Hashimoto, M. 1996, ApJ, 460, 408

Umeda, H., Nomoto, K. & Nakamura, W. 2000, to appear in the proceedings of the MPA/ESO conference "The first stars" eds A.Wess et al, Springer , astro-ph/9912248

Vangioni-Flam, E., Cassé, M., Fields, B. & Olive, K. 1996, ApJ, 468, 199

Vangioni-Flam, E., Cassé, M. & Ramaty, R. 1997, in " The Transparent Universe", 2nd INTEGRAL Workshop, ESA, SP382, 123

Vangioni-Flam E., Ramaty, R., Olive, K.A. & Cassé, M. 1998, A&A, 337, 714

Vangioni-Flam, E. Cassé, M., Cayrel, R., Audouze, J., Spite, M. & Spite F. 1999a, New Astronomy, vol. 4, no 4, 245

Vangioni-Flam, E., Ramaty, R., Cassé, M. & Olive, K.A. 1999b, in "LiBeB, Cosmic-Ray and related X-and gamma-rays", Edts E. Ramaty et al, ASP Conf. Ser. vol. 171, 268

Vangioni-Flam, E. & Cassé, M. 2000a, in "Galaxy Evolution: Connecting the distant Universe with the local fossil Record" Edts M. Spite, F. Crifo, Ap&SS, in press

Vangioni-Flam, E. Cassé, & Audouze, J. 2000b, Physics Report, in press

Webber, W.R., Kish, J.C. & Schrier, D.A. 1990a, Phys. Rev. C 41, 520

Webber, W.R., Kish, J.C. & Schrier, D.A. 1990b, Phys. Rev. C 41, 547

Wiedenbeck, M.E. et al 1999, ApJ, 523, L61

Woosley, S.E., & Weaver, T.A. 1995, APJS, 101, 181

LiBeB Evolution: Three Models

Reuven Ramaty

NASA/GSFC, Greenbelt, MD 20771, USA

Richard E. Lingenfelter

CASS/UCSD, LaJolla, CA 92093, USA

Benzion Kozlovsky

Tel Aviv University, Israel

Abstract.
We consider the three principal LiBeB evolutionary models, CRI in which the cosmic-ray source at all epochs of Galactic evolution is the average ISM, CRI+LECR in which metal enriched low energy cosmic rays (LECRs) are superimposed onto the CRI cosmic rays, and CRS in which the cosmic-ray source, accelerated in superbubbles, is constant, independent of the ISM metallicity. By considering the evolutionary trend of log(Be/H) vs. both [Fe/H] and [O/H], we demonstrate that the CRI model is energetically untenable. We present evolutionary trends for $^{11}B/^{10}B$ and B/Be which, combined with future precision measurements, could distinguish between the CRS and CRI+LECR models. We show that delayed LiBeB synthesis in the CRS model, due to the transport of the cosmic rays, could explain why log(Be/H) is steeper vs. [O/H] than vs. [Fe/H]. We also show that delayed deposition of Fe into star forming regions, due to its incorporation into high velocity dust, could provide an explanation for the possible rise of [O/Fe] with decreasing [Fe/H]. Observations of refractory and volatile α-elements could test this scenario. There seems to be a need for pregalactic or extragalactic ^{6}Li sources.

1. Introduction

Cosmic-ray driven nucleosynthesis has been known to be important for the origin of the light elements Li, Be and B (LiBeB) for three decades (Reeves, Fowler, & Hoyle 1970). But constraints on the relevant evolutionary models could only be obtained after LiBeB abundances of low metallicity stars started to become available (e.g. Ryan et al. 1990). The principal three models currently considered are: (i) the cosmic-ray interstellar model (hereafter CRI), in which the cosmic-ray source composition at all epochs of Galactic evolution is assumed to be similar to that of the average ISM at that epoch (Vangioni-Flam et al. 1990; Fields & Olive 1999); (ii) the CRI+LECR model, in which metal enriched low energy cosmic rays (LECRs) are superimposed onto the CRI cosmic rays

(Cassé, Lehoucq, & Vangioni-Flam 1995; Vangioni-Flam et al. 1996; Ramaty, Kozlovsky, & Lingenfelter 1996); and (iii) the cosmic-ray superbubble model (hereafter CRS), in which the cosmic-ray source composition is taken to be constant, independent of the ISM metallicity (Ramaty et al. 1997;2000; Higdon, Lingenfelter, Ramaty 1998). Both the CRS cosmic rays and the LECRs are thought to be accelerated out of supernova enriched matter in superbubbles.

Because of the excess of the observed Be abundances in low metallicity stars over the predictions of the CRI model, and motivated by reports of the detection of C and O nuclear gamma-ray lines from the Orion star formation region (Bloemen et al. 1994), LECRs, enriched in C and O relative to protons and α particles, were superimposed on the CRI cosmic rays, hence the CRI+LECR model. These LECRs, with maximum energies not exceeding about 100 MeV/nucleon, were thought (e.g. Ramaty 1996) to be responsible for the gamma rays reported from Orion. It was suggested that such enriched LECRs might be accelerated out of metal-rich winds of massive stars and supernova ejecta (Bykov & Bloemen 1994; Ramaty et al. 1996; Parizot, Cassé, & Vangioni-Flam 1997) by an ensemble of shocks in superbubbles (Bykov & Fleishman 1992; Parizot Cassé, & Vangioni-Flam 1997). The Orion gamma-ray data, however, have been retracted (Bloemen et al. 1999). Nonetheless, as the possible existence of the postulated LECRs remains, new gamma-ray line data are needed to determine the role of LECRs in LiBeB production.

Recent O abundance data, which suggest that [O/Fe] increases with decreasing [Fe/H] at low metallicities (Israelian et al. 1998; Boesgaard et al. 1999), led Fields and Olive (1999) to reexamine the viability of the CRI model. More recent measurements (Fulbright & Kraft 1999; Westin et al. 1999) argue against such an [O/Fe] increase. But as demonstrated in (Ramaty et al. 2000), and also shown below, this model is inconsistent with cosmic-ray energetics, an [O/Fe] increase notwithstanding.

Alternatively, it was suggested (Lingenfelter, Ramaty, & Kozlovsky 1998; Higdon et al. 1998), that the Be evolution can be best understood in the CRS model, in which the cosmic-ray metals at all epochs of Galactic evolution are accelerated predominantly out of supernova ejecta. Lingenfelter et al. (1998) and Lingenfelter & Ramaty (1999) showed that the arguments (e.g. Meyer, Drury, & Ellison 1997) against the supernova ejecta origin of the current epoch cosmic rays can be answered, and Higdon et al. (1998) and Higdon, Lingenfelter, & Ramaty (1999) showed that the most likely scenario is collective acceleration by successive supernova shocks of ejecta-enriched matter in the interiors of superbubbles. This scenario is consistent with the delay between nucleosynthesis and acceleration (time scales $\sim 10^5$ yr), suggested by the ^{59}Co and ^{59}Ni observations (Wiedenbeck et al. 1999). In both the CRS and CRI+LECR models, the bulk of the Be in the early Galaxy is produced by accelerated C and O interacting with ambient H and He. That these "inverse reactions" are dominant in the early Galaxy was first suggested by Duncan, Lambert, & Lemke (1992).

In the present paper we present results from a complete set of LiBeB evolutionary calculations for all three models using our production code described in Ramaty et al. (1997) and evolutionary code detailed in Ramaty et al. (2000).

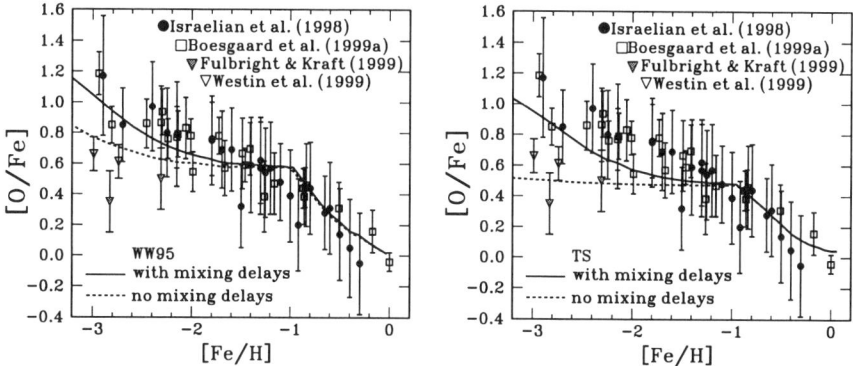

Figure 1. Galactic evolution of [O/Fe] vs. [Fe/H] for supernova O and Fe yields taken from minimum Fe yield models of Woosley & Weaver (1995, WW95 shown in the left panel) and Tsujimoto & Shigeyama (1998, TS shown in the right panel), assuming zero and finite Fe and O mixing delay times, $\tau_{Fe}(\text{mix}) = 30$Myr and $\tau_O(\text{mix}) = 1$Myr, and including contributions from Type Ia supernovae. The characteristic halo and disk infall times are 10Myr and 5Gyr, respectively, the star formation rate coefficient is 0.5Gyr^{-1}, and the ratio of halo-to-disk masses is 0.1. (See Ramaty et al. 2000 for more details).

2. Analysis

Figure 1 shows the evolution of [O/Fe] as a function of [Fe/H]. In order to account for the possible rise of [O/Fe] with decreasing [Fe/H], we introduced mixing delays, i. e. the delayed deposition of the synthesized products into the star forming regions due to differences in transport and mixing. We choose a short mixing time for oxygen because we expect the bulk of the O and other volatiles in the ejecta to mix with the ISM after the remnant slows down to local sound speeds. But we consider longer mixing times for Fe, assuming that the bulk of the ejected Fe is incorporated into high velocity refractory dust grains which continue moving for longer periods of time before they stop and can be incorporated into newly forming stars. The incorporation of a large fraction of the synthesized Fe into dust grains is supported by observations of both supernova 1987A and the Galactic 1.809 MeV gamma-ray line resulting from the decay of ^{26}Al (for more details and references see Ramaty et al. 2000). We see that with the mixing delays (solid curves in Figure 1) both the WW95 and TS cases (see figure caption) become consistent with the Israelian et al. (1998) and Boesgaard et al. (1999a) data, showing that delayed Fe deposition could indeed be the cause for the rise of [O/Fe]. In this connection, it is interesting to note that, unlike [O/Fe], the abundance ratios of the α-nuclei Mg, Si, Ca and Ti relative to Fe do not increase with decreasing [Fe/H] below [Fe/H] = −1 (Ryan, Norris, & Beers 1996). This may be consistent with the fact that these elements are also refractory, and thus are affected by mixing in the same way as is Fe. A test may be provided by sulfur, which is volatile, and thus should show

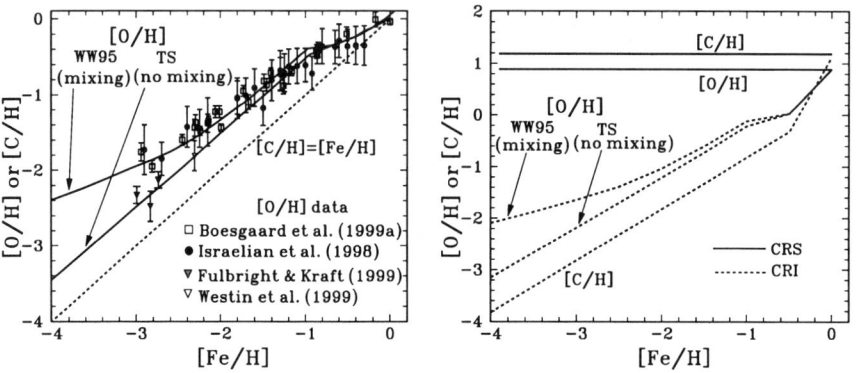

Figure 2. The employed ISM (left panel) and cosmic-ray source (right panel) C and O abundances as functions of [Fe/H]. For the ISM, [O/H] is obtained from the evolutionary calculation for the two limiting cases, WW95 with mixing delays and TS with no mixing delays (Figure 1). [C/Fe] is taken constant. For the cosmic rays the abundances are normalized to the solar values, i.e. $[X/Fe] \equiv \log(X/Fe) - \log(X/Fe)_\odot$ where X stands for C or O. For the CRS model [C/H] and [O/H] are independent of [Fe/H]. For the CRI model, [C/H] and [O/H] are based on the ISM values with enhancement factors consistent with shock acceleration theory (see text), except at [Fe/H]=0 where the cosmic ray CRS and CRI values are equal.

a rise similar to the rise of [O/Fe] vs. decreasing [Fe/H] (G. Israelian, private communication, 1999).

Figure 2 shows the employed C and O abundances of the ISM and the cosmic ray source. For the ISM (left panel) we take [C/H]≃[Fe/H], because at early times both the C and Fe come primarily from core collapse supernovae of massive stars, while at later times the increased C contribution from the winds of intermediate mass stars is compensated by the Fe contribution from the thermonuclear supernovae of the white dwarf remnants of such stars (see Timmes, Woosley & Weaver 1995). The O abundances follows from the results of evolutionary calculations shown in Figure 1. Unlike in Ramaty et al. (2000), where we used a constant He abundances, here we allow He/H (by number) to vary slowly from 0.08 at very low metallicities to 0.1 at [Fe/H]=0. For the cosmic-ray source (right panel), we define the logarithmic ratios [C/H] and [O/H] in the same way as is done for the corresponding ISM values, including normalization to solar (not current epoch cosmic ray source) abundances. As the CRS cosmic rays are accelerated primarily out of supernova ejecta enriched superbubbles, [C/H] and [O/H] are constant, set equal to current epoch cosmic-ray values. For the CRI model we scale [C/H] and [O/H] to the ISM values with enhancement factors of 1.5 and 2, consistent with the mass-to-charge dependent acceleration of volatiles (Ellison, Drury & Meyer 1997), except at [Fe/H]=0 where the cosmic ray CRS and CRI values are equal. For details on the rest of the employed cosmic-ray sources abundances see Ramaty et al. (2000). For the LECRs we adopt the CRS abundances. The CRS, CRI and LECR source energy spectra are

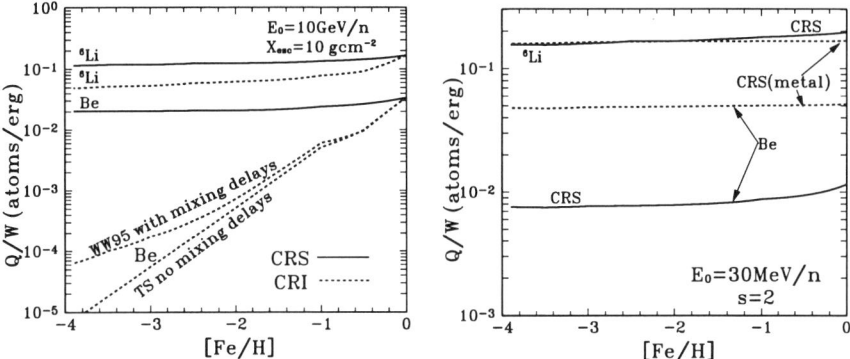

Figure 3. Number of Be and ^{6}Li atoms produced per unit cosmic-ray energy. Left panel – for the CRS and CRI models using a cosmic-ray source momentum power law spectral index of 2.5. For the CRI Be the two limiting mixing cases are shown. For ^{6}Li in both the CRS and CRI models, and for Be in the CRS model, the mixing cases do not affect Q/W. Right panel – for the LECR model with a turnover energy of 30 MeV/nucleon, spectral index of 2, CRS and CRS(metal) abundances, where the latter is identical to CRS, except that the proton and α particle abundances are set to zero. For these abundances the mixing cases have no effect.

power laws in momentum with high energy exponential cutoffs (characteristic energy E_0), which we set to an ultrarelativistic value for the CRS and CRI cosmic rays and to 30 MeV/nucleon for the LECRs.

Figure 3 shows the resultant Q/W's, the total number of nuclei Q produced by an accelerated particle distribution normalized to the integral cosmic-ray energy W, for a given source energy spectrum, and cosmic ray and ambient medium compositions as described above. We make the reasonable assumption that the accelerated particle source energy spectrum is independent of [Fe/H]. The resultant CRS and CRI Q/W's for Be and ^{6}Li (left panel) are for $X_{esc} = 10$ gcm^{-2}, typical of currently inferred values for "leaky box" cosmic-ray propagation models. While $\alpha\alpha$ dominated $Q(^6Li)/W$ is not very different for the CRS and CRI models, $Q(Be)/W$ is drastically different for the two models, reflecting the fact that efficient Be production in the early Galaxy can only result from C and O enriched accelerated particles. The different O abundances employed in the calculation of Q/W for the two mixing delays cases (see Figure 2) lead to significantly different $Q(Be)/W$'s for CRI model. The LECR Q/W's (right panel) show that, while removal of the protons and α particles (the CRS(metal) composition) significantly increases Q/W for Be, it essentially leaves $Q(^6Li)/W$ unchanged, because the lack of ^{6}Li production by α particles is compensated by a smaller W due to the absence of the protons and alphas.

Figure 4 shows the Be evolution for the CRS and CRI models. In the calculations, 10^{50} erg per supernova are imparted to the cosmic rays, a value in very good agreement with current epoch cosmic-ray energetics. We note that even though the overall slope of log(Be/H) vs. [Fe/H] is practically unity, while

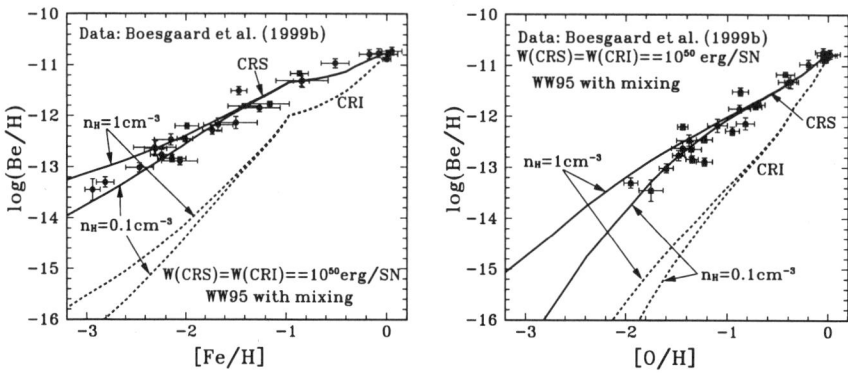

Figure 4. Be abundance evolution for the CRS and CRI models, as a function of [Fe/H] (left panel) and [O/H] (right panel). n_H is the density of ambient hydrogen which influences the propagation of the cosmic rays and determines the delay between the supernova explosion and the deposition of the synthesized Be. For both the CRS and CRI model, 10^{50} ergs per supernova are imparted to the cosmic rays. Results for only WW95 with mixing delays are shown.

that of log(Be/H) vs. [O/H] is significantly steeper (0.96±0.04 and 1.45±0.04, respectively, Boesgaard et al. 1999b), the CRS model provides a good fit to these evolutionary trends, particularly if n_H is near 0.1. Such a low value might not be unreasonable for an average halo hydrogen density if $10^{10} M_\odot$ are spread over a few kpc^3. The calculated log(Be/H) vs. [Fe/H] is flatter than log(Be/H) vs. [O/H] because the delayed deposition of the synthesized Be, caused by the low n_H, is compensated by the delayed Fe deposition, due to the incorporation of Fe in high velocity dust, but not compensated by the very short delay of the deposition of O, which is mostly volatile. As in Ramaty et al. (2000), we see that the CRI model, normalized to a reasonable energy in cosmic rays per supernova, severely underproduces the measured Be abundances. However, unlike in that paper where we showed the result only for log(Be/H) vs. [Fe/H], here we show that the same result also holds for log(Be/H) vs. [O/H]. This removes the remaining ambiguity concerning our argument against the result of Fields & Olive (1999), who claimed that the CRI model would be viable if instead of Fe ejecta per supernova based on calculations, which are somewhat uncertain, they used values based on their fit to the increasing [O/Fe] with decreasing [Fe/H]. As there is no such uncertainty concerning the O ejected masses, our present result unequivocally demonstrates that the CRI model is untenable.

Figure 5 shows the Be evolution for the CRI+LECR model. Here, as before, 10^{50} erg per supernova are imparted to the CRI cosmic rays, but in order to achieve a fit to the log(Be/H) vs. [Fe/H] data, we had to add more energy (1.5×10^{50} erg per supernova) in LECRs. The need for this increased energy can be seen in Figure 3, where the $Q(Be)/W$ for the LECR model with the CRS composition (right panel) is lower by about a factor of 2 than the corresponding value for the CRS model (left panel). Returning to Figure 5, we see that the CRI+LECR model leads to a log(Be/H) vs. [O/H] evolutionary curve with a

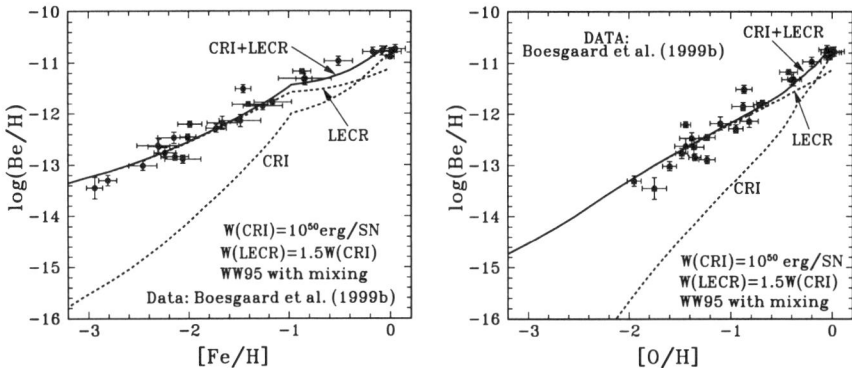

Figure 5. Be abundance evolution for the CRI+LECR model with a turnover energy of 30 MeV/nucleon, as a function of [Fe/H] (left panel) and [O/H] (right panel). LECRs slow down much faster than higher energy cosmic rays, hence there is no delay between the supernova explosion and the deposition of the Be. 10^{50} ergs per supernova are imparted to the CRI cosmic rays and 50% more to the LECRs.

slope which is only slightly steeper than 1, while, as mentioned above, the data indicate a slope of 1.45±0.04 (Boesgaard et al. 1999b). Indeed, the simplest evolutionary considerations for both the CRS and LECR models would predict that log(Be/H) vs. [O/H] should have a slope of 1. But, as shown above, the delay introduced by cosmic-ray transport with n_H=0.1 will steepen the slope. However, the LECRs slow down much faster than the higher energy CRS cosmic rays. Thus, the calculation of Figure 5 assigns no delay to the Be deposition, leading to a possible inconsistency between the predictions of the CRI+LECR model and the log(Be/H) vs. [O/H] data.

Figure 6 shows the boron isotopic ratio (left panel) and log(B/Be) (right panel) vs. Fe/H for the CRS and CRI+LECR models. In the evolutionary calculations, the production ratios $Q(^{11}B/^{10}B) = 2.4$ and $Q(B/Be) = 14$ for the CRS and CRI cosmic rays, and $Q(^{11}B/^{10}B) = 3.3$ and $Q(B/Be) = 22$ for the LECRs, were taken as independent of [Fe/H]. These numerical values are from Ramaty et al. (1997) and are valid for high energy CRS and CRI cosmic rays and LECRs with $E_0 = 30$ MeV/nucleon. In order to reproduce the meteoritic $^{11}B/^{10}B$ both the CRS and CRI+LECR models require the addition of ν-produced ^{11}B (Woosley & Weaver 1995). We take into account the metallicity dependence of this ^{11}B production, and the fact that only the core collapse supernovae produce ^{11}B by neutrinos. We find that for the CRS model the meteoritic data can be fit with f_ν=0.18, a lower value than we found in Ramaty et al. (2000) because of the lower cosmic-ray energy per supernova that we use here (10^{50} vs. 1.5×10^{50} erg that we used in that paper). The rise in $^{11}B/^{10}B$ and B/Be for $n_H = 0.1$ below [Fe/H] of about -2 in the CRS model is due to the delayed deposition of the cosmic-ray produced Be and B relative to the ν-produced ^{11}B for which we took a short delay time, the same as for O (1 Myr). For the CRI+LECR model, in which the LECRs employ a larger energy per supernova, $f_\nu = 0.2$ is required to fit the boron isotope data. The rise in $^{11}B/^{10}B$ and B/Be with decreasing

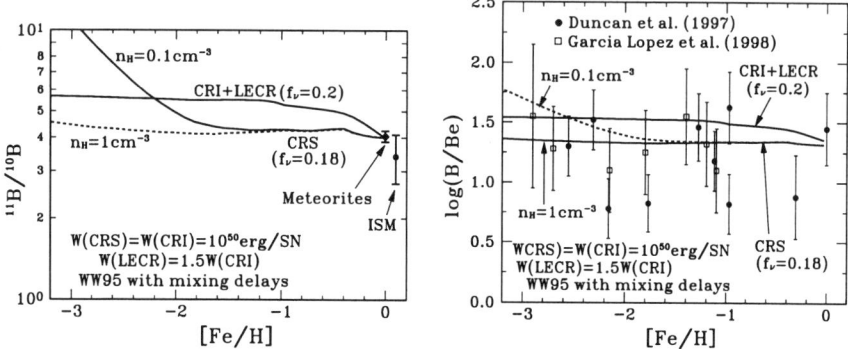

Figure 6. ^{11}B-to-^{10}B (left panel) and B-to-Be (right panel) abundance ratios vs. [Fe/H] for the CRS and CRI+LECR models. f_ν is the ratio of the employed ν-produced ^{11}B to the nominal Woosley & Weaver (1995) yields. n_H is defined in the caption of Figure 4. The B isotope data for meteorites and the ISM are from Chaussidon & Robert (1995) and Lambert et al. (1998), respectively.

[Fe/H] is mostly due to the larger value of the corresponding LECR production ratios. It is evident from Figure 6 that future high precision measurements of ^{11}B/^{10}B and B/Be as functions of [Fe/H] will distinguish between the models.

Figure 7 shows the Li evolution. The CRS model underproduces the ^{6}Li abundance for [Fe/H]< -2, suggesting the existence of pregalactic or extragalactic ^{6}Li sources. With the lower cosmic-ray normalization mentioned above, the CRS model also slightly underproduces the meteoritic ^{6}Li. With the discovery of solar flare produced ^{6}Li in the solar wind (via measurements in lunar soil, Chaussidon & Robert 1999), the possibility of some locally produced ^{6}Li in the solar system must be considered. The CRI+LECR model produces more ^{6}Li relative to Be, simply because the $\alpha\alpha$ cross section for ^{6}Li production peaks in the nonrelativistic region.

3. Discussion and Conclusions

We have summarized a complete set of O, Fe and LiBeB evolutionary calculation. We have considered the three principal evolutionary models, CRI in which the cosmic-ray source composition at all epochs of Galactic evolution is similar to that of the average ISM at that epoch, CRI+LECR in which metal enriched low energy cosmic rays (LECRs) are superimposed onto the CRI cosmic rays, and CRS in which the cosmic-ray source, accelerated in superbubbles, has a constant composition, independent of the ISM metallicity. By considering the evolutionary trend of log(Be/H) vs. both [Fe/H] and [O/H], we demonstrated that the CRI model is energetically untenable. Although the CRI+LECR mix considered here is consistent with the Be evolution vs. Fe, a plausible scenario for producing the required mix has yet to be proposed.

For Fe, our code allows for a delay between nucleosynthesis and deposition into star forming regions due to the incorporation of the synthesized Fe into

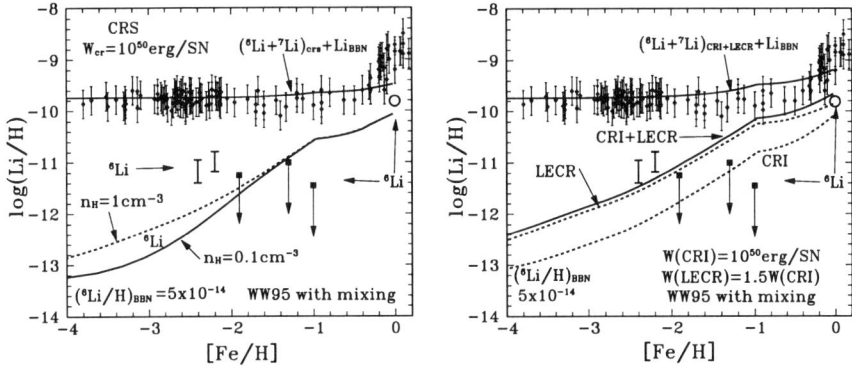

Figure 7. The evolution of the total Li and ^{6}Li for the CRS (left panel) and the CRI+LECR models (right panel). The ^{6}Li data at [Fe/H] of -2.4, -2.2, -1.9, -1.3 and -1 are from the summary of Hobbs (1999), and the meteoritic value at [Fe/H]=0 is from Grevesse et al. (1996). The total Li data is from a compilation by M. Lemoine (private communication 1997). n_H is defined in the caption of Figure 4. We took the primordial ^{7}Li/H=1.8×10^{-10} (Molaro 1999).

high velocity dust. This delay could provide an explanation for the possible rise of [O/Fe] with decreasing [Fe/H] indicated by some of the data. A test for this scenario would be the demonstration that the abundances of refractory α-elements Mg, Si, Ca and Ti relative to Fe do not increase with decreasing [Fe/H] below [Fe/H]= -1, but that volatile sulfur does rise.

For the LiBeB there is also a delay. Due to the transport of the cosmic rays, LiBeB synthesis lags behind the explosion of the supernova responsible for accelerating the cosmic rays by as much a hundred million years, depending on the average gas density in the halo of the early Galaxy. We show that this delay, combined with the delayed Fe deposition, could provide an explanation for the steeper evolutionary trend of log(Be/H) vs. [O/H] than vs. [Fe/H], as indicated by the recent data of Boesgaard et al. (1999b).

The trends of ^{11}B/^{10}B and B/Be vs. [Fe/H] show structure resulting from the above mentioned delays, as well as from the hybrid nature of the CRI+LECR model. Future observations of these ratios may distinguish between the models. The fact that the ^{6}Li abundances at [Fe/H] below -2 seem inconsistent with all the models suggests the existence of pregalactic or extragalactic ^{6}Li sources.

References

Bloemen, H. et al. 1994, A&A, 281, L5

Bloemen, H. et al. 1999, ApJ, 521, L137

Boesgaard, A., King, J., Deliyannis, C., & Vogt, S. 1999a, AJ, 117, 492

Boesgaard, A., Deliyannis, C., King, J., Ryan, S., Vogt, S., & Beers, T. 1999b, AJ, 117, 1548

Bykov, A., & Bloemen, H. 1994, A&A, 283, L1
Bykov, A. M., & Fleishman, G. D. 1992, MNRAS, 15, 269
Cassé, M., Lehoucq, R., & Vangioni-Flam, E. 1995, Nature, 373, 318
Chaussidon, M., & Robert, F. 1995, Nature, 374, 337
Chaussidon, M., & Robert, F. 1999, Nature, 402, 270
Duncan, D. K., Lambert, D. L., & Lemke, M., 1992, ApJ, 401, 584
Duncan, D. K. et al. 1997, ApJ, 488, 338
Ellison, D. C., Drury, L.O'C., & Meyer, J-P. 1997, ApJ, 487, 197
Fields, B. D., & Olive, K. A. 1999, ApJ, 516, 797
Fulbright, J. P, & Kraft, R. P. 1999, AJ, 118, 527
Garcia Lopez, R. J. et al. 1998, ApJ, 500, 241
Grevesse, N., Noels, A., & Sauval, A. J. 1996, in: Cosmic Abundances, eds. S. S. Holt, & G. Sonneborn, ASP Conf. Ser., 99, (San Francisco: ASP), 117
Higdon, H. C., Lingenfelter, R. E., & Ramaty, R., 1998, ApJ, 509, L33
Higdon, H. C., Lingenfelter, R. E., & Ramaty, R., 1999, 26th Internat. Cosmic Ray Conf., eds. D. Kieda et al. (Salt Lake City), 4, 144
Hobbs, L. M. 1999, in: LiBeB, Cosmic Rays, and Related X- and Gamma-Rays, eds. R. Ramaty et al., ASP Conf. Ser., 171, (San Francisco: ASP), 23
Israelian, G., Garcia Lopez, R. J., & Rebolo, R. 1998, ApJ, 507, 805
Lambert, D. L. et al. 1998, ApJ, 494, 614
Lingenfelter, R. E., & Ramaty, R., 1999, 26th Internat. Cosmic Ray Conf., eds. D. Kieda et al. (Salt Lake City), 4, 148
Lingenfelter, R. E., Ramaty, R., & Kozlovsky, B. 1998, ApJ, 500, L153
Meyer, J-P., Drury, L. O'C., & Ellison, D. C. 1997, ApJ, 487, 182
Molaro, P. 1999, in: LiBeB, Cosmic Rays, and Related X- and Gamma-Rays, eds. R. Ramaty et al. , ASP Conf. Ser., 171, (San Francisco: ASP), 6
Parizot, E. M. G., Cassé, M., & Vangioni-Flam, E. 1997, A&A, 328,107
Ramaty, R. 1996, A&A(Suppl.), 120, C373
Ramaty, R., Kozlovsky, B., & Lingenfelter, R. E. 1996, ApJ, 456, 525
Ramaty, R., Kozlovsky, B., Lingenfelter, R., & Reeves, H. 1997, ApJ, 488, 730
Ramaty, R., Scully, S., Lingenfelter, R., & Kozlovsky, B. 2000, ApJ, in press
Reeves, H., Fowler, W. A., & Hoyle, F. 1970, Nature, 226, 727
Ryan, S. G., Norris, J. E., & Beers, T. C. 1996, ApJ, 471, 254
Ryan, S., Bessell, M., Sutherland, R., & Norris, J. 1990, ApJ, 348, L57
Timmes, F. X., Woosley, S. E., & Weaver, T. A., 1995. ApJ(suppl), 98, 617
Tsujimoto, T., & Shigeyama, T. 1998, ApJ, 508, L151
Vangioni-Flam E., Cassé, M., Audouze, J., & Oberto, Y. 1990, ApJ, 364, 568
Vangioni-Flam, E., Cassé, M., Fields, B., & Olive, K. 1996, A&A, 468, 199
Westin, J. et al. 1999, ApJ, in press (astro-ph/9910376)
Wiedenbeck, M. E. et al. 1999, ApJ(Letters), 523, L61
Woosley, S. E. & Weaver, T. A. 1995, ApJS, 101, 181

Sinks of Light Elements in Stars – Part I

Constantine P. Deliyannis

Indiana University, Astronomy Department, 319 Swain Hall West, 727 E. 3rd Street, Bloomington, IN 47405-7105, USA

Marc H. Pinsonneault

The Ohio State University, Department of Astronomy, 140 W. 18th Ave. Columbus, Ohio 43210, USA

Corinne Charbonnel

Laboratoire d'Astrophysique de l'Observatoire Midi-Pyrénées, CNRS-UMR 5572, 14, av.E.Belin, F-31400 Toulouse, France

Abstract. The fragile light elements lithium, beryllium, and boron (Li, Be, B) are easily destroyed in stellar interiors, and are thus superb probes of physical processes occuring in the outer stellar layers. The light elements are also excellent tracers of the chemical evolution of the Galaxy, and can test big bang nucleosynthesis (BBN). These inter-related topics are reviewed with an emphasis on stellar physics.

In Part I (presented by CPD), an overview is given of the physical processes which can modify the surface abundances of the light elements, with emphasis on Population I dwarfs - convection; gravitational settling, thermal diffusion, and radiative levitation; slow mixing induced by gravity waves or rotation. We will discuss the increasingly large body of data which begin to enable us to discern the relative importance of these mechanisms in Population I main sequence stars. In Part II (presented by MHP), discussion is extended to the issue of whether or not the halo Li plateau is depleted, and includes the following topics: Li dispersion in field and globular cluster stars, Li production vs. destruction in Li-rich halo stars, and constraints from ^{6}Li. Also discussed are trends with metal abundance and T_{eff} and implications for chemical evolution and BBN. In Part III (presented by CC), evidence is reviewed that suggests that in situ mixing occurs in evolved low mass Population I and Population II stars. Theoretical mechanisms that can create such mixing are discussed, as well as their implications in stellar yields.

1. Introduction

In this three-part review we cover some of the most basic and important current topics of the subject; however, even in such an expanded review, it is not possible to do justice to all interesting topics. Part I complements Deliyannis (2000);

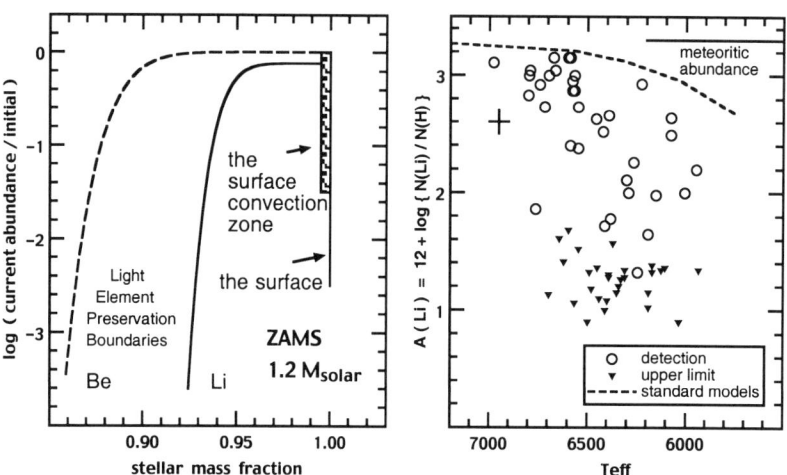

Figure 1. Left: The Li and Be preservation regions in a zams F-star. Right: Predictions of the classical theory compared to field star data. It is clear that additional physics is occuring inside these stars.

other recent reviews and points of view include Jeffries (2000), Pasquini (2000, these proceedings), Cayrel (1999), Martin (1997), and Pinsonneault (1997).

The light element tracers lithium, beryllium, and boron (hereafter Li, Be, and B) are excellent probes of physical processes occuring in stellar interiors, the chemical evolution of our Galaxy, and BBN. Within standard BBN, knowledge of the primordial abundances of these light elements (and also deuterium and the two stable helium isotopes) would constrain the Universal baryonic density (see Part II), the nature of dark matter, and ultimately, perhaps even some of the fundamental laws of nature. Delineation of the light element evolution with metal abundance can teach us about processes related to the general chemical evolution of the Galaxy, such as halo mass outflow and disk mass inflow, history of the cosmic ray spectrum, and the initial mass function. In stars, it is mainly the fragility of the light elements that makes them excellent tracers of physical processes occuring in stellar interiors. In Part I we focus on what Li, Be, and B tell us about Population I dwarfs and stars near the turnoff.

2. The Observed Lithium Morphology

The classical theory of stellar evolution, which assumes spherical symmetry and ignores possible effects due to rotation, diffusion, mass loss, magnetic fields, and so on, fails to explain nearly all observed Population I Li, Be, and B abundance patterns. Perhaps the most striking example is the Li gap in F stars, as illustrated in Figure 1. Inside stars, Li, Be, and B are destroyed by energetic protons

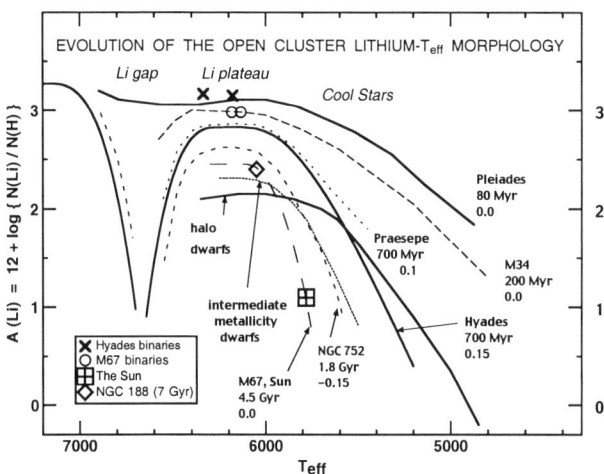

Figure 2. The main sequence **Lithium Morphology** of open clusters of different ages and composition, as indicated, of intermediate metallicity field stars, and of halo field stars.

at temperatures of only a few million K, and thus survive only in the outermost layers. The left panel of Figure 1 shows the ^{7}Li and ^{9}Be profiles in a classical model of an F star that has arrived on the ZAMS (adapted from Deliyannis & Pinsonneault 1997). ^{9}Be is sturdier than ^{7}Li, and thus survives to about twice the depth; the B isotopes are even sturdier and survive to a bit over twice the Be-survival depth; and ^{6}Li is the most fragile of these isotopes and survives to about only half the ^{7}Li- survival depth. The surface convection zone (hereafter, SCZ) is much shallower than any of the light element preservation regions, and so it is impossible to deplete the surface light element abundances during the main sequence (see §3; the small amount of ^{7}Li depletion shown was a result of pre-main sequence destruction). Yet, in stark contrast to the strictures of the classical theory, the vast majority of field F and early G stars deplete their surface Li abundances, often quite severely (right panel of Figure 1; data from Boesgaard & Tripicco 1986b). Clearly, interesting physics occurs in stars that is not included in the classical model.

An increasingly large body of data can help ascertain what the nature of this physics might be. Figure 2 (adapted from Deliyannis et al. 1998b) shows the Lithium Morphology (the mean trend of Li vs. T_{eff}) in open clusters of different ages, in intermediate metallicity field stars, and in halo field stars. This figure encapsulates a tremendous amount of work carried out over the past 15 years. A partial list includes the following. Pleiades: Pilachowski et al. (1987), Boesgaard et al. (1988), and Soderblom et al. (1993a); M34: Jones et al. (1997); Hyades: Thorburn et al. (1993), and Boesgaard & Tripicco (1986a); Praesepe: Soderblom et al. (1993b); NGC 752 and M67: *preliminary* analysis of the lead author's WIYN/Hydra data, for M67 see also Pasquini et al. (1997); intermediate metallicity and halo: see Ryan & Deliyannis (1995). A more complete list of Li cluster work can be found in the excellent review by Jeffries (2000). The

WIYN Open Cluster Study (Mathieu 2000) and several other groups are busily working at increasing the amount of information contained in this figure. It is clear that there are advantages to working with cluster data over working with field star data alone: whereas field stars have different compositions, ages, and evolutionary states, and probably as a result exhibit rather chaotic abundances (Figure 1), by contrast, a sample of cluster stars have the same (knowable) composition and age, and evolutionary state can be determined precisely. Thus, one can hope to find a well-defined pattern in cluster stars, as is indeed illustrated by Figure 2. Furthermore, one can study the evolution of the Lithium Morphology by studying clusters of different ages, and the dependence of the Lithium Morphology on composition. It should be noted that only the main trends for each cluster are shown here; in addition, there is evidence that spreads of Li exist at a given T_{eff}. It has been argued that part of the spread (in the Pleiades, particularly) may be due to atmospheric effects, as opposed to real abundance differences (King et al. 1999), but evidence remains that at least some of the spread is intrinsic.

On the basis of the empirical content of Figure 2, Deliyannis (2000) proposed that mass (more or less T_{eff}) is the first parameter that governs Li depletion in solar-type stars. Essentially, this reflects the striking Lithium Morphology exhibited in any given cluster. It was proposed that the second parameter is age, reflecting the apparent evolution of the Lithium Morphology, particularly main sequence Li depletion. Third parameters might include metallicity, initial rotation rates, etc. Theory (below) predicts metallicity-dependence of the Li depletion, but observations have not yet clearly established such a connection. However, some additional parameter may be required to explain the observed spreads of Li at a given T_{eff}. Three main regions are evident in Figure 2: the Li Gap, the Li Plateau, and the Cool Stars. The challenge for theory is to explain these various empirical Li trends for dwarfs, as well as Li turnoff-subgiant observations (§5; see also Part III), and Be and B data (§6). We consider each of these regions in more detail in §4; but first, we introduce various classes of light element depletion mechanisms in §3.

3. Classical Models and Possible Missing Physics

We argue that classical models fail in that they do not include certain physics that are clearly operating in real stars. The challenge is to identify, understand and properly describe this physics. Nonetheless, the classical theory provides the basic background. We thus review the fundamental aspects of Li depletion in classical models (see Deliyannis et al. 1990 for a detailed description of halo star models; low mass disk star models behave in a very similar way; see also Proffitt & Michaud 1989, Pinsonneault et al. 1990, Swenson & Faulkner 1992, Forestini 1994, D'Antona & Mazzitelli 1994). In an evolutionary sequence of a one-solar mass model, it is necessary to begin evolution at the fully convective deuterium birthline. As models contract on the Hayashi track, the core becomes hot enough to destroy light elements (first ^{6}Li, then ^{7}Li, and then perhaps a tiny amount of ^{9}Be). Convective mixing depletes the surface abundance. (To generalize, *ANY* physical mechanism that affects light element abundances at the base of the SCZ is nearly instantly evident at the surface as well. For example, diffusion out of or into the base of the SCZ decreases or increases, respectively, the surface abundance.) At about the same time, a radiative core develops, but light element destruction continues at the base of the SCZ (and thus surface) until the models reach the radiative Henyey track. The SCZ continues to get shallower (and cooler) essentially throughout the rest of the pre-main sequence (pre-MS) and main sequence, and there is no more surface Li depletion throughout this time. After the main sequence turnoff, the SCZ deepens and eventually sweeps past the boundary of Li preservation that was established during the main sequence. At this point, the fixed amount of Li mixes with material that has no Li, and so the surface abundance is diluted. Note that the temperature at the base of the (now much deeper) SCZ is not high enough to destroy Li, so dilution is the only Li depletion mechanism (see Part III). These are the only two ways to deplete the surface Li abundance in a classical sequence of F and G star models from the deuterium birthline up to the RGB tip: pre-MS destruction, and subgiant (and giant) dilution. There is a strong mass dependence: a pre-MS model of a later spectral type has a deeper convection zone that is hotter at its base, and will thus destroy more Li and for a longer period of time. This is the reason for the shape of the Li isochrone in Figure 1. For sufficiently cooler spectral types, main sequence Li depletion is also a possibility (but not in published F and G dwarf models).

In perfectly stable radiative layers, atomic diffusion of a given species must take place against the hydrogen background. Types of diffusion include gravitational settling (downward), thermal diffusion (downward), and radiative acceleration (upward; see, e.g., Michaud 1986 and 2000, these proceedings). Gravitational settling results from the fact that hydrostatic equilibrium varies slightly from species to species; thermal diffusion results from the dependence of atomic motion on atomic mass; and radiative acceleration is related to the different effective cross sections that result from different stages of ionization. However, even a small amount of turbulence can make the diffusion timescales extremely long and thus effectively decrease or even erase its signatures. Thus, no diffusion takes place within SCZ's. Furthermore, diffusion becomes increasingly efficient with decreasing density, i.e., increasing radius; thus, while little diffusion is predicted and indeed inferred from helioseismology in the Sun (Bahcall & Pinsonneault 1996; Richard et al. 1996; Guenther et al. 1996; Vauclair 1998; Basu et al. 2000), more is predicted for hotter dwarfs. Diffusion clearly plays a role in creating peculiar abundances in some A stars and earlier types (Stephens & Deliyannis 1999; Michaud et al. 1976, Michaud 1970, 2000, these proceedings, and references therein).

Mass loss has been invoked as a Li depletion mechanism in low mass PopI stars (Schramm et al. 1990; Guzik et al. 1987; Hobbs et al. 1989). The main idea is that the Li preservation region (Figure 1) is lost, so the surface abundance goes down (this is different from the suggestion by Vauclair & Charbonnel 1995 for Pop II stars as discussed in Part II). In G stars the SCZ occupies a significant fraction of the Li preservation region, so a necessary complication to this scenario is the continuous dilution of Li as mass is lost. The required mass loss rates are orders of magnitude larger than what is observed in the Sun.

At least two types of slow mixing are possible. Gravity waves (Garcia Lopez & Spruit 1991, Schatzman 1993, Montalban & Schatzman 1994, Kumar & Quartart 1997, Zahn et al. 1997, Kumar et al. 1999) can cause slow mixing below the SCZ when convective motions pound against the base of the SCZ. Slow mixing can also be rotationally-induced (Pinsonneault et al. 1990; Chaboyer & Zahn 1992; Chaboyer et al. 1995; Vauclair et al. 1978, Vauclair 1988, Charbonnel et al. 1992, 1994, Charbonnel & Vauclair 1992, Zahn 1992, Maeder 1995, Talon & Zahn 1997, Talon & Charbonnel 1998, Charbonnel & Talon 1999, Brun et al. 1999) as stars losing angular momentum from their surfaces trigger instabilities in the interior that cause mixing. Two schools have been constructing detailed models that include rotational mixing of light elements, the Yale school and the French-Swiss school. Although some details differ, from a broad perspective, the two schools are much more similar than they are different (see Maeder & Meynet 2000 for an indispensable and enlightening review).

Of course, it is also possible that other types of mechanisms, not discussed here, may cause depletion of light element abundances.

4. The Cool Stars

Like the Solar Neutrino problem, the Solar Li problem has been with us for over three decades. The problem is that the Solar Li abundance lies well below the Hyades Li morphology (Figure 2), yet classical models predict no main sequence Li depletion. Other old stars near the solar T_{eff} of 5780 K, particularly 16 Cyg A and B, and α Cen A show similarly low Li abundances (King et al. 1997, and references therein). The accumulating cluster evidence (Figure 2, but also additional clusters discussed by Jeffries 2000) argues that G stars experience main sequence Li depletion: there is a good correlation between mean cluster Li and age. These data contradict one possible and often-visited solution, namely that of ad hoc convective overshoot (e.g. Straus et al. 1976). This effectively makes the SCZ deeper, allowing for more Li depletion. The problem is that convective overshoot preserves the basic time dependence of light element depletion in classical models, e.g. rapid pre-MS depletion and a sharp boundary between main sequence depletion and non-depletion; it does not permit main sequence depletion across a wide range of SCZ depths. Thus, if the solar Li is now matched, the predicted Hyades Li depletion (and that of most other clusters) is now too great. This issue also relates to the historical debate of the Hyades vs. Pleiades Li morphologies, dating back to when only these two clusters had good G dwarf Li data. Proponents of the classical model pointed out that the Hyades is metal-rich compared to the Pleiades, and should thus experience more pre-MS Li depletion. However, the preponderance of data in Figure 2 argues that the Li depletion is predominantly a main sequence phenomenon, in contrast to the classical theory. Specific examples can be cited. The Pleiades, M34, UMa, and M67 are all solar metallicity clusters and show an age-Li relation. NGC 752 is a factor of 2 more metal-poor than the Hyades, yet its Li Morphology lies below that of the Hyades. Regardless of the true explanation for Li depletion, this suggests that age is a more important Li depletion parameter than metallicity; it remains to be seen how much of a role (if any) metallicity plays in G dwarf Li depletion, at least near solar metallicity. At intermediate and halo metallicity, it is clear that Li depletion does depend on the metal content because then structural effects are sufficiently important that the mass scale has shifted to higher T_{eff}.

Atomic diffusion at the 10 % level is inferred from a comparison of solar models and helioseismic data, close to the magnitude expected from theoretical calculations. However, the atomic diffusion time scales are too long to explain the bulk of the G dwarf Li depletion, and the timing is inverted (diffusion rates should increase with age as the convective envelope gets thinner, but the rate of Li depletion depletion seems to decline with age). To explain the Hyades Li morphology using mass loss requires an unrealistic initial mass function (Swenson & Faulkner 1992).

The most promising mechanism is slow mixing which is inferred from helioseismology (e.g., Christensen-Dalsgaard et al. 1993). Since there exist rotation-age, activity-age, and Xray-age correlations, it is tempting to prefer rotational mixing over gravity wave mixing. Furthermore, higher Li abundances in short period binaries also points to the action of rotational mixing (Deliyannis et al. 1990, Deliyannis 1990, Zahn 1992, 1994, Cayrel de Strobel et al. 1994, Ryan & Deliyannis 1995, Barrado y Navascues & Stauffer 1996). However, both mechanisms require further theoretical investigation (see Talon & Charbonnel 1998 for a discussion on the present description of the transport of angular momentum in low mass stars) and particularly, predictions should be made that might distinguish between the two types of slow mixing (see also §6).

5. The Li Plateau

The level of the Li abundance of the open cluster Li plateau appears to be lower with age, but the effect is not as dramatic as in the cooler G stars. In principle, this could reflect one of two things: Li depletion with age, or Galactic Li production (or both). Short period tidally locked binaries in the Hyades (Deliyannis *et al.* 1990; Soderblom *et al.* 1990; Thorburn *et al.* 1993; Ryan & Deliyannis 1995) and in M67 (Deliyannis *et al.* 1994; Pasquini *et al.* 1997) have Li abundances that lie above the mean Li Morphology of those clusters. This argues that, at least in those clusters, the Li plateaus have been depleted, and supports rotationally-induced mixing as the cause of the Li depletion (see Deliyannis et al. 1990, Deliyannis 1990, Zahn 1992, 1994, Ryan & Deliyannis 1995 for the detailed discussion of how such binaries can protect against rotational mixing).

The possible connection to the halo Li plateau is intriguing. Rotational models that reproduce the open cluster Li plateau depletion also suggest that the halo Li plateau is depleted. It is becoming increasingly clear that the depletion needs not be very large, but it also seems that if near-zero Li depletion has occured, halo dwarfs would have to be radically different than their disk counterparts. We refer to Part II for more discussion of this issue.

6. The Li Gap

The sharp decline of Li abundances in F stars offers, perhaps, the most striking evidence for the inadequacy of classical models. The Li Gap was first discovered in the Hyades (Boesgaard & Tripicco 1986), and its absence in the Pleiades (Boesgaard *et al.* 1988) demonstrated that it is a main sequence phenomenon.

At least three classes of mechanisms purport to explain it. Finely tuned (large) mass loss rates can, in principle, create the gap. Downward diffusion can create the red (cool) side of the gap; early models indicated that radiative acceleration might create the blue (hot) side of the gap, though improved later models did not show this (Michaud 1986; Richer & Michaud 1993). Slow mixing, induced either by gravity waves or rotation, can also create the gap, by mixing the Li preservation region with material below that is devoid of Li. Combinations of these mechanisms have also been proposed (e.g. Charbonneau & Michaud 1988, Talon & Charbonnel 1998). There are several ways to distinguish between these Li gap-producing mechanisms. Here, we focus on two methods: subgiants evolving out of the Li Gap, and information provided by additional elements, particularly Be and B.

The subgiant method (see also Part III and Talon & Charbonnel 1998 and in these proceedings) relies on determining the Li profile. Whereas all mechanisms predict a Li gap, the degeneracy is broken in the different signatures each mechanism leaves on the Li profile (Deliyannis et al. 1997). To deplete the surface Li abundance (in the presence of a very shallow SCZ), mass loss must get rid of essentially all of the Li preservation region (Figure 2). This leaves a tiny sliver of a Li preservation region, in which the Li abundance declines steeply with depth. For the red side of the gap, increasing diffusion rates with depth means Li piles up below the SCZ, and forms a peak. So Li increases with depth, but then decreases again at the Li preservation boundary. Rotationally-induced mixing generally flattens the Li profile (e.g. Deliyannis & Pinsonneault 1997, Talon & Charbonnel 1998), depending on the details of how the mixing occurs. These different profiles could be determined if a cluster existed with a turnoff in the Li gap: its subgiants would be evolving out of the gap, and as their convection zones deepen and dredge up progressively deeper material, they reveal the main sequence Li profile. Particularly, for mass loss, the subgiant surface Li abundances would decrease very sharply with decreasing T_{eff}. For diffusion, the Li abundances would increase at first, and then decrease at cooler T_{eff}. The case of slow mixing would lie in between the other two cases.

Fortuitously, nature provides a nearby rich cluster with turnoff on the cool side of the Li gap, namely M67. WIYN/Hydra observations of numerous stars near the turnoff at resolution R $\sim$ 15,000 = 3 pixels and S/N (of order 200–300 per pixel) are able to address this issue; models were also specifically constructed using the Yale/Ohio State code to interpret the observations (Deliyannis et al. 1997; Sills & Deliyannis 2000). These models were calibrated to create the Hyades Li gap, and took into account the metallicity difference between the two clusters. The subgiant abundances drop quickly from their turnoff value of A(Li) = 2.4, and also exhibit a significant spread. Mass loss is generally argued against in that straightforward scenarios that explain one cluster fail to explain the other; however, some scenarios with time-dependent mass loss rates might, conceivably, still be viable. Even these, though, are ruled out by the Li–Be depletion correlation (below). Pure diffusion is clearly argued against, except possibly for the intriguing case of one star. Rotational mixing might possibly be consistent with the observations if the observations are exhibiting large Li spreads, rather than a steeply declining Li abundance. In the case of the latter, however, the models do not have a sufficiently steep Li profile. We refer to Part III for further discussion concerning field subgiants.

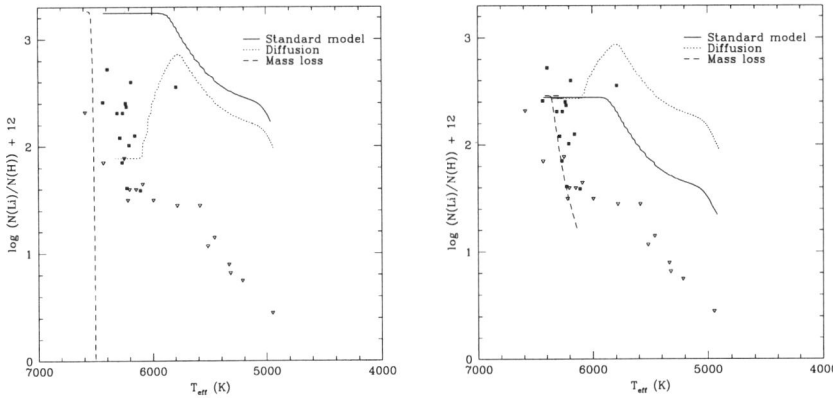

Figure 3. WIYN/Hydra MOS Li observations in M67 subgiants near the turnoff, compared to various models. Left: the models are calibrated to the Hyades Li gap depletion. Right: the models are calibrated to the M67 abundance just below the turnoff.

Very powerful tests of the models are realized when one uses more than one light element simultaneously. Consider using Li and Be. Whereas all classes of mechanisms predict a Li Gap, the Be signatures vary from mechanism to mechanism (Deliyannis & Pinsonneault 1997; Deliyannis et al. 1998a, Talon & Charbonnel 1998). In the case of mass loss, there is no change in Be as the surface Li goes down (Figure 1). All detectable Li must long be gone before any Be depletion occurs. In the case of pure diffusion, all the action takes place at the base of the SCZ: both elements diffuse downwards at similar rates. In the case of slow mixing, both the Li and Be profiles are flattened, and the surface abundances of both elements decline simultaneously. Li goes down faster than Be, depending on how the mixing occurs. In one extreme case, Li goes down as little as twice as fast as Be – the ratio of the sizes of their preservation regions. In models with rotational mixing, the efficiency of mixing increases with radius, and as a result Li goes down a little more than 4 times faster than Be does; both the Yale and French models agree on this (Deliyannis & Pinsonneault 1997; Charbonnel et al. 1994, Talon & Charbonnel 1998). In models with gravity waves (Garcia Lopez & Spruit 1991) the mixing is confined to smaller regions near the SCZ, so Li goes down at an even faster rate relative to Be. Thus, in a plot of A(Be) vs. A(Li), mass loss would be a horizontal line, diffusion would be a 45-degree line, rotational slow mixing would be shallower, and wave-driven slow mixing would be even shallower.

Figure 4 shows data (left panel) of a Li and Be survey of field F stars aimed specifically at identifying which mechanism creates the Li Gap (left panel; Deliyannis et al. 1998a). Great care was taken to define empirically the undepleted Li and Be abundances, by observing Be in stars that have A(Li) *greater than* 3.0 (rather than assume, as is often done, that stars with lower Li will have undepleted Be), and to get detections in Li at low A(Li) ~ 1. The data show a correlation between the depletion of Li and Be. This is remarkable in view of the

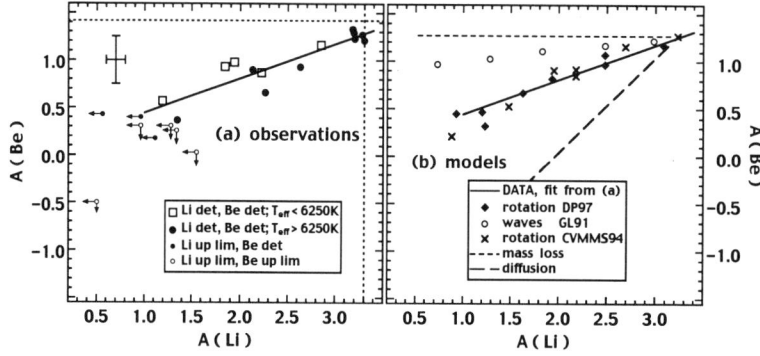

Figure 4. Keck I/HIRES, CFHT/Gecko, and UH 2.2m/Coudé observations of Li and Be in field F and G stars, illustrating a correlated depletion of surface Li and Be. *This argues strongly against mass loss and diffusion, and in favor of slow mixing as the dominant physical mechanism that creates the Li gap.* Rotationally-induced mixing is favored over mixing induced by gravity waves.

fact that many of the stars in the sample are the same stars that show chaotic Li abundances in Figure 1, with a variety of different ages. Figure 4b shows that this Li–Be depletion correlation argues strongly against mass loss and pure diffusion. *Rotational mixing is strongly supported.* Wave driven mixing does not do too badly, but the data favor rotational mixing. Note that the slope is slightly shallower for cooler stars, as predicted by the rotational models: this is due to the fact that cooler stars have deeper SCZ's which thus do a greater share of the mixing, leaving rotational instabilities free to deplete a larger amount of Li for a given Be depletion (since Be is preserved to a deeper level, convection does a smaller fraction of Be mixing than Li mixing). It is worth emphasizing the power of this method: not only does it identify the general class of model responsible for the Li gap, but it also seems able to distinguish between two types of slow mixing. It should be noted that we have identified the *dominant* mechanism; it is still possible that other mechanisms also occur to a more modest degree. In view of these findings, it is understandable why pure diffusion is not favored as the dominant mechanism, if mixing is favored: the mixing probably renders diffusion much less efficient. Future studies utilizing both Li and Be should be able to provide further constraints on slow mixing, especially if these studies include cluster stars; there is hope that the dominant mechanism can be understood in detail. Finally, a limited number of B observations show an essentially constant B abundance for highly varying degrees of Li and Be depletion (Boesgaard *et al.*

1998); this provides yet another independent argument against diffusion acting alone, and restricts the depth to which material can be mixed[1].

7. Conclusions

Classical stellar evolution theory fails to explain nearly all patterns of Li, Be, and B abundances in Population I low mass stars. Although several physical mechanisms have been proposed, the preponderance of data favor slow mixing, probably induced by rotation, that must compete with atomic diffusion. Future work should improve understanding of this slow mixing and evaluate what secondary role (if any) is played by wave-induced mixing, mass loss, and magnetic fields.

8. Acknowledgements

C.P.D. acknowledges support from the United States National Science Foundation under grant AST-9812735. M.H.P. acknowledges support from the United States National Science Foundation under grant AST-9731621 and from NASA under grant NAG5-7150. C.C. acknowledges support from the Action Spécifique de Physique Stellaire and the Conseil National Français d'Astronomie.

References

Barrado y Navascues, D. & Stauffer, J. R. 1996, A&A, 310, 879
Basu, S., Bahcall, J.N., & Pinsonneault, M.H. 2000, ApJ, 529, 1084
Boesgaard, A. M., & Tripicco, M. J. 1986a, ApJ, 302, L49
Boesgaard, A. M., & Tripicco, M. J. 1986b, ApJ, 303, 724
Boesgaard, A. M., Budge, K. G., & Ramsay, M. E. 1988, ApJ, 327, 389
Boesgaard, A. M., Deliyannis, C. P., Stephens, A., & Lambert, D. L. 1998, ApJ, 492, 727.
Brun, A.S., Turck-Chièze, S., Zahn, J.P., 1999, ApJ525, 1032
Cayrel, R. 1999, in "LiBeB, Cosmic Rays, and Related X- and Gamma-Rays,'", eds. R. Ramaty, E. Vangioni-Flam, M. Casse, and K. Olive, ASPCS 171, p. 268
Cayrel de Strobel, G., Cayrel, R., Friel, E., Zahn, J.P., Bentolila, C., 1994, A&A, 291, 505
Chaboyer, B., Zahn, J.P., 1992, A&A253, 173
Chaboyer, B., Demarque, P., & Pinsonneault, M. H. 1995, ApJ, 441, 876.
Charbonnel, C. & Talon, S., 1999, A&A, 351, 635
Charbonnel, C. & Vauclair, S. 1992, A&A, 265, 55

---

[1] In addition, the carbon and ozygen underabundances which are expected in the case of pure diffusion was not found in the Hyades F stars (Friel & Boesgaard 1990, Garcia Lopez et al. 1993)

Charbonnel, C., Vauclair, S. & Zahn, J.-P. 1992, A&A, 255, 191

Charbonnel, C., Vauclair, S., Maeder, A., Meynet, G., & Shaller, G. 1994, A&A, 283, 155

Christensen-Dalsgaard, J., Proffitt, C.R., Thompson, M.J., 1993, ApJ403, L75

Charbonneau, P., & Michaud, G. 1988, ApJ, 334, 746

D'Antona, F. & Mazzitelli, I. 1994 ApJS, 90, 467

Deliyannis, C. P., Ph.D. Dissertation, Yale University

Deliyannis, C. P., Demarque, P., & Kawaler, S. 1990, ApJS, 73, 21

Deliyannis, C. P., King, J., Boesgaard, A.M, & Ryan, S.G. 1994, ApJ, 434, L71

Deliyannis, C. P., & Pinsonneault, M. H. 1997, ApJ, 488, 836

Deliyannis, C. P., King, J., & Boesgaard, A. M. 1997, in "Wide Field Spectroscopy", eds. E. Kontizas et al., (Kluwer), Ast. & SSL 212, p. 201

Deliyannis, C. P., Boesgaard, A. M., Stephens, A., King, J. R., Vogt, S. S., & Keane, M. J. 1998a, ApJ, 498, L147

Deliyannis, C. P., Boesgaard, A. M., Stephens, A., King, J. R., Vogt, S. S., & Keane, M. J. 1998b, in "Nuclei in the Cosmos V", eds. N. Prantzos & S. Harissopoulos, Editions Frontieres, p. 41

Deliyannis, C. P. 2000, in "Stellar Clusters and Associations: Convection, Rotation, and Dynamos", eds. R. Pallavicini, G. Micela, & S. Sciortino, ASPCS 198, p. 235

Forestini, M 1994, A&A, 285, 473

Friel, E.D., Boesgaard, A.M., 1990, ApJ351, 480

Garcia López, R. J., Rebolo, R., Herrero, A., Beckman, J.E., 1993, ApJ412, 173

Garcia López, R. J., & Spruit, H. C. 1991, ApJ, 377, 268

Guenther, D. B., Kim, Y.-C., & Demarque, P. 1996, ApJ, 463, 382

Guzik, J.A., Willson, L.A., Brunish, W.M., 1987, ApJ, 319, 957

Hobbs, L. M., Iben, I., & Pilachowski, C. 1989, ApJ, 347, 817

Jeffries, R. D. 2000, in "Stellar Clusters and Associations: Convection, Rotation, and Dynamos", eds. R. Pallavicini, G. Micela, & S. Sciortino, ASPCS 198, p. 245

Jones, B. F., Fischer, D., Shetrone, M., Soderblom, D. R., 1997, AJ, 114, 352

King, J. R., Deliyannis, C. P., Hiltgen, D. D., Stephens, A., Cunha, K., & Boesgaard, A. M. 1997, AJ, 113, 1871

Kumar, P., Quartaert, E.J., 1997, ApJ, 475, L143

Kumar, P., Talon, S., Zahn, J.P., 1999, ApJ, 520, 859

Maeder, A., 1995, A&A, 299, 84

Maeder, A., Meynet, G., 2000, ARAA

Martin, E. L. 1997, Mem.S.Ast.It. 68, p. 905

Mathieu, R. 2000, in "Stellar Clusters and Associations: Convection, Rotation, and Dynamos", eds. R. Pallavicini, G. Micela, & S. Sciortino, ASPCS 198, p. 517

Michaud G., 1970, ApJ, 160, 641

Michaud, G. 1986, ApJ, 302, 650

Michaud, G., Charland, Y., Vauclair, S., Vauclair, G., 1976, ApJ, 210, 447
Montalban, J., Schatzman, E., 1994, A&A, 305, 513
Pasquini, L., Randich, S., & Pallavicini, R. 1997, A&A, 325, 535
Pilachowski, C. A., Booth, J., & Hobbs, L. M. 1987, PASP, 99, 1288
Pinsonneault, M. H. 1997, ARA&A, 35, 557
Pinsonneault, M. H., Kawaler, S. D., & Demarque, P. 1990, ApJS, 74, 501
Proffitt, C. R., & Michaud, G. 1989, ApJ, 346, 976
Richard, O., Vauclair, S., Charbonnel, C., Dziembowski, W.A. 1996, A&A, 312, 1000
Richer, J., & Michaud, G. 1993, ApJ, 416, 312
Ryan, S. G., & Deliyannis, C. P. 1995, ApJ, 453, 819
Schatzman, E., 1993, A&A279, 431
Schramm, D. N., Steigman, G., & Dearborn, D. S. P. 1990, ApJ, 259, L55
Sills, A., & Deliyannis, C. P. 2000, ApJ, submitted.
Soderblom, D., Oey, M., Johnson, D., & Stone, R., 1990, AJ, 99, 595
Soderblom, D., Jones, B., Balachandran, S., Stauffer, J., Duncan, D., Fedele, S., & Hudon, J. 1993a, AJ, 106, 1059.
Soderblom, D., Fedele, S., Jones, B., Stauffer, J., & Prosser, C. 1993b, AJ, 106, 1080.
Stephens, A., & Deliyannis, C. P. 1999, PASP, 758, 482
Straus, J. M., Blake, J. B., & Schramm, D. N. 1976, ApJ, 204, 481
Swenson, F.J. & Faulkner, J. 1992, ApJ, 395, 654
Talon, S., Charbonnel, C., 1998, A&A, 335, 959
Talon, S., Zahn, J.P. 1997, A&A, 317, 749
Thorburn, J. A., Hobbs, L. M., Deliyannis, C. P., & Pinsonneault, M. H. 1993, ApJ, 415, 150
Vauclair, G., Vauclair, S., Michaud, G., 1978, ApJ223, 920
Vauclair, S. 1988, ApJ335, 971
Vauclair, S. 1998, Space Sci.Rev., 85, 71
Zahn, J.-P., 1992, A&A, 265, 115
Zahn, J.-P., 1994, A&A, 288, 829
Zahn, J.-P., Talon, S., Matias, J., 1997, A&A, 332, 320

Sinks of Light Elements in Stars - Part II

Marc H. Pinsonneault

The Ohio State University, Department of Astronomy, 140 W. 18th Ave. Columbus, Ohio 43210, USA

Corinne Charbonnel

Laboratoire d'Astrophysique de l'Observatoire Midi-Pyrénées, CNRS-UMR 5572, 14, av.E.Belin, F-31400 Toulouse, France

Constantine P. Deliyannis

Indiana University, Astronomy Department, 319 Swain Hall West, 727 E. 3rd Street, Bloomington, IN 47405-7105, USA

Abstract.

See the abstract given in Part I (Deliyannis, Pinsonneault, Charbonnel, hereafter DPC, in these proceedings). In Part II we discuss the lithium data for metal-poor stars and the constraints it places on stellar depletion. There are a variety of indicators that place interesting bounds on the degree of stellar depletion, and in contrast to the other two sections the data for Population II stars provides weaker evidence for stellar lithium depletion. We review both the theoretical studies and the observational data, and critically evaluate the degree of stellar depletion consistent with the data.

1. Introduction

The study of lithium in metal-poor stars has implications for stellar structure, galactic chemical evolution, and Big Bang nucleosynthesis (BBN). There has thus been considerable observational and theoretical effort on this question, and we will therefore attempt to draw together the main themes and outstanding questions rather than undertake a detailed analysis of the fine points. This paper will be organized as follows. In section 2, we recall the main trends of the theoretical expectations. In section 3, we discuss the major features of the observational data; in particular we stress the similarities and differences with the population I pattern. Section 4 is devoted to the comparison of different classes of theoretical models with the data, and our conclusions are summarized in section 5.

2. Main Trends of the Theoretical Expectations

Let us first briefly recall that the so-called standard case refers to models which exclude any kind of transport processes of the chemicals in the radiative zones, and thus consider only convection as a mixing mechanism. Lithium is destroyed at moderate temperature by stellar interior standards; for typical main sequence (MS) and pre-MS densities, Li^6 burns at a time scale comparable to or shorter than the evolutionary time scale at around 2 million K and Li^7 burns at around 2.6 million K. Both beryllium and boron are less fragile, with characteristic burning temperatures of order 3.5 million K and 5 million K respectively. Because the observational data points to modest lithium depletion in halo stars (and even this conclusion is controversial!) we will restrict our discussion of light element depletion to lithium.

The major predictions of classical (sometimes referred to as standard) stellar evolution models for halo stars are summarized in Deliyannis et al. (1990). They depend mainly on the variations of the depth of the stellar convective envelope (and thus of the temperature at its base) with the stellar mass, metallicity and evolutionary stage. Pre-MS depletion increases with decreased mass, and there will therefore be a strong decrease of lithium with decreased T_{eff} for cool stars. Main sequence depletion is predicted to be minimal for all but the coolest stars; the absolute degree of lithium depletion decreases with decreased metal abundance. The net effect predicted is that hot halo dwarfs should exhibit little or no dependence of their surface lithium abundance on effective temperature, and for the lowest metal abundances there should be a minimal dependence on [Fe/H], in the sense that lower abundances would be predicted for higher metallicities. This implies a small dispersion in lithium at fixed effective temperature; in the Population I case, it also implies a weak dependence of lithium on age which can be tested in open clusters.

Classical stellar models neglect some physical processes which are known to be important for interpreting the surface lithium abundances of stars. The linked phenomena of gravitational settling and thermal diffusion, which are solidly based in our knowledge of plasma physics, are among the most important. Atomic diffusion is a fundamental process which must occur in the stellar gas unless some macroscopic motions counteract it, causing heavy elements to sink with respect to light ones under the conditions applicable for halo dwarfs. The time scale for this process decreases as the depth of the surface convection zone decreases.

Theoretical models which include pure atomic diffusion[1] (see Michaud et al. 1984 for the first computations for halo stars) therefore predict that lithium sinks below the surface convection zone for the conditions appropriate for sub-dwarfs, and furthermore that the degree of diffusion increases with increased T_{eff}. Microscopic diffusion will not generate a dispersion in abundance at fixed T_{eff}, and the effects at a given surface temperature are not strongly metallicity dependent.

---

[1] By "pure atomic diffusion" we mean that it is not counteracted by any macroscopic process in the stellar radiative zones.

Stellar winds can counteract and prevent atomic diffusion without leading to nuclear destruction (Vauclair & Charbonnel 1995). The corresponding mass loss necessary in the hottest halo stars is in excess of the value inferred from an extrapolation of the solar Mdot to halo stars by a factor of about 10 to 30, but not by a degree that can be ruled out observationally. For even larger mass loss rates the outer layers containing lithium can be removed; as noted by Swenson & Faulkner (1992) the finite depth of the surface convection zone must be accounted for and models with strong (stronger than inferred by Vauclair & Charbonnel 1995) mass loss alone are incompatible with the Population I lithium pattern. For hot halo stars the combined effect of diffusion and mass loss produce both lithium depletion and a small dispersion in abundance.

There are known mechanisms for mild mixing in the radiative envelopes of low mass stars. The two most frequently studied are rotationally-induced mixing and turbulence induced by gravity waves (see Pinsonneault 1997 for a review and DPC). Gravity waves can be produced by turbulence in the surface convection zone; because the convection zone depth is a strong function of mass, this can produce mass-dependent lithium depletion that is a function of time on the main sequence. It would not produce a dispersion in abundance for a sample of uniform age and composition, but abundance differences could be generated by a range of age and composition as seen in field halo stars.

The degree of rotational mixing depends on several major factors. Low mass Population I stars are observed to have a range of surface rotation rates and stars of the same mass, composition, and age could therefore have different initial angular momenta, different rotation rates as a function of time, and different degrees of rotational mixing. A dispersion in surface lithium abundance, even at fixed effective temperature in clusters, is therefore expected. However, the observed distribution of rotation rates in young clusters is nongaussian which strongly affects the expected lithium depletion pattern. We also cannot directly observe the initial conditions for Population II stars. To predict the detailed distribution of abundances the best we can to is to infer the distribution of initial conditions from young open cluster stars (see Pinsonneault et al. 1999, hereafter PWSN). The degree of mixing is also directly linked to the internal transport of angular momentum (e.g., Zahn 1992, Maeder 1995). Rotational mixing can also be inhibited by gradients in mean molecular weight - induced by nuclear burning in the cores of stars and possibly also by gravitational settling of helium in their outer layers (see also Michaud and Vauclair in these proceedings). In contrast with classical models, more modern models including mild mixing below the envelope on the main sequence can simultaneously produce modest depletions of species, such as lithium and beryllium, that burn at very different temperatures. Rotational mixing will also be a function of age, and some classes of models predict trends with [Fe/H] and effective temperature that can be tested in halo stars.

Realistic models should include the possible interactions between the above, since mass loss can counteract atomic diffusion and diffusion can interact with mixing (also see Michaud, these proceedings). We note that Vauclair (these proceedings) has proposed a nonlinear interaction between mixing and microscopic diffusion which would permit negligible halo star lithium depletion; note, however, that the details of such an interaction need to be computed and that such a cancellation would still have to be consistent with the globular cluster and Population I star data.

3. Observational Pattern

As we have just seen the surface lithium abundances of stars are sensitive to a variety of effects, both those accounted for in classical stellar models and those caused by physically well-motivated but still so-called non-standard effects. Uniqueness is thus a real issue when interpreting the observational data. We will therefore begin with the overall conclusions from studies of Population I stars, and then proceed to the current status of the observational data for Population II stars.

3.1. Population I Properties

The properties of Population I stars are summarized in DPC. Here we briefly recall those that are important for the problem of lithium depletion in Population II stars. In progressively older open cluster stars there is clear evidence for increased lithium depletion with age and a dispersion in abundance at fixed T_{eff} which is inconsistent with classical models. The predicted dropoff of lithium for cool stars from pre-MS burning is clearly seen. The overall properties favor mild mixing below the envelope on the MS; there is also evidence for microscopic diffusion playing a role for F stars and in the helioseismic inversions of the solar sound speed relative to theoretical models (see Guzik & Cox 1993, Richard et al. 1996, Basu et al. 2000). Large enough amounts of mass loss to directly cause lithium depletion are inconsistent with the observed population I pattern (Swenson & Faulkner 1992). As of this time a single theoretical model capable of explaining all of the Population I data has not yet been found (Talon & Charbonnel 1998, PWSN).

3.2. Overall Population II Properties

Beginning with the pioneering work of Spite & Spite (1982), there have been a series of progressively more sophisticated observational studies of lithium in halo field stars; the largest sample is that of Thorburn (1994). Ryan et al. (1999, hereafter RNB) obtained a smaller sample with a lower formal error ($\sigma \sim 0.036$ dex) than the errors in earlier studies ($\sigma \sim 0.07$-0.09 dex). There have also been preliminary studies of small samples of stars near the turnoff in globular clusters. Different investigators of field stars agree on some general properties:

 1. Halo stars hotter than 5800 K exhibit a weaker dependence on T_{eff}, [Fe/H], and a smaller dispersion than seen in Population I stars. There is vigorous debate about the existence and magnitude of any dispersion in the field star case, and there are active controversies about trends with T_{eff} and [Fe/H].

 2. Turnoff globular cluster stars were first studied by Molaro & Pasquini (1994), who reported a Li abundance for a turnoff star in NGC 6397 consistent with the halo plateau abundances. These are technically challenging observations owing to the faintness of the stars and the need for high resolution spectroscopy. A subsequent Keck study of the globular cluster M92 (Deliyannis et al. 1995, Boesgaard et al. 1998) revealed a large scatter, similar in morphology to the old open cluster M67. The sample, however, is small (seven stars,

including only three with S/N greater than 40). Pasquini & Molaro (1997) also observed a range of Li abundances in three turnoff stars in the intermediate metal abundance globular cluster 47 Tuc. Any successful theory must explain both the cluster and field star patterns; more data is clearly needed for the globular cluster stars (see also Part III, Charbonnel, Deliyannis & Pinsonneault).

We now turn to a summary of the most recent data on the important global features in field halo stars.

3.3. Trends with T_{eff}

There has been a spirited debate about the existence and magnitude of trends in the halo star data with metallicity and effective temperature. The slope with metal abundance is important for constraining the galactic chemical evolution contribution to the observed lithium abundances, and it may also contribute to the small scatter in the data at fixed effective temperature. Trends with effective temperature are important as a diagnostic of the mass dependence of any physical processes which affect the surface abundances. Thorburn (1994) reported evidence for a positive slope of lithium with respect to T_{eff}; this conclusion was challenged by Molaro et al. (1995). Subsequently Ryan et al. (1996) reanalyzed their data, claiming confirmation of the original results; see also Bonifacio & Molaro (1997). The existence of a modest rising trend with increased T_{eff} is only predicted in the models including atomic diffusion and stellar winds (Vauclair & Charbonnel 1995). However, what is even more important than the existence of a mild mean trend is the thing which is *not* seen : any evidence for a decline in lithium among the hottest stars.

As discussed above, models which include only microscopic diffusion predict a decline in surface lithium for the hottest halo stars; models with strong depletion from mixing also predict a downwards trend in lithium for the hottest stars (Chaboyer & Demarque 1994). This observational fact therefore constitutes an important limit on lithium depletion in halo stars.

3.4. Trends with [Fe/H]

The existence of trends of lithium with metallicity is a signature of galactic chemical evolution, and it could also contribute to the dispersion observed in halo stars. The majority of the observational investigations have looked for a correlation between [Li] and [Fe/H]; in parallel to the controversy over trends with effective temperature, there have been conflicting results on metallicity trends. Both Thorburn (1994) and Ryan et al. (1996) found some evidence for an increase in [Li] with [Fe/H] with a slope of order 0.1. This was disputed by Molaro et al. (1995) and Bonifacio & Molaro (1997). RNB obtained a sample with a small intrinsic range in effective temperature, but a wider range in metallicity. They could therefore evaluate metallicity but not effective temperature trends, and found a slope of [Li] with respect to [Fe/H] consistent with the Thorburn (1994) level. Chemical evolution trends should most logically be evaluated in the linear Li - linear Fe/H plane (see Olive and Matteucci, these proceedings).

A general feature of the derived chemical evolution trends is that they are sensitive to the source used for the metallicity, the subset of the data which is used, and the treatment of outliers in the fit. Ryan et al. (1996) also noted that the evidence for trends in the data with T_{eff} and [Fe/H] is more convincing in a bivariate analysis than when either variable is treated separately. The existence and magnitude of metallicity trends is also important for the interpretation of the dispersion in abundance (see below). The metallicity dependence of any rotational mixing is small (PWSN) and would be difficult to disentangle from chemical evolution effects.

3.5. Dispersion

The dispersion in the lithium abundances of halo plateau stars has been the subject of a number of studies. This is largely because the existence or absence of a detectable range in abundance at fixed metallicity and T_{eff} is the best direct test and constraint on the transport processes of chemicals in these stars. Lithium abundances can be studied as a function of age in the Population I case, which makes it easier to unambiguously distinguish between different classes of theoretical models (or at least rule bad models out). All of the metal-poor stars that we observe are old, and we therefore cannot directly reconstruct the depletion of lithium by sorting stars of progressively increased age into an evolutionary sequence.

Studies of the dispersion tend to fall into two groups. Some investigators (Deliyannis et al. 1993, Thorburn 1994) found evidence for a dispersion at a low level; others (Spite et al. 1996, Bonifacio & Molaro 1997) placed bounds on the dispersion consistent with their observational errors. The level of dispersion inferred by Deliyannis et al. (1993) is not inconsistent with the latter two studies (greater than 0.04 dex as compared with less than 0.08 and 0.07 dex respectively). In a recent paper, RNB have claimed a more stringent constraint on the overall dispersion. We examine this most recent data set below; a comparison of models including rotational mixing with the Thorburn (1994) data set was performed by PWSN.

The formal dispersion of the RNB data set is 0.053 dex, greater than their observational error of 0.036[2]. They attribute this to a correlation between metallicity and lithium abundance, e.g. chemical evolution rather than stellar depletion. There is a substantial overlap between the RNB and Thorburn (1994) data sets, and the markedly lower dispersion inferred by RNB can be traced directly to differences in equivalent width measurements. In Figure 1 we illustrate and compare the properties of the RNB sample (excluding one upper limit) with the stars in common as measured by Thorburn (1994); both have been shifted to the same effective temperature scale. There are both significant zero-point shifts and a marked difference in the overall dispersion of the sample. RNB attribute this to possible scattered light and sky subtraction issues in the Thorburn (1994) data set. We note, however, that similar differences appear in samples in common with other investigators[3]. The systematic differences between various observational data sets therefore require more scrutiny, especially given the relatively small sample size of the RNB data set and the small number of overdepleted stars expected for modest stellar depletion.

---

[2] Not 0.033 as claimed by RNB.

[3] e.g. compare G64-12 and CD -33 1173 as measured by both Thorburn 1994 and Spite & Spite 1993 to their relative abundances in RNB.

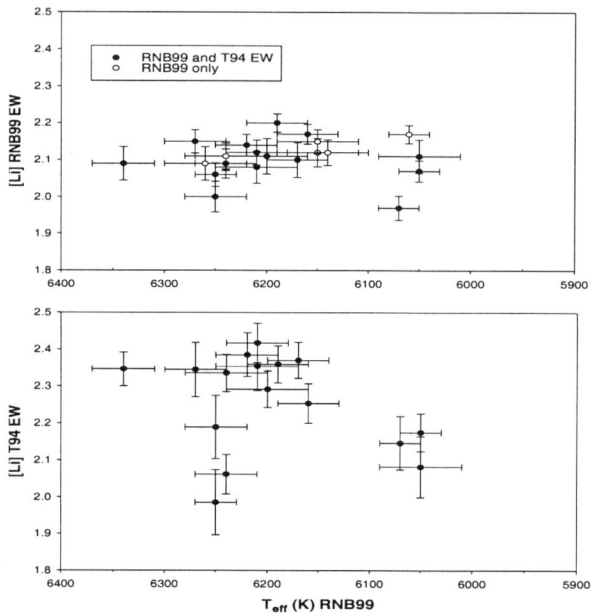

Figure 1. Data from RNB (top panel) compared with data for stars in common with Thorburn (1994) corrected to the same effective temperature (bottom panel).

We compare the theoretical distribution of the lowest depletion case of PWSN to the RNB data in Figure 2. The majority of young low mass stars in open clusters are slow rotators, which implies that they should have experienced similar rotational histories and similar degrees of rotational mixing. However about 1/5 of the stars in open clusters are observed to be rapid rotators and these should manifest themselves as overdepleted objects, producing an excess dispersion which is measureable. The existence of a core in the sample with minimal internal dispersion therefore does not by itself rule out more modest stellar destruction (as noted by RNB). The existence and number of outliers is a more stringent test. In the raw RNB sample there are three stars more than 0.1 dex below the median, one of which has an upper limit of 1.36 for its abundance; this simulation would predict ~ 4 depending on the criterion for defining what constitutes an overdepleted star. It is legitimate to question whether the highly overdepleted star is produced by the same mechanism as the other stars, but in any case the sample size is small and it is certainly difficult to make a persuasive case against modest depletion factors based on the data without chemical evolution corrections.

RNB placed more stringent constraints than the above based upon attributing some of their small dispersion to galactic chemical evolution. RNB fitted the data for a trend with [Fe/H] and concluded that there was a 10 % probability that as few outliers (one) as observed would be present by chance. This conclusion depends on the usage of a logarithmic, rather than a linear, rela-

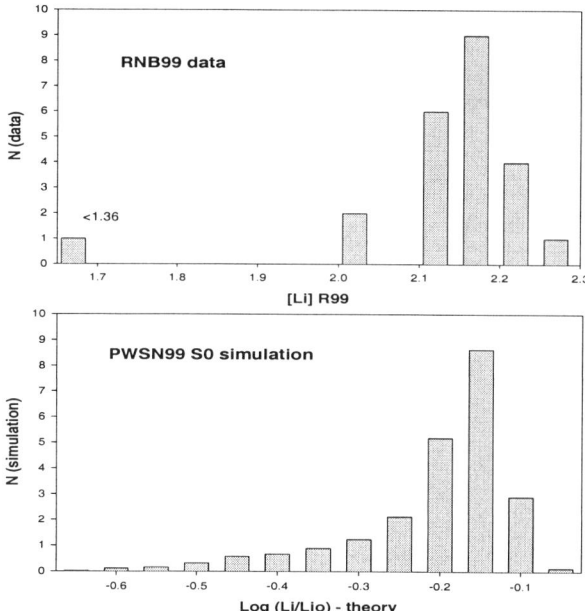

Figure 2. Data from RNB (top panel) compared with the theoretical simulation of PWSN for their low depletion case. The theoretical distribution has been convolved with an observational error of 0.03 dex.

tionship between lithium abundance and metallicity (Pinsonneault et al. 2000). In conclusion, the RNB data places more severe constraints on the dispersion in abundance than previous studies, and depending on the treatment of trends with metal abundance it may either be consistent with modest stellar depletion factors or places a bound of order 0.1 dex on the absolute depletion from the class of rotational mixing models considered by PWSN.

The existence of stars above the plateau may also provide some important clues; they could either be underdepleted or they could have experienced lithium production. Stars with very low initial angular momentum would experience much less rotational mixing than the norm and would therefore appear as underdepleted. However, there are strong observational selection effects against detecting very slow rotators in open clusters, and it is therefore difficult to estimate the fraction of such objects that would be expected in rotational mixing models. An alternative explanation would be differential lithium production. King et al. (1996) examined the most prominent such star, BD+23:3912, and found no evidence for lithium production in the abundances of other elements that would be affected by the main mechanisms (see King et al. 1996 for a detailed discussion and caveats).

3.6. Li^6/Li^7

Li^6 is more fragile than Li^7 and it is not produced in significant quantities in standard BBN models. The detection of Li^6 in halo stars can therefore be used to set powerful constraints on the absolute depletion of Li^7, with the caveat that the initial Li^6 abundance must be inferred from chemical evolution models. Smith et al. (1993) first claimed a detection of Li^6 in the halo star HD 84937. This was confirmed in subsequent studies by different investigators who added two more possible detections and a number of upper limits (see Cayrel et al. 1999, Hobbs et al. 1999, Nissen et al. 1999 for recent work on the subject and Nissen in these proceedings). The detected amount of Li^6 is small, but it appears to be secure. The amount is greater than would be expected from the beryllium and boron data, suggesting that alpha-alpha fusion may contribute to the production of Li^6.

One important uncertainty in the usage of Li^6 data is therefore what the initial abundance of the species could be; for example, Lemoine et al. (1996) and Cayrel et al. (1999) obtained bounds of a factor of four and three respectively on the absolute depletion of Li^6 in HD 84937. PWSN argued that an even higher initial abundance could not be excluded, and considered the extreme limiting case where the halo Li^6 abundance could have been as high as the solar system value.

The second uncertainty is the ratio of Li^6 to Li^7 depletion. Nuclear burning in the convective envelope in standard models would produce strong Li^6 depletion before any Li^7 depletion occurred; this has been used to argue that any detected Li^6 implies negligible Li^7 depletion (e.g. Lemoine et al. 1996). Both models with microscopic diffusion and models with mild envelope mixing, however, predict simultaneous detectable depletion of both isotopes to varying degrees, with Li^6 being more sensitive but not infinitely so (see PWSN). Nonetheless, the Li^6 data does provide one of the best independent checks on any stellar depletion of Li^7, and it indicates that large depletion factors are very unlikely to be consistent with the observed detections (see below.)

4. Theoretical Constraints on Lithium Depletion

In light of the observational data above, what can we infer about the depletion of lithium in halo stars? The first and most generally agreed-upon conclusion is that a variety of observational tests make a large depletion factor unlikely. There have been three major features of the halo data which have been used to constrain the absolute depletion: the degree of dispersion in the halo plateau, the detection of Li^6 in some halo stars, and the absence of a decline in surface Li for hotter halo stars.

4.1. Bounds from the Dispersion in the Plateau

PWSN inferred a range of 0.2-0.4 dex depletion factors from models including rotational mixing; the lower end of the range was more consistent with the dispersion inferred from the Thorburn (1994) data set, while the upper end of the range permitted the rare highly overdepleted stars to be explained within the framework of rotational mixing. As discussed above, the most recent data set of RNB is marginally consistent with the lower end of the depletion range in PWSN (of order 0.2 dex.) Other investigators (e.g. Bonifacio & Molaro 1997, RNB) have claimed more stringent limits of order 0.1 dex on the absolute depletion based upon the small (or, in their view, nonexistent!) dispersion in the halo Li data. We will return to this claim after reviewing other measures based upon the detection of Li^6 and the absence of the observed signature of pure microscopic diffusion in halo stars.

4.2. Bounds from Li^6 / Li^7 Measurements

PWSN set a less severe, but firm, limit of 0.5-0.6 dex Li^7 depletion from the measured Li^6 / Li^7 abundance ratios in halo stars under the assumption that the halo stars did not have a Li^6 abundance higher than the solar system value. Lemoine et al. (1996) derived a bound of a factor of 4 on the absolute Li^6 depletion of HD 84937, which in the rotationally mixed models of PWSN would imply a bound of 0.25 dex on the Li^7 depletion, while Cayrel et al. (1999) used the lithium data in the same star to set a bound of 0.1 dex on its Li^7 depletion; both of these calculations, however, are dependent on the chemical evolution model which is used. We note that if the Cayrel et al. (1999) bound of a factor of three Li^6 depletion is used in conjunction with the PWSN models, an absolute Li^7 depletion of 0.15 dex is inferred for HD 84937 rather than an upper limit of 0.1 dex.

4.3. Bounds from the Flatness of the Plateau

Vauclair & Charbonnel (1998) used a different set of properties to infer a stellar depletion factor. Pure microscopic diffusion in halo models would produce a strong decrease in surface lithium with increased T_{eff} which is not observed. It is therefore clear that something must be inhibiting microscopic diffusion, especially given the improved agreement with helioseismology from the inclusion of micrscopic diffusion in solar model calculations. They noted that there is a subsurface peak in the Li^7 abundance of the pure diffusion models which does not vary greatly across the plateau.

Vauclair & Charbonnel (1995) (see also Swenson 1995) argued that mass loss at a rate 10-30 times greater than the solar value could counteract the effects of diffusion if the rate was tuned across the lithium plateau. This would have the effect of exposing a uniform abundance across the plateau, with an absolute depletion of 0.15 dex. Vauclair & Charbonnel (1998) noted that in the presence of sufficiently mild mixing the height of this peak could be preserved, implying that a uniform depletion of order 0.15 dex could apply if the absence of a measurable surface signature of diffusion arose from either the competing effects of mass loss or the interaction of diffusion and mild mixing. As noted by Chaboyer & Demarque (1994), sufficiently strong mixing can cancel the effects of diffusion while not preserving the height of the peak; however, models with the high degree of depletion inferred by that latter paper are difficult to reconcile with the other observational tests of lithium depletion.

4.4. Is There Any Depletion?

In light of the remarkable observed properties of the halo lithium plateau, it is reasonable to ask whether there is in fact any depletion at all. In other words, are we directly seeing the primordial lithium abundance (e.g. Bonifacio & Molaro 1997) or is the primordial lithium abundance in fact *lower* than the observed values because of a significant contribution from galactic chemical evolution (RNB)?

"Standard" stellar models are sometimes invoked as evidence against significant depletion. These classical models, however, achieve this prediction by simply neglecting known physics rather than by demonstrating that it is unimportant. For example, one could construct stellar models which ignore the CNO cycle, and they might even agree with some data, but it does not follow that these should be placed on an equal physical basis with models that include the known nuclear physics. In particular, atomic diffusion cannot be excluded arbitrarily from the computations, on the pretext that it produces unobserved features. This disagreement is a simple signature of some macroscopic motions (mass loss or rotation-induced motions) which counteract the diffusion process.

Any model predicting zero stellar depletion in halo field stars must be reconciled with the apparent scatter in the globular cluster turnoff stars. If both the halo stars and the globular cluster stars are depleted by (say) mild mixing, it is possible to explain the difference in the abundance patterns by a different set of initial angular momenta. Such a difference could arise in the context of the currently popular model for the origin of the range of rotation rates, namely that the lifetime of accretion disks determines the rotation rate, if globular cluster stars experienced more frequent interactions which disrupted their accretion disks early in their lifetimes relative to the lower density systems that the halo stars arose from. It is more challenging, however, to explain why stars with similar thermal structures should experience completely different depletion histories.

The *complete* absence of depletion in Population II stars would also have to be reconciled with the strong evidence for depletion in Population I stars which is not predicted by standard models.

At the same time, it is also clear that none of the existing theoretical models provide a complete description of the complex lithium abundance pattern seen in stars. Although the most recent classes of rotational models are reasonably successful at reproducing the observed angular momentum evolution of low mass Population I stars, they do not reproduce the solar rotation profile as inferred from helioseismology. They also require an extrapolation of the initial conditions from Population I to Population II stars, which may introduce systematic errors in the calculations. Further theoretical work is clearly needed, and a more refined set of models could potentially alter the inferred degree of stellar depletion.

5. Summary

We have compared theoretical models with the observational Population II lithium data to obtain bounds on stellar lithium depletion. From a combination of the dispersion in the data, the detection of Li^6, and the flatness of the halo plateau interesting bounds can be set on the stellar depletion of lithium in Population II stars. The majority of tests are roughly consistent with depletion at the 0.15-0.2 dex level, with a firm upper bound of 0.5-0.6 dex from a combination of the detected Li^6 abundance in HD 84937 and an extreme chemical evolution model. The most recent data set of RNB places the most severe observational constraints on the dispersion of lithium in halo stars. It is marginally consistent with the least depleted set of models of PWSN including rotational mixing; a larger statistical sample would permit a more definitive test of the limits on depletion from mixing in halo stars. There are unexplained differences between the equivalent width measurements of Thorburn (1994) and RNB which need to be understood; indeed, the systematic differences between investigators on the absolute level of the plateau are approaching the uncertainty in the inferred degree of stellar depletion.

A new generation of theoretical models is needed to further refine our understanding of light element depletion in stars. The interaction between different physical mechanisms, such as microscopic diffusion, mass loss, and rotational mixing, and a better physical description of each of them, may prove important in this context.

Finally, any stellar lithium depletion has implications for BBN. PWSN discussed the implications of a higher primordial lithium abundance; see Olive (these proceedings) for the case of negligible stellar depletion. If the preliminary data from the BOOMERANG mission is confirmed (Lange et al. 2000), we note that there may be a disagreement between the stellar lithium abundances and the predictions of standard BBN as well as the low deuterium results of Tytler (these proceedings). This is significantly relaxed if stellar lithium depletion has occured.

6. Acknowledgements

M.P. would like to acknowledge support from NASA grant NAG5-7150 and NSF grant AST-9731621. C.C. thanks the Action Spécifique de Physique Stellaire and the Conseil National Français d'Astronomie for support. C.P.D. acknowledges support from the United States National Science Foundation under grant AST-9812735.

References

Basu, S., Bahcall, J.N., & Pinsonncault, M.H. 2000, ApJ, 529, 1084

Boesgaard, A.M., Deliyannis, C.P., Stephens, A., & King, J.R. 1998,ApJ, 492, 727

Bonifacio, P., & Molaro, P. 1997, MNRAS, 285, 847

Cayrel, R., Spite, M., Spite, F., Vangioni-Flam, E., Casse, M., & Audouze, J. 1999, A&A, 343, 923

Chaboyer, B. & Demarque, P. 1994, ApJ, 433, 519

Deliyannis, C.P., Demarque, P., & Kawaler, S.D. 1990, ApJS, 73, 21

Deliyannis, C.P., Pinsonneault, M.H., & Duncan, D.K. 1993, ApJ, 414, 740

Guzik, J.A., Cox, A.N. 1993, ApJ, 411, 394

Hobbs, L.M., Thorburn, J.A., & Rebull, L.M. 1999, ApJ, 523, 797

King, J.R., Deliyannis, C.P., & Boesgaard, A.M. 1996, AJ, 112, 2839

Lange et al. 2000, astro-ph/000504

Lemoine, M., Schramm, D.N., Truran, J.W., & Copi, C.J. 1996, ApJ, 478, 554

Maeder, A. 1995 A&A299, 84

Michaud, G., Fontaine, G., Beaudet, G., 1984, ApJ, 282, 206

Molaro, P. & Pasquini, L. 1994, A&A, 281, L77
Molaro, P., Primas, F., & Bonifacio, P. 1995, A&A, 295, L47
Pasquini, L. & Molaro, P. 1997, A&A, 322, 109
Nissen, P.E., Lambert, D.L., Primas, F., & Smith, V.V. 1999, A&A, 348, 211
Pinsonneault, M.H. 1997, ARA&A, 35, 557
Pinsonneault, M.H., Walker, T.P., Steigman, G., & Narayanan, V.K. 1999, ApJ, 527, 180 (PWSN)
Pinsonneault, M.H., Walker, T.P., Steigman, G., & Narayanan, V.K. 2000, in preparation
Richard, O., Vauclair, S., Charbonnel, C., Dziembowski, W.A. 1996, A&A, 312, 1000
Ryan, S.G., Beers, T.C., Deliyannis, C.P., & Thorburn, J.A. 1996, ApJ, 458, 543
Ryan, S.G., Norris, J.E., & Beers, T.C. 1999, ApJ, 523, 654 (RNB)
Smith, V.V., Lambert, D.L., & Nissen, P.E. 1993, ApJ, 408, 262
Spite, F. & Spite, M. 1982, A&A, 115, 357
Spite, F. & Spite, M. 1993, A&A, 279, L9
Spite, M., Francois, P. Nissen, P.E., & Spite, F. 1996, A&A, 307, 172
Swenson, F. 1995, ApJ, 438, L87
Swenson, F.J. & Faulkner, J. 1992, ApJ, 395, 654
Talon, S. & Charbonnel, C. 1998, A&A, 335, 959
Thorburn, J.A. 1994, ApJ, 421, 318
Vauclair, S. & Charbonnel, C. 1995, A&A, 295, 715
Vauclair, S. & Charbonnel, C. 1998, ApJ, 502, 372
Zahn, J.-P. 1992 A&A, 265, 115

The Light Elements and Their Evolution
IAU Symposium, Vol. 198, 2000
L. da Silva, M. Spite, J. R. de Medeiros, eds.

Sinks of Light Elements in Stars – Part III

Corinne Charbonnel

Laboratoire d'Astrophysique de l'Observatoire Midi-Pyrénées, CNRS-UMR 5572, 14, av.E.Belin, F-31400 Toulouse, France

Constantine P. Deliyannis

Indiana University, Astronomy Department, 319 Swain Hall West, 727 E. 3rd Street, Bloomington, IN 47405-7105, USA

Marc H. Pinsonneault

The Ohio State University, Department of Astronomy, 140 W. 18th Ave., Columbus, OH 43210, USA

Abstract. See the abstract given in Part I (Deliyannis, Pinsonneault, Charbonnel, hereafter DPC, this volume). In Part III, we first discuss the LiBeB observations in subgiant stars and the constraints they bring on the transport processes occuring on the main sequence. Evidence is then reviewed that suggests that in situ mixing occurs in evolved low mass Population I and Population II stars. Theoretical mechanisms that can create such mixing are discussed, as well as their implications for the evolution of the LiBeB and ^{3}He.

1. Introduction

Evolved stars can be important sinks for light elements. In Parts I and II we reviewed the observational and theoretical situation for Population I and Population II main sequence stars. The structural evolution of evolved low mass stars is different than their main sequence precursors. The surface abundances are diluted as stars travel from the main sequence across the subgiant branch to the red giant branch. This first dredge-up is completed when the surface convection zone reaches its maximum depth in mass (below the luminosity of the horizontal branch). The properties of stars as they undergo the first dredge-up can provide clues about the internal abundance profiles of their main sequence precursors; we review the conclusions that can be drawn from subgiants in §2. On the upper giant branch standard models - which predict constant surface abundances after the completion of the first dregdge-up - are in serious conflict with the observational data, which exhibits strong trends with increased luminosity; we discuss first-ascent red giants in §3.

2. LiBeB on the subgiant branch - Constraints for the processes occuring on the main sequence

2.1. Population I subgiants

LiBeB abundances in subgiants are *a posteriori* tracers of the hydrodynamical processes that affect these elements during the previous evolutionary phases. Indeed, when the first dredge-up occurs, the convective dilution of the external stellar layers with the internal LiBeB free regions induces a decrease of the surface abundances down to values that depend on the stellar mass and metallicity (which dictate the dredge-up efficiency) and on the total LiBeB content in the star at the turnoff. One thus expects very low post-dilution abundances for stars which significantly destroyed these elements during the pre-main sequence and the main sequence. This is the case in particular for Li in Pop I evolved stars with initial masses lower than about $1.4 M_\odot$ for which the large Li dispersion reflects the distribution on the main sequence and at the turnoff (see DPC).

Until recently, the situation was less clear for the more massive stars (A and early-F types) which spend their main sequence on the hot side of the Li-dip (note that there are strong observational selection effects for the hotter stars which are rapidly rotating and for which lithium abundances can not be measured). In open clusters like Coma, Praesepe and the Hyades, these stars show Li abundances close to the galactic value, except for some Li deficient Am stars (Boesgaard 1987; Burkhart & Coupry 1989, 1998, 2000). However, important Li underabundances are exhibited by some of their field main sequence or slightly evolved counterparts before the dilution starts (Alschuler 1975, Brown et al. 1989, Balachandran 1990, Wallerstein et al. 1994). This, in addition to the fact that dilution alone can not explain the low lithium values shown by the giants in the open clusters with turnoff masses higher than $\sim 1.5 M_\odot$ (Gilroy 1989, Charbonneau et al. 1989; see Figure 1) suggested that some Li depletion occurs inside these stars while they are on the main sequence but shows up at the surface relatively late (i.e., still on the main sequence but after the age of the Hyades) compared to cooler dwarfs (Vauclair 1991, Charbonnel & Vauclair 1992). This was confirmed by Randich et al. (1999) and do Nascimento et al. (2000, using data by Lèbre et al. 1999) on the basis of the spectroscopic analysis of large samples of field Pop I subgiants for which Hipparcos data allowed the precise determination of both the mass and evolutionary status (see also Mallik 1999 and in this volume)[1]. Observations of Be and B in a few Hyades giants brought additional constraints on the processes that occur in the external layers of the A and early-F main sequence stars : In these giants indeed, Be is moderately underabundant (Boesgaard et al. 1977) while B is almost normal (Duncan et al. 1998) compared to the dilution predictions.

In A and F main sequence stars, atomic diffusion is known to play an important role (mainly in the very outermost stellar layers; see Michaud in this volume). However, diffusion theory alone predicts abundance anomalies much larger than the observed ones (Richer et al. 2000 and references therein). In particular at the age of the Hyades, radiative diffusion in stars with $T_{\text{eff}} \simeq 7000K$

---

[1]Let us note that the very low Li values found for the most massive subgiants are confirmed even when non-LTE effects are taken into account (do Nascimento et al. 2000)

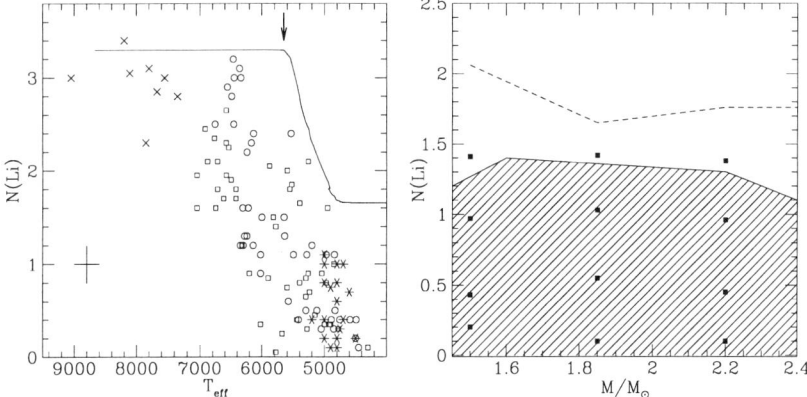

Figure 1. Li in stars with masses higher than 1.5M$_\odot$. **(left)** Hyades main sequence stars (crosses, Burkhart & Coupry 2000), field subgiants (circles, Lèbre et al. 1999; squares, Wallerstein et al. 1994), giants of open clusters (stars, Gilroy 1989). The solid line shows the evolution of the surface lithium with T$_{\text{eff}}$ for a standard 1.85M$_\odot$ model (Charbonnel & Talon 1999, CT99); the arrow points the theoretical start of dilution. **(right)** Evolved stars of open clusters (shaded area). The dotted curve shows the Li values expected with dilution alone, i.e. no additional internal mixing on the main sequence. The squares give the Li values predicted when rotation induced mixing is taken into account as in CT99 for different rotation velocities of the main sequence star ranging between 50 and 150 km.sec^{-1}

should lead to important Li overabundances, while at T_{eff} >7200K strong Li underabundances should be due to settling. This is not observed. Some macroscopic processes thus occur, that both decrease the efficiency of the atomic diffusion and lead to non standard Li depletion in evolved stars as we discussed previously (see also Part II, Pinsonneault, Charbonnel & Deliyannis in this volume)

Charbonnel & Talon (1999, hereafter CT99) studied the combined effect of atomic diffusion and rotational mixing on the LiBeB in these stars up to the completion of the first dredge-up, in the framework of the transport of matter and of angular momentum by wind-driven meridional circulation and shear turbulence (Zahn 1992, Maeder 1995, Talon & Zahn 1997, Maeder & Zahn 1998). Their models were computed for rotation velocities covering the large Vsini range observed at these spectral types. While lithium is found not to vary much at the stellar surface at the age of the Hyades, more destruction occurs inside the rotating models compared to the classical ones and its signature appears at the surface before the onset of the dilution. The post dredge-up Li values are much lower than predicted classically and agree with the observations both in evolved stars belonging to open clusters and in the field, as can be seen in Fig.1. The less fragile Be and B are less affected than Li by the rotation-induced mixing, and the corresponding predictions also reproduce the observations in the Hyades giants (Boesgaard et al. 1977, Duncan et al. 1998). The main success of CT99 models which include the most complete description currently available for rotation-induced mixing is their ability to reproduce abundance anomalies of various elements over a large domain of stellar masses and evolutionary phases. Indeed the same treatment of the hydrodynamical process which can account for the C and N anomalies in B type stars (Talon et al. 1997) also shapes the hot side of the Li dip in the open clusters (Talon & Charbonnel 1998) and explains the LiBeB observations in main sequence F and A stars as well as in their evolved counterparts (CT99; see also Talon & Charbonnel in this volume).

2.2. Population II subgiants

The behavior of lithium in metal-poor field subgiants appeared clearly in the large sample of Pilachowski et al. (1993) and was confirmed recently by Gratton et al. (2000; see also Carretta et al. 1998). As can be seen in Fig.2, the observed trend (steady decline of lithium abundance with temperature decreasing between ~ 5600 and 4900K on the subgiant branch) is well explained by the theoretical dilution up to the completion of the first dredge-up (e.g. Deliyannis et al. 1990, Proffitt & Michaud 1991, Charbonnel 1995). The precise temperature at which theoretical dilution begins is somewhat model dependent, but nonetheless, given current lingering uncertainties in the temperature scales, the agreement between theory and observation is impressive.

Pilachowski et al. (1993) noted that the lithium abundances in their subgiant sample showed more scatter than do the turnoff stars; they interpreted this as an indication for variations in the main sequence Li destruction below the observable surface layers. Reconsidering Pilachowski's sample to which they added some more stars, Ryan & Deliyannis (1995) showed however that much of the scatter disappears with a careful and self-consistent treatment of the reddening. Note that this difference does not show up in Gratton's sample. However, the

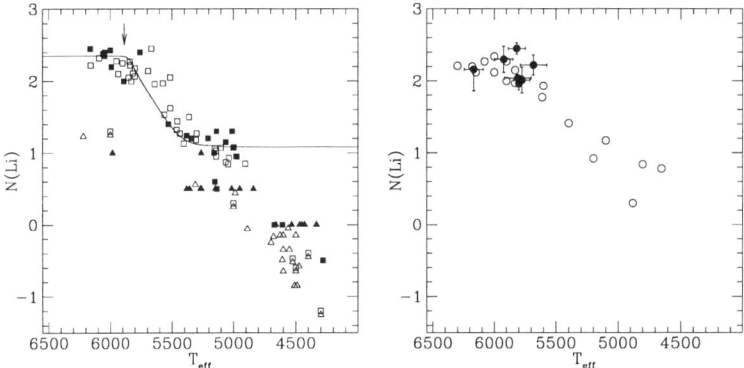

Figure 2. Li in metal-poor subgiants. **(left)** Field stars with $-2 \leq$[Fe/H]≤ -1 (Ryan & Deliyannis 1995, open squares and triangles for real detection and upper limits respectively; Gratton et al. 2000, black squares and triangles). The solid line shows the prediction of the Li variations due to dilution alone (which starts as indicated by the arrow) at the surface of a 0.8M$_\odot$, Z=10^{-4} star (Palacios et al. 2000). **(right)** Stars in globular clusters M92 (Deliyannis et al. 1995, Boesgaard et al. 1998, black points) and NGC 6397 (Molaro & Pasquini 1994, Pasquini & Molaro 1996); the values presented here were derived within Carney's temperature scale (Spite, private com.)

PopII field subgiants with Li upper limits are presumably the counterparts of the few plateau stars with no Li detection (see Part II in this volume); both samples should be analysed simultaneously to understand and quantify a possible Li depletion in halo stars.

Very few data exist up to now for LiBeB in globular clusters. Lithium abundances have been determined for slightly evolved stars which have not reached the onset of dilution in NGC 6397 (Molaro & Pasquini 1994, Pasquini & Molaro 1996) and M92[2] (Deliyannis et al. 1995, Boesgaard et al. 1998). These turnoff stars with very similar effective temperature show a Li dispersion of a factor about two and three respectively in NGC 6397 and M92 (see Fig.2). Boesgaard et al.(1998) favor the explanation of differential depletion in M92 by rotation induced mixing in objects with different stellar rotational histories. It is worth knowing however that the M92 subgiants show other "surprising abundances" (King et al. 1998), i.e., under and overabundances respectively of Mg and Na compared to field stars with the same metallicity. While the observational data need to be confirmed, these anomalies reveal stricking field to cluster differences (see also §3) which could reflect environmental effects (pollution of the intracluster gas, distribution of the initial rotation velocities, ...) which have to be disentangled. This is of particular importance if one wants to use and com-

---

[2]The globular cluster stars were chosen to be pre-dilution stars, relative to the empirical commencement of the Li dilution from field stars

pare the Li data in globular cluster and halo stars to constrain the primordial abundance of this element and its evolution in the early epochs (see Part II).

3. LiBeB and ^{3}He on the Red Giant Branch

3.1. Evidences for extra-mixing in low-mass RGB stars

Observational evidences have accumulated during the last years of a second and distinct mixing episode that occurs in low mass stars when they climb the red giant branch (RGB; see Kraft 1994, and more recently Charbonnel et al. 1998, Sneden 1999 and Gratton et al. 2000 for references). The signatures of this non-standard process in terms of abundance anomalies are numerous. In metal-poor field giants, it leads to a further major decrease of the Li abundance (around 4800K as can be seen in Fig.2). By reaching the regions of incomplete CNO burning inside the RGB stars, it induces a decrease of the carbon abundance and of the carbon isotopic ratio, and a corresponding increase of the N. In most of the metal-deficient field and globular cluster stars, the surface $^{12}C/^{13}C$ ratio even approaches the equilibrium value; this anomaly also appears, while at a lower extent, in evolved stars belonging to open clusters with turnoff masses lower than $\sim 2M_\odot$ (Gilroy 1989, Gilroy & Brown 1991). This extra-mixing is also frequently invoqued to explain the global O versus Na anticorrelation observed in globular cluster red giants (see Weiss et al. 2000 for references).

All the relevant data clearly indicate that the extra-mixing starts acting when the stars pass the so-called RGB bump in the luminosity function. At this point, the hydrogen burning shell (HBS) crosses the discontinuity in molecular weight built by the convective envelope during the dredge-up. Before this evolutionary point, the mean molecular weight gradient probably acts as a barrier to the mixing between the convective envelope and the HBS (Sweigart & Mengel 1979, Charbonnel 1994, 1995, Deliyannis 1995, Charbonnel et al. 1998). Above this point, this barrier disappears and the extra-mixing, whatever its nature, is free to act.

3.2. The origin of the extra-mixing ...

Several attempts have been made to simulate this extra-mixing in order to reproduce the abundance anomalies in RGB stars. Denissenkov & Weiss (1995) and Weiss et al. (2000) modelled this deep mixing by adjusting both the mixing depth and rate in their diffusion procedure, but focussed on the O-Na anticorrelation (see also Cavallo et al. 1998). Wasserburg et al. (1995), Boothroyd & Sackmann (1999) and Sackmann & Boothroyd (1999; see also Sackmann in this volume) used an ad-hoc "conveyor-belt" circulation model, where the depth of the "extra-mixed region" is related to a parametrized temperature difference up to the bottom of the HBS and is adjusted to reach the observed carbon isotopic ratios as a function of stellar mass and metallicity.

Some studies attempted to relate the RGB extra-mixing with physical processes, among which rotation seems to be the most promising. Sweigart & Mengel (1979) suggested that meridional circulation on the RGB could lead to the low $^{12}C/^{13}C$ observed in field giants. Charbonnel (1995) investigated the influence of such a process by taking into account more recent progress in the description of the transport of chemicals and angular momentum in stellar interiors : Zahn's (1992 and subsequent developments) consistent theory which describes the interaction between meridional circulation and turbulence induced by rotation (as already discussed in §2.1). This framework is appealing because it takes advantage of some particularities of the non-homologous RGB evolution. In particular, some mixing is expected to take place wherever the rotation profile presents steep vertical gradients, and near nuclear burning shells. Moreover due to the stabilizing effect of the composition gradients, the mixing is expected to be efficient on the RGB only when the hydrogen-burning-shell crosses the chemical discontinuity created by the convective envelope during the first dredge-up. Using a simplified version of this description Charbonnel (1995) showed that the rotation-induced mixing can indeed account for the observed behavior of carbon isotopic ratios and for the Li abundances in Population II low mass giants. Simultaneously, when this extra-mixing begins to act, ^{3}He is rapidly transported down to the regions where it burns by the ^{3}He$(\alpha, \gamma)^7$Be reaction. This leads to a decrease of the surface value of ^{3}He/H (see also Deliyannis 1995, Hogan 1995 and Sackmann & Boothroyd 1999).

The study of horizontal branch stars also provides some intriguing clues about angular momentum evolution on the RGB. Peterson (1983) first discovered that some blue horizontal branch stars are rapid rotators (see Behr et al. 2000 for more recent data and a discussion of the observational situation). Pinsonneault et al. (1991) noted that the combination of RGB mass loss, high horizontal branch rotation, and low main sequence rotation required strong differential rotation with depth in giants. If the convection zone of RGB stars had solid body rotation, differential rotation with depth in their MS precursors was also required. Behr et al. (2000) found a break in the rotational properties of HB stars, in the sense that very blue HB stars both exhibited the surface signature of atomic diffusion and rotated more slowly than slightly cooler stars. Sills & Pinsonneault (2000) interpreted this as evidence that mean molecular weight gradients caused by atomic diffusion inhibit angular momentum transport in hot horizontal branch stars; this is an independent test of the impact of composition gradients in a different evolutionary phase. In addition, Sills & Pinsonneault (2000) found that uniform rotation at the main sequence turnoff was only compatible with rapid horizontal branch rotation under the following conditions: (1)turnoff rotation of order 4 km/s rather than the 1 km/s inferred from an extrapolation of the Population I angular momentum loss law to Population II stars; (2)constant specific angular momentum in the convective envelopes of giants; (3)strong differential rotation with depth in the radiative cores of giants. All of these are radically different from the expectations from main sequence angular momentum evolution models, and they are an indication that further theoretical work is needed in physical models of RGB rotational mixing. It is encouraging, however, that all of the above properties favor more vigorous rotational mixing on the RGB than would be expected from the opposite conclusions.

3.3. ... and its consequences for ^{3}He

While there is a consensus on the fact that the mechanism which is responsible for the chemical anomalies on the RGB also affects the ^{3}He abundance (as first suggested by Rood et al. 1984), large uncertainties remain on the quantitative extent of this ^{3}He depletion. In the PopII models of Charbonnel (1995) ^{3}He decreases by a large factor in the ejected envelope material but low mass stars remain net producers (while far much less efficient than in the case of models without RGB extra-mixing) of ^{3}He. On the other hand the models of Sackmann & Boothroyd (1999) predict a net destruction of ^{3}He by low mass stars. Both studies also differ on the predictions for the evolution of the ^{12}C/^{13}C ratio at the stellar surface. While the rotation-induced mixing by Charbonnel reproduces the observed sharp drop of the ^{12}C/^{13}C just beyond the RGB bump and its constancy at higher luminosity, the cool bottom processing model by Sackmann & Boothroyd predicts a smooth decrease of the surface ^{12}C/^{13}C all along the RGB up to the tip where the low values are finally (but too belatedly compared to the data) reached.

As is customary, we can say that at this stage additional studies are needed. In particular, a consistent treatment of the transport of chemicals and angular momentum as decribed in §2.1 will prove most useful to quantify with confidence the impact of the rotation-induced mixing on the RGB. Work is in progress in this direction (Palacios et al. 2000).

The stake of this problem for what concerns the primordial nucleosynthesis and the galactic evolution of ^{3}He is discussed by Tosi and by Rood & Bania in this volume. Let us note that in a statistical study of the carbon isotopic ratio observed in post-bump RGB stars, Charbonnel & do Nascimento (1998) showed that at least 95 % of the low mass stars do undergo the extra-mixing. This thus leads to a strong revision of the actual contribution of low mass stars to ^{3}He evolution, and should account for both the measurements of ^{3}He/H in galactic H$_{II}$ regions (Balser et al. 1994) and in the planetary nebulae (Rood et al. 1992, Balser et al. 1997).

3.4. The Li rich giants

If the ^{7}Be produced deep inside the RGB stars by ^{3}He burning could be rapidly transported to cooler regions before its electron capture to ^{7}Li can take place, fresh ^{7}Li could show up at the stellar surface. The so-called Cameron-Fowler (1971) mechanism due to the extra-mixing would thus have a chance to produce the so-called Li rich RGB stars (de la Reza and Charbonnel & Balachandran, this volume). Sackmann & Boothroyd (1999; see also Sackmann in this volume and Denissenkov & Weiss 2000) showed that certain assumptions, which depend critically on the speed, geometry and episodicity of their parametrized mixing, can indeed lead to important Li creation along the RGB. In particular high Li enrichment (up to logN(Li)=4) is obtained by Sackmann & Boothroyd (1999) with a continuous mixing which simultaneously induces a smooth decrease of the carbon isotopic ratio up to the RGB tip. As we discussed in the previous section this prediction for the ^{12}C/^{13}C ratio is not sustained by the observations and cast doubt on the underlying assumptions of the model. Moreover Charbonnel & Balachandran (2000; see also this volume) showed that the field Li rich stars are not observed all along the RGB, but that they do clump at the bump phase. This result fits well to the observations of the ^{12}C/^{13}C ratio which reveal a very fast mixing episode at the RGB bump (Palacios et al. 2000). Thus the Li-rich phase is extremely short, and its contribution to the Li enrichment of the ISM is certainly negligible.

3.5. Field-to-cluster differences

At this point, it is important to remember some striking differences which distinguish field and globular cluster red giants. In particular, while field giants show C anomalies but no O nor Na variations (indicating that the extra mixing does not reach the deep internal regions where the complete CNO cycling and the NeNa cycle occur), globular cluster bright giants exhibit the so-called universal O-Na anticorrelation (e.g., Kraft et al. 1993, Kraft 1994, Denissenkov et al. 1998) which could be of primordial origin or due to a deeper extra-mixing than in field stars. Up to now, observations in globular clusters are available only for some very bright stars close to the RGB tip (except in M13 for which very sparse data exist for stars close to the RGB bump) in very few globular clusters. Observations are crucially needed in less evolved stars down to the main sequence turnoff in order to quantify the primordial contamination of the intracluster gas by an earlier generation of more massive stars and to disentangle it from the real impact of in situ extra-mixing. In other words, the field-to-cluster differences have to be observationaly quantified and understood to constrain the physics and the efficiency of the extra-mixing in the advanced evolutionary phases.

4. Acknowledgements

C.C. thanks the Action Spécifique de Physique Stellaire and the Conseil National Français d'Astronomie for support. C.P.D. acknowledges support from the United States National Science Foundation under grant AST-9812735. M.P. would like to acknowledge support from NASA grant NAG5-7150 and NSF grant AST-9731621.

References

Alschuler, W.R. 1975, ApJ, 195, 649

Balachandran, S. 1990, ApJ, 354, 310

Balser, D.S., Bania, T.M., Brockway, C.J., Rood, R.T., Wilson, T.L. 1994, ApJ, 430, 667

Balser, D.S., Bania, T.M., Rood, R.T., Wilson, T.L. 1997, ApJ, 483, 320

Behr, B.B., Cohen, J.G., McCarthy, J.K. 2000, ApJ, 531, L37

Boesgaard, A.M. 1987, PASP, 99, 1067

Boesgaard, A.M., Deliyannis, C.P., Stephens, A., King, J.R. 1998, ApJ, 493, 206

Boesgaard, A.M., Heacox, W.D., Conti, P.S. 1977, ApJ, 214, 124

Boothroyd, A.I., Sackmann, I.J. 1999, ApJ510, 232

Brown, J.A., Sneden, C., Lambert, D.A., Dutchover, E. 1989, ApJS, 71, 293

Burkhart, C., Coupry, M.F. 1989, A&A, 220, 197

Burkhart, C., Coupry, M.F. 1998, A&A, 338, 1073

Burkhart, C., Coupry, M.F. 1999, private communication

Burkhart, C., Coupry, M.F. 2000, A&A, 354, 216
Cameron, A.G.W., Fowler, W.A., 1971, ApJ, 164, 111
Caretta, E., Gratton, R.G., Sneden, C., E., Bragaglia, A. 1998, in Paris-Meudon Observatory meeting, Galaxy Evolution : Connecting the Distant Universe with the Local Fossil Record, in press
Cavallo, R., Sweigart, A., Bell, R., 1998, ApJ, 492, 575
Charbonneau, P., Michaud, G., Proffitt, C.R. 1989, ApJ, 347, 821
Charbonnel, C. 1994, A&A, 282, 811
Charbonnel, C. 1995, ApJ, 453, L41
Charbonnel, C., Balachandran, S., 2000, A&A, in press
Charbonnel, C., Brown, J.A., Wallerstein, G. 1998, A&A, 332, 204
Charbonnel, C., do Nascimento, J.D. 1998, A&A, 336, 915
Charbonnel, C., Talon, S. 1999, A&A, 351, 635, CT99
Charbonnel, C., Vauclair, S. 1992, A&A, 265, 55
Denissenkov, P.A., Weiss, A. 1996, A&A308, 773
Denissenkov, P.A., Weiss, A. 2000, A&A, accepted
Deliyannis, C.P., 1995, in the ESO/EIPC workshop, The Light Element Abundances, Elba, 395
Deliyannis, C.P., Boesgaard, A.M., King, J.R., 1995, ApJ, 452, L13
Deliyannis, C.P., Demarque, P., Kawaler, S.D., 1990, ApJS, 73, 21
Denissenkov, P.A, Da Costa, G.S., Norris, J.E., Weiss, A., 1998, A&A, 333. 926
do Nascimento, J.D., Charbonnel, C., Lèbre, A., de Laverny, P., de Medeiros, J.R. 2000, A&A, 357, 931
Duncan, D.K., Peterson, R.C., Thorburn, J.A., Pinsonneault, M.H. 1998, ApJ, 499, 871
Gilroy, K.K. 1989, ApJ, 347, 835
Gilroy, K.K., Brown, J., A. 1991, ApJ, 371, 578
Gratton, R.G., Sneden, C., Caretta, E., Bragaglia, A. 2000, A&A, 354, 169
Hogan, C.J. 1995, ApJ, 441, L17
King, J.R., Stephens, A., Boesgaard, A.M., Deliyannis, C.P. 1998, ApJ, 115, 666
Kraft, R.P. 1994, PASP, 106, 553
Kraft, R.P., Sneden, C., Langer, G.E., Shetrone, M.D., 1993, AJ, 106, 1490
Lèbre, A., de Laverny, P., de Medeiros, J.R., Charbonnel, C., da Silva, L. 1999, A&A, 345, 936
Maeder, A. 1995, A&A, 299, 84
Maeder, A., Zahn, J.P. 1998, A&A, 334, 1000
Mallik, S.V. 1999, A&A, 352, 495
Molaro, P., Pasquini, L., 1994, A&A, 281, L77
Palacios, A., Charbonnel, C., Forestini, M. 2000, in preparation
Pasquini, L., Molaro, P. 1996, A&A, 307, 761
Pilachowski, C.A., Sneden, C., Booth, J. 1993 ApJ, 407, 699
Proffitt, C.P., Michaud, G. 1991, ApJ, 371, 584

Randich, S., Gratton, R., Pallavicini, R., Pasquini, L., Carretta, E. 1999, A&A, 348, 487
Richer, J., Michaud, G., Turcotte, S. 2000, ApJ, 529, 338
Rood, R.T., Bania, T.M., Wilson, T.L. 1992, Nature, 555, 618
Ryan, S.G., Deliyannis, C.P. 1995, ApJ, 453, 819
Sackmann, I.J., Boothroyd, A.I. 1999, ApJ510, 217
Sneden, C., 1999, 35rd Liège Int. workshop on The Galactic Halo - From Globular Clusters to Field Stars, in press
Sweigart, A.W., Mengel, K.G. 1979, ApJ, 229, 624
Talon, S., Charbonnel, C. 1998, A&A, 335, 959
Talon, S., Zahn, J.P., Maeder, A., Meynet, G. 1997, A&A, 322, 209
Vauclair, S. 1991, Evolution of Stars : The Photospheric Abundance Connection, IAU Symp. 145 (eds G.Michaud & A.Tutukov), 327
Talon, S., Zahn, J.P. 1997, A&A, 317, 749
Wallerstein, G., Böhm-Vitense, E., Vanture, A.D., Gonzalez, H. 1994, AJ, 107, 2111
Wasserburg, G.J., Boothroyd, A.I., Sackmann, I.J., 1995, ApJ, 447, L37
Weiss, A., Denissenkov, P.A., Charbonnel, C., 2000, A&A, 356, 181
Zahn, J.-P., 1992, A&A, 265, 115

Creation and Destruction of ^{7}Li and ^{3}He in RGB and AGB Stars

I.-Juliana Sackmann and Arnold I. Boothroyd

W. K. Kellogg Radiation Laboratory 106-38, California Institute of Technology, Pasadena, CA 91125, U.S.A.

Abstract. Early in this decade our theoretical work demonstrated that all AGB stars in the mass range ~ 4 to $\sim 7 M_\odot$ pass through a stage when a tremendous amount of lithium [up to $\log \varepsilon(^7\text{Li}) \sim 4.5$] is created and transported to the surface. These lithium-rich AGB stars are predicted to occupy a narrow luminosity range between $M_{bol} = -6$ and -7, in excellent agreement with the observations of Smith & Lambert (1989), and might be useful as approximate standard candles. Recently, we found that even low mass stars (~ 1 to ~ 2 $M_\odot$) on the RGB could create a tremendous amount of surface lithium. In both the AGB and RGB cases, it is the Cameron-Fowler mechanism that is responsible for the lithium creation.

In the AGB stars, it is *hot bottom burning* (nuclear burning at the base of the convective envelope) that produces the lithium. In the RGB stars, it is "*cool bottom processing*" that can lead to either lithium production or destruction. Cool bottom processing results when extra mixing (presumably rotation-induced) transfers material from the cool convective envelope down to the outer wing of the hydrogen-burning shell (where nuclear reactions can take place) and back out to the envelope. If the extra mixing is slow, ^{7}Li is destroyed; if it is fast enough, then ^{7}Li is created — for sufficiently fast and deep extra mixing, $\log \varepsilon(^7\text{Li}) \sim 4$ is possible.

Unlike ^{7}Li, the ^{3}He abundance is almost independent of the mixing speed, and is constrained by observations of ^{12}C/^{13}C or [C/Fe] on the RGB. Cool bottom processing causes low mass stars of sub-solar metallicity to be net destroyers of ^{3}He, rather than net producers. This is in contrast to previous theoretical predictions, and has a far-reaching effect on our understanding of galactic chemical evolution of ^{3}He.

1. Introduction

Lithium burns at only a few million degrees K; thus the story of lithium in stars has traditionally been one of destruction. Lithium is destroyed during the pre-main sequence evolution by some low mass stars ($M \lesssim 1.2$ $M_\odot$ for Population I); it is also observed to be destroyed during the main sequence evolution of low mass stars ($M \lesssim 2$ $M_\odot$). Surface lithium abundances are diluted by two orders of magnitude in all low and intermediate mass stars when they reach the lower RGB (due to first dredge-up). According to classical stellar evolution theory, these reduced lithium abundances were not expected to change much subsequent to the RGB stage. On the other hand, as early as 1940, the AGB star WZ Cas was discovered to be superrich in lithium (McKellar 1940). In the next three decades, a handful of other superrich lithium stars were discovered in our galaxy. Some of these stars have lithium abundances orders of magnitude above that of the interstellar medium from which they were born (see, e.g., Wallerstein & Conti 1969; Boesgaard 1970; Abia et al. 1991). All of the superrich lithium stars discovered during this period were AGB stars. Furthermore, in the Magellanic Clouds, lithium enrichment has been discovered in the AGB stars in the magnitude range $-6 \gtrsim M_{bol} \gtrsim -7$ (Smith & Lambert 1989, 1990). Such lithium creation in AGB stars can be understood in terms of "hot bottom burning" (HBB) in intermediate mass stars of $4-7 M_\odot$ (Sackmann, Smith, & Despain 1974; Scalo, Despain, & Ulrich 1975; Sackmann & Boothroyd 1992), as will be discussed in § 3.

In the last two decades, stars rich in lithium have also been discovered on the RGB, starting with the work of Wallerstein & Sneden (1982). In fact, observations indicate that a few percent of all RGB stars are unusually lithium-rich (Brown et al. 1989). A few RGB stars have even been discovered to be superrich in lithium, with abundances above that of the interstellar medium (see, e.g., da Silva, de la Reza, & Barbuy 1995). About 20 lithium-rich RGB stars were discovered by the Pico dos Dias (PDS) Survey, which was searching for T Tauri candidates selected by means of the IRAS Point Source Catalog (Gregorio-Hetem et al. 1992); the fact that most lithium-rich RGB stars show a far-infrared excess (Gregorio-Hetem et al. 1993) suggests a connection between lithium enrichment and mass loss (de la Reza, Drake, & da Silva 1996). These lithium-rich RGB stars appear to have relatively low mass, of order $1-2.5\ M_\odot$ (in contrast to the intermediate masses, $4-7 M_\odot$, of the lithium-rich AGB stars). Such lithium creation in RGB stars can be understood in terms of "cool bottom processing" (CBP) in low mass stars (Sackmann & Boothroyd 1999), as will be discussed in § 4.

As far as ^{3}He is concerned, stars were traditionally considered to be net sources of this isotope. Rich pockets of ^{3}He are built up outside the cores of low and intermediate mass stars during their main sequence evolution; subsequently, during the early RGB stage, first dredge-up mixes ^{3}He from this pocket to the surface. High mass stars do not have time to build up a significant ^{3}He-pocket, and a reduction in surface ^{3}He results as first dredge-up reaches into ^{3}He-depleted layers further in. However, the ejecta from low and intermediate mass stars far outweighs the ejecta from the less common high mass stars. Thus, in classical stellar evolution, stars were predicted to strongly enrich the interstellar medium in ^{3}He. This conclusion is dramatically altered when cool bottom processing on the RGB is taken into account, as will be shown in § 4.

2. The Cameron-Fowler Mechanism

Cameron (1955) noted that a significant amount of ^{7}Be could be produced by the ^{3}He + α reaction in the p-p chain; he also noted that the half-life for electron capture of ^{7}Be is greatly lengthened (to of order 100 years) under stellar interior conditions due to ^{7}Be being almost completely ionized, compared to the 53-day half-life from K-capture under laboratory conditions. Therefore he reasoned that with the deep convective envelopes of red giants the ^{7}Be would be able to be carried out from the interior regions where it was created to the outer layers before electron capture would take place. The resulting ^{7}Li would thus be created under cool outer-layer conditions, where nuclear burning could not destroy it; it would be observable until the convective currents brought it down to interior layers hot enough for its burning, while at the same time more ^{7}Be was brought up to create new ^{7}Li.

At a jolly party of the Kellogg Radiation Laboratory of Caltech in 1970, Cameron took Willy Fowler aside and told him about the above idea. Willy then added that he had a clue where such a scenario could actually take place. Helium shell flashes (thermal pulses) had just been discovered to take place in AGB stars. Willy suggested that helium shell flashes might occasionally induce complete convection of the outer envelope down to the helium-burning shell. This back-of-the-envelope idea became known thereafter as the Cameron-Fowler mechanism (Cameron & Fowler 1971).

3. Hot Bottom Burning on the AGB

Classical stellar evolution calculations were unable to produce lithium in the surface layers of stars. Classical stellar evolution models did not provide convective envelopes reaching down into high-temperature regions where the ^{3}He+$\alpha \to$ ^{7}Be reaction could take place; certainly the "entropy barrier" prevents helium-shell flash convection from joining with the envelope convection (as had been proposed by Cameron & Fowler [1971]). However, Iben (1975) was able to reach high temperatures (60 million K) at the base of the conventional convective envelope of a 7 $M_\odot$ AGB star — hot enough for ^{7}Be production. Despite this fact, he found no lithium production, because he made the classical assumption of *instantaneous* convective mixing (Iben 1973).

Since the early observations of the superrich lithium stars clearly demonstrated that lithium creation did indeed take place in some stars, Sackmann et al. (1974) introduced models with a novel (non-classical) feature: an improved description of convective mixing was introduced, namely, time-dependent "convective diffusion" coupled to nuclear burning, discretized over many layers from the surface to the base of convection. In addition, the assumption was made that the convective envelope penetrated right down into the center of the hydrogen-burning shell, at 50 million K; such deep convective envelopes were not achievable in a self-consistent way at the time. On the other hand, this non-classical calculation *was* able to account for the superrich lithium abundances for the first time. Scalo et al. (1975), using the Sackmann et al. (1974) convective diffusion code in AGB envelope models (with inner boundary conditions estimated from the core mass–luminosity relation), again found high temperatures at the base of the convective envelope (which they christened "hot bottom burning"), together with the lithium production.

It became clear in the 1980's, when improved opacities became available, that the canonical value of unity for the mixing length parameter "α" (the mixing length relative to the pressure scale height) was incorrect; it had to be increased significantly (by about 50 to 100%, depending on the opacities used),

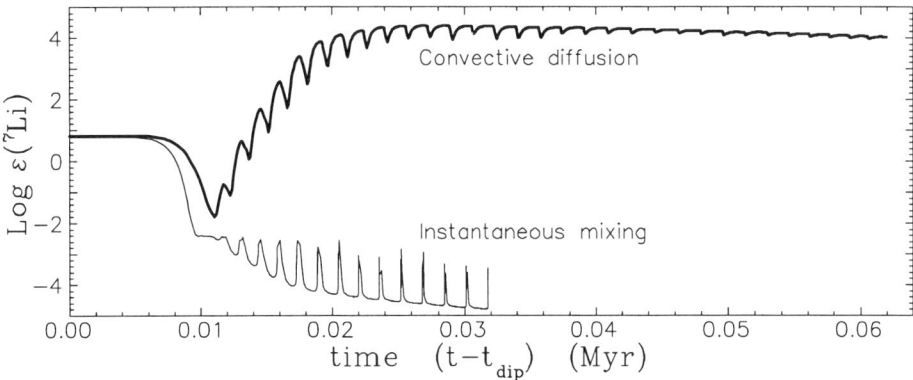

Figure 1. Lithium production in a 6 $M_\odot$, $Z = 0.02$ AGB star, comparing the time-dependent "convective diffusion" mixing algorithm with the "instantaneous mixing" approximation ($t_{\rm dip}$ is the time of the pre-flash luminosity "dip").

e.g., to obtain a correct solar model (Sackmann, Boothroyd, & Fowler 1990) and the correct position of the base of the RGB. Sackmann & Boothroyd (1991) demonstrated that the temperature at the base of the convective envelope on the AGB increases steeply as a function of α, for $1 \lesssim \alpha \lesssim 2$. For AGB stars between about 4 and 7 $M_\odot$, these new opacities and the corresponding larger α values yielded hot bottom burning in the envelopes of stellar models, i.e., in *self-consistent* models (rather than merely being assumed, as in the earlier lithium-production models).

Sackmann & Boothroyd (1992) applied the time-dependent "convective diffusion" algorithm of Sackmann et al. (1974) to the new hot bottom burning AGB models, producing for the first time self-consistent models for ^{7}Li production (shown as a function of time in Figure 1; the wiggles correspond to successive helium shell flashes. Also demonstrated in Figure 1 is the fact that the classical "instantaneous mixing" approximation leads to lithium destruction, rather than creation. Figure 2 is a key diagram, showing the predicted ^{7}Li abundances as a function of AGB luminosity for various stellar masses and metallicities. The peak lithium abundance reached, of $\log \varepsilon(^7{\rm Li}) \sim 4.5$, is consistent with the highest lithium abundances observed in galactic superrich lithium stars (Abia et al. 1991; Denn, Luck, & Lambert 1991). Figure 2 illustrates the theoretical prediction that the high ^{7}Li abundances occur in the magnitude range $-6 \gtrsim M_{bol} \gtrsim -7$, in excellent agreement with the Magellanic Cloud observations of Smith & Lambert (1989, 1990). Figure 2 also demonstrates that the peak ^{7}Li abundances are roughly independent of stellar mass and metallicity (for stars in the range $4 - 7$ $M_\odot$, where hot bottom burning takes place). Since the magnitude range (where ^{7}Li abundance is high) is relatively narrow, these superrich lithium stars on the AGB can in principal be used as approximate standard candles to yield distances; Ventura, D'Antona, & Mazzitelli (1999) pointed out that considering only the lithium-rich C-stars would work even better, as they have a narrower luminosity range and are less prone to confusion with lithium-rich RGB stars. Figure 2 also illustrates that the ^{7}Li abundance

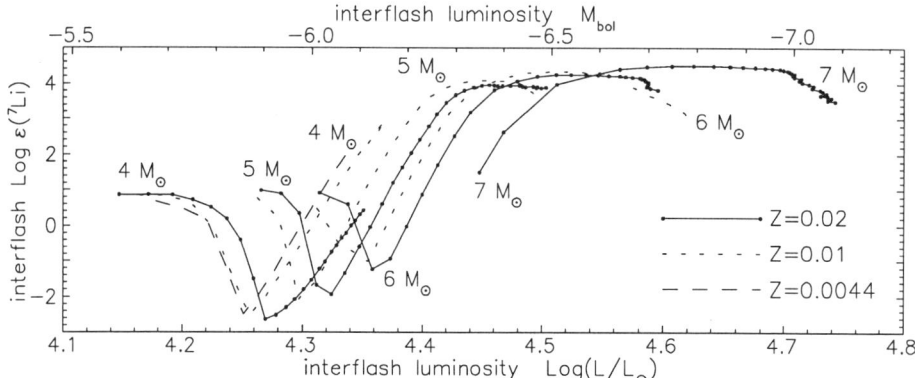

Figure 2. Lithium abundances in hot bottom burning AGB stars as a function of (interflash) luminosity, for various stellar masses at solar metallicity ($Z = 0.02$) and at approximate Large and Small Magellanic Cloud metallicities ($Z = 0.01$ and 0.0044, respectively).

declines as the ^{3}He fuel is used up; this decline continues much further than shown by the truncated runs of Figure 2, and thus these models are consistent with the observed range of abundances in lithium-rich Small Magellanic Cloud AGB stars, namely, $1.9 \lesssim \log \varepsilon(^7\text{Li}) \lesssim 3.5$ (Plez, Smith, & Lambert 1993).

4. Cool Bottom Processing on the RGB

After the excellent agreement between our theoretical predictions of superrich lithium stars on the AGB and the observations of such stars, we were made aware of recent observational discoveries of lithium-rich low-mass stars on the RGB (as summarized, unfortunately too briefly, in § ??). These lithium-rich RGB stars could *not* be accounted for by our theoretical models, since hot bottom burning does not occur on the RGB (nor does it occur even on the AGB in such low mass stars). To attempt to explain these mysterious lithium-rich RGB stars, we turned to the RGB extra mixing phenomenon that had long been invoked to explain RGB carbon observations. It has been suggested since the 1970's that the anomalously low $^{12}\text{C}/^{13}\text{C}$ ratios observed in low mass RGB stars could be accounted for by extra mixing below the convective envelope, which could convert ^{12}C into ^{13}C (see, e.g., Dearborn, Eggleton, & Schramm 1976; Sweigart & Mengel 1979). The term "cool bottom processing" (CBP) for this extra mixing with nuclear processing was coined by Wasserburg, Boothroyd, & Sackmann (1995); the difference between hot bottom burning and cool bottom processing is illustrated schematically in Figure 3.

First dredge-up leaves behind a composition discontinuity in the star, with a steep molecular weight gradient, frequently referred to as the "μ-barrier"; this μ-barrier is strongly stable against mixing, acting like a wall. RGB observations of Population I stars and field Population II stars indicate that the extra mixing region cannot reach from the convective envelope into the neighborhood of the hydrogen-burning shell until the hydrogen-burning shell has reached and de-

Creation and Destruction of ^{7}Li and ^{3}He in RGB and AGB

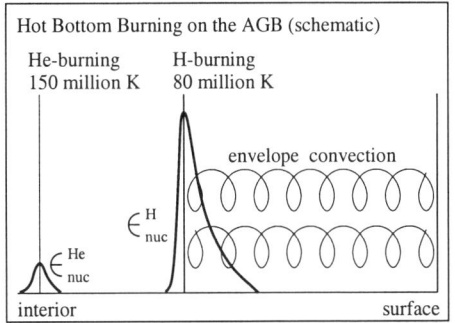 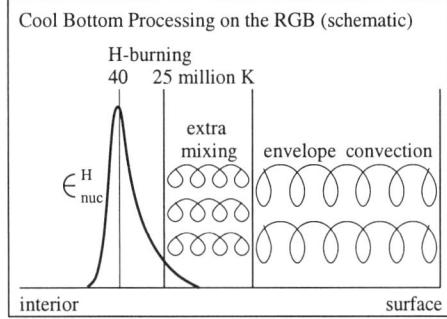

Figure 3. Schematic illustration of the difference between *hot bottom burning* and *cool bottom processing*; typical temperatures are shown.

stroyed this intervening μ-barrier (see, e.g., Charbonnel 1994; Charbonnel 1995; Boothroyd & Sackmann 1999). On the other hand, it is noteworthy that the extra mixing does appear to be able to penetrate this μ-barrier in globular cluster RGB stars, as discussed in Boothroyd & Sackmann (1999).

Charbonnel (1995) showed that extra mixing could lead to lithium *destruction* in low-mass RGB stars of low metallicity, explaining the anomalously *low* lithium abundances that are observed late on the RGB in in field Population II stars. Sackmann & Boothroyd (1999) demonstrated for the first time that cool bottom processing on the RGB could also lead to lithium *creation*. The extra mixing was assumed to reach into the outer wing of the hydrogen-burning shell, and was modelled as a "conveyer-belt" type circulation. As shown in Figure 4, a relatively long-lived mixing "episode" was modelled in a solar-mass Population I star at an RGB luminosity $\log L = 1.5$ (i.e., at the stage on the RGB where the hydrogen-burning shell has just reached and destroyed the μ-barrier): the extra mixing was assumed to reach down to within $\Delta \log T = 0.17$ of the bottom of the hydrogen-burning shell, and was assumed to last long enough to reproduce the observed RGB ^{12}C/^{13}C ratio of ~ 11 (namely, lasting 12.5 million years — short compared to the RGB evolutionary timescale). The left-hand panel of Figure 4 illustrates, for a circulation speed of 10^{-3} $M_\odot$/yr, the lithium production reaching as high a value as $\log \varepsilon(^7\mathrm{Li}) \sim 3$ for a period of $\sim 10^6$ yr, after which the lithium abundance declines due to its ^{3}He fuel being used up, as shown in the figure. Different speeds of circulation were considered; the right-hand panel of Figure 4 illustrates the strong sensitivity of the lithium abundance to the mixing speed. It demonstrates that rapid mixing speeds ($\gtrsim 10^{-4}$ $M_\odot$/yr, in this model) lead to lithium creation, while slower mixing leads to lithium destruction. The upper limit to the speed of this extra mixing is that it should be much slower than convective mixing, which has a speed of order 1 $M_\odot$/yr in RGB envelopes; the lower limit is the requirement that the mixing be faster than the speed with which the hydrogen shell burns its way outwards (of order 10^{-8} $M_\odot$/yr for the above model).

Figure 5 illustrates a different ("continuous") extra mixing scenario, where circulation was assumed to start when the μ-barrier was destroyed and to continue until the tip of the RGB was reached, but always reaching the same

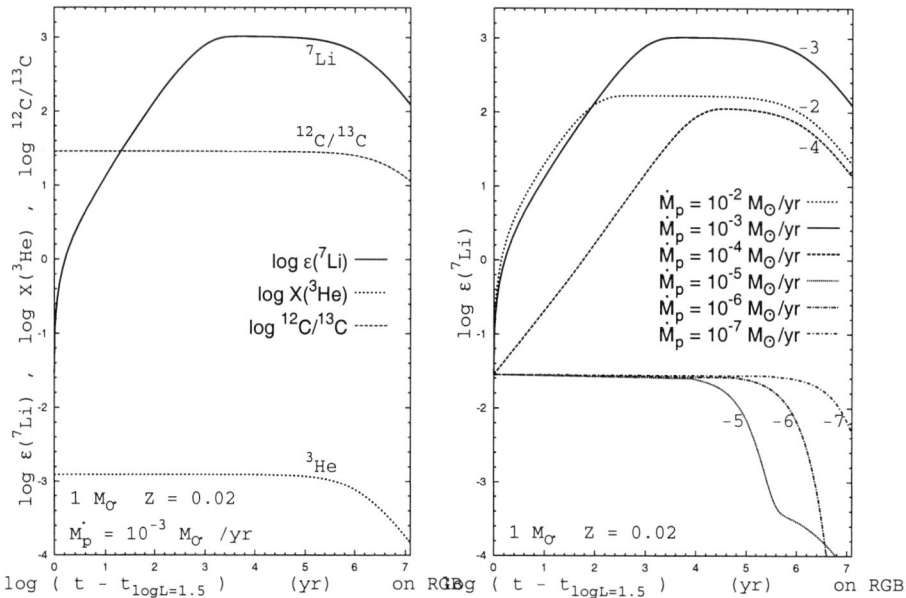

Figure 4. Lithium production as a function of time on the RGB, for a circulation model of a long-lasting extra mixing episode.

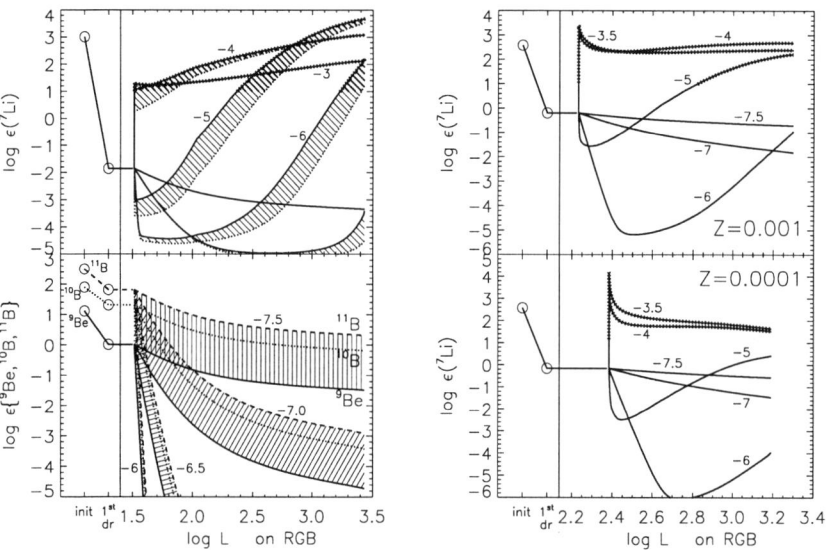

Figure 5. Lithium production as a function of RGB luminosity, for a "continuous" model where extra mixing continues to the end of the RGB.

distance in temperature from the base of the hydrogen-burning shell, namely, $\Delta \log T = 0.26$, in order to reproduce the observed RGB $^{12}C/^{13}C$ ratio of ~ 11 by the time the star reached the tip of the RGB. Figure 5 shows the lithium creation and destruction from such a model in a $1\,M_\odot$ star as it climbs the RGB. The upper left-hand panel illustrates that, for most mixing speeds, such continuous mixing leads to the highest lithium abundances near the tip of the RGB; for a mixing speed of $\sim 10^{-4}\,M_\odot/\mathrm{yr}$, $\log \varepsilon(^7\mathrm{Li}) \sim 4$ can be attained. Note the switchover between lithium creation and destruction as one changes the mixing speed. The shaded areas show the effect of changing the circulation geometry, as discussed in detail in Sackmann & Boothroyd (1999). The lower left-hand panel illustrates that the stable beryllium and boron isotopes are rapidly depleted, except at the very lowest mixing speeds; shaded areas tie together abundance curves for a given mixing speed. The right-hand panel of Figure 5 illustrates this "continuous" circulation model (with the same $\Delta \log T = 0.26$) for Population II objects. At all but the very slowest mixing speeds, the ^{3}He is largely destroyed soon after extra mixing begins, due to the higher temperature of the hydrogen-burning shell in Population II RGB stars. As a consequence, the highest attainable lithium abundances $\log \varepsilon(^7\mathrm{Li}) \sim 4$ are attained shortly after extra mixing starts on the RGB (rather than at the tip of the RGB), but only for rapid mixing speeds $\gtrsim 10^{-4}\,M_\odot/\mathrm{yr}$; lower but still observable amounts of lithium are maintained thereafter. Mixing speeds of $\sim 10^{-5}\,M_\odot/\mathrm{yr}$ lead first to destruction, then to creation of observable amounts of lithium near the RGB tip; slower mixing speeds lead only to lithium destruction.

The RGB carbon observations suggest an extra mixing scenario in between the above two mixing scenarios; presently, lithium observations may point towards a number of short successive mixing episodes (de la Reza et al. 1996).

5. Creation and Destruction of ^{3}He

Figures 6 and 7 illustrate the creation and destruction of ^{3}He as a function of stellar mass, for Population I and II stars, respectively. First dredge-up leads to considerable envelope enrichment of ^{3}He in low mass stars, and slight depletion in higher mass stars ($\gtrsim 5M_\odot$); see also § ??. In low mass stars ($\lesssim 2\,M_\odot$), cool bottom processing subsequently on the RGB destroys ^{3}He, and thus low mass stars of sub-solar metallicity are *not* sources of ^{3}He enrichment in the interstellar medium; Population II stars destroy much more ^{3}He than Population I stars. The amount of ^{3}He destruction due to CBP is uncertain by a factor of about 2; the strongest constraint is provided by observations of $^{12}C/^{13}C$ and [C/Fe] on the RGB, which constrain the total amount of processing, but the ^{3}He-burning reactions have a different temperature dependence from the CN-cycle rates, and thus the temperature at which the processing takes place has some effect on the amount of ^{3}He-burning associated with a given (observed) amount of CN-cycle processing. However, it is clear from Figure 7 that CBP in low mass Population II stars destroys essentially all their ^{3}He. Second dredge-up on the early AGB has only a relatively minor effect. For stars between ~ 4 and $7\,M_\odot$, hot bottom burning on the AGB will destroy essentially all the ^{3}He that is present.

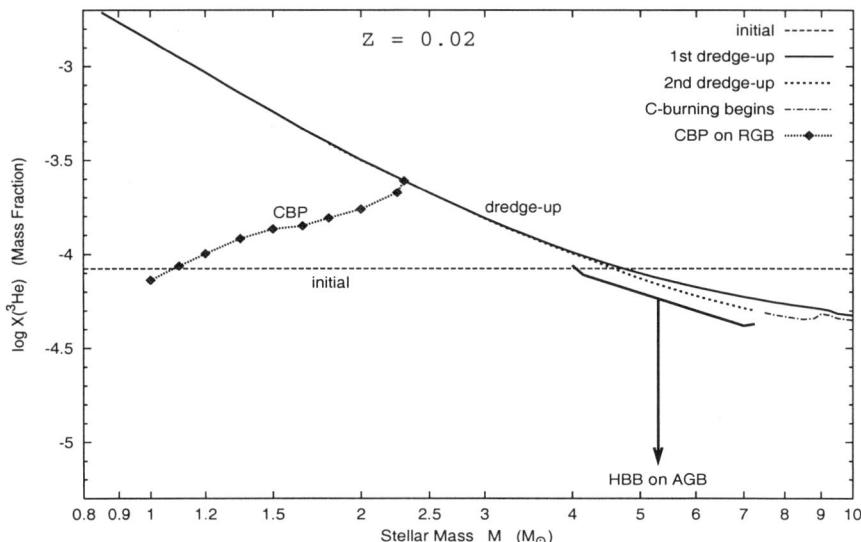

Figure 6. Production and depletion of ^{3}He in Population I stars. Heavy black diamonds show an estimate of the destruction of ^{3}He due to cool bottom processing (CBP), relative to the first dredge-up value. Bracket with arrow shows the mass region where hot bottom burning (HBB) on the AGB destroys the ^{3}He.

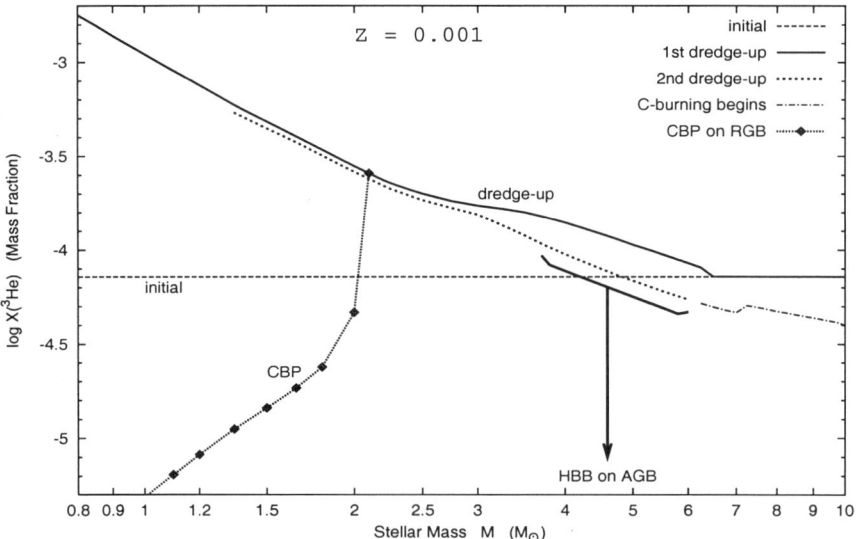

Figure 7. Production and depletion of ^{3}He in Population II stars; symbols as in Fig. 6.

Acknowledgments. We wish to acknowledge helpful discussions with C. A. Barnes and the support of R. D. McKeown of the Kellogg Radiation Laboratory. This work was partially supported by a grant from NSF PHY-9420470. One of us (I.-J. S.) wishes to acknowledge partial travel support to the Natal meeting from T. A. Tombrello, Division Chairman of Physics, Mathematics, and Astronomy of the California Institute of Technology.

References

Abia, C., Boffin, H. M. J., Isern, J., & Rebolo, R. 1991, ApJ, 245, L1
Brown, J. A., Sneden, C., Lambert, D. L., & Dutchover, E., Jr. 1989, ApJS, 71, 293
Boesgaard, A. M. 1970, ApJ, 161, 1003
Boothroyd, A. I., & Sackmann, I.-J. 1999, ApJ, 510, 232
Cameron, A.G.W. 1955, ApJ, 121, 144
Cameron, A.G.W., & Fowler, W. A. 1971, ApJ, 164, 111
Charbonnel, C. 1994, A&A, 282, 811
Charbonnel, C. 1995, ApJ, 453, L41
Dearborn, D., Eggleton, P. P., & Schramm, D. N. 1976, ApJ, 203, 455
da Silva, L., de la Reza, R., & Barbuy B. 1995, ApJ, 448, L41
de la Reza, R., Drake, N. A., & da Silva, L. 1996, ApJ, 456, L115
Denn, G. R., Luck, R. E., & Lambert, D. L. 1991, ApJ, 377, 657
Gregorio-Hetem, J., Castilho, B. V., & Barbuy, B. 1993, A&A, 268, L25
Gregorio-Hetem, J., Lépine, J. R. D., Quast, G. R., Torres, C. A. O., & de la Reza, R. 1992, AJ, 103, 549
McKellar, A. 1940, PASP, 52, 407
Iben, I., Jr. 1973, ApJ, 185, 209
Iben, I., Jr. 1975, ApJ, 196, 525
Plez, B., Smith, V. V., & Lambert, D. L. 1993, ApJ, 418, 812
Sackmann, I.-J., & Boothroyd, A. I. 1991, ApJ, 366, 529
Sackmann, I.-J., & Boothroyd, A. I. 1992, ApJ, 392, L71
Sackmann, I.-J., & Boothroyd, A. I. 1999, ApJ, 510, 217
Sackmann, I.-J., Boothroyd, A. I., & Fowler, W. A. 1990, ApJ, 360, 727
Sackmann, I.-J., Smith, R. L., & Despain, K. H. 1974, ApJ, 187, 555
Scalo, J. M., Despain, K. H., & Ulrich R. K. 1975, ApJ, 196, 805
Smith, V. V., & Lambert, D. L. 1989, ApJ, 345, L75
Smith, V. V., & Lambert, D. L. 1990, ApJ, 361, L69
Sweigart, A. V., & Mengel, J. G. 1979, ApJ, 229, 624
Ventura, P., D'Antona, F., & Mazzitelli, I. 1999, ApJ, 524, L111
Wallerstein, G. & Conti, P. S. 1969, ARA&A, 7, 99
Wallerstein, G. & Sneden, C. 1982, ApJ, 255, 577
Wasserburg, G. J., Boothroyd, A. I., & Sackmann, I.-J. 1995, ApJ, 447, L37

The Light Elements and Their Evolution
IAU Symposium, Vol. 198, 2000
L. da Silva, M. Spite, J. R. de Medeiros, eds.

Cosmological gravitons back reaction and the primordial nucleosynthesis

M. R. G. Maia[1], J. C. Carvalho[1], J. S. Alcaniz[1] and J. M. F. Maia[1,2]

[1] Departamento de Física-UFRN, CP 1641, 59072-970 Natal, RN-Brasil
[2] Instituto de Física-USP, CP 66318, 05389-970 São Paulo, SP-Brasil

Abstract. The back reaction of effective gravitons created during noninflationary epochs due to the inequivalence of vacuum states at different eras is examined in the context of primordial nucleosynthesis. Our final purpose is to obtain limits on the model employed to study such a process.

The inequivalence of vacuum states at different eras of the history of the universe determines the production of gravitational waves in scales larger than the Hubble lengh. During noninflationary periods of expansion, these very long tensor perturbations (VLTP's) enter the Hubble radius, thus becoming effective gravitational waves (EGW's) (Allen 1988). Such an effect adds new contributions to the gravitons energy density ρ_g within the horizon, the modes energetically meningful, and can be described by using a macroscopic approach to matter creation (see de Garcia Maia et al., 1997 an references therein). This is done by introducing a creation pressure term in the balance equation for ρ_g and the EWG's back reaction leads to the following dynamical equation for the scale factor $a(t)$ during the radiation era (for the flat case):

$$\frac{\ddot{a}}{a} + \frac{3\gamma_r - 2}{2} \frac{\dot{a}^2}{a^2} = 4\pi G \left(\gamma_r \rho_g + \frac{\dot{\rho}_g}{3H} \right), \qquad (1)$$

where H is the Hubble parameter, γ_r is the barotropic parameter related to the dominant component of the era denote by the subscript r and the whole R.H.S. is due to the effective gravitons creation process. It can be shown (de Garcia Maia et al. 1997) that given γ_{r-1}, γ_r and $H(t_r)$, the explicit expression for ρ_g is univocally determined in terms of $a(t)$ and $\dot{a}(t)$, so that $a(t)$ can be determined.

Here, we will consider that a transition occurred between a phase denoted by $r = 0$, for example an inflationary period, to the radiation era ($r = 1$). In such a case,

$$\rho_g(t) = \frac{1}{\pi^2} \left[\frac{(3\gamma_0 - 2)H_1}{2} \right]^4 \left[\frac{a_1}{a(t)} \right]^4 I(t), \qquad (2)$$

where $I(t)$ accounts for the graviton creation and is given by (Carvalho et al. 2000)

$$I(t) = \frac{(2m_0 + 1)^2 \Gamma^2(-m_0)}{\pi(2m_0 + 3)2^{2m_0+5}} \left[1 - \left(\frac{4\pi}{a_1 H_1 (3\gamma_0 - 2)} \dot{a}(t) \right)^{2m_0+3} \right], \qquad (3)$$

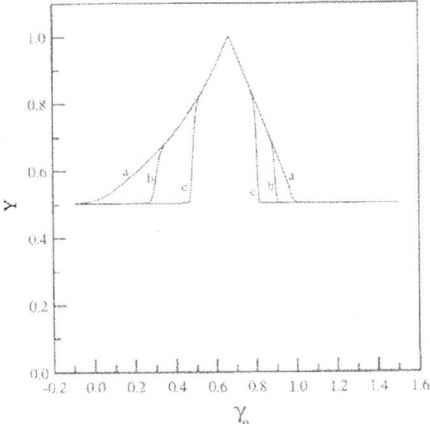

Figure 1. The asymptotic value of the slope $Y = [d\log a(t)/a_1][d\log t]$ as a function of the barotropic parameter γ_o. Curves (a), (b), and (c) correspond to $(H_1/m_{\text{planck}}) = 1$, $(H_1/m_{\text{planck}}) = 10^{-5}$, $(H_1/m_{\text{planck}}) = 10^{-10}$, repectively. As can be seen, as γ_o gets close to $2/3$, the slope Y changes drastically from the standard value $1/2$ to peak at the value 1 for $\gamma_o = 2/3$.

with $m_r \equiv 2/(3\gamma_r - 2) - 1/2$. For details on obtaining the above equations, see de Garcia Maia et al. 1997.

We should mention that the production of VLTP's is a natural outcome of quantum mechanics and general relativity in the cosmological context. This phenomenon may predict observational effects, depending on the values of free parameters of the model (de Garcia Maia et al. 1999). Here, we are interested in constraining these parameters from a primordial nucleosynthesis analysis.

Since only effective gravitons are produced, and they are not coupled to the other particles at almost all times of the thermal history of the universe, the radiation expands adiabatically, as in the standard picture. The main modification on primordial nucleosynthesis comes from the time-temperature relation, due to the back reaction effect on the expansion rate (Carvalho et al. 2000):

$$t = 2(10.4)^2 \int_T^\infty \frac{dT}{T^3\sqrt{1 + \alpha I(T)}}, \qquad (4)$$

where $\alpha = 4(10.4)^4 \frac{8GH_1}{3\pi}(\frac{3\gamma_0 - 2}{2T_1})^4$. We are presently implementing the numerical code to solve the above equation as well as making the necessary modifications on Kawano's code (Kawano 1992) in order to determine the bounds on the parameter γ_0 from nucleosynthesis. Because the dynamics of the Universe is not very sensitive to the value of H_1 (de Garcia Maia et al. 1997), we expect that for $\gamma_0 \sim 2/3$ the standard results will be considerably affected (see Fig. 1). In particular, the present estimates of the ^{4}He abundance could impose constraints on the possible values of γ_0 (Carvalho et al. 2000).

References

Allen, B. 1988, Phys. Rev. D, 37, 2078.

Kawano, L. 1992, Fermilab preprint PUB-92/04-A (unpublished)

de Garcia Maia, M. R., Carvalho, J. C. and Alcaniz, J. S. 1997, Phys. Rev. D, 50, 6351.

de Garcia Maia, M. R., Carvalho, J. C. and Alcaniz, J. S. 1999, Phys. Rev. D, 60, 123510.

Carvalho, J. C., de Garcia Maia, M. R., Alcaniz, J. S. and Maia, J. M. F. 2000, in preparation.

Constraints from big bang nucleosynthesis on a time-varying cosmological constant

J. A. S. Lima[1], J. M. F. Maia[1,2] and N. Pires[1]

[1] *Departamento de Física-UFRN, CP 1641, 59072-970 Natal, RN-Brasil*
[2] *Instituto de Física-USP, CP 66318, 05389-970 São Paulo, SP-Brasil*

Abstract. The limits from big bang nucleosynthesis on a class of models with time-varying Λ-term are reexamined. In particular, it is discussed how the stringent bounds previously derived may be relaxed.

Recently, many authors have proposed cosmological models where the Λ-term decays continously during the evolution of the Universe (Overduin & Cooperstock 1998 and references therein). In this approach, the Λ-term depends on the variables dynamically relevant and also on some phenomenological parameters, which should be constrained from observations.

Here, we rediscuss the bounds inferred from primordial nucleosynthesis on the parameter β arising in the phenomenological law (Lima & Maia 1994, Lima & Trodden 1996)

$$\Lambda = 3\beta H^2 + 3(1-\beta)\frac{H^3}{H_I},$$

where H is the Hubble parameter and H_I^{-1} is an arbitrary time scale related to an initial inflationary stage. This model presents a smooth transition between the inflationary epoch and the FRW-like radiation era. In particular, if $H_I \sim M_{\rm pl}^2$, it explains the discrepancy between the expected and the effectively observed values of the cosmological term. At the time of nucleosynthesis, the second term in the r.h.s. of the above equation is negligible, resulting in a Universe with an effective age of $t_0 = 2H_0^{-1}/3(1-\beta)$. Using the estimates of H_0 and t_{gc} (the age of globular clusters) available at that time, it was found that $\beta \geq 0.21$. Later on, by analysing the constraints from nucleosynthesis, Birkel and Sarkar (1997, hereafter BS) derived that $\beta \leq 0.13$, thereby concluding that the model was ruled out by these combined data. Here we argue that both results are not so definitive and that the limits on β still deserve a closer escrutiny.

Recent measurements of the Hubble parameter yields $H_0 = 70 \pm 10 km/s/Mpc$ (Freedman 2000) and $t_{gc} = 13 \pm 1 Gyr$, thus decreasing the age lower bound on β to 0.06, if the minimal values of H_0 and t_{gc} are considered. This lower bound is much smaller than the one of our previous analysis, and also well below the upper limit of BS. In addition, using updated values of element abundances in an adapted version of Kawano's nucleosynthesis code (Kawano, 1992), we have found a slightly greater upper bound, $\beta \leq 0.16$), which relax even further the age problem. Our main results are summarized in Fig. 1 below.

Finally, we stress that the most attractive feature of the model is its non-singular and deflationary evolution. Since both aspects are independent of β, the scenario is by no means ruled out from observational constraints based on

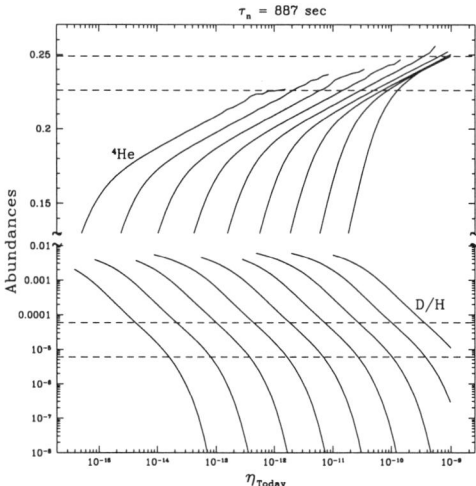

Figure 1. Abundances of primordial ^4He and D/H as functions of η_{today}. Dashed lines correspond to observational values (Molaro 1997). From right to left, β ranges from 0 to 0.16 (step 0.02). We assumed 3 light neutrino species and the neutron lifetime $\tau_n = 887 \pm 2s$.

this parameter. These analyses are also rested on the assumption that Λ decays only into the dominant component of the cosmic fluid. However, after the end of the radiation era, a decay into matter driven by an effective parameter (β_m) is not forbidden from first principles. In this case, bounds for this free parameter derived in the radiation epoch are not valid to the matter dominated phase. Still following this reasoning, one may ask why to neglect the decay into neutrinos after their decoupling from radiation, as have been done here and in BS. In the same vien, without a more careful study, the chemical potential associated with the products of the decaying vacuum cannot be neglected, as usually done in the standard nucleosynthesis treatment. In principle, all these potentially important issues should be addressed before a more definitive conclusion.

Acknowledgments This work was partially supported by CAPES, CNPq and the project PRONEX-FINEP (No. 41.96.0908.00).

References

Birkel, M. & Sarkar, S. 1997, Astropart. Phys., 6, 197.

Freedman, W. L. 2000, Phys. Rep., in press.

Kawano, L. 1992, Preprint FERMILAB-PUB-92/04-A.

Lima J. A. S. & Maia, J. M. F. 1994, Phys. Rev. D, 49, 5597.

Lima J. A. S. & Trodden M. 1996, Phys. Rev. D, 53, 4280.

Molaro, P. 1997, P. A. S. P. Conference Series, 126, 103.

Overduin, F. M. & Cooperstock, F. I. 1998, Phys. Rev. D, 58, 043506.

Photon Creation in the Universe and Primordial Nucleosynthesis

J. A. S. Lima, J. S. Alcaniz, J. Santos and R. Silva Jr.

Departamento de Física - UFRN, CP 1641, 59072-970 Natal, RN - Brasil

Abstract. In hot big bang cosmologies, the irreversible process of continous photon creation may phenomenologically be described through a thermodynamic approach. In these models, the radiation temperature law depends on a phenomenological parameter β which is closely related to the photon creation rate. It is shown that a stringent constraint on the value of this parameter is imposed from primordial nucleosynthesis.

1. Cosmology with photon creation

The increasing difficulties of the standard CDM cosmology are now compelling the investigation of alternative big-bang mo- dels. One of the main observational motivations is the conflict between the expanding age of the universe, as inferred from recent measurements of the Hubble parameter (Friedmann 2000) and the age of the oldest stars in globular clusters (Pont et al. 1998).

It is also widely believed that matter and radiation need to be created in order to overcome some conceptual problems of the standard hot big bang model. In this concern, a thermodynamic description for gravitational creation of matter and radiation has been proposed in the literature (Prigogine et al. 1989; Calvão et al. 1992). The crucial ingredient of this formulation is the explicit use of a balance equation for the number density of the created particles in addition to Einstein field equations. In this framework, the thermodynamic second law leads naturally to a reinterpretation of the energy momentum tensor corresponding to an additional stress term (creation pressure), which in turn depends on the photon creation rate, and alters considerably the observational predictions of the CDM model (Lima and Alcaniz 1999; Alcaniz and Lima 1999).

In this context, by addopting a photon creation scenario recently discussed (Lima and Abramo 1999; Wichoski and Lima 1999), an upper limit to the creation parameter is derived from primordial nucleosynthesis constraints.

The Einstein field equations for a fluid endowed with "adiabatic" photon creation and the balance equation for the particle number density are (Calvão et al. 1992)

$$8\pi G \rho = 3\frac{\dot{R}^2}{R^2} , \tag{1}$$

$$8\pi G(p + p_c) = -2\frac{\ddot{R}}{R} - \frac{\dot{R}^2}{R^2} , \tag{2}$$

Table 1.   Limits to β

η	^{4}He Abundance	β
10^{-12}	≤ 0.25	≤ 0.15
10^{-11}	≤ 0.25	≤ 0.12
10^{-10}	≤ 0.25	≤ 0.06

$$\frac{\dot{n}}{n} + 3\frac{\dot{R}}{R} = \frac{\psi}{n} \quad , \qquad (3)$$

where an overdot means time derivative and ρ, p, n and ψ are the energy density, pressure, particle number density and photon creation rate, respectively. The creation pressure p_c depends on the photon creation rate, and for "adiabatic" photon creation takes the form $p_c = -\frac{\rho+p}{3nH}\psi$ (Calvão et al. 1992).

Let us now assume a physically reasonable expression to the photon creation rate ψ. In this work we confine our attention to the simplest phenomenological expression $\psi = 3\beta nH$ (Lima et al. 1996), where β is smaller than unity, and given by some particular microscopic model for gravitational creation. This parameter must be a function of the cosmic era, however, in what follows we assume only photon creation (β constant).

2. Limits to β

Modifying the Wagoner-Kawano code (Kawano 1992), we have put limits on the photon creation rate β. In virtue of the photon creation process, the baryon-to-photon ratio is a function of time, which is now expressed as

$$\eta = \eta_{today}\left(\frac{T_9}{T_o \times 10^{-9}}\right)^{\frac{3\beta}{(1-\beta)}} . \qquad (4)$$

The ^{4}He abundance as a function of the baryon-to-photon ratio for some selected values of β parameter is shown in Fig. 1. The limits derived for the β parameter have been obtained assuming a fixed maximum abundance for ^{4}He (≤ 0.25), $N_\nu = 3$, $\tau_n = 887 \pm 2$, and several lower bounds for η. The limits on the β parameter obtained from ^{4}He abundance are summarized in Table 1.

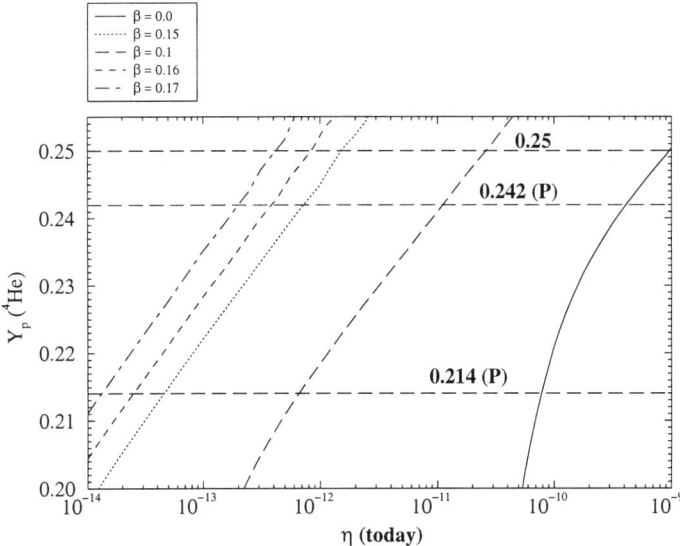

Figure 1. The ^{4}He abundance as a function of the baryon-to-photon ratio for some selected values of β parameter. Dashed lines labelled as 0.214 (P) and 0.242 (P) stand for the ^{4}He abundance (95% cl) obtained by Pagel et al (1992).

Acknowledgments

This work was partially supported by CNPq and the project PRONEX-FINEP (No. 41.96.0908.00).

References

Freedman, W. L. 2000, Phys. Rep., in press.

Pont, F., Mayor, M., Turon, C., & Vandenberg, D. A. 1998, A&A, 329, 87

Prigogine, I., Geheniau, J., Gunzig, E., Nardone, P., et al. 1989, Gen. Rel. Grav., 21, 767

Calvão, M. O., Lima, J. A. S. & Waga, I. 1992, Phys. Lett. A, 162, 233

Lima, J. A. S. & Alcaniz J. S. 1999, A&A, 348, 1

Alcaniz, J. S. & Lima, J. A. S. 1999, A&A, 349, 729

Lima, J. A. S. & Abramo, L. R. W. 1999, Phys. Lett. A, 257, 123

Wichoski, U. F. & Lima, J. A. S. Phys. Lett. A, 262, 103

Lima, J. A. S., Germano, A. S., & Abramo, L. R. W. 1996, Phys.Rev.D, 53, 4287

Kawano, L. 1992, in Let's Go: Early Universe II, Fermilab preprint PUB-92/04-A (unpublished)

Pagel, B. E. J., Simonson, E. A., Terlevich, R. J., and Edmunds, M. G. 1992, MNRAS, 255, 325

Change in Primordial Abundances Due to a Change in the Primordial Plasma Energy Density

M. Opher[1]

Dept. of Physics, UCLA, Los Angeles, Calif., USA

R. Opher[2]

Instituto Astronômico e Geofísico - IAG/USP, Av. Miguel Stéfano, 4200 CEP 04301-904 São Paulo, S.P., Brazil

Abstract. We recently showed that the energy density of a plasma is appreciably different than previously thought when high frequency plasma fluctuations, $\omega \geq k_B T/\hbar$, are taken into account (M. Opher & R. Opher, 1999). A change in the primordial plasma energy density changes the primordial expansion rate of the universe, the neutron temperature freeze-out, and the primordial nucleosynthesis abundances. The change in the primordial abundances due to the change in the primordial plasma energy density is evaluated, taking into account the high frequency, as well as the low frequency fluctuations of the plasma.

The universe at the epoch of primordial nucleosynthesis was an electron positron plasma at the beginning and, after the electron positron annihilation, an electron-proton-helium plasma. The energy estimated in the primordial nucleosynthesis epoch is the sum of the energies of all of the particles present in the primordial soup. In the standard calculation, the energy density of the particles is obtained by treating the particles as non-interacting, as in an ideal gas. The photon energy density is estimated to be the energy density of the blackbody in vacuum (Kolb & Turner 1990).

Recently, we showed that the correction to an ideal gas is greater than the Debye-Hückel correction, in the low temperature classical limit (Opher & Opher 1999). This investigation was made by studying the electromagnetic fluctuations present in a plasma, where all of the frequencies were included. The Debye-Hückel theory is recovered when only fluctuations with frequencies $\omega < T$ are included ($\hbar = k_B = 1$). We studied plasmas with temperatures $T = 10^3 - 10^5$ K and densities $n = 10^{13} - 10^{19}$ cm^{-3}, obtaining corrections on the order of $10^2 - 10^{10}$ times greater than those given by the Debye-Hückel theory. We may, therefore, expect that in the epoch of primordial nucleosynthesis, a similar important correction exists to the energy, as that found for classical low temperature plasmas (Opher & Opher 1999).

By integrating the spectra in wave number and frequency, we obtain the energy densities of the magnetic field, ρ_B, and of the transverse and longitudinal electric fields, ρ_{E_T} and ρ_L. We perform the integration of the spectra over frequency and wave number, without assuming that $\omega < T$. For details of the calculation, see our previous investigation (Opher & Opher 1999).

To obtain the interaction energy, ρ_{int}, we subtract the energy of the particles due to their own fields from the longitudinal energy density, ρ_L. We previously found that the transverse energy (summing the transverse electric and magnetic field energies, ρ_{E_T} and ρ_B) has an additional energy, compared to the blackbody energy density in vacuum. The additional transverse energy is $\Delta\rho_\gamma = \rho_B + \rho_{E_T} - \rho_\gamma$, where ρ_γ is the photon energy density, estimated to be the blackbody energy density in vacuum.

Adding the interaction energy ρ_{int} to $\Delta\rho_\gamma$, we obtain the total change in the energy density due to the transverse and longitudinal components, $\rho_{new} = \Delta\rho_\gamma + \rho_{int}$. We calculate ρ_{new} for the primordial electron-positron plasma for $3 \times 10^9 \ K < T < 1.3 \times 10^{10} \ K$. ρ_{new} is also calculated for low temperatures, after the electron-positron annihilation, when we have an electron-proton-helium plasma. After annihilation, the density drops by a factor of 10^{10}. Therefore, ρ_{new} will be negligble for low temperatures. We find that $\rho_{new} \sim -10\%\rho_\gamma$ and goes to zero at lower temperatures. This is consistent with our previous analysis at low temperatures (Opher & Opher 1999), where we found that ρ_{new} is positive.

It is interesting to note that the ratio ρ_{new}/ρ_{part} over this range of temperatures and densities (i.e., between $T = 3.5 \times 10^9 \ K$ and $T = 1.3 \times 10^{10} \ K$) is approximately constant (~ -0.33), where ρ_{part} is the energy density of the particles, $\rho_{part} = (3/2)nT$. The maximum variation is $\sim 2\%$. Thus, we can write $\rho_{new} \cong -0.33((3/2)nT)$, valid for all temperatures.

We found that the energy content is $\sim -10\%$ less than previously thought. A decrease of ρ_γ by 10% acts like a decrease in the number of relativistic neutrinos from $N_\nu = 3.0$ to $N_\nu = 2.6$, which predicts a decrease in the theoretical abundance of helium Y (Theor.). This is in agreement with the results presented at this meeting. Whereas the standard model (with low deuterium abundance) predicts Y (Theor.)=0.245, Viegas and Gruenwald (2000) report at this meeting that observations of HII extragalactic regions (taking into account their inhomogeneity) indicate that Y (Observ.)=0.241 and Peimbert (2000) reports at this meeting, as well, that observations of the Magellanic Clouds indicate that Y(Observ.)=0.236

Acknowledgments. M.O. thanks the Brazilian agency FAPESP for support (no. 97/13427-8) and R.O. thanks the Brazilian agency CNPq for partial support. Both authors thank the Brazilian project Pronex/FINEP (no. 41.96.0908.00) for support. The authors are also grateful to the late David Schramm for sending us the primordial nucleosynthesis code used in this paper.

References

Kolb, E. W. & Turner, M. S. 1990, The Early Universe
(Redwood City: Addison-Wesley)
Opher, M. & Opher, R. 1999, Phys.Rev.Lett, 82, 4835
Viegas, S. M. & Gruenwald, R. 2000,
(see article in these proceedings)
Peimbert, M. 2000, (see article in these proceedings)

The Light Elements and Their Evolution
IAU Symposium, Vol. 198, 2000
L. da Silva, M. Spite, J. R. de Medeiros, eds.

The Recombination in a FRW Universe with a Variable Cosmological Term

N. Pires and J. A. S. Lima

Departamento de Física - UFRN, CP 1641, 59072-970 Natal, RN - Brasil

Abstract. The effects of a time-dependent Λ term in the recombination epoch are analysed taking into account the main physical processes occurring in the primordial plasma. For a vacuum decaying into photons, we show that the recombination begins when the Universe was smaller and denser, leading to a greater recombination rate. These results may have several implications to the large scale structure. In particular, the earlier recombination means that more time is available to the evolution of density perturbations.

1. The Model

We consider an adiabatic model with a Λ–term decaying into photons (Lima 1996 and Lima and Trodden 1996), given by $c^2 \Lambda_v(t) = 8\pi G \beta \rho_T$, where ρ_T is the total energy density (radiation, matter and vacuum). The parameter $\beta < 0.16$ is constrained from nucleosynthesis studies (Maia et al. 2000). In this model the temperature law scales as $T_\gamma \propto a^{1-\beta}$, where a is the scale factor. For FRW models with a Λ-term, the age of the Universe reads

$$H_0 t_0 = \int_0^1 \left(\frac{(1-\beta)x^{(1-3\beta)}}{\Omega_{m0} + (1 - \beta - \Omega_{m0} - \lambda_0)x^{(1-3\beta)} + \lambda_0 x^{3(1-\beta)}} \right)^{1/2} dx, \quad (1)$$

where $H_0 = 100\ h \text{Mpc}^{-1}\text{km s}^{-1}$ is the Hubble constant, $\lambda_0 = \Lambda_c/3H_0^2$ is the cosmological constant and $\Omega_{m0} = (8\pi G/3H_0^2)\rho_{m0}$ is the matter density parameter. The index "0" denotes the present day values. Assuming that the Universe is a mixture of protons, electrons and hydrogen, we consider a variety of physical processes envolving matter and radiation during and after recombination: photon drag, ionization due to radiation and electronic collisions, recombination cooling and Compton heating–cooling (Opher et al. 1998).

The results have been obtained using a hydrodynamical code (including the above processes), and are summarized in the table below. All models are parametrized by Ω_{m0}, h, β, and λ_0. This table yields the temperature of the radiation, $T_{\gamma rec}$ and the redshift, z_{rec}, at the recombination epoch, which is assumed to occur when 50% of the electrons have been captured. The last column of this table yields the age of the Universe as given by (1).

Model	Ω_{m0}	h	β	λ_0	$T_{\gamma rec}(K)$	z_{rec}	$Age(Gyr)$
A1	0,024	0,73	0,1	0,0	3900	3201	13,1
A2	"	"	0,05	0,0	3740	2003	13,0
A3	"	"	0,0	0,876	3540	1296	20,3
A4	"	"	0,1	0,876	3920	3220	25,1
A5	"	"	0,0	0,0	3540	1296	12,9
B1	0,05	0,5	0,1	0,0	3900	3201	18,8
B2	"	"	0,05	0,0	3740	2003	18,6
B3	"	"	0,0	0,85	3540	1296	26,7
B4	"	"	0,1	0,85	3920	3220	31,7
B5	"	"	0,0	0,0	3540	1296	18,3
C1	0,1	0,73	0,1	0,0	4070	3357	12,5
C2	"	"	0,05	0,0	3900	2093	12,3
C3	"	"	0,0	0,8	3720	1362	16,1
C4	"	"	0,1	0,8	4070	3357	18,6
C5	"	"	0,0	0,0	3720	1362	12,0
D1	0,1	0,5	0,1	0,0	3990	3283	18,3
D2	"	"	0,05	0,0	3820	2048	17,9
D3	"	"	0,0	0,8	3620	1325	23,5
D4	"	"	0,1	0,8	3990	3283	27,1
D5	"	"	0,0	0,0	3620	1325	17,6

2. CONCLUSIONS

For models with cosmological constant (or standard CDM), same results already established are also diplayed: The recombination begins in a higher redshift if the Universe is denser. The Universe is older if the cosmological constant increases (for the same h). For models with a decaying vacuum into photons the main results are:
1) $\beta \neq 0$ always increases the age of the Universe.
2) The recombination temperature ($T_{\gamma rec}$) is greater and the same happens with the redshift (z_{rec}).
3) The β parameter has also a deep influence on the late time evolution of fluctuations. In particular, comparing several models at equal redshift, the Jeans mass is smaller if the β parameter increases.

References

Lima J. A. S. 1996, Phy. Rev. D., 54, 2571
Lima J. A. S. & Trodden M. 1996, Phy. Rev. D., 53, 4280
Maia J., Lima J. A. S. & Pires N., Constraints from Big Bang Nucleosynthesis on a Time Varying Cosmological "Constant", in this book.
Opher R., Pires N. & de Araujo J.C.N. 1997, MNRAS, 285, 811
Pires N. 1999, PhD Thesis, IAG-USP

Improved Use of Inputs to Primordial Nucleosynthesis

Kenneth M. Nollett[1,2] and Scott Burles[2]

[1]Department of Physics and [2]Department of Astronomy and Astrophysics, The University of Chicago, 5640 S. Ellis Ave., Chicago, IL 60637, USA

Abstract. We present a new method to incorporate laboratory nuclear data directly into the standard big-bang nucleosynthesis calculation. Using the quoted uncertainties on the data in a Monte Carlo procedure, we estimate likelihood distributions of the predicted abundances. Our results indicate significantly narrower confidence limits for the predictions than those presently in use. This technique provides error estimates that become smaller as more input data are added, and it allows new nuclear data to be incorporated easily as they become available.

Calculations of the primordial abundances in the standard big bang nucleosynthesis (BBN) framework require as inputs the thermal reaction rates for twelve key nuclear reactions. The results of the standard BBN calculation are limited in accuracy almost exclusively by the accuracy of these rates. The quoted uncertainties on the observed D and ^{7}Li abundances are now smaller than the uncertainties quoted on most standard BBN calculations.

Our new method is intended to make the most direct possible use of the nuclear data and their published uncertainties, to be explicitly less conservative and less subjective than the previous evaluations (while still being reasonable), and to take correlated uncertainties into account. Perhaps most importantly, we wanted our method to reduce the size of the formal abundance uncertainties as newer, more accurate laboratory data are added. It is important to note that the current standard evaluation of the uncertainties in ? set many of its error estimates by requiring that all published data lie within the formal two-sigma errors on the cross section. While this is fine for a one-time conservative error estimate, it can not be re-applied to a growing database in a way that reduces error estimates as data are added, unless some old data are discarded.

We examined the entire experimental literature on the key reactions from approximately 1945 onward, to ensure that we were using the right data, and that we were not mis-using them. We then used the data to obtain abundance predictions by a Monte Carlo process. Each abundance calculation contained the following steps:

1) For each measured cross section (data point), a random number was drawn from a Gaussian distribution whose mean was the measured cross section and whose standard deviation was the (independent) uncertainty for that point. For each set of points sharing a common normalization, a normalization was drawn from a Gaussian distribution with standard deviation given by the normalization error for that set of points.

2) An integrable function was generated from these data by a weighted least-squares fit of the S-factor to a piecewise polynomial. Thermally averaged reaction rates were then generated.

3) The thermal reaction rates were used as inputs to a standard BBN code, which evolved the reaction network to produce primordial abundances of ^{4}He, D, ^{3}He, and ^{7}Li.

After 25,000 repetitions of this procedure, we examined the distribution of output abundances and extracted the 90% confidence limits corresponding to the input cross section distributions. For more thorough descriptions of our procedure, see ? and ?.

The median abundances generated by our procedure differ little from the previous standard, although there are cases where new measurements tend to produce some deviation ($d(p,\gamma)^3$He, ^{3}He$(n,p)^3$H), as well as cases where existing data sets receive different weighting than in the previous evaluation (the $d+d$ reactions, ^{3}He$(\alpha,\gamma)^7$Be, and ^{3}He$(n,p)^3$H). These changes tend to cancel each other in the cases of D and Li abundances, but not in the case of ^{3}He. More notably, we arrive at narrower limits on the D, ^{3}He and ^{7}Li predictions, with 2σ errors estimated to be smaller by up to a factor of three than in the previous, conservative, estimate.

The greatest strength of our method is its close coupling to the nuclear data. It uses only the most relevant data for a given temperature, and it allows new data to be incorporated easily into the calculation. To identify which reactions are contributing to the uncertainties and to the differences from the previous standard rates, we ran the Monte Carlo code several more times. For each of these runs, we set ten of the key reaction rates to their values in ? while we treated only one reaction with our Monte Carlo technique.

Combining the reaction-by-reaction analysis with the study of the reaction rate sensitivitites in ?, we conclude that the places where very precise cross sections would be most useful are as follows: For deuterium, $p(n,\gamma)d$ at 20–300 keV, $d(d,n)^3$He above 100 keV, $d(d,p)^3$H above 100 keV, and $d(p,\gamma)^3$He everywhere, while relative importance varies with baryon density. For lithium at $\Omega_B h^2 = 0.019$, the most important reactions, in order, are $p(n,\gamma)d$ at 20–150 keV, ^{3}He$(\alpha,\gamma)^7$Be at 150–380 keV (and overall normalization), $d(p,\gamma)^3$He everywhere, and $d(d,n)^3$He above 100 keV. For lithium at $\Omega_B h^2 = 0.009$, the most important contributions are from ^{3}H$(\alpha,\gamma)^7$Li at 40–300 keV and ^{7}Li$(p,\alpha)^4$He at 40–300 keV.

References

Burles, S., Nollett, K. M., Truran, J. W., & Turner, M. S. 1999, Phys.Rev.Lett, 82, 4176

Nollett, K. M. & Burles, S. 1999, submitted to Phys.Rev.D

Smith, M. S., Kawano, L. H., & Malaney, R. A. 1993, ApJS, 85, 219

The welcome reception,
in a pleasant Springtime evening, typical of Natal.

ABUNDANCES OF D, ^{3}He AND ^{4}He

Hebert Reeves making a pertinent and humorous comment, in the good company of Gary Steigman, Francesca Matteucci, Sylvie Vauclair, and Etienne Parizot.

Measurements of The Primordial D/H Abundance Towards Quasars

David Tytler, John M. O'Meara, Nao Suzuki and Dan Lubin

Center for Astrophysics and Space Sciences; University of California, San Diego; MS 0424; La Jolla; CA 92093-0424

Scott Burles

Department of Astronomy and Astrophysics, University of Chicago, Chicago, IL 60637

David Kirkman

Bell Laboratories, Lucent Technologies, Murray Hill, NJ 07574

Abstract.

Big Bang Nucleosynthesis (BBN) is the synthesis of the light nuclei, Deuterium (D or ^{2}H), ^{3}He, ^{4}He and ^{7}Li during the first few minutes of the universe. In this review we concentrate on recent data which give the primordial deuterium (D) abundance.

We have measured the primordial D/H in gas with very nearly primordial abundances. We use the Lyman series absorption lines seen in the spectra of quasars. We have measured D/H towards three QSOs, while a fourth gives a consistent upper limit. All QSO spectra are consistent with a single value for D/H: $3.325^{+0.22}_{-0.25} \times 10^{-5}$. From about 1994 – 1996, there was much discussion of the possibility that some QSOs show much higher D/H, but the best such example was shown to be contaminated by H, and no other no convincing examples have been found. Since high D/H should be much easier to detect, and hence it must be extremely rare or non-existent.

The new D/H measurements give the most accurate value for the baryon to photon ratio, η, and hence the cosmological baryon density: $\Omega_b = 0.0190 \pm 0.0009$ (1σ) A similar density is required to explain the amount of Lyα absorption from neutral Hydrogen in the intergalactic medium (IGM) at redshift $z \simeq 3$, and to explain the fraction of baryons in local clusters of galaxies. The D/H measurements lead to predictions for the abundances of the other light nuclei, which generally agree with measurements. The remaining differences with some measurements can be explained by a combination of measurement and analysis errors or changes in the abundances after BBN. The measurements do not require physics beyond the standard BBN model. Instead, the agreement between the abundances is used to limit the non-standard physics.

1. Introduction

There are now four main observations which validate the Big Bang theory: the expansion of the universe, the Planck spectrum of the Cosmic Microwave Background (CMB), the density fluctuations seen in the slight CMB anisotropy and in the local galaxy distribution, and BBN. Together, they show that the universe began hot and dense (Turner 1999).

BBN occurs at the earliest times at which we have a detailed understanding of physical processes. It makes predictions which are relatively precise (10% – 0.1%), and which have been verified with a variety of data. It is critically important that the standard theory (SBBN) predicts the abundances of several light nuclei (H, D, ^{3}He ^{4}He and ^{7}Li) as a function of a single cosmological parameter, the baryon to photon ratio, $\eta \equiv n_b/n_\gamma$ (Kolb & Turner 1990). The ratio of any two primordial abundances should give η, and the measurement of the other three tests the theory.

The abundances of all the light elements have been measured in a number of terrestrial and astrophysical environments. Although it has often been hard to decide when these abundances are close to primordial, it has been clear for decades (e.g. Reeves et al 1973; Wagoner 1973) that there is general agreement with the BBN predictions for all the light nuclei. The main development in recent years has been the increased accuracy of measurement. In 1995 a factor of three range in the baryon density was considered $\Omega_b = 0.007 - 0.024$. The low end of this range allowed no significant dark baryonic matter. Now the new D/H measurements towards quasars give $\Omega_b = 0.0190 \pm 0.0009$ (1σ) – a 5% error, and there have been improved measurements of the other nuclei.

2. Measurement of Primordial Abundances

The goal is to measure the primordial abundance ratios of the light nuclei made in BBN. We normally measure the ratios of the abundances of two nuclei in the same gas, one of which is typically H, because it is the easiest to measure.

The two main difficulties are the accuracy of the measurement and departures from primordial abundances. The state of the art today (1 σ) is about 3% for Y_p, 10% for D/H and 8% for ^{7}Li, for each object observed. These are random errors. The systematic errors are hard to estimate, usually unreliable, and potentially much larger.

By the earliest time at which we can observe objects, redshifts $z \simeq 6$, we find heavy elements from stars in most gas. Although we expect that large volumes of the intergalactic medium (IGM) remain primordial today (Gnedin & Ostriker 1997), we do not know how to obtain accurate abundances in this gas. Hence we must consider possible modifications of abundances. This is best done in gas with the lowest abundances of heavy elements, since this gas should have the least deviations caused by stars.

The nuclei D, ^{3}He, ^{6}Li and ^{7}Li are all fragile and readily burned inside stars at relatively low temperatures of a few 10^6 K. They may appear depleted in the atmosphere of a star because the gas in the star has been above the critical temperature, and they will be depleted in the gas returned to the interstellar medium (ISM). Nuclei ^{3}He, ^{7}Li and especially ^{4}He are also made in stars.

2.1. From Observed to Primordial Abundances

Even when heavy element abundances are low, it is difficult to prove that abundances are primordial. For deuterium, we can make the following argument. The observations are made in gas with two distinct metal abundances. The quasar absorbers have from 0.01 to 0.001 of the solar C/H, while the ISM and pre-solar observations are near solar. Since D/H towards quasars is twice that in the local ISM, 50% of the D is destroyed when abundances rise to near the solar level, and less than 1% of D is expected to be destroyed in the quasar absorbers, much less than the random errors in individual measurements of D/H. Since there are no other known processes which destroy or make significant D (e.g. Reeves et al. 1973; Jedamzik & Fuller 1995), we should be observing primordial D/H in the quasar absorbers.

Since we are now obtaining "precision" measurements, it now seems best to make a few measurements with the highest possible accuracy and controls, in places with the least stellar processing, rather than multiple measurements of lower accuracy. We will now discuss observations of each of the nuclei, and especially D, in more detail.

3. Deuterium in quasar spectra

The abundance of deuterium (D or ^{2}H) is the most sensitive measure of the baryon density (Wagoner 1973). No known processes make significant D, because it is so fragile (Reeves et al. 1973; Epstein et al. 1976; Boyd et al. 1989; Jedamzik & Fuller 1997). Gas ejected by stars should contain zero D, but substantial H, thus D/H decreases over time as more stars evolve and die.

We can measure the primordial abundance in quasar spectra. The measurement is direct and accurate, and with one exception, simple. The complication is that the absorption by D is often contaminated or completely obscured by the absorption from H, and even in the rare cases when contamination is small, superb spectra are required to distinguish D from H.

Contamination by H has about 1000 times the effect of the destruction of D in stars. If stellar processing were the main uncertainty, then we would use the highest measured D/H as the best indication of the primordial value. However, contamination by H is extremely common, and has a much larger effect. We expect that stellar processing has reduced D/H by <1% in the quasar absorbers with abundances below 0.01 solar, while contamination of the D lines by H can make D/H appear >10 times too large.

Prior to the first detection of D in quasar spectra (Tytler, Fan, & Burles 1996), D/H was measured in the ISM and the solar system. The primordial abundance is larger, because D has been destroyed in stars. Though generally considered a factor of a few, some papers considered a factor of ten destruction (Audouze 1986). At that time, most measurements of ^{4}He gave low abundances, which predict a high primordial D/H, which would need to be depleted by a large factor to reach ISM values (Vidal-Madjar & Gry 1984).

Reeves, Audouze, Fowler & Schramm (1973) noted that the measurement of primordial D/H could provide an excellent estimate of the cosmological baryon density, and they used the ISM ^{3}He +D to conclude, with great caution, that primordial D/H was plausibly $7 \pm 3 \times 10^{-5}$.

Deuterium has been measured in the ISM for many years, but only recently have we been able to measure primordial D/H, in QSO spectra. The first measurement of D in interstellar gas was reported in 1973 using DCN (Jefferts, Penzias, & Wilson 1973). In 1973 Vidal-Madjar and colleagues proposed using the Lyman series absorption lines (Vidal-Madjar & Gry 1984) to measure the the column density of neutral atomic D I, and this was done with great success in the ISM using ultraviolet spectra from the Copernicus satellite from 1973 – 1982. Adams (1976) suggested that it might be possible to measure primordial D/H towards low metallicity absorption line systems in the spectra of high redshift quasars. This gas is in the outer regions of galaxies or in the IGM, and it is not connected to the quasars. The importance of such measurements was well known in the field in the late 1970s (Webb et al. 1991), but the task proved too difficult for 4-m class telescopes (Chaffee et al. 1985; Chaffee et al. 1986; Carswell et al. 1994). The high SNR QSO spectra obtained with the HIRES echelle spectrograph (Vogt et al. 1994) on the W.M. Keck 10-m telescope provided the breakthrough.

The primordial D/H is now well established from the spectra of four QSOs obtained by our group using the HIRES spectrograph: first, 1937–1009 (Tytler, Fan, & Burles 1996; Burles & Tytler 1998a); second, 1009+2956 (Burles & Tytler 1998b); third, 0130–4021 (Kirkman et al. 2000), and fourth, HS 0105+1619 (O'Meara et al. 2000). We give some of the parameters associated with these measurements in Table 1.

Table 1. Parameters for the D/H Measurements

QSO	z_{dh}	$\log N_{HI}$ (cm^{-2})	$\log n_{H\,I}/n_H$	$b(D)$ (km s^{-1})	[C/H]
Q1937-1009 [a]	3.572	17.86 ± 0.02	$-2.35, -2.29$	14.0 ± 1.0	$-3.0, -2.1$
Q1009+2956 [a]	2.504	17.39 ± 0.06	$-2.97, -2.84$	15.7 ± 2.1	$-2.8, -3.0$
Q0130-4021	2.799	16.66 ± 0.02	-3.4	$16 - 23$	≤ -2.6 [b]
Q0105+1619	2.536	19.41 ± 0.02	0.8	9.4 ± 0.4	-2.0 [c]

[a] We list the parameters for each of the two components, where available.
[b] Abundances are for Si.
[c] Abundance is a measurement for O, and an upper limit for C, Al, Fe and Si.

The third case, Q0130–4021, is secure because the entire Lyman series is well fit by a single velocity component. The velocity of this component and its column density are well determined because many of its Lyman lines are unsaturated. Its Lyα line is simple and symmetric, and can be fit using the H parameters determined by the other Lyman series lines, with no additional adjustments for the Lyα absorption line. There is barely enough absorption at the expected position of D to allow low values of D/H, and there appears to be no possibility of high D/H. Indeed, the spectra of all three QSOs are inconsistent with high D/H.

The fourth quasar, HS 0105+1619, gives the most secure measurement of D/H to date, because the absorption system has a simple, one component, velocity structure, and the N(H I) is high, giving strong absorption lines, including four lines of D and numerous very narrow metal lines which show that the absorption is produced by D and is not significantly contaminated with H.

The result agrees with our prior measurements in three other QSOs, and reduces the best value for the primordial D/H, from all QSOs, by 2.0%. The best estimate of the baryon density is then also almost unchanged, and becomes much more secure because these QSOs sample different directions in the universe, and the measurement conditions differ substantially, because the H I column densities differ by a factor of 240, and the fraction of the gas which is neutral varies over 3 orders of magnitude.

There remains uncertainty over a case at $z_{abs} = 0.701$ towards quasar PG 1718+4807, because we lack spectra of the Lyman series lines which are needed to determine the velocity distribution of the Hydrogen, and these spectra are of unusually low signal to noise, with about 200 times fewer photons per kms^{-1} than those from Keck. Webb et al. (1997a, 1997b) assumed a single hydrogen component and found D/H = $25 \pm 5 \times 10^{-5}$. This is the best case for a significantly different value for D/H, but it is not convincing. Levshakov et al. (1998) allow for non-Gaussian velocities and find D/H $\sim 4.4 \times 10^{-5}$, while Tytler et al. (1999) find $8 \times 10^{-5} <$ D/H $< 57 \times 10^{-5}$ (95%) for a single Gaussian component, or D/H as low as zero if there are two hydrogen components, which is not unlikely. This quasar is fully consistent with the usual D/H.

Recently Molaro et al. (1999) claimed that D/H might be measured and low in an absorber at $z = 3.514$ towards quasar APM 08279+5255, though they noted that higher D/H was also possible. Only one H I line, Lyα, was used to estimate the hydrogen column density N_{HI} (measured in H I atoms per cm^{-2} along the line of sight) and we know that in such cases the column density can be highly uncertain. Their Figure 1 (panels a and b) shows that there is a tiny difference between D/H = 1.5×10^{-5} and 21×10^{-5}, and it is clear that much lower D is also acceptable because there can be H additional contamination in the D region of the spectrum. Levshakov et al. (2000) show that $\log N_{HI} = 15.7$ (too low to show D) gives an excellent fit to these spectra, and they argue that this is a more realistic result because the metal abundances and temperatures are then normal, rather than being anomalously low with the high N_{HI} preferred by Molaro et al.

The first to publish a D/H estimate using high signal to noise spectra from the Keck telescope with the HIRES spectrograph were Songaila et al. (1994), who reported an upper limit of D/H $< 25 \times 10^{-5}$ in the $z_{abs} = 3.32$ Lyman limit system (LLS) towards quasar 0014+813. Using different spectra, Carswell et al. (1994) reported $< 60 \times 10^{-5}$ in the same object, and they found no reason to think that the deuterium abundance might be as high as their limit. Improved spectra (Burles, Kirkman, & Tytler 1999) support the early conclusions: D/H $< 35 \times 10^{-5}$ for this quasar. High D/H is allowed, but is highly unlikely because the absorption near D is at the wrong velocity, by 17 ± 2 km s^{-1}, it is too wide, and it does not have the expected distribution of absorption in velocity, which is given by the H absorption. Instead this absorption is readily explained entirely by H (D/H $\simeq 0$) at a different redshift.

Very few LLS have a velocity structure simple enough to show deuterium. Absorption by H usually absorbs most of the quasar flux near where the D line is expected, and hence we obtain no information of the column density of D. In these extremely common cases, very high D/H is allowed, but only because we have essentially no information.

All quasar spectra are consistent with the current best value for the primordial: D/H $\sim 3.3 \pm 0.2 \times 10^{-5}$. Three quasars (1937–1009, 1009+2956 & HS 0105+1619) are inconsistent with D/H $\geq 5 \times 10^{-5}$, and the third (0130–4021) is inconsistent with D/H $\geq 6.7 \times 10^{-5}$. Most quasar spectra allow much higher D/H, because we lack spectra of sufficient quality to distinguish D from H, or because the D is contaminated by H and the spectra provide no useful information.

3.1. Galactic Chemical Evolution of D

Numerical models are constructed to follow the evolution of the abundances of the elements in the ISM of our Galaxy.

The main parameters of the model include the yields of different stars, the distribution of stellar masses, the star formation rate, and the infall and outflow of gas. These parameters are adjusted to fit many different data. These Galactic chemical evolution models are especially useful to compare abundances at different epochs, for example, D/H today, in the ISM when the solar system formed, and primordially.

In an analysis of a variety of different models, Tosi et al. (1998) concluded that the destruction of D in our Galaxy was at most a factor of a few, consistent with low but not high primordial D. They find that all models, which are consistent with all Galactic data, destroy D in the ISM today by less than a factor of three. Such chemical evolution will destroy an insignificant amount of D when metal abundances are as low as seen in the quasar absorbers.

Others have designed models which do destroy more D (Vangiono-Flam & Cassé 1995; Timmes et al. 1997; Scully et al. 1997; Olive 1999b), for example, by cycling most gas through low mass stars and removing the metals made by the accompanying high mass stars from the Galaxy. These models were designed to reduce high primordial D/H, expected from the low Y_p values prevalent at that time, to the low ISM values. Tosi et al. (1998) describe the generic difficulties with these models. To destroy 90% of the D, 90% of the gas must have been processed in and ejected from stars. These stars would then release more metals than are seen. If the gas is removed (e.g. expelled from the galaxy) to hide the metals, then the ratio of the mass in gas to that in remnants is would be lower than observed. Infall of primordial gas does not help, because this brings in excess D. These models also fail to deplete the D in quasar absorbers, because the stars which deplete the D, by ejecting gas without D, also eject carbon. The low abundance of carbon in the absorbers limits the destruction of D to <1% (Jedamzik & Fuller 1997).

4. Is there spatial variation in D/H towards quasars?

It seems highly likely that we are measuring the primordial D/H in the QSO spectra.

Are there other places where D is much larger than we observe? All quasar spectra are consistent with a single D/H value. The cases which are also consistent with high D are readily explained by the expected H contamination. We now explain why we have enough data to show that high D must be rare, if it occurs at all.

High D should be much easier to find than the usual D. Since we have not found any convincing examples, high D must be very rare. If D were ten times the usual value, the D line would be ten times stronger for a given N_{HI}, and could be seen in spectra with ten times lower signal to noise, or 100 times fewer photons recorded per Å. If such high D/H were common, it would have been seen many times in the high resolution, but low signal to noise, spectra taken in the 1980's, when the community was well aware of the importance of D/H. High D would also have been seen frequently in the spectra of about 100 quasars taken with the HIRES spectrograph on the Keck telescope. In these spectra, which have relatively high signal to noise, high D could be detected in absorption systems which have 0.1 of the N_{HI} needed to detect the usual D/H. Such absorbers are about 40 – 60 times more common than those needed to show the usual D/H, and hence we should have found tens of excellent examples.

4.1. Conclusions from D/H from quasars

Most agree that D is providing the most accurate and reliable value for η (Schramm & Turner 1998). The prior concern, (Olive 1999b; Audouze 1998), that some quasar absorbers might show much higher values of D/H, is now assuaged by the continuing lack of any secure examples, and the measurement of the usual D/H towards HS 0105+1619.

The weighted mean D/H from three QSOs 1937–1009, 1009+2956 and HS 0105+1619 measured by our group, including Scott Burles & Xiao-Ming Fan: (Burles & Tytler 1998a; Burles & Tytler 1998b; Kirkman et al. 2000; Burles & Tytler 1997; O'Meara et al. 2000)

- log D/H = -4.478 ± 0.031 (7% error)
- D/H = $3.325^{+0.22}_{-0.25} \times 10^{-5}$,

which is 2.0% lower than our previous best estimate prior to HS 0105+1619. The value and its error are still dominated by the measurement from Q1937–1009, which gave half the errors of Q1009+2956 & HS 0105+1619.

This best primordial D/H value, together with over 50 years of theoretical work and laboratory measurements of reaction rates, leads to the following values for cosmological parameters:

- D/H = $3.325^{+0.22}_{-0.25} \times 10^{-5}$ (from the spectra of three quasars)
- $\eta = 5.2 \pm 0.3 \times 10^{-10}$ (from SBBN and D/H)
- $Y_p = 0.2464 \pm 0.0007$ (mass fraction of ^{4}He, from SBBN and D/H)

- $^7\text{Li/H} = 3.46^{+0.55}_{-0.48} \times 10^{-10}$ (from SBBN and D/H)

- $n_\gamma = 411.7 \pm 1.8$ photons cm^{-3} (photon density, from the CMB temperature)

- $\rho_b = 3.57 \pm 0.17 \times 10^{-31}$ gcm^{-3} (baryon density, from the photon density and η, 0.22 ± 0.01 H atoms m^{-3} today)

- $\Omega_b h^2 = 0.01897 \pm 0.00091$ (baryon density, in units of the critical density ρ_c, 4.8% error)

- $\Omega_b = 0.039 \pm 0.002$ for $h = 0.7$

- $N_\nu < 3.20$ (number of neutrino species, from SBBN, D/H and Y_p data).

Here the photon density is from the COBE (Fixen et al. 1996) measurement of the CMB temperature, $T = 2.728 \pm 0.004$ K, and the neutrino limit from Burles et al. (1999). In the past we have often quoted 95% confidence intervals, but now, the agreement between the D/H obtained for three QSOs gives us more confidence that we have approximately the right magnitude for the errors, and hence, in this paper, we quote 1σ errors.

The η and Ω_b from D/H are considered both the most reliable and the most accurate, because the observation of D is straightforward, given sufficient telescope time. The measurement of D/H from spectra is relatively simple, again provided the data are adequate, and the D/H observed is likely the primordial value.

This work was funded in part by grant G-NASA/NAG5-3237 and by NSF grants AST-9420443 and AST-9900842, and is based in part on data obtained from the Lick Observatory, and the W.M. Keck observatory, which is managed by a partnership among Caltech, the University of California and NASA. We are grateful to Steve Vogt, the PI for the Keck HIRES instrument which enabled our work on D/H, and to the W.M. Keck Observatory staff; the observing assistants Joel Aycock, Teresa Chelminiak, Gary Puniwai, Ron Quick, Barbara Schaefer, Chuck Sorenson, Terry Stickel and Wayne Wack, and the instrument specialists Tom Bida, Randy Campbell, Bob Goodrich, David Sprayberry and Greg Wirth. It is a pleasure to thank Constantine Deliyannis, Carlos Frenk, George Fuller, Yuri Izotov, Hannu Kurki-Suonio, Sergei Levshakov, Keith Olive, Jerry Ostriker, Evan Skillman, Gary Steigman and Trinh Xuan Thuan for suggestions and many helpful and enjoyable discussions.

References

Turner, M.S., in "The Proc. of Particle Phys. and the Universe" ed. D.O. Caldwell astro-ph/9904359 (AIP, Woodbury, NY, 1999).

Kolb, E.W., and Turner, M.S., "The Early Universe", (Addison Wesley 1990).

Reeves, H., Audouze, J., Fowler, W. & Schramm, D. N., Astrophys.J. **179**, 909 (1973).

Wagoner, R. V., Astrophys. J. **179**, 343 (1973).

Gnedin, N. Y. & Ostriker, J. P., Astrophys. J. **486**, 581 (1997).

Jedamzik, K. & Fuller, G.M., Astrophys. J. **452**, 33 (1995).
Epstein, R.I., Lattimer, J.M. & Schramm, D.N., Nature **263**, 198 (1976).
Boyd, R.N., Ferland, G.J. & Schramm, D.N., Astrophys. J. **336**, L1 (1989).
Jedamzik, K. & Fuller, G.M., Astrophys. J. **483**, 560 astro-ph/9609103(1997).
Tytler, D., Fan, X. M., & Burles, S., Nature **381**, 207 (1996).
Audouze, J., in "Nucleosynthesis & Chemical Evolution" Saas-Fee, p.433 (Geneva Observatory, 1986).
Vidal-Madjar, A. & Gry, C. Astron. Astrophys. **138**, 285 (1984)
179, L57 (1973).
Adams, T.F., Astron. Astrophys. **50**, 461 (1976).
Webb, J. K., Carswell, R.F., Irwin, M.J. & Penston, M.V. Mon.Not.Roy.Astr.Soc. **250**, 657 (1991)
Chaffee, F. H., Foltz, C. B., Roser, H.-J., Weymann, R. J., & Latham, D. W., Astrophys. J. **292**, 362 (1985).
Chaffee, F. H., Foltz, C. B., Bechtold, J. & Weymann, R. J. Astrophys. J. **301**, 116 (1986).
Carswell, R. F., Rauch, M., Weymann, R. J., Cooke, A. J.,& Webb, J. K., Mon. Notes R. Astron. Soc. **268**, L1 (1994).
Vogt, S. S., et al., Proc. SPIE, **2198**, 362 (1994).
Burles, S. & Tytler, D., Astrophys. J. **499**, 699 astro-ph/9712108 (1998a).
Burles, S. & Tytler, D., Astrophys. J. **507**, 732 astro-ph/9712109 (1998b).
Kirkman, D., Tytler, D., Burles, S., Lubin, D., & O'Meara, J., Astrophys. J., **529**, 655, astro-ph/9907128 (2000).
O'Meara, J.M., Tytler, D., Prochaska, J.X., Burles, S., Kirkman, D., Lubin, D., Suzuki, N. Wolfe, A.M. in preparation
Webb, J. K., Carswell, R. F., Lanzetta, K. M., Ferlet, R., Lemoine, M., Vidal-Madjar, A., & Bowen, D. V. Nature **388**, 250 (1997a).
Webb, J. K., et al., preprint, astroph 9710089 (1997b).
Levshakov, S.A., Kegel, W.H., & Takahara, F., astro-ph/9812114 (1998).
Tytler, D., et al., Astrophys. J. **117**, 63 astro-ph/9810217 (1999).
Molaro, P., Bonifacio, P., Centurion, M., Vladilo, G., Astron. Astrophys. **349**, no. 1, L13-16 (1999).
Levshakov, S.A., Agafonova, I.I. & Kegel, W.H. 2000b, Astron. Astrophys. in preparation astro-ph/9911261.
Songaila, A., Cowie, L. L., Hogan, C. J. & Rugers, M., Nature **368**, 599 (1994).
Burles, S., Kirkman, D.& Tytler, D., Astrophys. J., **519**, 18 (1999c).
Tosi, M., Steigman, G., Matteucci, F. & Chiappini, C., Astrophys. J. **498**, 226 (1998).
Vangiono-Flam, E. & Cassé, M., Astrophys. J. **441**, 729 (1995).
Timmes, F. X., Truran, J. W., Lauroesch, J. T. & York, D.G., Astrophys.J. **476**, 464 (1997).
Scully, S. Cassé, M., Olive, K. A. & Vangioni-Flam, E., Astrophys. J. **476**, 521 astroph/9607106 (1997).

Olive, K.A., 19th Texas Symposium on Relativistic Astrophysics and Cosmology, Paris 1998, astro-ph/9903309 (1999b).

Schramm, D. N. & Turner, M. S., Rev. Mod. Phys **70**, 303-318 (1998).

Audouze, J., in conference "Galactic Evolution..." Meudon, (1998).

Burles, S. & Tytler, D., Astron. J. **114**, 1330 astro-ph/9707176 (1997).

Fixen, D.J. et al. ApJ **473**, 576 (1996)

Burles, S., Nollett, K.M., Truran, J.N. & Turner, M.S., Phys. Rev. Lett. **82**, 4176 (1999) astro-ph/9901157 (1999b).

The Deuterium Abundance in QSO Absorption Systems: A Mesoturbulent Approach[1]

Sergei A. Levshakov

Department of Theoretical Astrophysics, Ioffe Physico-Technical Institute, 194021 St. Petersburg, Russia

Abstract. A new method, based on simulated annealing technique and aimed at the inverse problem in the analysis of intergalactic or interstellar complex spectra of hydrogen and metal lines, is outlined. We consider the process of line formation in clumpy stochastic media accounting for fluctuating velocity and density fields self-consistently. Two examples of the analysis of 'H+D'-like absorptions seen at $z_a = 3.514$ and 3.378 towards APM 08279+5255 are presented.

1. Introduction

The cosmological significance of the deuterium abundance measurements in metal-deficient QSO absorption systems has been widely discussed in the literature (see e.g. the review by Lemoine et al. 1999). Practical applications of such measurements were clearly outlined in Tytler & Burles (1997) : (i) the primordial D/H value gives the density of baryons Ω_b at the time of big bang nucleosynthesis (BBN), a precise value for Ω_b might be used (ii) to determine the fraction of baryons which are missing, (iii) to specify Galactic chemical evolution, and (iv) to test models of high energy physics. A measurement of the D/H ratio, together with other three light element abundances (^{3}He/H, ^{4}He/H, and ^{7}Li/H), provides the complete test of the standard BBN model.

Deuterium has been reported up-to-now in a few QSO Lyman limit systems (LLS), – the systems with the neutral hydrogen column densities of $N_{\rm H\,I} \sim 10^{17} - 10^{18}$ cm^{-2} (Kirkman et al. 1999). The difficulties inherent to measurements of D/H in QSO spectra are mainly caused by the confusion between the D I line (which is always partially blended by the blue wing of the saturated hydrogen line) and the numerous neighboring weak lines of H I observed in the Lyα forest at redshifts $z > 2$ (Burles & Tytler 1998).

Currently, there are two methods to analyse the absorption spectra : (i) a conventional Voigt-profile fitting (VPF) procedure which usually assumes several subcomponents with their own physical parameters to describe a complex absorption profile, and (ii) a mesoturbulent approach which describes the line formation process in a continuous medium with fluctuating physical characteris-

---

[1] Based on data obtained at the W. M. Keck Observatory, which is jointly operated by the University of California, the California Institute of Technology, and the National Aeronautics and Space Administration.

tics. It is hard to favor this or that method if both of them provide good fitting. But the observed increasing complexity of the line profiles with increasing spectral resolution gives some preference to the model of the fluctuating continuous medium.

Here, we set forward a mesoturbulent approach to measure D/H and metal abundances, which has many advantages over the standard VPF procedures. A brief description of a new Monte Carlo inversion (MCI) method is given in this report. For more details, the reader is referred to Levshakov, Agafonova, & Kegel (2000b).

An example of the MCI analysis of two 'H+D'-like profiles with accompanying metal lines observed at $z_a = 3.514$ and 3.378 towards the quasar APM 08279+5255 is described. The high quality spectral data have been obtained with the Keck-I telescope and the HIRES spectrograph by Ellison et al. (1999).

2. The MCI method and results

The MCI method is based on simulated annealing technique and aimed at the evaluation both the physical parameters of the gas cloud and the corresponding velocity and density distributions along the line of sight. We consider the line formation process in clumpy stochastic media with fluctuating velocity and density fields (*mesoturbulence*). The new approach generalizes our previous Reverse Monte Carlo (Levshakov, Kegel, & Takahara 1999) and Entropy-Regularized Minimization (Levshakov, Takahara, & Agafonova 1999) methods dealing with incompressible turbulence (i.e. the case of random bulk motions with homogeneous gas density n_H and kinetic temperature T).

The main goal is to solve the *inverse problem*, i.e. the problem to deduce physical parameters from a QSO absorption system. The inversion is always an optimization problem in which an objective function is minimized. To estimate a goodness of the minimization we used a χ^2 function augmented by a regularization term (a penalty function) to stabilize the MCI solutions. The MCI is a stochastic optimization procedure and one does not know in advance if the global minimum of the objective function is reached in a single run. Therefore to check the convergency, several runs are executed for a given data set with every calculation starting from a random point in the simulation parameter box and from completely random configurations of the velocity and density fields. After these runs, the median estimation of the model parameters is performed.

Our model supposes a continuous absorbing gas slab of a thickness L. The velocity component along a given line of sight is described by a random function in which the velocities in neighboring volume elements are correlated with each other. The gas is optically thin in the Lyman continuum. We are considering a compressible gas, i.e. n_H is also a random function of the space coordinate, x. Following Donahue & Shull (1991) and assuming that the ionizing radiation field is constant, the ionization of different elements can be described by one parameter only – the ionization parameter $U \propto 1/n_H$. Furthermore, for gas in thermal equilibrium, Donahue & Shull give an explicit relation between U and T. The background ionizing spectrum is taken from Mathews & Ferland (1987).

In our computations, the continuous random functions $v(x)$ and the normalized density $y(x) = n_H(x)/n_0$, n_0 being the mean hydrogen density, are represented by their sampled values at equally spaced intervals Δx, i.e. by the vectors $\{v_1, ..., v_k\}$ and $\{y_1, ..., y_k\}$ with k large enough to describe the narrowest components of complex spectral lines. For the ionization parameter as a function of x, we have $U(x) = \hat{U}_0/y(x)$, with $\hat{U}_0$ being the reduced mean ionization parameter defined below.

Absorption system at $z_a = 3.514$. A measurement of the primordial D/H in the $z_a = 3.514$ system has been recently made by Molaro et al. (1999). They suggested that the blue wing of H I Lyα is contaminated by D I and evaluated a very low deuterium abundance of D/H$\simeq 1.5 \times 10^{-5}$ in the cloud with $N_{\mathrm{H\,I}} = (1.23^{+0.09}_{-0.08}) \times 10^{18}$ cm^{-2}. They considered, however, the derived D abundance as a lower limit because their analysis was based on a simplified one-component VPF model which failed to fit the red wing of the Lyα line as well as the profiles of Si III, Si IV, and C IV lines exhibiting complex structures over approximately 100 km s^{-1} velocity range. They further assumed that additional components would decrease the H I column density for the major component and, thus, would yield a higher deuterium abundance. Given the MCI method, we can test this assumption since the MCI accounts self-consistently for the velocity and density fluctuations.

Our aim is to fit the model spectra simultaneously to the observed H I, C II, C IV, Si III, and Si IV profiles. In this case the mesoturbulent model requires the definition of a simulation box for the six parameters : the carbon and silicon abundances, Z_{C} and Z_{Si}, respectively, the rms velocity σ_{v} and density dispersion σ_{y}, the reduced total hydrogen column density $\hat{N}_{\mathrm{H}} = N_{\mathrm{H}}/(1+\sigma_y^2)^{1/2}$, and the reduced mean ionization parameter $\hat{U}_0 = U_0/(1+\sigma_y^2)^{1/2}$. For the model parameters the following boundaries were adopted : Z_{C} ranges from 10^{-6} to 4×10^{-4}, Z_{Si} from 10^{-6} to 3×10^{-5}, σ_{v} from 25 to 80 km s^{-1}, σ_{y} from 0.5 to 2.2, $\hat{N}_{\mathrm{H}}$ from 5×10^{17} to 8×10^{19} cm^{-2}, and $\hat{U}_0$ ranges from 5×10^{-4} to 5×10^{-2}. We fix $z_a = 3.51374$ (the value adopted by Molaro et al.) as a more or less arbitrary reference velocity at which $v_j = 0$.

Having specified the parameter space, we minimize the χ^2 value. The objective function includes those pixels which are critical to the fit. In Fig. 1, these pixels are marked by shaded areas. In Fig. 1 (panels **f** and **c**), the observed profiles of C IVλ1550 and, respectively, Si IVλ1402 are shown together with the model spectra computed with the parameters derived from Lyα, C IIλ1334, C IVλ1548, Si IIIλ1206, and Si IVλ1393 fitting to illustrate the consistency. For the same reason the Lyβ model spectrum is shown in panel **b** at the expected position. All model spectra in Fig. 1 are drawn by continuous curves, whereas filled circles represent observations (normalized fluxes). The corresponding distributions of $v(x)$, $y(x)$, and $T(x)$ are shown in panels **i**, **j**, and **k**. The restored velocity field reveals a complex structure which is manifested in non-Gaussian density-weighted velocity distribution as shown in panels **l** and **m** for the total hydrogen as well as for the individual ions. We found that the radial velocity distribution of H I in the vicinity of $\Delta v \simeq -100$ km s^{-1} may mimic the deuterium absorption and, thus, the asymmetric blue wing of the hydrogen Lyα absorption may be readily explained by H I alone.

The median estimation of the model parameters gives $N_{\mathrm{H}} = 5.9 \times 10^{18}$ cm^{-2}, $N_{\mathrm{H\,I}} = 5.3 \times 10^{15}$ cm^{-2}, $U_0 = 1.6 \times 10^{-2}$, $\sigma_{\mathrm{v}} = 51$ km s^{-1}, and $\sigma_{\mathrm{y}} = 1.1$.

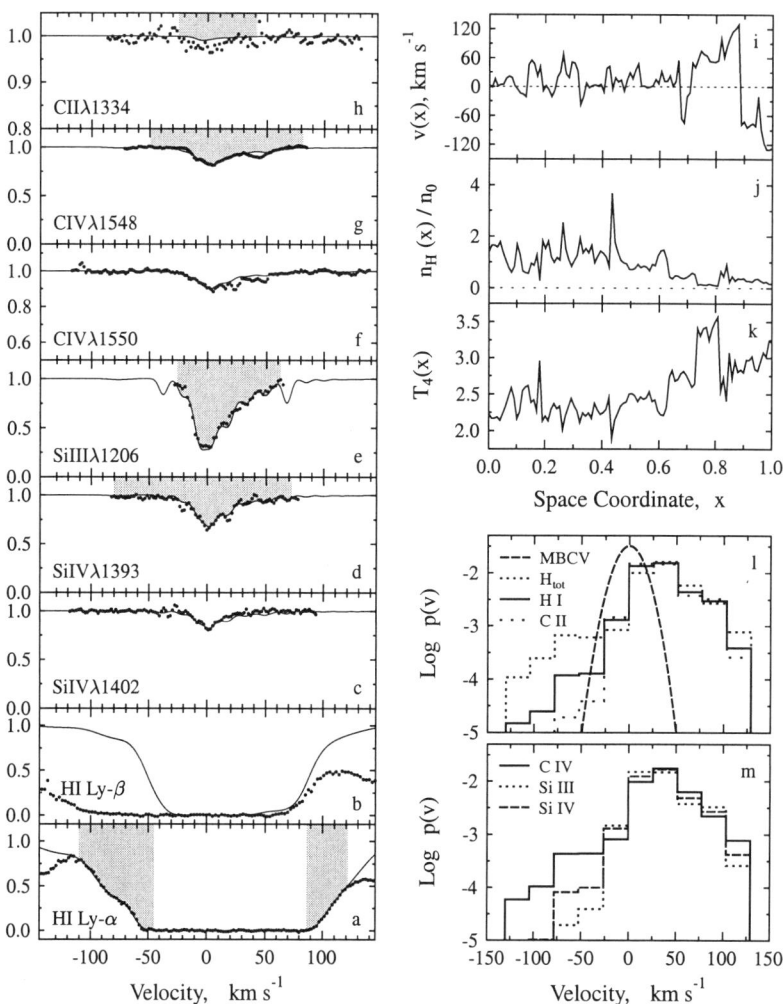

Figure 1. (a–h). Observed normalized intensities (dots) and theoretical (solid curves) profiles of the absorption lines from the $z_a = 3.514$ system (note different scales in panels **h** and **f**). Shaded areas mark pixels which are critical to the MCI fit. (**i–k**). The corresponding MCI reconstraction of the radial velocity, density, and kinetic temperature T_4 which is given in units of 10^4 K. (**l,m**). The corresponding density-weighted radial velocity distribution functions, $p(v)$, for the total H_{tot} and neutral H I hydrogen, C II, C IV, Si III, and Si IV as restored by the MCI. For comparison, in panel **l** the short dashed curve shows $p(v)$ adopted by Molaro et al.

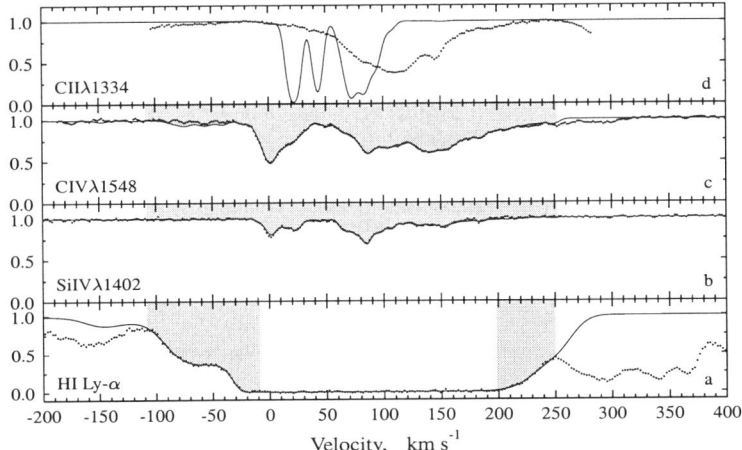

Figure 2. (a–d). Observed normalized intensities (dots) and theoretical (solid curves) profiles of the absorption lines from the $z_a = 3.378$ system (shaded areas mark pixels used by the MCI). The 'H+D'-like absorption is explained by H I alone. An excellent common fit to H I, Si IV, and C IV should, however, be rejected since the upper panel demonstrates the inconsistency between the synthetic spectrum of C II computed with the parameters derived from the foregoing lines and the observed intensities.

The results were obtained with $k = 100$ and the correlation coefficients $f_v = f_y = 0.95$ (for more details, see Levshakov, Agafonova, & Kegel 2000a).

The MCI allowed us to fit precisely not only the blue wing of the saturated Lyα line but the red one as well. We found that the actual neutral hydrogen column density may be a factor of 250 lower than the value obtained by Molaro et al. if one accounts for the velocity field structure. Besides we did not confirm the extremely low metallicity of [C/H] $\simeq -4.0$, and [Si/H] $\simeq -3.5$ reported by Molaro et al. Our analysis yields [C/H] $\simeq -1.8$, and [Si/H] $\simeq -0.7$. A similar silicon overabundance has also been observed in halo (population II) stars (Henry & Worthey 1999).

Absorption system at $z_a = 3.378$. The following example illustrates how the realibility of the inversion procedure can be controlled. We have chosen the $z_a = 3.378$ system since at the position of the narrowest C IV subcomponent with $z_{C\,IV} = 3.37757$ and $b_{C\,IV} = 6.5$ km s^{-1} (see Ellison et al.) one can see an 'H+D'-like absorption in the blue wing of the saturated Lyα line (Fig. 2a). The C IV, and Si IV profiles from this system were treated by Ellison et al. separately. They found $N^{tot}_{C\,IV} = (9.12^{+0.65}_{-0.61}) \times 10^{13}$ cm^{-2} and $N^{tot}_{Si\,IV} = (1.70 \pm 0.08) \times 10^{13}$ cm^{-2}. For the neutral hydrogen, they estimated $N_{H\,I} = 2.8 \times 10^{15}$ cm^{-2} and $b_{H\,I} = 78.6$ km s^{-1} (errors for both quantities are greater than 30%).

In this example, we assumed no deuterium absorption and tried to force a common fit to the lines shown in panels **a–c** (Fig. 2). Pixels used in the fitting are labeled by shaded areas. The MCI fit, shown by solid curves, looks excellent, and gives $N_{\rm H\,I} = 2.1 \times 10^{17}$ cm^{-2}, $N_{\rm C\,IV} = 1.0 \times 10^{14}$ cm^{-2}, $N_{\rm Si\,IV} = 2.6 \times 10^{13}$ cm^{-2}, $\sigma_{\rm v} = 41.6$ km s^{-1}, and $\sigma_{\rm y} = 1.3$. The results were obtained with $k = 150$ and the correlation coefficients $f_{\rm v} = f_{\rm y} = 0.97$.

The obtained MCI solution should, however, be rejected because the synthetic spectrum of C II (solid curve in Fig. 2d), computed with the parameters derived from the Lyα, C IV, and Si IV fitting, differs significantly from the observed intensities (dots in Fig. 2d). This example shows that we can always control the MCI results using additional portions of the analysed spectrum.

Another issue of this example is that we can, in principle, fit an 'H+D'-like absorption by H I alone even for the systems with $N_{\rm H\,I} \simeq 10^{17}$ cm^{-2} and accompanying metal lines. Examples of false deuterium identifications in systems with 10 times lower neutral hydrogen column densities and without supporting metal lines have been discussed in Tytler & Burles. Both cases stress the importance of the comprehensive approach to the analysis of each individual QSO system showing possible D absorption.

We may conclude that up-to-now deuterium was detected in only four QSO spectra (Q 1937-1009, Q 1009+2956, Q 0130-4021, and Q 1718+4807) where $N_{\rm H\,I}$ was measured with a sufficiently high accuracy. These measurements are in concordance with D/H = $(3-4) \times 10^{-5}$.

Acknowledgements. This paper includes results obtained in collaboration with I. I. Agafonova and W. H. Kegel. The author is grateful to Ellison et al. for making their data available. I would also like to thank the conference organizers for financial assistance.

References

Burles, S., & Tytler, D. 1998, in Primordial Nuclei and Their Galactic Evolution, ed. N. Prantzos, M. Tosi & R. von Steiger (Dordrecht : Kluwer Academic Publishes), 65

Donahue, M., & Shull, J. M. 1991, ApJ, 383, 511

Ellison, S. L., Lewis, G. F., Pettini, M., Sargent, W. L. W., Chaffee, F. H., Foltz, C. B., Rauch, M., & Irwin, M. J. 1999, PASP, 111, 919

Henry, R. B. C., & Worthey, G. 1999, PASP, 111, 919

Kirkman, D., Tytler, D., Burles, S., Lubin, D., & O'Meara, J. M. 1999, AAS, 194, 3001 (astro-ph/9907128)

Lemoine, M., Audouze, J., Jaffel, L. B., Feldman, P., Ferlet, R., Hébrard, G., Jenkins, E. B., Mallouris, C., Moos, W., Sembach, K., Sonneborn, G., Vidal-Madjar, A., & York, D. C. 1999, New Astronomy, 4, 231

Levshakov, S. A., Kegel, W. H., & Takahara, F. 1999, MNRAS, 302, 707

Levshakov, S. A., Takahara, F., & Agafonova, I. I. 1999, ApJ, 517, 609

Levshakov, S. A., Agafonova, I. I., & Kegel, W. H. 2000a, A&A, in press

Levshakov, S. A., Agafonova, I. I., & Kegel, W. H. 2000b, A&A., submit.

Mathews, W. D., & Ferland, G. 1987, ApJ, 323, 456

Molaro, P., Bonifacio, P., Centurion, M., & Vladilo, G. 1999, A&A, 349, L13

Tytler, D., & Burles, S. 1997, in Origin of Matter and Evolution of Galaxies, ed. T. Kajino, Y. Yoshii & S. Kubono (Singapore : World Scientific), 37

Deuterium Observations in our Galaxy - View A

Jeffrey L. Linsky and Brian E. Wood

JILA, University of Colorado, & NIST, Boulder CO 80309-0440 USA

Abstract. Accurate measurements of the D/H ratio in our Galaxy provide critical tests of Galactic chemical evolution and constrain the primordial value of D/H. Very high quality ultraviolet spectra from the GHRS and STIS instruments on HST have been analyzed for lines of sight toward both early and late-type stars and hot white dwarfs. We will summarize the results that are being obtained for D/H along sightlines through the Local Interstellar Cloud (LIC) and other nearby warm clouds. All sightlines through the LIC are consistent with D/H $= (1.53 \pm 0.18) \times 10^{-5}$. Whether or not significantly different values of D/H are present in other clouds within 100 pc of the Sun is not yet settled, but there is evidence that D/H is significantly lower in Orion (500 pc). We will describe the likely sources of systematic errors in determining D/H that must be understood and quanitified when analyzing such ultraviolet spectra.

1. Introduction

Accurate measurements of the ratio by number of deuterium to hydrogen (hereafter called D/H) in different environments are critically needed to address two fundamental problems in astrophysics. First, measurements of D/H in environments where there has been very little processing of gas through stars and the metal abundance is small is essential to infer the primordial D/H ratio that determines the baryonic density of the universe. Second, measurements of the D/H ratio in interstellar gas located throughout the Galactic disk and halo will allow us to track the end result of chemical evolution resulting from nuclear burning of D inside stars (astration), transfer of D-depleted gas into the ISM, formation of new generations of stars from this D-depleted interstellar gas, mixing of the gas with other environments in the ISM, and infusion of gas with presumably primordial D/H from the halo. This complex set of processes is modelled by Galactic chemical evolution codes (e.g., Tosi *et al.* 1998). Accurate D/H measurements can test the predictions of these codes and thereby provide insight into the rates for these processes over the lifetime of the Galaxy.

Lemoine *et al.* (1999) have written an excellent review of D/H measurements in quasar spectra, the ISM, and the solar system (reflecting the D/H ratio in the protosolar nebula). Although the D/H ratio can be inferred from radio frequency hyperfine transitions and from deuterated molecules, the most accurate method today is to measure the interstellar column densities of the H and D Lyman lines seen in absorption against the spectra of normal stars and hot white dwarfs. In the last few years measurements of D/H along many lines of sight have been determined from the analysis of high resolution spectra from the GHRS and STIS instruments on HST, but the IMAPS experiment on the Astro-Spas platform has also provided data for D/H measurements. We will summarize all of these results below. Careful analysis of the excellent quality spectra now available has led to a better appreciation of the systematic errors that likely are more important than the published random errors and, if not included in the analysis, could lead to factor of 2 or larger underestimates in D/H. For this reason we start with a discussion of systematic and other errors.

2. Systematic and Other Measurement Errors

Since the D/H ratio in the ISM is roughly 1.5×10^{-5}, detection of a D Lyman line requires that the H Lyman line be very optically thick. This simple fact can lead to large systematic errors in the D/H ratios derived by fitting Lyman line spectra unless one includes in the analysis all sources of H absorption. The magnitude of this problem was not appreciated until recently. Observations of the higher Lyman lines with FUSE and IMAPS will lessen the impact of this large ratio of the D and H opacities as one can compare D absorption in a more opaque Lyman line with H absorption in a less opaque higher Lyman line. Even so, the problems described here will still be important and must be addressed in order to obtain credible values for the D/H ratio.

Horizon: Since the D line is located at -81 km s^{-1} relative to the H line, saturated H Lyman line absorption can cover up the D line creating a visibility horizon. For Lyman-α the horizon sets in at about $\log N_{\rm HI} = 18.8$, while for Lyman-γ it begins at about 20.0. The presence of a horizon is important because the large H column densities to hot stars prevent visibility of the D Lyman-α line and often the Lyman-β line, but the sensitivity of the FUSE spectrograph is a factor of 3 or smaller at the higher Lyman lines compared to Lyman-β and much lower than the sensitivity of the HST spectrographs to Lyman-α.

Confusion: Only a few lines of sight to nearby stars appear to have only one velocity component at a resolution of 2.6–3.6 km s^{-1} (GHRS and STIS). Even the short lines of sight to Sirius (2.63 pc, Hébrard et al. 1999) and 61 Cyg A (3.48 pc, Wood & Linsky 1998) often show two or more velocity components, and lines of sight to more distant targets can be even more complex. Higher resolution studies of interstellar Na I absorption led Welty, Morton, & Hobbs (1996) to estimate that only 10% of the interstellar velocity components are resolved for unsaturated lines at a resolution of 3.6 km s^{-1}. This result may overestimate the number of velocity components expected for the H and D lines since neutral H and D are present in warm clouds whereas Na I is also formed in cold clouds. Unresolved velocity components can lead to errors in derived H column densities and thus D/H, since $N_{\rm HI}$ is often derived from the shape of the H absorption just outside of the saturated line core where all of the velocity components contribute and the opacity is a nonlinear function of $N_{\rm HI}$ and the broadening parameter b.

Since the H and D lines are broad, closely spaced velocity components identified in the heavier (and thus narrower) metal lines cannot be separated in the H and D lines. The abundance ratio of H to a metal depends on the relative ionization of H and the metal and the depletion of metal atoms onto grains.

Both of these effects are likely different in each cloud, making it difficult to infer H and D column densities by extrapolating from the metal lines. The best metal lines to use as surrogates for H are N and O because they have nearly the same ionization potential as H, but the uncertain depletion of N and O implies an uncertain distribution in the H column densities in the different clouds. A second problem is uniqueness. In a multicomponent fit, one generally cannot solve for N_{HI} and b, and D/H, for each of several clouds along a line of sight from the analysis of only a few line profiles without making some assumptions, such as assuming common b or D/H values in all clouds, or that the distribution of N_{HI} among the several clouds is the same as is seen in the metal lines. It is difficult to assess the systematic errors than result from such assumptions.

Uncertain background: Absorption by interstellar H and D is measured against the stellar emission or absorption line that serves as the background "continuum." For late-type stars this "continuum" is the chromospheric Lyman emission line, whereas for hot stars, including white dwarfs, the "continuum" is the photospheric absorption line. Unless the star has an extremely high radial velocity, the stellar line is not separable from the interstellar absorption feature and must be estimated or computed. On the other hand, for large column desities ($\log N_{HI} > 20.0$), for example the line of sight to δ Ori A analyzed by Jenkins et al. (1999), the interstellar Lyman-α line is so broad that the shape of the background stellar absorption line is irrelevant.

For late-type stars, one approach is to scale the solar Lyman-α line in width depending on the stellar luminosity. This scaling, which is analogous to the Wilson-Bappu effect for the Ca II H and K lines (cf. Linsky 1999), appears to work for late-type dwarfs but not for luminous stars with massive winds, which have P-Cygni emission lines with blue-shifted absorption. An alternative approach is to assume that the stellar Lyman-α line has the same shape (on a velocity scale) as the Mg II k line. This should be valid at least for late-type dwarfs as both lines are optically thick and are formed in about the same layers in a chromosphere. Also, these two lines have the same shape in the solar spectrum. When there is a significant radial velocity separation (> 40 km s^{-1}) between the star and the ISM absorption, it is possible to reconstruct the stellar emission line empirically. Wood, Alexander, & Linsky (1996) developed a technique that assumes that the far wings of the stellar emission line are symmetric about the photospheric radial velocity, although this assumption is probably only valid for solar-like main sequence stars. This technique has recently been applied to the analysis of the line of sight to 36 Oph A. Another very powerful technique is to analyze spectra of RS CVn systems obtained at opposite orbital velocity quadratures. When applied to Capella, for which the component stars have a maximum radial velocity separation of 53 km s^{-1}, and HR 1099, for which the radial velocity separation is 111 km s^{-1}, Linsky et al. (1995) and Piskunov et al. (1997) could reconstruct the emission line profiles for each component star in the system and thereby obtain very accurate values of D/H.

When analyzing hot star spectra, one must assume a shape for the photospheric Lyman-α absorption line. For hot main sequence stars, the stellar absorption line may be asymmetric due to absorption from a wind. For hot white dwarfs like Sirius B, it is important to compute the Lyman-α line profile using a non-LTE model with accurately determined T_{eff} and $\log g$ (e.g., Hébrard *et al.* 1999). Uncertainties in the photospheric line profile, especially near zero flux where the scattered light correction is important, can lead to errors in the derived D/H ratio.

Scattered light: A little known fact concerning the GHRS instrument is that the echelle grating could not be built to specifications. Instead an off-the-shelf echelle grating with greater scattered light was substituted at the last minute. Unfortunately, the Digicon detector had tall pixels (in the direction perpendicular to the echelle dispersion) and could not sample the spectral shape of the scattered light with high precision. By contrast, the STIS echelle gratings have lower amounts of scattered light that can be measured accurately by the two-dimensional MAMA detector with small pixels. Sahu *et al.* (1999) showed a comparison of the Lyman-α spectra of the hot white dwarf G191-B2B obtained with the echelle gratings on GHRS and STIS. The profiles differ in the core of the D line, which led Sahu *et al.* to conclude that the D/H ratio in the non-LIC cloud toward the star is consistent with the LIC value, whereas Vidal-Madjar *et al.* (1998) had concluded from the GHRS data alone that the D/H ratio in the non-LIC components are significantly lower than in the LIC. The conclusion to this controversy may require additional analysis and new observations, but the important point is that accurate D/H ratios require spectra with an accurate correction for scattered light.

Heliospheric hydrogen absorption: In their analysis of the lines of sight to α Cen A and α Cen B, Linsky & Wood (1996) found two problems: (a) the central velocity of the H absorption is redshifted by $+3.2$ km s^{-1} relative to the velocity of the D, Mg II, and Fe II lines; (b) the broadening of H is larger than expected from the T and turbulent velocity values derived from the other lines. These difficulties imply the existence of extra H absorption on the red side of the interstellar absorption feature. Older *Copernicus* spectra also suggested a redshift of H relative to D in α Cen data and several explanations were proposed (cf. Lemoine *et al.* 1999). The GHRS spectra confirmed the missing H opacity and provided important clues as to its properties. A two-component fit to the α Cen Lyman-α line profiles showed that the extra H has a temperature of about 27,000 K and a redshift of at least 2 km s^{-1} relative to the main interstellar absorber.

The H responsible for this absorption is believed to be located near the heliopause where the inflowing partially ionized interstellar gas interacts with the outflowing ionized solar wind by charge exchange reactions of the protons and hydrogen atoms (cf. review by Zank 1999). Named the "hydrogen wall" by Baranov and Malama (1995), the computed pileup of decelerated and heated neutral H in the wall has properties consistent with those inferred from the additional H opacity toward α Cen. Even though $N_{\rm HI}$ in the H wall is only 10^{-3} that of the interstellar gas, the wall provides measurable absorption on the red side of the interstellar line because the H gas is redshifted, hot, and optically thick. Recent calculations by Müller, Zank, & Lipatov (1999) show that the H wall absorption is redshifted (relative to the interstellar flow) in all directions, because downstream the solar wind accelerates the heliospheric H. Heliospheric absorption has been detected toward α Cen, 36 Oph A (Wood, Linsky, & Zank 1999), and probably also Sirius (Izmodenov, Lallement, & Malama 1999).

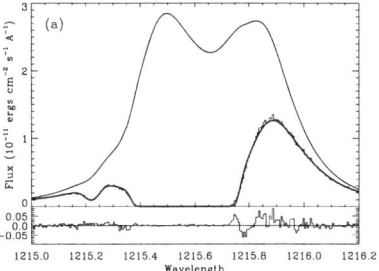 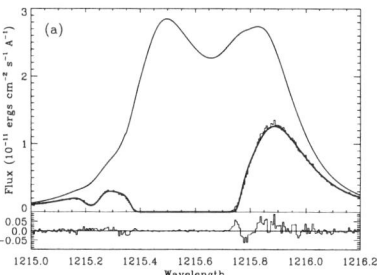

Figure 1. (a) Fitting of the observed Lyman-α line toward 36 Oph A (histogram) with only interstellar absorption. (b) Fitting with three components (interstellar, heliosphere, and astrosphere). The dotted line indicates absorption due only to the ISM, the dashed line is absorption due to the heliosphere, and the dot-dashed line is absorption due to the astrosphere. The thick solid line is the combined absorption of all three components. From Wood et al. (1999).

Astrospheric hydrogen absorption: Wood et al. (1996) noticed that H absorption is blueshifted relative to D in the spectrum of ε Ind. They interpreted this as absorption in the H wall around the target star. At least 9 stars (see Tables 1 and 2) now have identified astrospheric absorption, although many of these detections must be considered to be tentative. The combination of heliospheric absorption on the red side and astrospheric absorption on the blue side of the interstellar absorption means that model fits to the Lyman line that ignore this excess absorption will overestimate $\log N_{HI}$, and thus underestimate D/H. For the nearby stars α Cen, 36 Oph A, and Sirius, fitting the Lyman-α line without absorption in the heliosphere and astrosphere underestimates D/H by about a factor of 2. Figure 1 shows a comparison of the fits of the Lyman-α line toward 36 Oph A, which is located only 12° from the direction of inflowing LIC gas, with only interstellar absorption ($\log N_{HI} = 18.18$) and with all three components ($\log N_{HI} = 17.85$ in the ISM). Since the analysis of the Lyman-α lines toward many nearby stars have not included heliospheric and astrospheric absorption, more refined analyses could potentially lead to an upward revision of some of the reported D/H values.

3. Summary of D/H Measurements

3.1. D/H in the LIC

Table 1 summarizes the published D/H measurements toward nearby stars for which the interstellar gas has a velocity consistent with the LIC velocity vector (Lallement & Bertin 1992). In those cases where the authors derived a D/H ratio by assuming a value for N_{HI}, we cite their conclusions that the data are consistent ("con") with a given value of D/H. Figure 2 shows a plot of the D/H measurements with respect to the distance of the background star and the Galactic longitude. There is no trend with either variable, and we conclude that

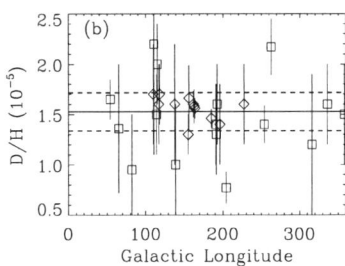 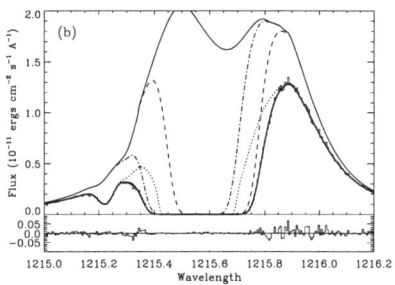

Figure 2. (a) D/H values from Tables 1 and 2 plotted vs. distance to the target star. Diamonds show the LIC values and boxes show the non-LIC values. The horizontal solid line is the weghted mean value of D/H for the LIC measurements and the dashed lines are $\pm 2\sigma$ about the mean value. (b) Similar to (a), but D/H values are plotted versus Galactic longitude.

$D/H = (1.53 \pm 0.18) \times 10^{-5}$ for gas in the LIC, based on the weighted mean and 2σ dispersion of the measurements in Table 1.

The largest potential source of systematic error in the LIC value of D/H is probably heliospheric and/or astrospheric absorption that has not been taken into account in the analyses of many of the target stars. In the analysis of data for 9 of the lines of sight through the LIC, hydrogen absorption in the heliosphere and/or astrosphere was included in the analysis. This is indicated by the symbols A and H in the Helio/Astro column of the tables. For the other lines of sight no inclusion of either absorber was attempted, as these data did not clearly show the presence of non-LISM H absorption, although heliospheric/astrospheric absorption might conceivably be present. The effect of not including heliospheric and astrospheric absorption is to ascribe all of the hydrogen absorption to the LIC and thereby potentially overestimate $N_{\rm HI}$ and underestimate D/H. Calculations of the heliospheric and astrospheric absorption for the lines of sight to Sirius (Izmodenov, Lallement, & Malama 1999) and α Cen A (Gayley et al. 1997) indicate that the changes in the derived D/H values could be significant. A reanalysis of many of these data sets is underway by Wood, Müller, & Zank (2000), which should better quantify this source of systematic error. The results summarized in Table 1 suggest, however, that the increase in the mean value of D/H in the LIC that will come from this reanalysis may not be very large, as the D/H values for those lines of sight analyzed including heliospheric and/or astrospheric absorption are not significantly larger than for the lines of sight analyzed without the corrections.

3.2. D/H in other clouds within 500 pc

Two fundamental issues in the study of Galactic chemical evolution are the distance scale over which significant variations in D/H occur and the chemical history of the environments in which different values of D/H are measured. While the value of D/H in the LIC may now be accurately measured and can serve as a reference by which D/H values in different environments can be com-

Table 1. Summary of D/H Measurements in the LIC

Star	d (pc)	l (°)	b (°)	Instrument	Helio/ Astro	D/H (10^{-5})	Ref.
Sirius A	2.63	227	−09	GHRS/M	H,A	con. with 1.6	L
Sirius A,B	2.63	227	−09	GHRS/M		1.6 ± 0.4	H
ε Eri	3.22	196	−48	GHRS/E	A	1.4 ± 0.4	F
61 Cyg A	3.48	082	−06	GHRS/E	A	con. with 1.5	G
Procyon	3.50	214	+13	GHRS/M		con. with 1.6	B
40 Eri A	5.04	201	−38	GHRS/E	A	con. with 1.5	G
Capella	12.9	163	+05	GHRS/E		1.47–1.72	A
Capella	12.9	163	+05	GHRS/M		1.41–1.74	B
Capella	12.9	163	+05	GHRS/E	A	1.56 ± 0.1	I
β Cas	16.7	118	−03	GHRS/M		1.6 ± 0.4	D
β Cas	16.7	118	−03	GHRS/E		1.7 ± 0.3	F
α Tri	19.7	139	−31	GHRS/E		1.6 ± 0.6	F
λ And	25.8	110	−15	GHRS/E	A	1.7 ± 0.5	C
HR 1099	29.0	185	−42	GHRS/E		1.46 ± 0.09	D
G191-B2B	68.8	156	+07	GHRS/E		1.1–1.5	E
G191-B2B	68.8	156	+07	GHRS/E		con. with 1.6	I
G191-B2B	68.8	156	+07	STIS/H		1.33–1.99	J
ε CMa	187.	240	−11	GHRS/M		con. with 1.6	K

References: A: Linsky et al. (1993); B: Linsky et al. (1995); C: Wood, Alexander, & Linsky (1996); D: Piskunov et al. (1997); E: Lemoine et al. (1996); F: Dring et al. (1997); G: Wood & Linsky (1998); H: Hébrard et al. (1999); I: Vidal-Madjar et al. (1998); J: Sahu et al. (1999); K: Gry et al. (1995); L: Izmodenov et al. (1999).

pared, the LIC value of D/H may or may not be representative of D/H in the disk at the galactocentric distance of the Sun. Future observations, especially with the FUSE spacecraft, are needed to settle this question.

Table 2 summarizes the published D/H measurements for interstellar gas with line of sight velocities different from the LIC velocity vector. The first five stars in the table show absorption at velocities corresponding to the velocity vector of the G cloud (Lallement & Bertin 1992). This cloud, centered in roughly the Galactic center direction, begins very close to the Sun in the direction toward α Cen and 36 Oph, because no LIC absorption is detected along these lines of sight. We include ε Ind in this group even though its velocity is the same as the predictions of both the LIC and G vectors, because the LIC model proposed by Redfield & Linsky (2000) predicts very little LIC absorption for this line of sight. Dring et al. (1997) proposed that the main absorption feature toward β Gem (22.0 ± 1.8 km s^{-1}) and σ Gem (21.5 ± 1.8 km s^{-1}) are consistent with the projected LIC velocity (19.6 km s^{-1}), but we find that the projected G cloud velocity (21.2 km s^{-1}) is a better match to the data. The present data indicate that D/H for the G cloud is consistent with the LIC value.

Three other clouds can be identified by their respective absorption velocities. The North Galactic Pole (NGP) cloud is seen toward HZ 43 and 31 Com. The South Galactic Pole (SGP) cloud is seen toward β Cet. A third cloud identified by its radial velocity of about +33 km s^{-1} is seen toward β Gem and σ Gem. The D/H ratios for these three clouds, all located within about 30 pc of the Sun, appear to be consistent with the LIC value.

Table 2 includes another 4 stars (Sirius, 61 Cyg A, α Tri, and G191-B2B) located within 100 pc of the Sun for which there are interstellar velocity components different from the LIC, G, NGP, SGP, and +33 clouds. Observations of other lines of sight are needed to determine the size and locations of the clouds responsible for the observed absorption components. Hébrard et al. (1999) argue that the +11.7 km s^{-1} absorption component toward Sirius A and B has a D/H ratio that is barely consistent with the LIC value but is more likely much smaller. However, their analysis does not include heliospheric or astrospheric absorption, which Izmodenov et al. (1999) argue are important and must increase the interstellar D/H ratio. We await a reanalysis of this line of sight before concluding that the D/H ratio is low in this cloud. The line of sight to the hot white dwarf G191-B2B has been studied by Vidal-Madjar et al. (1998) using GHRS spectra and by Sahu et al. (1999) using STIS spectra. These analyses arrive at very different conclusions concerning the D/H ratio in non-LIC clouds detected along this line of sight: Vidal-Madjar et al. concluded that two clouds are present (+8.2 and +13.2 km s^{-1}) with average D/H $\sim 0.9 \times 10^{-5}$, while Sahu et al. concluded that there is only one cloud at +8.6 km s^{-1} with D/H consistent with the LIC value. Sahu et al. argued that scattered light in the GHRS echelle spectra is responsible for the very different conclusions. Given this disagreement and no inclusion of heliospheric and astrospheric absorption in either analysis, we believe that it is premature to conclude that D/H has a lower value in these clouds.

The last four lines of sight listed in Table 2 are for stars located beyond 100 pc from the Sun. The observations of the higher Lyman lines were obtained with the high resolution IMAPS spectrograph ($\lambda/\Delta\lambda = 75,000$) or the lower resolution ORFEUS spectrograph ($\lambda/\Delta\lambda = 10,000$). For γ^2 Vel, δ Ori A, and ζ Pup, the D column densities were obtained by fitting the higher Lyman lines, but the H column densities were obtained by fitting IUE spectra of the Lyman-α line. The very broad interstellar absorption in the Lyman-α line means that H absorption in the heliosphere and astrosphere occur where the interstellar absorption is saturated and thus do not need to be included. For these lines of sight there may be several clouds responsible for the absorption, in which case the derived D/H value would represent an average through these different clouds.

The most interesting of these lines of sight is toward δ Ori A, for which Jenkins et al. (1999) find that D/H lies in the range (0.61–0.93) $\times 10^{-5}$ in what appears to be a single absorbing cloud. We believe that this is the most reliable measurement so far of a D/H ratio in the Galactic disk that is significantly different from the LIC value. If analysis of other lines of sight in star forming regions in Orion confirm this result, then Galactic chemical evolution models are faced with an interesting test. These models (e.g., Tosi et al. 1998) typically show an inverse correlation of D/H with metal abundance as chemical evolution in stars inevitably proceeds in this direction. Jenkins et al. (1999), however, find no overabundance of metals in the δ Ori A line of sight, but rather a low abundance of N and a marginally low abundance of O. Since Orion is not a region of enhanced metal abundance, why is D/H so low?

Table 2. Summary of D/H Measurements within 500 pc

Star	d (pc)	l (°)	b (°)	Instrument	Helio/ Astro	Cloud	D/H (10^{-5})	Ref.
α Cen A,B	1.35	316	−01	GHRS/E	H,A	G	1.2 ± 0.7	A
ε Ind	3.63	336	−48	GHRS/E	A	G?	1.6 ± 0.4	B
36 Oph A	5.46	358	+07	STIS/H	H,A	G	1.5 ± 0.5	F
β Gem	10.3	192	+23	GHRS/E		G?	1.4 ± 0.4	D
σ Gem	37.5	191	+23	GHRS/E		G?	1.4 ± 0.4	D
HZ 43	32.0	054	+84	GHRS/E		NGP	1.65 ± 0.2	G
31 Com	94.2	115	+89	GHRS/M		NGP	1.5 ± 0.4	C
31 Com	94.2	115	+89	GHRS/E		NGP	2.0 ± 0.4	D
β Cet	29.4	111	−81	GHRS/M		SGP	2.2 ± 1.1	C
β Gem	10.3	192	+23	GHRS/E		+33.2	1.6 ± 0.4	D
σ Gem	37.5	191	+23	GHRS/E		+32.0	1.3 ± 0.4	D
Sirius A	2.63	227	−09	GHRS/M	H,A	+11.7	con. 1.6	O
Sirius A, B	2.63	227	−09	GHRS/M		+11.7	≤ 1.6	H
61 Cyg A	3.48	082	−06	GHRS/E	A	−9.0	con. 1.5	E
α Tri	19.7	139	−31	GHRS/E		+13.1	1.0 ± 0.6	D
G191-B2B	68.8	156	+07	GHRS/E		+8.2	≈ 0.9	K
G191-B2B	68.8	156	+07	STIS/H		+8.6	> 1.26	L
G191-B2B	68.8	156	+07	GHRS/E		+13.2	≈ 0.9	K
BD+28°4211	104.	082	−19	ORFEUS		−4	0.4–1.5	N
γ² Vel	260.	263	−08	IMAPS			1.89–2.44	J
BD+39°3226	270.	065	+29	ORFEUS		−24	0.72–1.99	M
δ Ori A	400.	204	−18	IMAPS		+25	0.61–0.93	I
ζ Pup	430.	254	−05	IMAPS			1.21–1.59	J

References: A: Linsky & Wood (1996); B: Wood, Alexander, & Linsky (1996); C: Piskunov et al. (1997); D: Dring et al. (1997); E: Wood & Linsky (1998); F: Wood, Linsky, & Zank (1999); G: Landsman et al. (1999); H: Hébrard et al (1999); I: Jenkins et al. (1999); J: Sonneborn et al. (1999); K: Vidal-Madjar et al. (1998); L: Sahu et al. (1999); M: Bluhm et al. (1999); N: Gölz et al. (1998); O: Izmodenov et al. (1999).

Acknowledgments. This work is supported by NASA grant S-56500-D to the University of Colorado and NIST.

References

Baranov, V.B., & Malama, Y.G. 1995, *J. Geophys. Res.*, 100, 14755

Bluhm, H., Marggref, O., de Boer, K.S., Richter, P., Heber, U. 1999, A&A, 352, 287

Dring, A., Linsky, J., Murthy, J., Henry, R.C., Moos, H., Vidal-Madjar, A., Audouze, J., Landsman, W. 1997, ApJ, 488, 760

Gayley, K.G., Zank, G.P., Pauls, H.L., Frisch, P.C., & Welty, D.E. 1997, ApJ, 487, 259

Gölz, M. et al. 1998, in Proc. IAU Colloq. 166, The Local Bubble and Beyond, eds. D. Breitschwerdt, M.J. Freyberg, & J. Trumper, p. 75

Gry, C., Lemonon, L., Vidal-Madjar, A., Lemoine, M., & Ferlet, F. 1995, A&A, 302, 497

Hébrard, G., Mallouris, C., Ferlet, R., Koester, D., Lemoine, M., Vidal-Madjar, A., & York, D. 1999, A&A, 350, 643

Izmodenov, V.V., Lallement, R., & Malama, Y.G. 1999, A&A, 342, L13

Jenkins, E.B., Tripp, T. M., Wozniak, P. R., Sofia, U. J., & Sonneborn, G. 1999, ApJ, 520, 182

Lallement, R., & Bertin, P. 1992, A&A, 266, 479

Landsman, W., Sahu, M., Bruhweiler, F.C., Gull, T.R., Barstow, M.A., Hubeny, I., & Holberg, J.B. 1999, poster presented at IAU Symp 198

Lemoine, M. et al. 1999, New Astronomy, 4, 231

Lemoine, M., Vidal-Madjar, A., Bertin, P., Ferlet, R., Gry, C., & Lallement, R. 1996, A&A, 308, 601

Linsky, J.L. 1999, ApJ, 525, 776

Linsky, J.L., Brown, A., Gayley, K., Diplas, A., Savage, B.D., Ayres, T.R., Landsman, W., Shore, S.N., & Heap, S.R. 1993, ApJ, 402, 694

Linsky, J.L., Diplas, A., Wood, B.E., Brown, A., Ayres, T.R., & Savage, B.D. 1995, ApJ, 451, 335

Linsky, J.L., & Wood, B.E. 1996, ApJ, 463, 254

Müller, H.R., Zank, G.P., & Lipatov, A.S. 1999, to appear in J. Geophys. Res.

Piskunov, N., Wood, B.E., Linsky, J.L., Dempsey, R.C., & Ayres, T.R. 1997, ApJ, 474, 315

Redfield, S. & Linsky, J.L. 2000, ApJ, to appear May 10

Sahu, M.S., Landsman, W., Bruhweiler, F.C., Gull, T.R., Bowers, C.A., Lindler, D., Feggans, K., Barstow, M.A., Hubeny, I., & Holberg, J.B., 1999, ApJ, 523, L159

Sonneborn, G., Jenkins, E.B., Tripp, T., Vidal-Madjar, A., Ferlet, R., & Sofia, U.J. 1999, poster presented at IAU Symp 198

Tosi, M., Steigman, G., Matteucci, F., & Chiappini, C. 1998, ApJ, 498, 226

Vidal-Madjar, A., Lemoine, M., Ferlet, R., Hébrard, G., Koester, D., Audouze, J., Cassé, M., Vangioni-Flam, E., & Webb, J. 1998, A&A, 338, 694

Welty, D.E., Morton, D.C., & Hobbs, L.M. 1996, ApJS, 106, 533

Wood, B.E., Alexander, W.R., & Linsky, J.L. 1996, ApJ, 470, 1157

Wood, B.E., & Linsky, J.L. 1998, ApJ, 492, 788

Wood, B.E., Linsky, J.L., & Zank, G.P. 1999, submitted to ApJ

Wood, B.E., Müller, H.R., & Zank, G.P. 2000, in preparation

Zank, G.P. 1999, Space Sci.Rev., 89, 1

Deuterium observation in our Galaxy - View B

Alfred Vidal-Madjar

Institut d'Astrophysique de Paris, 98bis Boulevard Arago, F-75014 Paris, FRANCE

Abstract.
Galactic D/H evaluations from observations completed in the far UV with first the Copernicus satellite then followed by IUE and the GHRS on the HST studies already suggest that D/H variations may be present in the interstellar medium. More recent IMAPS (the Interstellar Medium Absorption Profile Spectrograph) observations confirm that conclusion while new STIS (which replaced the GHRS on board HST) studies seem to indicate the contrary. The situation is discussed here.

Hopefully FUSE (the Far Ultraviolet Spectroscopic Explorer, launched the 24th of June 1999) will give access to the D/H ratio in many different galactic sites. This will help us reach a better global view of the evolution of that key element, and thus better constrain any evaluation of its primordial abundance.

1. Introduction

Deuterium is understood to be only produced in significant amount during primordial Big Bang nucleosynthesis (BBN) and thoroughly destroyed in stellar interiors. Deuterium is thus a key element in cosmology and in galactic chemical evolution (see e.g. Audouze & Tinsley 1976; Gautier & Owen 1983; Vidal-Madjar & Gry 1984; Boesgaard & Steigman 1985; Olive et al. 1990; Pagel 1992; Vangioni-Flam & Cassé 1994; Prantzos 1996; Scully et al. 1997). Indeed, its primordial abundance is the best tracer of the baryonic density parameter of the Universe Ω_B, and the decrease of its abundance along the galactic evolution should trace the amount of star formation (among others).

The first, although indirect, measurement of the deuterium abundance of astrophysical significance was carried out through ^{3}He evaluation in the solar wind, leading to D/H$\simeq 2.5 \pm 1.0 \times 10^{-5}$ (Geiss & Reeves 1972), a value representative of 4.5 Gyrs ago. The first measurements in the interstellar medium (ISM) of the D/H ratio, representative of the present epoch, were reported shortly thereafter (Rogerson & York 1973). Their value of D/H$\simeq 1.4 \pm 0.2 \times 10^{-5}$ has, as a representative value, since then nearly not changed. For nearly three decades, these interstellar abundances have been used to constrain BBN in a direct way.

In the following, we discuss the different measurements of the deuterium abundance in the ISM within our galaxy and try to show that the situation is not that simple.

2. ISM observations (various approaches)

There are several methods to measure the interstellar abundance of deuterium (see Vidal-Madjar 1991; Ferlet 1992). One of them is to observe deuterated molecules such as HD, DCN, etc... and to form the ratio of the deuterated molecule column density to its non-deuterated counterpart (H_2, HCN, etc....). More than twenty different deuterated species have been identified in the ISM, with abundances relative to the non-deuterated counterpart ranging from 10^{-2} to 10^{-6}. This means that fractionation effects are important, and that, as a consequence, this method cannot provide a precise estimate of the true interstellar D/H ratio.

However, if the molecular form is HD in a place where all or nearly all the deuterium atoms are trapped in that molecule, then D/H could be deduced reasonably well from an evaluation of the HD/H_2 ratio. In such cases it may be possible to evaluate the D/H ratio as recently shown in the IR through ISO observations of the HD molecule in Orion by Bertoldi et al. (1999). They evaluated: $(D/H)_{Orion} = 7.6 \pm 2.9 \times 10^{-6}$.

More recently, it was also possible to observe with FUSE the HD molecule in the direction of a reddened star where possibly most of the deuterium is also under the HD molecular form. This observation is still very preliminary (Ferlet et al. 2000) but could lead in a near future to new estimates of the D/H ratio in denser parts of the ISM.

Another way to derive the D/H ratio comes through radio observations of the hyperfine line of DI at 92cm (Cesarsky, Moffet & Pasachoff 1973). The detection of this line is however extremely difficult. An interresting upper limit was derived toward Cas A (Heiles et al. 1993): $D/H \leq 2.1 \times 10^{-6}$ which results from a large differential fractionation of atomic D in molecular form.

In the galactic anticenter direction one could expect a higher D/H values due to less stellar processing and from the searches made in that direction, the most recent one from Chengalur et al. (1997) leads to a possible detection corresponding to $(D/H)_{Gal.Anticenter} = 3.9 \pm 1.0 \times 10^{-5}$.

Finally, a detection of the Balmer $D\alpha$ and $D\beta$ lines in the direction of Orion was recently reporteded by Hébrard et al. (2000; see also these proceedings) showing that a new technique for evaluating D/H may become available soon. Similar observations made in the direction of planetary nebulae lead to a new strong upper limit on the deuterium abundance: $(D/H)_{NGC\ 6572} \leq 1. \times 10^{-7}$. This shows, as expected, that deuterium is burned in stars and gives a direct observational confirmation of such a fact. This is one of the possible explanations for deuterium abundance variations.

3. ISM observations (Lyman lines)

Another way to derive the ISM D/H ratio is to observe the atomic transitions of the Lyman series against the background continuum of cool or hot stars.

3.1. Cool stars

The main advantage of observing cool stars is that they can be selected in the vicinity of the Sun. This results in low HI column densities, and "trivial" or at least "simpler" lines of sight. However even with low HI column densities of the order of 10^{19} cm^{-2} the presence of several interstellar components with different b-values implies large errors on the HI column density, in particular when very low column density "hot" HI gas is detected as the so called circumstellar "hydrogen walls" (see Linsky 2000 in these proceedings). Moreover, the chromospheric Lyman α emission line of the target cool star has to be modeled to set the continuum for the interstellar absorption. Such a procedure necessarily introduces systematic errors along with all the other unknown instrumental systematics. In addition, in the "cool stars" approach, the detailed structure of the line of sight could be found only through the observation of the FeII and the MgII ions, which are unfortunately present in both HI and HII regions and thus may not trace properly the HI and DI gas. In particular, species like NI and OI could not be observed. This could lead to additional uncertainties.

For these reasons, deriving the HI column density has always been the limiting factor of accurate D/H ratios measurements. Nevertheless, this method has provided the most precise measurement of the local D/H ratio in the direction of Capella, using HST-GHRS:
$(D/H)_{Capella} = 1.60 \pm 0.09^{+0.05}_{-0.10} \times 10^{-5}$ (Linsky et al., 1993; 1995).

This result was further confirmed by the re-analysis of Vidal-Madjar et al. (1998). Such an agreement and is probably due to the *very high quality of the GHRS data*, the high signal to noise ratio, the binary star motion giving access to the stellar Lyα profiles, the remarquably simple line of sight with apparently only one component and negligible perturbation by "hot" HI gas.

From several several additional cool stars observed with HST (see Linsky 2000, these proceedings) it is however only possible to conclude that the deduced D/H evaluations are compatible with the Capella one since none of these results are precise enough to place new constraints.

3.2. Hot stars

Hot stars are located further away from the Sun, so that one always has to face a higher HI column density and a non-trivial line of sight structure. In these cases, DI could not be detected at Lyα, and one has to observe higher order lines, *e.g.* Lyγ, Lyδ, Lyϵ; hence these measurements have primarily come through Copernicus observations. The stellar continuum is however smooth at the location of the interstellar absorption and, moreover, NI and OI lines are available to probe the velocity structure of the line of sight. They were shown to be good tracers of HI in the ISM (Ferlet 1981; York et al. 1983) although NI may present some difficulties (Vidal-Madjar et al. 1998).

The D/H ratios evaluated in the direction of hot stars by Copernicus range from $\sim 5\times10^{-6}$ to $\sim 2.5\times10^{-5}$. A large scatter is clearly detected and represents differences of the *average* D/H ratio in the nearby galactic ISM (within 1kpc), that may be as large as a factor $\simeq 4$. The essential question is : do these variations really exist?

Unfortunately from the Copernicus data alone, no one has been able to answer this question. To progress, one may have either to re-analyze all these data in a consistent way, looking for possible undetected systematics, or complete new observations in the direction of a great variety of targets. In effect each type of target will generate its own type of problems and systematics.

As an example, one should recall that time variations of the D/H ratio have already been reported toward ϵ Per (Gry et al. 1983), which were interpreted as due to the ejection of high velocity hydrogen atoms from the star. But since this perturbation can only enhance the D/H ratio, it is worth noting that in at least four cases the D/H ratio was found to be really low: $0.7 \pm 0.2 \times 10^{-5}$ and $0.65 \pm 0.3 \times 10^{-5}$ toward δ and ϵ Ori (Laurent et al. 1979); $0.8 \pm 0.2 \times 10^{-5}$ toward λ Sco (York 1983) and $0.5 \pm 0.3 \times 10^{-5}$ toward θ Car (Allen et al. 1992). In each case, authors discussed in details possible systematics but concluded that none of the identified ones could explain such values. These low values thus seems to be real.

To make more progresses, observations were made with new space instrumentations. One with Orfeus II (Bluhm et al. 1999), in the direction of a galactic halo star lead to D/H=$1.2^{+0.5}_{-0.4} \times 10^{-5}$, an interresting value since it concerns a new part of the ISM where no evaluation was made before. The result is not however significatively different from the local ISM evaluation.

Other observations were made with the IMAPS instrument in the direction of stars already observed with Copernicus. The combination of these observations with a new technique to evaluate the total HI content on the line of sight through the use of many IUE observations of the Lyα line, lead to three new evaluations of the *average* D/H ratio in the nearby ISM (see also Sonneborn et al. in these proceedings). They are, $(D/H)_{\delta Ori\ A} = 0.74^{+0.19}_{-0.13} \times 10^{-5}$ (Jenkins et al. 1999), $(D/H)_{\gamma^2 Vel} = 2.14^{+0.30}_{-0.25} \times 10^{-5}$ (Sonneborn et al. 2000), $(D/H)_{\zeta Pup} = 1.39^{+0.20}_{-0.18} \times 10^{-5}$ (Sonneborn et al. 2000).

The interresting point is that the δOri A result confirms the Copernicus evalutaion made by Laurent et al. (1979) showing that this low *average* D/H value is estimated through two different studies using different instruments. This low D/H is also compatible with the ISO, Bertoldi et al. (1999) evaluation made in the Orion region. This seems to make that case of a low D/H value, very strong. This *average* evaluation is already incompatible with the "supposed uniform" value found in the local ISM.

Note that with the same IMAPS and IUE instruments and approach, a high D/H value is evaluated in the direction of γ^2Vel. This is discussed in detail by Sonneborn et al. (in these proceedings). This strengthens the case of D/H variations.

The *average* D/H ratio seems thus to possibly vary by at least a factor 3 in the nearby ISM.

3.3. White dwarfs

Observing white dwarfs has many advantages. Such targets can be chosen near to the Sun, circumventing the main disadvantage of hot stars, and they can also be chosen in the high temperature range, so as to provide a smooth stellar profile at Lyα. At the same time, the NI triplet at 1200 Å as well as the OI line at 1302 Å would be available, allowing thus an accurate sampling of the HI part of the line of sight. Such observations have now been conducted using HST toward three white dwarfs: G191-B2B (Lemoine et al. 1996; Vidal-Madjar et al. 1998), Hz43 (Landsman et al. 1996) and Sirius B (Hébrard et al. 1999; see also these proceedings). Although both Hz43 and Sirius B D/H *average* evaluations are compatible with the local ISM value evaluated toward Capella, in the case of Sirius B, this compatibility is marginal since two components are detected on that line of sight and, imposing the "local" D/H value to the component identified with the Local Interstellar Cloud (LIC) in which the sun is embedded, leads to a very low value in the other component. This observation may confirm that D/H possibly varies from cloud to cloud even over few parsecs.

In the case of G191-B2B, from Lemoine et al. (1996) and Vidal-Madjar et al. (1998) studies of two independent sets of GHRS data, the line of sight velocity structure comprises one HI region, identified with the LIC observed toward Capella, together with two HII regions. If the D/H ratio for the LIC is common to both the G191-B2B and Capella sight-lines (these stars are separated by only 8° on the sky) and equal to $(D/H)_{LIC} = 1.6 \times 10^{-5}$, then the D/H ratio for the other components appears to be significantly lower and of the order of $(D/H)_{HII} = 0.9 \times 10^{-5}$.

This is a very strong case for D/H variations since even the *average* value found on that line of sight is (Vidal-Madjar et al. 1998): $(D/H)_{G191-B2B} = 1.12 \pm 0.08 \times 10^{-5}$.

However since this result is contested by the recent study of Sahu et al. (1999 ; see also these proceedings) which claims from new STIS observations, that "D/H values in both components are consistent with $(D/H)_{LIC} = (1.5 \pm 0.1) \times 10^{-5}$", I will discuss below in detail their arguments and underline the possibble cause of such a disagreement.

4. The case of G191-B2B

This case is interresting because the line of sight was observed with similar high signal to noise ratio with two different and excellent instruments on board HST, GHRS and STIS. Both instruments present a very high spectral resolution.

Furthermore these data sets were treated independently by three different groups using different approaches leading to different spectra:
- GHRS data were analysed by Vidal-Madjar et al. (1998) (V98) with different procedures described there in, as well as by Sahu et al. (1999) (S99) and Howk & Sembach (2000) (H00) using each their own GHRS package;
- STIS data were analysed by S99 using a complete package developed by the STIS Instrument Team and by H00 who used their own package to correct for the stray light in the STIS echelle, leading to two other STIS spectra kindly forwarded to me by the two groups in order to complete a precise analysis.

4.1. The number of components and the zero level

S99 claimed that only two components are needed on that line of sight arguing that the STIS SiIII line is shifted by 4km/s relative to the GHRS one. Although this point is correct, other arguments justify the need for three components along this line of sight. They are : i) the separation between the "two clearly seen" components in the SiIII lines is of 10.25 km/s while 11.00 km/s in the NI triplet, ii) the bottom of the saturated OI line is more easily represented, close to the zero level, by the presence of a third component and iii) the width of the strongest SiIII component is larger than the corresponding SiII one. All these arguments concern relative widths and spectral separations and are thus strong observational evidences.

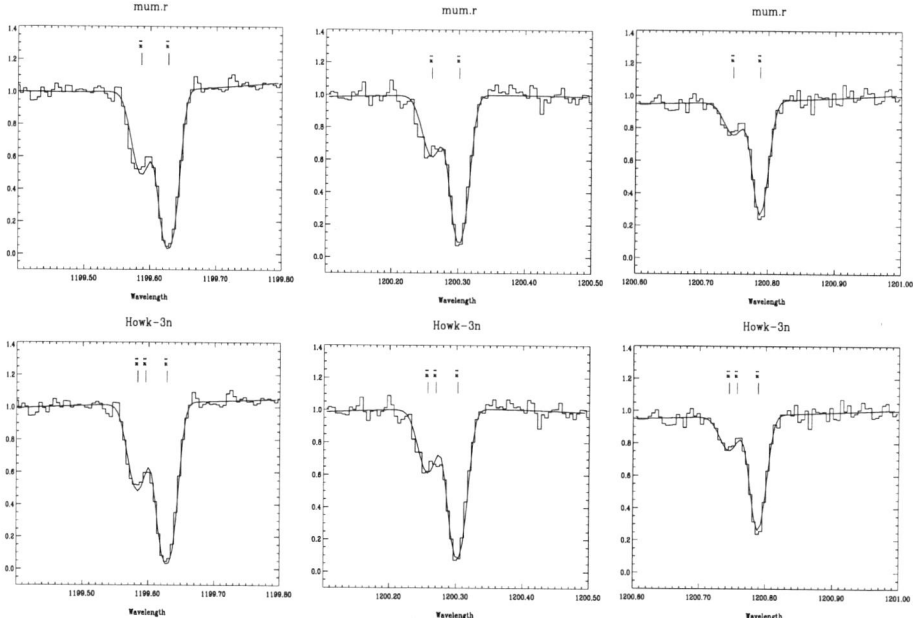

Figure 1. The STIS NI triplet. Top, 2 components, bottom, 3 components. Fits made along with the Si lines where the "two" apparent components are closer. The 2-components fit is significantly poorer. This is not seen if one looks only at the weaker NI line (right pannels).

Figure 1 shows that the STIS data also ask for a third component if *all* NI line are used.

However, as shown in Figure 2 (left), the study of S99 STIS data shows that the D/H evaluation is insensitive to the number of components supposed, as long as more than one is assumed. It is interesting to note however that if only one component is taken into account, then D/H is significantly different, a warning for more complex lines of sight studies. Thus having fitted the data with two or three components is not the cause of the discrepancy.

It was also argued that the zero level in an echelle spectrograph is not well controlled and that is may be the cause of the disagreement. Using the set of STIS data from S99 and varying arbitrarily the zero level by few percent of the nearby continuum level, it is shown in Figure 2 (right) through the $\Delta\chi^2$ method (see e.g. V98) that within $\pm 2\sigma$, i.e. conservatively when χ^2 varies by less than 10, this induces at most a variation on D/H of $\pm 0.1 \times 10^{-5}$. This is also found with the other data sets and is thus not the explanation for the discrepancy.

4.2. The stellar Lyα continuum

In both V98 and S99, the study of the stellar continuum was made carefuly. S99 argued that they used the best available non-LTE model (NLTE) and thus that their correction has to be the right one. V98 however tried to analyse the effect of the assumption of different types of continuum and showed that it can be adjusted in any case by the fitting procedure. So the precise knowledge of the

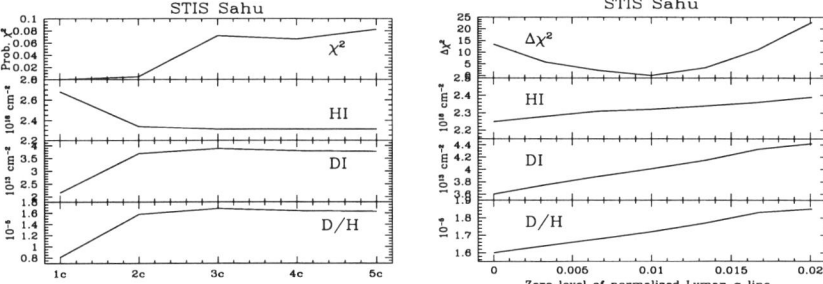

Figure 2. Left: Variation of the STIS Sahu et al. (1999) data fit solutions as a function of the number of components (c) assumed. The probability of obtaining each χ^2 is low due to hidden systematics. However a significative jump is observed when going from 2 (2c) to 3 (3c) components. Evaluated total HI and DI column densities are shown. The average D/H is insensitive to the number of components above 2c. This is because the H and D lines are intrinsically broad.

Right: Variation of the fitting parameters as a function of the zero level assumed at the bottom of the Lyman α H and D lines. The "real" zero is found to be about 1% higher than the instrument corrected zero (according to Sahu et al. 1999 treatment). The $\Delta\chi^2$ variation relative to the best χ^2 is shown and indicates that a poor estimate of the zero level cannot induce an error on D/H larger than about $\pm 0.1 \times 10^{-5}$ ($\sim 2\sigma$).

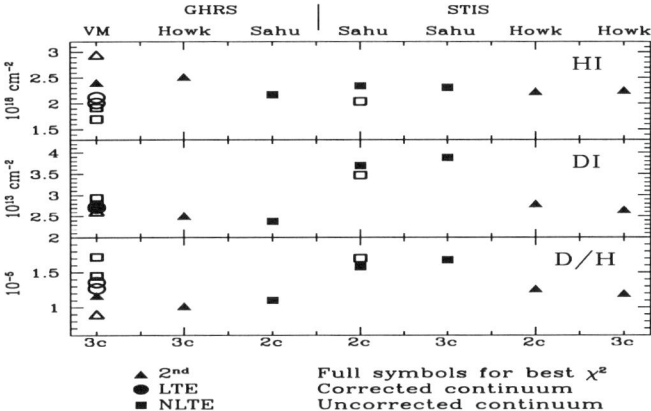

Figure 3. The HI and DI total column density on the G191-B2B line of sight as a function of the data treated : GHRS (left) from V98 (VM), H00 (Howk) and S99 (Sahu) and STIS (right) from S99 (Sahu) or H00 (Howk) (see text). The number of components assumed is indicated in the abscissa (2c or 3c). The filled symbols correspond to the best fits, in most cases obtained with corrected stellar continuum when available (see V98 and text for more details). The total HI is very stable. The disagreement comes from the DI column density deduced from the STIS data as treated by S99. All other evaluations agree together within $1.15 \pm 0.15 \times 10^{-5}$.

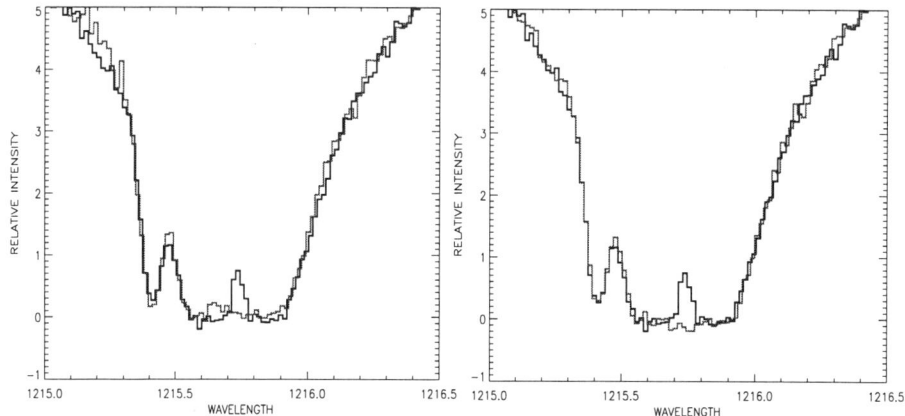

Figure 4. The comparison of the S99 (left) and the H00 (right) STIS profiles (thin histogram, binned 3 times) with the GHRS one as processed by H00 (since all GHRS studies agree together, thick histogram in both, binned 5 times) are shown (see text). In both pannels the central emmission is due to GHRS data and corresponds to the earth geocoronal Lyman α emission. Note the remarquable conspiracy to produce similar line shapes from two independant sets of data in the right pannel while on the left hand side, deviations could be seen in many places at 1215.2Å, 1215.4Å (bottom of D line), 1215.5Å (top of D line), 1215.65Å (position of earth geocorona during STIS observations), 1216.1Å, 1216.2Å. The deviations near the D line are the cause of the discrepency between the different studies.

continuum is not too critical. To illustrate this, in Figure 3 are presented the different data sets analysed by assuming three types of stellar continua (2nd order polynomial, LTE or NLTE) fixed or fitted simultaneously with the interstellar lines (see V98, for more details). The result is striking: the HI column density is not sensitive to such changes and is thus very precisely evaluated. The HI total content of the line of sight is very stable with any data set used or whatever stellar continuum assumed as long as the fitting procedure is able to adjust it (by the way one can note that the stellar continuum used by S99, is quite good since an additional adjustment of the fitting procedure does not change significantly their result).

On that same Figure 3 however, one can note that on the contrary it is the DI column density that seems to vary! It varies, not in relation with that study, but only in the case of the S 99 STIS data set. All other sets, GHRS or STIS from V98 or H00 agree very well together. The difference thus clearly comes from the detailed D line profile as in fact explained by S99.

4.3. The D line profile

In Figure 4 the GHRS and STIS line profiles are compared by binning the GHRS data five times and the STIS data three times in order to have a similar sampling. It is striking to see how the GHRS and STIS, H00 data look similar while the GHRS and STIS, S99 data are discrepent, particularly over the deuterium line (see also comparison figures presented in both S99 and H00 papers). Contrary to what S99 claimed, this is not a background issue as discussed earlier, but a line profile issue since in the two different treated STIS data one (H00) looks exactly like the GHRS one while the other (S99) does not. How could a process correct the STIS data to look like the GHRS one ? Could it be a strange *conspiracy* or is it, more simply, that one set was corrected properly and the other not?

To strengthen that argument, note that the GHRS instrument had a very efficient procedure called FP-SPLIT able to correct for detector defects, while in these STIS data a similar procedure was not used. Also note that along the H I Lyα wings, which should be smooth by definition, other fluctuations could be seen between the STIS S99 profile and the GHRS one while everywhere the GHRS and STIS H00 data nicely match together. Even in the core of the Lyα line, supposed to be flat, fluctuations are seen in one set and less in the other.

In conclusion, out of these four data sets, the disagreement is coming from only one set and concerns solely the D profile, leading to a different D/H value. From that discussion it seems that one set of data contains probably some uncorrected instrumental fluctuations that led S99 to erroneous conclusions.

5. Conclusion

The different D/H evaluations made within the galaxy lead to the fact that the D/H ratio seems to vary from region to region for reasons still unknown as discussed in e.g. Lemoine et al. (1999).

The present observations lead to the following conclusions:
– in the local ISM (<100 pc) average variations are of the order of 30%
– in the local ISM component to component variations may reach a factor of 2
– in the nearby ISM (<1000 pc) average variations may reach a factor of 3
– in the nearby ISM component to component variations may be even larger

This is today's situation, but now with FUSE in orbit we should have soon many more evaluations of the D/H ratio in all the galaxy, within the halo and even in some extra galactic lines of sights.

Acknowledgments. I thank M.S. Sahu for providing me with her STIS data and J.C. Howk & K.R. Sembach for giving me access to the same STIS data as treated with their procedure prior to publication.

References

Audouze, J., & Tinsley, B.M. 1976, ARA&A, 14, 43

Allen, M., Jenkins, E.B., & Snow T.P. 1992, ApJS, 83, 261

Bertoldi, F., Timmermann, R., Rosenthal, D., Drapatz, S., & Wright, C.M. 1999, A&A, 346, 267

Bluhm, H., Marggraf, O., de Boer, K.S., Richter, P., & Heber, U. 1999, A&A, 352, 287

Boesgaard, A.M., & Steigman, G. 1985, ARA&A, 23, 319 ApJ, 243, 161

Cesarsky, D.A., Moffet, A.T. & Pasachoff, J.M. 1973, ApJ, 180, L1

Chengalur, J.N., Braun, R., & Burton, W.B. 1997, A&A, 318, L35
Ferlet, R. 1981, A&A, 98, L1
Ferlet, R. 1992, in IAU Symposium 150, 85
Ferlet, R., André, M., Hébrard, G., Lecavelier, A., Lemoine, M., Pineau des Forêts, G., Roueff, E., Vidal-Madjar, A., & The FUSE Team 2000, ApJ, submitted
Gautier, D., & Owen, T. 1983, Nature, 304, 691
Geiss, J., & Reeves, H. 1972, A&A, 18, 126
Gry, C., Laurent, C., & Vidal-Madjar, A. 1983, A&A, 124, 99
Howk, J.C. & Sembach, K.R. 2000, AJ, in press
Hébrard, G., Mallouris, C., Ferlet, R., Koester, D., Lemoine, M., Vidal-Madjar, A., & York, D. 1999, A&A 350, 643
Hébrard, G., Péquignot, D., Vidal-Madjar, A., Walsh, J.R., & Ferlet, R. 2000, accepted for publication in A&A *Letters*
Heiles, C., McCullough, P., & Glassgold, A. 1993, ApJS, 89, 271
Jenkins, E.B., Tripp, T.M., Woźniak, P.R., Sofia, U.J. & Sonneborn, G. 1999, ApJ, 520, 182
Landsman, W., Sofia, U.J., & Bergeron, P. 1996, in Science with the Hubble Space Telescope - II, STScI, 454
Laurent, C., Vidal-Madjar, A., & York, D.G.: 1979, ApJ, 229, 923
Lemoine, M., et al. 1996, A&A, 308, 601
Lemoine, M., et al. 1999, New Astronomy, 4, 231
Linsky, J., et al. 1993, ApJ, 402, 694
Linsky, J., et al. 1995, ApJ, 451, 335
Olive, K., Schramm, D., Steigman, G., & Walker, T. 1990, Phys.Rev.Lett, B236, 454
Pagel, B., et al. 1992, MNRAS, 255, 325
Prantzos, N. 1996, A&A, 310, 106
Rogerson, J., & York, D. 1973, ApJ, 186, L95
Sahu, M.S., Landsman, W., Bruhweiler, F.C., Gull, T.R., Bowers, C.A., Lindler, D., Feggans, K., Barstow, M.A., Hubeny, I., & Holberg, J.B. 1999 ApJ, 523, L159
Scully, S.T., Cassé, M., Olive, K.A., & Vangioni-Flam, E. 1997, ApJ, 476, 521
Sonneborn, G., Tripp, T.M., Ferlet, R., Jenkins, E.B., Sofia, U.J., Vidal-Madjar, A. & Woźniak, P.R., 2000, ApJ, submitted
Vangioni-Flam, E., & Cassé, M. 1994, ApJ, 427, 618
Vidal-Madjar, A., Laurent, C., Bruston, P., & Audouze, J. 1978, ApJ, 223, 589
Vidal-Madjar, A., & Gry, C.: 1984, A&A, 138, 285
Vidal-Madjar, A.: 1991, in Adv. Space Res., 11, 97
Vidal-Madjar, A., Lemoine, M., Ferlet, R., Hébrard, G., Koester, D., Audouze, J., Cassé, M., Vangioni-Flam, E., & Webb, J., 1998, A&A, 338, 694
York, D.G. 1983, ApJ, 264, 172
York, D.G., et al. 1983, ApJ, 266, L55

The D/H Ratio in Interstellar Gas Toward the Hot, White Dwarf G191-B2B

M. S. Sahu

NASA/Goddard Space Flight Center, Code 681, Greenbelt, MD 20771 and National Optical Astronomy Observatories, 950 North Cherry Avenue, Tucson, AZ 87519-4933

Abstract.

Space Telescope Imaging Spectrograph (STIS) observations of the D/H ratio in the two velocity components towards G191-B2B are consistent with $1.5 \pm 0.1 \times 10^{-5}$ and do not agree with the values derived using the Goddard High Resolution Spectrograph (GHRS) data. We present some new work on the G191-B2B sightline, and the results we obtain are consistent with those of Sahu et al. (1999).

1. Introduction

The fraction of deuterium not processed into helium during primordial nucleosynthesis is a sensitive function of the cosmological baryon density (η) and can be used to probe conditions in the earliest times of the Universe. Deuterium is destroyed when it is cycled through thermonuclear processes in stars and there is no reasonable way other than BBN to produce any significant quantity of deuterium in the Universe (Epstein, Lattimer & Schramm, 1976). Since the amount of deuterium in the Universe decreases monotonically with time, the D/H abundance ratio in the Local Interstellar Medium (LISM) places a lower limit on the density of baryons in the Universe.

In the early 1970's, observations of interstellar deuterium toward early-type stars with the *Copernicus* satellite provided the first strong evidence that the Universe is not closed by baryons and that a significant amount of non-baryonic dark matter should be present (Reeves et al. 1973). More recently, deuterium abundance measurements in metal-poor Lyman-α clouds toward high-redshift QSOs have provided more direct measurements of the primordial deuterium abundance (e.g. Burles & Tytler, 1998). The more detailed D/H studies possible in the LISM now focus on testing the Galactic chemical evolution models of deuterium, and, in particular, on testing the assumption that there is no significant non-cosmological deuterium production.

Hubble Space Telescope (HST) – Goddard High Resolution Spectrograph (GHRS) measurements of the D/H ratio in the Local Interstellar Cloud (LIC), using nearby, late-type stars and WDs as background sources are all consistent with a uniform D/H value of $1.5 \pm 0.1 \times 10^{-5}$ (e.g. Linsky et al. 1993, 1995, hereafter L93 & L95; Linsky 1998, Landsman et al. 1996, Dring et al. 1998). This view of the constancy of the D/H ratio in the LISM was questioned by the results of Vidal-Madjar et al. (1998, hereafter VM98) who used GHRS echelle data. VM98 reported a value of D/H = 0.9×10^{-5} for two non-LIC interstellar components observed toward the white dwarf G191-B2B (d=69 pc), implying $\sim$ 30% local variation in the D/H ratio. Variations of the D/H ratio in the LISM would complicate the use of deuterium as a cosmological probe and affect Galactic chemical evolution model calculations.

We re-examined the D/H ratio toward G191-B2B using newer HST-Space Telescope Imaging Spectrograph (STIS) data along with the archival HST-GHRS echelle data (Sahu et al. 1999; hereafter S99). The STIS data of other interstellar species indicate the presence of only two velocity components and the D/H ratios derived for the two velocity components are consistent with previous determinations (L93 & L95). For the second (non-LIC) component, the STIS data yield D/H $> 1.26 \times 10^{-5}$, which although poorly constrained, is consistent with the LIC value. The STIS echelle data provide no evidence for local or component-to-component variation of the D/H ratio in the LISM. We found a clear disagreement between the D/H ratios derived from GHRS and STIS data.

In this paper, we present some of the results from S99 together with a brief report on recent progress in the G191-B2B D/H study.

2. Overview of the G191 B2B Sightline

The Sun is embedded within the Local Interstellar Cloud (LIC), a warm (T $\sim$ 7,000K), low-density (n_e $\sim$0.1), partially ionized region, which is observed in projection toward most, but not all, nearby stars (Figure 1). Models of the LIC (Redfield & Linsky, 2000) show the Sun is located just inside the edge of the LIC in the direction of the Galactic Center and toward the North Galactic Pole (NGP). The Hipparcos based distance to G191-B2B is 69 pc (Vauclair et al, 1997). The position of G191-B2B [$(l,b) = (155°.9, +7°.1)$] is relatively close to Capella [$(l,b) = (162°.6, +4°.6)$] in the sky although Capella at $d = 12.9$ pc, is much closer to the Sun. The results obtained for Capella [L93, L95] are relevant for interpreting the G191-B2B data. The 19.5 km s^{-1} component present in the G191-B2B data (§3.3) is at the projected velocity of the LIC and also seen in the Capella data (L93), suggesting that both the G191-B2B and Capella sightlines intercept the LIC.

3. Our Approach

Our analysis differs from the VM98 analysis in three main aspects which are listed below.

3.1. The Radial Velocity of G191-B2B

The radial velocity of G191-B2B that we use in our analysis is estimated from STIS data of other WD photospheric lines (Bruhweiler et al. 2000, in prepara-

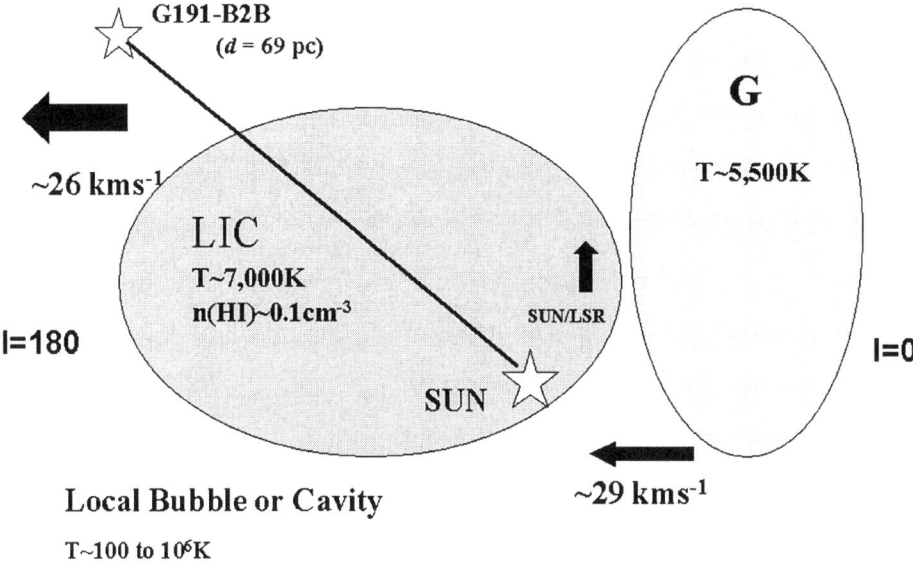

Figure 1. *A schematic view of the immediate vicinity of our Sun as viewed from above the Galaxy, shows the relative motions of the Local Standard of Rest (LSR), the Sun, LIC and G clouds. The shapes of the LIC and G clouds as shown here are only for the purpose of illustration and do not represent their true shapes.*

tion) and is 24.6 ± 0.4 km s^{-1} (including gravitational redshift). This differs by ~ 4.4 km s^{-1} from the value of 29 km s^{-1} derived by VM98. In their analysis, the Si III 1206.5Å stellar feature, which has low S/N (~ 10) in the GHRS data and is blended with the interstellar contribution has been used to get a radial velocity of 29 km s^{-1} (see § 3.1 of VM98). The STIS wavelength coverage for G191-B2B is ~ 20 times more than the GHRS coverage and Bruhweiler et al. use several photospheric lines (mainly C,N,O,Si, Fe and Ni) in the ~ 1150 - 1700Å range. They obtain a peak for the velocity distribution of 24.6 km s^{-1} with a FWHM of 1.3 km s^{-1}. This value is consistent with the radial velocity of 22 ± 2 km s^{-1} obtained for G191-B2B by Reid & Wegner (1988) from the H-alpha emission core. Furthermore, Holberg et al. (1998) report an average radial velocity of 21.98 ± 0.54 km s^{-1} obtained from high-dispersion IUE spectra of metal lines. This average value and the range of velocities quoted in their Table 4 from the different metal lines are consistent with the Bruhweiler et al. (2000) values.

3.2. Use of NLTE Stellar Atmosphere Models

Using a physically realistic model to predict the intrinsic stellar Lyman-α profile for G191-B2B is crucial for the D/H ratio measurement because the model calculations provide the continuum against which the interstellar D I and H II absorptions are measured. In their analysis VM98 used a pure hydrogen LTE model with T_{eff} ~ 61,400 K and log$g$ ~ 7.55 and no photospheric metals. The

GHRS data used by VM98 to determine the Lyman-α photospheric line shape covers only a few Å of data which do not extend out to the continuum. In contrast, in our analysis we have used stratified, line-blanketed NLTE model calculations by Barstow, Hubeny and co-workers fitted to the STIS data covering the 1140 to 1700Å region. The best-fit NLTE model atmosphere [T_{eff} = 54,000 ± 2000 K and log g = 7.5± 0.03] is adopted in our analysis to predict the intrinsic WD Lyman-α profile and to check for possible contamination of the interstellar lines by narrow WD photospheric absorption lines.

3.3. The Number of Velocity Components

In addition to the interstellar D I and H I absorption lines, the STIS echelle spectra show interstellar absorption due to N I ($\lambda\lambda$1199.5, 1200.2 and 1200.7), C II λ1334.5, C* II λ1335.7, O I λ1302, Si II ($\lambda\lambda$1190, 1193, 1260, 1304 and 1526), Si III λ1206.5, Al II λ1670.8, S II λ1259.5 and Fe II λ1608.5. As mentioned in §3.2, the stratified NLTE calculations predict weak contaminations in several of the interstellar lines mainly by WD photospheric Fe and Ni absorption lines. The wavelength regions near the interstellar N I λ1200.7, Si II $\lambda\lambda$1193 & 1304 and Fe II lines show no contamination by WD photospheric lines. The number and relative velocity spacing of individual components that make up an absorption line profile were estimated by an initial inspection of the STIS spectra.

All velocities quoted in this paper are in the heliocentric frame. One velocity component is at $\sim$ 8.6 km s^{-1} (hereafter referred to as comp 1) and the other is at 19.5 km s^{-1}, the projected velocity of the LIC (Lallement et al. 1995) in the line-of-sight to G191-B2B (see §2). Our analysis of the STIS and GHRS data explicitly assumes the existence of two distinct absorption components.

4. Profile fitting of the interstellar D I and H I lines

Each velocity component is assumed to be homogeneous and characterized by a column density N, radial velocity v and a line-of-sight velocity dispersion defined by $b = (2kT/m + \xi^2)^{1/2}$ where ξ is the turbulent velocity parameter along the line-of-sight, T is the kinetic temperature and m is the ion mass. In our analysis, we have included the fact that the D I and H I lines are doublets and the two components of each doublet are separated by 0.0054Å. We have simultaneously fitted both the D I and H I interstellar lines since they are separated by only 0.33Å and the D I absorption is located on the wing of the broad H I absorption. For each interstellar species, intrinsic line profiles were computed assuming two absorption components and then convolved with either the STIS instrumental LSF for the 0.2 × 0.2 arcsec slit given by the STIS Handbook (Sahu, 1999) or the two-component Gaussian LSF for GHRS given by Spitzer & Fitzpatrick (1993). Three parameters (N, v, b) were determined for each component by an iterative least-squares fit. The turbulent velocity parameters for the two components were determined by plotting the velocity dispersions (b values) for the various atomic species (D, N, O, C, Si, S, Al and Fe) as a function of ion mass m and performing a least-squares fit to the data points. The best-fit value of ξ for the LIC component is 1.7 km s^{-1} (consistent with the value derived for this component by L93, L95) while for comp 1, ξ is 2.5 km s^{-1}.

For modeling of the Lyman-α profile, the velocities of the two components are kept fixed at 8.6 (comp 1) and 19.5 km s^{-1} (LIC) and the ξ values of two components are fixed at 1.7 (LIC) and 2.5 km s^{-1} (comp 1). The results of the profile fitting analysis are listed in S99.

5. Some new results

Subsequent to the S99 analysis, several refinements have been made to the IDL-based CALSTIS data reduction package developed by the STIS Instrument Development Team (IDT) which is used to process the STIS data. These refinements include updated dispersion coefficients and improved on-board Doppler corrections. The use of these updated dispersion coefficients has resulted in velocity shifts for the various interstellar absorption lines of up to ± 0.3 km s^{-1} for G191-B2B as compared to the results published in S99. The velocity shifts introduced due to improved on-board Doppler corrections for the G191-B2B data is insignificant, 0.006 km s^{-1} and 0.002 km s^{-1} for the $\lambda_c = 1234$ and 1598Å settings respectively and 0.2 km s^{-1} for the 1426Å setting. The version of IDT-CALSTIS used here is as of 17 December 1999. The derived D/H ratios parameters using the new reduction are not significantly different from the values published in S99 and do not affect our earlier conclusions.

In addition to this new analysis, we have also re-analyzed the STIS data using the alternate scattering algorithm provide by Howk & Sembach (2000). The D/H ratios derived using the Howk & Sembach algorithm is consistent with the D/H ratio derived using the IDT reduction. The STIS derived D/H ratios *do not agree* with the GHRS-derived values irrespective of the scattering algorithm used (Sahu et al. 2000, in preparation).

Acknowledgments. L. da Silva, M. Spite and the Scientific and Local Organizing committees deserve great credit for their organization of this successful meeting.

References

Burles, S., & Tytler, D. 1998, ApJ, 499, 699
Dring, A.R. et al. 1997, ApJ, 488, 760
Epstein, R. Lattimer, J., & Schramm, D. N. 1976, Nature, 263, 198
Holberg, J.B., Barstow, M.A. & Sion E.M. 1998, ApJS, 119, 207
Howk,C.J. & Sembach, K.R. 2000, AJ, 119, 248
Lallement, R. et al. 1995, A&A, 286, 898
Landsman, W. et al. 1996, Science with the Hubble Space Telescope-II, 454
Linsky, J. L. 1998, Space Sci.Rev., 84, 285
Linsky, J.L. et al. 1993, ApJ, 402, 694
Linsky, J.L. et al. 1995, ApJ, 451, 335
Redfield, S. & Linsky, J.S. 1999, ApJ(in press)
Reid, N. & Wegner, G. 1988, ApJ, 335, 953

Reeves, H. et al. 1973, ApJ179, 909
Sahu, M.S. et al. 1999, ApJ, 523, L159 (S99)
Spitzer, L. & Fitzpatrick,E.L. 1993, ApJ, 409, 299
Vauclair, S. et al. 1997, A&A325, 1055
Vidal-Madjar, A. et al. 1998, ApJ, 338, 694 (VM98)

The Light Elements and Their Evolution
IAU Symposium, Vol. 198, 2000
L. da Silva, M. Spite, J. R. de Medeiros, eds.

The Deuterium Abundance In The Galactic Center 50 km/s Molecular Cloud: Evidence For A Cosmological Origin Of D

D. A. Lubowich

Dept. of Physics, Hofstra U., Hempstead, NY & AIP, Melville, NY

Jay M. Pasachoff and Robert P. Galloway

Dept. of Astronomy, Williams College, Williamstown, MA

Thomas J. Balonek and Christy Tremonti

Dept. of Physics, Colgate University, Hamilton, NY

Tom Millar and Helen Roberts

Dept. of Physics, Univ. Manchester Institute of Technology, Manchester, UK

Abstract. We confirm that deuterium exists in the Galactic Center (GC) and estimate that D/H = 3×10^{-6} using a new 5192-chemical reaction model. This is the lowest D/H ratio observed in the Galaxy, five times lower than the local ISM D/H = 1.5×10^{-5} but $10^6 \times$ larger than D/H predicted by GC models. We detected DCN in the GC Sgr A 50 km/s molecular cloud located 10 pc from the GC with the NRAO 12m telescope and obtained $T_R^* = 0.061 \pm 0.007$ K and 0.04 ± 0.02 K for the J = 1-0 and 2-1 lines. The most likely source of the GC D is continuous injection from the infall of primordial matter with D/H = 5×10^{-5} with the D/H determined by astration and mixing. Thus there are no significant Galactic sources of D and no recent quasar or AGN activity in the GC. This primordial D/H implies that the baryon density is less than the density necessary to close the Universe; most of the baryons are in dark matter; and there are fewer than four ν families.

1. Introduction and observations

Because D is not produced in stars, the abundance of D will decrease with time unless there are any additional sources of D. The D/H ratio is an important prediction of big-bang nucleosynthesis(Schramm & Turner, 1998) because D/H depends on the T and baryonic density during the first 1000 seconds and might determine if the density is sufficient to close the universe. Recent observations of D including QSO absorption spectra are reviewed by Vidal-Madjar (1999).

The GC is the most active and heavily processed region of the Galaxy and the Sgr A 50 km/s cloud (M-0.02-0.07), 10 pc from the GC, is related to the

GC activity and is the best place to search for Galactic D. If D is produced by any stellar or Galactic process, then it should be more abundant in the GC and there should be a corresponding gradient in the D abundance (Pasachoff & Vidal-Madjar 1989). Chemical models of the GC with no sources of D predict D/H = 5×10^{-12} at 12 Gyr (Audouze et al. 1976).

We used the NRAO 12-m telescope during May 16-18, 1993, and June 29, 1993, and observed the DCN J=1-0 and J=2-1 lines at 72.404 GHz and 144.83 GHz in total-power mode using position switching with the 3-mm and 2-mm SIS receivers centered at $\alpha = 17^h 42^m 42^s$ $\delta = -28°58'00"$ (the peak CS J= 7-6 and J=5-4 emission; Serabyn, Lacy, & Achtermann 1992) insuring that we observed the densest part of this cloud (n= 10^6 cm^{-3} We analyzed the data with the UNIPOPS program. We detected both the J= 1-0 and J= 2-1 lines of DCN and obtained $T_R^* = 0.061 \pm 0.007$ K and $T_R^* = 0.042 \pm$K+/-0.02 K, respectively, where T_R^* is the source antenna temperature corrected for atmospheric attenuation and all telescope losses (ohmic and spillover) except for coupling of the source and beam. We detected DCN at 4/5 additional positions offset by 1' and 2' S from our center position and also observed the J=1-0 lines of DNC,HC^{15}N, H^{13}CN, HCN, HCO$^+$, and HNC. As a check we observed the known DCN lines in the Sgr B2 molecular at the (OH) position ($\alpha = 17^h 44^m 11^s$; $\delta = -28°22'30"$) where the DCN 1-0 line is blended with the H$_2$CO 5(1,4)-5(1,5) transition.

The spectra are shown in figures 1 and 2 and our results are given in tables 1, 2, and 3 where the lines were identified from the Lovas (1992) rest frequencies (unknown lines listed with U) and fit by Gaussian profiles.

Table 1. Sgr A 50 km/s Cloud, Center Position

Molecule	Transition	νMHz)	T_r^*(mK)	rms(mK)	ΔV(km/s)
		U72323.9	68	7.1	22.7
		U72344.2	18	7.1	30.6
DCN	1-0	72414.9	61	7.1	30.9
HC^{13}CCN	8-7	72475.1	145	7.1	21.9
HCC^{13}CN	8-7	72482.1	120	7.1	27.9
HC^{15}N	1-0	86055.0	192	27	26.8
SO	2(2)-1(1)	86093.5	226	27	19.8
H^{13}CN	1-0	86340.2	1510	20	35.2
HCN	1-0	88631.8	5408	48	27.1
HCO$^+$		89188.5	3070	23	25.7
		U89204.3	532	23	11.7
		U89215.5	250	23	11.7
		U89221.8	130	23	7.8
HC^{13}CCN	10-9	90593.1	130	23	18.5
HCC^{13}CN	10-9	90601.8	120	23	25
HNC	1-0	90663.5	2198	23	41.3
C$_2$S	7,7-6,6	90686.4	167	23	12.4
		U144735.3	67.6	15	12.6
DCN	2-1	144828.0	38.9	15	25.2

We used 1-MHz filters; a 256 MHz bandwidth; 4.14 km/s resolution and 86 " beam at 72 GHz, 2.07 km/s resolution and 43 " beam at 144 GHz. Pointing

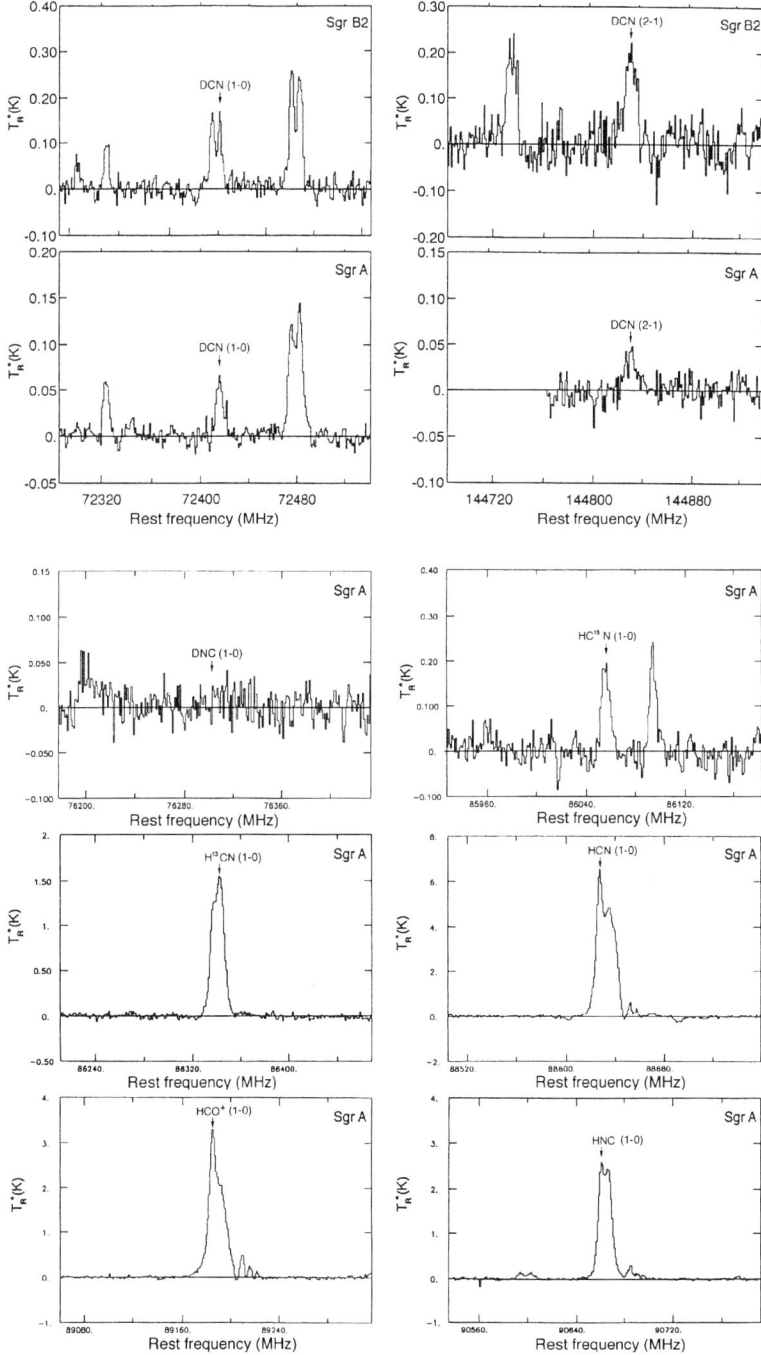

Figure 1. Spectra of the GC Sgr A 50 km/s and Sgr B2 molecular clouds. The J = 1-0 lines of DCN, DNC, $HC^{15}N$, $HC^{13}N$, HCN, HCO^+, and HNC are labeled.

Table 2. Sgr A 50 km/s Cloud, Offset Positions

Molecule	Transition	νMHz)	T_r^*(mK)	rms(mK)	ΔV(km/s)
1' West					
		U72324.9	31	16	25.2
DCN	1-0	72414.9	30	16	41.5
HC^{13}CCN	8-7	72475.1	102	16	21.4
HCC^{13}CN	8-7	72482.1	91	16	21.4
1' North					
		U72323.1	23	14	62.9
DCN	1-0	72414.9	11	14	59.1
HC^{13}CCN	8-7	72475.1	60	14	15.1
HCC^{13}CN	8-7	72482.2	58	14	22.8
1' East					
DCN	1-0	72419.0	40	19	33.1
HC^{13}CCN	8-7	72475.1	100	19	27.3
HCC^{13}CN	8-7	72482.1	100	19	27.3
1' South					
		U72324.5	64	16	19.7
DCN	1-0	72414.9	55	16	33.9
HC^{13}CCN	8-7	72475.1	135	16	18.8
HCC^{13}CN	8-7	72482.1	118	16	20.1
2' South					
DCN	1-0	72414.9	32	15	27.1
HC^{13}CCN	8-7	72475.1	80	15	18.6
HCC^{13}CN	8-7	72482.1	79	15	27.5

Table 3. Sgr B2(OH)

Molecule	Transition	νMHz)	T_r^*(mK)	rms(mK)	ΔV(km/s)
CH^{13}CCH$_3$	10(1,9)-10(0,10)	72300.2	61	16	18.1
		U72324.0	106	16	17.3
H$_2$CO	5(1,4)-5(1,5)	72404.4	167	16	18.6
DCN	1-0	72414.9	167	16	16.6
HC^{13}CCN	8-7	72475.1	226	16	20.0
HCC^{13}CN	8-7	72482.1	260	16	20.0
		U144734.8	217	38	16.9
DCN	2-1	U144828.0	196	38	21.5

was checked using Jupiter and Uranus. The double sideband T_{sys} was 350 K at 72 GHz and 400 K at 144 GHz. We could not resolve the hyperfine splitting for the DCN, HCN, $H^{13}CN$, and $HC^{15}N$ lines because of our use of 1 MHz filters and the large line widths of $\approx$ 30 km/s.

Because the HCN J=1-0 line is optically thick, we used the optically thin J=1-0 line of $HC^{15}N$ to estimate $DCN/HC^{15}N$ and DCN/HCN (Hatchell, Millar & Rodgers 1998). DCN/HCN = $T_{DCN}\Delta V / T_{HC^{15}N}\Delta V$ $(^{15}N/^{14}N)_e$-$(\Delta E/kT_{ex})$, where (ΔE is difference between the energies of DCN and $HC^{15}N$ transitions, T_{ex} = 75 K, and sources are extended relative to our beam for the J = 1-0 lines. We obtained $DCN/HC^{15}N$ = 0.36 and DCN/HCN = 4.0×10^{-4} using $(^{15}N/^{14}N)$ = 900 (Güsten & Ungerechts 1985) which are the lowest $DCN/HC^{15}N$ and DCN/HCN ratios observed in any molecular cloud.

2. Discussion

Chemical fractionation resulting from the lower zero-point energy for deuterated molecules due to its larger mass significantly increases these abundances. Our chemical model is an updated version of Rodgers & Millar (1996) containing 165 non-deuterated, 122 deuterated species, 5192 gas-phase reactions, recent data on the dissociative recombination of H_2D^+, S chemistry, and grain formation of H_2 and HD. For a model with the physical parameters typical of the 50 km/s cloud (T = 75 K; n(H) = 10^6 cm-3; an ionization rate of 1.3×10^{17} sec^{-1}; fractional abundances of C, N, O equal to $9.9 \times 10^{-4}, 3.6 \times 10^{-4}$, and 1.98×10^{-3}, respectively, approximately 3× their solar abundances; Minh, Irving, & Friberg (1992); Serabyn, Lacy, & Actermann (1992); Poglitsch et al. (1991),Simpson et al. (1995)), we obtained a best agreement with the observed DCN/HCN for D/H = 3×10^{-6} (a degree of fractionation of 133).

Although the column densities of DCN and HCN will be changed by varying the physical conditions, the fractionation is independent of density, metallicity, and ionization rate where a faster ionization rate resulted a shorter time to reach the steady-state values. DCN/HCN as a function of T and n is shown in figure 3. Our D/H is also consistent with the upper limits of D/H $< 8.3 \times 10^{-5}$ (from the 92-cm hfs line; Lubowich, Anantharamaiah, & Pasachoff 1989)and DCN/HCN $< 6 \times 10^{-4}$ (Jacq et al. 1999)in GC 50 km/s molecular cloud. There is one reported marginal 1σ detection of D from the J = 1-0 line of DCN in the 50 km/s Sgr A molecular cloud core (Penzias 1979) with T_a^*= 0.02±0.015 K.

Using a closed-box model and a time scale for GC astration of 2×10^8 yr Audouze et al. (1976) calculate D/H = 5×10^{-12}. If there were no additional sources of D, the GC molecular clouds should be composed of astrated material completely depleted in D and DCN should not be detectable. If D is produced via any mechanism related to the nucleosynthesis of O (massive stars or type-II SN), then the D and O abundances will be positively correlated. Because the GC O abundance is enhanced by 3× while the D abundance is reduced by 4.5×, D is anticorrelated with the O abundance. Thus if any D nucleosynthesis exists, it is not correlated with O nucleosynthesis or massive stars.

Combining our result of the GC D/H = 3.0×10^{-6}, the Sgr B2(OH) D/H = 5.0×10^{-6} (Jacq et al. 1999), the local ISM D/H = 1.5×10^{-5} (Linsky 1998), and the possible detection of DI with D/H = 3.9×10^{-4} in the Galactic anticenter

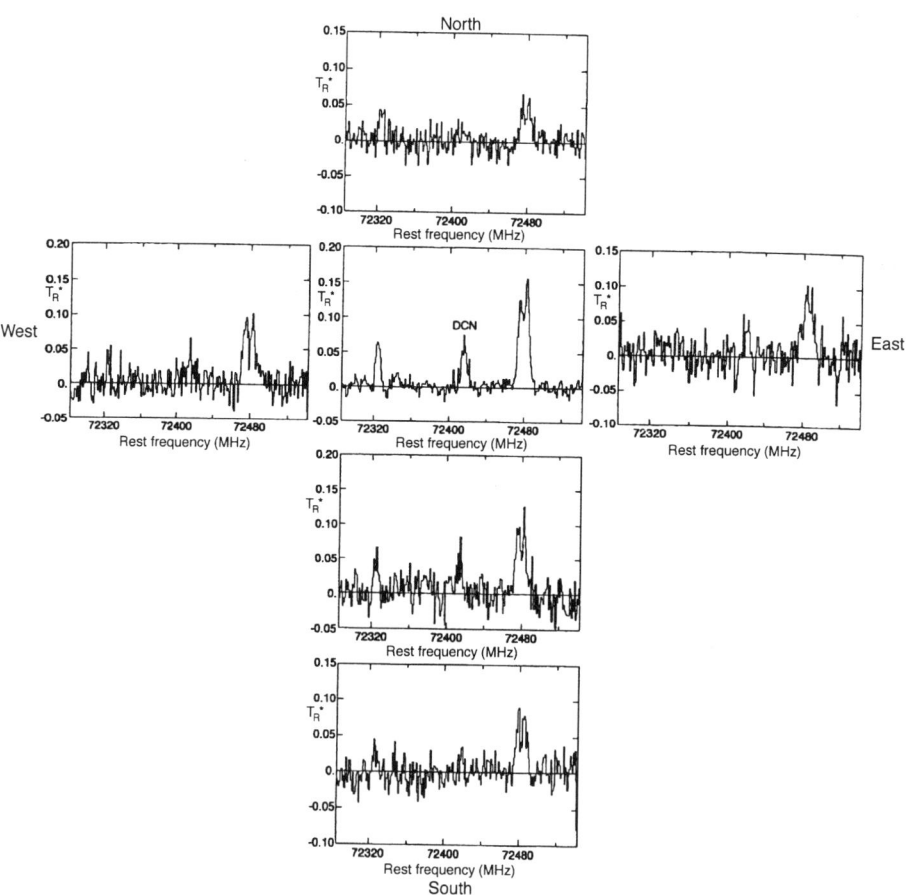

Figure 2. Spectra of the GC 50 km/s molecular cloud offset by 1' North, South, East, West and by 2' South of our center position.

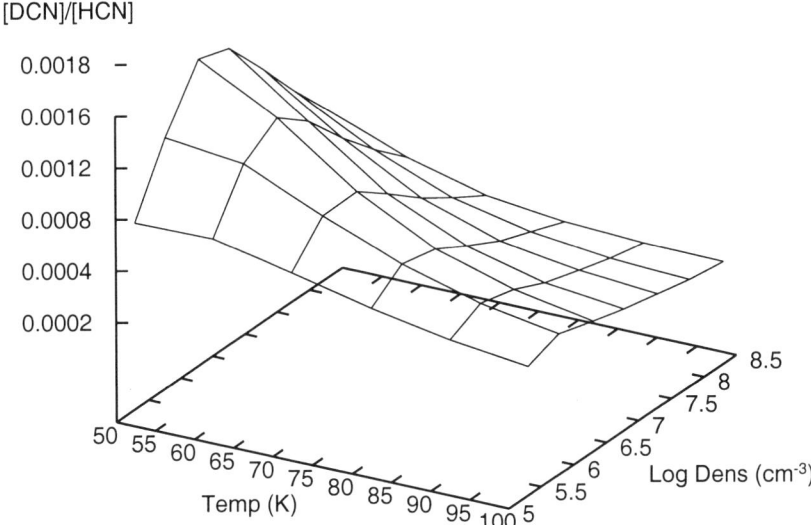

Figure 3. DCN/HCN as a function of temperature of temperature and density

(Chengalur, Braun, & Burton, 1997), we obtain a positive abundance gradient of D in the Galaxy indicating that there is no significant Galactic production of D (Pasachoff & Vidal-Madjar 1988). The most likely source of the GC D is continuous infall of primordial matter with D/H $\approx 5 \times 10^{-5}$ where the resultant D/H ratio is determined by the astration and mixing which always reduces the D/H. This replenishment would negate much of the effects of astration and inject gas enhanced in D but deficient in O gas into the GC. This is in agreement with the analysis of the Galactic D/O ratio (Prantzos 1996) where the ISM D is probably the result of infall plus astration.

Our results constrain models of GC activity that predict a GC D/H = 10^{-4} (10 pc from the GC) from γ-ray photospallation reactions (Boyd, Ferland, & Schramm 1989) for an AGN or Seyfert γ-ray luminosity $L_\gamma = 10^{42}$ erg/s; D/H = 10^{-2} and $L_\gamma = 10^{44}$ erg/s for a quasar; and D/H = 2×10^{-5} for an AGN cosmic-ray (CR) proton luminosity $L_p = 10^{43}$ erg/s for 1 Gyr (Ozernoi & Chernomordik 1975). The GC production of D by large fluxes of CRs or γ-rays in the early Galaxy is possible only if there was rapid astration so that any D produced was destroyed in the GC but there was little astration in the local ISM. Because the current GC D/H << D/H predicted by quasar or AGN activity, there has not been recent AGN activity in the GC. The luminosities required to produce the GC D/H are still many orders of magnitude larger than the observed GC $L\gamma$ or $L_p = 2 \times 10^{37}$ erg/s (Mayer-Hasselwander et al. 1998; Mastichades & Ozernoy 1994). The γ-rays from CR spallation reactions were not detected during 9 years of GC observations (Harris, Share, & Messina, 1995).

Thus AGN activity, (γ-ray photodisintegration reactions, or CR spallation reactions are not significant sources of Galactic deuterium. Because almost all nucleosynthesis processes that can produce a significant abundance of D always overproduce Li or B by 10^3 - 10^5 times, the observed upper limit on the GC Li of $(Li/H)_{GC} < 3.9 \times 10^{-8}$ or $(Li/H)_{GC} < 20\ (Li/H)_{disk}$, (Lubowich, Turner, & Hobbs 1998) further implies that there are no Galactic Center sources of deuterium. Although weak AGN activity or periodic bursts of star formation may occur, the Milky Way has not had a recent active phase nor an early active phase for any period longer than 1 Gyr. Although unlikely, we cannot exclude low astration models with no infall or a continuous production of deuterium combined with the exact astration necessary to produce the Galactic Center, the local, and the anticenter D/H values.

If all the deuterium is primordial and the astration models Prantzos (1996) are correct, then the primordial or early Galactic D/H $\approx 5 \times 10^{-5}$. For this D/H big-bang nucleosynthesis models imply that the baryon-to-photon ratio ratio $\eta_b = 3 \times 10^{-10}$, there are less than four neutrino families, the baryon density of the Universe $\rho_b = 3 \times 10^{-31}$ gm cm^{-3} < critical density of $\rho_c = 3H_o^2/8\pi G = 1.88 H_o^2 \times 10^{-29}$ gm cm^{-3} = 9.2×10^{-30} gm cm^{-3} (for a Hubble constant H_o = 70 km/s/Mpc) necessary to close the Universe for a flat Einstein-de Sitter Universe, and $\Omega_b = \rho_b/\rho_c = 0.04$ (Copi, Schramm, & Turner, 1995). Thus the fraction of the critical density contributed by baryons (Ω_b) in a closed Universe requires that most of the baryons are in the form of dark matter.

Note added in proof: Based on additional modelling D/H = 1.7×10^{-6}, $9\times$ lower than the local D/H which strengthens our conclusion that GC D is the result of infalling matter.

We thank Ann Mancuso (Hofstra) Sebastian Diaz (Williams), Matthew Pickard (Keck Northeast Astronomy Consortium Summer Fellow), Ken Pagliuca (AIP), & M. L. Kutner. We acknowledge a Hofstra Faculty Research and Development Grant, a Bronfman Science Center Grant from Williams, and a PPARC grant at UMIST.

References

Audouze, J., Lequeux, J., Reeves, H., & Vigroux, L. 1976, ApJ, 208, L51
Boyd, R.N., Ferland, G.J., & Schramm, D.N. 1989, ApJ, 336, L1
Chengalur, J. N., Braun, R., & Butler, Burton W. 1997, A&A, 318, L35.
Copi, C. J., Schramm, D.N, & Turner, M.S. 1995, Phys. Rev. Letters, 75, 3981
Genzel, R., Hollenbach, D., & Townes, C.H. 1994, Reports Prog. Phys., 57, 417
Güsten, R. & Ungerechts, H. 1985, A&A, 145, 241
Harris, Michael J., Share, Gerald H., & Messina, Daniel C. 1995, ApJ, 448, 157
Hatchell, J. , Millar, T.J., & Rodgers, S. D. 1998, A&A, 332, 695.
Jacq, T., Baudry, A., Walmsley, C.M., & Caselli, P., 1999, A&A, 347, 957
Linsky, J. L. 1998, Space Sci. Rev. 84, 285
Lovas, F. J. 1992. J. Phys. & Chem. Reference Data 21, 181
Lubowich, D. A., Anantharamaiah, K.R., & Pasachoff, J.M. 1989, ApJ, 345, 770
Lubowich, D. A., Turner, B.E., & Hobbs, L.M. 1998 ApJ, 508, 729
Mastichades, A. & Ozernoy, L. M. 1994, ApJ, 426, 599
Mayer-Hasselwander et al. 1998, A&A, 335, 161

Minh, Y.C., Irvine, W.M., & Friberg, P. 1992, A&A, 258, 489

Ozernoi, L., & Chernomordik, V.V. 1975, Sov. Astron. 19, 693

Pasachoff, J. M, & Vidal Madjar, A. 1989, Comments in Astrophysics, 14, 61

Penzias, A.A. 1979, ApJ, 228, 430

Poglitsch, A., Stacey, G. J., Geis, N., Haggerty, M., Jackson, J., Rumitz, M., Genzel, R., Townes, C. H. 1991, ApJ, L33

Prantzos, N. 1996, A&A, 310, 106

Rodgers, S.D. & Millar, T.J. 1996, MNRAS, 280, 1046

Schramm, D. N. , & Turner, M. S. 1998, Rev. Mod. Physics, 70, 303

Serabyn, E., Lacy, J.H., & Actermann, J.M. 1992, ApJ395, 166

Simpson, J.P., Colgan, S.W.J., Rubin, R.H., Erickson, E.F., & Haas, M.R. 1995, ApJ, 444, 721

Vidal-Madjar, A. 2000, Nucl. Phys. B (Proc. Suppl.), 80 119

Blue compact galaxies and the primordial ^{4}Helium abundance

Trinh Xuan Thuan

Astronomy Department, University of Virginia, P.O. Box 3818, University Station, Charlottesville, VA 22903, USA

Yuri I. Izotov

Main Astronomical Observatory, Ukrainian National Academy of Sciences, Goloseevo, Kiev 03680, Ukraine

Abstract.
Blue compact galaxies (BCG) are ideal objects in which to derive the primordial ^{4}He abundance because they are chemically young and have not had a significant stellar He contribution. We discuss a self-consistent method which makes use of all the brightest He I emission lines in the optical range and solves consistently for the electron density of the He II zone. We pay particular attention to electron collision and radiative transfer as well as underlying stellar absorption effects which may make the He I emission lines deviate from their recombination values. Using a large homogeneous sample of 45 low-metallicity H II regions in BCGs, and extrapolating the Y-O/H and Y-N/H linear regressions to O/H = N/H = 0, we obtain Y_p = 0.2443±0.0015, in excellent agreement with the weighted mean value Y_p = 0.2452±0.0015 obtained from the detailed analysis of the two most metal-deficient BCGs known, I Zw 18 and SBS 0335–052. The derived slope dY/dZ = 2.4±1.0 is in agreement with the value derived for the Milky Way and with simple chemical evolution models with homogeneous outflows. Adopting Y_p = 0.2452±0.0015 leads to a baryon-to-photon ratio of $(4.7^{+1.0}_{-0.8}) \times 10^{-10}$ and to a baryon mass fraction in the Universe $\Omega_b h_{50}^2 = 0.068^{+0.015}_{-0.012}$, consistent with the value derived from the primordial D abundance of Burles & Tytler (1998).

1. Introduction

Blue compact galaxies (BCG) are low-luminosity ($M_B \gtrsim -18$) systems which are undergoing an intense burst of star formation in a very compact region (less than 1 kpc) which dominates the light of the galaxy (Figure 1) and which shows blue colors and a HII region-like emission-line optical spectrum (Figure 2). BCGs are ideal laboratories in which to measure the primordial ^{4}Helium abundance because of several reasons:

1) With an oxygen abundance O/H ranging between 1/50 and 1/3 that of the Sun, BCGs are among the most metal-deficient gas-rich galaxies known. Their gas has not been processed through many generations of stars, and thus

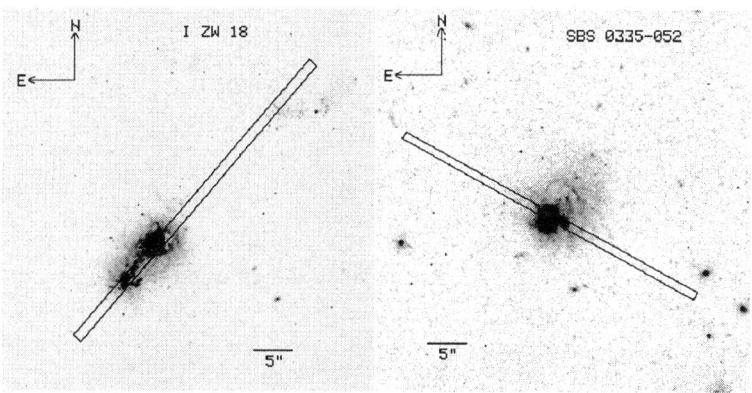

Figure 1. Slit orientations superposed on *HST* archival V images of I Zw 18 and SBS 0335–052. The slit orientation of I Zw 18 is chosen in such a way as to get spectra of the SE and NW components as well as of the C component to the NW of the main body of the galaxy. The spatial scale is $1'' = 49$ pc in the case of I Zw 18 and is $1'' = 257$ pc in the case of SBS 0335–052.

best approximates the pristine primordial gas. Izotov & Thuan (1999) have argued that BCGs with O/H less than $\sim 1/20$ that of the Sun may be genuine young galaxies. Their argument is based on the observed constancy and very small scatter of the C/O and N/O ratios in extremely metal-deficient BCGs with $12 + \log O/H \lesssim 7.6$, which they interpret as implying that the C and N in these galaxies have been made in the same massive stars (M $\gtrsim 9$ M$_\odot$) which manufactured O, and that intermediate-mass stars (3 M$_\odot$ $\lesssim$ M $\lesssim 9$ M$_\odot$) have not had time to release their nucleosynthetic products. Since the main-sequence lifetime of a 9 M$_\odot$ star is ~ 40 Myr, Izotov & Thuan (1999) suggest that very metal-deficient BCGs are younger than ~ 100 Myr. Thus the primordial Helium mass fraction Y_p can be derived accurately in very metal-deficient BCGs with only a small correction for Helium made in stars.

2) Because of the relative insensitivity of ^{4}He production to the baryonic density of matter, Y_p needs to be determined to a precision better than 5% to provide useful cosmological constraints. This precision can in principle be achieved by using BCGs because their optical spectra show several He I recombination emission lines and very high signal-to-noise ratio emission-line spectra with moderate spectral resolution of BCGs can be obtained at large telescospes (4 m class or larger) coupled with efficient and linear CCD detectors with a relatively modest investment of telescope time. The theory of nebular emission is well understood and the theoretical He I recombination coefficients calculated by Brocklehurst (1972) and Smits (1996) are well known enough to allow to convert He emission-line strengths into abundances with the desired accuracy.

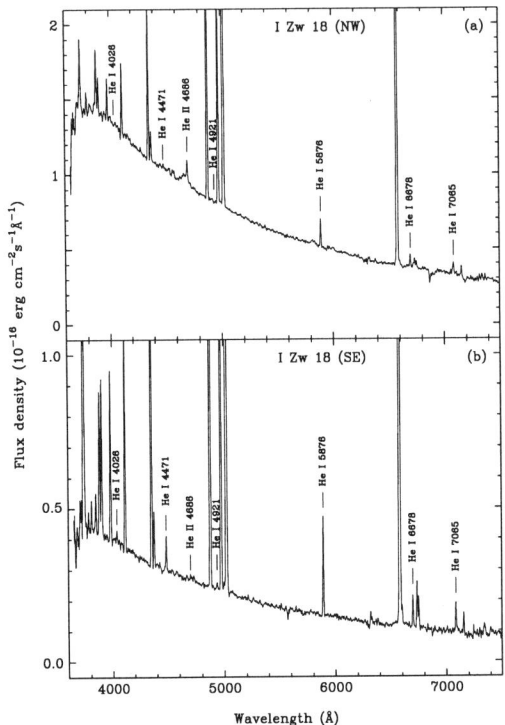

Figure 2. The 0".6×1".5 aperture MMT spectra of the brightest parts of the NW and the SE components of I Zw 18. The spectrum of the SE component is extracted at the angular distance of 5".4 from the NW component. The positions of He I lines are marked. Note that all marked He I lines in the spectrum of the SE component are in emission while the two He I λ4026 and λ4921 lines are in absorption and the He I λ4471 emission line is barely detected in the spectrum of the NW component.

2. The primordial He abundance from extrapolation of the Y-O/H and Y-N/H linear regressions

2.1. A new large sample of Blue Compact Galaxies

Peimbert & Torres-Peimbert (1974, 1976) first noted the correlation between He and O abundances in a small sample of dwarf magellanic irregulars and BCGs, and they proposed to determine Y_p by linear extrapolation of the correlation to O/H = 0. Later, Pagel, Terlevich & Melnick (1986) proposed to use also the Y-N/H correlation for the determination of Y_p, to take into account the temporary local excess of helium and nitrogen due to pollution by winds from massive stars. Many attempts at determining Y_p have been made, using the Y versus O/H and Y versus N/H correlations on various samples of dwarf irregulars and BCGs (e.g. Pagel et al. 1992, Izotov, Thuan & Lipovetsky (1994, 1997, hereafter ITL94 and

ITL97; Olive, Steigman & SKillman 1997, hereafter OSS97; Izotov & Thuan 1998ab, hereafter IT98ab).

Before our work, the largest, most accurate and consistent published data set was by Pagel et al. (1992). Their observations were reduced in a uniform manner and they paid careful attention to such points as the correction for the unseen neutral helium and electron collisional effects which may make some He I lines deviate from their recombination values. Pagel et al. (1992) obtained Y_p = 0.228±0.005, below the limit set by the standard hot big bang model of nucleosynthesis (SBBN) and consistent with it only at the 2σ level. This prompted us to consider obtaining another measurement of Y_p from an independent data set with as high or better precision to test SBBN.

Starting in 1993, we embarked on a large-scale program to obtain high signal-to-noise ratio spectra for a relatively large sample of BCGs assembled from several objective-prism surveys: the First Byurakan or Markarian survey (Markarian et al. 1989), the Second Byurakan Survey (SBS, Izotov et al. 1993) and the University of Michigan survey (Salzer, MacAlpine & Boroson 1989). The SBS sample was particularly interesting because it contained about a dozen BCGs with O/H less than 1/15 of $(O/H)_\odot$, more than doubling the number of such known low-metallicity BCGs. The total sample consists of 45 H II regions in 42 BCGs. The data have been published in a series of papers in the Astrophysical Journal: ITL94, ITL97 and IT98ab.

2.2. Methodology

There are a number of features which distinguish our work from previous efforts in determining the primordial He abundance. Our methodology is described in detail in ITL94, ITL97, and IT98ab.

1) We have observed all the galaxies in our sample with the same telescopes (the Kitt Peak 4 m and 2.1 m telescopes) and instrumental set up, and the data were all reduced in a homogeneous way. This differs from OSS97, for example, which used a more heterogeneous sample of BCGs observed by different observers on different telescopes, with some of the data obtained many years ago with nonlinear detectors. A uniform sample is essential to minimize as much as possible the artificial scatter introduced by assembling different data sets reduced in different ways.

2) To derive the He mass fraction, previous authors use mainly one He emission line, He I 6678. Correction to this line's intensity is usually made only for one effect, electron collisional enhancement. This correction is usually carried out adopting the electron density derived from the [S II] 6717/6731 emission-line ratio. The approach just described has several shortcomings. The metastability of the 2^3S state of He I can also lead to possible radiative transfer effects (also called fluorescence effects) in the triplet lines which may be enhanced at the expense of the He I 3889 line (Robbins 1968). When a single He I emission line is used, one cannot distinguish between between electron collisional and radiative transfer effects. Thus, fluorescent enhancement is neglected, while it may be important. Furthermore, at the low electron number densities N_e which often characterize the HII regions in BCGs, the determination of N_e from [S II] emission lines is very uncertain. In the majority of cases, N_e is arbitrarily set to 100 cm^{-3}. This assumption can lead to artificially low He abundance, as in

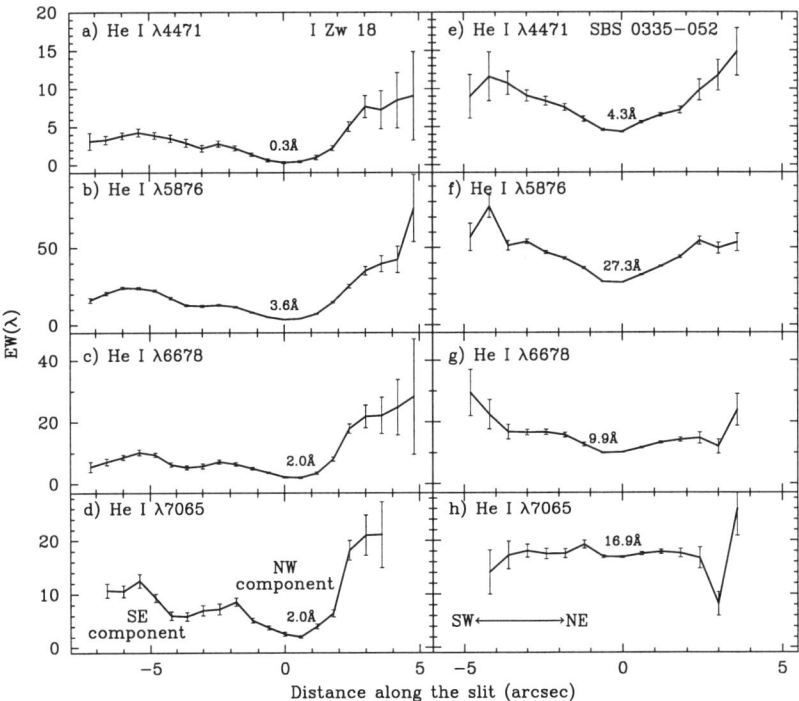

Figure 3. The spatial distributions of the He I nebular emission line equivalent widths in I Zw 18 (left panel) and in SBS 0335–052 (right panel). The error bars are 1σ deviations. The value of the minimum equivalent width for each He I emission line is given.

the case of the southeast component of I Zw 18 where the true N_e is only ~ 10 cm^{-3} (Izotov et al. 1999). More importantly, setting N_e(He II) equal to N_e(S II) is not physically reasonable as the S$^+$ and He$^+$ regions are not expected to coincide, given the large difference in the S I and He I ionization potentials.

To remedy these problems, we have proposed a self-consistent method in which we use all five brightest He I emission-lines in the optical range (the He I 3889, 4471, 5876, 6678 and 7065 lines) and solve simultaneously for N_e(He II) and the optical depth in the He I 3889 line so that the He I 3889/4471, 5876/4471, 6678/4471 and 7065/4471 line ratios have their recombination values, after correction for both collisional (Kingdon & Ferland 1995) and fluorescent (Robbins 1968) enhancements. The He I 3889 and 7065 lines play an important role because they are particularly sensitive to both optical depth and electron number density.

2.3. Underlying stellar absorption

Effects other than collisional and fluorescent enhancements can also change He I line intensities. An important effect is the underlying stellar absorption in He I lines caused by hot OB stars which decreases the intensities of nebular He

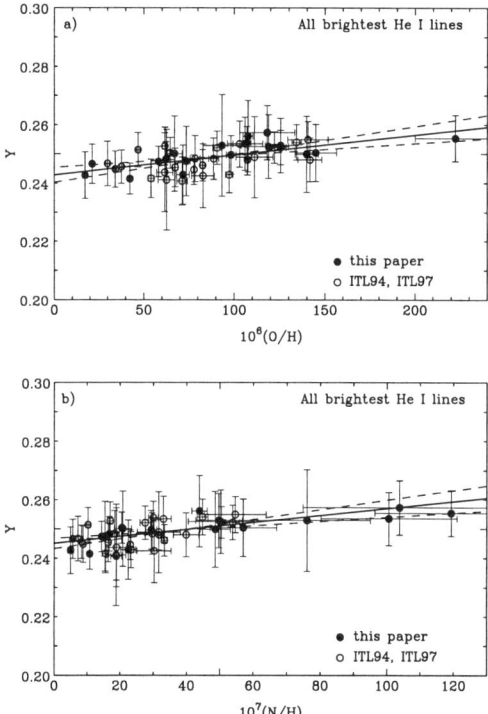

Figure 4. Linear regressions of (a) the helium mass fraction Y vs. oxygen abundance O/H and (b) the helium mass fraction Y vs. nitrogen abundance for our sample of 45 H II regions. The Ys are derived self-consistently by using the 5 brightest He I emission lines in the optical range. Collisional and fluorescent enhancements, underlying He I stellar absorption and Galactic Na I interstellar absorption are taken into account. Open circles denote data from ITL94 and ITL97 and filled circles are data from IT98ab. 1σ alternatives are shown by dashed lines.

I lines. This effect is most important for the emission lines with the smallest equivalent widths.

The neglect of He I underlying stellar absorption can lead to a severe underestimate of the He mass fraction. One of the most spectacular examples is that of the BCG I Zw18. This object plays a key role in the determination of the primordial He abundance because, with an O/H only 1/50 that of the Sun, it is the most metal-deficient BCG known and has great influence on the derived slopes and intercepts of the Y-O/H and Y-N/H linear regression lines. Figure 2 shows very high signal-to-noise ratio spectrophotometric observations of I Zw 18 obtained with the Multiple Mirror Telescope (MMT), with the slit oriented so as to go through the two main centers of star formation in the BCG, the so-called NW and SE components (Figure 1, left). Comparison of the spectrum of the NW component (Figure 2a) with that of the SE component (Figure 2b)

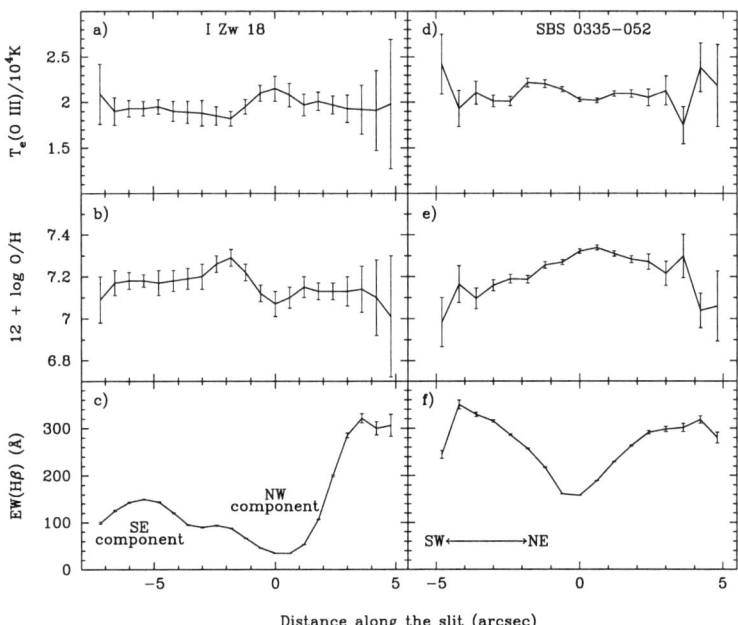

Figure 5. The spatial distributions of the electron temperature T_e(O III), oxygen abundance 12 + log (O/H), and the equivalent width EW(Hβ) of the Hβ emission line in I Zw 18 (left panel) and in SBS 0335–052 (right panel). The error bars are 1σ deviations.

shows clearly that underlying stellar absorption is much more important in the NW than in the SE component: all marked He I lines in the spectrum of the SE component are in emission while the two He I 4026 and 4921 lines are in absorption and the He I 4471 emission line is barely visible in the spectrum of the NW component. Prior to our work, most of the He work relied on measurements of Y in the NW component because of its larger brightness as compared to the SE component (Figure 1, left). This led to a systematic underestimate of Y in I Zw 18 . Izotov et al. (1999) derive the implausibly small Y(4471)= 0.169±0.023 and Y(5876)= 0.192±0.007 from the He I 4471 and 5876 lines respectively. In addition to underlying stellar absorption, the 5876 line intensity in I Zw 18 is reduced further by absorption from the Galactic interstellar 5890 and 5896 Na I lines.

While the NW component cannot be used for He determination, the SE component is better suited as the influence of underlying stellar absorption on the He I emission line intensities is significantly smaller in this component. This can be seen in the left panel of Figure 3 which shows the spatial distributions of the He I nebular emission-line equivalent widths in I Zw 18: they are systematically larger in the SE component than in the NW one, implying less absorption.

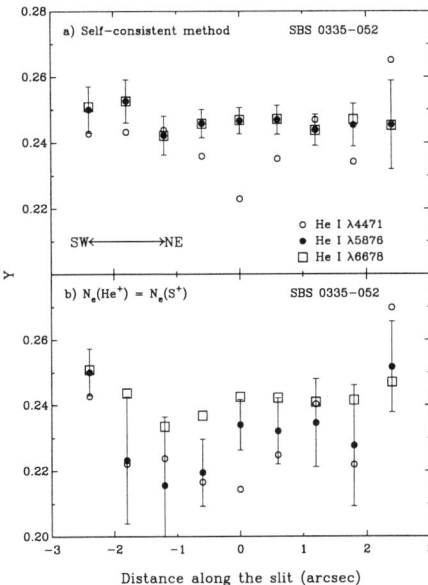

Figure 6. The spatial distributions of the helium mass fractions in SBS 0335–052 derived from the He I $\lambda4471$, $\lambda5876$ and $\lambda6678$ emission line intensities. The intensities of the He I emission lines in Figure 6a are corrected for fluorescent and collisional enhancement with an electron number density $N_e(\text{He II})$ and an optical depth $\tau(\lambda3889)$ derived self-consistently from the observed He I $\lambda3889$, $\lambda4471$, $\lambda5876$, $\lambda6678$ and $\lambda7065$ emission line intensities. The intensities of the He I emission lines in Figure 6b are corrected only for collisional enhancement with an electron number density $N_e(\text{S II})$. The 1σ error bars are shown only for the He mass fraction derived from the He I $\lambda5876$ emission line. They are larger in Figure 6b because of the large uncertainties in the determination of $N_e(\text{S II})$.

2.4. Results

Figure 4 shows the Y-O/H and Y-N/H linear regressions for the whole sample of 45 H II regions in BCGs. The sample includes most of the very metal-deficient BCGs known, including the two most extreme ones, the SE component of I Zw 18 and SBS 0335–052 with O/H about 1/43 that of the Sun. We obtain $Y_p = 0.2443 \pm 0.0015$ with $dY/dZ = 2.4 \pm 1.0$ (IT98b). Our Y_p is considerably higher than those derived by other groups which range from 0.228 ± 0.005 (Pagel et al. 1992) to 0.234 ± 0.002 (OSS97). At the same time, our derived slope is significantly smaller than those of other authors, $dY/dZ = 6.7 \pm 2.3$ for Pagel et al. (1992) and $dY/dZ = 6.9 \pm 1.5$ for OSS97. This shallower slope is in good agreement with the value derived from stellar data for the Milky Way's disk and with simple models of galactic evolution of BCGs with well-mixed homogeneous outflows.

3. He abundance in the two most metal-deficient blue compact galaxies known

3.1. I Zw 18 and SBS 0335−052

Instead of the statistical approach described above, we can also derive the primordial He abundance from accurate measurements of the He abundance in a few objects selected to have very low O/H to minimize the amount of He manufactured in stars.

Izotov et al. (1999) have carried out such a study for the two most metal-deficient BCGs known. I Zw 18 and SBS 0335−052 provide a study in contrast concerning the different physical mechanisms which may modify the He I emission-line intensities. While in I Zw 18, the electron number density is small ($N_e \lesssim 100$ cm^{-3}) and collisional enhancement has a minor effect on the derived helium abundance, N_e is much higher in SBS 0335−052 ($N_e \sim 500$ cm^{-3} in the central part of the H II region). Additionally, the linear size of the H II region in SBS 0335−052 is ~ 5 times larger than in I Zw 18, suggesting that it may be optically thick for some He I transitions. In fact, both collisional and fluorescent enhancements of He I emission lines play an important role in this galaxy. By contrast, underlying stellar He I absorption is much less important in SBS 0335−052 than in I Zw 18. Since the equivalent widths (EW) of the He I emission lines scale roughly as the Hβ EWs, it is evident that this is the case from Figures 5c and 5f which show the spatial distribution of EW(Hβ) in both BCGs. In SBS 0335−052, EW(Hβ) has a lowest value of 160 Å and increases to ~ 300 Å in the outer parts, while in I Zw 18, EW(Hβ) is only 34 Å in the center of the NW component. Given equal EWs for He I absorption lines in both BCGs, we may expect the effect of underlying stellar absorption to be ~ 5 times smaller in SBS 0335−052 than in I Zw 18.

To disentangle the various effects which may make the He I emission-line intensities deviate from their recombination values, it is thus essential to use as many He I lines as possible in a self-consistent method as decribed above. An important and essential check that all corrections have been properly applied is the agreement between the He mass fraction Y derived independently from each He line. Figure 6 shows the Ys derived from the 4471, 5876 and 6678 He I emission lines in SBS 0335−052 at different spatial locations. It is clear that the self-consistent method (Figure 6a) gives much better agreement between the different lines. The Ys derived from the 4471 line are systematically below because only collisional and fluorescence effects have been taken into account and not underlying stellar absorption, and because the 4471 line is more subject to the latter effect. By contrast, there is not very good agreement between the Ys from different lines when N_e(He II) is set equal to N_e(S II) and only collisional enhancement is taken into account (Figure 6b).

3.2. Results

Izotov et al. (1999) derive Y = 0.243±0.007 for the SE component of I Zw 18 in very good agreement wth the value found by IT98a and Y = 0.2463±0.0015 for SBS 0335−052, excluding the He I 4471 line. The weighted mean is then Y = 0.2462±0.0015. Using dY/dZ = 2.4 (IT98b), the stellar He contribution is 0.0010, giving a primordial value Y_p = 0.2452±0.0015, in excellent agreement

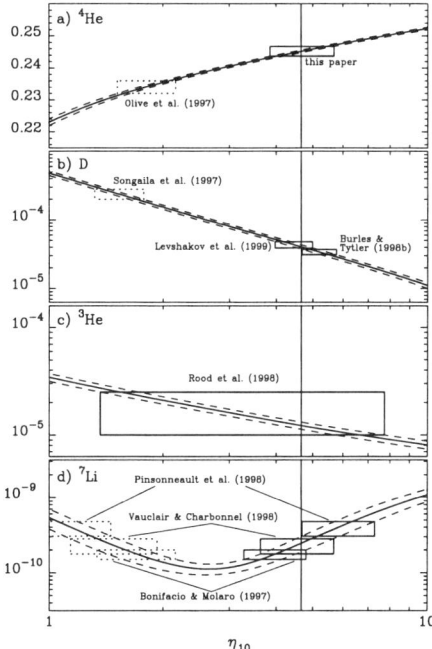

Figure 7. The abundance of (a) ^{4}He, (b) D, (c) ^{3}He and (d) ^{7}Li as a function of $\eta_{10} \equiv 10^{10}\,\eta$, where η is the baryon-to-photon number ratio, as given by the standard hot big bang nucleosynthesis model. The abundances of D, ^{3}He and ^{7}Li are number ratios relative to H. For ^{4}He, the mass fraction Y is shown. Our value $Y_p = 0.2452\pm0.0015$ gives $\eta = (4.7^{+1.0}_{-0.8})\times 10^{-10}$ as shown by the solid vertical line. We show other data with 1σ boxes.

with the value 0.2443±0.0015 derived from extrapolation of the Y-O/H and Y-N/H regression lines for our large BCG sample. It is, however, higher than the value $Y_p = 0.2345\pm0.0030$ derived by Peimbert & Peimbert (2000) from Magellanic Clouds H II regions. These authors suggest that two systematic effects may cause the disagreement: the presence of neutral hydrogen inside the helium Stromgren sphere and temperature fluctuations in our BCGs.

There is no evidence that the first effect is important in our objects. In our work, we have used the 'radiation softness parameter' of Vilchez & Pagel (1988) to estimate the correction factor for neutral helium and found the fraction of neutral helium to be insignificant ($\lesssim 2\%$) in all our objects, i.e their HII and He II Stromgren spheres are coincident to a very good approximation. This conclusion is corroborated by the detailed modeling of I Zw 18 by Stasinska & Schaerer (1999) who found the amount of neutral Helium to be negligible. On the other hand, they did find the observed T_e(OIII) to be $\sim 15\%$ higher than predicted by the photoionization model. Future progress in the determination of the primordial ^{4}He abundance using BCGs will rely on the discovery of more I Zw 18-like objects and on detailed modeling of very high signal-to-noise ratio and high-spectral resolution spectra of a few very metal-deficient BCGs (those with O/H less than 1/20 of solar) to look into such systematic effects as those mentioned above.

4. Cosmological implications

Figure 7 shows the primordial abundances of ^{4}He, D, ^{3}He, and ^{7}Li predicted by standard big bang nucleosynthesis theory as a function of the baryon-to photon number ratio η. The dashed lines are 1 σ uncertainties in model calculations. The solid boxes show the 1 σ predictions of η as inferred from our derived primordial abundance of ^{4}He, and the primordial abundances of D (Levshakov, Kegel & Takahara 1999; Burles & Tytler 1998), ^{3}He (Rood et al. 1998) and ^{7}Li (Bonifacio & Molaro 1997; Vauclair & Charbonnel 1998; Pinsonneault et al. 1999). All these determinations are consistent to within 1 σ, although the most stringent constraint is provided by D. For comparison, we have also plotted the Y_p derived by OSS97 which is low, partly because underlying stellar absorption was not taken into account in I Zw 18. Their $Y_p = 0.234$ happens also to be the value obtained by Peimbert & Peimbert (2000) in the Magellanic Clouds. This low value would have been consistent with the primordial D abundance obtained by Songaila et al (1997) which is one order of magnitude higher than the value obtained by Burles & Tytler (1998), except that this high value is now believed to be erroneous.

Our $Y_p = 0.2452 \pm 0.0015$ value implies a baryon-to-photon number ratio $\eta = 4.7^{+1.0}_{-0.8} \times 10^{-10}$. This translates to a baryon mass fraction $\Omega_b h_{50}^2 = 0.068^{+0.015}_{-0.012}$ where h_{50} is the Hubble constant in units of 50 km s^{-1}Mpc^{-1}. For a Hubble constant equal to 65 km s^{-1}Mpc^{-1}, $\Omega_b = 0.040^{+0.009}_{-0.007}$. Our derived baryonic mass fraction is consistent with the one obtained by analysis of the Lyα forest in a cold dark matter cosmology. Depending on the intensity of diffuse UV radiation, the inferred lower limit is $\Omega h_{50}^2 = 0.05 - 0.10$ (Weinberg et al. 1997; Bi & Davidsen 1997; Rauch et al. 1997), while Zhang et al. (1998) have derived $0.03 \leq \Omega h_{50}^2 \leq 0.08$.

Finally, for the most consistent set of primordial abundances – D from Levshakov et al. (1999), our above value for ^{4}He, and ^{7}Li from Vauclair & Charbonnel (1998) – we derive an equivalent number of light neutrino species $N_\nu = 3.0 \pm 0.3$ (2σ) (Izotov et al. 1999).

Acknowledgments. We thank the partial financial support of NSF grant AST-96-16863 and an IAU Travel grant. We acknowledge useful conversations with M. Peimbert and B. Pagel and thank the organizers for a stimulating meeting in a superb locale.

References

Bi, H., & Davidsen, A. F. 1997, ApJ, 479, 523
Bonifacio, P. & Molaro, P. 1997, MNRAS, 285, 847
Brocklehurst, M. 1972, MNRAS, 157, 211
Burles, S., & Tytler, D. 1998a, ApJ, 507, 732 1998, Phys. Rev. D, 58, 063506

Izotov, Y. I., & Thuan, T. X. 1998a, ApJ, 500, 188 (IT98a)
———. 1998b, ApJ, 497, 227 (IT98b)
———. 1999, ApJ, 511, 639
Izotov, Y. I., Thuan, T. X., & Lipovetsky, V. A. 1994, ApJ, 435, 647 (ITL94)
———. 1997, ApJS, 108, 1 (ITL97)
Izotov, Y. I., Lipovetsky, V.A., Guseva, N.G., Kniazev, A. Y., Neizvestny, S. I. & Stepanian, J. A. 1993, Astron. Astrophys. Trans., 3, 193
Izotov, Y. I., Chaffee, F. H., Foltz, C. B., Green, R. F., Guseva, N. G., Thuan, T. X. 1999, ApJ, 527, 757
Kingdon, J., & Ferland, G. J. 1995, ApJ, 442, 714
Levshakov, S. A., Kegel, W. H., & Takahara, F. 1998, MNRAS, 302, 707
Markarian, B. E., Lipovetsky, V.A., Stepanian, J.A., Erastova, L. K., & Shapovalova, A. I. 1989, Commun. Special Astrophys. Obs., 62, 5
Olive, K. A., Skillman, E. D., & Steigman, G. 1997, ApJ, 483, 788 (OSS97)
Pagel, B. E. J., Simonson, E. A., Terlevich, R. J., & Edmunds, M. G. 1992, MNRAS, 255, 325
Pagel, B. E. J., Terlevich, R. J., & Melnick, J. 1986, PASP, 98, 1005
Peimbert, M., & Torres-Peimbert, S. 1974, ApJ, 193, 327
———. 1976, ApJ, 203, 581
Peimbert, M. & Peimbert, A. 2000, this volume
Pinsonneault, M. H., Walker, T. P., Steigman, G., & Naranyanan, V. K. 1999, ApJ, 527, 180
Rauch, M., Miralda-Escudé, J., Sargent, W. L. W., Barlow, T. A., Weinberg, D. H., Hernquist, L., Katz, N., Cen, R., & Ostriker, J. P. 1997, ApJ, 489, 7
Robbins, R. R. 1968, ApJ, 151, 511
Rood, R. T., Bania, T. M., Balser, D. S., & Wilson, T. L. 1998, Space Sci. Rev., 84, 185
Salzer, J. J., MacAlpine, G. M., & Boroson, T. A. 1989, ApJS, 70, 447
Smits, D. P. 1996, MNRAS, 278, 683
Songaila, A., Wampler, E. J., & Cowie, L. L. 1997, Nature, 385, 137
Stasińska, G., & Schaerer, D. 1999, A&A, 351, 72
Vauclair, S., & Charbonnel, C. 1998, ApJ, 502, 372
Vílchez, J. M., & Pagel, B. E. J. 1988, MNRAS, 231, 257
Weinberg, D. H., Miralda-Escudé, J., Hernquist, L., & Katz, N. 1997, ApJ, 490, 564
Zhang, Y., Meiksin, A., Anninos, P., & Norman, M. L. 1998, ApJ, 495, 63

The Light Elements and Their Evolution
IAU Symposium, Vol. 198, 2000
L. da Silva, M. Spite, J. R. de Medeiros, eds.

Inhomogeneous H II Regions and the Helium Abundance

S. M. Viegas and R. Gruenwald

IAG-USP; Av. Miguel Stefano, 4200; 04301-904 São Paulo, SP, Brazil

Abstract. When calculating the helium (^{4}He) abundance in low metallicity H II regions, the ionization correction factor (icf) for unseen neutral helium (and hydrogen) is usually assumed to be unity. In this paper, we explore this factor for H II regions ionized by young stellar clusters. Our main result is that $icf < 1$ for homogeneous H II regions and, that the effect of density condensations in the H II regions is to further *reduce* the icf. For $icf < 1$, the primordial helium abundance inferred from observations of low-metallicity, extragalactic H II regions is *decreased*.

1. Introduction

The importance of the primordial helium abundance as a key test of the standard hot big bang model and the necessity of determining an accurate value are discussed in another paper in this volume (Steigman 2000). The usual method to derive the primordial ^{4}He abundance from very low-metallicity H II regions has achieved very small statistical uncertainties ($\sim 1\%$) for Y_P, the primordial helium mass fraction (Olive & Steigman 1995, Olive, Skillman & Steigman 1997, Izotov, Thuan & Lipovetsky 1994, 1997 (ITL), Izotov & Thuan 1998 (IT)). However, the derived value may be contaminated by unrecognized systematic uncertainties (Davidson & Kinman 1985, Pagel et al. 1992, Skillman et al. 1994, ITL, IT, Skillman, Terlevich & Terlevich 1998). Here we discuss one potential source of systematic error – the ionization correction for unseen neutral hydrogen and/or helium, both for homogeneous and inhomogeneous H II regions.

The empirical method usually used to derive the chemical abundances from the observed emission-line intensities from an assumed homogeneous nebula was first proposed by Peimbert & Costero (1969). The fractional abundances of the ions present in the gas which produce observable emission-lines are obtained and combined to obtain the element abundances. In order to account for the unseen ions and neutral atoms, an ionization correction factor (icf), derived from considerations of the ionization potential or from photoionization models (Peimbert & Torres-Peimbert 1977, Stasinska, 1980, 1982, Mathis 1982, Peña 1986) is used. However, these ionization correction factors were obtained while assuming a homogenous nebula. In contrast, recent HST imaging reveals that real H II regions are far from homogeneous, showing many different features such as condensations, filaments and voids. Thus, the true icf may differ from calculated values, introducing a systematic error in the calculation of the element abundances. In particular, an error in the helium abundance determination will reflect directly on the inferred primordial helium abundance. $\equiv$ valleys) as shown

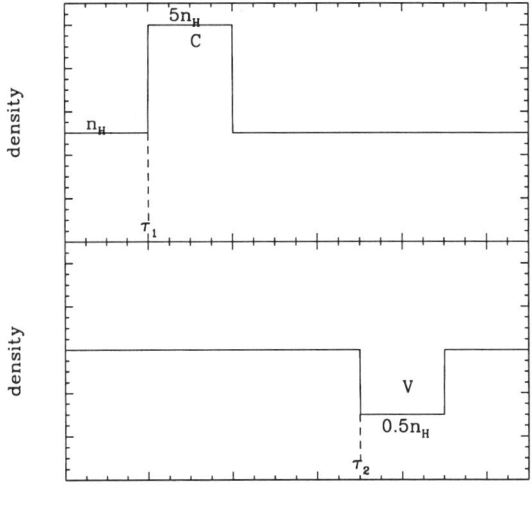

Figure 1. Illustration of an enhancement (C) and a deficit (V) of the density, where n_H is the density of the corresponding homogeneous model. In our models, the density increases by a factor of 5, or decreases by a factor of 2 from the homogeneous case extending over a distance which is 10% of the thickness of the H II region for the corresponding homogeneous model. The position of the condensation (C) or valley (V) is fixed by the choice of the optical depth (τ_1, τ_2) at that location.

in Figure 1. The location of the condensations and valleys are chosen by fixing the corresponding optical depth at the Lyman limit, τ_H, in the homogeneous model (H); the following values were chosen: τ_H = 0.02, 0.03, 0.04, 0.10, and 0.40. Using a Monte Carlo method on a variety of H, C and V models, we mimic the physical conditions in more realistic H II regions.

2. Simulating an Inhomogeneous H II Region

The available 1-D photoionization codes have been widely used to analyse observed emission-line spectra and several of them have been intercompared showing good agreement (Péquignot 1986, Ferland et al. 1995). The input parameters are the ionizing radiation spectrum, the gas density and the chemical abundances in the gas, in addition to the assumption of spherical or plane-parallel symmetry in order to account for the diffuse radiation.

Here, in this analysis, the photoionization code AANGABA is used to build spherically symmetric homogeneous models. Further, in order to mimic the presence of condensations and voids revealed by H II region imaging, several models with different choices of the input parameters are combined. The ionizing radiation of a young stellar cluster (Cid-Fernandes et al. 1992) is adopted for two evolutionary phases of the stellar cluster: the initial phase (t = 0), when the spectrum is dominated by the hottest, most massive stars (appropriate for low-metallicity, high-excitation H II regions), and a later phase (t= 2.5 Myr), when the massive stars have evolved and there are fewer He$^+$ ionizing photons. The stellar cluster is characterized by the number of ionizing photons above the hydrogen Lyman limit, Q_H, which, along with a choice of the gas density defines the H II region model. Regarding the ionization parameter U, commonly used to define photoionization models, for a given density our models correspond to a fixed value of UR_i^2 (where R_i is the inner radius of the H II region). More details on this point can be found in Viegas, Gruenwald & Steigman (2000). Because we are interested in low-metallicity H II regions, 0.1 solar composition is chosen.

Based on the homogeneous models described above, models mimicking condensations and voids are built with spatially bounded density enhancements (C ≡ condensations) and density deficits (V

3. The Helium Correction Factor and the Helium Abundance

Pagel et al. (1992) proposed a method to estimate the ionization correction factor icf, based on the "radiation softness parameter" $\eta = (O^+/S^+)(S^{++}/O^{++})$ defined by Vilchez & Pagel (1988). Comparing with photoionization models, they concluded that $icf = 1$ for $\log \eta < 0.9$, corresponding to models with effective temperature higher than 37 000 K, while for $\log \eta > 0.9$ the icf may differ from unity. In the following we will analyse the behaviour of icf as a function of η from H II region models using the stellar cluster radiation spectra at t = 0 and t = 2.5 Myr.

In order to account for the presence of unseen He^{++}, and H^0, the icf is defined as

$$icf = [1 + \frac{(n(He^0) + n(He^{++}))}{n(He^+)}]/[1 + \frac{n(H^0)}{n(H^+)}]. \tag{1}$$

The results for the icf are shown in Figure 2. For the homogeneous models we fixed the density at $n_H = 10$ cm^{-3} and varied Q_H by nine orders of magnitude from 7.5×10^{44} s^{-1} to 7.5×10^{53} s^{-1}. All the models were calculated assuming a filling factor equal to unity, although the results obtained with a lower filling factor all lie along the same curves (trading Q_H for filling factor).

Models with the radiation spectrum of a young (t = 0) stellar cluster predict $icf \leq 1$, indicating that neutral hydrogen is present where the helium is still ionized. On the other hand, a different result is obtained with an evolved stellar cluster (t = 2.5 Myr). In this case, because the ionizing radiation lacks photons beyond the He$^+$ Lyman limit, there is neutral He inside the H$^+$ zone and the ionization correction factor exceeds unity.

Regarding the inhomogeneous models, the results for "valleys" are indistinguishable from the corresponding homogeneous models, while those for condensations tend to lead to a bigger downward correction (reduction in the predicted icf) than for the homogeneous cases, with the greatest deviation occurring in models where the condensation is located at $\tau = 0.04$. Since a real H II region must be a mix of H, C and V regions located at different τ, a single composite model can be created assuming that the weighted contribution of each region to the composite H II region is proportional to the solid angle occupied by the H, C or V model. The solid angles are chosen randomly, with the constraint that their sum is 4π. The Monte Carlo method is used for different choices of Q_H

Inhomogeneous H II Regions and the Helium Abundance

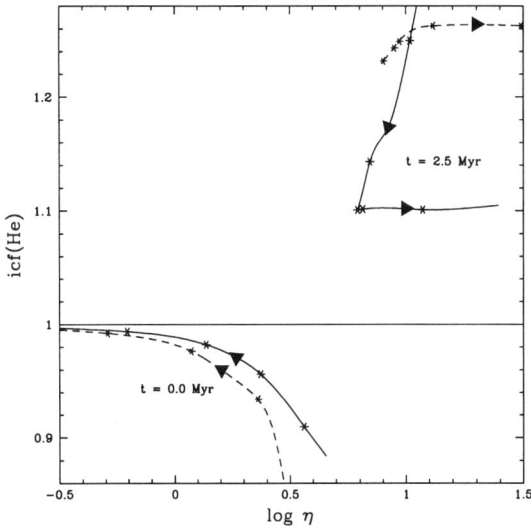

Figure 2. The $icf - \log \eta$ relation for H II regions ionized by a young (t = 0, left bottom curves) and an evolved (t = 2.5 Myr, top right curves) stellar cluster. The solid lines correspond to homogeneous models, and the dashed lines to models with condensation located at $\tau = 0.04$. The tickmarks correspond to models with $n_H = 10$ cm^{-3}, $\epsilon = 1.$, and $Q_H/(7.5 \times 10^{45} = 1., 10^2, 10^4, 10^6, 10^8$. Arrows indicate the effect of increasing values of Q_H.

and n. As expected, the results for these composite models lie between those for the homogenous and the "extreme" condensation models.

Another kind of composite model has been investigated by Dinerstein & Shields (1986), who considered the case where the spectrum of an H II region ionized by a young stellar cluster could be contaminated by emission from a region powered by an older cluster lying along the same line of sight. As seen in Figure 2, the $\log \eta$ values for models with young and old stellar clusters are very different. Thus, for most combinations of the two cases, the resulting H II region would show the $\log \eta$ and icf characteristic of the dominant H II region (i.e., the one with the largest Q_H). However, models with similar Q_H would generally result in $\log \eta$ larger than 1, and would not be appropriate to describe the H II regions used for deriving the primordial helium abundance which are selected to have $\log \eta$ less than unity. However, the combination of an H II region powered by a young stellar cluster and low Q_H, with an older cluster with high Q_H, could result in a $\log \eta \leq 0.8$ but with icf larger than unity. In fact, combining the results of the model where t = 0 and $Q_H = 7.5 \times 10^{47}$ s^{-1}, for which $\log \eta = 0.37$, with the models for t = 2.5 Myr and $Q_H = 7.5 \times 10^{49}$ s^{-1}, 7.5×10^{51} s^{-1}, and 7.5×10^{53} s^{-1}, for which $\log \eta \approx 0.8$, we obtain $\log \eta \approx 0.4$ and $icf \approx 1.04$. However, for these models the emission-line ratios differ noticeably from the case of a single H II region with the same value of $\log \eta$ as can be seen for the key line ratios which are shown in Table 1, where the results for a young, less luminous H II region are compared to those in the composite models described above (characterized by the Q_H value of the older H II region).

Table 1. Table1 - Emission-line intensities relative to Hβ

Emission-line	t = 0	Composite model		
	7.5E47	7.5E59	7.5E51	7.5E53
[O II]3727	2.65	1.76	0.86	0.48
[Ne III]3868+	0.43	0.39	0.53	0.63
[O III]5007+	3.88	5.48	7.58	8.81
He I 5876	0.11	0.099	0.098	0.098
[N II]6584+	0.68	0.41	0.20	0.11
[S II]6717+	0.32	0.13	0.041	0.011

4. Concluding Remarks

As described above, the helium ionization correction factor for homogeneous H II regions ionized by a young stellar cluster is ≤ 1. The icf may be further reduced by the presence of condensations, which are likely to be present in real H II regions as indicated by HST imaging. Composite models which combine the results of a young (t = 0), less luminous stellar cluster with an evolved (t = 2.5 Myr), more luminous stellar cluster, may mimic a single H II region with log η ≈ 0.4, but having icf larger than unity. Since the composite models will have different emission-line intensities, relative to Hβ, only by comparison to all the observational data can composite H II regions mimicking a single H II region be unmasked, allowing us to choose the appropriate icf.

Finally, in order to quantify our results we used the data from a sample of low-metallicity H II regions observed by Izotov & Thuan (1998) in a Monte Carlo simulation (see Viegas, Gruenwald & Steigman 2000 for details), to calculate the systematic error in the inferred primordial helium abundance due to realistic values of the icf. The results indicate that the systematic error in the mass fraction, Y_P, could be as large as -0.004, which is significant when compared to the quoted statistical errors on Y_P. Recent observations of Magellanic Cloud H II regions, which may be less affected by systematic errors than the distant low-metallicity H II regions, provide independent support for this conclusion by also indicating a lower value for the primordial helium abundance (Peimbert 2000, this volume).

Acknowledgments. We are indebted to G. Steigman for many fruitful discussions. This work has been partially supported by FAPESP, CNPq and PRONEX/Finep.

References

Cid-Fernandes, R., Dottori, H., Gruenwald, R., & Viegas, S.M. 1992, MNRAS, 255, 165
Davidson, K., & Kinman, T.D. 1985, ApJS, 58, 321
Dinerstein, H.L., & Shields, G.A. 1986, ApJ, 311, 45

Ferland et al. 1995, in The Analysis of Emission Lines, STScI Symp. Ser. 8,ed. R.E. Williams & M. Livio (Cambridge: Cambridge University Press), 143

Izotov, Y.I., Thuan, T.X., & Lipovetsky, V.A. 1994, ApJ, 435, 647

Izotov, Y.I., Thuan, T.X., & Lipovetsky, V.A. 1997, ApJS, 108, 1

Izotov, Y.I. & Thuan, T.X. 1998, ApJ, 500, 188

Mathis, J.S. 1982, ApJ, 261, 195

Olive, K.A., Skillman, E., & Steigman, G. 1997, ApJ, 483, 788

Olive, K. A., &, Steigman, G. 1995, ApJS, 97, 49

Pagel, B.E.J., Somonson, E.A., Terlevich, R.J. & Edmunds, M. 1992, MNRAS, 255, 325

Peimbert, M. & Costero, R. 1969, Bol. Obs. Tonantziltla y Tacubaya, 5, 3

Peimbert, M. & Torres-Peimbert, S. 1977, MNRAS, 179, 217

Pena, M. 1986, PASP, 98, 1061

Pequignot 1986, Workshop on Model Nebulae, ed. D. Pequignot (Meudon, Observatoire de Paris-Meudon)

Skillman, E., Terlevich, R.J., Kennicutt, R.C., Garnett, D.R., & Terlevich, E. 1994, ApJ, 431, 172

Skillman, E., Terlevich, E., & Terlevich, R.J. 1998, Sp. Sci. Rev, 84, 105

Stasinska, G. 1980, A&A, 84, 320

Stasinska, G. 1982, A&AS., 41, 513

Viegas, S.M., Gruenwald, R., & Steigman, G. 2000, ApJ (in press)

Vilchez, J.M. & Pagel, B.E.J. 1988, MNRAS, 231, 257

The Magellanic Clouds and the Primordial Helium Abundance

Manuel Peimbert and Antonio Peimbert

Instituto de Astronomía, Universidad Nacional Autónoma de México. Apdo. Postal 70-264, México, D.F. 04510

Abstract. A new determination of the pregalactic helium abundance based on the Magellanic Clouds H II regions is discussed. This determination amounts to $Y_p = 0.2345 \pm 0.0030$ and is compared with those derived from giant extragalactic H II regions in systems with extremely low heavy elements content. It is suggested that the higher primordial value derived by other authors from giant H II region complexes could be due to two systematic effects: the presence of neutral hydrogen inside the helium Strömgren sphere and the presence of temperature variations inside the observed volume.

1. Introduction

The determination of the pregalactic, or primordial, helium abundance by mass Y_p is paramount for the study of cosmology, the physics of elementary particles, and the chemical evolution of galaxies (e. g. Fields & Olive 1998, Izotov et al. 1999, Peimbert & Torres-Peimbert 1999, and references therein). In this review we briefly discuss the method used to derive Y_p and its main sources of error as well as a new determination based on observations of the SMC. This determination is compared with those carried out earlier based on extremely metal poor extragalactic H II regions.

The Magellanic Clouds determination of Y_p can have at least four significant advantages and one disadvantage with respect to those based on distant H II region complexes: a) no underlying absorption correction for the helium lines is needed because the ionizing stars can be excluded from the observing slit, b) the determination of the helium ionization correction factor can be estimated by observing different lines of sight of a given H II region, c) the accuracy of the determination can be estimated by comparing the results derived from different points in a given H II region, d) the electron temperature is generally smaller than those of metal poorer H II regions reducing the effect of collisional excitation from the metastable 2^3 S level of He I, and e) the disadvantage is that the correction due to the chemical evolution of the SMC is in general larger than for the other systems.

2. He$^+$/H$^+$ Determinations

To derive accurate He$^+$/H$^+$ values we need very accurate N_e (He II), T_e (He II), and $\tau(3889,$ He I) values. Good approximations to determine He$^+$/H$^+$ for $N_e <$ 300 cm^{-3}, 13 000 K $< T_e <$ 20 000 K and $\tau(3889) = 0.0$, have been presented for the main helium lines by Benjamin, Skillman and Smits (1999):

$$\frac{N(\text{He}^+)}{N(\text{H}^+)} = \frac{I(6678)}{I(\text{H}\beta)} 2.58 T_4^{0.249 - 2.0 \times 10^{-4} N_e}, \tag{1}$$

$$= \frac{I(4471)}{I(\text{H}\beta)} 2.01 T_4^{0.127 - 4.1 \times 10^{-4} N_e}, \tag{2}$$

$$= \frac{I(5876)}{I(\text{H}\beta)} 0.735 T_4^{0.230 - 6.3 \times 10^{-4} N_e}. \tag{3}$$

Fortunately for giant extragalactic H II regions $\tau(3889)$ is very small and frequently close to zero. $\tau(3889)$ can be estimated together with N_e (He II) from the 3889/4471 and the 7065/4471 ratios computed by Robbins (1968).

Most authors assume that $T_e(\text{O III}) = T_e(\text{He II})$, there are two reasons for this assumption the He$^+$ and O^{++} emission regions occupy similar volumes (but not identical ones) and $T_e(\text{O III})$ is easy to measure. Nevertheless the assumption is correct only for isothermal nebulae, and not for real nebulae if very accurate abundances are needed. In the presence of temperature variations along the line of sight the [O III] lines originate preferentially in the high temperature zones and the helium and hydrogen lines in the low temperature zones (e.g. Peimbert 1967, 1995). It can be shown that the temperature derived from the ratio of the Balmer continuum to a Balmer emission line, $T_e(\text{Bac})$, is similar to $T_e(\text{He II})$, and that both temperatures are smaller than $T_e(\text{O III})$.

From models computed with CLOUDY (Ferland 1996) it is found that $T_e(\text{Bac})$ —labeled $T_e(\text{Hth})$ by CLOUDY— is about 5% smaller than $T_e(\text{O III})$. Moreover in the model with a homogeneous sphere of I Zw 18 by Stasinska and Schaerer (1999) it is found that $\langle T_e(\text{Ar III})\rangle = 16300$ K (for 63% of the volume) and $\langle T_e(\text{Ar IV})\rangle = 18300$ K (for 36% of the volume) indicating the presence of temperature variations. Notice that the average temperature has to be weighted by the emissivities strengthening the effect of the temperature variations. Moreover there is additional evidence that indicates that the temperature variations are even higher than those predicted by photoionization models (e. g. Peimbert 1995; Luridiana, Peimbert, & Leitherer 1999; Stasinska and Schaerer 1999). From equations (1–3) it follows that the smaller the adopted temperature the smaller the derived He$^+$/H$^+$ value.

Due to the presence of very strong density variations inside gaseous nebulae the different methods to derive the density yield very different values. The root mean square density, $N_e(\text{rms})$, is usually obtained from the observed flux in a Balmer line and by assuming a spherical geometry; $N_e(\text{rms})$ provides a minimum value for $N_e(\text{He II})$, the local density needed to derive the helium abundance. Forbidden line ratios of lines of similar excitation energy give us an average density for cases where the density is similar or smaller than the critical density for collisional deexcitation; available line ratios in the visual region are those of

[S II], [O II], [Cl III], and [Ar IV], unfortunately giant extragalactic H II regions are close to the low density limit of these ratios, the line intensities of [Cl III] are very faint, for most observations of the [O II] lines the resolution is not high enough to separate $\lambda 3726$ from $\lambda 3729$ nor $\lambda 4711$ of [Ar IV] from $\lambda 4713$ of He I. Consequently most of the densities in the literature are those derived from the [S II] lines, but as rightly mentioned by Izotov, Thuan, & Lipovetsky (1994, 1997) they are not representative of the regions where the He I lines originate. The self-consistent method, advocated by Izotov and collaborators, is based only on line ratios of helium I and is the best method to derive the density of the He II zone: $N_e(\mathrm{He\ II})$.

The stellar underlying absorption can affect the derived He I and H I emission line intensities and has to be taken into account. The best way to reduce this effect is to avoid the presence of bright early type stars in the observed slit, this can only be done for objects inside the Galaxy and the Magellanic Clouds. To minimize the effect of the underlying absorption it is recommended to use only objects where the line intensities show very large equivalent widths in emission, and to increase the spectral resolution. The best way to correct for underlying absorption is to use starburst models that predict the underlying stellar spectrum. The correction for underlying absorption can be tested by comparing the higher order Balmer lines, that are most affected by this effect, with the brightest Balmer lines that are the least affected. Similarly the underlying absorption effect is larger for $\lambda 4471$ and smaller for $\lambda\lambda 5876, 6678$ and 7065; the effect for these three lines is in general negligible due to a combination of causes (mainly their large equivalent widths in emission).

3. Ionization Structure

The total He/H value is given by:

$$\frac{N(\mathrm{He})}{N(\mathrm{H})} = \frac{N(\mathrm{He}^0) + N(\mathrm{He}^+) + N(\mathrm{He}^{++})}{N(\mathrm{H}^0) + N(\mathrm{H}^+)}, \quad (4)$$

$$= ICF(\mathrm{He})\frac{N(\mathrm{He}^+) + N(\mathrm{He}^{++})}{N(\mathrm{H}^+)}. \quad (5)$$

The $\mathrm{He}^{++}/\mathrm{H}^+$ ratio can be obtained directly from the $4686/\mathrm{H}\beta$ intensity ratio. In objects of low degree of ionization the presence of neutral helium inside the H II region is important and $ICF(\mathrm{He})$ becomes larger than 1. The $ICF(\mathrm{He})$ can be estimated by observing a given nebula at different lines of sight since He^0 is expected to be located in the outer regions. Another way to deal with this problem is to observe H II regions of high degree of ionization where the He^0 amount is expected to be negligible. Vílchez & Pagel (1988) (see also Pagel et al. 1992) defined a radiation softness parameter given by

$$\eta = \frac{N(\mathrm{O}^+)N(\mathrm{S}^{++})}{N(\mathrm{S}^+)N(\mathrm{O}^{++})}; \quad (6)$$

for large values of η the amount of neutral helium is significant, while for low values of η it is negligible.

The Magellanic Clouds and the Primordial He Abundance

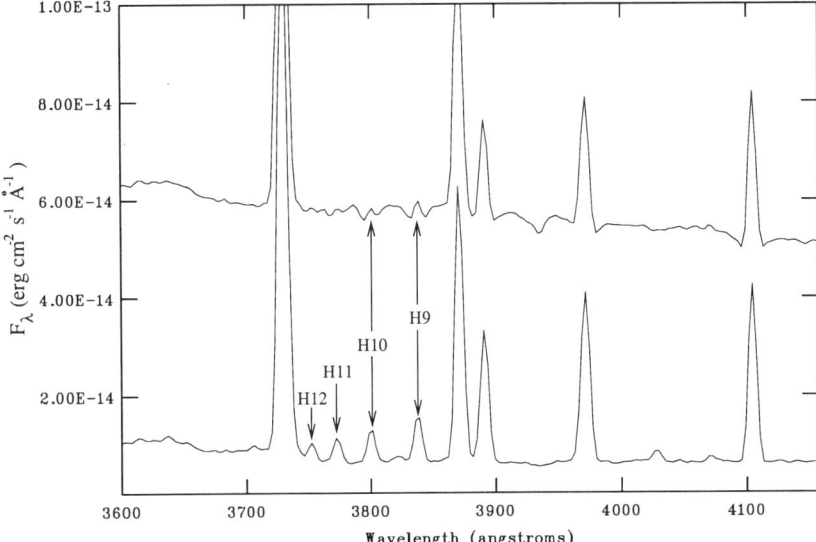

Figure 1. Spectra of NGC 346 with and without underlying absorption. The vertical scale is for the lower spectrum (region 346–A). The flux of the upper spectrum (region 346–B) was normalized to the Hα emission line flux of the lower spectrum.

On the other hand for ionization bounded objects of very high degree of ionization the amount of H^0 inside the He^+ Strömgren sphere becomes significant and the ICF(He) can become smaller than 1. This possibility was firstly mentioned by Shields (1974) and studied extensively by Armour et al. (1999) for constant density models. Since H^0 will be located in a thin shell in the border of the nebula it can be shown that for ionization bounded nebulae of homogeneous density the fraction of H^0/He^+ will be three times larger for an observation of the whole object than for an observation of a line of sight that includes the center. Similarly this fraction becomes higher than a factor of three for models with decreasing density from the center outwards. Alternatively for density bounded nebulae this effect can be neglected.

4. SMC

Peimbert, Peimbert, & Ruiz (2000a) presented long slit observations of the most luminous H II region in the SMC: NGC 346. They divided the two long slit positions into thirteen areas, four including the brightest stars ($m \sim 14$) and 9 without stars brighter than $m = 17$. In the upper part of Figure 1 we present a spectrum that includes all the observed areas (region 346–B); while in the lower part of Figure 1 we present a spectrum made by the seven brightest areas that do not include the brightest stars (region 346–A).

After correcting region 346–A for extinction, based on the four brightest Balmer lines, it is found that the weaker Balmer lines (H9 to H12) are not affected by stellar underlying absorption (see Figure 1), therefore the He lines are not expected to be affected by underlying absorption. The extinction derived from the Balmer lines is in very good agreement with the extinction derived from the stellar cluster by Massey, Parker, & Garmany (1989), another indication that the stellar underlying absorption of the Balmer lines is negligible.

Table 1. $N(\mathrm{He}^+)/N(\mathrm{H}^+)^a$ and χ^2 for NGC 346

T_e(K)	N_e(cm^{-3})				
	53	100	143	162	247
11200	805 (83.2)	798 (47.7)	793 (26.4)	791 (20.0)	**781**[b] *(8.24)*[c]
11800	806 (38.6)	799 (15.9)	793 (7.37)	**790** *(6.59)*	780 (20.4)
11950	806 (30.8)	799 (11.7)	**793** **(6.53)**[d]	790 (7.25)	779 (27.7)
12400	807 (15.0)	**799** *(7.17)*	793 (12.5)	790 (17.9)	778 (58.6)
13000	**809** *(9.72)*	800 (18.2)	793 (38.4)	790 (50.2)	777 (118)

[a] Given in units of 10^{-4}, χ^2 values in parenthesis.

[b] The He$^+$/H$^+$ values in boldface correspond to the minimum χ^2 values at a given temperature.

[c] The minimum χ^2 value at a given temperature is presented in italics.

[d] The smallest χ^2 value for all temperatures and densities is presented in boldface, thus defining T_e(He II) and N_e(He II).

To derive the He$^+$/H$^+$ value, in addition to the Balmer lines, we made use of nine He I lines, $\lambda\lambda$ 3889, 4026, 4387, 4471, 4921, 5876, 6678, 7065, and 7281 to determine N_e(He II) and T_e(He II) self-consistently. In Table 1 we present He$^+$/H$^+$ values for different temperatures and densities; the temperatures were selected to include T_e(O III), T_e(Bac), T_e(He II), and two representative temperatures; the densities were selected to include the minimum χ^2 at each one of the five temperatures. The temperature with the minimum χ^2 is the self-consistent T_e(He II) and amounts to 11950 ± 560 K; this temperature is in excellent agreement with the temperature derived from the Balmer continuum that amounts to 11800 ± 500 K, alternatively T_e(O III) amounts to 13070 ± 100 K. Notice that the χ^2 test requires a higher density for a lower temperature, increasing the dependence on the temperature of the He^+/H^+ ratio. The values in Table 1 correspond to the case where τ (3889) equals zero, for higher values of τ (3889) the χ^2 values increase. In Table 1 the He^+/H^+ values in boldface and the χ^2 values in italics correspond to the minimum value of χ^2 at a given T_e, the χ^2 value in boldface is the smallest value for all temperatures and all densities.

From Table 1 we obtain that $T_e(\text{He II}) = 11950$ K and $N_e(\text{He II}) = 143$ cm^{-3}, which correspond to $He^+/H^+ = 0.0793 \pm 0.007$. By comparing the He/H values for lines of sight with different ionization degree it is found that $ICF(\text{He}) = 1.00$. To obtain the total He/H value we have added the contribution of He^{++}/H^+ that amounts to 2.2×10^{-4}.

In Table 2 we present the helium abundance by mass $Y(\text{SMC})$, derived from NGC 346. The $Y(\text{SMC})$ values were derived from Table 1, the He^{++}/H^+ values and the Z values determined by Peimbert et al. (2000a).

Table 2. $Y(\text{SMC})$

T_e(K)	N_e(cm^{-3})				
	53	100	143	162	247
11200	0.2431	0.2416	0.2404	0.2399	**0.2377**[a]
11800	0.2435	0.2419	0.2405	**0.2399**	0.2375
11950	0.2436	0.2420	**0.2405**	0.2399	0.2374
12400	0.2439	**0.2421**	0.2406	0.2399	0.2372
13000	**0.2443**	0.2423	0.2407	0.2400	0.2370

[a] Boldface values correspond to minimum χ^2 values, see Table 1.

5. Y_p

To determine the Y_p value from the SMC it is necessary to estimate the fraction of helium present in the interstellar medium produced by galactic chemical evolution. We will assume that

$$Y_p = Y(\text{SMC}) - Z(\text{SMC}) \frac{\Delta Y}{\Delta Z}. \quad (7)$$

Peimbert et al. (2000a) find that for $T_e(\text{He II}) = 11950 \pm 500$ K, $Y(\text{SMC}) = 0.2405 \pm 0.0018$ and $Z(\text{SMC}) = 0.00315 \pm 0.00029$. To estimate $\Delta Y/\Delta Z$ we will consider three observational determinations and a few determinations predicted by chemical evolution models.

Peimbert, Torres-Peimbert, & Ruiz (1992) and Esteban et al. (1999) found that $Y = 0.2797 \pm 0.006$ and $Z = 0.0212 \pm 0.003$ for the Galactic H II region M17, where we have added 0.10dex and 0.08dex to the carbon and oxygen gaseous abundances to take into account the fraction of these elements embedded in dust grains (Esteban et al. 1998). By comparing the Y and Z values of M17 with those of NGC 346 (Peimbert et al. 2000a) we obtain $\Delta Y/\Delta Z = 2.17 \pm$

0.4. M17 is the best H II region to determine the helium abundance because among the brightest galactic H II regions it is the one with the highest degree of ionization and consequently with the smallest correction for the presence of He^0 (i.e. ICF(He) is very close to unity). It can be argued that the M17 values are not representative of irregular galaxies, on the other hand they provide the most accurate observational determination. From a group of 10 irregular and blue compact galaxies, that includes the LMC and the SMC, Carigi et al. (1995) found $\Delta Y/\Delta Z = 2.4\pm0.6$, where they added 0.2dex to the O/H abundance ratios derived from the nebular data to take into account the temperature structure of the H II regions and the fraction of O embedded in dust, moreover they also estimated that O constitutes 54% of the Z value. Izotov & Thuan (1998) from a group of 45 supergiant H II regions of low metallicity derived a $\Delta Y/\Delta Z = 2.3 \pm 1.0$; we find from their data that $\Delta Y/\Delta Z = 1.5 \pm 0.6$ by adding 0.2dex to the O abundances to take into account the temperature structure of the H II regions and the fraction of O embedded in dust.

Based on their two-infall model for the chemical evolution of the Galaxy Chiappini, Matteucci, & Gratton (1997) find $\Delta Y/\Delta Z = 1.6$ for the solar vicinity. Carigi (2000) computed chemical evolution models for the Galactic disk, under an inside-out formation scenario, based on different combinations of seven sets of stellar yields by different authors; the $\Delta Y/\Delta Z$ spread predicted by her models is in the 1.2 to 1.9 range for the Galactocentric distance of M17 (5.9 kpc).

Carigi et al. (1995), based on yields by Maeder (1992), computed closed box models adequate for irregular galaxies, like the SMC, and obtained $\Delta Y/\Delta Z = 1.52$. They also computed models with galactic outflows of well mixed material, that yielded $\Delta Y/\Delta Z$ values similar to those of the closed box models, and models with galactic outflows of O-rich material that yielded values higher than 1.52. The maximum $\Delta Y/\Delta Z$ value that can be obtained with models of O-rich outflows, without entering into contradiction with the C/O and $(Z-C-O)/O$ observational constraints, amounts to 1.69.

Carigi, Colín, & Peimbert, (1999), based on yields by Woosley, Langer, & Weaver (1993) and Woosley & Weaver (1995), computed chemical evolution models for irregular galaxies also, like the SMC, and found very similar values for closed box models with bursting star formation and constant star formation rates that amounted to $\Delta Y/\Delta Z = 1.71$. The models with O-rich outflows can increase the $\Delta Y/\Delta Z$, but they predict higher C/O ratios than observed.

From the previous discussion it follows that $\Delta Y/\Delta Z = 1.9 \pm 0.5$ is a representative value for models and observations of irregular galaxies. Moreover, this value is in good agreement with the models and observed values of the disk of the Galaxy.

The Y_p values in Table 3 were computed by adopting $\Delta Y/\Delta Z = 1.9 \pm 0.5$. The differences between Tables 2 and 3 depend on T_e because the lower the T_e value the higher the Z value for the SMC.

6. Discussion

The Y_p value derived by us is significantly smaller than the value derived by Izotov & Thuan (1998) from the $Y - $ O/H linear regression for a sample of 45 BCGs, and by Izotov et al. (1999) from the average for the two most metal

Table 3. Y_p Derived from the SMC

$T_e(K)$	$N_e(cm^{-3})$				
	53	100	143	162	247
11200	0.2363	0.2348	0.2336	0.2331	**0.2309**[a]
11800	0.2373	0.2357	0.2343	**0.2337**	0.2313
11950	0.2376	0.2360	**0.2345**	0.2339	0.2314
12400	0.2384	**0.2366**	0.2351	0.2344	0.2317
13000	**0.2395**	0.2375	0.2359	0.2352	0.2322

[a] Boldface values correspond to minimum χ^2 values, see Table 1.

deficient galaxies known (I Zw 18 and SBS 0335-052), that amount to 0.2443 ± 0.0015 and 0.2452 ± 0.0015 respectively.

The difference could be due to systematic effects in the abundance determinations. There are two systematic effects not considered by Izotov and collaborators that we did take into account, the presence of H^0 inside the He^+ region and the use of a lower temperature than that provided by the [O III] lines. We consider the first effect to be a minor one and the second to be a mayor one but both should be estimated for each object.

From constant density chemicaly homogeneous models computed with CLOUDY we estimate that the maximum temperature that should be used to determine the helium abundance should be 5% smaller than T_e(O III). Moreover, if there is additional energy injected to the H II region T_e(He II) should be even smaller.

Luridiana, Peimbert, & Leitherer (1999) produced a detailed photoionized model of NGC 2363. For the slit used by Izotov, Thuan, & Lipovetsky (1997) they find an ICF(Hc) $= 0.993$; moreover they also find that the T_e(O III) predicted by the model is considerably smaller than observed. From the data of Izotov et al. (1997) for NGC 2363, adopting a T_e(He II) 10% smaller than T_e(O III) and $\Delta Y/\Delta Z = 1.9 \pm 0.5$ we find that $Y_p = 0.234 \pm 0.006$.

Similarly, Stasinska & Schaerer (1999) produced a detailed model of I Zw 18 and find that the photoionized model predicts a T_e(O III) value 15% smaller than observed, on the other hand their model predicts an ICF(He) $= 1.00$. From the observations of $\lambda\lambda$ 5876 and 6678 by Izotov et al. (1999) of I Zw 18, and adopting a T_e(He II) 10% smaller than T_e(O III) we obtain $Y_p = 0.237 \pm 0.007$; for a T_e(He II) 15% smaller than T_e(O III) we obtain $Y_p = 0.234 \pm 0.007$, both results in good agreement with our determination based on the SMC. Further discussion of these issues is presented elsewhere (Peimbert, Peimbert, & Luridiana 2000b).

The primordial helium abundance by mass of $0.2345 \pm 0.0030(1\sigma)$ — based on the SMC — combined with the computations by Copi, Schramm, & Turner (1995) for three light neutrino species implies that, at the 95 percent confidence level, $\Omega_b h^2$ is in the 0.0046 to 0.0103 range. For $h = 0.65$ the Y_p value corresponds to $0.011 < \Omega_b < 0.024$, a value considerably smaller than that derived from the pregalactic deuterium abundance, D_p, determined by Burles & Tytler (1998) that corresponds to $0.040 < \Omega_b < 0.050$ for $h = 0.65$, but in very good agreement with the observational estimate of the global budget of baryons by Fukugita, Hogan, & Peebles (1998) who find $0.007 < \Omega_b < 0.038$ for $h = 0.65$. The discrepancy between Y_p and D_p needs to be studied further.

To increase the accuracy of the Y_p determinations we need observations of very high quality of as many He I lines as possible to derive T_e(He II), N_e(He II), and $\tau(3889)$ self-consistently. We also need observations with high spatial resolution to estimate the ICF(He) along different lines of sight.

It is a pleasure to acknowledge several fruitful discussions on this subject with: L. Carigi, V. Luridiana, B. E. J. Pagel, M. T. Ruiz, E. Skillman, G. Steigman, S. Torres-Peimbert, and S. Viegas.

References

Armour, M. H., Ballantyne, D. R., Ferland, G. F., Karr, J., & Martin, P.G. 1999, PASP, 111, 1251
Benjamin, R. A., Skillman, E. D., & Smits, D. P. 1999, ApJ, 514, 307
Burles, S., & Tytler, D. 1998, ApJ, 507, 732
Carigi, L. 2000, ApJ, submitted
Carigi, L., Colín, P., Peimbert, M., & Sarmiento, A. 1995, ApJ, 445, 98
Carigi, L., Colín, P., & Peimbert, M. 1999, ApJ, 514, 787
Chiappini, C., Matteucci, F., & Gratton, R. 1997, ApJ, 477, 765
Copi, C. J., Schramm, D. N., & Turner, M. S. 1995, Science, 267, 192
Esteban, C., Peimbert, M., Torres-Peimbert, S., & Escalante, V. 1998, MNRAS, 295, 401
Esteban, C., Peimbert, M., Torres-Peimbert, S., & García-Rojas, J. 1999, RevMexAA, 35, 85
Ferland, G. J. 1996, *Hazy, a Brief Introduction to CLOUDY*, Univ. of Kentucky Dept. of Phys. & Astron. Internal Report
Fields, B. D., & Olive, K. A. 1998, ApJ, 506, 177
Fukugita, M., Hogan, C. J., & Peebles, P. J. E. 1998, ApJ, 503, 518
Izotov, Y. I., Chaffee, F. H., Foltz, C. B., Green, R. F., Guseva, N. G., & Thuan, T. X. 1999, ApJ, 527, 757
Izotov, Y. I., & Thuan, T. X. 1998, ApJ, 500, 188
Izotov, Y. I., Thuan, T. X., & Lipovetsky, V. A. 1994, ApJ, 435, 647
Izotov, Y. I., Thuan, T. X., & Lipovetsky, V. A. 1997, ApJS, 108, 1
Luridiana, V., Peimbert, M., & Leitherer, C. 1999, ApJ, 527, 110
Maeder, A. 1992, A&A, 264, 105
Massey, P., Parker, J. W., & Garmany, C. D. 1989, AJ, 98, 1305
Pagel, B. E. J., Simonson, E. A., Terlevich, R. J., & Edmonds, M. G. 1992, MNRAS, 255, 325
Peimbert, A., Peimbert, M., & Luridiana, V. 2000b, RevMexAA, submitted
Peimbert, M. 1967, ApJ, 150, 825

Peimbert, M. 1995, in *The Analysis of Emission Lines.*, ed. R.E. Williams & M. Livio (Cambridge University Press), p. 165
Peimbert, M., Peimbert, A., & Ruiz, M.T. 2000a, ApJ, submitted
Peimbert, M., & Torres-Peimbert, S. 1999, ApJ, 525 (Part 3), 1143
Peimbert, M., Torres-Peimbert, S., & Ruiz, M.T. 1992, RevMexAA, 24, 155
Robbins, R. R. 1968, ApJ, 151, 511
Shields, G. A. 1974, ApJ, 191, 309
Stasinska, G., & Schaerer, D. 1999, A&A, 351, 72
Vílchez, J. M., & Pagel, B. E. J. 1988, MNRAS, 231, 257
Woosley, S. E., Langer, N., & Weaver, T. A. 1993, ApJ, 411, 823
Woosley, S. E., & Weaver, T. A. 1995, ApJS, 101, 181

Some aspects of the chemical evolution of ^{4}He in the Galaxy: the He/H radial gradient and the $\Delta Y/\Delta Z$ enrichment ratio

W. J. Maciel

IAG/USP - Av. Miguel Stefano 4200 - CEP 04301-904 São Paulo SP, Brazil

Abstract. Two aspects of the chemical evolution of ^{4}He in the Galaxy are considered on the basis of a sample of disk planetary nebulae by the application of corrections due to the contamination of ^{4}He from the progenitor stars. First, the He/H radial gradient is analyzed, and then, the helium to heavy element enrichment ratio is determined for metallicities up to the solar value.

1. Introduction

The study of radial abundance gradients and of the enrichment ratio between helium and the heavy elements ($\Delta Y/\Delta Z$) can be performed on the basis of photoionized nebulae, comprising HII regions and planetary nebulae (PN, see for example the reviews by Peimbert 1990 and Maciel 1997). The former have the advantage of representing the interstellar composition directly, while for the latter the influence of the contamination from the progenitor stars should be taken into account, particularly for those elements that are dredged up to the outer layers of the progenitor stars, such as helium and nitrogen.

In the case of helium, the contamination from the evolution of the progenitor stars has been recently determined on the basis of different theoretical models (cf. van den Hoek & Groenewegen 1997, Boothroyd & Sackmann 1999), so that it is now possible to apply individual corrections to the observed PN abundances in order to obtain improved values of the He/H radial gradient and of the $\Delta Y/\Delta Z$ ratio. In the present work, these recent calculations are used in order to estimate the He contamination for a sample of disk planetary nebulae recently studied by Maciel & Quireza (1999). The application of corrections to the observed abundances leads to a new determination of the He/H radial gradient d(He/H)$/dR$ and of the slope $\Delta Y/\Delta Z$.

2. The He/H radial gradient and the $\Delta Y/\Delta Z$ ratio

2.1. The He/H radial gradient

The O/H radial gradient is now well known, and can be derived for HII regions, planetary nebulae, hot stars, etc. It amounts roughly to $d\log$(O/H)$/dR \simeq -0.07$ dex/kpc, and similar values have been obtained for other elements, such as S and Ar (Maciel 1997, Maciel & Quireza 1999). The existence of a He/H

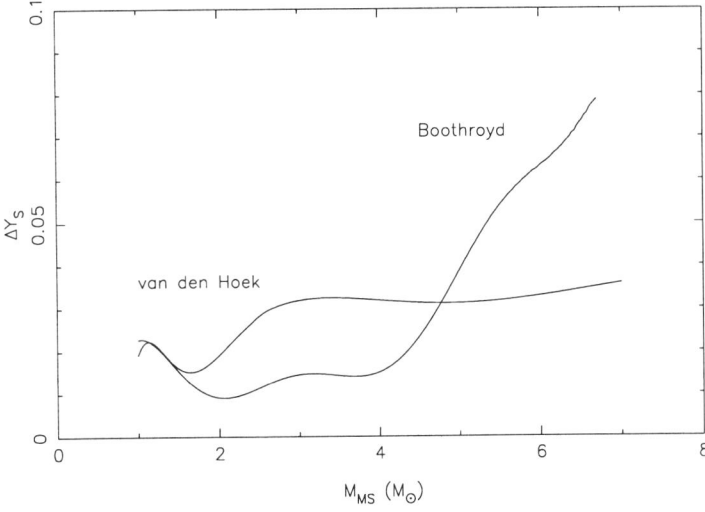

Figure 1. He contamination from PN progenitor stars.

gradient is much more uncertain, and results obtained both from HII regions and planetary nebulae show a very large uncertainty, as seen in Table 1 (see Esteban & Peimbert 1995 for a review).

Table 1. Determinations of the He/H radial gradient

Reference	PN/HII	$d(\text{He}/\text{H})/dR$ (kpc^{-1})	$d\log(\text{He}/\text{H})/dR$ (dex/kpc)
D'Odorico et al. (1976)	PN	−0.007	−0.03
Peimbert & Serrano (1980)	PN	−0.005	−0.02
Shaver et al. (1983)	HII	−0.0003	−0.001
Faúndez-Abans & Maciel (1986)	PN	−0.005	−0.02
Pasquali & Perinotto (1993)	PN	−0.003	−0.009
Amnuel (1993) [min]	PN	−0.0005	−0.002
Amnuel (1993) [max]	PN	−0.006	−0.026
Maciel & Chiappini (1994)	PN	−0.0004	−0.002
Esteban et al. (1999)	HII	−0.001	−0.004

All determinations involving PN until now have *not* taken into account the He contamination by the progenitor stars, so that it is interesting to revise the values of the He/H gradient by considering the amount of He produced and dredged up to the outer layers of the stars and eventually shed into the nebulae.

2.2. The $\Delta Y/\Delta Z$ enrichment ratio

The helium to metals enrichment ratio $\Delta Y/\Delta Z$ is generally determined adopting a linear variation for the helium abundance by mass with the metallicity Z of the form $Y(Z) = Y_p + (\Delta Y/\Delta Z) Z$, where Y_p is the pregalactic helium abundance, as proposed by Peimbert and Torres-Peimbert (1974, 1976). Photoionized nebulae such as HII regions and blue compact galaxies are generally used (cf. Izotov et al. 1997, Thuan & Izotov 1998, Esteban et al. 1999), and recent work has also taken into account the fine structure in the main sequence of nearby stars (Pagel & Portinari 1998). Results are generally in the range $2 \leq \Delta Y/\Delta Z \leq 6$, as shown in Table 2.

Table 2. Determinations of the $\Delta Y/\Delta Z$ ratio

Reference	object	$\Delta Y/\Delta Z$
D'Odorico et al. (1976)	PN	2.95
Peimbert & Serrano (1980)	PN	2.2 – 3.6
Maciel (1988)	PN	3.5 ± 0.3
Pagel et al. (1992)	HII	6.1 ± 2.1
Chiappini & Maciel (1994)	PN/HII	3.4 – 5.6
Pagel & Portinari (1998)	MS stars	3 ± 2
Thuan & Izotov (1998)	HII	2.3 ± 1.0
Esteban et al. (1999)	HII	1.9 – 3.9

Maciel (1988) determined Y_p and $\Delta Y/\Delta Z$ using a sample of type II PN (Peimbert 1978), and assuming that the He contamination from the central star was negligible. Chiappini & Maciel (1994) made a first attempt at including this contamination in a systematic way, and added a term ΔY_s to the $Y(Z)$ relation, which can be written as

$$Y = Y_p + \frac{\Delta Y}{\Delta Z} Z + \Delta Y_s \ . \qquad (1)$$

The stellar contribution ΔY_s was obtained on the basis of some calculations by Boothroyd (private communication). The adopted values were $\Delta Y_s = 0.0, 0.008, 0.015$ and 0.022, which were applied to all nebulae in the sample, so that any differences in their *individual* behaviour were lost.

3. He contamination from the PN progenitor stars

The He excess by mass ΔY_s can be estimated from the yields of the intermediate mass stars as a function of the stellar mass on the main sequence M_{MS} and total metallicity Z, adopting $Z = 0.020$, which is appropriate for type II PN. The stellar mass on the main sequence can be obtained by an initial mass-final mass relation as a function of the PN core mass M_c, which can in principle be determined from the observed nebular abundances, particularly the N/O ratio.

In this work, we have adopted the recent yields by van den Hoek & Groenewegen (1997), using as comparison the calculations by Boothroyd & Sackmann

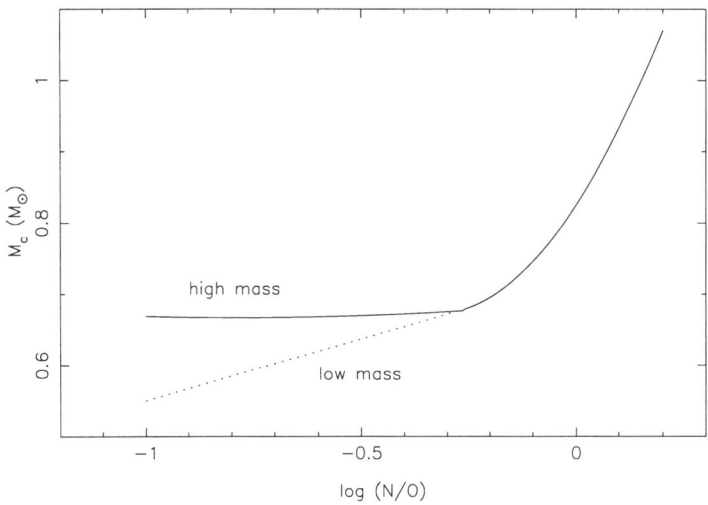

Figure 2. The adopted $M_c \times$ N/O calibrations.

(1999). The former includes all three dredge up processes that occur in the late stages of the intermediate mass stars, and generally produce larger yields than the latter, as can be seen in Figure 1. We have assumed that most of the He excess that are dredged up to the surface is mixed up in the outer layers and ejected into the nebulae, so that our derived ΔY_s is the largest possible correction for a given mass.

In view of the uncertainties on the PN central star masses, we have adopted two different calibrations, referred to as *high mass* and *low mass* calibrations, respectively. For the *high mass calibration*, we adopted the $M_c \times$ N/O relation recently proposed by Cazetta & Maciel (2000),

$$M_c = a + b \, \log(\text{N/O}) + c \, [\log(\text{N/O})]^2 \qquad (2)$$

where $a = 0.689$, $b = 0.056$ and $c = 0.036$ for $-1.2 \leq \log(\text{N/O}) \leq -0.26$ and $a = 0.825$, $b = 0.936$ and $c = 1.439$ for $\log(\text{N/O}) > -0.26$. The average initial mass–final mass relation was taken from the gravity distance work of Maciel & Cazetta (1997), and can be written as

$$M_c = a_0 + a_1 \, M_{MS} + a_2 \, M_{MS}^2 + a_3 \, M_{MS}^3 + a_4 \, M_{MS}^4 \qquad (3)$$

where $a_0 = 0.5426$, $a_1 = 0.02093$, $a_2 = -0.01122$, $a_3 = 0.00447$ and $a_4 = -0.0003119$. This calibration leads to core masses $M_c \geq 0.67 M_\odot$ and main sequence masses $M_{MS} \geq 3 \, M_\odot$, showing a good agreement with the results from NLTE model atmospheres of Méndez et al. (1988, 1992).

For the *low-mass calibration* we can still use equation 2, replacing the coefficients by $a = 0.7242$, $b = 0.1742$ and $c = 0$ for $\log(\text{N/O}) \leq -0.26$. For the initial mass–final mass relation we have the coefficients $a_0 = 0.4877$, $a_1 = 0.0623$, $a_2 = a_3 = a_4 = 0$. This calibration produces masses in the ranges $M_c \geq 0.55 \, M_\odot$ and $M_{MS} \geq 1 M_\odot$, which agree with the recent determinations of PN central star masses of Stasińska et al. (1997) and with the masses of type II PN originally proposed by Peimbert (1978). Both $M_c \times$ N/O calibrations are shown in figure 2, and the initial mass-final mass relation for the high-mass calibration can be seen in the figure 1 of Maciel & Cazetta (1997).

4. Results and discussion

We have applied the corrections outlined in the previous section to a sample of disk planetary nebulae in order to derive the He/H radial gradient and the $\Delta Y/\Delta Z$ enrichment ratio. For details on the objects and abundances, the reader is referred to Maciel & Quireza (1999) and Maciel (2000).

4.1. The He/H radial gradient

The average He/H abundances by number of atoms are shown in Table 3, using the van den Hoek & Groenewegen (1997) and Boothroyd & Sackmann (1999) data, both for the high- and low-mass calibrations.

Table 3. The He/H average abundances

	He/H
Uncorrected abundances	0.106 ± 0.003
Boothroyd, low mass calibration	0.100 ± 0.003
Boothroyd, high mass calibration	0.100 ± 0.003
van den Hoek, low mass calibration	0.094 ± 0.003
van den Hoek, high mass calibration	0.091 ± 0.003

The derived gradients are negligible, and can be written as $d(\text{He/H})/dR = 0.0000 \pm 0.0004$, irrespective of the calibration used, as shown for example in Figure 3 for the van den Hoek & Groenewegen (1997) data with the low mass calibration. We have adopted $R_0 = 7.6$ kpc as in Maciel & Quireza (1999).

It can be seen that the correction procedure simply reduces the average He/H abundances, so that stars with different masses are probably scattered homogeneously in the whole range of galactocentric distances, thus destroying any systematic variations. In view of the uncertainties involved in the abundances (and distances), it is unlikely that any He/H radial gradient could be presently detected from planetary nebulae, and previous determinations were probably affected by the use of small samples.

On the other hand, an upper limit to the He/H gradient can be derived on the basis of the total dispersion σ_d observed. Taking $\sigma_d \simeq 0.04$, we have $|d(\text{He/H})/dR| < \sigma_d/\Delta R \simeq 0.004$ kpc^{-1}, or $d\log(\text{He/H})/dR \simeq -0.02$ dex/kpc. Therefore, any existing He/H radial gradient should be lower than the O/H gradient by at least a factor of 3. This conclusion is in agreement with the small gradients derived for galactic HII regions by Esteban et al. (1999) and with some recent chemical evolution models for the Galaxy (Chiappini et al. 1997, Chiappini, this conference).

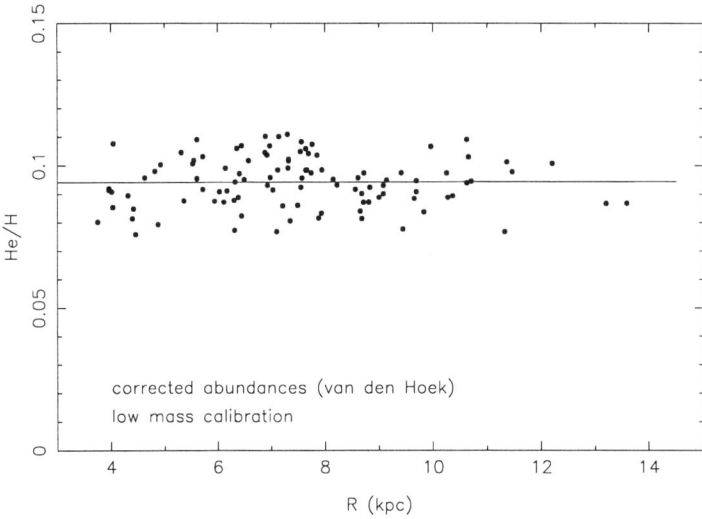

Figure 3. He/H abundances as a function of the galactocentric distance for the van den Hoek and Groenewegen (1997) data with the low mass calibration.

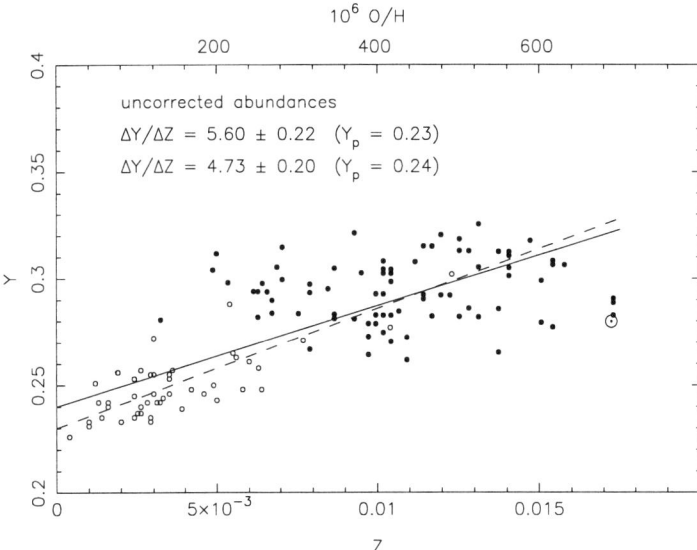

Figure 4. Uncorrected He abundances by mass Y for PN (solid circles) as a function of O/H (top axis) and Z (bottom axis). Also shown are the Sun ($\odot$) and HII regions (empty circles). The straight lines are least squares fits for $Y_p = 0.23$ (dashed line) and $Y_p = 0.24$ (solid line).

4.2. The $\Delta Y/\Delta Z$ enrichment ratio

In order to apply the correction procedure to the PN sample, we have considered the O/H abundances as representative of the total metallicity, adopting $Z \simeq 25$ O/H (Peimbert 1990, Chiappini & Maciel 1994). Since the pregalactic abundance can be better determined on the basis of very low metallicity objects, we have taken Y_p as a parameter, with the values $Y_p = 0.23$ and $Y_p = 0.24$ (see for example Olive et al. 1999, Izotov et al. 1999, Steigman, this conference).

We have taken into account all PN in our sample having metal abundances up to 10^6 O/H $\simeq 700$, which corresponds approximately to the solar value, $\epsilon(O) = \log(O/H) + 12 = 8.83$ (Grevesse & Sauval 1998), or $Z \simeq 0.017$. As pointed out by Chiappini & Maciel (1994), He abundances of PN show some tendency to flatten out for very large Z, where a more sophisticated relation than eq. (1) would be needed. Moreover, some large metallicity PN have larger than average N/O ratios, so that the corrections to the He abundance are also larger and more uncertain.

The main results are shown in Table 4 and in figures 4 and 5, where the PN are shown as filled circles. Figure 4 shows the uncorrected abundances and fits, and figure 5 shows the results using the corrections according to the van den Hoek and Groenewegen data both for the high- and low-mass calibrations. In both figures, the straight lines are least squares fits using $Y_p = 0.23$ (dashed lines) and $Y_p = 0.24$ (solid lines).

Table 4. The $\Delta Y/\Delta Z$ ratio

	$\Delta Y/\Delta Z$
$Y_p = 0.23$	
uncorrected	5.60 ± 0.22
Boothroyd low mass	4.42 ± 0.20
Boothroyd high mass	3.95 ± 0.19
van den Hoek low mass	3.59 ± 0.19
van den Hoek high mass	2.87 ± 0.17
$Y_p = 0.24$	
uncorrected	4.73 ± 0.20
Boothroyd low mass	3.55 ± 0.19
Boothroyd high mass	3.08 ± 0.17
van den Hoek low mass	2.73 ± 0.17
van den Hoek high mass	2.01 ± 0.16

For comparison purposes, figures 4 and 5 also include the Sun ($\odot$) and the HII regions and metal poor blue compact galaxies (empty circles) from the compilation of Chiappini & Maciel (1994). These objects have *not* been taken into account in the determination of the linear fits, and are included for comparison only.

It can be seen that the scatter in the PN data is somewhat higher, which is partially due to the higher uncertainties in the abundances and also to the

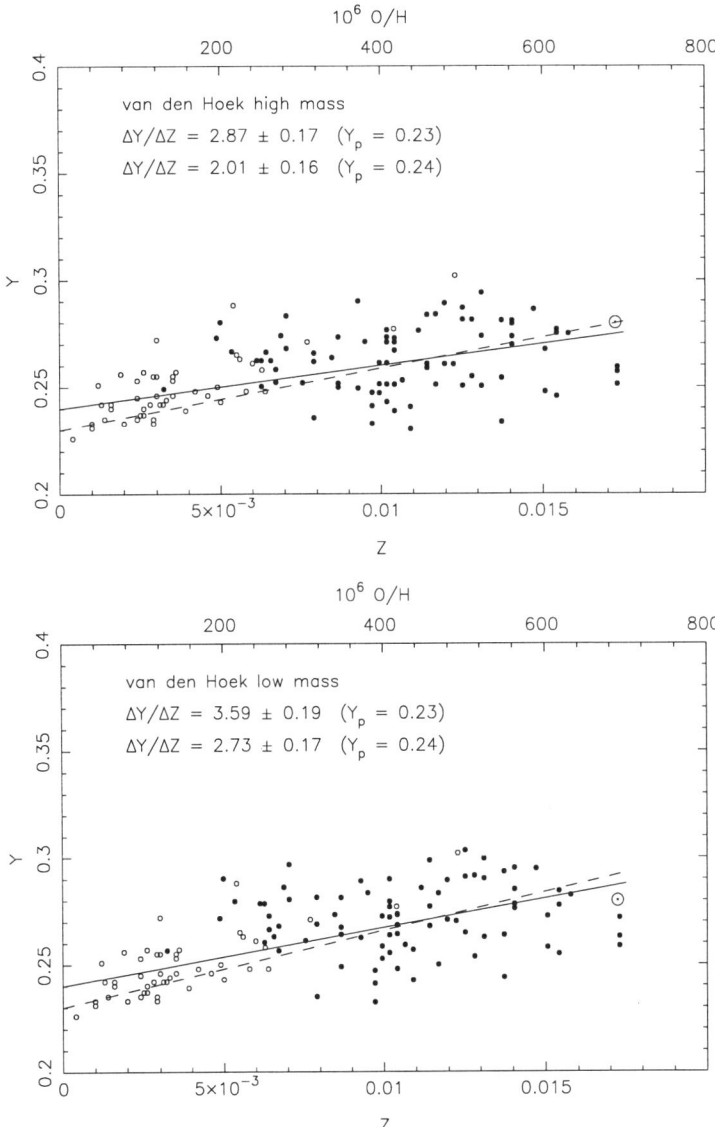

Figure 5. The same as figure 4 using corrected abundances according to the van den Hoek & Groenewegen (1997) data for the high mass calibration (top panel) and low mass calibration (bottom panel).

adopted correction procedure, so that future improvements along these lines would be desirable. The higher uncertainties in the PN abundances and the lower He/H gradient are probably responsible for the lack of a He/H $\times$ R correlation, while some correlation between Y and Z (or O/H) can be observed.

The application of the correction procedure reduces the average He abundances and the $\Delta Y/\Delta Z$ ratio, and also decreases the uncertainty of the derived slopes. The $\Delta Y/\Delta Z$ ratio decreases from the range 4.7 – 5.6 to values in the range 2.8 – 3.6 for $Y_p = 0.23$ and 2.0 – 2.8 for $Y_p = 0.24$. These results are closer to recent independently derived ratios, as seen in Table 2 (see also Thuan, this conference), and to the predictions of theoretical models (Allen et al. 1998, Chiappini et al. 1997), particularly for the high mass calibration using the van den Hoek & Groenewegen (1997) data.

The average uncertainties in the derived slopes are smaller as a consequence of the fact that a large number of objects has been included, with a larger metallicity spread. An average including both calibrations would give $\Delta Y/\Delta Z = 3.2 \pm 0.5$ for $Y_p = 0.23$ and $\Delta Y/\Delta Z = 2.4 \pm 0.5$ for $Y_p = 0.24$, which can be compared with the uncertainties shown in Table 2.

Finally, the apparent continuity between the low metallicity and high metallicity objects of figure 5 suggests that their chemical evolution may not have been very different. In fact, the chemical evolution of a system is basically defined by its initial mass function (IMF) and star formation history (SFH). The IMF is now believed to be universal (Maciel & Rocha-Pinto 1998, Padoan et al. 1997). Blue compact galaxies have bursts of star formation, a feature that has been recently reinforced in our own Galaxy, as shown by Rocha-Pinto et al. (2000) on the basis of chromospheric ages. Therefore, the similarity in the chemical evolution of the different systems shown in figure 5 is probably not surprising.

Acknowledgements. This work was partially supported by CNPq and FAPESP.

References

Allen, C., Carigi, L., & Peimbert, M. 1998, ApJ, 494, 247

Amnuel, P.R. 1993, MNRAS, 261, 263

Boothroyd, A.I., & Sackmann, I.-J. 1999, ApJ, 510, 232

Cazetta, J.O., & Maciel, W.J. 2000, Rev. Mex. Astron. Astrofis., in press

Chiappini, C., Maciel, W.J., 1994 A&A, 288, 921

Chiappini, C., Matteucci, F., & Gratton, R. 1997, ApJ, 477, 765

D'Odorico, S., Peimbert, M., & Sabbadin, F. 1976, A&A, 47, 341

Esteban, C., & Peimbert, M. 1995, Rev. Mex. Astron. Astrofis. SC, 3, 133

Esteban, C., Peimbert, M., Torres-Peimbert, S., & Garcia-Rojas, J. 1999, Rev. Mex. Astron. Astrofis., 35, 65

Faúndez-Abans, M., & Maciel, W.J. 1986, A&A, 158, 228

Grevesse, N., & Sauval, A.J. 1998, Space Sci. Rev., 85, 161

Izotov, Y.I., Thuan, T.X., & Lipovetsky, V.A. 1997, ApJS, 108, 1

Izotov, Y.I., Chaffee, F.H., Foltz, C.B., Green, R.F., Guseva, N.G., & Thuan, T.X. 1999, ApJ, in press

Maciel, W.J. 1988, A&A, 200, 178
Maciel, W.J. 1997, in IAU Symp. 180, ed. H.J. Habing & H.J.G.L.M. Lamers (Dordrecht: Kluwer), 397
Maciel, W.J. 2000, A&A, submitted
Maciel, W.J., & Cazetta, J.O. 1997, Ap&SS, 249, 341
Maciel, W.J., & Chiappini, C. 1994, Ap&SS, 219, 231
Maciel, W.J., & Quireza, C. 1999, A&A, 345, 629
Maciel, W.J., & Rocha-Pinto, H.J. 1998, MNRAS, 299, 889
Méndez, R.H., Kudritzki, R.P., Herrero, A., Husfeld, D., & Groth, H.G. 1988, A&A, 190, 113
Méndez, R.H., Kudritzki, R.P., & Herrero, A. 1992, A&A, 260, 329
Olive, K.A., Steigman, G., & Walker, T.P. 1999, Phys. Rep., in press
Padoan, P., Nordlund, Å., & Jones, B.J.T. 1997, MNRAS, 288, 145
Pagel, B.E.J., & Portinari, L. 1998, MNRAS, 298, 747
Pagel, B.E.J., Simonson, E.A., Terlevich, R.J., & Edmunds, M.C. 1992, MNRAS, 255, 325
Pasquali, A., & Perinotto, M. 1993, A&A, 280, 581
Peimbert, M. 1978, in IAU Symp. 76, ed. Y. Terzian (Dordrecht: Reidel), 215
Peimbert, M. 1990, Rep. Prog. Phys., 53, 1559
Peimbert, M., & Serrano, A. 1980, Rev. Mex. Astron. Astrofis., 5, 9
Peimbert, M., & Torres-Peimbert, S. 1974, ApJ, 193, 327
Peimbert, M., & Torres-Peimbert, S. 1976, ApJ, 203, 581
Shaver, P.A., McGee, R.X., Newton, L.M., Danks, A.C., & Pottasch, S.R. 1983, MNRAS, 204, 53
Rocha-Pinto, H.J., Scalo, J., Maciel, W.J., & Flynn, C. 2000, ApJ, in press
Stasińska, G., Gorny, S.K., & Tylenda, R. 1997, A&A, 327, 736
Thuan, T.X., & Izotov, Y.I. 1998, Space Sci. Rev., 84, 83
van den Hoek, L.B., & Groenewegen, M.A.T. 1997, A&AS, 123, 305

The Primordial 3-Helium Abundance At Last?

T. M. Bania

Institute for Astrophysical Research, Boston University, Boston, MA

Robert T. Rood

Department of Astronomy, University of Virginia, Charlottesville, VA

Dana S. Balser

National Radio Astronomy Observatory, Green Bank, WV

Abstract. We summarize the past 17 years of our efforts to determine the cosmic abundance of the ^{3}He isotope. The vast majority of our ^{3}He$^+$ observations were made with the NRAO 140 Foot telescope in Green Bank, WV. The 140 Foot ceased operations in July 1999 so that NRAO could prepare to commission its replacement, the Green Bank Telescope (GBT). Our ^{3}He experiment was the last scientific program at the 140 Foot. It is thus poignant and timely for us to reassess the astrophysical context of our ^{3}He results. Here we argue that the existence of "The ^{3}He Plateau" for our sample of simple sources and recent advances in the understanding of the evolution of solar analog stars together suggest that we can finally estimate the primordial abundance of ^{3}He. Our current best estimate for the primordial abundance is ^{3}He/H $= (1.5 \pm 0.6) \times 10^{-5}$ (s.e.) by number.

1. Introduction

3-Helium is one of the isotopes produced in the Big Bang. Standard models for stellar evolution predict that it is also produced, perhaps in prodigious amounts, by Solar-type stars. Thus measurements of the ^{3}He abundance are of interest both for cosmology and the chemical evolution of the Galaxy. For almost two decades we have been determining the ^{3}He abundance using measurements of the 3.5 cm hyperfine transition of ^{3}He$^+$. Our 66 source observational sample includes evolved stars (planetary nebulae) and star forming (H II) regions which are located throughout the Milky Way's gaseous disk, from the Galactic Center to the far reaches of the outer Galaxy. Although the observations are technically quite challenging, we have detected ^{3}He over a larger fraction of the Galaxy than any other light isotope. We are finding that Milky Way H II regions are not ^{3}He enriched and, furthermore, that the distribution of ^{3}He in the Galaxy cannot be easily explained by chemical evolution models. One thing is clear: real stars haven't heard about standard stellar models—there is no evidence that Solar-type stars are enriching the Galaxy in ^{3}He. On the other hand we

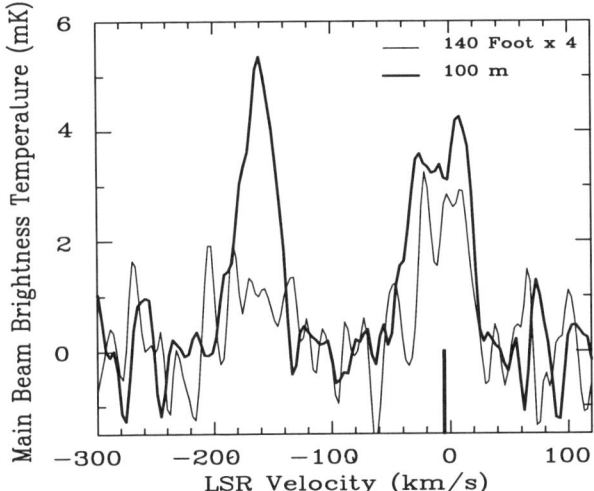

Figure 1. Confirmation of the detection of ^{3}He$^+$ emission from the planetary nebula NGC 3242. The thick curve is the spectrum taken with the MPIfR 100 meter telescope. The thin curve is the spectrum taken with the NRAO 140 foot telescope with the observed intensity multiplied by a factor of 4. The vertical line at -5.3 kms^{-1} flags the ^{3}He$^+$ emission at the LSR velocity of NGC 3242. The prominent feature in the MPIfR spectrum is the H171η recombination line.

find lots of ^{3}He in some planetary nebulae—the ejecta of Solar-type stars. This seemingly contradictory situation has led to "The ^{3}He Problem" (Galli et al. 1997). Here we review the problem as a whole and reflect upon how cosmological interpretations of light element abundances are made so very much more difficult when one is burdened by a surfeit of sources.

2. The 3-Helium Experiment

We derive the ^{3}He abundance in Milky Way sources using measurements of the hyperfine transition of ^{3}He$^+$ at 3.46 cm wavelength together with numerical models for the density and ionization structure of each source (see, e.g., Balser et al. 1997 [BBRW97], 1999 [BBRW99]; Bania et al. [BBRWW97]). Because the Milky Way interstellar medium (ISM) is optically thin at centimeter wavelengths, our source sample probes a larger volume of the Galactic disk than does any other light element tracer of Galactic chemical evolution. The source sample is currently comprised of 6 planetary nebulae (PNe) and 60 H II regions. For technical reasons the bulk of the PNe observations were made using the 100 m telescope of the MPIfR in Effelsberg, Germany, whereas the majority of the H II observations were made using the 140 Foot telescope of the NRAO in Green Bank, WV, USA.

2.1. Planetary Nebulae

Standard models for the evolution of solar analog stars predict (1) that these objects should produce copious amounts of ^{3}He during their main sequence lifetimes, (2) that the ^{3}He is mixed to the surface during the first dredge-up low on the red giant branch, and (3) that ^{3}He rich material flows into the ISM in stellar winds along the first and second red giant branches and in the final planetary nebula. The subsequent ^{3}He enrichment of the ISM should be significant. If we are to understand the chemical evolution of the Galaxy, it is important to pin down definitively the contribution of PNe to the ^{3}He abundance. Galli et al. (1997) show how the PNe ^{3}He abundances and estimates of the PN progenitor masses together provide a direct test of stellar evolution theory.

It is technically quite challenging to detect ^{3}He$^+$ emission from PNe even though the theoretically expected abundances, ^{3}He/H $\sim$ 1–10 $\times$ 10^{-4}, are a factor of ten or more larger than that measured for Galactic H II regions. Unfortunately typical PNe angular sizes of $\sim 20''$ are much smaller than the beamsizes of even the largest centimeter wave telescopes, so the ^{3}He$^+$ emission from these objects is weakened by geometrical dilution. For the MPIfR 100 m telescope the typical dilution factor is ~ 12 and for the NRAO 140 Foot it is ~ 92.

Because of this geometrical dilution we used the MPIfR 100 m telescope to search for ^{3}He$^+$ emission from PNe (Rood, Bania, & Wilson 1992; BBRW97). Even so, we were working at the sensitivity limit of the 100 m telescope. Our survey sample of six PNe was therefore chosen to maximize the likelihood of ^{3}He$^+$ detections, and so it was highly biased. For example, we chose sources that were ~ 500 pc from the Galactic plane. This criterion increased the likelihood that a given PN was part of the old/thick disk or halo population. On the average, such PNe would have low progenitor masses, longer main-sequence lifetimes, and, thus, presumably higher ^{3}He abundances. Furthermore, our PNe sample showed no evidence in the nebular gas of nuclear processing on the RGB or AGB. This excluded objects where the gas had undergone CNO processing since ^{3}He would have also been depleted in such objects. Specifically, our PNe sample had low He, N, and ^{13}C abundances.

This PNe survey detected ^{3}He emission in NGC 3242 and probably also detected it in NGC 6543 and NGC 7009. These detections showed that there is *some* stellar production of ^{3}He. The abundances derived for the detected sources are consistent with the idea that ^{3}He is produced in significant quantities by at least some stars with masses in the range 1–2 $M_\odot$. Finally, in BBRW97 we argued that the ^{3}He$^+$ emission was just below the detection threshold for many of the sources.

Because we were working at the sensitivity limit of the 100 m telescope it was important that we confirm these results. Since the NRAO 140 Foot telescope had an additional factor of ten geometrical dilution above that of the 100 m, we concentrated our efforts on the PN NGC 3242. Figure 1 shows that after a 270 hour integration we have confirmed the presence of ^{3}He$^+$ emission from the PN NGC 3242 (Balser, Rood, & Bania 1999). Both spectrometers show that the observed ^{3}He$^+$ hyperfine emission has a double-peaked line profile structure. This line shape naturally arises if NGC 3242 is modeled as a dense core surrounded by a very low density expanding shell. After modeling this source we conclude that a ^{3}He/H abundance ratio by number of (2–5) $\times$ 10^{-4} is

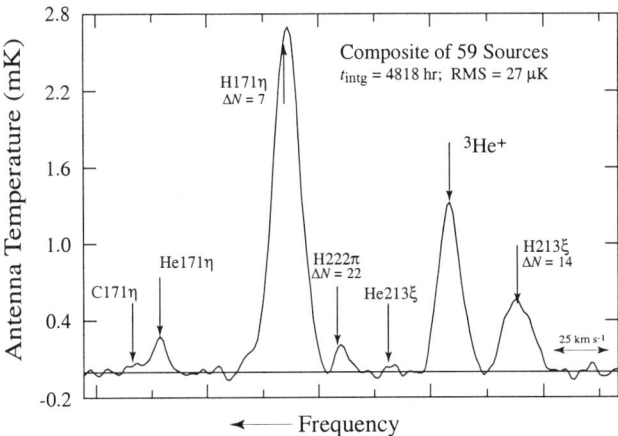

Figure 2. The grand average of ^{3}He observations obtained with the Green Bank 140 Foot Telescope from 1982 through 1999. The spectrum is a composite of 59 Galactic H II regions. The observed RMS noise after almost 5000 hours of integration is just as predicted by the radiometer equation. This shows that we are correctly dealing with possible sources of systematic error like spectral baseline removal.

consistent with both observed spectra. This result confirms stellar production of ^{3}He by some yet to be determined fraction of the Galactic PN population.

2.2. H II Regions

Because H II regions are zero-age objects when compared to the age of the Galaxy, the elemental abundances of these nebulae chronicle the results of billions of years Galactic chemical evolution. Measurement of the present ^{3}He abundance should therefore be an important diagnostic of chemical evolution in the Galaxy. Standard stellar ^{3}He nucleosynthesis predicts that the ^{3}He/H abundance ratio should grow with time and be higher in those parts of the Galaxy where there has been substantial stellar processing. Specifically: (1) the protosolar value should be less than that found in the present ISM; and (2) there should be a ^{3}He/H abundance gradient across the Galactic disk with the highest abundances occurring in the highly-processed inner Galaxy.

We have now accumulated a sample of 60 H II regions that sample the entire Galactic disk, from the Galactic Center to the far reaches of the outer Galaxy. During the past few years at Green Bank we have dramatically increased the number of H II regions with ^{3}He detections due to a combination of technical improvements and better astrophysical insight in our selection of candidate sources. We have benefited from the new generation of high sensitivity receivers constructed by the NRAO for the GBT. We have also learned something about source selection: in the past few years nearly all of our candidate H II regions have yielded ^{3}He detections.

The selection of ^{3}He targets is counter-intuitive because the ^{3}He$^+$ hyperfine line strength is proportional to the source density, while one usually thinks of H II regions in terms of radio continuum or recombination line strength both of

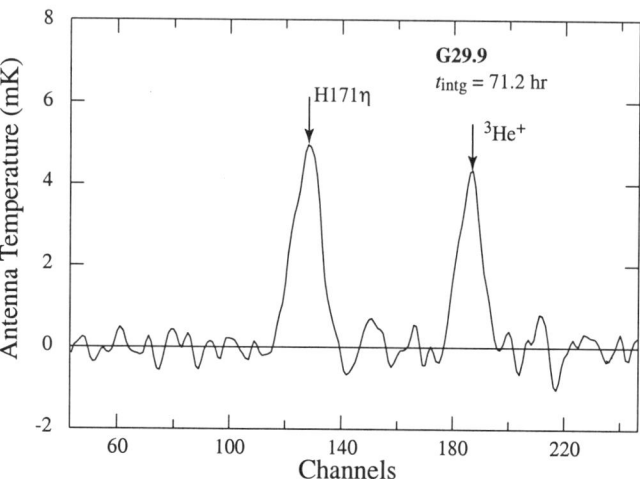

Figure 3. ^{3}He$^+$ spectrum of the H II region G29.9. The spectrum has a velocity resolution of 8.1 km s^{-1}. Vertical arrows flag the expected positions of the detected lines. Although relatively obscure as H II regions go, G29.9 has one of the strongest ^{3}He$^+$ lines we have observed.

which depend on the square of the density. The ^{3}He$^+$ line strength depends on the ^{3}He$^+$ abundance ratio and a number of other factors:

$$T_L^A(^3\text{He}^+) \propto \frac{N(^3\text{He}^+)}{N(\text{H}^+)} \frac{(T_C^A D)^{1/2} T_e^{1/4} (\theta_{\text{obs}}^2 - \theta_{\text{beam}}^2)^{3/4}}{\Delta v(^3\text{He}^+)[\ln(5.717 \times 10^{-3} T_e^{3/2})]^{1/2} \theta_{\text{obs}}}$$

where T_L^A and Δv are the antenna temperature and FWHM of the ^{3}He$^+$ line, D is the nebular distance, T_C^A and θ_{obs} are the antenna temperature and observed FWHM angular size of the continuum emission, θ_{beam} is the telescope's FWHM beam, and T_e is the nebular electron temperature. For H II regions much larger than the telescope beam we can select targets using the criterion: $T_L^A(^3\text{He}^+) \sim \sqrt{T_C^A D \theta_{\text{obs}}}$. This is the case since we can neglect the weak dependence on T_e and because we do not know either ^{3}He$^+$/H$^+$ or Δv. Thus big, distant H II regions could be potential ^{3}He$^+$ targets even if their continuum emission is weak.

Armed with this knowledge we included H II regions like S209 in our early observing list along with more famous sources like W43. Still we did not have the temerity to push this reasoning to the limit. We have now found, however, that this selection criterion is valid for even the wimpiest known H II regions. It is this realization that has led to the dramatic increase in the number of ^{3}He$^+$ sources discovered at Green Bank during the last several years. As expected, these are all large, low density, diffuse H II regions.

2.3. The 3-Helium Plateau

We have found source-to-source variations in the ^{3}He/H abundance of at least a factor of five for Galactic H II regions (e.g. BBRW99). The observed abundance

The Primordial ^{3}He Abundance At Last?

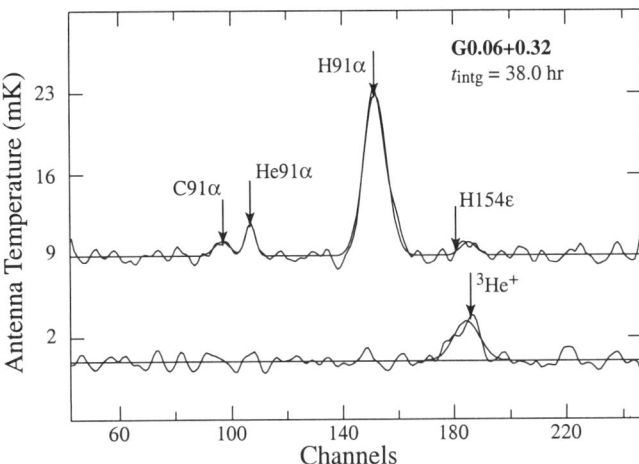

Figure 4. Spectrum of the H II region G0.06+0.32. This is an example of the large, diffuse sources detected during the past few years with the NRAO 140 Foot telescope. The top figure is the H91α spectrum; the bottom figure is the ^{3}He$^+$ spectrum. The expected centers of detected transitions are flagged. The H91α line is only 4 times stronger than the ^{3}He$^+$ line; in tradional H II regions like W43 the H91α line is several hundred times stronger than the ^{3}He$^+$ line. Even though G0.06+0.32 is roughly in the direction of the Galactic center, it is actually 11.5 kpc *on the other side of the center.*

pattern, however, cannot be easily explained by existing chemical evolution models. There is no evidence for a gradient in the Galaxy. Nor do the H II regions show any evidence for systematic stellar ^{3}He enrichment during the last 4.5 Gyr. Indeed, we have made little progress in understanding how the few H II regions with high ^{3}He got that way.

Our goal is to derive the most accurate possible ^{3}He/H abundances, developing in the process a theoretical framework that will allow us to understand quantitatively the inherent systematic errors. As is the case for any cosmic abundance determination, converting the measured ^{3}He$^+$ column density into an abundance ratio relative to hydrogen, ^{3}He/H, is nontrivial. For the case of ^{3}He this conversion depends on the density and ionization structure of each nebula. In BBRW99 we developed numerical models for our ^{3}He$^+$ sources which account for density structure. We discovered that a major advantage of diffuse H II regions is that they tend to be "simple," i.e., the H II region has a relatively simple density structure. This allows us to derive abundances from the observed ^{3}He$^+$ line parameters without recourse to the rather complex modeling required for most classic H II regions (see BBRW99). Because the abundance corrections for simple sources are negligible, these sources can in principle yield the most accurate ^{3}He/H abundance ratios attainable.

At present our H II region sample has 19 sources that are "simple" in the sense just described. As was the case for the full H II region sample, our sample of simple sources shows no evidence for an abundance gradient in the Galaxy, no evidence for any stellar ^{3}He enrichment during the last 4.5 Gyr, and no trend

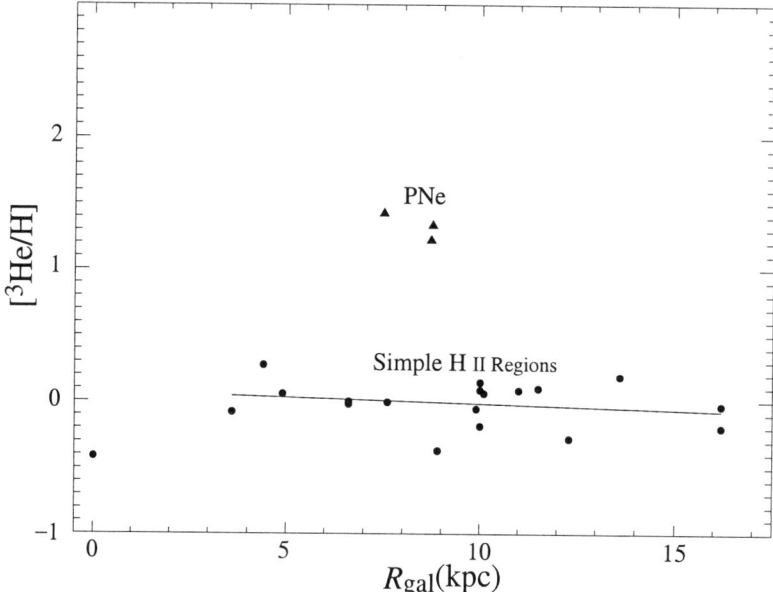

Figure 5. [^{3}He/H] (= $\log(^3\mathrm{He/H}) - \log(^3\mathrm{He/H})_\odot$) abundances plotted as a function of galactocentric radius for the sample of "simple" Galactic H II regions (see text). Also shown are the abundances derived for the planetary nebulae NGC 3242, NGC 6543, and NGC 7009. The best linear fit to the data is shown. The solar abundance $(^3\mathrm{He/H})_\odot$ was taken to be 1.5×10^{-5}. The point near $R_{\mathrm{gal}} = 0$ is the H II region G1.1. It is not included in the averages or fits because the Galactic center region is thought to have a different chemical enrichment history than the disk.

of abundance with metallicity. (We can derive an [O/H] abundance from our nebular models using the relationship between nebular electron temperature and [O/H] derived by Shaver et al. 1983.)

But this result *is* a trend: *simple ^{3}He sources show absolutely no abundance gradient with respect to either galactocentric radius or metallicity.* The dispersion of ^{3}He abundances, moreover, is very small—comparable to that of the famous "Lithium Plateau". The "^{3}He Plateau" defined by our sample of simple sources has an average ^{3}He$^+$/H$^+$ abundance ratio by number of $(1.37 \pm 0.54) \times 10^{-5}$ (s.e.) by number.

Here we make a preliminary ionization correction to these abundances using the observed source ^{4}He$^+$/H$^+$ abundance ratio, y_4, measured via the H and He 91α recombination transitions. Since we have never found ^{4}He^{++} emission in any of our H II regions, we assume that y_4 is a reasonably accurate estimate of a source's helium ionization state. To correct for source ionization we demand that $y_4 = 0.10$. That is, for sources with $y_4 < 0.10$, we scale the ^{3}He$^+$/H$^+$ abundance by the ratio $0.10/y_4$, otherwise we set ^{3}He/H = ^{3}He$^+$/H$^+$. Nine of the simple sources need to be ionization corrected. The "^{3}He Plateau" defined by our ionization corrected sample of simple sources has an average ^{3}He/H abundance ratio of $(1.50 \pm 0.58) \times 10^{-5}$ (s.e.) by number. This value and the absence of a gradient are completely consistent with the proto-solar system and local ISM values of ^{3}He (Gloeckner 2000).

3. Astrophysical Implications: "The 3-Helium Problem"

We have shown that at least *some* low-mass stars have produced significant amounts of ^{3}He$^+$. Yet our observations of ^{3}He in zero-age H II regions imply that essentially no stellar enrichment has occurred over the last 4.5 Gyr (BB-BRW99). Chemical evolution models of the Galaxy indicate that to account for the observed abundances in H II regions, no more than 15 or 20% of low-mass stars can produce ^{3}He like the progenitor of NGC 3242 did (Galli et al. 1997).

3.1. Stellar Production of 3-Helium

As early as 1984 (Rood, Bania, & Wilson 1984, §V.c) we suggested that the low abundances of ^{3}He which we were just then finding indicated a breakdown in standard stellar models. We also pointed out that there might be some connection with observed Li abundances and CNO anomalies. In the intervening years the case for low abundances became stronger and chemical evolution models became far more sophisticated. Chemical evolution models that adopted standard stellar ^{3}He nucleosynthesis overproduced ^{3}He. All Galactic evolution models which match the other observational constraints (e.g., star formation rate, gas and total mass density, mass infall rate, IMF, etc.) predict ^{3}He abundances that are inconsistent with those observed both locally and globally in the Milky Way unless they adopt alternative nucleosyntheses with a strongly reduced ^{3}He contribution from low- and intermediate-mass stars (Tosi 2000).

In the past few years some stellar models with non-standard assumptions have been shown to be ^{3}He non-producers. In particular, Charbonnel, et al. (1995, 1998, 1998a, 1998b) investigate rotationally driven mixing processes in an effort to explain various observed abundance anomalies. They show that while observed ^{12}C/^{13}C ratios agree well with standard models for stars less luminous than the red giant luminosity function bump, the ratios above the bump drop indicating non-canonical mixing. They attribute this continued mixing to rotationally driven turbulence. Their models nicely account for a number of observables like the ^{12}C/^{13}C ratios in red giants in open clusters and with Hipparchos distances. Stars with low ^{12}C/^{13}C ratios would have a correspondingly low ^{3}He. A few stars, presumably slow rotators, preserve a high value for ^{12}C/^{13}C. In these the ^{3}He would also survive allowing for high ^{3}He in a few PNe. The fraction of stars with high ^{12}C/^{13}C is approximately that allowed by the ^{3}He constraint on chemical evolution models. The Charbonnel models illustrate a subtle but important point: the stars with low ^{12}C/^{13}C have indeed destroyed most of the ^{3}He *produced on the main sequence*. Still the resulting ^{3}He is comparable to or higher than the initial value. As far a chemical evolution goes such stars are *non-producers* rather than *destroyers* of ^{3}He.

Sackmann & Boothroyd (1999 and references therein) have also explored non-standard mixing processes. Their parametric approach is not as tightly tied to a physical mechanism as is that of Charbonnel et al. It does not appear to be as tightly constrained by observations. They do, however, find a part of parameter space in which stars do indeed destroy ^{3}He. If operative in all stars this mechanism could lead to net destruction of ^{3}He. Additional observations of ^{12}C/^{13}C and other elements are crucial to delineate the character and statistics of these non-standard mixing processes.

3.2. The Primordial 3-Helium Abundance

The existence of the ^{3}He Plateau suggests that there is a floor to the ^{3}He/H abundance. If, as seems plausible, stellar processing enriches the ISM ^{3}He abundance by a positive-definite (but perhaps nearly zero) amount, this floor value provides an upper limit to the primordial ^{3}He abundance produced by Big Bang Nucleosynthesis. That is, BBNS must not overproduce ^{3}He beyond the floor abundance of ^{3}He/H $= (1.5 \pm 0.6) \times 10^{-5}$ (s.e.) by number.

4. Summary and Future Prospects

During the past 20 years much progress has been made in understanding the Galactic abundance of ^{3}He and its temporal evolution. Although there are still outstanding issues regarding "The ^{3}He Problem," recent progress in non-standard stellar mixing has provided new insight into the chemical evolution of ^{3}He. When combined with the existence of "The ^{3}He Plateau" for our sample of "simple" Galactic H II regions, we now believe that we can estimate the primordial ^{3}He abundance.

There are still many more H II regions that can be studied. We were still detecting new ^{3}He sources with the NRAO 140 Foot telescope until the bitter end at 08:12 (EDT) 19 July, 1999. Indeed, we have already measured the ^{3}He$^+$ line in many H II regions not included in Figure 5. It is the lack of adequate continuum measurements and even simple source modeling that prevent us from giving abundances now. The GBT and the newly upgraded Arecibo telescope will greatly enhance the sensitivity of the ^{3}He experiment. We are especially excited by the potential of the GBT as its unblocked-aperture design is optimal for the ^{3}He experiment.

Ours in the only project in the world making a systematic study of ^{3}He throughout the Galaxy. While our project has already continued for many years, we can continue to improve our understanding of the origin and evolution of ^{3}He by: (1) observing a larger sample of PNe to reduce the observational bias in our current sample; (2) improving the accuracy of our current PNe sample; (3) observing new H II regions chosen to test specific questions which have risen from current results; (4) obtaining improved spectra for a few crucial H II regions; (5) refining the emission line models of detected ^{3}He$^+$ sources; (6) making further improvements to the ^{3}He abundance determinations by constructing more sophisticated source models; and (7) searching for ^{3}He$^+$ emission from extra-galactic H II regions.

After spending so long on a project it is amusing to reflect on how it may have come out differently. If we were starting the project now and applying for telescope time on a highly oversubscribed telescope, like the GBT, we might be forced to submit the minimum proposal to prove our point. We likely would propose observations of W43 and W51 in the inner Galaxy and the remote outer Galaxy H II region S209. We would make easy detections and find a somewhat lower ^{3}He/H ratio than expected; there would be a gradient in ^{3}He consistent with our naive expectations. We would never get any more telescope time, because the TAC would respond "Nothing to be learned; standard model confirmed." Unfortunately the early 1980's Green Bank TACs were generous. We observed too many H II regions in the wrong order, and the situation was very confusing from our first observing run. We have learned many things since 1982: (1) too much data can be confusing; (2) The Green Bank kitchen staff can be enticed to cook spaghetti carbonara (but the raw egg part causes panic); (3+) no more space; deferred to a future paper.

Acknowledgments. We thank Tom Wilson for his contributions to the ^{3}He project over the years and for making this paper necessary. The research was supported by a NATO Travel Grant and the U.S. National Science Foundation (AST 97-31484).

References

Balser, D. S., Bania, T. M., Rood, R. T., & Wilson, T. L. 1997, ApJ, 483, 320 (BBRW97)

Balser, D. S., Bania, T. M., Rood, R. T., & Wilson, T. L. 1999, ApJ, 510, 759 (BBRW99)

Balser, D. S., Rood, R. T., & Bania, T. M. 1999, ApJ, 522, L73

Bania, T. M., Balser, D. S., Rood, R. T., Wilson, T. L., & Wilson, T. A. 1997, ApJS, 113, 353 (BBRWW97)

Charbonnel, C. 1995, ApJ, 453, L41

Charbonnel, C. 1998, Space Science Reviews, 84, 199

Charbonnel, C., Brown, J. A., Wallerstein, G. 1998a, A&A, 332, 204

Charbonnel, C., & do Nascimento Jr, J. D. 1998b, A&A, 336, 915

Galli, D., Stanghellini, L., Tosi, M., Palla, F. 1997, ApJ, 477, 218

Gloeckler, G. 2000, in this volume

Rood, R. T., Bania, T. M., and Wilson, T. L. 1984, ApJ, 280, 629

Rood, R. T., Bania, T. M., & Wilson, T. L. 1992, Nature, 355, 618

Rood, R. T., Bania, T. M., Balser, D. S., & Wilson, T. L. 1998, Space Science Reviews, 84, 185

Rood, R. T., Steigman, G., & Tinsley, B. M. 1976, ApJ, 207, L57

Sackmann, I.-J. & Boothroyd, A. I. 1999, ApJ, 510, 217

Shaver, P. A., McGee, R. X., Newton, L. M., Danks, A. C., & Pottasch, S. R. 1983, MNRAS, 204, 53

Tosi, M. 2000, in this volume

The Light Elements and Their Evolution
IAU Symposium, Vol. 198, 2000
L. da Silva, M. Spite, J. R. de Medeiros, eds.

Deuterium and Helium-3 in the Protosolar Cloud

George Gloeckler

Department of Physics and IPST, University of Maryland, College Park, Maryland 20742, and Department of Atmospheric, Oceanic and Space Sciences, University of Michigan, Ann Arbor, Michigan 48109, USA

Johannes Geiss

International Space Science Institute, Hallerstrasse 6, CH-3012 Bern, Switzerland

Abstract. New measurements of the isotopic composition of helium in the solar wind obtained from the Solar Wind Ion Composition Spectrometer (SWICS) on Ulysses are presented and compared with earlier SWICS results and previous mass spectrometric determinations of this ratio with the Apollo Solar Wind Composition (SWC) experiment and the Ion Composition Instrument (ICI) on the International Sun Earth Explorer 3 (ISEE 3). The new SWICS data from both the fast and slow solar wind are extrapolated to the photosphere to obtain a representative value of the present-day ratio of ^{3}He/^{4}He $= (3.75 \pm 0.27) \times 10^{-4}$ in the Outer Convective Zone (OCZ) of the Sun. After corrections of this ratio for secular changes caused by diffusion, mixing and ^{3}He production by incomplete H-burning (Vauclair 1998), we obtain (D + ^{3}He)/H $= (3.6 \pm 0.38) \times 10^{-5}$ for the Protosolar Cloud (PSC). Adopting the Jovian ^{3}He/^{4}He ratio $= (1.66 \pm 0.05) \times 10^{-4}$ measured by the Galileo Probe mass spectrometer (Mahaffy et al. 1998) as representative for the PSC, we obtain (D/H)$_{\text{protosolar}} = (1.94 \pm 0.39) \times 10^{-5}$. Using results of galactic evolution studies (Tosi 1998, 2000) and the D and ^{3}He abundances in the Protosolar Cloud and the Local Interstellar Cloud (Linsky 1998; Gloeckler & Geiss 1998), we estimate (D/H)$_{\text{primordial}} = (2.4 - 4.2) \times 10^{-5}$. This range corresponds to a universal baryon/photon ratio of $(5.9-4.2) \times 10^{-10}$.

1. Introduction

At present, primordial abundances of deuterium and helium-3 are deduced from observations and measurements in three samples of cosmic material. (1) The present-day galaxy, (2) the Protosolar Cloud (PSC), and (3) clouds with very low contamination from stellar nucleosynthesis, using metallicity as an indicator. Here, we present estimates for the abundances of D and ^{3}He in the Protosolar Cloud that represents a sample of galactic material with a nucleosynthetic age of approximately 4.6 Gyr. Since D was converted into ^{3}He in the early Sun, the abundance of ^{3}He in the Outer Convective Zone (OCZ) of the Sun corresponds approximately to the abundance of D + ^{3}He in the PSC. We summarize and

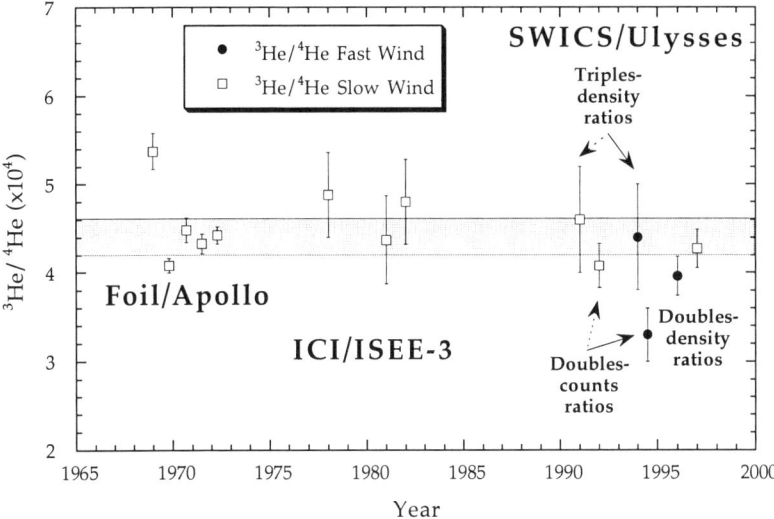

Figure 1. Plot of the average ^{3}He/^{4}He ratio measured with space-borne mass spectrometers (Apollo: Geiss et al. 1970, 1972; ISEE-3: Ogilvie et al. 1980, Coplan et al. 1984, Bochsler 1984; Triples — density ratios: Bodmer et al. 1995; Doubles — counts ratios: Gloeckler & Geiss 1998; Doubles — density ratios: present work) as a function of the approximate time when the measurements were made. Except for some of the recent Ulysses measurements, all observations were made in the in-ecliptic, slow solar wind. The errors bars shown are the result of statistical uncertainties and the real variability of the solar wind ratio during the averaging period.

review the various ^{3}He/^{4}He abundance measurements in the solar wind. Then we obtain an estimate of the ^{3}He/^{4}He ratio in the present-day OCZ by extrapolating our new long-time averages of the ^{3}He/^{4}He ratio in both the slow and fast stream solar wind using a new method. After a discussion of the processes that have changed the He/H and ^{3}He/^{4}He ratios in the OCZ over the lifetime of the Sun, we give our best estimate for $(D + {}^3He)/H$ in the PSC. The protosolar D/H ratio is then obtained by combining the solar wind ^{3}He/^{4}He data with the recently determined ^{3}He/^{4}He ratio in Jupiter (Mahaffy et al. 1998). Finally, by comparing the D and ^{3}He abundances in the PSC and in the Local Interstellar Cloud (LIC) we discuss constraints on the evolution of these two nuclei in the galaxy and on their primordial abundances.

2. The ^{3}He/^{4}He Ratio in the Solar Wind

The ^{3}He/^{4}He ratio in the solar wind has been measured by several space-borne instruments. Since the results clearly show that this ratio is not constant, a comprehensive database and at least some understanding of the causes for the changes in ^{3}He/^{4}He is needed for obtaining the best estimate for the ^{3}He/^{4}He ratio in the present-day OCZ. Comprehensive results have been obtained from three investigations: (1) the Apollo Solar Wind Composition (SWC) experiments, using solar wind collection in foils with subsequent analysis by laboratory mass spectrometry, (2) the Ion Composition Instrument (ICI) on the International Sun Earth Explorer 3 (ISEE 3) using an electromagnetic mass spectrometer allowing unambiguous measurement of the mass/charge ratio of the ions, and (3) the Solar Wind Composition Spectrometer (SWICS) on Ulysses, a time-of-flight system giving the mass/charge ratio as well as the mass of the ions. The results obtained with these three techniques are shown in Figure 1.

True variations in the ^{3}He/^{4}He ratio were found by all three experiments. However, for comparable solar wind flows (cf. the averages in the slow wind) the agreement between the results obtained over 25 years by these three completely different techniques is very remarkable. Prior to the Ulysses mission, all ^{3}He/^{4}He data were taken in the ecliptic plane, where the low speed solar wind dominates, although some data were obtained in fast streams and during CME events. The SWC-Apollo foils collected solar wind particles mainly during slow wind conditions. Because of the polar orbit of Ulysses and the large energy range of the SWICS instrument, it became possible for the first time to investigate systematically the helium isotopes in the high speed streams coming out of large coronal holes. Below we present the latest SWICS results based on measurements of the phase-space density distributions of ^{3}He and ^{4}He using the double coincidence technique (see Gloeckler and Geiss 1998). The final results of this analysis for the fast and slow wind are shown in Figure 1 as the two respective ratios labeled 'Doubles-Density Ratios'. It is evident that ^{3}He/^{4}He is lower in the fast streams than it is in the average slow solar wind (shaded region of Figure 1). This is consistent with previous results based on other methods of analysis of the SWICS data.

3. SWICS measurements of the ^{3}He/^{4}He Ratio in the Slow and Fast Solar Wind

For the present analysis we have chosen four 300-day long time periods, the same time periods during which von Steiger et al. (2000) computed average solar wind elemental abundances. In two of the periods we sampled the in-ecliptic, slow wind and in the other two the high-speed wind. The solar wind ^{3}He^{++}/^{4}He^{++} ratios were derived from the phase-space densities of ^{3}He^{++} and ^{4}He^{++}. The ^{3}He^{++} spectra were computed from the SWICS pulse-height (PHA), double-coincidence (i.e. time-of-flight) data (see Gloeckler & Geiss 1998 and Gloeckler et al. 1992 for details). A background correction amounting to less than 6% of the peak ^{3}He^{++} density was made to eliminate the spill-over (mostly at $W < {\sim}0.9$ and $W > 1.1$) from the many orders of magnitude more abundant solar wind protons and ^{4}He^{++}. The ^{4}He^{++} distributions were derived from the unsaturated triple-coincidence rate MR1 which was then multiplied by the ratio of double-coincidence to triple-coincidence PHA counts selected for ^{4}He^{++}. The spectral shapes of ^{3}He^{++} and ^{4}He^{++} were found to be identical within experimental uncertainties, and the ^{3}He/^{4}He ratio in the fast solar wind (3.83×10^{-4}) was lower than in the slow solar wind (4.34×10^{-4}).

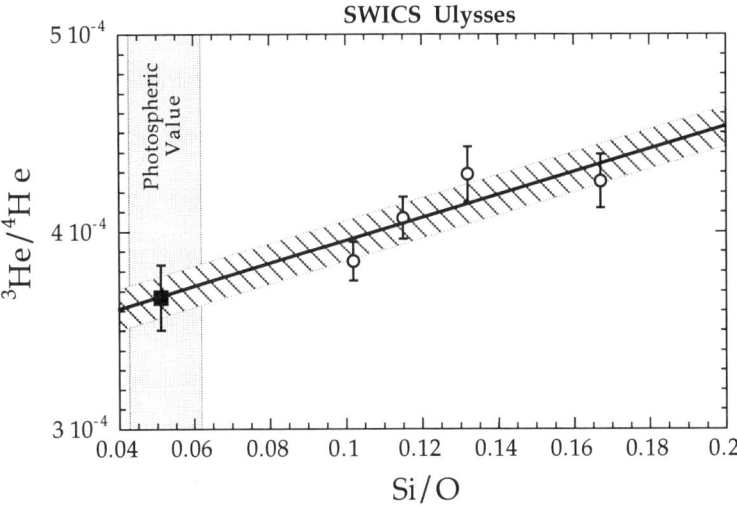

Figure 2. Correlation of the 300-day average ^{3}He/^{4}He and the Si/O ratios measured in the two fast wind and the two slow wind intervals. The Si/O ratios are taken from von Steiger et al. (2000). Using linear regression (solid line) and the photospheric abundance of Si/O (Grevesse & Sauval 1998) we estimate the ^{3}He/^{4}He ratio (solid square) in the OCZ.

4. Derivation of the Present-day ^{3}He/^{4}He Ratio in the Outer Convective Zone

It is now well established that the abundance of elements with low First Ionization Potential (FIP) is higher in the solar wind than in the photosphere of the Sun (e.g. Geiss 1982). This so called 'FIP bias' is found to be smaller in the fast solar wind than in the slow wind (e.g. von Steiger et al. 2000). Furthermore, the solar wind H$^+$/He^{++} ratio is lower in the fast wind than the slow wind. We use the FIP-dependent compositional bias measured for Si/O, H/He and ^{3}He/^{4}He in the slow and fast solar wind to derive the ^{3}He/^{4}He abundance ratio in the present-day Outer Convective Zone (OCZ) of the Sun. Figure 2 shows the dependence of the average solar wind ^{3}He/^{4}He on the Si/O ratios for the four 300-day intervals. We notice that the two ratios are correlated, and using linear regression and the photospheric value of Si/O we estimate the ^{3}He/^{4}He ratio in the OCZ to be $(3.67 \pm 0.17[\text{stat.} + \text{var.}] \pm 0.20[\text{sys.}]) \times 10^{-4}$. The 1-$\sigma$ errors include statistical and systematic instrumental uncertainties as well as the spread due to solar wind variability.

In Figure 3 we plot the average ^{3}He/^{4}He ratios in the polar coronal hole solar wind (average of the two high-speed wind periods) and the in-ecliptic solar wind (average of the two low-speed wind periods) against the ^{1}H/^{4}He ratios for the corresponding periods. Using a linear extrapolation and the value of 11.9 (Pérez Hernández & Christensen-Dalsgaard 1994) for the OCZ H/He, we obtain an OCZ ^{3}He/^{4}He ratio of $(3.82 \pm 0.14[\text{stat.} + \text{var.}] \pm 0.20[\text{sys.}]) \times 10^{-4}$. The two extrapolation methods give essentially the same result. The average of the two

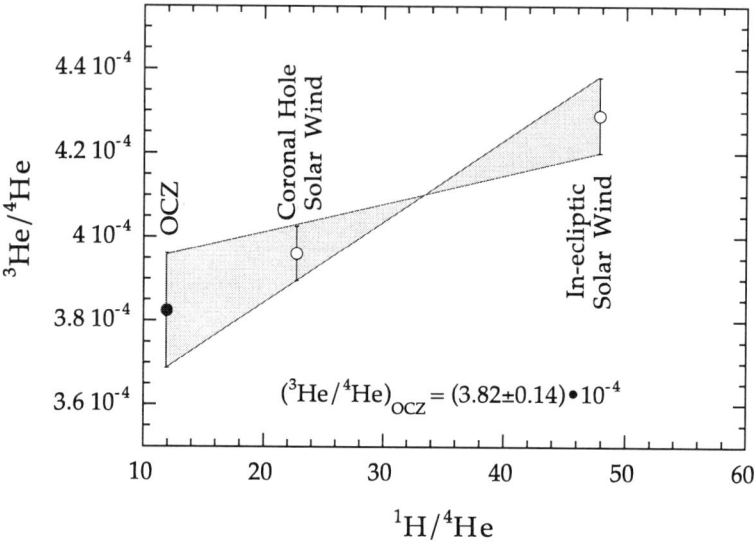

Figure 3. Correlation of the average ^{3}He/^{4}He and the ^{1}H/^{4}He ratios measured in the coronal hole wind and the in-ecliptic wind, respectively. The H/He ratios are measured by the SWOOPS instruments on Ulysses (Bame et al. 1992).

values gives $(^3$He/^{4}He$)_{OCZ} = (3.75 \pm 0.11[\text{stat.} + \text{var.}] \pm 0.20[\text{sys.}]) \times 10^{-4}$, which is in remarkable agreement with the value $(3.8 \pm 0.5) \times 10^{-4}$ obtained by Geiss & Gloeckler (1998).

5. (D+^{3}He)/H, D/H and ^{3}He/H in the Protosolar Cloud

When material from an interstellar cloud collapsed to form the solar system 4.6 Gy ago, the Sun was largely formed by direct infall (Tscharnuter 1987) implying that the material going into the Sun was representative of the Protosolar Cloud (PSC). In the early Sun, D was converted into ^{3}He which has not been further processed in the OCZ, as can be surmised from the existence there of beryllium (Geiss & Reeves 1972). Thus, the ^{3}He/^{4}He ratio in the OCZ basically represents the protosolar (D + ^{3}He)/H ratio. There are, however, two processes that could have changed the ^{3}He/^{4}He ratio in the OCZ during the lifetime of the Sun.

Solar seismic data and solar models show that He/H in the OCZ is 16% lower than it was in the PSC (e.g. Bahcall & Pinsonneault 1995). The difference is interpreted as being due to settling of helium out of the OCZ into deeper layers of the Sun. ^{3}He settles more slowly than ^{4}He resulting in an increase in the present-day OCZ ^{3}He/^{4}He ratio of a few percent (Gautier & Morel 1997).

The second possible change of $(^3$He/^{4}He$)_{OCZ}$ over solar history is due to solar mixing. Using mixing models to various solar depths, Vauclair (1998) has shown that in order to deplete lithium by two orders of magnitude, as is required, some p-p-produced ^{3}He will be added to the OCZ, possibly increasing ^{3}He /^{4}He there by several percent.

A record of solar wind irradiation on the lunar surface goes back to about 4 Gy. While in most lunar materials, the ^{3}He/^{4}He ratio of the old solar wind samples has been affected by strong diffusive losses of helium, the loss is least severe in ilmenite. Using an on-line etching technique applied to this mineral, Wieler et al. (1992) have deduced that the change in the solar wind ^{3}He/^{4}He ratio over the last 3 Gy was less than 10 percent. Since the peak of ^{3}He increases over solar history and moves slowly outward, we expect a large fraction of the contamination of the OCZ with ^{3}He from incomplete H-burning to have occurred during the last 3 billion years of solar history. Thus settling of helium out of the OCZ and solar mixing could not have increased the ^{3}He/^{4}He ratio in the OCZ by much more than 10%. We thus adopt a correction of $-(4 \pm 2)$% and apply this to the ^{3}He/^{4}He ratio in the present-day OCZ to obtain $[(D + {}^3He)/H]_{PSC} = (3.6 \pm 0.38) \times 10^{-4}$.

The solar wind data give only the protosolar abundance of the sum of D and ^{3}He. We use the Jovian value of ^{3}He/^{4}He $= (1.66 \pm 0.05) \times 10^{-4}$, determined by the mass spectrometer on the Galileo Probe (Mahaffy et al. 1998), and (He/H)$_{PSC} = 0.10$ (Bahcall & Pinsonneault 1995) to obtain the Protosolar value of (D/H)$_{PSC} = (1.94 \pm 0.39) \times 10^{-5}$. The error limits are 1-σ uncertainties. They include statistical and systematic (instrumental) errors, broadening due to solar wind variability, uncertainties in the correction for chromospheric and coronal effects on the solar wind composition, and the uncertainty resulting from helium settling and admixture of p-p produced ^{3}He to the OCZ. We have not included an error for a possible difference between the ^{3}He/^{4}He ratio in Jupiter's atmosphere and in the PSC. We note, however, that a small decrease (probably less than 10%) in ^{3}He/^{4}He by gravitational escape from the protoplanetary disc or the Jovian sub-nebula cannot be excluded.

The method of deriving the protosolar D abundance from the ^{3}He/^{4}He ratio in the solar wind has been used for over 25 years (Geiss & Reeves 1972: (D/H)$_{PSC} = (2.5 \pm 1.0) \times 10^{-5}$; Reeves et al. 1973: $(2.6 \pm 1.0) \times 10^{-5}$). As additional solar wind data, new estimates of He/H in the OCZ and PSC, better data for the protosolar ^{3}He/^{4}He ratio, and revised assumptions or model results on solar mixing and fractionation processes in the solar atmosphere became available, several authors have used this method as well as remote and in situ measurements of the Jovian atmosphere (e.g. see Mahaffy et al. (1998) for a summary of in situ and spectroscopic measurement techniques) for deriving (D/H)$_{PSC}$. The more recent results (since 1985) are summarized in Figure 4 along with the present determination of (D/H)$_{PSC}$. The average of the four data points after 1997 gives (D/H)$_{PSC} = (2.11 \pm 0.18) \times 10^{-5}$. The errors are standard errors of the average values. It is interesting to note that both averages, lie within the error bars given in 1972 and 1993, indicating the robustness of the solar wind method to derive the protosolar D/H ratio.

6. Implications for Galactic Evolution and Cosmology

With the Protosolar Cloud and the Local Interstellar Cloud (LIC) we have two galactic samples, differing in nucleosynthetic age by 4.6 Gy, for which we have

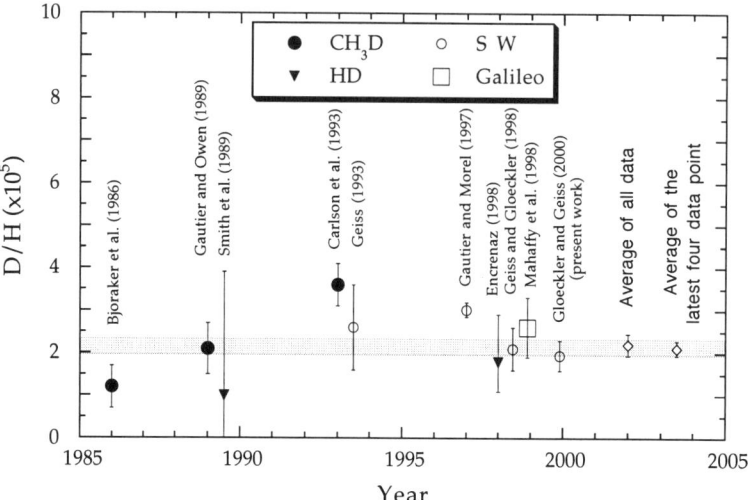

Figure 4. Measurements or estimates of the protosolar D/H ratio as a function of the approximate time when the measurements were made or reported. D/H ratios based on solar wind ^{3}He/^{4}He are indicated by open circles. The in situ mass spectrometric measurement with the Galileo Probe (Mahaffy et al. 1998) is shown as the open square. Spectroscopic determinations of D/H derived from CH$_3$D and HD measurements in Jupiter's atmosphere are shown as solid circles and solid triangles respectively.

reliable data on the isotopic abundances of both hydrogen and helium. D/H in the LIC (Linsky 1998) is lower than it was in the PSC while ^{3}He/^{4}He in the LIC (Gloeckler & Geiss 1998) is higher than it was in the PSC. The direction of these changes is as expected. Because D is destroyed but not produced by stars, the D/H ratio in the galactic interstellar medium ought to decrease continuously with time. ^{3}He, on the other hand, is both destroyed and produced by stars. The observed increase in ^{3}He from the PSC to the LIC value is mainly due to p-p production in small stars (cf. Tosi 1998) that began to leave the main sequence and to lose material only relatively late in galactic history.

The LIC is of course not a direct descendent of the PSC. However, since (a) we have good D and ^{3}He abundance data for both, (b) they evolved at roughly the same distance from the galactic center and (c) their age difference is not small, but > 30% of the age of the universe, a comparison of the two clouds provides us with unique information on galactic evolution. The PSC is the sample with the best defined nucleosynthetic status. Our knowledge on elemental and isotopic abundances in the LIC is still scarce, but will be growing, thanks to refined spectroscopic methods, and to direct measurements on interstellar grains and gas components that pass through the heliosphere. At the present time, the data do not allow a determination of the difference in the He/H ratio or in metallicity between LIC and PSC.

In Figure 5 we plot the abundance ratios as a function of nucleosynthetic age, with the age of the universe taken as $\sim$14 Gy. (cf. Tammann 1998). The

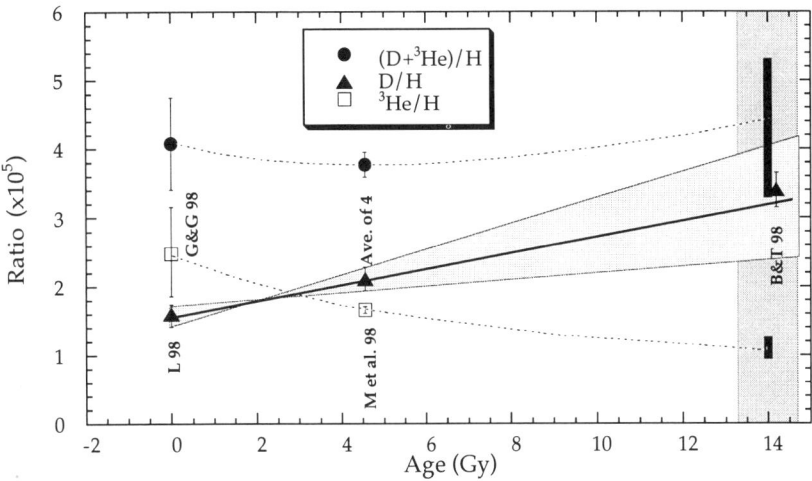

Figure 5. Illustration of the evolution of D and ^{3}He in the "solar ring" of the galaxy. The measured H/D and ^{3}He/^{4}He ratios for the Local Interstellar Cloud (LIC) at age 0 are from Linsky (1998) and Gloeckler & Geiss (1998), those for the Protosolar Cloud (PSC) at age 4.6 Gyr are from this work and from Mahaffy et al. (1998). The shaded region between the lines through the 1-σ limits of the LIC and PSC D/H ratios indicates a range of primordial D/H at 14 Gyr. From the range of primordial deuterium (D/H $\sim(2.4$–$4.2) \times 10^{-5}$) given here, the theory of Standard Big Bang Nucleosynthesis (Walker et al. 1991) allows the determination of the ranges for primordial abundance ratios of (D+^{3}He)/H and ^{3}He/H shown as the vertical solid bars, respectively, from the PSC data. The dotted curves are drawn through the mean values of the (D+^{3}He)/H and ^{3}He/H ratios.

(D/H)$_{\rm LIC}$ is from Linsky (1998) and the (D/H)$_{\rm PSC}$ is the average of the latest four measurements from Figure 4. Since interstellar deuterium is destroyed and not produced by stars and since production of D by cosmic rays is minor, the line going through the lower 1-σ limit of (D/H)$_{\rm PSC}$ and the upper 1-σ limit of (D/H)$_{\rm LIC}$ gives the lower limit of primordial D/H ($\sim 2.3 \times 10^{-5}$). The other line going through the upper 1-σ limit of (D/H)$_{\rm PSC}$ and the lower 1-σ limit of (D/H)$_{\rm LIC}$ gives an approximate upper limit of primordial D/H of $\sim 4 \times 10^{-5}$. Predictions of galactic evolution models (Tosi 1998; Tosi 2000, and references therein) fall well within the range of the two lines (shaded region). We note that the D/H = 3.4 $\pm$ 0.25) $\times 10^{-5}$ obtained in high-z clouds by Burles and Tytler (1998) is consistent with this range of primordial D/H ratios. Higher values are difficult to reconcile with the D/H and ^{3}He/^{4}He abundance ratios measured in the solar ring of the galaxy.

Galactic evolution models (Tosi 1998) show that the slopes of the D/H and ^{3}He/H curves given in Figure 5 can only be reproduced if an infall into the galactic disc of relatively unprocessed material is postulated. Tosi et al. (1998) give an infall rate for the solar ring of 4×10^{-9} M$_{\rm Sun}$ pc^{-2} yr^{-1}. A radial motion of the Sun by a few kpc during its lifetime would not affect the infall rate very much, as the radial evolution calculations of D/H by Tosi et al. (1998) indicate. Comparison of the infall rate given above with the gas density in the solar ring of the galactic disc of 6 M_{Sun} pc^{-2} lead to a replenishment time of 1.5×10^9 yr. This implies that most of the gas presently in the solar ring of the disc had not yet been accreted into the galaxy at the time of birth of the solar system.

Probably, gas does not fall into the galaxy just as a steady drizzle. Observations of high velocity clouds approaching the galaxy (de Boer & Savage 1983) indicated significant variations of the infall rate in space and time. Accretion of dwarf galaxies into our galaxy may contribute significantly, as the observations of the Sagittarius dwarf galaxy indicate. Thus, spatial variations in D/H and in metallicity are produced by inhomogenous infall as well as inhomogenous nucleosynthesis.

Recent microlensing observations indicate that gravitational wells produced by spiral galaxies have dimensions of up to several hundred kpc (Fischer et al. 2000). If this result is applicable to our galaxy, this would certainly very much influence the dynamics of infall, and it would make a difference whether the gravitational well is mainly produced by dispersed baryonic matter or by non-baryonic matter. Measurement of metallicities and deuterium abundances and their variations could help to understand the dynamics of infall and the evolution of our galaxy.

Acknowledgments. The SWICS instrument was developed by a collaboration of the universities of Maryland, Bern and Braunschweig, and the Max-Planck-Institut für Aeronomie. We thank C. Gloeckler for her help with data reduction, and M. Tosi for discussions. This work was supported by NASA/JPL contract 955460 (GG) and by the International Space Science Institute and the Swiss National Science Foundation (JG).

References

Bahcall, J. N., & Pinsonneault, M. H. 1995, Rev. Modern Phy., 67, 781

Bame, S. J., McComas, D. J., Barraclough, B.L., et al. 1992, ApJS, 92, 239

Bjoraker, G. L., Larson, H. P., & Kunde, V.G. 1986, Icarus, 66, 579

Bochsler, P. 1984, Helium and Oxygen in the Solar Wind, University of Bern Habilitation Thesis

Bodmer, R., Bochsler, P., Geiss, J., von Steiger, R., & Gloeckler, G. 1995, Space Sci.Rev., 72, 61

Burles, S., & Tytler, D. 1998, Space Sci.Rev., 84, 65

Carlson, B. E., Lacis, A. A., & Rossow, W. B. 1993, J. Geophys. Res. 98, 5251

Coplan, M. A., Ogilvie, K. W., Bochsler, P., & Geiss, J. 1984, Solar Phys. 93, 415

De Boer, K. S., & Savage, B. D. 1993, ApJ265, 210-215

Encrenaz, T., de Graauw, T., Schaeidt, S., et al. 1996, A&A, 315, L397

Fischer, Ph., McKay, T. A., Sheldon, E., et al. 2000, Astron. J., in press

Gautier, D., & Morel, P. 1997, A&A, 323, L9

Gautier, D., & Owen, T. 1989, in Origin and Evolution of Planetary and Satellite Atmospheres, eds. S. K. Atreya et al., (Cambridge: Univ. Arizona Press), 487

Geiss, J. 1982, Space Sci.Rev., 33, 201

Geiss, J. 1993, in Origin and Evolution of the Elements, eds. N. Prantzos, E. Vangioni-Flam, & M. Cass Cambridge University Press, 89

Geiss, J., & Reeves, H. 1972, A&A, 18, 126

Geiss, J., & Gloeckler, G. 1998, Space Sci.Rev., 84, 275

Geiss, J., Eberhardt, P., Bühler, F., Meister, J., & Signer, P. 1970, J. Geophys. Res., 75, 5972

Geiss, J., Bühler, F., Cerutti, H., Eberhardt, P., & Filleux, Ch. 1972, Apollo 16 Preliminary Science Report, NASA SP-315, section 14

Gloeckler, G., & Geiss, J. 1998, Space Sci.Rev., 84, 239

Gloeckler, G., Geiss, J., Balsiger, H., et al. 1992, A & A Suppl. Ser., 92, 267

Grevesse, N., & Sauval, A. J. 1998, Space Sci.Rev., 85, 161

Linsky, J. L. 1998, Space Sci.Rev., 84, 285

Mahaffy, P. R., Donahue, T. M., Atreya, S. K., Owen, T. C., & Niemann, H. B. 1998, Space Sci.Rev., 84, 251

Ogilvie, K. W., Coplan, M. A., Bochsler, P., & Geiss, J. 1980, J. Geophys. Res., 85, 6021

Pérez Hernández, F., & Christensen-Dalsgaard, J. 1994, MNRAS, 269, 475

Reeves, H., Audouze, J., Fowler, W. A., & Schramm, D. N. 1993, ApJ, 179, 909

von Steiger, R., Schwadron, N. A., Fisk, L. A., et al. 2000, J. Geophys. Res. (in press)

Smith, W. H., Schempp, W. V., & Baines, K. H. 1989, ApJ, 336, 967

Tammann, G. A. 1998, Space Sci.Rev., 84, 15

Tosi, M. 1998, Space Sci.Rev., 84, 207-218

Tosi, M., Steigman G., Matteucci, F., & Chiappini C. 1998, ApJ, 498, 226-235

Tosi, M., 2000, in The Light Elements and Their Evolution, eds. L. da Silva, M. Spite, & J. R. de Medeiros, ASP Conf. Ser. 3

Tscharnuter, W. M. 1987, A&A, 188, 55

Vauclair, S. 1998, Space Sci.Rev., 84, 265

Walker, T. P., Steigman G., Schramm D. N., Olive K. A., & Kang H. S. 1991, ApJ, 376, 51

Wieler, R., Baur, H., & Signer, P. 1992, Lunar Plantet Sci., XXIII, 1525

The Light Elements and Their Evolution
IAU Symposium, Vol. 198, 2000
L. da Silva, M. Spite, J. R. de Medeiros, eds.

Helium and Oxygen Abundances in SMC Planetary Nebulae

R. D. D. Costa, J. A. de Freitas Pacheco and T. P. Idiart

Instituto Astronômico e Geofísico/USP, Av. Miguel Stéfano 4200, 04301-904 - São Paulo - SP, Brazil

Abstract. In this work we report new high quality spectroscopic data for a sample of PNe in the SMC, aiming to derive physical parameters and chemical abundances, in particular to settle the question concerning the oxygen discrepancy found for type I planetaries with respect to stars and HII regions.

1. Introduction

Planetary nebulae can provide valuable clues to understand the evolution of stellar systems like the Magellanic Clouds. In particular, they give informations on intermediate mass stars from which they are originated. While helium and nitrogen abundances are affected by dredge-up episodes along the progenitor life, abundances of other elements like oxygen, neon and argon trace the chemical history of the interstellar medium. Low mass (old) stars are supposed to suffer a low contamination from the first dredge-up, and these objects have been used by several authors to estimate the primordial He abundance.

In this work we report new data for a sample of 18 PNe in the SMC, and we show that there is an anti-correlation between the nitrogen enrichment and the oxygen abundance (supposed to represent the arrow of time). Combining our previous LMC data with existing data in the literature, we found a similar anti-correlation, also observed for galactic type-I PNe.

Face to these unexpected results, the use of He/H vs O/H diagrams for low mass planetaries, aiming to derive primordial He values, should be taken cautiously.

2. Observations and Data Reduction

Observations for this work were made with two different telescopes and instrumental configurations: LNA 1.60 m telescope, in Brazil, using a Boller & Chivens Cassegrain spectrograph, grating of 300 lines/mm, allowing a 4.4 Å/pixel dispersion in an observational run from 18 to 21/Jul/1999, and ESO-Chile 1.52 m telescope, with a Boller & Chivens Cassegrain spectrograph, grating of 400 l/mm allowing a 2.2 Å/pixel dispersion, in an observational run from 15 to 20/Aug/1999.

Line fluxes were calculated by gaussian profiles and reddening was derived from the Balmer $H\alpha/H\beta$ ratio. Electron density was calculated from [SII]

6716/6730 line ratio and electron temperatures from [OIII] and [NII] lines. Ionic abundances were derived from a 3-level atom model and then Elemental abundances were calculated through IC-factors. Typical uncertainties in abundances are of 0.1 to 0.2 dex.

3. Discussion

From our sample we derive the following mean abundances:

$He/H = 0.113 \pm 0.019$
$\varepsilon(O) = 8.22 \pm 0.24$
$\varepsilon(N) = 7.11 \pm 0.44$
$\varepsilon(Ne) = 7.47 \pm 0.23$
$\varepsilon(S) = 6.34 \pm 0.18$
$\varepsilon(Ar) = 5.79 \pm 0.28$

These values for SMC planetary nebulae are compatible with those from their galactic counterparts, however, to classify the objects among the Peimbert (1983) types, we took into account the smaller enrichment of the SMC interstellar medium with respect the Galaxy, as already made to the LMC (see de Freitas Pacheco et al. 1993 for further details). Following this scheme, we classified 5 objects as type-I PNe.

Abundances from our sample correlate as expected: helium vs. log(N/O) reflect the mass spectrum of the progenitors, with objects of higher masses, which experienced more dredge-up episodes, displaying higher abundances. Oxygen and neon, elements not produced by the progenitors have also correlated abundances, reflecting the interstellar medium at the moment of each progenitor's formation.

A clear anti-correlation appeared from our data, between $\varepsilon(O)$ and log(N/O) values, indicating that in the timescale of intermediate mass stars evolution in the Magellanic Clouds, oxygen abundance cannot be seen as a steadily increasing function of time. Two possible interpretations arise to explain this behavior: oxygen conversion into nitrogen, or a relationship between the efficiency of the dredge-up process with the metallicity of the progenitors. As a consequence, one shoud be cautious when classifying extragalactic planetaries on the basis of chemical abundances only. These problems are examined in more detail by Costa et al. (2000).

References

Costa, R.D.D., de Freitas Pacheco, J.A. & Idiart, T.P. 2000, A&A, submitted

de Freitas Pacheco, J.A., Costa, R.D.D. Barbuy, B. & Idiart, T.P. 1993, A&A 271, 429

Peimbert, M. in IAU Symposium 103, Planetary Nebulae, ed. R.D. Flower (Dordrecht: Reidel), 233

The Light Elements and Their Evolution
IAU Symposium, Vol. 198, 2000
L. da Silva, M. Spite, J. R. de Medeiros, eds.

Interstellar D/H on the sightline of Sirius[1]

G. Hébrard, A. Vidal-Madjar and R. Ferlet

Institut d'Astrophysique de Paris, CNRS, 98 bis boulevard Arago, F-75014 Paris, France

C. Mallouris, D. York

Department of Astronomy and Astrophysics, University of Chicago, 5640 South Ellis Avenue, Chicago, IL 60637, USA

M. Lemoine

DARC, Observatoire de Paris-Meudon, UPR-176 CNRS, F-92195 Meudon Cédex, France

Abstract. We present observations of the binary Sirius A / Sirius B performed with HST-GHRS. Two interstellar clouds are detected on this sightline, one of them being identified as the Local Interstellar Cloud (LIC). Lyman α interstellar lines are also observed, but whereas the deuterium Lyman α line is well detected in the LIC with an abundance in agreement with that obtained by Linsky et al. (1995), no significant D I line is detected in the second cloud. The deuterium abundance which we measured in this cloud is $0 < (D/H)_{ISM} < 1.6 \times 10^{-5}$. Despite the large error bar, this sightline appears consequently as a good candidate for a low $(D/H)_{ISM}$.

Deuterium is a key element in cosmology and in galactic chemical evolution. The decrease of its abundance all along galactic evolution is a function, amongst other things, of the star formation rate. In order to measure the interstellar D/H ratio, representative of the present epoch, we introduced in Cycle 1 of HST a new type of target, white dwarfs in the high temperature range, for which the depth of the Lyman α photospheric absorption line is reduced, and whose stellar continuum remains smooth. These targets also allow the study of lines of other species, such as N I and O I, which are shown to be reliable tracers of H I in the ISM. We have already observed the white dwarf G191-B2B in HST Cycle 1 (Lemoine et al. 1996) and Cycle 5 (Vidal-Madjar et al. 1998). Continuing that program, we present here ultraviolet observations of Sirius A and its white dwarf companion Sirius B performed with HST-GHRS in the frame of Cycle 6.

Our observations were performed in November 1996, using G140M and Echelle-A gratings. In our high and medium spectral resolution data, we have detected 10 interstellar lines toward Sirius A and/or Sirius B: N I 1200 Å triplet, O I 1302 Å, C II 1334 Å, Si II 1190 Å, 1193 Å and 1304 Å, D I 1215 Å and H I

---

[1]Based on observations with the NASA/ESA *Hubble Space Telescope*.

Lyman α. We completed our data set with Sirius A spectrum including Mg II and Fe II lines from HST-GHRS archives (Lallement et al. 1994).

The main results of our observations are the following:

• Two interstellar components are detected on the short (2.6 pc) Sirius line of sight, in agreement with previous study by Lallement et al. (1994). One is the Local Interstellar Cloud and we named the second the Blue Component (BC).

• The Lyman α lines do not present the same profile toward Sirius A and Sirius B, an extra absorption being observed in the blue wing of the Sirius A Lyman α line, and an extra absorption being observed in the red wing of the Sirius B Lyman α line. We interpreted these excesses respectively as the signatures of the wind from Sirius A and of the core of the Sirius B photospheric Lyman α line.

• A composite Lyman α profile was constructed from these two lines and fitted in order to measure the $(D/H)_{ISM}$ ratio in the two components. Our data are compatible with D/H ratio found in the LIC by Linsky et al. (1995), *i.e.* $(D/H)_{LIC} = 1.60 \pm 0.09^{+0.05}_{-0.10} \times 10^{-5}$. Our result is $(D/H)_{LIC} = 1.6 \pm 0.4 \times 10^{-5}$. In the other component, BC, we did not detected a significant D I line. The ratio we derived is $0 < (D/H)_{BC} < 1.6 \times 10^{-5}$.

• We did not detect the interstellar absorption of Si III at 1206.5 Å and C II* at 1335.7 Å. This implies a low electron density n_e, for which we found the upper limit $n_e \leq 0.05$ cm^{-3} in the LIC, assuming equilibrium between collisional excitation to excited-state C II* and radiative de-excitation to ground-state C II. Since measured values of the electron density are higher in the LIC toward other sightlines, the new value of n_e toward Sirius could point to inhomogeneities in the Local Interstellar Cloud (Ferlet 1999).

The data are thus consistent with $(D/H)_{BC}$ in the range 0 to 1.6×10^{-5}. The BC cloud is a candidate region for low $(D/H)_{ISM}$, but no definite conclusion about D/H can be made at this time. We intend to continue the study of the Sirius system until we can come to a definitive conclusion as to whether or not a low $(D/H)_{ISM}$ is present in the BC. In particular, it is critically important that this experiment be done again with deep HST-STIS observations to study the Lyman α line and FUSE observations to study the higher Lyman lines.

Details on these observations can be found in Hébrard et al. (1999).

References

Ferlet, R. 1999, A&AR 9, 153

Hébrard, G., Mallouris, C., Ferlet, R., Koester, D., Lemoine, M., Vidal-Madjar, A., & York, D. 1999, A&A 350, 643

Lallement, R., Bertin, P., Ferlet, R., Vidal-Madjar, A., & Bertaux, J.L. 1994, A&A 286, 898

Lemoine, M., Vidal-Madjar, A., Bertin, P., et al. 1996, A&A 308, 601

Linsky, J., Diplas, A., Wood, B.E., Brown, A., Ayres, T.R., & Savage, B.D. 1995, ApJ 451, 335

Vidal-Madjar, A., Lemoine, M., Ferlet, R., Hébrard, G., Koester, D., Audouze, J., Cassé, M., Vangioni-Flam, E., & Webb, J. 1998, A&A 338, 694

Deuterium Balmer emission from nebulae[1]

Guillaume Hébrard

Institut d'Astrophysique de Paris, CNRS, 98 bis boulevard Arago, F-75014 Paris, France

Daniel Péquignot

DAEC, Observatoire de Paris-Meudon, France

Alfred Vidal-Madjar

Institut d'Astrophysique de Paris, France

Jeremy R. Walsh

European Southern Observatory, Garching, Germany

Roger Ferlet

Institut d'Astrophysique de Paris, France

Abstract. We report on the detection and first identification of the deuterium Balmer lines Dα and Dβ, observed in emission in the Orion Nebula (M 42). The excitation mechanism is UV fluorescence from the Lyman(D I) lines at the interface between the H II region and the molecular cloud. These lines may open the possibility to measure D/H in galactic H II regions and, *e.g.*, low-metallicity extragalactic H II regions, using optical spectroscopy. Fluorescence provides an extremely sensitive way to *detect* deuterium. Thus, the non-detection of Dα and Dβ in the planetary nebula NGC 6572 leads to the stringent upper limit (D/H)$_{\text{NGC 6572}}$ less than the order of 1×10^{-7}.

A detailed appraisal of the evolution of deuterium is crucial for cosmology. New methods to determine D/H are of interest [see *e.g.* Lemoine et al. (1999) for a review]. So far, attempts to identify the Balmer lines of deuterium have been unsuccessful in any astrophysical site. We have performed a dedicated search for these lines in H II regions.

We have secured deep spectra of the Orion Nebula at the Canada-France-Hawaii Telescope using the Gecko spectrograph in October 1997 and September 1999. Two very narrow emission lines (FWHM $\simeq 8$ km s^{-1}) were detected in the blue wings of both Hα and Hβ, ~ 11 km s^{-1} redward of the positions expected for the deuterium Balmer lines Dα and Dβ in the H I emission line frame

---

[1]Based on observations collected at the Canada-France-Hawaii Telescope, Hawaii, USA.

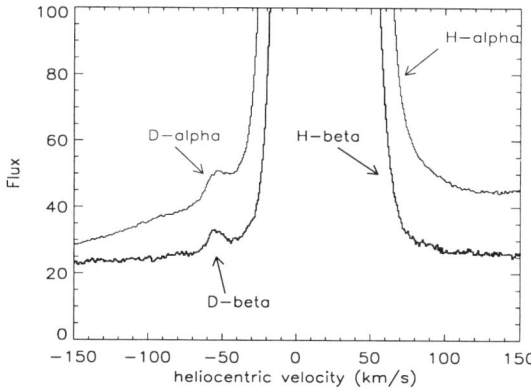

Figure 1. Detection of Dα and Dβ in Orion. The vertical scale corresponds to peak fluxes 7250 and 2500 for Hα and Hβ respectively.

(see Figure 1). It is argued (Hébrard et al. 2000) that: (1) the narrowness of the lines excludes emission from the hot H^+ region and therefore either H I or D I recombination excitation; (2) H I emission is highly unlikely considering the very high velocity (70 km s^{-1}) *and* velocity coherence over a large area (5'); (3) however, assuming D I emission, the radial velocity of these lines closely coincides with that of the H^0 interface between the H^+ and H_2 regions (the Photon Dominated Region); (4) then, the D I lines are likely to be excited by fluorescence following Ly(D I) radiative pumping by the stellar UV continuum, as confirmed by order of magnitude estimates. The fluxes of Dα and Dβ being quite similar, the next deuterium Balmer lines are likely to be detectable. A careful study of the deuterium Balmer decrement may lead to a new determination of D/H.

Using the same configuration, we obtained spectra of six planetary nebulae (PNe). We did not detect any line at the Dα and Dβ wavelengths. Deuterium is believed to be much depleted in PNe owing to its destruction in the parent stars. Indeed, deuterium has never been observed in PNe by whatever means. Nevertheless, our data can lead to interesting upper limits to D/H. In the case of NGC 6572, we provisionally obtain $(D/H)_{NGC\ 6572} < 1 \times 10^{-7}$. To our knowledge, this is the most stringent upper limit to D/H ever found for a PN.

We suspect that D I Balmer lines have already been detected in H II regions but have never been recognized as such. The present results suggest that Balmer deuteriun fluorescence is a very sensitive way to *detect* deuterium. This is due to the large photoexcitation cross sections involved. With the new generation of very large telescopes, it is foreseen that these lines will become detectable in low-metallicity extragalactic H II regions, hence possibly leading to a new determination of the primordial abundance of deuterium.

References

Hébrard, G., Péquignot, D., Vidal-Madjar, A., Walsh, J. R., & Ferlet, R. 2000, accepted for publication in A&A *Letters* (astro-ph/0002141)

Lemoine, M., Audouze, J., Ben Jaffel, L., et al. 1999, *New Astronomy* 4, 231

The Light Elements and Their Evolution
IAU Symposium, Vol. 198, 2000
L. da Silva, M. Spite, J. R. de Medeiros, eds.

FUSE Spectra of Sk 80, an O7 Supergiant in the Small Magellanic Cloud

R. C. Iping, G. Sonneborn

NASA/GSFC, Code 681, Greenbelt, MD 20771

D. Massa

NASA/GSFC and Raytheon ITSS

A. W. Fullerton

Johns Hopkins Univ., Baltimore, MD 21218 and Univ. of Victoria

J. B. Hutchings

National Research Council of Canada

and the FUSE Science Team

Abstract. We present far-UV spectra (910-1190Å) of Sk 80, an O7 supergiant in the Small Magellanic Cloud, observed by the FUSE satellite during its In-Orbit Checkout phase. The spectra reveal many interstellar absorption lines, including H_2, and O VI, and several key stellar wind features.

1. Observations and Results

The FUSE satellite was launched in June 1999 and is described by Sonneborn et al. (this volume). Sk 80 (=AV 232, spectral type O7 Iaf) was observed by FUSE in September and October 1999. The FUSE spectra of SK 80 cover the full wavelength range of the FUSE instrument. The stellar wind lines present in the FUV spectrum of Sk 80 include S VI 933-944, C III 977, N III 990, S IV 1062-1073, and P V 1118-1128. Interstellar absorption lines are present from Milky Way gas as well as from the SMC. Species detected include HI, NI, OI, C II, CIII, Fe II, Ar I, P II, SiII, O VI, and H_2. The H I Lyman series is present up to Ly-8 923.15 Å before overlapping of higher series lines obliterates discernable features. Molecular hydrogen is very strong in the spectrum of Sk 80. The FUSE spectra of Sk 80 are shown in Fig. 1

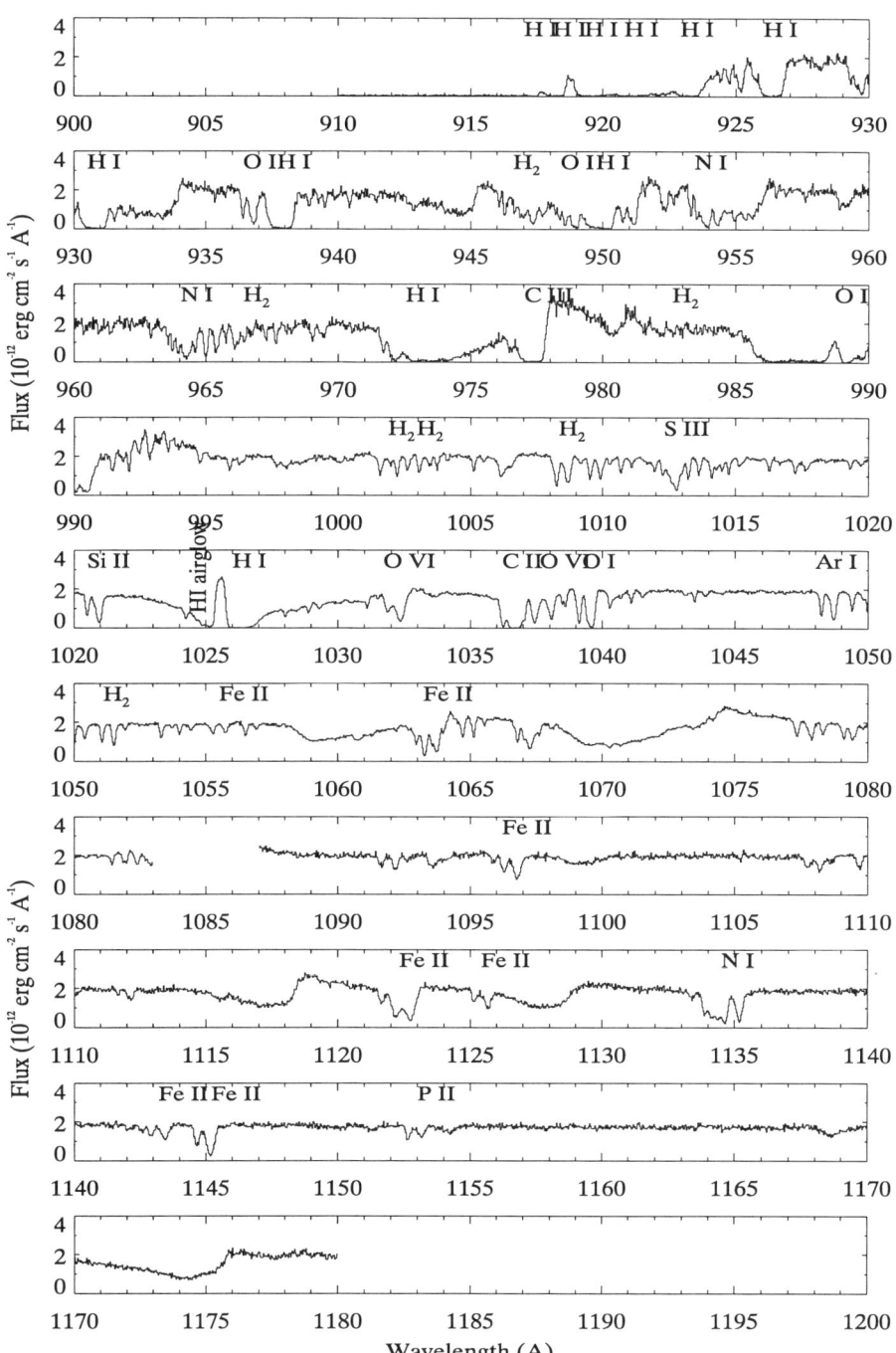

Figure 1. FUSE spectra of Sk 80. Some of the principal interstellar lines are indicated.

The Light Elements and Their Evolution
IAU Symposium, Vol. 198, 2000
L. da Silva, M. Spite, J. R. de Medeiros, eds.

Spatial Variations in the Atomic D/H Ratio in the ISM

G. Sonneborn

NASA/Goddard Space Flight Center, Code 681, Greenbelt MD 20771 USA

E. B. Jenkins, T. Tripp and P. Wozniak

Princeton University Observatory, Princeton, NJ 08544, USA

R. Ferlet and A. Vidal-Madjar

IAP/CNRS, 98bis Blvd. Arago, 75014 Paris, France

U.J. Sofia

Dept. of Astronomy, Whitman College, Walla Walla, WA 99362, USA

Abstract. Observations with IMAPS demonstrate that interstellar atomic D/H abundance ratios differ by a factor of 3 on the sightlines toward δ Ori A, ζ Pup, and γ^2 Vel, early-type stars within ~ 500 pc of the Sun. The observed D/H differences are not inversely correlated with N/H, as would be expected if injection of CNO-processed material was the primary mechanism for the D/H abundance variations in the ISM.

1. Introduction

Observations made with the Interstellar Medium Absorption Profile Spectrograph (IMAPS, see Jenkins et al. 1996) on the US-German ORFEUS-SPAS II mission in late 1996 provide the first new measurements of Galactic interstellar atomic deuterium beyond the local ISM ($d \lesssim 80$ pc) since the *Copernicus* mission in the 1970s. IMAPS is an objective grating echelle spectrograph designed for high spectral resolution ($\lambda/\Delta\lambda \sim 80,000$) in the far UV (930-1160Å). IMAPS observed δ Orionis A (O9.5 II), γ^2 Velorum (WC8+O8), and ζ Puppis (O4 Iaf). H I Lyδ (949.485Å) and Lyϵ (937.548Å). In this contribution we summarize results of Jenkins et al. (1999) for δ Ori A and by Sonneborn et al. (2000) for γ^2 Vel and ζ Pup.

2. D, H, and N Analyses

The total D I column density ($N_{\rm DI}$) toward each star was determined by simultaneously modelling multiple D I profiles (Lyδ 949.485Å and Lyϵ 937.548Å, plus Lyγ 972.272Å for ζ Pup). The parameters defining the profiles are $N_{\rm DI}$, temperature T, the continuum fit near the D I and H I features, and the background (zero intensity) level near each line. The velocity profile of neutral gas toward each star was defined by absorption features from several multiplets of N I recorded in the IMAPS spectra, covering a range of 1.80 in log(fλ). The N I velocity profiles ($N_a(v)$) were used as templates for modelling the D I profiles. Analysis of O I 1039, 1302, and 1356 confirms that N I accurately traces H I on these sightlines. This approach tightens the error limits for N_{DI}. For each star, the optimum solution for N_{DI} was determined by minimizing χ^2 for the multi-parameter fit. The details of this analysis technique are given in Jenkins et al. (1999) and Sonneborn et al. (2000). N_{NI} was determined by integrating $N_a(v)$.

Values of the total H I column density (N_{HI}) for these stars in the literature have sufficiently large uncertainties that new evaluations were made. N_{HI} was measured by performing a χ^2 analysis of Lyα using all the available *IUE* SWP high-dispersion spectra for each star. The large number of *IUE* spectra significantly reduces statistical errors in N_{HI} so that systematic errors (e.g. binary phase, stellar variability) could be assessed and incorporated into the error estimates. In this N_{HI} range ($\sim 10^{20}$cm^{-2}), the same gas produces the broad H I Lyα wings and the D I Lyδ and Lyε features.

The D/H and N/H abundance ratios given in Table 1 (errors are 90% confidence limits, 1.65σ) conclusively demonstrate that D/H in the ISM can vary substantially from that measured in the local ISM (D/H = $1.60 \pm 0.17 \times 10^{-5}$, Piskunov et al. 1997). We find a difference of 3X in D/H between δ Ori and γ^2 Vel. There is no trend of lower D/H with higher metallicity, as would be expected if the injection of CNO-processed (i.e. D-depleted) material into the ISM was the primary mechanism responsible for the D/H variability. (O/H is also low toward δ Ori A– Meyer et al. 1998) Measuring D, N, and O abundances on many sightlines in the Milky Way, understanding the distribution function for D/H ratios, and its correlation with Galactic and stellar environments are major objectives of the FUSE mission (Sonneborn et al., this volume).

Table 1. IMAPS Deuterium and Nitrogen Abundance Ratios

Star	l^{II}	b^{II}	d (pc)	N/H (10^{-5})	D/H (10^{-5})
δ Ori A	203°.9	-17°.7	372	3.97 ± 0.32	0.74 ± 0.16
ζ Pup	256°.0	-4°.7	429	8.30 ± 1.22	1.42 ± 0.24
γ^2 Vel	262°.8	-7°.7	258	7.99 ± 1.02	2.18 ± 0.33

References

Jenkins, E. B., et al. (1996), Ap&SS, 239, 315
Jenkins, E. B., et al. (1999), ApJ, 520, 182
Meyer, D. M. Jura, M., & Cardelli, J. A. 1998, ApJ, 493, 222
Piskunov, N. et al. (1997), ApJ, 474, 315
Sonneborn, G., et al. (2000), ApJ, in press

In-Orbit Performance of the Far Ultraviolet Spectroscopic Explorer

G. Sonneborn

NASA/GSFC, Code 681, Greenbelt MD 20771 USA

H. W. Moos and K. R. Sembach

Dept. of Physics & Astronomy, Johns Hopkins Univ., Baltimore, MD 21218 USA

The FUSE Science Team

Abstract. The Far Ultraviolet Spectroscopic Explorer (FUSE) satellite was launched by a Delta II rocket on 24 June 1999 into a ~768 km, 25° orbit. Following the In-Orbit Checkout phase, FUSE is obtaining high-resolution spectra ($\lambda/\Delta\lambda \gtrsim 20,000$) in the spectral range 905 – 1187 Å. The sensitivity meets or exceeds the pre-launch predictions. Many QSOs and AGNs have already been observed.

1. FUSE In-Orbit Performance

The FUSE satellite is fully operational and has been obtaining scientific data since 1999 Sept. Normal science operations for PI Team and Guest Investigator programs started on 1999 Dec. 1. Pointing performance is excellent ($<0.5''$). The far-UV effective area meets or exceeds pre-flight predictions over the entire bandpass ($A_{eff}(\lambda) \sim 22\text{cm}^2$ to $\sim 70\text{cm}^2$ when all four channels are combined). FUSE has four spectroscopic channels (pairs of mirrors and gratings), two coated with SiC, two with LiF+Al. The background of the cross-delay-line microchannel-plate detectors are below the expected 1 ct cm^{-2} s^{-1}. Scattered far-UV light in the instrument is extremely low. Several anomalies that effect observing efficiency, but not scientific data quality, have been encountered; several of these have been solved and remaining issues are being addressed. The grating and mirror positions in-orbit are not as stable as expected. Small alignment motions related to changing thermal environment of the satellite are primarily a function of solar orientation, but secondary thermal effects within a single orbit have also been observed. Nevertheless, scientific on-target observing efficiency was 25-35% in late 1999. Examples of FUSE spectra are shown in Fig. 1 and 2.

Spectral resolving power is currently (Jan. 2000) $\sim 20,000$. Further improvements are expected as instrument optimization continues. For more information on all aspects of the FUSE mission, please consult the FUSE web site http://fuse.pha.jhu.edu.

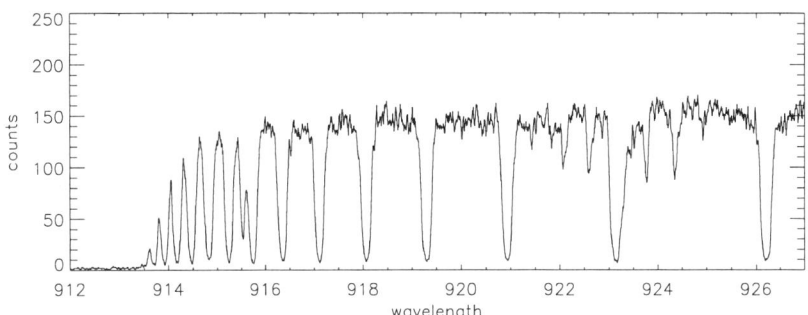

Figure 1. FUSE Commissioning Phase spectrum (resolution $\sim$20 km s^{-1}) of the white dwarf WD2211-495 showing the interstellar H I Lyman series in absorption down to Ly21 913.3Å.

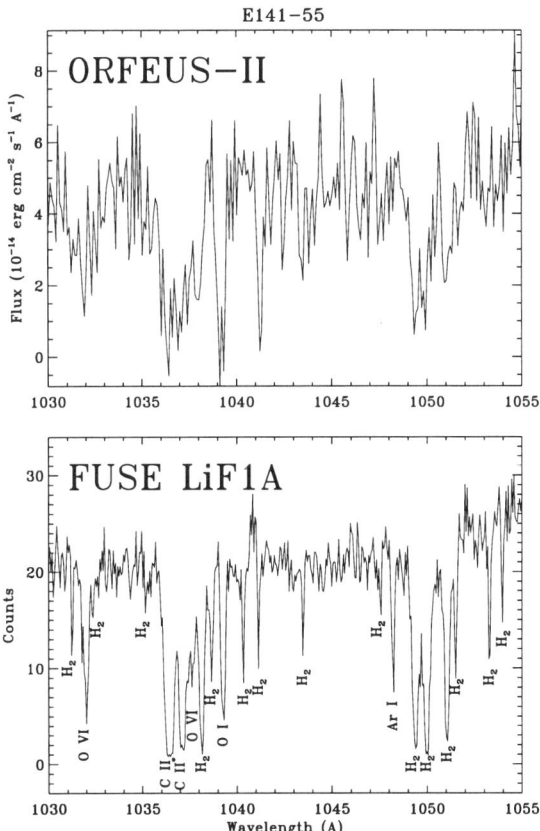

Figure 2. Comparison between FUSE and ORFEUS-II spectra of the Seyfert 1 galaxy ESO 141-55. The significant improvement in spectral resolution and sensitivity over previous missions is evident in the clarity with which the FUSE spectrum separates the H$_2$ features from adjacent atomic lines.

Coffee break

LITHIUM ABUNDANCES

Licio da Silva, Renan de Medeiros, and Roger Cayrel

The Light Elements and Their Evolution
IAU Symposium, Vol. 198, 2000
L. da Silva, M. Spite, J. R. de Medeiros, eds.

^{7}Li in Metal-Poor Stars: The Spread of the Li Plateau

Sean G. Ryan

Dept of Physics and Astronomy, The Open University, Walton Hall, Milton Keynes MK7 6AA, UK

Abstract. A highly homogeneous study of 23 halo field dwarf stars has achieved a Li abundance accuracy of 0.033 dex per star. The work shows that the intrinsic spread of the Li abundances of these stars *at a given metallicity* is < 0.02 dex, and consistent with zero. That is, the Spite Li plateau for halo field dwarfs is incredibly thin. The thinness rules out depletion by more than 0.1 dex by a rotational-induced extra-mixing mechanism. Despite the thinness of the plateau, an increase of Li with [Fe/H] is seen, interpreted as evidence of Galactic chemical evolution (GCE) of Li, primarily due to Galactic cosmic ray (GCR) spallation reactions in the era of halo formation. The rate of Li evolution is concordant with: (1) observations of spallative ^{6}Li in halo dwarfs; (2) GCE models; and (3) data on Li in higher metallicity halo stars. New data have also revealed four new ultra-Li-deficient halo dwarfs, doubling the number known. Based on their propensity to cluster at the halo main sequence turnoff *and also* to exist redward of the turnoff, we hypothesise that they are the products of binary mergers that ultimately will become blue stragglers. We explain their low Li abundances by normal pre-main-sequence (and possibly main-sequence) destruction in the low mass stars prior to their merging. If this explanation is correct, then such stars need no longer be considered an embarrassment to the existence of negligible Li destruction in the majority of field halo dwarfs.

1. Introduction

The first indication that the old stars of the Galaxy exhibited an almost uniform Li abundance emerged at IAU Coll. 68, when Spite & Spite (1981, 1982) presented their first observations of warm halo dwarfs. Almost two decades later, IAU Symp. 198 met to consider progress in studies of this and other light elements.

Studies by many workers in the decade following the Spite & Spite discoveries resulted in mounting evidence that the warm halo dwarfs exhibited a unique Li abundance (e.g. Rebolo, Molaro, & Beckman 1988). The interpretation of this abundance as the primordial one reflecting big bang nucleosynthesis, at worst "hardly altered" (Spite & Spite 1982), hinged on the importance of possible depletion of Li from a higher initial abundance. While ample evidence existed of Li destruction in some stars, the lack of a significant spread in halo dwarf Li abundances provided empirical evidence that destruction may have been minimal in these objects. (See Boesgaard & Steigman 1985 for a review of that period.) Classical stellar evolution models (e.g. Deliyannis, Demarque, & Kawaler 1990) fitted this interpretation, showing that Li destruction in metal-deficient dwarfs with shallow surface convective zones would be minimal ($\lesssim 0.05$ dex). However, this same class of stellar models failed to explain numerous Population I star observations, and an alternative class of models invoking extra mixing implied that considerable Li depletion (as high as 1 dex; Pinsonneault, Deliyannis, & Demarque 1992) could have occurred in the halo stars.

In the next decade, several dissenting voices were heard. Deliyannis, Pinsonneault, & Duncan (1993) argued that there was a non-negligible spread in the Li abundances of the halo dwarfs that would not be consistent with a perfectly primordial composition. Depending on the sample selected, they found a Li spread of $\sigma \geq 0.04$ dex. Thorburn (1994) found an even greater intrinsic spread $\sigma \simeq 0.10$ dex, and moreover claimed, as did Norris, Ryan, & Stringfellow (1994), that the abundances depended on both $T_{\rm eff}$ and [Fe/H]. Such dependences were contrary to the notion of a unique Li abundance in the halo stars, and thus undermined the association of the observed Li abundance(s) with the primordial one. The efforts of Ryan et al. (1996) to bring all previous observations onto a uniform temperature and abundance scale did not eliminate the cited dependences.

One of the largest uncertainties in abundance analyses is errors in the effective temperature scales. Spite & Spite (1993) and Bonifacio & Molaro (1997) discussed the possible role of such errors in distorting an otherwise uniform Li abundance, the latter work finding the previously reported trends to become insignificant when a more recent temperature calibration based on the infra-red flux method (IRFM) was applied. Although the IRFM scale might be expected to provide better systematics, the large individual uncertainties attached to each temperature determination by this method limit the scale's ability to distinguish between effects at the level of those claimed for Li. The existence of some large errors even in the metallicity estimates for program stars also hampers the efforts. In particular, the literature data utilised by Bonifacio & Molaro (1997) includes several poorly determined values which, in hindsight, frustrated their analysis by smearing out the data (Ryan, Norris, & Beers 1999, §7.3.3).

2. The Intrinsic Spread of ^{7}Li

In an effort to avoid the impact of undesirable errors, Ryan, Norris, & Beers (1999) set out to obtain a highly homogeneous data set on a sample occupying only a narrow range of $T_{\rm eff}$, [Fe/H], and evolutionary type. Restricting their sample to 6000 K $\lesssim T_{\rm eff} \lesssim$ 6400 K and $-3.5 \lesssim$ [Fe/H] $\lesssim -2.5$, applying double-blind data analysis techniques, obtaining multiple high-resolution, high-S/N observations of the targets, and using multiple temperatures indicators to minimise random errors, they achieved a formal abundance error as low as $\sigma_{\rm err} = 0.033$ dex per star. These results are at considerably higher precision than most previous Li measurements (typically having $\sigma_{\rm err} \simeq 0.06 - 0.08$ dex).

The sample was known to contain one previously known ultra-Li-deficient star, G186-26, which was excluded from the analysis. Remaining objects exhibited a total observed spread $\sigma_{\rm obs} = 0.053$ dex, considerably less than that

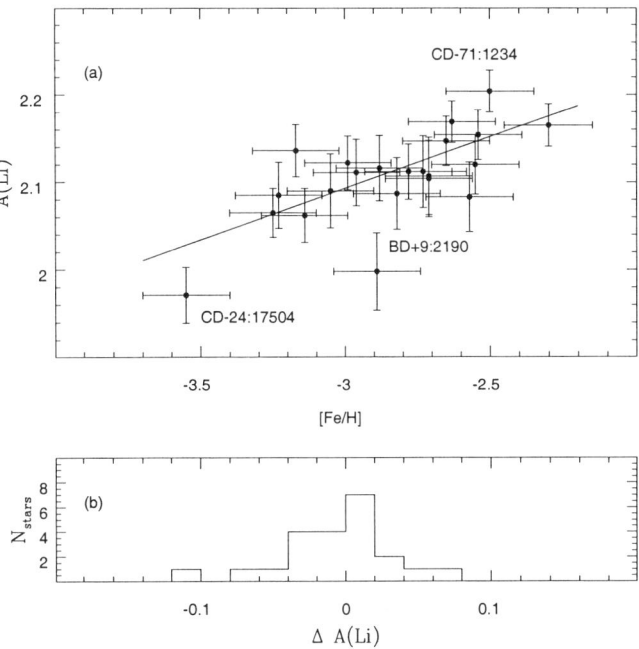

Figure 1. Metallicity-dependence of turnoff Li abundances, and residual observed spread with $\sigma_{\rm obs} = 0.031$ dex.

found by Thorburn (1994). However, this 0.053 dex was found to be dominated by an underlying metallicity dependence, and the spread of the Li abundances about this trend is a mere $\sigma_{\rm obs} = 0.031$ dex (see Figure 1), and Gaussian in form. This corresponds to the spread in Li abundance at a given metallicity. Comparing this with the formal measurement errors of $\sigma_{\rm err} = 0.033$ dex leads to the conclusion that the intrinsic spread in the stars must be negligible. We state a generous upper-limit on the intrinsic spread as being $\sigma_{\rm int} < 0.02$ dex.

An important consequence of the very narrow spread of Li abundances is its ability to constrain the impact of possible extra-mixing in so far as extra-mixing models predict a spread in the final Li abundances of a population of stars. The rotationally-induced mixing models of Pinsonneault et al. (1993) suggested that Li depletion by as much as an order of magnitude could have occurred in halo turnoff dwarfs. The more recent work of Pinsonneault et al. (1999), in concert with the observational data of Thorburn (1994), revised downward the depletion level to $\simeq 0.2 - 0.4$ dex. As Figure 2 shows, the data from Ryan et al. (1999) with their narrower spread (at a given metallicity) rule out rotationally-induced mixing models that exhibit even 0.1 dex median depletion. Considering the size of the observed sample and the absence of stars in the tail of the theoretical distribution, Poisson statistics provide only 10% chance that the observed and theoretical curves are compatible, or 90% probability that the median depletion is less than 0.1 dex. Other statistical tests discriminate the models even more significantly.

We seek now to explain previous results that yielded contrary conclusions. Thorburn's (1994) data reduction process explicitly excluded sky background

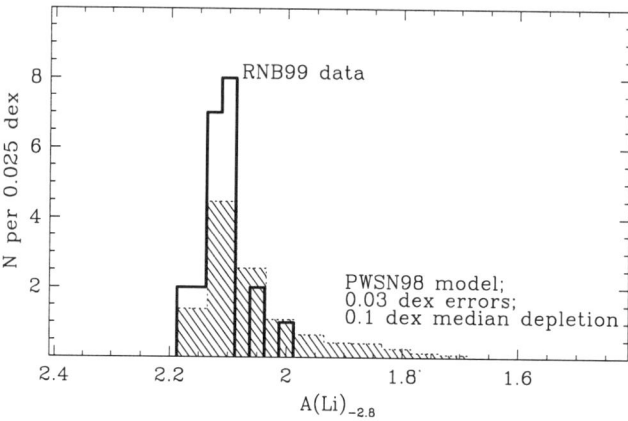

Figure 2. Spread in observations (at a given metallicity after compensation for the [Fe/H] dependence of Li), compared with predictions for a rotationally-induced mixing model exhibiting a median depletion of 0.1 dex.

and scattered light subtractions, but this shortcut was not reflected in the formal error estimates. Incorporation of the errors introduced by this procedure are enough, on average, to inflate the error estimates to the size required by the observed scatter. That is, the scatter observed by Thorburn is almost certainly consistent with that resulting from data acquisition and analysis. Bonifacio & Molaro (1997) found no significant metallicity dependence in their analysis, but as discussed above, certain metallicities they adopted from the literature were found subsequently to be unreliable. This, and the large random errors inherent in the IRFM temperature scale, resulted in the weak Li evolution being washed out. (See Ryan et al. 1999, §7.3.3, for a detailed analysis.) Finally, we note that the small spread of abundances found by Deliyannis et al. (1993) is consistent with the observed spread in our sample if the underlying metallicity trend is overlooked. In fact, the Deliyannis et al. study pre-dated any claims of a metallicity dependence, and their result was probably driven by the large metallicity range in their sample.

3. The Underlying Li vs [Fe/H] Trend

Although Li GCE during the halo-forming era has often been ignored, we should not be surprised that it exists. If recent detections of ^{6}Li in halo stars (Smith, Lambert, & Nissen 1993,1998; Hobbs & Thorburn 1994,1997; Cayrel et al. 1999, Deliyannis & Ryan 2000) are correct, then we would be surprised *not* to see ^{7}Li GCE. With the measurement precision attainable using modern CCDs and large aperture telescopes, even small levels of ^{7}Li enrichment can be measured, and it is consistent with the measured ^{6}Li abundances.

To see whether the observed trend was compatible with GCE, Ryan et al. (2000a) examined the Fields & Olive (1999a,b) model. For halo stars, the ν-

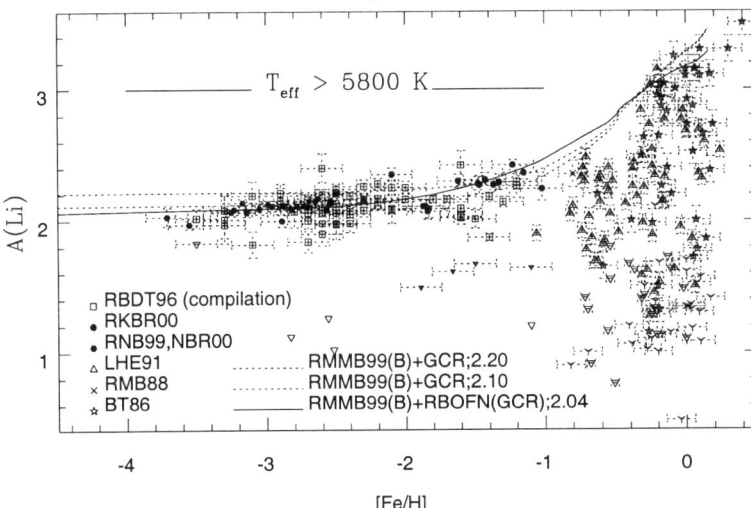

Figure 3. Evolution of Li with metallicity. Observations are for halo stars having $T_{\rm eff} > 5800$ K, to avoid lower-mass stars with Li depletion and to reduce the heterogeneity of the sample, and Population I stars from sources indicated. Models are (dashed curves) from Romano et al. (1999) for two different primordial values ($A({\rm Li})_p = 2.10$ and 2.20), and (solid curve) a hybrid model using the GCR contribution of Fields & Olive (1999a,b; Ryan et al. 2000a) with Population I evolution from Romano et al.

process and GCR spallation are the most likely sources of ^{7}Li. The Fields & Olive model normalises the GCR contribution to meteoritic Be and ^{6}Li abundances, and normalises the ν-process to the otherwise unaccounted for ^{11}B. The model does not include stellar ^{7}Li sources acting in the later stages of Galactic evolution, and hence does not model the Population I abundance. The models of Romano et al. (1999) incorporate many Population I sources (primarily the ν-process, AGB stars, and novae). Figure 3 shows two variants of Romano et al's models and a hybrid using the GCR predictions of Fields & Olive from Ryan et al. (2000a). The model reproduces not only the halo star Li evolution discussed above, but also fits new data around [Fe/H] ~ -1.5 (Ryan et al. 2000b; see below) and the lowest metallicity datum at [Fe/H] $= -3.7$ (Norris, Beers, & Ryan 2000) which were added *after* the Fields & Olive model was produced.

Ryan et al. (1996) indicated that a selection bias existed in the Li data available in the literature, in that most studies centred on more-metal-poor objects. Few more-metal-rich halo stars had been examined, and those which had were on the whole cooler than the metal-poor ones. Given the difficulties with temperature scales for stars, the comparing of warmer metal-poor stars with cooler metal-rich stars was clearly undesirable. In an effort to address this bias, Ryan et al. (2000b) obtained data on 18 more-metal-rich halo stars, with $-2.0 \lesssim$ [Fe/H] $\lesssim -1.0$, but in the warm temperature range $T_{\rm eff} \gtrsim 6000$ K as for the most metal-poor samples. The sample was found to contain four ultra-Li-deficient halo stars, which will be discussed separately in the next section. The remaining stars, shown as solid circles in Figure 3, were found to sit exactly where the Fields & Olive model predicts. We emphasise that the model was completed prior to the reduction and analysis of the metal-rich halo sample, so the agreement between the two is a genuine accomplishment, not something achieved artificially in the model. This is viewed as additional evidence that the trend of Li with [Fe/H] evidenced in Figure 1 is a result of natural GCE of the element during formation of the Galactic halo.

4. The Primordial Li Abundance and Uncertainties

We combine these measurements in Table 1 to present a new accounting for the primordial Li abundance. Beginning with the observed abundance at the mean metallicity of our sample, we apply corrections for the inferred GCE contribution (with uncertainties) and for stellar depletion. For the latter we take the value implied by classical models, but in the uncertainties allow for additional depletion up to the 0.1 dex limit of the rotationally-induced mixing models. Temperature scale uncertainties remain one of the largest sources of error, and in this analysis we apply an offset of 0.08 dex to the Li abundance, corresponding to a change from the temperature scale adopted in our original analysis (based on the Bell & Oke (1986) and Magain (1987) b-y scales) to the systematically hotter IRFM scale of Alonso, Arribas, & Martínez–Roger (1996). However, we associate an uncertainty of ±0.08 dex with this process, in recognition of the remaining difficulties in the temperature scales for halo dwarfs. These and the other affects tabulated lead us to infer a primordial abundance $A(\text{Li})_p = 2.09\,^{+0.19}_{-0.13}$ dex, where the uncertainties resemble 2σ limits (Ryan et al. 2000a).

Table 1.   Transforming the observed halo Li abundance into the primordial abundance, accounting for random and systematic errors.

Systematic Effects Influencing Inferred Primordial Lithium Abundance		
Observed:		
$\langle A(\text{Li})_{-2.8}\rangle =$ 2.12		± 0.02
Corrections to apply:		
GCE/GCR	-0.11	$^{+0.07}_{-0.09}$
Stellar depletion	$+0.02$	$^{+0.08}_{-0.02}$
T_{eff} scale zeropoint	$+0.08$	± 0.08
1-D atmosphere models	$+0.00$	$^{+0.10}_{-0.00}$
Model temperature gradient	$+0.00$	$^{+0.08}_{-0.00}$
NLTE	-0.02	± 0.01
gf-values	$+0.00$	± 0.04
Anomalous/pathological objects	$+0.00$	± 0.01
Total	-0.03	$^{+0.19}_{-0.13}$
Inferred:		
$A(\text{Li})_p =$ 2.09		$^{+0.19}_{-0.13}$

5. The Ultra-Li-Deficient Halo Dwarfs: Blue Stragglers After All?

Boesgaard & Tripicco (1986a) showed that Hyades stars with 6400 K $< T_{\text{eff}} <$ 6900 K exhibit extremely low surface Li abundances. These and similar stars in other Population I clusters became known as "Li-dip stars". They showed that a second mechanism, besides convection on the pre-main sequence (and possibly main sequence) for lower mass stars, could greatly deplete surface Li abundances. Lambert, Heath, & Edvardsson (1991) showed that *most* of the strongly Li-depleted Population I field stars, shown for example in Figure 3, could be explained as either having evolved from the Li dip or being low mass convectively-depleted stars. However, they also noted that a small number of stars did not share these histories, and proposed that perhaps 10% of stars had experienced additional severe Li depletion through unknown causes. This work preceded the discovery of halo dwarfs whose temperatures and metallicities coincided with the Spite plateau but which were ultra-Li-deficient by more than an order of magnitude or so (Hobbs, Welty & Thorburn 1991; Thorburn 1992; Spite et al. 1993). The halo examples have been estimated at perhaps 3–5% of the Population, and may result from the same process as the Population I class proposed by Lambert et al. The nature of the process resulting in their Li-depletion has remained unclear, and our inability to explain them has been given as a reason to mistrust the entire Population II interpretation (e.g. Thorburn 1994). Whether or not such a view is held (cf. Ryan et al. 1999), they identify an embarrassing deficit in our knowledge of stellar processing of this important element. Efforts to identify common chemical signatures other than Li deficiency proved impossible; instead considerable diversity and heterogeneity was found amongst the complete sample (four) known at the end of 1998 (Norris et al. 1997; Ryan, Norris, & Beers 1998).

The study of 18 more-metal-rich halo stars by Ryan et al. (2000b) discussed above resulted in the discovery of four new ultra-Li-deficient halo stars; see Figure 4 (Ryan et al. 2000c). The discovery rate in that study, 22%, contrasts greatly with the previous Population estimate of 3–5%, and indicates that they are preferentially clustered in the stellar parameter range singled out in that investigation, namely warm, more-metal-rich halo stars. The stars are therefore seen to be grouped preferentially towards the main sequence turnoff of the halo, but not exclusively so. The clustering near the turnoff is reminiscent of blue stragglers, but the hypothesis that they were redward-evolving blue stragglers had already been ruled out for the previously known examples (Thorburn 1994; Norris et al. 1997). However, the discovery of four more such stars preferentially close to the main sequence turnoff resulted in the re-examination of the blue-straggler hypothesis, with the distinction that main-sequence blue-stragglers-to-be are now considered. Halo stars initially cooler than about 5700 K, corresponding to a mass of $\sim 0.7\ M_\odot$, deplete their surface Li during their pre-main-sequence evolution. When two low-mass stars merge to become a single object higher up the main sequence near the halo turnoff with a total mass around 0.7-0.85 $M_\odot$, they will in most cases form from stars which have already destroyed their Li. They will appear, then, as normal halo main-sequence stars *except* with respect to two parameters: (1) their Li abundances will be extremely low and hence appear abnormal, and (2) their main sequence lifetimes will be extended due to the delayed onset of nuclear burning

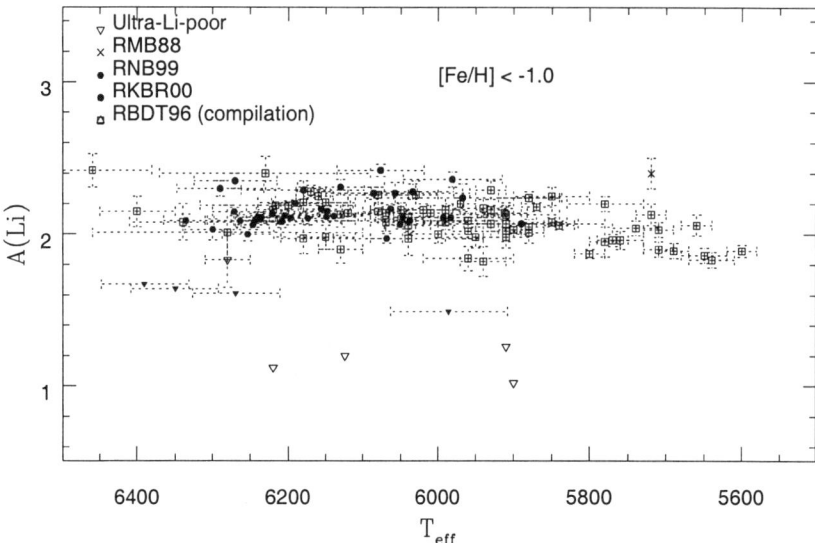

Figure 4. Recent observations doubling the number of known ultra-Li-deficient halo stars. Their location close to the main sequence turnoff leads us to the hypothesis that they are the progeny of low-mass binary-star mergers, destined to become blue-stragglers.

at a rate expected of stars of their (combined) mass. The stars we now observe as ultra-Li-deficient halo stars, preferentially but not uniquely clustered towards the turnoff, may indeed be the progeny of such mergers and the progenitors of future blue-stragglers. Indeed, extreme Li deficiency may be the only common signal of future Pop II blue-stragglers-to-be.

If this hypothesis is correct, then we may finally remove such stars with confidence from discussions of the spread about the Spite plateau, and consider them as a truly distinct class of stars whose evolutionary history explains their abnormal Li abundances. Whether this mechanism can also explain the heterogeneity found for the abundances of their other elements remains to be seen.

6. Differences Between Halo Field and Globular Cluster Stars?

Although the halo field stars discussed above have minimal intrinsic spread about the Li Spite plateau (at a given abundance), data for globular cluster samples show a different picture. Figure 5 compares the very metal-poor field turnoff dwarf data for stars spanning a dex in [Fe/H] with observations of subgiants in M92 (Boesgaard et al. 1998). The two groups exhibit quite different Li characteristics! The considerable spread in the globular cluster sample prompted Boesgaard et al. to favour a mechanism in which a higher pre-stellar abundance has been depleted by varying degrees in the stars, possibly by the rotationally-induced mixing mechanism discussed earlier. Why this mechanism should differ for the globular cluster and field star samples is unclear. Differences in angular momentum evolution in the two environments may be responsible, but the

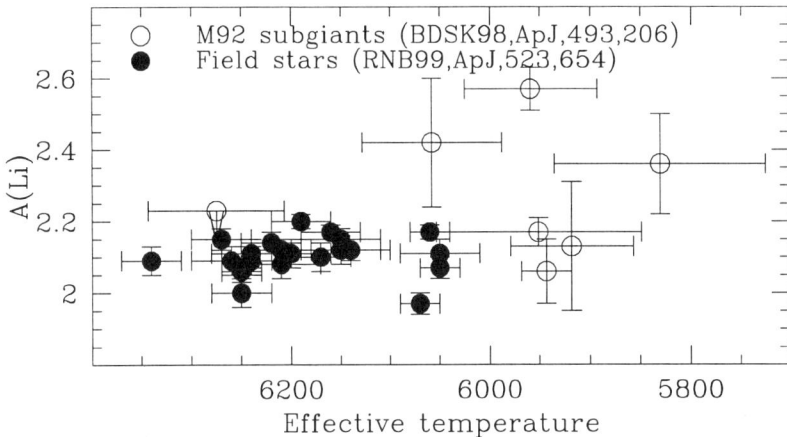

Figure 5. Contrast between the tight distribution of Li for halo field dwarfs (having a range of [Fe/H]), compared with the broad spread in globular cluster subgiants. See text for discussion.

details have yet to be proposed. Other examples of differences in the mixing of stellar envelopes in field star and globular cluster samples has been forthcoming in recent years (e.g. Hanson et al. 1998), adding to previous evidence of field-vs-cluster differences in CNO element ratios.

Until the cause of the difference is understood, one must choose whether to use the field star or the globular cluster data to interpret GCE. I would argue that the large Galactic volume sampled by field stars in contrast to the small total volume of globular clusters, and the greatly increased possibility of star-to-star interactions in the high stellar densities of the latter, would render field star samples more representative of the evolution of the Galaxy as a whole. Of course, this in no way reduces the importance of understanding the globular cluster element abundance patterns for what they may tell us about the evolution of the Galaxy and stellar processes in dense environments.

Acknowledgments. This work represents the outcomes of collaborations involving Prof. J. E. Norris (Australian National University), Dr T. C. Beers (Michigan State University), Dr K. A. Olive (University of Minnesota), Dr B. D. Fields (University of Illinois at Urbana-Champaign), Dr T. Kajino (National Astronomical Observatory of Japan), Ms. D. Romano (SISSA, Trieste), and Ms K. Rosolankova (The Open University, & St Hilda's College, Oxford), all of whose contributions are gratefully acknowledged. The author is likewise grateful to the IAU for partial support to attend this meeting.

References

Alonso, A., Arribas, S., & Martínez–Roger, C. 1996, A&AS, 117, 227
Bell, R. A. & Oke, J. B. 1986, ApJ, 307, 253
Boesgaard, A. M., Deliyannis, C. P., Stephens, A., & King, J. R. 1998, ApJ, 493, 206
Boesgaard, A. M. & Steigman, G. 1985, ARAA, 23, 319

Boesgaard, A. M. & Tripicco, M. J. 1986a, ApJ, 302, L49
Boesgaard, A. M. & Tripicco, M. J. 1986b, ApJ, 303, 724
Bonifacio, P. & Molaro, P. 1997, MNRAS, 285, 847
Cayrel, R., Spite, M., Spite, F., Vangioni-Flam, E., Cassé, M., & Audouze, J. 1999, A&A, 343, 923
Deliyannis, C. P., Demarque, P., & Kawaler, S. D. 1990, ApJS, 73, 21
Deliyannis, C. P., Pinsonneault, M. H., & Duncan, D. K. 1993, ApJ, 414, 740
Deliyannis, C. P. & Ryan, S. G. 2000, in prep
Fields, B. D. & Olive, K. A. 1999, New Astronomy, 4, 255
Fields, B. D. & Olive, K. A. 1999, ApJ, 516, 797
Hanson, R. B., Sneden, C., Kraft, R. P., & Fulbright, J. 1998, AJ, 116, 1286
Hobbs, L. M., & Thorburn, J. A. 1994, ApJ, 428, L25-L28
Hobbs, L. M., Welty, D. E., & Thorburn, J. A. 1991, ApJ, 373, L47
Magain, P. 1987, A&A, 181, 323
Hobbs, L. M. & Thorburn, J. A. 1997, ApJ, 491, 772
Lambert, D. L., Heath, J. E., & Edvardsson, B. 1991, MNRAS, 253, 610
Norris, J. E., Ryan, S. G., & Stringfellow, G. S. 1994, ApJ, 423, 386
Norris, J. E., Ryan, S. G., Beers, T. C., & Deliyannis, C. P. 1997, ApJ, 485, 370
Pinsonneault, M. H., Deliyannis, C. P., and Demarque, P. 1992, ApJS, 78, 179
Pinsonneault, M. H., Walker, T. P., Steigman, G., & Narayanan, V. K. 1999, ApJ, 527, 180
Rebolo, R., Molaro, P., & Beckman, J. E. 1988, A&A, 192, 192
Romano, D., Matteucci, F., Molaro, P., & Bonifacio, P. 1999, A&A, in press
Ryan, S. G., Beers, T. C., Deliyannis, C. P., & Thorburn, J. A. 1996, ApJ, 458, 543
Ryan, S. G., Beers, T. C., Olive, K. A., Fields, B. D., & Norris, J. E. 2000a, ApJL, in press
Ryan, S. G., Beers, T. C., Kajino, T., & Rosolankova, K. 2000c, in prep
Ryan, S. G., Kajino, T., Beers, T. C., Suzuki, T., Rosolankova, K., & Romano, D. 2000b, in prep
Ryan, S. G., Norris, J. E., & Beers, T. C. 1998, ApJ, 506, 892
Ryan, S. G., Norris, J. E., & Beers, T. C. 1999, ApJ, 523, 654
Smith, V. V., Lambert, D. L., & Nissen, P.-E. 1993, ApJ, 408, 262
Smith, V. V., Lambert, D. L., & Nissen, P. E. 1998, ApJ, 506, 405
Spite, F. & Spite, M. 1982, A&A, 115, 357
Spite, F. & Spite, M. 1993, A&A, 279, L9
Spite, M., Molaro, P., Francois, P., & Spite, F. 1993, A&A, 271, L1
Spite, M., & Spite, F. 1981, IAU Coll. 68, Astrophysical Parameters for Globular Clusters, A. G. David Phillip & D. S. Haynes eds (Kluwer, Dordrecht)
Thorburn, J. A. 1992, ApJ, 399, L83
Thorburn, J. A. 1994, ApJ, 421, 318

Observations of ^{6}Li in metal poor stars

P. E. Nissen

Institute of Physics and Astronomy, University of Aarhus, DK–8000 Aarhus C, Denmark

Abstract.
Methods and accuracies of determining the lithium isotope ratio in stellar atmospheres from high resolution observations of the profile of the $\lambda 6708$ Li I doublet are discussed, and recent results for metal-poor disk stars and halo stars are reviewed. The data are compared to models for the Galactic evolution of the ^{6}Li abundance. It is concluded that a much larger data set is needed to understand properly the formation of ^{6}Li in the interstellar medium as well as the depletion in stars.

1. Introduction

Measurements of the abundance of the ^{6}Li isotope in stellar atmospheres are of considerable interest and have attracted much attention since the first probable detection of ^{6}Li in the metal-poor turnoff star HD 84937 by Smith, Lambert, & Nissen (1993). The reason for this interest is threefold:

i) Detection of ^{6}Li in halo turnoff stars puts strong limits on the possible depletion of ^{7}Li, and thus allows better determination of the primordial ^{7}Li abundance from the observed Li abundance of stars on the 'Spite plateau'.

ii) ^{6}Li abundances as a function of [Fe/H] provide an additional test of theories for the production of the light elements Li, Be and B by interactions between cosmic ray nuclei and ambient ones.

iii) Information on depletion of ^{6}Li as a function of stellar mass and metallicity puts new constraints on stellar models in addition to those set by ^{7}Li depletion. This is so because the proton capture cross section of ^{6}Li is much larger than that of ^{7}Li. Hence, at a given metallicity there will be a mass interval, where ^{6}Li but not ^{7}Li is being destroyed according to standard stellar models.

Altogether, ^{6}Li abundances may contribute to the study of such different fields as Big Bang nucleosynthesis, cosmic ray physics and stellar structure. It will, however, require a rather large data set of ^{6}Li abundances to get information in all these areas. The most metal-poor stars around the turnoff are of particular interest in connection with the determination of the primordial ^{7}Li abundance, whereas more metal-rich halo stars and disk stars are of interest for the study of the formation and astration of the light elements.

2. Methods to determine the Li isotope ratio

The ^{6}Li/^{7}Li ratio in stars can be determined by two methods: From the center-of-gravity (*cog*) of the Li I 6708 Å line or from a detailed model atmosphere synthesis of the profile. The isotopic shift of the ^{6}Li doublet is $+0.158$ Å relative to the ^{7}Li doublet. Addition of ^{6}Li therefore shifts the Li I line to longer wavelengths and increases the width. The *cog*-method relies in principle on a very simple and straightforward measurement, but its accuracy is limited by possible errors in the laboratory wavelengths of the lithium line and the reference lines needed to correct for the radial velocity shift of the star. Furthermore, differences in convective blueshifts of the lines may be a problem. Hence, the profile method is superior to the *cog*-method and the exact wavelength of the Li I line should be considered as a free parameter in the comparison between synthetic and observed profiles.

In order to determine the Li isotopic ratio with good accuracy, the profile of the Li I 6708Å line should be observed at high resolution and very high S/N. Experience shows that $R \simeq 100\,000$ and $S/N > 400$ are required to obtain errors less than ± 0.02 in ^{6}Li/^{7}Li.

The synthesis of the Li I line has to be based on a model atmosphere with the same effective temperature, surface gravity and composition as the star. As discussed by e.g. Smith et al. (1993) the errors of these parameters do not add significantly to the error of the ^{6}Li/^{7}Li determination. Furthermore, data for the fine structure and hyper-fine structure splitting of the Li line as well as the isotopic shift between the ^{6}Li and ^{7}Li components is known with superior accuracy; see e.g. Table 3 of Smith, Lambert, & Nissen (1998).

Due to the weakness of the lithium line in the solar spectrum one has a good possibility to see if it is blended by other lines. As discussed by Nissen et al. (1999), the blue wing of the Li line is contaminated by a weak Fe I line ($\lambda 6707.43$) and a few very weak CN lines. In the red wing there are two unidentified lines with equivalent widths of 0.6 and 1.1 mÅ. For more metal rich stars these blends are a problem and limits the obtainable accuracy of the ^{6}Li/^{7}Li determination. For metal-poor disk stars with $-0.8 <$ [Fe/H] < -0.5 only the Fe I line in the blue wing plays a role, and for halo stars with [Fe/H] below say -1.5 this line has also disappeared.

The basic problem in the synthesis is to determine the broadening of the lithium line due to rotation and turbulent motions in the stellar atmosphere. In the studies performed so far a homogenous, plane parallel model has been adopted and a symmetric velocity broadening profile has been determined from other lines in the spectrum with about the same strength as the lithium line. Due to convective motions in the atmosphere a slight asymmetry is expected, but as discussed by Smith et al. (1998) the asymmetry is probably too small to affect the derived ^{6}Li/^{7}Li ratio significantly. This has recently been confirmed by Asplund (2000) from an analysis of the Li line in HD 84937 based on inhomogeneous 3D hydrodynamical model atmospheres (Asplund et al. 1999) for which the only free parameter is the rotation velocity of the star.

As an example of the method of determining ^{6}Li/^{7}Li, Figs. 1 and 2 show the synthesis of the lithium line for two metal-poor disk stars, HR 8181 and HD 130551, as performed by Nissen et al. (1999). The atmospheric velocity

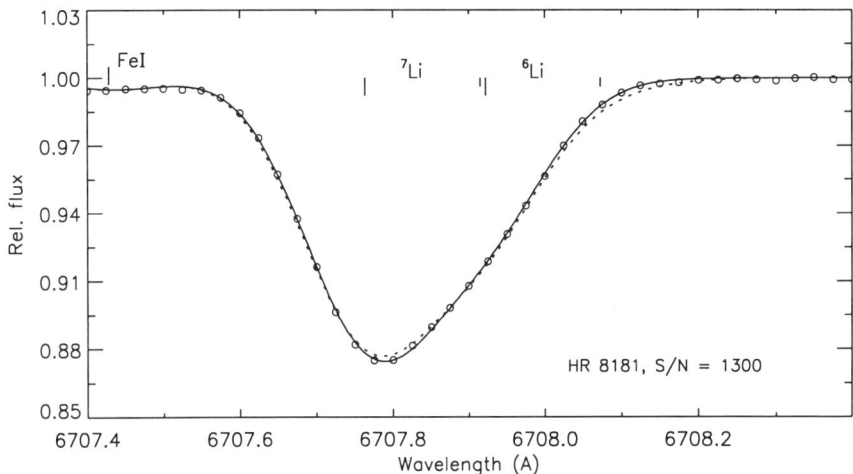

Figure 1. The model atmosphere synthesis of the the Li I 6707.8 Å line in the spectrum of HR 8181. The datapoints are shown with open circles. The full drawn line corresponds to ^{6}Li/^{7}Li = 0.0 (the best fit) and the dotted line to ^{6}Li/^{7}Li = 0.05. Note, that when ^{6}Li/^{7}Li is varied the other free parameters in the fit, the wavelength and the equivalent width of the line, are optimized to get the best possible fit

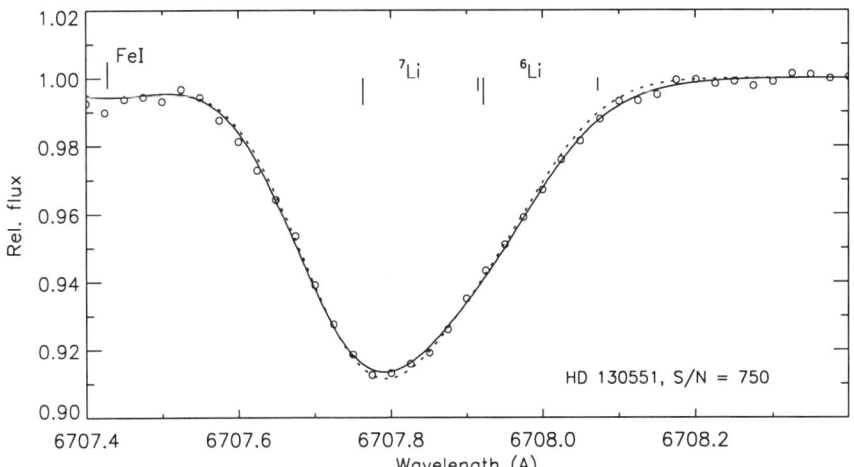

Figure 2. Same as Fig. 2 for HD 130551. Here the full drawn line corresponds to ^{6}Li/^{7}Li = 0.06 (the best fit) and the dotted line to ^{6}Li/^{7}Li = 0.0

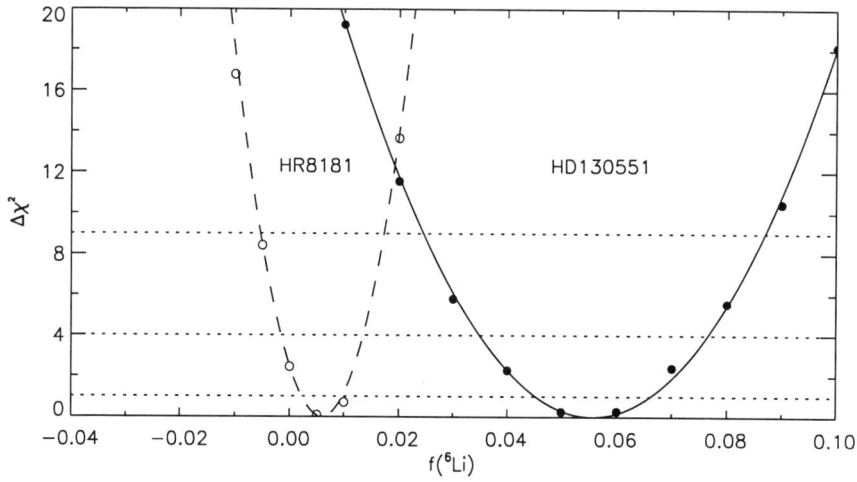

Figure 3. Variation of the χ^2 of the fit to the Li I 6707.8 Å line as a function of the relative abundance of ^{6}Li for two stars, HR 8181 and HD 130551

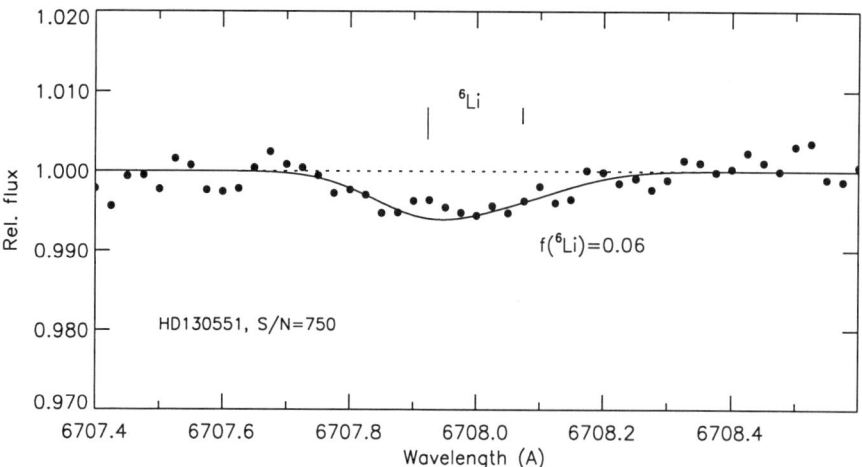

Figure 4. The residuals of the observations of HD 130551 after subtraction of the ^{7}Li and Fe I 6707.43 Å part of the synthesis of the Li I line. For comparison the synthesis of the ^{6}Li doublet is shown with a full drawn line

broadening was determined from two nearby Fe I lines ($\lambda 6703.6$ and $\lambda 6705.1$) of about the same strength as the Li line. A good fit ($\chi^2_{red} \simeq 1$) to these lines was obtained for a Gaussian broadening profile with FWHM parameters of 5.5 and 6.6 km s^{-1} for HR 8181 and HD 130551, respectively. These parameters have then been applied in a χ^2 fit of the Li line using the relative abundance of ^{6}Li, the total Li abundance and the wavelength of the line as free parameters. The χ^2 variation as a function of $f(^6\text{Li}) = {}^6\text{Li}/({}^6\text{Li}+{}^7\text{Li})$ is shown in Fig. 3. Note, that the steeper variation of χ^2 for HR 8181 is due to the higher S/N (1300) of the observations than in the case of HD 130551 ($S/N = 750$). It should also be stressed that for each value of $f(^6\text{Li})$, the other free parameters are optimized to get the lowest possible value of χ^2. $\Delta\chi^2 = 1$, 4 and 9 then correspond to the 1, 2, and 3σ confidence limits of $f(^6\text{Li})$ (Bevington & Robinson 1992). The setting of the continuum may also be considered as a free parameter in the fit, but normally it will be well constrained by the continuum on each side of the Li line.

As seen from Fig. 3, ^{6}Li is not present in HR 8181 but is detected in HD 130551 at about the 3σ confidence level. The presence of ^{6}Li can be seen from Fig. 2, and more clearly from Fig. 4, which shows a plot of the residuals in the observations after subtracting the ^{7}Li and Fe I part of the synthesis. A residual absorption at the wavelength of the ^{6}Li doublet is present.

3. Metal-poor disk stars

Nissen et al. (1999) have studied the Li isotope ratio in 5 metal-poor disk stars in the turnoff region of the HR diagram. The observations were carried out with the ESO 1.4m CAT and 3.6m telescopes using the CES instrument to obtain $R \simeq 110\,000$ spectra in the lithium line region. The derived ^{6}Li/^{7}Li ratio is given in Table 1 together with T_{eff} and [Fe/H] as well as the absolute magnitude derived from the Hipparcos parallax and the mass derived from comparing the position of the star in the T_{eff}-M_V diagram with evolutionary tracks from VandenBerg et al. (2000) based on stellar models with enhanced abundances of the α-elements, [α/Fe] $= 0.3$. In estimating the error of the mass, both the error of T_{eff} (± 70 K) and the error of M_V have been taken into account.

Table 1. Effective temperatures, absolute magnitudes, metallicities, masses and Li isotope ratios for 5 metal-poor disk stars. The quoted errors and upper limits are one sigma values.

ID	T_{eff}	M_V	[Fe/H]	$\mathcal{M}/\mathcal{M}_\odot$	^{6}Li/^{7}Li
HR 2883	5980 K	$3.55 \pm .06$	-0.75	$1.02 \pm .02$	< 0.02
HR 3578	5970	$4.16 \pm .04$	-0.82	$0.88 \pm .02$	< 0.01
HR 8181	6140	$4.41 \pm .01$	-0.67	$0.96 \pm .02$	< 0.01
HD 68284	5880	$3.41 \pm .19$	-0.59	$1.07 \pm .04$	$0.04 \pm .01$
HD 130551	6240	$3.77 \pm .09$	-0.62	$1.06 \pm .02$	$0.06 \pm .01$

As seen from Table 1, the two stars with ^{6}Li present have significantly higher masses than the three with non-detections. This makes sense, because the depth of the convection zone in a star on the main sequence decreases rapidly as a function of increasing mass. Hence, according to standard stellar models without mixing, the depletion of ^{6}Li is less severe in the more massive stars. In this connection we note that although HD 68284 is the coolest of the stars, it is a subgiant that has spent most of its life as a main sequence star at $T_{\rm eff} \simeq 6300$ K.

Hobbs & Thorburn (1997) have also studied a few metal-poor disk stars based on observations with the 2.7m reflector and coudé spectrograph at McDonald Observatory. No detections of ^{6}Li was obtained. The tightest upper limit, ^{6}Li/^{7}Li <0.02, was found for HD 134169 a subgiant with $T_{\rm eff} \simeq 5800$ K, [Fe/H] $\simeq -1.0$ and $M_V = 3.82 \pm 0.14$. From the evolutionary tracks of VandenBerg et al. (2000) we derive a mass $\mathcal{M}/\mathcal{M}_\odot = 0.90 \pm 0.03$, again significantly lower than the mass of the two stars with ^{6}Li detections.

4. Halo stars

The primary candidate in the search for ^{6}Li has been HD 84937 – the brightest metal-poor halo star near the turnoff. A probable detection of ^{6}Li was obtained by Smith et al. (1993) based on observations with the coudé spectrograph at McDonald Observatory. A later refinement of the analysis of these observations (Smith et al. 1998) resulted in ^{6}Li/^{7}Li $= 0.06 \pm 0.03$. Independent observations and analysis by Hobbs & Thorburn (1994, 1997) led to ^{6}Li/^{7}Li $= 0.08 \pm 0.04$. Recently, Cayrel et al. (1999) have observed HD 84937 with the GECKO spectrograph at the CFHT at a resolution of 10^5 and a S/N as high as 650 yielding ^{6}Li/^{7}Li $= 0.052 \pm 0.019$. The weighted mean of these values is ^{6}Li/^{7}Li $= 0.059 \pm 0.016$, where all errors quoted are one sigma values.

The paper of Cayrel et al. (1999) is particular interesting. Due to the very high S/N obtained, the χ^2 fitting of the Li line was carried out by including the FWHM of the Gaussian broadening function as a free parameter in addition to ^{6}Li/^{7}Li, the wavelength zero-point, the total abundance of Li, and the continuum setting. This is possible, because the slope of the blue wing of the Li line primarily depends on the FWHM of the broadening function, whereas the red wing depends both on the FWHM and ^{6}Li/^{7}Li. After having determined the FWHM parameter from the Li line it was checked if the Ca I line at 6162 Å could be fit with the same value, which is indeed the case. Cayrel et al. also show that the extra absorption seen in HD 84937 at the position of the ^{6}Li doublet is very unlikely to be due to a binary component. Furthermore, there is no sign of radial velocity variations of the star based on 30 years CORAVEL and ELODIE observations although such variations have been suspected by Carney et al. (1994).

Apart from HD 84937, ^{6}Li seems to have been detected in BD +26 3578 by Smith et al. (1998) based on a spectrum with $S/N \simeq 400$. Hobbs & Thorburn (1994, 1997) also studied this star, but due do a lower S/N they were only able to place an upper (3σ) limit, ^{6}Li/^{7}Li < 0.09. Another 15 halo stars have been searched for ^{6}Li by Smith et al. (1998), Cayrel et al. (1999), and Hobbs, Thorburn & Rebull (1999) but without detections. Table 2 lists the results for a group of [Fe/H] $\simeq -2.0$ halo stars with detections or tight upper limits of ^{6}Li/^{7}Li. The absolute magnitudes are based on Hipparcos parallaxes or (for the more distant stars) on the Strömgren c_1 index, and masses are derived from a comparison with the α-enhanced evolutionary tracks of VandenBerg et al. (2000). As in the case of the metal-poor disk stars, the two stars for which ^{6}Li is detected are among those having the highest masses. It is, however, puzzling that several other stars (HD 74000, HD 160617 and BD +20 2603) have equally high masses within the errors. Hence, it cannot be excluded that there is a real dispersion in the ^{6}Li abundance at a given mass and metallicity.

Table 2. Same as Table 1 for 7 halo turnoff stars with metallicities [Fe/H] $\simeq -2$

ID	T_{eff}	M_V	[Fe/H]	$\mathcal{M}/\mathcal{M}_\odot$	^{6}Li/^{7}Li
HD 19445	5870 K	5.12 ± .10	−2.00	0.65 ±.03	< 0.03
HR 74000	6190	4.47 ± .25	−1.80	0.70 ±.04	< 0.03
HR 84937	6310	3.81 ± .19	−2.20	0.75 ±.03	0.059 ± 0.016
HD 160617	5960	3.49 ± .25	−1.90	0.79 ±.06	< 0.03
HD 218502	6000	4.03 ± .18	−2.00	0.69 ±.04	< 0.02
BD +20 2603	6210	3.90 ± .25	−2.20	0.73 ±.04	< 0.02
BD +26 3578	6310	3.08 ± .25	−2.40	0.85 ±.07	0.05 ± .03

5. Discussion and conclusions

Fig. 5 shows a comparison of the ^{6}Li data from Smith et al. (1998) and Nissen et al. (1999) with three recent Galactic evolutionary models. In the model of Fields & Olive (1999) the 'standard' picture of Galactic cosmic ray nucleosynthesis of Li by spallation of C,N,O nuclei and $\alpha + \alpha$ fusion is adopted. The cosmic rays are assumed to have the composition of the average ISM at a given epoch, and the energy spectrum is that of present days cosmic rays in the solar neighborhood i.e. relativistic energies are dominant. A novel ingredient in the model of Fields & Olive is the incorporation of recent measurements of the oxygen abundance in halo stars indicating that [O/Fe] increases steply with decreasing [Fe/H] (Israelian et al. 1998, Boesgaard et al. 1999). Furthermore, the time integrated Li,Be,B outputs are normalized to the ^{6}Li, Be and ^{10}B abundances in meteorites. Vangioni-Flam et al. (1999) invoke an additional low energy cosmic ray component (LECR) which they associate with the acceleration of supernovae ejecta in superbubbles created collectively by winds from massive stars in OB associations. A key point about this component is that the He, C, and O abundances of the ejecta are considered to be much higher than in the halo interstellar medium and, then, the dominant spallation process is between O in the ejecta and protons in the interstellar gas, whereas in the standard picture the leading process is between protons in the cosmic rays and O in the interstellar gas.

As seen from Fig. 5, the models of Fields & Olive (1999) and Vangioni-Flam et al. (1999) reproduce the observed ^{6}Li abundances of the two halo stars very well. Ramaty et al. (2000) have, however, criticized these models for requiring an unrealistic high energy input of the supernovae into cosmic ray acceleration. Another problem with the two models is the high ^{6}Li abundance predicted in the

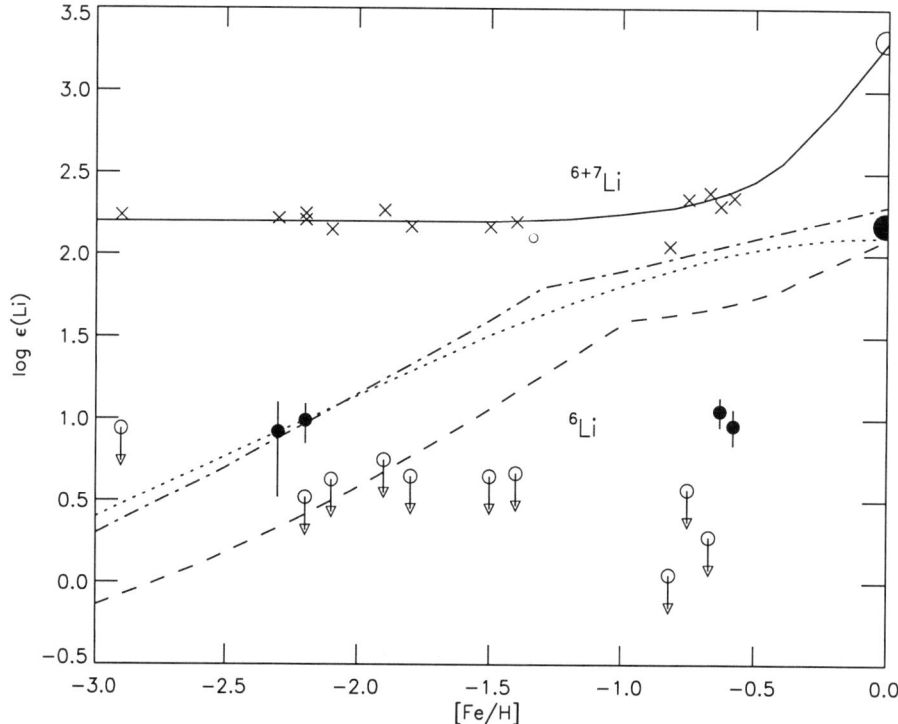

Figure 5. Abundances of the lithium isotopes as a function of [Fe/H] for 9 halo stars from Smith et al. (1998) and 5 disk stars from Nissen et al. (1999). Crosses indicate the total Li abundance, filled circles ^{6}Li abundances with one-sigma error bars, and open circles one-sigma upper limits of the ^{6}Li abundance. The big symbols indicate meteoritic abundances from Anders & Grevesse (1989). The upper full drawn line is a fit to the 'Spite plateau' of lithium abundances for [Fe/H] < -1.5 and to the upper envelope of the Li abundance distribution for disk stars according to Lambert et al. (1991). The dotted line represents the evolution of ^{6}Li in the model of Fields & Olive (1999), the dashed-dotted line the model of Vangioni-Flam et al. (1999), and the dashed line the model of Ramaty et al. (2000)

metallicity range $-1.0 < [\text{Fe/H}] < -0.5$ corresponding to $^6\text{Li}/^7\text{Li}$ of the order of 0.3 to 0.4. The measured ^6Li abundances in HD 68284 and HD 130551 are about a factor of 10 lower than predicted requiring a very large ^6Li depletion factor for the stars. Furthermore, the ^6Li production is accompanied by a ^7Li production in the ratio $(^7\text{Li}/^6\text{Li})_{CR} \simeq 1.5$ according to the well known cross sections for cosmic ray production of the two isotopes. Added to the primordial ^7Li value this gives an expected ^7Li abundance in HD 68284 and HD 130551 of about $\log \epsilon (^7\text{Li}) = 2.50$, considerably higher than the observed value of 2.30. Hence, a substantial ^7Li depletion in the stars is also required, if the two models are correct.

Ramaty et al. (2000) have studied the evolution of ^6Li in two models, the CRI model in which the cosmic ray composition is similar to that of the ISM at all epochs, and the CRS model in which the cosmic rays are accelerated out of the supernovae ejecta in superbubbles and hence always have a composition similar to that of present days cosmic rays. The energy spectrum is given by an expression appropriate for shock acceleration implying a spectrum extending to ultrarelativistic energies. The evolution of Li, Be and B is computed together with the evolution of O and Fe, and it is shown that the CRI model cannot reproduce the observed Be abundances as a function of O or Fe, whereas the CRS model is doing well. Both models fails, however, to account for the observed ^6Li abundance of the two halo stars. As seen from Fig. 5, the CRS model underpredicts ^6Li with about a factor of 4, and the CRI model fails with a still larger margin. The models require, on the other hand, only a moderate depletion of ^6Li in HD 68284 and HD 130551 and no depletion of ^7Li to fit the observations.

The discrepancy between the model predictions of Ramaty et al. (2000) and the observed values of ^6Li in HD 84937 and BD +26 3578 raises the interesting question, if there are other (pregalactic?) sources of ^6Li. Before answering this question a much larger data set ^6Li abundances must be obtained. We need detections of ^6Li in several additional metal-poor halo turnoff stars, before we can be confident that the abundance found in HD 84937 and BD +26 3578 represents the interstellar value at $[\text{Fe/H}] \simeq -2.3$, especially because other stars with about the same mass and metallicity do not have measurable abundances of ^6Li in their atmospheres. Furthermore, ^6Li data for more metal rich stars are needed to be able to study the evolution and astration of lithium in the Galaxy.

Up to now the search for ^6Li has been carried out with high resolution ($R \simeq 100\,000$) coudé spectrographs attached to 3-4 meter class telescopes. The limiting magnitude has been around $V \simeq 9.5$ meaning that only a few halo turnoff stars could be reached. With the advent of 8 meter class telescopes and new more efficient high resolution echelle spectrographs like UVES at the ESO VLT, HRS at the Hobby Eberly Telescope, and HDS at Subaru, the limit can be extended to $V \simeq 12$. This opens the possibility for a large ^6Li survey of halo turnoff stars with hopefully new rewarding results.

References

Anders E., Grevesse N. 1989, Geochim. Cosmochim. Acta 53, 197

Asplund M. 2000, (This volume)

Asplund M., Nordlund Å, Trampedach R., Stein R.F. 1999, A&A 346, L17
Bevington P.R., Robinson D.K. 1992, *Data reduction and error analysis for the physical sciences*, McGraw-Hill, p.212
Boesgaard A.M., King J.R., Deliyannis C.P., Vogt S.S. 1999, AJ 117, 492
Carney B.W., Latham D.W., Laird J.B., Aguilar L.A. 1994, AJ 107, 2240
Cayrel R., Spite M., Spite F., E. Vangioni-Flam, Cassé M., Audouze J. 1999, A&A 343, 923
Fields B.D., Olive K.A. 1999, New Astronomy 4, 255
Hobbs L.M., Thorburn J.A. 1994, ApJ 428, L25
Hobbs L.M., Thorburn J.A. 1997, ApJ 491, 772
Hobbs L.M., Thorburn J.A., Rebull L.M. 1999, ApJ 523, 797
Israelian G., García López R.J., Rebolo R. 1998, ApJ 507, 805
Lambert D.L., Heath J.E., Edvardsson B. 1991, MNRAS 253, 610
Ramaty R., Scully S.T., Lingenfelter R.E., Kozlovsky B. 2000, ApJ (in press)
Smith V.V., Lambert D.L., Nissen P.E. 1993, ApJ 408, 262
Smith V.V., Lambert D.L., Nissen P.E. 1998, ApJ 506, 405
VandenBerg D.A., Swenson F.J., Rogers F.J., Iglesias C.A., Alexander D.R. 2000, ApJ (in press)
Vangioni-Flam E., Cassé M., Cayrel R., Audouze J., Spite M., Spite F. 1999, New Astronomy 4, 245

Li Abundance in Pop I Stars

Luca Pasquini

*European Southern Observatory, Karl Schwarzschild Strasse 2
85748 Garching bei München, Germany*

Abstract.
The study of Li in Pop I stars has focussed in the last years on observations of open clusters, spanning a large range of ages and metallicities. So far the observational picture is quite complex: the data indicate several phenomena which are not well understood, from the 'Li dip', to the scatter in Li abundances found among (otherwise similar) stars both in young and old clusters. Models fail in reproducing most of the observed features; in particular the almost total lack of dependence of Li depletion with metallicity. The comparison between clusters of different ages show that no PMS depletion occurs, and that for a large fraction of old star no additional depletion occurs after the first ~1.5 Gyrs, while the depletion is strong at earlier ages. These observations are crucial for our understanding of the relationship between the Pop II Li Plateau and the Li primordial abundance.

1. Introduction

After the pioneering work of Spite and Spite (1982), the current paradigm is that while Pop II (metal poor) stars give (or are close to) the primordial Li abundance, Li evolution in Pop I stars can be used to trace the Li galactic enrichment and the phenomena occurring in the stellar interiors. It is however clear that it is definitely uncomfortable to use the Pop II (Spite) Li plateau as a signature of the primordial Li abundance as long as we do not understand how Li is produced in the Galaxy and depleted in stars. The study of Li in Pop I stars is therefore also relevant to the measurement of its primordial abundance.

In this review I will concentrate on *observations* of solar MS stars i.e. stars cooler than 6200 K and warmer than 5000 K, therefore not discussing either the whole 'Li dip' problem (Boesgaard and Tripicco 1986) or the interesting lower MS stars (Martin 2000, this volume). Although with a somewhat different approach, excellent recent reviews can be found in Jeffries 2000, Delyiannis 2000. In recent years most observations have been carried out among open cluster stars. The main reason for studying in detail open clusters is very simple: they represent the best approximation of a simple, homogeneous population: by studying them and covering an as wide as possible stellar parameter space (temperature, metallicity, age) we will be able to disentangle the dependence of Li evolution on stellar parameters. The optimization of high resolution spectrographs coupled to 4 and 8 meter telescopes has contributed to the observations of the often faint

Table 1. Summary of recent open clusters Li observations. For an explanation of the columns see the text.

OC	Age	Fe/H	Nstar	Sp. Ty	Ref.
NGC 2264	6.6*	-0.15	6,28	FG, GK	K1998, S99*
IC2602	7.36	-0.1*	25,26	FGKM,GKM	R97a,M99*
IC4665	7.58	?	14	GKM	MM97
IC2391	7.64	-0.1*	10,22	GKM, GKM	S89,M99*
Blanco 1	7.70	0.12*	39,17	FGK, GK	PO97,JJ99*
NGC2516	7.79	0.06	24	FGK	J98
NGC2547	7.88	-0.16	34	KM	J99
α Per	7.90	0.1	5,3,29,18	F,M,FGK,KM	B88,GL94,B96,R98
Pleiades	7.92	-0.034*	17,95,13,15,8	F,FGK,K,KM,GK	B88,S93,GL94,J96,R96
M35	8.0	-0.17	39	GK	BN99
NGC6475	8.11	+0.08	35,49	F,G,K	JJ97,R99a
NGC1039	8.26	-0.26	34	FGK	J97
Coma	8.69	-0.03	16,5,15,11	F,FG,FGK,GK	B87,S90,J99a,F99
NGC6633	8.66	0.0	21	FGK	Je97
Hyades a	8.92	0.12	32,14,23,68,12	F,F,FGK,K	B87,BB88,S90,T93,S95
Praesepe a	8.92	0.07	63	FG	S93
NGC752 a	9.15	-0.16	19,6	FG,F	HP86,HP88
NGC3680 a	9.18	-0.16	11	FGK	P98,R00
IC4651 a	9.25	-0.16	14	FGK	R00
NGC2243 a	9.75	-0.56	11	FG	H99
M67 a	9.80	-0.09	7,6,14,25	FG,F,FG,FG	HP86a,S87,P97,J99b
NGC188 a	9.86	-0.05	7	F	HP88a

open cluster stars. The results are impressive: from a total of 17 studies on 9 Clusters, covering less than 60 stars in clusters in addition to the well studied Hyades, Praesepe and Pleiades published prior to 1994, we have reached the present situation, summarized in Table 1, where more than 40 studies have been carried out on a total of 22 Clusters, with observations of more than 400 stars in the 19 clusters in addition to the 3 classic ones. In Table 1 a summary is given: in addition to the cluster name, age is given (following Mermilliod scale for the young clusters, the Friel scale for the old ones) metallicity (when spectroscopic determination is available, this has been preferred and they are marked with * in Table 1), number of stars observed/work, their spectral type and an abbreviation for the work, while full references are given in the reference section. Although an effort for completeness has been performed, I apologize if references are missing. In the following of the paper the references given in Table 1 are not repeated when used; note that all the figures show data brought to the same temperature and Li abundance scales.

By using this impressive set of data (which is likely to grow rapidly in the next years, thanks to the use of the new generation of multi object high resolution spectrographs, see e.g. Pasquini 2000) we can start to answer to the following questions:

1. Does Pre Main Sequence (PMS) depletion occur?

2. Does Main Sequence (MS) depletion occur? How large is this MS depletion? On which timescales does it occur? Does it dependence on metallicity? On other parameters?

3. Is the Li abundance uniquely determined by stellar macroscopic parameters like T_{eff}, Mass, Age, Composition ?

2. PMS depletion

To know if strong PMS depletion occurs among solar stars is relevant because most evolutionary models predict a strong Li PMS depletion, and if this would occur, it should be taken into account when interpreting the Plateau as primordial Li. Figure 1 shows Li vs. effective temperature for three young clusters, together with the slightly older Pleiades. IC2602, IC2391, IC4665 span a range of a few million years in age; their G stars have just arrived on the MS, while the cooler K stars are at the end of the PMS phase, approaching the MS. They have different metallicity, although the range spanned is admittedly small. Considering the original solar system value N(Li)=3.3 and that the same value is found today in PMS stars after proper NLTE correction, it is clear from Figure 1 that the G stars in these clusters, with N(Li)~3.2, have experienced hardly any PMS depletion, with no dependence on metallicity. The situation is different for K Stars with effective temperatures below ~ 5000 K: among them in fact a rather strong depletion is observed: since they have not yet reached the MS this depletion occurred in the PMS phase. It is finally worth noticing that almost no differences in Li abundance exist between these clusters and the older Pleiades despite their 60-90 Myr younger age.

3. Main Sequence Depletion

The question of whether MS Li depletion occurs and how, is intimately linked to the physical mechanisms present in the stellar interior and to the mixing acting at the basis of the convective zone. It is worth remembering indeed that standard models predict for the Sun that the bottom of the convective zone is not deep enough to produce any Li burning; in addition, still according to these models, a Li depletion should strongly depend on stellar metallicity: more metal poor stars should suffer a much lower depletion (Pinsonneault et al. 1989, Swensson 1995). It is also worth pointing out that in all non-standard models (i.e. those including interior extra mixing, either produced by diffusion, rotation or gravity waves) the extra depletion predicted is really 'extra', that is, it should apply in addition to the depletion predicted by the standard ones; this implies that some of the basic features of the standard models should still hold (Jeffries 2000). In Figure 2 the data for 4 clusters of different ages are shown: Hyades (600 Myr), IC4651 and NGC3680 (1.6 Gyrs) and the 4 Gyrs old M67.

It is clear that, YES, main sequence Li depletion is indeed occurring. In particular, when comparing the Hyades with the younger Pleiades (cfr. Figure 1) in the ~500 Myrs between the two clusters a depletion of ~ 0.4 dex had occurred for the G stars. This MS depletion, on the other hand, tends to become much smaller with an increase in stellar ages: the comparison of the two intermediate

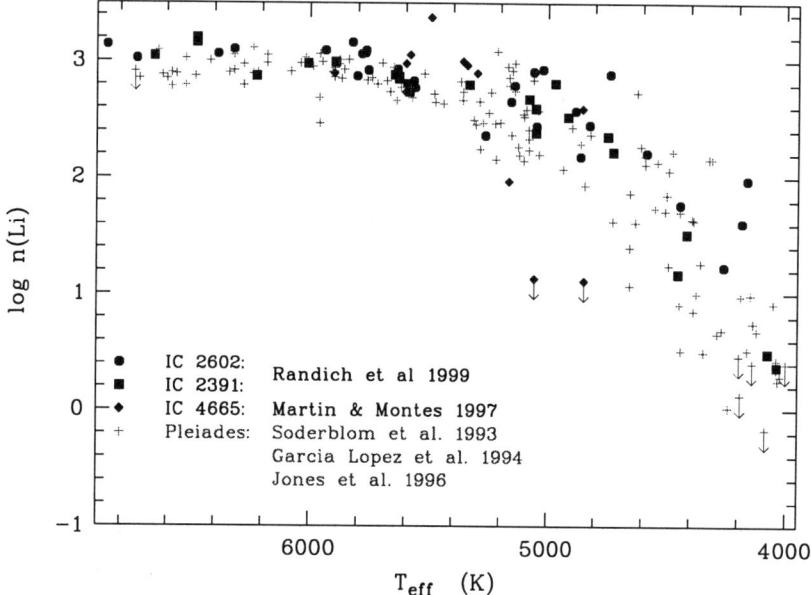

Figure 1. Li vs. effective temperature for three very young open clusters and the Pleiades

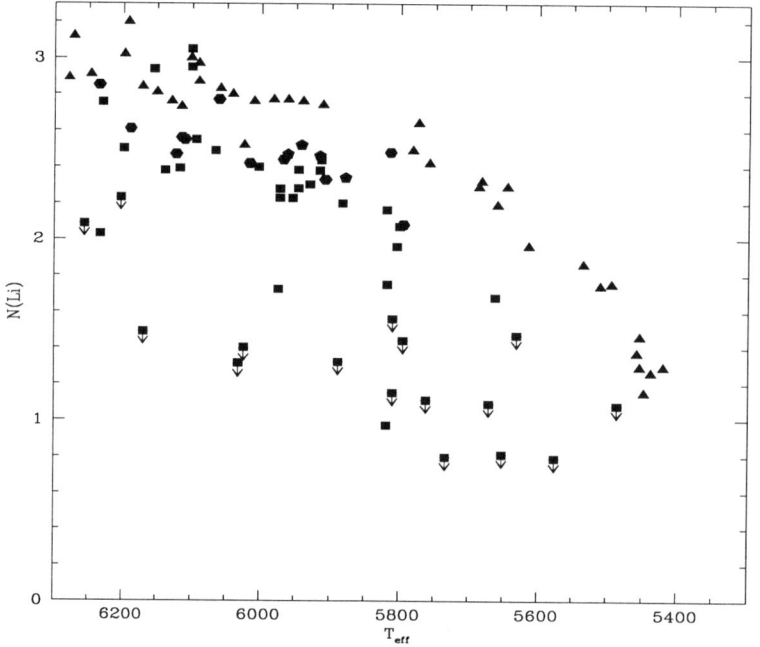

Figure 2. Li vs. effective temperature for Hyades (600 Myr, triangles), IC4651 and NGC3680 (1.6 Gyrs, pentagons and hexagons), M67 (4 gyrs, squares)

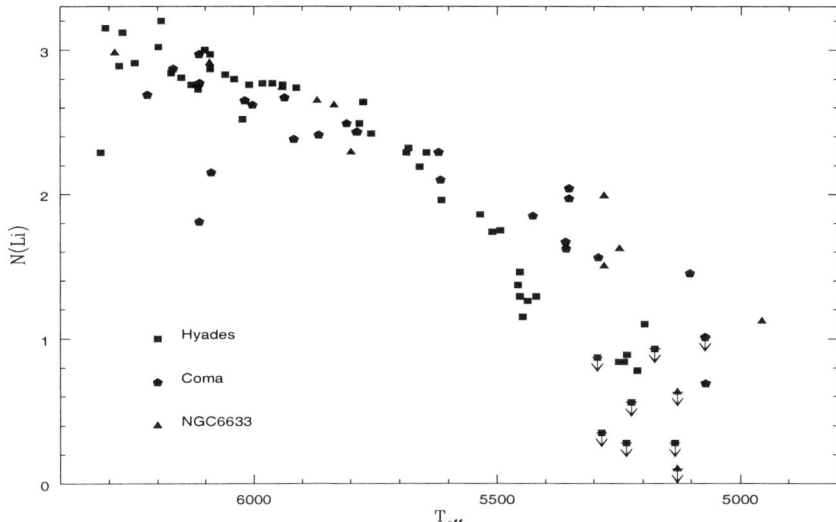

Figure 3. Li vs. effective temperature for three clusters with same age but different metallicity: Hyades ([Fe/H]=0.15), NGC6633([Fe/H]=0, Coma([Fe/H]=-0.03); no appreciable difference is observed among G stars; the effect of metallicity may be present among the K stars

age clusters IC4651 and NGC3680 with the Hyades shows that in the subsequent 1 Gyr only a very small amount of Li depletion occurs, of the order of 0.2 dex. The direct comparison between the intermediate age clusters and the upper envelope M67 stars, on the other hand, shows that for a large fraction of stars NO additional depletion between 1.6 and 4 Gyrs occurred.

3.1. Dependence on Metallicity?

The most relevant stellar parameter expected to determine depletion is stellar metallicity; to investigate this point in Figure 3 the Li data for 3 clusters with similar age but substantially different metallicity are shown: Coma, NGC6633 and the Hyades. Since, as shown above, most main sequence depletion has occurred at their age, this should be a rather stringent test. From Figure 3 it is evident that for G stars very little dependence exists on metallicity, if at all.

The situation could be different for the cooler K stars: pending more and more accurate data, for these stars a slight dependence on metallicity seems to be present, although at a much lower level than expected by theory.

One has to caveat, on the other hand, that according to several models stellar metallicity (which is here considered as given by Fe and a solar element pattern is assumed) is not the only relevant abundance, but Li depletion may be very sensitive to the abundance of elements like O, Mg, Si (Swenson et al. 1994). Unfortunately for most clusters these abundances are unknown and it would be extremely important to gather them soon.

3.2. Dependence on Rotation?

Stellar rotation is one of the parameters typically ignored in evolutionary models of low main sequence stars, but for the specific case of Li evolution, it may have a substantial role, either through meridional circulation (e.g. Vauclair, these proceedings) or rotationally induced mixing (e.g. Pinsonneault, these proceedings). The assessment of the relevance of rotation is not an easy task, because the driving parameter is expected to be the rotational history of the star, rather than its present, measured Vsini value.

The strongest case for the case of stellar rotation is given by short period binaries in open clusters (Delyiannis, these proceedings); in Figure 2, for instance, the M67 stars with the highest Li abundance refer to a short period binary system. Short period binaries are typically tidally locked, therefore their rotational period is much shorter than the corresponding one of single stars, which spin down during their main sequence lifetime.

Another strong case for rotation was presented by Soderblom et al. 1993, in their study of the Pleiades: among Pleiades K stars they found a large spread in Li abundances; and they showed that statistically the fast rotator K stars in the Pleiades have higher Li abundances than the slow rotating ones.

Unfortunately I believe that the arguments so far presented are not conclusive, on purely observational grounds.

Definitely more spectra of cluster short period binaries are required to ensure a proper sample; in M67, for instance, many exist and the case presented so far is not completely clear (cfr. the discussion in Pasquini et al. 1997).

Concerning the Pleiades K stars, the case of dependence on rotation is not at all clear: for these stars Soderblom et al. (1993) and Jeffries (1999) acquired potassium spectra. The potassium resonance lines are an important crosscheck, because these lines are formed in the same layers of the Li resonance doublet. In Figure 4 the equivalent widths of the potassium lines vs. the Li lines are shown for the early K stars in the Pleiades. If the spread of Li abundance could be explained with differential depletion, for the potassium lines this argument does not apply, since potassium is not burnt at low temperatures. Some of the scatter can be given by measurement errors (the equivalent widths measured by Jeffries and Soderblom for stars in common may vary by more than 100 mÅ), but it may also be that for these rapid rotators our models are inadequate to represent their atmospheres, or for instance, that the effective temperatures derived by colors are not appropriate. I believe that this point should be addressed as early as possible. Similar conclusions were reached by King et al (2000), who also found the same problem in the analysis of the potassium and Li lines, but in addition, found a possible dependence of the potassium line strength with a chromospheric index derived from the Ca II triplet.

Finally, the 4 G stars belonging to NGC3680 have rather different measured Vsini: 1.9,1.6,1.0 and 8 Km/sec but same Li; also, the median Hyades G stars Vsini is 6 Km/sec and as shown in Figure 3, the difference in Li between these clusters is very small. I am not claiming that rotational velocity is not relevant in Li evolution, rather that its relevance is not yet well established and understood on observational grounds.

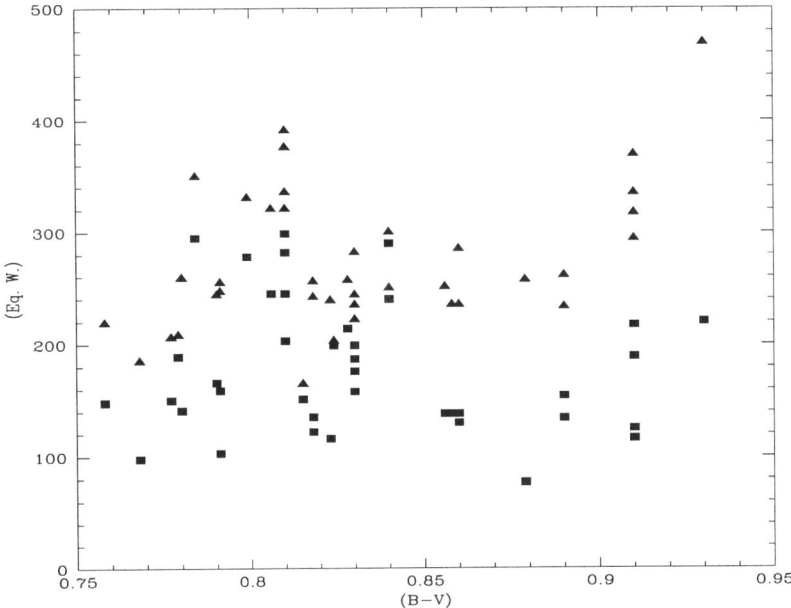

Figure 4. K equivalent widths vs. Li equivalent widths for Pleiades K stars, as measured by Jeffries 1999 and Soderblom et al. 1993

4. Li Scatter?

One advantage of studying stars in clusters is that we can assume a very simple population approximation: all the stars in a cluster share age, history and metallicity; they are perfect targets to investigate whether, at a given age and mass, all stars share the same Li abundance or not.

From Figures 1, 2 and 3 it is clear that almost *no scatter* is present (the definitive assessment on the presence of a small scatter would require a much better quality data sample) among G stars belonging to one cluster for ages comprised between 30 Myr and 1.8 Gyrs.

Yes, scatter is present among the G stars belonging to the 4 gyrs old M67. This fact is evident and it has been discussed by Pasquini et al. 1997. A couple of interesting points seem to emerge from the M67 data set: in 30-40% of the G stars a very strong depletion occurs, similar to that observed in the sun, and their Li content is at least 10 times lower than that of the other, otherwise similar objects. The second point is that almost no stars appear at intermediate Li values, suggesting that a bimodal distribution may be present and that the mechanism responsible for this extra depletion could be rather fast: an on-off type which occurs in some of the stars but not in the others. Note that this distribution is likely not peculiar to the M67 clusters, i.e. due to the cluster environment, because the same behavior is observed among G stars in the field, although these stars represent a much less homogeneous sample (Pasquini et al. 1994).

As far as K stars are concerned, it is clear that a strong scatter is present in all clusters younger than the Hyades, therefore younger than ~ 600 Myr. For older stars the Li abundance is so low that to obtain proper observations may become prohibitive.

5. Pop I - Pop II Plateau Stars

In the introduction I anticipated that the study of the Li evolution in Pop I stars has an impact on our understanding of the Li Plateau and the primordial nucleosynthesis. In this section I would like therefore to summarize how the above observations relate to this topic.

The fact that no PMS depletion is observed among metal rich stars is a strong support to the primordial interpretation of the Plateau, because PMS depletion could represent a strong Li sink, as well as very difficult to measure.

As far as MS depletion is concerned, the fact that ~30% of the stars show extra depletion in M67, while a few 'outlayers' are found among the Pop II stars could also support the Plateau as signature of primordial Li. In fact this suggests that the depletion mechanism at its origin is strongly metallicity dependent, and therefore much less effective among the Plateau stars.

One result which I find rather annoying is that, once we have accepted that a number of 'outlayers' exists (even if their percentage is decreasing with metallicity), the upper envelope of the intermediate age and old clusters tends to flatten out for temperatures higher than ~5800K, or, in other words, Li depletion tends to 'saturate' among old G stars. The early (and so far only!) observations of the very old NGC188 show stars with Li abundances comparable to the younger M67. I expect that a considerable fraction of very old, metal rich G stars still have still a Li abundance around N(Li)=2.1-2.2.

The fact that 'standard' main sequence depletion is not strongly metallicity dependent, per se does not affect much the Pop II issue; however it shows clearly that current theories, which are the same predicting no Li depletion in Plateau stars, have some fundamental problem (see also Randich et al. 2000), and that the Li problem is not really understood. In order to advance, a more detailed work, similar to that performed for the Pop II stars, should be carried out, in order to define for the cluster stars accurate T_{eff}, and other elemental abundances like, e.g. O, Mg, Si. Very few data are available on old open clusters and on metal poor clusters! They are fundamental because they constitute the link between the Pop I and the Pop II stars. To understand if model atmospheres and stellar parameter determination are adequate is also fundamental, and, for instance Potassium spectra for K stars in young clusters could be essential. Finally, measurements of rotational velocities and magnetic fields proxies will help to shine some light on the role played by these parameters.

6. Summary of Evidence

Observations of G and K stars in Open Clusters start to provide basic constraints on Li evolution. While until a few years ago different sets of (mostly field) data could provide different answers, now a quantity of well established observational evidence exists, and it should be considered when elaborating Li evolution theories.

- No strong PMS Depletion occurs among G stars; apparently with no strong dependence with metallicity.

- A Main Sequence 'normal depletion' exists: most of the action takes place between ~100 and ~600 Myr, while it smooths considerably for older stars and may tend to saturation. No dependence of this depletion on metallicity is observed among G stars, while this may be present among the K stars.

- A Main Sequence 'extra depletion' is present among G stars: it affects up to 40% of the stars and takes place after ~ 1.6 Gyrs. It decreases the Li abundance of a factor ~10 or more. The distribution of Li abundance among old stars may be bimodal, indicating that this extra depletion occurs in a rather short period.

- For a given cluster, no large spread is detected for G stars younger than 1.6 Gyrs, but it is present at 4 Gyrs.

- Binarity (or something connected; magnetic activity?) may slow down considerably Li depletion.

Acknowledgments. I am indebted to V. Hill, R. Jeffries, R. Pallavicini and S. Randich for many discussions and for providing some of the data; to P. Bristow for a careful reading of the manuscript.

References

Balachandran, S., Lambert, D., Stauffer, J. 1996, ApJ 470, 1243 (B96)
Barrado Navasquez, D. et al. 1999, in "Cool stars stellar systems and the Sun XI ", Rebolo R. and Garcia-Lopez R. eds. in press. (BN99)
Boesgaard A.M., Tripicco, M.J. 1986, ApJ 303, 724
Boesgaard, A.M., 1987, ApJ 321, 967
Boesgaard, A.M. 1987, PASP 99, 1067 (B87)
Boesgaard, A.M. et al. 1988, ApJ 327, 389 (B88)
Boesgaard, A.M., Budge, K.G. 1988, ApJ 332, 410 (BB88)
Delyiannis, C. 2000, in " Stellar Clusters and Associations " R. Pallavicini et al. eds., ASP Conf. Ser. in press
Ford, A. et al. 1999, in "Cool Stars Stellar Systems and the Sun XI", Rebolo R. and Garcia-Lopez R. eds., in press. (F99)
Garcia Lopez, R., Rebolo, R, Martin, E. 1994, A&A 282, 518 (GL94)
Hill, V., Pasquini, L. 1999, These proceedings (H99)
Hobbs, L.M.., Pilachowski, C. 1986, ApJ 309, L17 (HP86)
Hobbs, L.M. Pilachowski, C. 1986, ApJ 311, L37 (HP86a)
Hobbs, L.M., Pilachowski, C. 1988, PASP 100, 336 (HP88)
Hobbs, L.M. Pilachowski, C. 1988, ApJ , 334, 734 (HP88a)

James, D.J., Jeffries, R.D. 1997, MNRAS 292, 252 (JJ97)
Jeffries, R.D. 1997, MNRAS 292, 177 (Je97)
Jeffries, R.D. 1999, in " Stellar Clusters and Associations " R. Pallavicini et al. eds., ASP Conf. Ser. in press
Jeffries, R.D., James, D.J. 1999, ApJ, 511, 218
Jeffries, R.D., James, D.J., Thurston, M.R. 1998, MNRAS 300, 550 (J98)
Jeffries, R.D. et al. 2000, in " Stellar Clusters and Associations " R. Pallavicini et al. eds., ASP Conf. Ser. in press (J99)
Jeffries, R.D. 1999, MNRAS 304, 821 (J99)
Jeffries, R.D. 1999, MNRAS 309, 189
Jones, et al. 1996, AJ 114, 352 (J96)
Jones B.F., et al. 1997, AJ 114, 352 (J97)
Jones, B.F, Fischer, D. Soderblom, D.R. 1999, AJ 117, 330 (J99b)
King, J.R. 1998, AJ 116, 254
King, J.R. Krishnamurthi, A., Pinsonneault, M.H. 2000, AJ 119, 859
Martin, E., Montes, L., 1997, A&A 318, 805 (MM97)
Meola et al. 99, in " Stellar Clusters and Associations " R. Pallavicini et al. eds., ASP Conf. Ser. in press (M99)
Panagi, P.M., O'Dell, M.A. 1997, A&AS 121, 213 (PO97)
Pasquini, L., Liu, Q., Pallavicini, R. 1994, A&A 287, 191
Pasquini, L., Pallavicini, R., Randich, S. 1997, A&A 325, 535 (P97)
Pasquini, L. 1998, in "Cool Stars Stellar Systems and the Sun X" A.K. Dupree and J. Bookbinder eds. CD-947 (P98)
Pasquini, L. 2000, in "VLT Opening Symposium" J. Bergeron ed., in press.
Pinsonneault, M.H. et al. 1989, ApJ 338, 424
Randich et al. 1997, A&A 323, 86 (R97a)
Randich, S. et al. 1998, A&A 333, 591 (R98)
Randich, S., Pallavicini, R., Mermilliod, J.C. 1999 These proceedings (R99a)
Randich, S., Pasquini, L., Pallavicini, R. 2000, A&A, in press (R00)
Russell, S.C. 1996, ApJ, 463, 593 (R96)
Soderblom, D.R. et al. 1990, AJ 99, 595 (S90)
Soderblom, D.R. et al. 1993, AJ 106, 1080 (S93)
Soderblom, D.R. et al. 1995, AJ 110, 729 (S95)
Soderblom, D.R. et al. 1999, AJ 118, 1301 (S99)
Spite, M., Spite F. et al. 1987, A&A 171, L1 (S87)
Stauffer, J. et al. 1989, ApJ 342, 285 (S89)
Swenson, F.J. et al. 1994, ApJ 425, 286
Thorburn, J. et al. 1993, ApJ 415, 150 (T93)

The Light Elements and Their Evolution
IAU Symposium, Vol. 198, 2000
L. da Silva, M. Spite, J. R. de Medeiros, eds.

Evolution of lithium abundance in Pop I giants

Sushma V. Mallik

Indian Institute of Astrophysics, Bangalore 560034, India

Abstract. Recent observations of cool giants have shown that there exists a large range in their lithium abundances even for apparently similar stars. In order to explore whether this large spread could be interpreted in terms of mass, the Li I line at 6707.8 Å has been observed in more than 100 stars and these data have been combined with the already known lithium abundance data in literature. Absolute magnitudes of these stars have been estimated from the Hipparcos data and an attempt has been made to understand the Li evolution of a star as a function of its mass and the evolutionary status. There are strong evidences for extra-mixing taking place on the red giant branch which explain the unusually low abundances observed.

1. Introduction

Observations of cool giants have indicated severe depletions in their lithium abundances, far in excess of the predictions of the standard stellar model calculations of *e.g.* Iben (1967). According to the standard stellar models, the surface dilution of Li for a $1M_\odot$ star at the tip of the RGB is 28 whereas for a $5M_\odot$ star it is 60. On the main sequence (MS), while the early F stars and possibly the still hotter stars appear to retain their initial Li log N(Li)=3.1 ± 0.2, it is seen to decrease towards the later spectral types. Consequently, low mass giants ($\lesssim 1.5\text{-}2.0 M_\odot$) are expected to have surface Li much smaller than and up to their maximum predicted value of 1.8 ± 0.3 and higher mass giants ($\sim 2\text{-}5 M_\odot$) near their maximum predicted value of 1.5 ± 0.3. Observations, however, reveal a large majority of giants having $-1.5\leq$log N(Li)$\leq+1.0$. These observations have often been interpreted in terms of the Li abundance being primarily controlled by the stellar mass. With the Hipparcos data (ESA 1997) it is possible to obtain absolute magnitudes of these stars to a high accuracy and infer their masses on an evolutionary diagram which would give a clue to what progenitors they have evolved from. It is worth exploring then how the Li abundance is linked with mass. With this in mind, a study of the Li I line at 6707.78Å has been undertaken in a sample of 109 stars. These are described in the next section followed by the interpretation of the observations in Section 3.

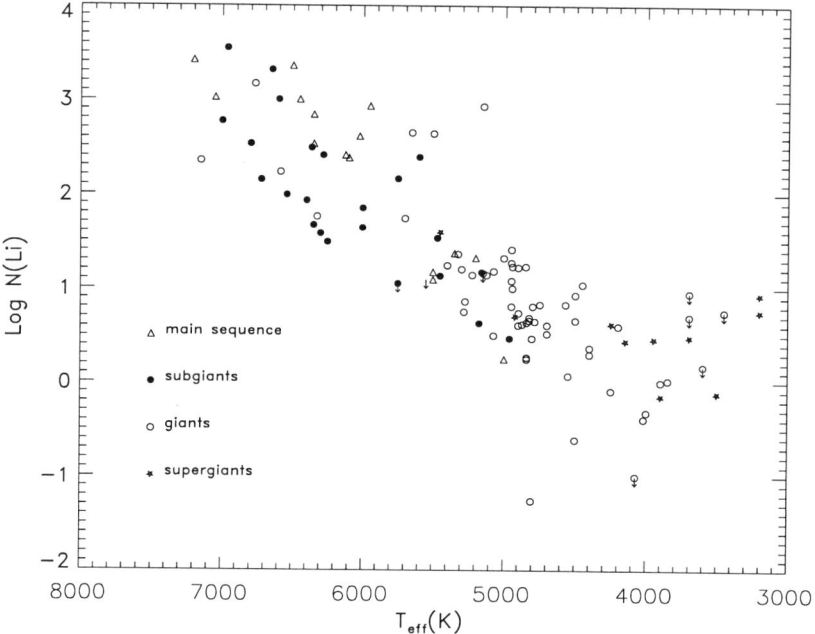

Figure 1. Lithium abundance vs. effective temperature for the 109 observed stars

2. Observations and Analysis

109 stars have been sampled from the Bright Star Catalogue (Hoffleit 1982) and the [Fe/H] Catalogue of Cayrel de Strobel et al. (1997) spanning spectral types from early F to late K mostly comprising of subgiants and giants. A few MS stars and supergiants have also been selected. CCD spectra of these stars have been obtained in the Li I region at a spectral resolution of ~0.35Å using the coudé echelle spectrograph at the 1m telescope at the Vainu Bappu Observatory at Kavalur, India. Standard procedure using the IRAF software has been adopted for reducing the data. With the measured EQWs as the input (after correcting for the Fe I blend at 6707.445Å) and an appropriate choice of the models based on the stellar parameters from the grid of model atmospheres due to Gustafsson et al. (1975, upgraded 1992) generated by Luck (1992), Li abundances have been determined using LINES (Sneden 1973, upgraded 1997). The Li abundance depends extremely sensitively, almost exclusively on T_{eff}. T_{eff} in the present study has been derived from the recent T_{eff}-(B−V) calibration of Flower (1996).

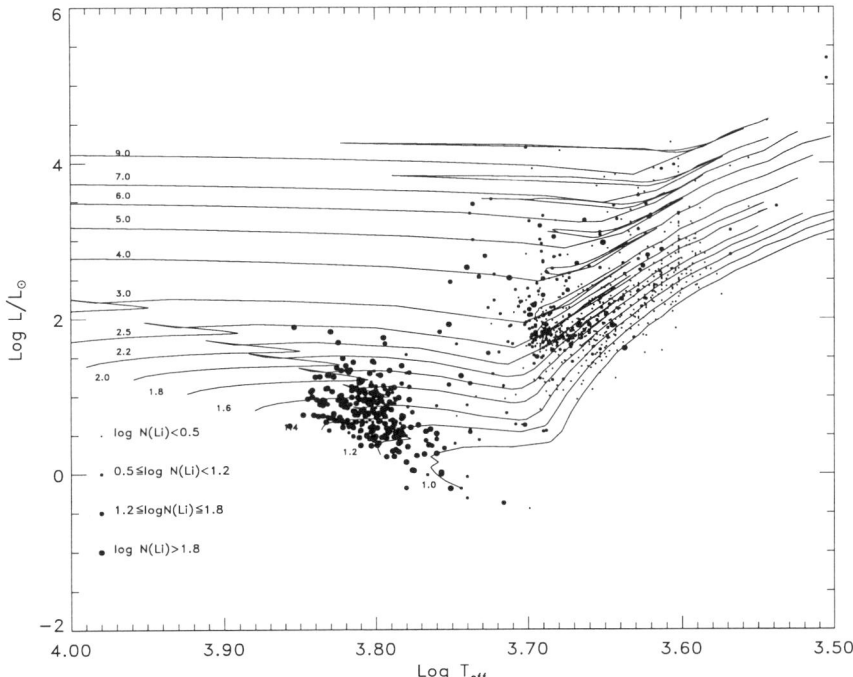

Figure 2. HR diagram of the entire sample of 1024 stars. The log N(Li) bins are indicated in the key. The theoretical evolutionary tracks are those of Bressan et al. for masses 1.0 to 9.0 $M_\odot$ for X=0.70, Y=0.28 and Z=0.02

3. Interpretation

3.1. Evolution of lithium up the red giant branch

Fig. 1 shows a gradual decline in the Li abundance for the stars observed as a function of T_{eff} - an evidence of the increasing dilution due to the deepening of the convective envelope. There is a large spread observed at any given T_{eff} which is perhaps due to the range in masses and evolutionary ages. But by and large, a large number of MS stars and the hotter subgiants have log N(Li) close to 3.0, supergiants appear to be heavily depleted and giants lie in between exhibiting a large range in abundances.

We have combined our data with the already known Li abundance data of another 915 stars from Balachandran (1990):113 MS stars and 49 subgiants and giants; Lèbre et al.(1999):104 subgiants; Pallavicini et al.(1987):6 subgiants; Brown et al.(1989):593 giants and supergiants; Lambert et al.(1980):38 giants and Luck (1997):12 supergiants. From the parallaxes and the apparent visual magnitudes given in the Hipparcos Catalogue, luminosities have been obtained for all these stars using the bolometric corrections from the T_{eff}-B.C. calibration of Flower (1996). The resulting accuracy in log $L/L_\odot$ is better than 0.08. There

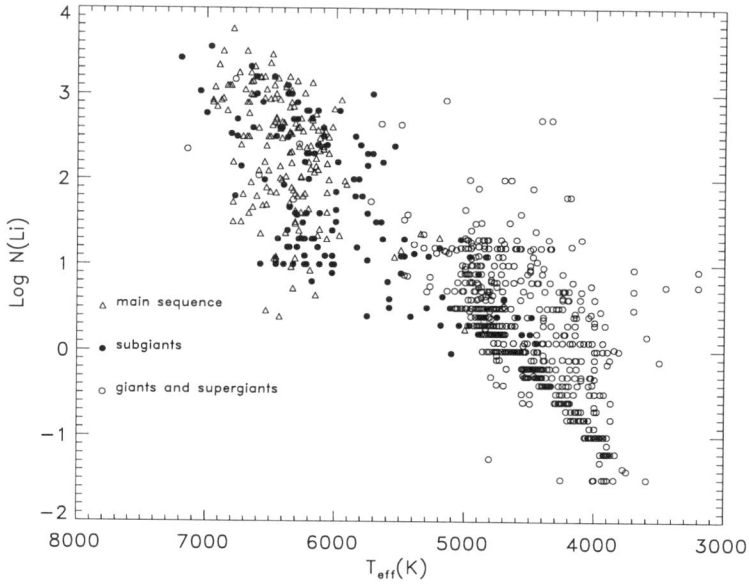

Figure 3. log N(Li) vs. T_{eff} for all the stars plotted in Fig.2

are a few supergiants whose distances exceed 100pc but the rest of the stars are all within 100pc, so the reddening effects are assumed negligible.

Fig.2 shows the positions of the 1024 stars on the HR diagram. Symbols of decreasing size indicate bins of decreasing values of the Li abundance. Superposed on these stars are the theoretical evolutionary tracks of Bressan et al. (1993) for masses ranging between 1 and $9 M_\odot$ with Pop I composition. Based on the location on the HR diagram, one finds that a fraction of subgiants lie very close to the turn-off point or on the MS. Similarly, several giants are actually subgiants and a few of them are supergiants. Although the sample is rather heterogeneous, its largeness aids in understanding the salient features of the evolution of the Li abundance of a star as a function of its mass and the evolutionary stage. Fig.2 shows a few very obvious trends :

1) Most of the stars around log T_{eff}=3.8 are either on the MS or evolving off it. Their MS Li has been largely preserved.

2) There is a paucity of stars in the range 3.78>logT_{eff}>3.70 corresponding to the Hertzsprung gap. Depletion as a result of convective dilution is seen in stars as they cross the subgiant branch.

3) The vast majority of the stars in the sample lie at log T_{eff} <3.7 (T_{eff} $\lesssim$5000K). These are all red giants and they appear to be severely depleted in Li, particularly those with masses less than $2 M_\odot$. Giants more massive than $2 M_\odot$ show a somewhat different behaviour. There are several of them which only show the effects of standard post-MS convective dilution. No MS depletion is anyway expected in these massive stars according to the standard stellar models. However, there are an equally large number of giants that have rather low abundances, as low as those in low mass giants.

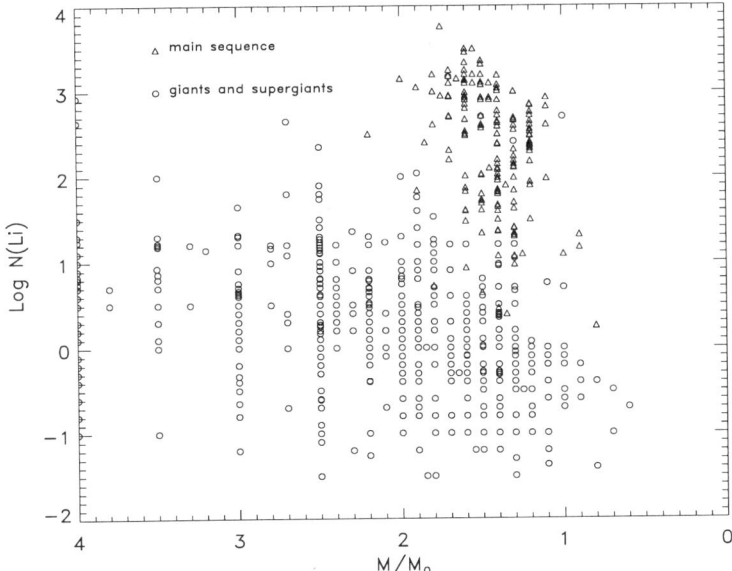

Figure 4. log N(Li) vs. $M/M_\odot$ for MS stars and giants of masses less than $4M_\odot$

Some of these trends are seen more clearly in Fig.3 where log N(Li) for all the 1024 stars is plotted against T_{eff}. This plot permits us to follow the Li evolution as the star evolves along the subgiant branch and up the red giant branch. The MS stars exhibit a large range in abundances even over their limited range in T_{eff}. A similar spread is seen in subgiants for $T_{eff} \geq 5400K$. The most remarkable feature is the extremely low values of log N(Li) observed in giants in most of them, much lower than what is expected from the standard model predictions after the first dredge up. A large fraction of them have abundances below +0.5 to as low as -1.5. The Li content goes lower and lower with cooler temperatures. All this implies there must be additional mixing and dilution of Li on the giant branch. It is clear from the sample of stars in Fig.2 with $T_{eff} >5000K$ that

1) For subgiants with $\log T_{eff} \geq 3.73$, dilution due to the deepening of the convective envelope has just about set in and one expects they display a range of abundances basically reflecting the observed spread in the MS stars. In particular, one notices that the Li spread is distinctly apparent between the tracks of 1.4 and $1.2 M_\odot$, exactly for stars corresponding to the Boesgaard-Tripicco dip on the MS. So the influence of the MS dip on Li in subgiants is evident.

2) The more evolved stars just to the left of $\log T_{eff}=3.7$ have undergone deeper mixing. A majority of them have Li already much less than 1.5 ± 0.3. There is a strong indication of extra mixing having already taken place even for these stars that are still on the horizontal trail and have not yet begun the ascent up the RGB.

On the other hand, the sample of stars with $T_{eff} <5000K$ reveals that

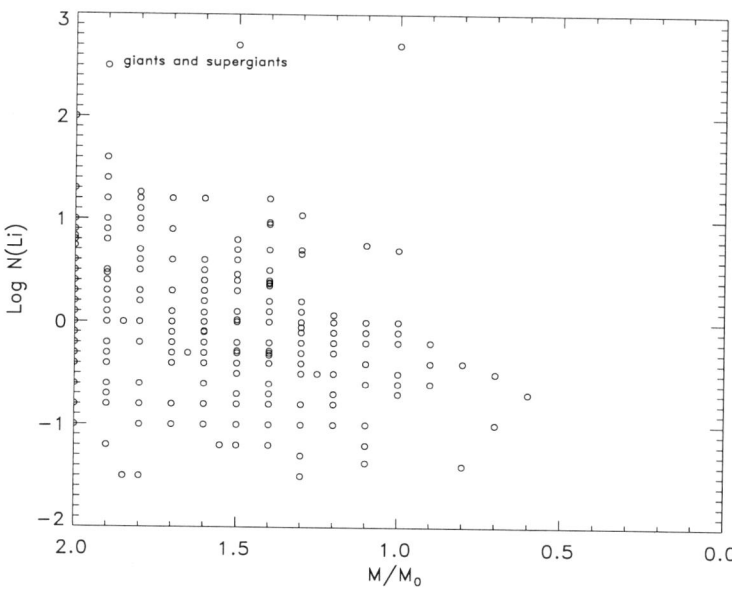

Figure 5. log N(Li) vs. $M/M_\odot$ for giants less massive than $2M_\odot$

1) The giants encompass a range of Li abundances. The scatter for a given mass is a consequence of the dilution effect; stars as they move more to the right on RGB are evidently lower in their Li content. Almost all giants of $M < 2M_\odot$ are heavily depleted independent of the fact that they were in or out of the B-T dip in their MS phase. A very large dilution possibly brings all low mass stars towards a low abundance, whatever their MS Li abundance is. The depletion in the MS phase observed for lower mass stars and the standard convective dilution are surely not sufficient enough to explain these unusually low abundances in them. It is imperative to invoke extra mixing on RGB to explain the observed values. Charbonnel and her collaborators (1994) have computed evolutionary models where they show that the standard convective dilution is essentially complete around $\log L/L_\odot = 1.5\text{-}2.0$ for the low mass stars, so any further dilution beyond this point is a result of extra mixing on the red giant branch. To be more specific, their calculations reveal that for stars between 1.0 and $1.7 M_\odot$ at luminosities between 1.5 and 2.0 respectively, the hydrogen burning shell reaches the chemical discontinuity created by the convective envelope at its maximum extent. The material from the convective envelope has thus access to the Li burning temperature regions. So there is a process identified in the giant phase that leads to fresh destruction of Li. It is worth noting that the point where $\log N(Li)$ reaches 0.5 and lower is at a higher luminosity for higher mass. A more systematic detailed study of such data could actually reveal exactly when Li destruction begins for a given mass.

2) Several giants more massive than $2M_\odot$ have also rather low abundances, much lower than the predicted maximum. Their MS progenitors are all earlier than F0, perhaps A and late B. Therefore all observed depletion of Li presumably takes

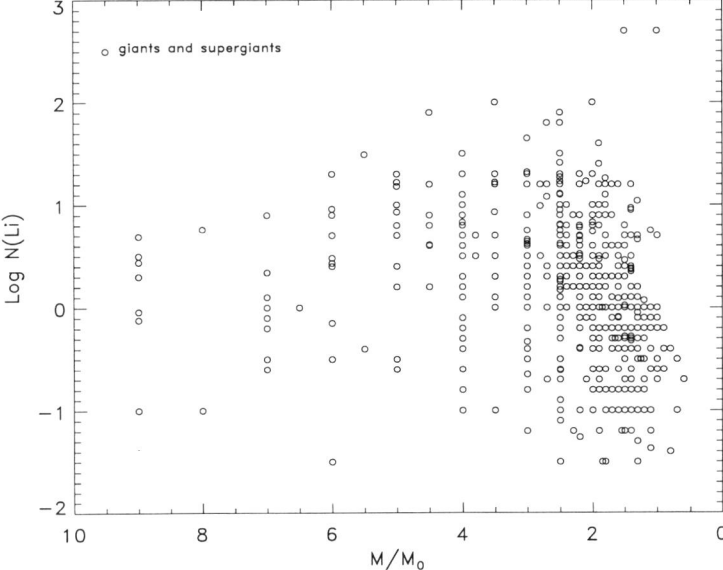

Figure 6. log N(Li) vs. $M/M_\odot$ for giants of all masses observed

place only in the post MS phase. One has to invoke non-standard mechanisms even for these more massive stars. Most of the stars more massive than $4M_\odot$ are supergiants and AGB stars and since these are expected to suffer more mass loss than the others, it is likely their Li depletion is further affected. Nevertheless, mass loss and the standard post MS depletion together may not still be adequate to explain the observed low abundances.

3.2. Lithium vs. Mass

Masses have been derived for all the stars using the same evolutionary tracks of Bressan et al. Fig. 4 plots log N(Li) vs. $M/M_\odot$ for the MS stars and the giants of masses less than $M_\odot$. There is a large scatter at all masses but it is hard to overlook a couple of key features. There is a suggestion of a trend among the MS stars of higher Li being associated with a higher mass. A more conspicuous trend seems to exist for giants $\leq 2M_\odot$, roughly over the same range of masses as the MS stars. This is seen better in Fig.5. The upper envelope shows a declining trend of log N(Li) for lower masses. One is tempted to interpret this as a consequence of the Li vs. mass spread present in the progenitors. One notes that the very low values are independent of mass. As pointed out earlier, the depletion is so large that it does not matter how much of it occured on the MS. Giants with masses $2.0 \leq M/M_\odot \leq 6.0$ essentially show a scatter diagram as seen in Fig.6. It could be because they evolve from progenitors which are expected to retain their initial lithium and not show any mass dependence. Lastly, it is curious to note that there is a distinct decline in the average Li abundance for masses $\geq 6M_\odot$. Most of them are supergiants and AGB stars and are likely to suffer more mass loss than the rest. Most possibly, mass loss combined with standard and non-standard mixing causes the lowering of the abundance in these stars.

Acknowledgments. It is a pleasure to acknowledge several stimulating discussions with D.C.V. Mallik

References

Balachandran, S. 1990, ApJ, 354, 310

Bressan, A. et al. 1993, A&AS, 100, 647

Brown, J.A. et al. 1989, ApJS, 71, 293

Cayrel de Strobel, G. et al. 1997, A&AS, 124, 299

Charbonnel, C. 1994, A&A, 282, 811

ESA 1997, The Hipparcos and Tycho Catalogues, ESA SP-1200

Flower, P.J. 1996, ApJ, 469, 355

Gustafsson, B. et al. 1975, A&A, 42, 407

Hoffleit, D. 1982, The Bright Star Catalogue, Yale University Observatory

Iben, I. Jr. 1967, ApJ, 147, 624

Lambert, D.L. et al. 1980, ApJ, 235, 114

Lèbre, A. et al. 1999, A&A, 345, 936

Luck, R.E. 1977, ApJ, 218, 752

Pallavicini, R. et al. 1987, A&A, 174, 116

Sneden, C.A. 1973, Ph.D. Thesis, The University of Texas, Austin

Lithium in the open cluster NGC 6475

S. Randich
Osservatorio Astrofisico di Arcetri, Firenze, Italy

R. Pallavicini
Osservatorio Astronomico di Palermo, Italy

J.-C. Mermilliod
University of Lausanne, Switzerland

Abstract. We present the results of a lithium study of the 220 Myr old cluster NGC 6475. Our data are merged with those from previous Li surveys of this cluster in order to construct a more statistically significant sample. The Li vs. effective temperature distribution of the merged sample is compared with the one of the co-eval, but more metal poor cluster M 34 and with the ones of the younger Pleiades and older Hyades. The results are discussed in the framework of non-standard Li depletion on the main-sequence between the age of the Pleiades ($\sim$ 120 Myr) and that of the Hyades ($\sim$ 600 Myr).

1. Introduction

NGC 6475 is a well populated southern hemisphere cluster located at the distance of $\sim$ 250 pc. Its estimated age of $\sim$ 220 Myr is intermediate between the age of the Pleiades ($\sim$ 120 Myr) and that of older clusters such as the Hyades, Praesepe, and Coma Berenices (ages between 500 and 700 Myr). A metallicity [Fe/H]= +0.11 has been determined spectroscopically for this cluster (James & Jeffries 1997). Given its distance and relatively small reddening, which allow reaching down to late–K stars with 4 m class telescopes, it represents a key cluster for the study of the main sequence (MS) evolution of stellar properties (e.g., lithium, rotation, chromospheric and coronal activity) of solar–type and lower mass stars. We mention in passing that NGC 6475 is indeed the closest and most compact cluster at its (or similar) age.

Focusing on lithium, the study of a cluster intermediate in age between the Pleiades and the Hyades allows addressing the question of MS Li depletion and its timescales, for both solar and later type stars. It is now clear that non–standard Li destruction and/or preservation processes (i.e., including rotation, magnetic fields, mass loss) must be at work during MS (and PMS) evolution as suggested by several observational evidences (Pinsonneault 1997; Jeffries 1999a; Deliyannis 1999, 2000; Pasquini 2000, and references therein). We mention, among them: *i)* the finding that, at variance with standard model predictions, solar-type stars deplete Li while on the MS, as shown by the comparison between the Pleiades and the Hyades; *ii)* the existence of a star-to-star scatter in Li abundances among stars cooler than ~ 5300 K as observed in the Pleiades and younger clusters such as α Persei and IC 2602 (see, e.g., Soderblom et al. 1993; Randich et al. 1997, 1998 and 2000; Jeffries 1999a). If Li depletion were driven only by convection, stars with the same age, mass, and chemical composition should have the same Li content. If the scatter reflects a real scatter in Li abundances and it is not caused by, e.g., the effect of chromospheric activity on the line formation process (e.g., Jeffries 1999b), its existence implies that additional mechanisms of Li destruction (or preservation) must be at work. In addition, standard models predict that Li depletion both on the MS and in the PMS should strongly depend on metallicity, a prediction that can be tested by observing clusters of similar age but different metallicities.

Li surveys of NGC 6475 were carried out by James & Jeffries (1997) and James et al. (1999). Both studies found that the Li vs. $T_{\rm eff}$ distribution of solar-type stars in NGC 6475 lie below that of similar stars in the Pleiades and above the Hyades; this in turn indicates that MS depletion does occur between 220 and 600 Myr. Not enough K-type stars were included in the two samples to allow firm conclusions about the existence of a star-to-star scatter.

Jones et al. (1997) carried out a Li study of M 34, a cluster that is about coeval to NGC 6475, but has a reported lower metallicity (either solar or slightly subsolar). They concluded that M 34 shows a Li depletion pattern which is intermediate between the Pleiades and the Hyades and that a dispersion in Li is present for stars cooler than 4700 K.

In this paper we present the results of additional Li observations of NGC 6475. Our data, merged with the ones of James & Jeffries (1997) and James et al. (1999) provides a larger (and thus more statistically significant) sample of stars to further address the issue of Li evolution between the Pleiades and the Hyades. In addition, our sample contains a few stars of later spectral-types than the previous surveys, allowing us to get some insight on MS Li depletion for very cool stars.

2. Sample, observations and data analysis

Our sample originally included 44 stars with $0.5 \leq (B-V)_0 \leq 1.44$: 10 of the stars were taken from the list of Koelbloed (1959) and are thus optically selected, while the remaining 34 are X-ray selected and come from Prosser et al. (1995). Observations were carried out at ESO in April 1994, April 1995, and July 1996; the CASPEC spectrograph at the 3.6 m telescope was used. The resolving powers in the different runs ranged from $R \sim 20,000$ to $R \sim 41,000$. 60% of the stars were also observed with Coravel which allowed inferring radial velocities and checking for binarity. 13 stars eventually turned out to be cluster-non members, 28 were confirmed as members, whereas for the remaining 3 a more accurate radial velocity analysis is needed.

Nine of the stars were in common with the sample of James & Jeffries (1997); for most of them we find a very good agreement between their and our Li equivalent widths. In the following we will use our own measurements.

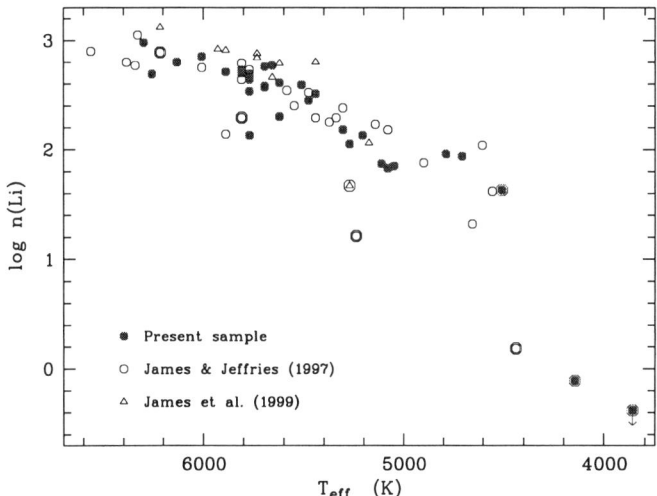

Figure 1. log n(Li) vs. T_{eff} for our sample stars (filled circles), James & Jeffries (1997) sample (open circles) and James et al. (1999) sample (open triangles). Circled symbols denote stars to be confirmed as members.

Li abundances were derived in the same fashion as in Randich et al. (1997) for their IC 2602 sample. Namely, effective temperatures were estimated from dereddened B−V colors using the calibration of Soderblom et al. (1993). LTE Li abundances were derived from the measured equivalent widths using Soderblom et al. COGs; abundances of stars warmer than 4500 K were then corrected for NLTE using the prescription by Carlsson et al. (1994). Both James & Jeffries (1997) and James et al. (1999) data were re-analyzed in the same way for consistency.

3. Results

In Figure 1 we compare the log n(Li) vs. T_{eff} distribution of our sample stars with those of James & Jeffries (1997) and James et al. (1999): the three distributions appear to be very similar (although the James et al. sample may lie slightly above the other two samples – see discussion in James et al.); in particular no systematic difference (due, e.g., to the use of different instruments or spectral resolutions) is present among the various samples. The three samples therefore can be safely merged into a single larger sample.

In Figures 2 and 3 the merged sample is compared to M 34 (Fig. 2) and to the Pleiades and Hyades (Fig. 3). Several conclusions can be drawn from these two figures. 1. No significant difference appears to exist between the mean log n(Li) vs. T_{eff} patterns of NGC 6475 and M 34, confirming the finding of Jeffries & James (1999) that metallicity (or at least the Fe abundance) does not affect Li depletion significantly. Note that, although based on very few

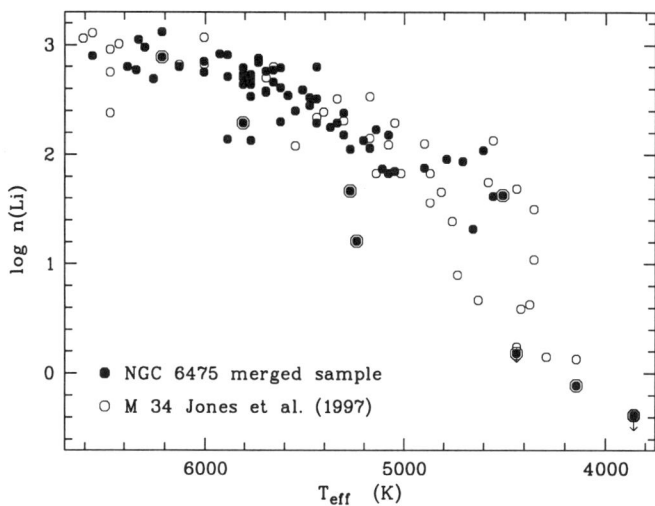

Figure 2. log n(Li) vs. $T_{\rm eff}$ for the merged NGC 6475 sample (filled circles) and the M 34 stars (from Jones et al 1997 – open symbols).

stars to be confirmed as members, this conclusion seems to hold also for stars considerably less massive than the Sun, for which convection should be the major Li destruction mechanism. For these stars the metallicity (which affects the depth of the convection zone) is expected by standard models to have a significant effect on Li depletion;

2. A few stars in NGC 6475 appear as depleted as (or more depleted than) the older Hyades. Whereas some of them are not confirmed as members, others seem to be *bona fide* cluster members (but one is probably a photometric binary from its position on the color-magnitude diagram). These stars more depleted than the Hyades should be further monitored; if additional observations confirm their cluster membership and the measured Li EWs, they will represent a very puzzling problem; **3.** The larger sample confirms that MS Li depletion occurs for stars cooler than $\sim$ 6200 K both between 120 and 220 Myr and between 220 and 600 Myr. Li destruction, however, appears to be slower between the Pleiades and NGC 6475 than between the latter one and the Hyades. If we consider, for example, the mean Li abundance of stars with $5500 \leq T_{\rm eff} \leq 5600$ K, we have a factor of $\sim$ 2 in Li depletion between the Pleiades and NGC 6475 (i.e., in a factor $\sim$ 1.9 in time) and of $\sim$ 5.4 between NGC 6475 and the Hyades (i.e., in a factor $\sim$ 2.7 in time). This is surprising and suggests that the MS Li depletion mechanism (which cannot be convection for solar–type stars that have not deep enough convective zones) becomes faster after 200–300 Myr; **4.** MS Li depletion between 220 and 600 Myr is more efficient for cooler/lower mass stars, probably witnessing the increasing contribution of convection to Li depletion. On the other hand, below 5400 K the NGC 6475 distribution lies on the lower envelope of the Pleiades distribution and, more specifically, several Pleiades stars exist that show the same amount of depletion as NGC 6475 stars; **5.** If we exclude the outliers discussed in point 2. above, no significant star-to-star scatter is present

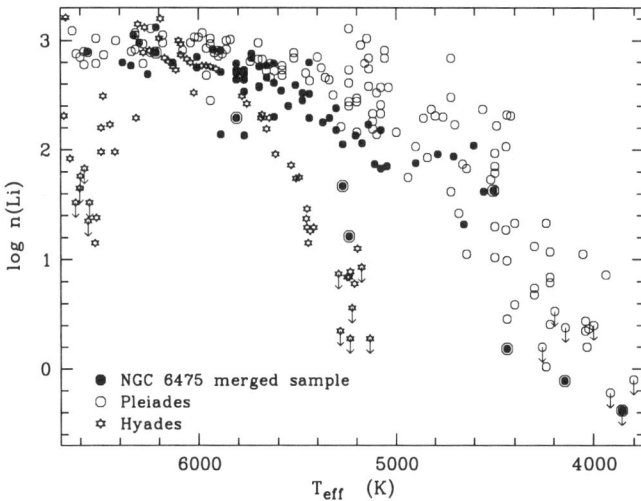

Figure 3. The NGC 6475 merged sample (filled circles) is compared with the Pleiades (open circles: data from Soderblom et al. 1993 and Jones et al. 1996) and the Hyades (star symbols – from Balachandran 1995).

down to ~ 4800 K. Among cooler stars, there might be a slight indication of the presence of a scatter, but this is evidenced only by two stars at present. Our sample contains too few cool stars to put sensible constraints on the existence of a scatter. Should further Li observations of NGC 6475 demonstrate that no spread is present for NGC 6475, while it does exist for M 34, this would suggest that, whereas metallicity does not affect the main Li depletion mechanism (i.e., the one determining the mean depletion pattern), it may indeed affect the mechanism causing the dispersion in Li.

4. Conclusions

The observations of the intermediate age cluster NGC 6475 confirm that MS Li depletion does occur for solar-type and lower-mass stars, both between the age of the Pleiades and that of NGC 6475 (and M 34) and between the latter age and that of the Hyades. The lack of any significant difference between the log n(Li) vs. T_{eff} distributions of NGC 6475 and M 34 indicates that metallicity does not play a significant role in Li depletion, if the reported metallicities for these two clusters are indeed correct (no spectroscopic determination of the metallicity of M 34 does yet exist, while the metallicity of NGC 6475 has been determined so far only by James & Jeffries 1997). Whereas a star-to-star scatter in Li abundances is clearly present for M 34, there is no significant scatter in NGC 6475 down to ~ 4800 K. Additional data on cooler stars are required to ascertain whether a scatter is present at lower temperatures. Finally, a few stars exist in NGC 6475 which are as depleted as, or more depleted than, stars in the older Hyades; conversely, stars exist in the younger Pleiades which are as depleted as the coolest stars in NGC 6475 and M 34.

References

Balachandran, S.C. 1995, ApJ 446, 203

Carlsson, M., Rutten, R.J., Bruls, J.H.M.J., & Shchukina, N.G. 1994, A&A 288, 860

Deliyannis, C. 1999, in Stellar Clusters and Associations: Convection, Rotation, and Dynamos, R. Pallavicini, G. Micela, and S. Sciortino (eds.), ASP Conf. Ser., in press

Deliyannis, C. 2000, these proceedings

James, D.J., and Jeffries, R.D. 1997, MNRAS 292, 252

James, D.J., Collier Cameron, A., Barnes, J.R., and Jeffries, R.D. 1999, in Stellar Clusters and Associations: Convection, Rotation, and Dynamos, R. Pallavicini, G. Micela, and S. Sciortino (eds.), ASP Conf. Ser., in press

Jeffries, R.J. 1999a, in Stellar Clusters and Associations: Convection, Rotation, and Dynamos, R. Pallavicini, G. Micela, and S. Sciortino (eds.), ASP Conf. Ser., in press

Jeffries, R.D. 1999b, MNRAS 309, 189

Jones, B.F., Shetrone, M., Fisher, D., and Soderblom, D.R. 1996, AJ 112, 186

Jones, B.F., Fisher, D., Shetrone, M., and Soderblom, D.R. 1997, AJ 114, 352

Koelbloed, D. 1959, Bull. Astron. Inst. Netherlands, 14, 265

Pasquini, L. 2000, these proceedings

Pinsonneault, M.H. 1997, ARA&A 35, 557

Prosser, C.F., Stauffer, J.R., Caillault, J.-P., Balachandran, S., Stern, R.A., and Randich, S. 1995, AJ 110, 1229

Randich, S., Aharpour, N., Pallavicini, R., Prosser, C.F., & Stauffer, J.R. 1997, A&A 323, 86

Randich, S., Martín, E.L., García López, R.J., and Pallavicini, R. 1998, A&A 333, 591

Randich, S., Pallavicini, R., Meola, G., Stauffer, J.R., and Balachandran, S.C. 2000, A&A, submitted

Soderblom, D.R., Jones, B.F., Balachandran, S., Stauffer, J.R., Duncan, D.K., Fedele, S.B., & Hudon, J.D. 1993, AJ 106, 1059

The Light Elements and Their Evolution
IAU Symposium, Vol. 198, 2000
L. da Silva, M. Spite, J. R. de Medeiros, eds.

Lithium in the Old Open Cluster NGC 2243

V. Hill and L. Pasquini

ESO, Karl Schwarzschild Strasse 2,
Garching bei München, D-85748, Germany

Abstract. We report observations of lithium in a sample of 11 stars in the metal-poor open cluster NGC 2243, that were obtained from high-resolution spectroscopy at CASPEC (ESO 3.6m telescope). The targets are located at the turnoff region, plus one red giant star.

NGC 2243 is one of the most metal-poor open cluster, almost as deficient as 47 Tuc, but substantially younger (∼4 Gyrs and [Fe/H]=-0.5 dex), which makes it a very interesting case to compare with more metal rich coeval clusters on the one hand, and old metal-rich globular clusters (47 Tuc) on the other hand. The preliminary Lithium abundances obtained are discussed in this framework.

1. Introduction

Lithium abundance of turnoff and dwarf stars in open clusters of various ages and metallicities will provide keys to the understanding of the lithium destruction mechanisms in main sequence stars. Being of known ages, metallicities, masses and evolutionary stages, cluster stars are the perfect way to disentangle the effect of each of these parameters on mixing phenomena (see Pasquini 1999, this volume).

NGC 2243 is an old and metal poor open cluster located towards the anti-center, at $R_g = 10.76$kpc and $z = 1.1$kpc below the plane of our Milky Way. With an age of 3 to 5 Gyrs (Bonifazzi et al. 1990, Bergbusch et al. 1991, Friel 1995), NGC 2243 has a metallicity of [Fe/H]∼ −0.5dex (Gratton et al. 1994), ie very close to that of the old metal-rich Pop II globular cluster 47 Tuc (age ∼ 13Gyrs and [Fe/H]∼ −0.7dex). It is therefore the perfect target to disentangle age from metallicity effects on lithium depletion. Moreover, with a low reddening (E($B-V$)=0.04 Bonifazzi et al. 1990) and a distance modulus of $(M-m)_V$=13.05 (Bonifazzi et al.1990), turnoff stars in NGC 2243 are of magnitudes between 15.5 and 17, just within reach of the high resolution spectrograph CASPEC mounted on the 3.6m telescope at ESO, La Silla.

2. Observations and analysis

Table 1 gives a summary of the basic parameters of the observed stars in NGC 2243, together with a summary of the observations. CASPEC was used for this purpose as a single-order spectrograph in long-slit mode, the order containing the Liλ6708Å line being selected by a filter. The long slit was oriented in a way to align as many as 4 targets along one single slit. The S/N reported in Table 1 refer to the measured S/N per pixel of the combined extracted spectra for each star. The achieved resolution is R≈19000 and the wavelength coverage λ6670-6740Å.

Table 1. Log book of the observations

Star	V	$(B-V)$	$(V-I)$	$(B-V)_o$	$(V-I)_o$	S/N	Exp. time
I1	14.90	0.91	0.91	0.87	1.10	80	2x1.5h
I14	15.66	0.50	0.59	0.46	0.69	45	2x1.5h+1.2h+3x1h*
I167	15.83	0.49	0.57	0.45	0.66	50	1.8h+1h+1.5h
I91	15.73	0.52	0.64	0.48	0.75	50	2x1.5h
I23	16.02	0.46	0.55	0.42	0.64	50	2x1.5+3x1h*
I33	16.30	0.46	0.57	0.42	0.67	35	1.8h+2x1.5h
I36	16.04	0.46	0.53	0.42	0.61	30	2x1.5h+1.2h
I31	16.35	0.47	0.57	0.43	0.66	25	2x1.5h
I25	16.05	0.49	0.60	0.45	0.70	35	1h+1.5h
I13	16.63	0.45	0.55	0.41	0.64	20	2x1.5h+1.2h
I90	17.77	0.63	0.80	0.59	0.96	25	2x1.5h

Identifications after Van den Bergh 1977.
$(B-V)$ from Bergbusch et al. 1991
$(V-I)$ from Kaluzny et al. 1996 $((V-I)_o$ transformed to Johnson colours)
* EMMI echelle spectra Dec96

The spectra were processed using standard MIDAS routines including flat-fielding, cosmic rejection, order extraction, wavelength calibration and sky subtraction. The radial velocity for each spectrum was then determined using the Ca Iλ6718Å line, and the spectra of each star were then coadded. Figure 1 displays an example of reduced coadded spectra for 3 stars in the sample.

The equivalent width of the lithium Li Iλ6708Å and Ca Iλ6718Å lines were measured by gaussian fitting and checked by straight integration of the combined spectra. Table 2 reports the measured radial velocities and equivalent width (gaussian fitting) for all stars.

The *atmospheric parameters* T_{eff} log g reported in Table 2 were determined for each star from photometry:

- T_{eff} was taken as the mean of the two temperature determinations using the Alonso et al. 1996 $(B-V)_o$ and $(V-I)_o$ calibrations, adopting a reddening of $E(B-V)$=0.04. The difference between the two indicators is small: temperatures deduced from $(B-V)$ are 36 ± 140K cooler than temperatures deduced from $(V-I)$. The estimated error on the adopted T_{eff} is therefore ±150K, which corresponds to a 0.11 dex uncertainty in lithium abundance.

- log g was determined from the bolometric magnitude of the stars, adopting the bolometric corrections of Alonso et al. 1996.

- the microturbulence velocity was assumed to be $vt = 1.5$km/s

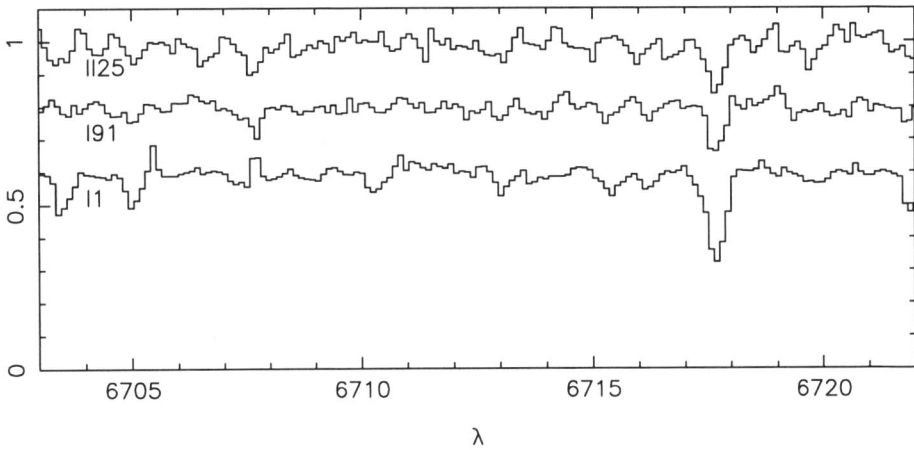

Figure 1. Example of a portion of three reduced spectra, containing the Li I and Ca I lines: I25, I91 and I1 (the spectra of I91 and I1 were shifted vertically by -0.2 and -0,4 for clarity). Lithium is detected in I25, I91, but not in I1, which is the most evolved star of our sample.

Finally, the Li and Ca abundances reported in Table 2 were computed, using Gustafsson et al. 1975 (and subsequent extensions) models.

The lithium line could be detected in 4 stars. In all cases where the lithium line was successfully detected, the Li abundance is reported in Table 2 together with its associated uncertainty (1σ); in all other cases, an upper limit for the equivalent width of the line was established via the Cayrel (1988) formula and translated into an upper limit on the Li abundance (denoted by a "<" sign).

Examination of the radial velocities of the sample of 11 stars reveal that only one star (I36) is a suspected non-member, with a radial velocity differing by more than 20 km/s from the mean value of the sample $Vr = 53.57 \pm 2.6$ km/s (10 stars). I36 was hence excluded from the following discussion, as probable non-member.

Table 2. Atmospheric parameters and Lithium abundances of the program stars

Star	T_{eff}	log g	M_V	W_{Ca}	W_{Li}	[Ca/H]	N(Li)	Vr(km/s)
I1	4987.	3.5	2.1	129	15	-0.27	<0.94	48.6
I14	6251.	4.0	2.9	61	20	-0.43	2.18±0.10	57.4
I167	6309.	4.0	3.0	52	15	-0.57	<2.09	55.7
I91	6101.	4.0	2.9	59	31	-0.55	2.26±0.05	54.5
I23	6441.	4.2	3.2	64	29	-0.29	2.48±0.15	52.6
I33	6384.	4.2	3.5	37	13	-0.75	<2.09	53.6
I36	*6489.*	*4.2*	*3.2*	*70*	*20*	*-0.17*	*<2.35*	*32.5*
I31	6364.	4.2	3.5	50	20	-0.56	<2.24	54.1
I25	6255.	4.2	3.2	62	41	-0.44	2.49±0.07	50.1
I13	6464.	4.4	3.8	69	25	-0.24	<2.41	53.8
I90	5578.	4.6	4.9	77	20	-0.68	<1.65	55.3

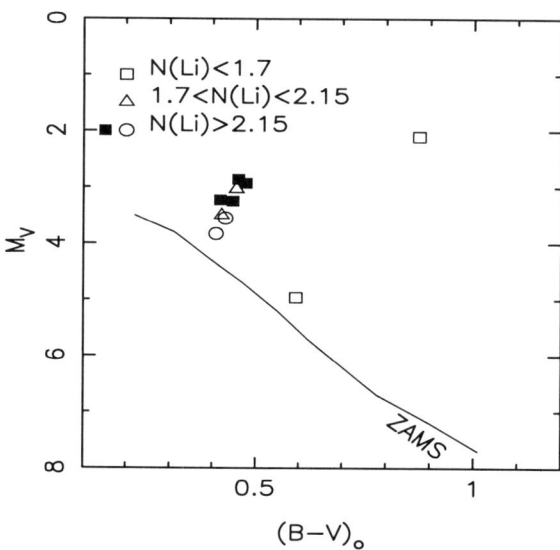

Figure 2. Color-magnitude diagram of the observed stars in NGC 2243. The lithium abundance determined is coded by symbols: open symbols are upper limits whereas filled symbols are detections. The ZAMS is a semi-empirical locus from Vandenberg 1989, for a metallicity of [Fe/H]=-0.65 and Y=0.24.

3. Discussion

Let us now examine the significance of the four Li detections and six upper limits that were established.

The status of the two coolest stars in our sample is rather straight-forward: the giant I1 is sufficiently evolved to have diluted its lithium, whereas the dwarf I90 is simply too cool (5600K) to have kept its lithium on the MS. Accordingly, no Li line was detected in either star, but stringent upper limits of N(Li)=0.94 and 1.67 were established, a factor respectively 35 an 6 times less than the maximum lithium abundance detected in this cluster : these stars have evidently undergone severe depletion.

3.1. Lithium *dip*

In open clusters such as the Hyades, it is well known that the lithium "plateau" that is observed for stars hotter than 5800K is interrupted by the so-called lithium *dip*, a narrow temperature (or mass) range where lithium is severely depleted by factors up to 100 (eg. Balachandran 1995 Fig.12), and interpreted as the signature of an extra-mixing mechanism. Moreover, Balachandran (1995) has shown that this *dip* appears at a *constant ZAMS temperature*, for clusters of a range of ages, whether the stars either side of the *dip* are still on the MS (for the younger clusters) of already evolved to the turnoff and subgiant region (for older clusters such as M67).

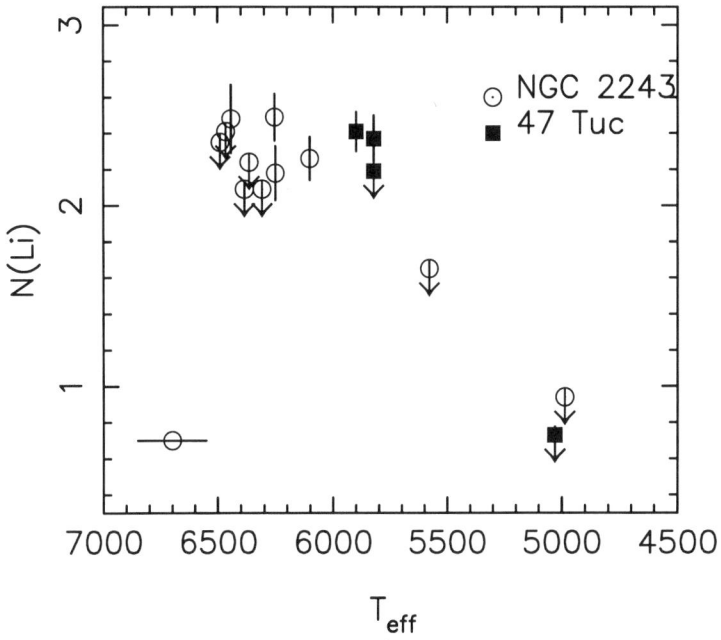

Figure 3. Li abundances versus effective temperature for the NGC 2243 stars (this paper), compared with literature data for stars in the globular cluster 47 Tuc (Pasquini & Molaro 1997)

Interestingly, Figure 2 indicates that in the case of NGC 2243, the stars in which we successfully detected the lithium line (filled symbols) are in fact those which have the hotter ZAMS temperatures (the larger masses) and are now evolved off the MS. On the other hand, stars closer to the MS have only upper limits for their lithium abundance. This fact could be interpreted as the signature of the *dip*: the stars on the blue side of the *dip* (hotter side) have already evolved from the MS, while the stars in the *dip*, which have destroyed lithium, are seen as hardly evolved above the MS.

This hypothesis could be tested by enlarging the sample towards the upper main sequence of the cluster: these stars should fall on the red side of the *dip* (cooler side) and show detectable lithium again. However, such observations require larger telescopes, since the magnitudes of the stars of interest will be V>17, and the signal to noise needed, larger than 30.

3.2. Lithium abundance in the Galaxy Evolution context

If this interpretation is correct, then the stars for which we could detect strong lithium should show no strong depletion, similarly to stars on the blue side of the *dip* in the Hyades. In this respect, the comparison to 47 Tuc is also interesting, and is shown in Figure 3, where the Lithium abundance of 5 stars in 47 Tuc (Pasquini & Molaro 1997) are shown together with the present determinations for NGC 2243. The "plateau" for temperatures hotter than 5800K is visible, and above this value, the two clusters show surprisingly similar lithium abundances: N(Li)=2.37 for 4 stars in 47 Tuc and 2.35 for the 4 detections in NGC 2243.

Considering that, with a similar metallicity, one would expect the two clusters to have started with a similar lithium abundance. The fact that the present day lithium abundance in the atmosphere of turnoff stars in the two clusters are so similar indicate that, if they have suffered MS lithium depletion, then the depletion has been of similar amplitude in the two clusters, despite the fact that NGC 2243 is almost 10 Gyrs younger than 47 Tuc.

Finally, we would like to note that, if the $N(Li)=2.35$ indeed represents the lithium abundance enrichment level reached by the galactic ISM at metallicities of $[Fe/H]=-0.5$ dex (ie if the NGC 2243 and 47 Tuc stars have not suffered significant MS Li depletion), then Galactic Chemical Evolution models predict too high lithium in this metallicity range. Moreover, the increase of lithium abundance between $[Fe/H]=-0.5$ and solar metallicities would be of a factor of 10, much more than predicted by Chemical Evolution models.

As compared to the standard Pop II field stars *lithium plateau* value of 2.15, or the recent lithium measurements of 2.18 ± 0.18 and 2.28 ± 0.10 in two metal poor old globular clusters M92 and NGC 6397 (Deliyannis et al. 1995, Boesgaard et al. 1998), the 47 Tuc and NGC 2243 stars are only slightly enriched in lithium: again, this very small enrichment is at variance from Chemical Evolution models predictions.

References

Alonso A., Arribas S., Martinez-Roger C., 1996 A&A 313, 873
Boesgaard A., Deliyannis C., Stephens A., King J., 1998, ApJ 493, 206
Balachandran S., 1995 ApJ 446, 203
Bergbusch P., Vandenberg D., Infante L., 1991, AJ 101.2102
Bonifazi A., Tosi M., Fusi Pecci F., Romeo G., 1990, MNRAS 245, 15
Cayrel R., 1988, IAU Symp. 132. p345
Delyannis C., Boesgaard A., King J., 1995, ApJ 452, L13
Friel E., 1995, ARA&A 33, 381
Gratton R., Contarini G., 1994, A&A 283, 911
Gustafsson B., Bell R., Eriksson K., Nordlund A., 1975, A&A 42, 407
Kaluzny J., Krzeminski W., Mazur B., 1996, A&AS 118, 303
Molaro P., Pasquini L., 1997, A&A 322, 109
Pasquini L., 1999 this volume
van den Bergh S., 1977, ApJ 215, 89
VandenBerg D, Poll H., 1989, AJ 98, 1451

Lithium in brown dwarfs

Rafael Rebolo
*Instituto de Astrofísica de Canarias and Consejo Superior de Investigaciones Científicas
E-38200 La Laguna, Tenerife, Spain*

Abstract. Lithium is a key element to establish the substellar nature of brown dwarf candidates. Theoretical calculations show that brown dwarfs with masses below ~ 0.065 $M_\odot$ preserve a significant fraction of their initial Li content while for higher masses total Li depletion occurs in short timescales. Lithium is preserved at masses well below the hydrogen burning mass. Strong lithium lines have been predicted and discovered in the spectra of brown dwarfs. Most of the bona fide brown dwarfs detected in stellar clusters and in the solar neighborhood have been confirmed to be substellar via detection of the lithium resonance doublet at 670.8 nm. I review these detections and the progress made in understanding the formation of Li lines in very cool high gravity dwarfs.

1. Introduction

Lithium is a well known tracer of stellar structure and a sensitive indicator of mixing processes in stellar interiors. As we move towards the bottom of the Main Sequence stars become more efficient destroyers of this fragile element. Observations in M-type stars of young clusters like α Persei and the Pleiades (see e.g. García López et al. 1994) show how the content of lithium is significantly reduced after several tens of Myr. This depletion is explained in terms of nuclear burning via the reaction ^{7}Li(p,α)^{4}He which is very efficient at temperatures above $\sim 2.4\ 10^6$ K. Such temperatures are easily attained in stellar interiors, but never reached in less massive objects like brown dwarfs or giant planets. Brown dwarfs are gaseous bodies with insufficient mass to produce stable hydrogen burning, they are less massive than 0.075 $M_\odot$ and are expected to form, as stars, from the direct collapse and fragmentation of molecular clouds. A comprehensive description of the theoretical properties of these fully convective objects can be found for instance in Burrows and Liebert (1993) and Chabrier et al. (2000). Brown dwarfs were first conceived by Kumar 1963, but their existence was not proved until 1995 with the discoveries of a young hot brown dwarf in the Pleiades star cluster (Teide 1, Rebolo et al. 1995) and a cool "methane" brown dwarf around a nearby M-type star (Gl 229 B, Nakajima et al. 1995). Since then, several hundreds of potential brown dwarfs have been identified in stellar clusters (see e.g. Zapatero-Osorio et al. 1997, Luhman et al. 1997, Bouvier et al. 1998, Béjar et al. 1999) and in the solar neighborhood (Ruiz et al. 1997, Delfosse et al.

1997, Kirkpatrick et al. 1999). Here I review the crucial role played by lithium in our understanding of this fascinating new class of astronomical objects.

2. The lithium test

Models describing the interiors of brown dwarfs predict the evolution of their core temperatures as a function of time. From the early work by D'Antona and Mazzitelli (1985), it was already possible to realize that below a certain mass, brown dwarfs will never reach a core temperature sufficiently high to produce lithium burning. Opposite to very low-mass stars, these brown dwarfs can preserve the original content of lithium (see e.g. Stringfellow 1989; Pozio 1991). Ten years ago, the effective atmospheric temperatures of objects close to the substellar limit were subject of controversy and difficult to predict on theoretical grounds. The available spectral type-$T_{\rm eff}$ calibrations appeared to indicate that brown dwarfs could have effective temperatures slightly lower than those of the coolest T Tauri stars with detected Li. These considerations prompted Rebolo Martín and Magazzú (1992) to compute the formation of the Li I resonance doublet at 670.8 nm using Allard (1990) model atmospheres with $T_{\rm eff}$ in the range 2000 to 2700 K. The computations showed the formation of a very strong line (equivalent width of several Å) in such cool atmospheres and led to propose a spectroscopic test based on its detection as a powerful tool to confirm the substellar nature of brown dwarf candidates: the Li test. This was presented in more detail by Magazzù, Martín and Rebolo (1993) who also reported on the first search for lithium in several of the best brown dwarf candidates known at that time. The evolution of the lithium abundance in very low-mass stars and brown dwarfs was later extensively considered in several works which reported remarkable agreement on the timescale for lithium depletion and the minimum mass for lithium preservation (see e.g. Nelson, Rappaport and Chiang 1993; D'Antona and Mazzitelli 1994; Chabrier, Baraffe and Plez 1996). This agreement is achieved in spite of the use of different interior models, equations of state, screening factors for nuclear reactions or atmospheric opacities, basically reflecting the little sensitivity to these parameters of the core temperatures as a function of time, and the robustness of the predictions regarding lithium. In Fig. 1 we plot lithium depletion curves obtained with the NextGen models by Chabrier et al. (1996). It is obvious from the figure that brown dwarfs with masses below 0.06 $M_\odot$ preserve a significant (detectable) amount of lithium, and that below 0.05 $M_\odot$ lithium is fully preserved. The mass limit for Li preservation is clearly below the substellar mass limit, usually accepted to lay between 0.08 and 0.07 $M_\odot$ for solar metallicity. The shape of the destruction curves also predict a sharp transition between Li-poor and Li-rich objects at the bottom of the main sequence of a cluster like the Pleiades.

3. Detection of lithium in brown dwarfs

The first searches for lihium in brown dwarf candidates gave negative results (Magazzù et al. 1993; Martín, Rebolo and Magazù 1994; Marcy, Basri and Graham 1994). Nearby late M-type dwarfs (GL 234 B, GL 473AB, GL 569B, LHS 2924, etc) and the faintest proper motion objects discovered in the Pleiades

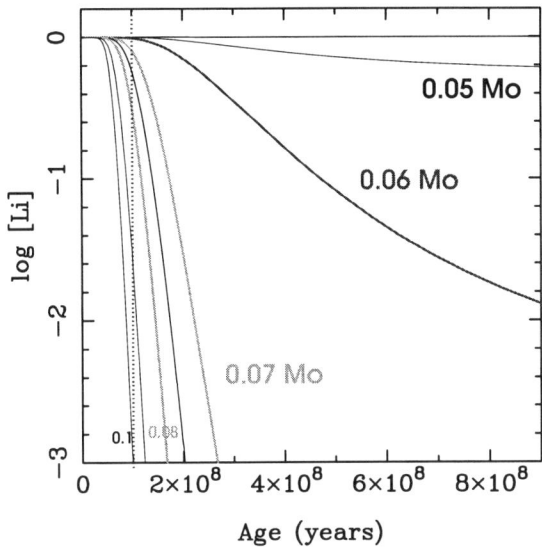

Figure 1. The evolution of lithium abundance as a function of time in very low mass stars and brown dwarfs.

cluster (like HHJ 10, HHJ 3 from Hambly et al. 1993) did not show lithium in their atmospheres. For the field objects with unknown ages the strong depletion of lithium just implied masses above 0.065 $M_\odot$, but the conclusions regarding substellar nature were much drastic for the Pleiades objects. Since at the age of the cluster ($\sim$ 100 Myr) a detectable amount of lithium was expected in objects with 0.08 $M_\odot$ or less (see Fig. 1), the depletion inferred from the observations clearly excluded the substellar nature of the examined Pleiads.

3.1. Brown dwarfs in stellar clusters

The first positive result of the lithium test was achieved by Basri et al. (1996) in PPl 15, an M6.5 dwarf in the Pleiades cluster discovered by Stauffer et al. (1994). The detection of lithium in PPl 15 put on empirical grounds the theoretical views on the reappearance of lithium at the bottom of the Main Sequence, but was not sufficient to establish the brown dwarf nature of this candidate which according to its effective temperature, luminosity and lithium abundance laid precisely on the boundary between stars and brown dwarfs. Uncertainties in these parameters, in the age of the cluster and in the theoretical timescales for lithium depletion prevented to conclude whether PPl 15 was stellar or substellar. The discovery of a fainter and cooler (M8 spectral type) proper motion member of the Pleiades (Teide 1) claimed to be a brown dwarf by Rebolo, Zapatero-Osorio and Martín (1995) offered a new extremely interesting opportunity to search for lithium in the substellar domain. Rebolo et al. (1996) reported the detection of a strong lithium doublet (EW$\sim$ 1 Å) in this brown dwarf and in the twin object Calar 3 (Martín, Rebolo and Zapatero-Osorio 1996). Estimates of the lithium abundances for these three objects can be seen in Fig. 2. After these discoveries many new brown dwarf candidates have been detected in the

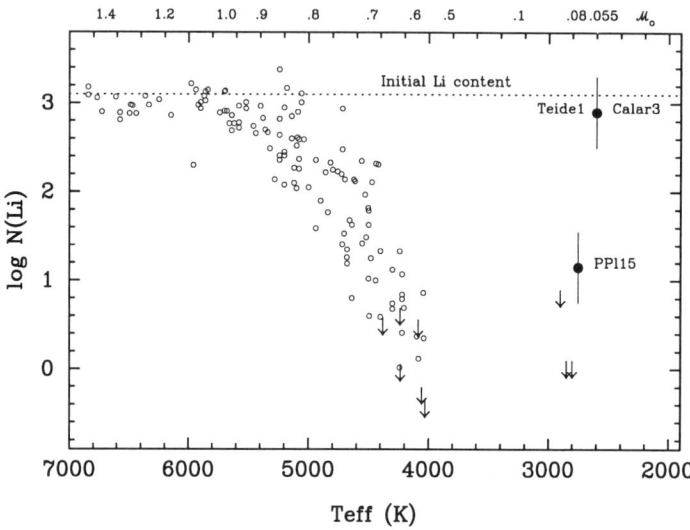

Figure 2. Lithium abundances versus effective temperatures for Pleiades F-M stars and first brown dwarfs (after Rebolo et al. 1996). Abundances are given in the usual scale log $N(H) = 12$.

Pleiades (Zapatero-Osorio et al. 1997, Bouvier et al. 1998) and new lithium detections have been performed beyond the stellar/substellar boundary (Martín et al. 1998, Stauffer, Schultz and Kirkpatrick 1998). These new observations confirm that there is an abrupt reappearance of lithium at the bottom of the Pleiades Main Sequence locating the edge of the lithium depletion at spectral type M6.5 and visual magnitude $I_c = 17.8 \pm 0.1$.

According to evolutionary models this depletion boundary is very sensitive to the age of the cluster and provides an effective method to estimate its age. The most accurate age determination for the Pleiades gives 125 ± 8 Myr (Stauffer et al. 1998), a value much higher than that derived from the upper main sequence turnoff of the cluster.

The lithium depletion boundary (LDB) has been determined in two more stellar clusters: αPersei (Stauffer et al. 1999, Basri and Martín 1999) and IC 2391 (Barrado y Navascués, Stauffer and Patten 1999). In both cases the ages resulting from the location of the LDB, 65-90 Myr and 53±5 Myr for α Persei and IC 2391, respectively, are several tens of Myr older than those determined from the turnoff. However, convective overshooting in massive stars may affect this age determination. The systematically older "lithium ages" could be an indication that indeed overshooting takes place in stars of the upper main sequence.

Brown dwarfs have also been discovered in very young star forming regions (Luhman et al. 1997, Newh'auser and Comerón 1998, Béjar, Zapatero-Osorio and Rebolo 1999). At very young ages, the lithium test cannot help as a substellar discriminator because there is no sufficient time for lithium to be depleted in very low-mass stars. However, the presence of lithium can give confidence on any brown dwarf candidate in star forming regions. An example of an extremely

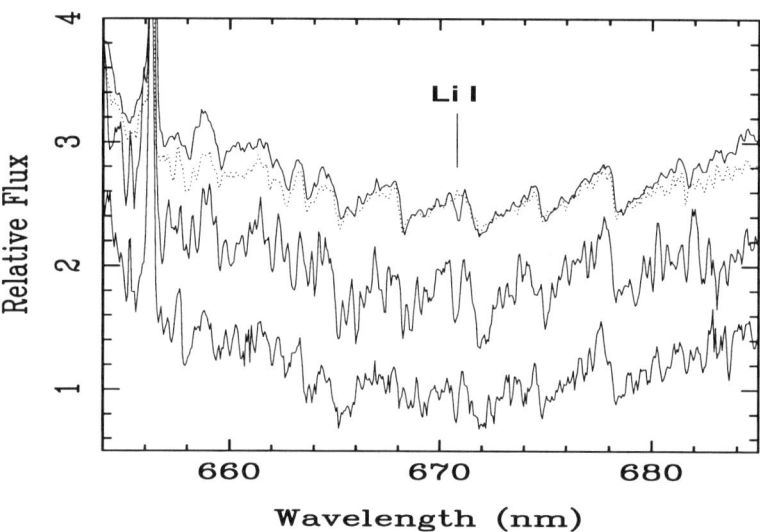

Figure 3. The lithium resonance doublet in spectra of several young brown dwarfs. From top to bottom: σOri 27, Calar 3 and Teide 1. The spectrum of the star vB 10, with no lithium has been overplotted (dotted line).

young brown dwarf discovered by Béjar et al. (1999) in the σ Orionis cluster (age 1-5 Myr) is displayed in Figure 3 where it is plotted next to older and more massive Pleiades brown dwarfs with similar spectral type. Hα can be seen in emission in the three objects.

3.2. Brown dwarf companions

Imaging techniques have revealed three brown dwarfs bounded to nearby stars. The first was Gl 229 B (Nakajima et al. 1995) a cool "methane" brown dwarf in orbit around a M1 V star at 5.7 pc from the Sun. Its effective temperature ($\sim$ 950 K) is so low that most Li atoms are forming part of molecules and consequently it is expected a very weak Li I resonance doublet. The second detected brown dwarf companion was G 196-3 B (Rebolo et al. 1998), it is in orbit around a very young, extremely active M3 dwarf star at a distance in the range 15 to 27 pc from the Sun. The brown dwarf which can be seen in Fig. 4 has an estimated mass of 25±10 Myr, and the spectral energy distribution of an early L-type dwarf according to the recent classification scheme proposed by Kirkpatrick et al. (1999) and Martín et al. (1999). In Fig. 4 it is also plotted the low resolution optical spectrum and the detection of lithium which guarantees that we are dealing with a bona fide brown dwarf. The estimated effective temperature of this object is $\sim$ 1800 K. More recently, Burgasser et al. (2000) have found the third companion brown dwarf, a cool "methane" dwarf in orbit around an M-type star.

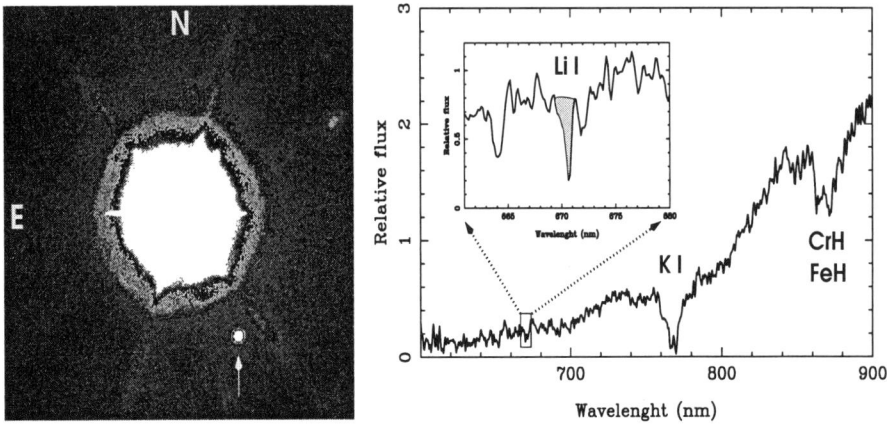

Figure 4. The brown dwarf companion G 196-3 B, identification image in the I band, low resolution optical spectroscopy and intermediate dispersion spectrum showing the presence of lithium (after Rebolo et al. 1998.

3.3. Free-floating brown dwarfs in the solar neighborhood

Nearby free-floating brown dwarfs have been discovered as a result of proper motion surveys (Kelu 1, Ruiz et al. 1997), large scale IR surveys like the DEep Near-Infrared Sky (DENIS, Delfosse et al. 1997) and 2-Micron All-Sky Survey (2MASS, Kirkpatrick et al. 1999), and more recently by the SLOAN Digital Sky survey (Strauss et al. 1999). As expected from the findings in stellar clusters, brown dwarfs are populating the solar neighborhood in significant numbers. Examination of the first 371 sq. deg of 2MASS data have produced 20 objects later than M9.5 V which can be classified as L dwarfs (Kirkpatrick et al. 1999). These objects can be either stellar or substellar and span the effective temperature range between 2000 and 1200 K. In their spectra, the characteristic metallic oxides of M dwarfs are replaced by metallic hydrides and neutral alkali metals as the most remarkable spectroscopic features. About one third of the L dwarfs show strong lithium lines in their spectra (typical equivalent widths are several Å, with a strongest reported detection of 15 Å). Most of these brown dwarfs with lithium will have masses below 0.065 $M_\odot$. The remaining L dwarfs where lithium has not been detected can either be more massive brown dwarfs or simply very low mass stars. Unfortunately, their distances are not sufficiently well known to determine accurate luminosities and subsequently constrain their masses using evolutionary models. It is quite likely that the early L dwarfs are indeed very low mass stars just above the hydrogen burning mass limit. Spectra of L dwarfs showing lithium absorption are plotted in Fig. 5.

Cooler free floating "methane" brown dwarfs whose spectra resemble that of Gl 229 B, have been discovered in the last two years (Strauss et al. 1999, Burgasser et al. 1999, Leggett et al. 2000). These intrinsically fainter objects have effective temperatures close or below 1000 K which do not favor the detection of lithium. Some of these objects could be old massive brown dwarfs (m ≥ 0.065 $M_\odot$) which have cooled significantly and have not retained any lithium

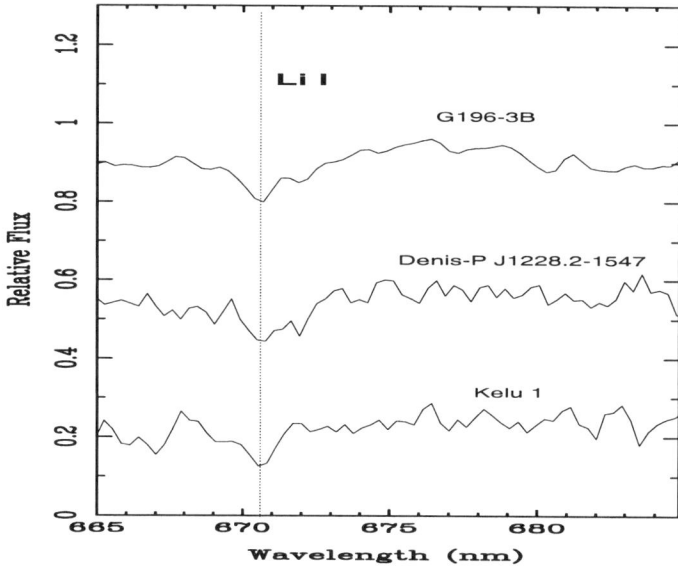

Figure 5. Spectra of several L dwarfs in the region of the lithium resonance doublet.

because burning took place in their interiors. Some other could be younger less massive brown dwarfs where lithium will be preserved for ever.

4. The formation of Li lines in brown dwarfs.

The computations by Rebolo et al. (1992), Magazzù et al. (1993), and Pavlenko et al. (1995) clearly revealed the formation of strong Li I λ670.8 nm lines in the spectra of cool dwarfs with temperatures above 1800 K showing that Li is detectable in the forest of molecular bands expected at these temperatures. Pavlenko et al. (1995) analyzed in detail the LTE and NLTE formation of Li I lines using Allard (1990) model atmospheres of T_{eff} 2000, 2500 and 3000 K. In these computations, the dissociation equilibria for seven Li molecules (LiH, LiO, LiCl, LiF, LiBr, LiI, LiOH) was taken into account. More than 20 atoms in two ionization states and 54 molecules were included in the state equation system. The synthetic spectra were able to reproduce the TiO bands in the region around the Li resonance line, correctly describing positions and intensities of the observed bands. The resulting LTE computations showed the formation of prominent Li I 670.8 nm lines with equivalent widths of several Å. The NLTE effects were found to be small, less than 0.1 dex in abundance. LTE and NLTE curves of growth for the weaker Li I lines at 610.3 and 812.6 nm. were also given. Very recently, Pavlenko, Zapatero-Osorio and Rebolo (2000) have produced synthetic spectra using cooler model atmospheres (effective temperatures in the range 2000-1000 K) suitable to describe the new L dwarfs. The synthesis code was based on that of Pavlenko et al. (1995) but extended the number of considered molecular species up to 100 and included detailed opacities for the most relevant bands. The new computations show that the alkali elements Li, Na, K, Rb and Cs govern the optical spectra of the cooler L dwarfs and that there is a need for implementation of additional opacity in the computations in order to reproduce the far red spectral energy distribution. This additional opacity could be associated to absorption or scattering processes in the atmospheres of cool dwarfs and its dependence with frequency can be simply described by a power law of the form $a_\nu = a_0(\nu/\nu_0)^N$ where the parameters N and a_0 are determined from the observations and ν_0 is arbitrarily adopted as the frequency of the KI resonance line at 769.9 nm. It was found that $N=4$, corresponding to the case of pure Rayleigh scattering, is adequated for all the objects examined so far. The effects of additional dust opacity on the formation of Li I lines (resonance and subordinate ones) also deserve detailed consideration. The computations show that both, the resonance line at 670.8 nm, and the subordinate lines at 601.3 nm and 812.6 nm are very sensitive to the additional opacity.

Table 1. Equivalent widths (Å) of the Li I resonance doublet at $\lambda 670.8$ nm computed for cosmic Li abundance ($\log N(\text{Li}) = 3.2$) and gravity $\log g = 5.0$ (after Pavlenko et al. 2000).

T_{eff}	a_o		
	0.00	0.01	0.10
1000 K	17	8	0.6
1200 K	30	12	0.7
1400 K	42	21	0.9

The predicted equivalent widths for the 812.6 nm line, assuming fully preserved lithium, are rather small, ranging from EW= 0.4 Å to 0.04 Å for effective temperatures in the range 2000 K to 1200 K. These EWs are significantly reduced by the inclusion of the additional opacity described above. In Table 1, equivalent widths of the Li resonance doublet at 670.8 nm are listed for several of the coolest model atmospheres (1400-1200 K) considered in Pavlenko et al. (2000). In the absence of any additional opacity, we would expect rather strong Li resonance lines in the spectra of objects as cool as Gl 229 B. The chemical equilibrium of Li species, still allow a sufficient number of Li atoms produce a rather strong resonance feature. However, it is expected a very dusty atmosphere in Gl 229 B, and for the best value of the opacity parameter a_0 derived for this object (a_0=0.1) there is little hope to detect any signature from atomic lithium in its spectrum. It is interesting to note, that in much less dusty atmospheres of similar effective temperature it could be possible to achieve the detection of lithium.

Another important issue is the variability of the lithium lines due to changes in the physical conditions of these atmospheres. In particular, changes in the dust content (condensation) can modify the atmospheric opacity and lead to detectable variations in the EWs of the lithium lines. As a consequence, weak lithium lines do not necessarily imply a depletion of this element. The observed LiI variability in Kelu 1 (with changes in EW by factor 5), could be an indication

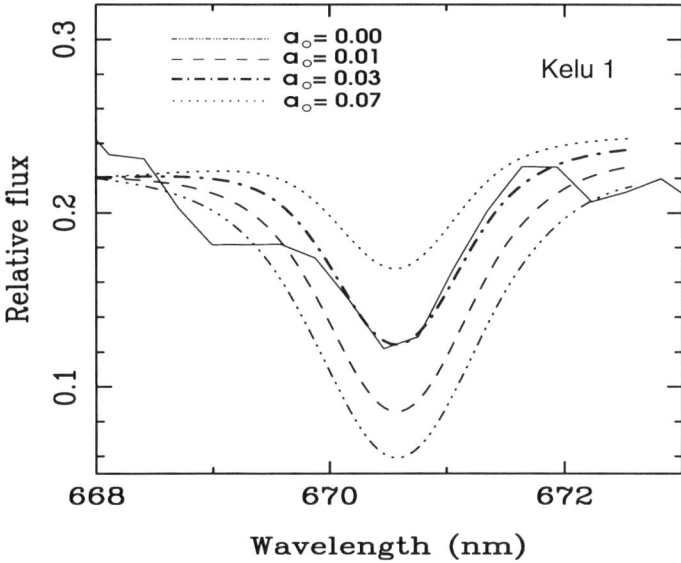

Figure 6. Spectrum of the brown dwarf Kelu 1 (T_{eff}=2000 K, log g=5.0) in the Li region (solid line) and computed spectra assuming a cosmic Li abundance but various atmospheric opacities.

of metheorological changes in the atmosphere of this rapidly rotating cool object. In Fig. 6 several spectral synthesis show the sensitivity of the lithium line to the extra opacity in the atmosphere. The additional opacity parameters which give the best fit for the observed lithium line in Kelu 1 (EW =6.5±1.0 Å) coincide with those also providing the best fit to the whole optical spectrum. The obtained lithium abundance is consistent with complete preservation.

5. Concluding remarks

Observations in very late type M dwarfs and L dwarfs have confirmed early claims that the resonance lithium doublet would be observable in very cool dwarfs. The lithium test is now widely used as a mean to confirm the substellar nature of brown dwarfs. Detection of lithium is rutinary achieved in brown dwarf candidates in stellar clusters and in the solar neighborhood and most bona fide brown dwarfs known at present have been recognized through the presence of lithium in their spectra. Another promising area of activity promoted by these lithium searches is the new method for datation of clusters based on the empirical determination of the end of the lithium depletion at the bottom of the main sequence.

While large progress has been achieved in understanding the formation of lithium lines in cool atmospheres, there are many aspects that require refinement to fully exploit the potential of this element. In particular, it is important to investigate the formation of lines from molecular species containing lithium, to refine the spectral synthesis computations and generate better model atmospheres. Future searches may provide detection of the more fragile ^{6}Li isotope in brown dwarfs. Detection of metal-poor halo brown dwarfs at suitable distances from the Sun will allow investigation of their lithium content which could open new ways to examine the early galactic evolution of this element.

Acknowledgments. I thank M. R. Zapatero-Osorio for comments on the paper and help with some figures.

References

Allard, F. 1990 PhD. Thesis, Univ. of Heidelberg
Barrado-Navascués, D., Stauffer, J.R., & Patten, B.M., 1999, ApJ, 522, 53
Basri, G., Marcy, G., & Graham, J. R. 1996, ApJ, 458, 600
Basri, G., & Martín, E.L., 1999, ApJ, 510, 266
Béjar, V.J.S., Zapatero Osorio, M.R., & Rebolo, R., 1999, ApJ, 521, 671
Bouvier, J., et al., 1998, A & A, 336, 490
Burgasser et al. 1999, ApJ, 522, 65
Burgasser et al. 2000, ApJ, 531, 57
Burrows, A., & Liebert, J. 1993, Rev. Mod. Phys, 65, 301
Chabrier, G., Baraffe, I., & Plez, B. 1996, ApJ, 459, L91
Chabrier, G., Baraffe, I., Allard, F., & Hauschildt, P.A., 2000, ApJ, in press
D'Antona, F., & Mazzitelli, I., 1985, ApJ, 296, 502
D'Antona, F., & Mazzitelli, I., 1994, ApJS, 90, 467
Delfosse et al., 1997, A & A Lett., 327, L25
García López, R.J., Rebolo, R., & Martín E.L. 1994, A&A282, 518
Hambly, N. C., Hawkins, M.R.S., & Jameson, R.F., 1993, A&AS, 100, 607
Kirkpatrick et al. 1999, ApJ, 519, 802
Leggett, et al. 2000, ApJ, 536, 35
Luhman, K.L., Liebert, J., & Rieke, G. H., 1997, ApJ, 489, L165
Magazzù, A., Martín, E. L., & Rebolo, R., 1993, ApJ, 404, L17
Marcy, G.W., Basri, G., & Graham, J.R. 1994, ApJ, 428, 57
Martín, E. L., Delfosse, X., Basri, G., Goldman, B.; Forveille, T., & Zapatero-Osorio, M.R. 1999, AJ, 118, 2466
Martín, E.L., Basri, G., Gallegos, J.E., Rebolo, R., Zapatero Osorio, M.R., & Béjar, V.J.S., 1998, ApJ, 499, L61
Martín, E.L., Rebolo, R., & Zapatero Osorio, M.R., 1996, ApJ, 469, 706
Nakajima, T., Oppenheimer, B. R., Kulkarni, S. R., Golimowski, D. A., Matthews, K. & Durrance, S. T., 1995, Nature, 378, 463
Nelson, L. A., Rappaport, S., & Chiang, E., 1993, ApJ, 413, 364
Neuhäuser, R., & Comerón, F., 1998, Science, 282, 83
Pavlenko, Y., Rebolo, R., Martín, E.L., García López, R.J. 1995, A&A, 303, 807
Pavlenko, Y., Zapatero-Osorio, M.R., & Rebolo, R. 2000, A&A, 355, 245
Pozio, F. 1991, Mem. Soc. Astr. It. 62, 171

Rebolo, R., Martín, E.L., Basri, G., Marcy, G.W., & Zapatero Osorio, M.R., 1996, ApJ, 469, L53
Rebolo, R., Martín E.L.,& Magazzú, A. 1992, ApJ389, 83
Rebolo, R., et al. 1998, Science, 282, 1309
Rebolo, R., Zapatero Osorio, M.R., & Martín, E.L., 1995, Nature, 377, 129
Ruiz, M.T., Leggett, S.K., & Allard, F., 1997, ApJ, 491, L107
Stauffer, J.R., Hamilton, D., & Probst, R.G., 1994, AJ, 108, 155
Stauffer, J., Schultz, G., & Kirkpatrick, J. D., 1998, ApJ, 499, L199
Strauss et al. 1999, ApJ, 522, 61
Stringfellow, G. 1989, Ph.D. Thesis, Univ. of Calif. Santa Cruz
Zapatero Osorio, M.R., Rebolo, R., Martín, E. L., Basri, G., Magazzù, A., Hogdkin, S. T., Jameson,R. F., & Cossburn, M. R., 1997, ApJ, 491, L81

Lithium in Giant Stars

Ramiro de la Reza

Observatório Nacional, Rio de Janeiro, 20921-400 Brazil

Abstract. Lithium continues to be a surprising element and more than ever the study of its creation, destruction and distribution is giving us tools to understand, not only the chemical evolution of this element, but also the nature of mass loss of evolved giants and nucleosynthesis of other elements in cases of very low-metal giants. It also helps to set constraints for cosmological models. The presence of very strong lithium lines in some giant stars of different spectral types and stages of evolution has been considered up to the end of the nineties as a puzzle. To solve this problem, non standard evolutionary mechanisms must be invoked. We review here all the mechanisms presented in the literature and which are divided into internal and external processes of lithium enrichment. We will also discuss the observational tests which are being performed in order to discard (or not) some of them. In any case, the more realistic values of the lithium abundances in giants are, as we will see, the main test of these proposed scenarios. Because of this importance we discuss here the state of art of the Non-LTE determinations of lithium abundances in strong lithium giants. Evolved giants, with lithium abundances larger than that of the interstellar medium and with their important mass losses can be considered the most realistic sources of lithium in the Galaxy. We believe that a complete physical picture of this problem will give a powerful tool to understand the chemical evolution of a large part of all light elements.

1. Introduction

Three discoveries, realized in very different years, 1940, 1982 and 1989 put in evidence the existence of very strong lithium giants. Mac Kellar (1940) detected a very strong Li I resonance line in the spectrum of WZ Cas, a cool carbon star located in the Asymptotic Giant Branch (AGB). Much later, Wallerstein & Sneden (1982) detected the first Li strong K giant (HD 112127), being a first ascending red giant branch star (RGB). In 1989, Smith & Lambert (1989) detected several Li strong S and MS type AGB giants in the Magellanic Clouds. Today we know several of these stars, nevertheless, their number in respect to the Li-poor giants is only known approximately. It depends on the kind of giant stars and on the definition of the minimum abundance defining a Li-rich star. The values oscillate between 2% and 8% according to different authors. As we will see, this proportion appears to increase to 60% among the fast rotating giants. In any case, the presence of these very strong and also moderately strong Li lines is opposed to what is expected from the first dredge-up standard evolutionary theory where, on the contrary, only Li-poor giants are expected to exist. This situation has received names such as '"the Lithium problem˘"' or "the 'puzzle of the Lithium˘" in the literature.

Even today, 60 and 18 years respectively after the discoveries of Li-strong giants among the carbon and K type stars, we have to decide between the internal and external scenarios for the Li enrichment in giants. The internal mechanisms are all based on the ^{7}Be mechanism (Cameron 1955; Cameron & Fowler 1971), whereas the external ones are based on the direct injection of ^{7}Li into the atmospheres of the giant produced by hot nova companions or by engulfing of planets or brown dwarf companions by the giant stars.

In the following, we will discuss the different scenarios that appeared in the literature. For each model the present or future observational tests that could confirm or discard these models will be indicated. Afterwards, we will present the state of art of the Non-LTE methods employed to determine the abundances of Li in giants. At the end, we will present some conclusions and perspectives concerning this problem.

2. Internal Lithium Enrichment Mechanisms

2.1. The Internal Mixing and Convective Scenarios

These are based in the efficiency of the ^{7}Be mechanism, in which ^{7}Be is produced in the hot internal layers corresponding to the H-burning zone and which are necessary to produce the reaction ^{3}He$(\alpha, \gamma)^7$Be. For giants with masses between 3.5 and $6M_\odot$ this reaction is produced very near the internal base of the convective envelope, and in this way, fresh ^{7}Be can be rapidly transported to the external layers where it can be transformed into ^{7}Li by ^{7}Be$(e^-, \nu)^7$Li before being destroyed by reactions as ^{7}Li$(p, \alpha)^4$He or at hotter temperatures by ^{7}Be$(p, \alpha)^4$He. These are the main ingredients of the scenario called 'Hot Bottom Burning˘ (HBB) which have been developed in detail by Sackmann & Boothroyd (1992), and Mazzitelli, D`'Antona & Ventura (1999). For giants with masses equal or larger than $7 - 8M_\odot$ the ^{7}Be mechanism is no more in action. This is because the internal temperatures in this case are hot enough to destroy all the ^{7}Be at the base of the convection zone. The HBB reproduces very well the existence of Li-rich AGB giants of types S and MS discovered by Smith & Lambert (1989) in the Magellanic Clouds.

Now, let us consider the situation of low mass giants with masses $< 2.5 M_\odot$. Because the temperatures at the base of the convective layer are not hot enough to produce ^{7}Be, an internal mixing mechanism has to be invoked in order to connect the base of the convective layer to the H-burning zone. For this purpose an *ad-hoc* conveyor belt has been introduced by Sackmann & Boothroyd (1999) which is able to produce ^{7}Li surface enrichments as large as $\log \varepsilon(\text{Li}) = 4.2$ for low-metal giants with [Fe/H] ~ -2.3 due to the hotter internal temperatures attained in these giants. This mechanism is called 'Cool Bottom Processing˘ (CBP) and has its efficiency increased due to the existence of internal excesses of ^{3}He in only this range of stellar masses.

Because of the efficiency of the CBP in very metal-poor giants all the ^{3}He can be consumed in the RGB and in this way no Li-rich AGB giants are expected to exist for this extreme case (Sackmann & Boothroyd 1999). Details of these mechanisms can also be found in Sackmann & Boothroyd in these proceedings. Which mechanism is acting in the intermediate range of masses between 2.5 and $3.5 M_\odot$? Probably the conveyor belt, reducing its importance and passing smoothly from the CBP to the HBB case.

Observational tests of this model can be made by observing the abundances of the primordial ^{9}Be and ^{10,11}B abundances in these Li-rich giants. CBP predicts that these elements, having existed in the atmospheres since the creation of these stars, will vanish rapidly by the action of the deep mixing process. In fact, the first ^{9}Be observations in Li-rich giants have shown that this is the case (Castilho et al. 1999, see also Castilho in these proceedings). Because ^{6}Li is the most fragile of the primordial elements we expect that a large depletion of this isotope will be produced in a continued action of the CBP.

2.2. The Internal Rotation Induced Mechanisms

Low-mass giants are in general low rotation giants with $v \sin i \sim 2$ km s^{-1}, however, the existence of some giants with large rotational velocities have suggested in the past that in some way the external transport of momentum of a high rotation core could be the cause of relatively fast external rotations and even of mass loss. Fekel (1988) first suggested that ^{7}Li could be the clue to understand this process. Later Fekel & Balachandran (1993) developed the idea in which ^{7}Be is transported to the exterior, helped by this transfer of momentum, but not exists a one to one relation between rotation and Li abundances, because ^{7}Li enrichment and rotation decay times are not necessary the same. More recently, Drake et al. (2000) have found a clear correlation existing between very fast rotation ($v \sin i \geq 8$ km^{-1}) and high and very high Li abundances. This correlation appears, however, only when the giants present clear evidences of mass loss through the presence of large excesses of far infrared (FIR) radiation and also through quite peculiar asymmetric Hα absorption lines. This is the case of giants HD 233517, HD 219025 and PDS 365. Interestingly, Drake et al. (2000) have also shown that considering moderately Li-strong giants together with the Li-strong ones, the proportion of Li-rich objects increases to 60%! among the high rotation stars. Somehow, rotation induces the reduction of Li depletion. This rotation induced mechanism has to explain the existence of very Li-rich K giants with very low rotation velocities ($v \sin i \sim 1$ km s^{-1}) as is the case of HD 19745 (de la Reza & da Silva 1995).

2.3. The Prompt ^{7}Li Enrichment - Mass Loss Scenario

The discovery of the fact that a very large part of the low-mass Li-rich giants are the optical counterparts of IRAS point sources (Gregório-Hetem et al. 1992; Gregório-Hetem, Castilho, & Barbuy 1993) led to a scenario in which all RGB stars pass by a short interval of time ($10^{3-} - 10^5$ yrs) during which these giants are Li-rich (de la Reza, Drake, & da Silva 1996; de la Reza et al. 1997; see also de la Reza et al. in these proceedings). This 'Li cycle ˜is produced by a sudden ^{7}Li enrichment of internal origin producing the formation of a circumstellar shell (CS). When this internal rapid mechanism ceases, the CS is detached from the star, transporting the new produced ^{7}Li into the interstellar matter contributing this way to the Galaxy enrichment with this element. The fresh ^{7}Li that remains in the photosphere of the stars is gradually depleted by convection in times of 10^3 to 10^5 yrs for expansion velocities of the CS of less than 5 km s^{-1} and episodic mass losses of the order of $\sim 5 \times 10^{-8} M_\odot$/yr. Being ^{3}He the 'fuel' by which the ^{7}Be mechanism came into action, the Li cycle can be repeated depending on the abundance of the internal remaining ^{3}He. The repeated action can produce eventual multiple CS. This Li cycle is represented in a color-color diagram containing the fluxes at 12, 25 and 60 microns, by closed curves representing the CS expansion and passing over the observed distribution of K giants in this diagram. Whereas the main bulk of K giants, which are Li-poor, is concentrated in a region of this diagram presenting no CS, the Li-rich and Li-poor giants with CS having FIR excesses are distributed in various regions of the diagram. Because the CS ejection and Li depletion in stars are time synchronized, the depletion times can be estimated by the position of the Li-poor K giants in this diagram. An alternative model of this scenario, in which the velocity of the CS depends on the mass of the CS, is presented by Torres et al. in these proceedings.

Another way to examine the Li enrichment in this scenario in the RGB stage, consists in relating it to the luminosity bump that appears in this evolutionary stage. This bump corresponds to the evolutionary stage when the hydrogen-burning shell erases the chemical discontinuity left behind by the first dredge-up at the moment when the convective layer was at its maximum extent or deepest penetration. Because the CBP is expected to start at this stage until the tip of the RGB is attained, we expect (if the CBP is responsible for the ^{7}Li enrichment) to observe Li-rich giants at luminosities equal or higher than the RGB bump. We verified that this is effectively the case, using all Li K giants for which we have Hipparcos distances and using the bumps proposed by Charbonnel (1994) for giants with masses between 1.0 to 1.5 $M_\odot$ (see also Charbonnel et al. in this proceedings).

What is the relation of the ^{7}Li production with stellar masses concerning this scenario? We do not have yet a Li production - fine tuning mass relation. This relation will be important for Li galactic evolutionary models. Nevertheless, Drake (1998) has found a relation of Li depletion with mass among the low-mass giant stars. Giants with $\sim 1 M_\odot$ have smaller Li depletion times of the order of 3×10^3 yrs whereas larger depletion times ($\geq 10^4$ yrs) appear to be common for giants with $\sim 2 M_\odot$. These considerations result for CS expansion velocities of the less than 5 km s^{-1} and mass losses between $10^{-8} - {}^-10^{-7} M_\odot$/yr.

There are several ways to test the prompt ^{7}Li enrichment - mass loss scenario. Fekel et al. (1996) have found that the star HDE 233517 is not a β Pictoris dwarf star as had been considered before (Skinner et al. 1995). On the contrary, this is a distant very Li-rich K giant. Fekel et al. also showed that the size of CS measured by Skinner et al. (1995) at 10 microns corresponds to what is expected by a Li-strong giant in the prompt ^{7}Li enrichment - mass loss scenario.

Detecting detached (or multiple detached) CS would be a strong indication of the validity of this scenario. Observations by means of CO lines have beautifully shown the existence of detached and almost perfect spherical CSs and even multiple CSs, among the carbon giant stars (Olofsson et al. 1999; Mauron & Huggins 1999). To detect detached CSs in RGB stars is, however, much more difficult. The main reason of this difficulty is due to the smaller involved shell masses during the RGB phase. One possibility could be the visualization of the dusty CS by FIR images. In the case of HDE 233517, the 10 microns radiation, produced nearby the star, is probably not the most indicated wavelength to detect detached shells. Longer wavelengths measured by ISO could be more convenient. Disappointedly, this satellite has not shown the same clear detached shells in C giants that have been measured with the CO lines, as far as we know. Other possibilities of exploration of the presence of detached shells can be made in the gas phase of the CS, using the strong Na D lines. Extra absorption features, especially blueward, will indicate the number of shell components, their opacities and their velocities (see Jasniewicz et al. in these proceedings). The appearance of rapid pulses of ^{7}Li enrichment can be detected by observing the simultaneous presence of a very strong ^{7}Li line together with Be and B lines and, depending on its position in the Li cycle, this would be an indication that the star is in the beginning of its ^{7}Li enrichment phase and that the mixing process has not yet destroyed the Be and B. In the next Li cycles the Be and B elements will be already destroyed and only the Li line will be observed (see also Sackmann & Boothroyd 1999).

3. External Enrichment Processes

3.1. The Nova Mechanism

In this case, the photosphere of the giant star is enriched with new ^{7}Li, produced by a nearby hot nova companion (Gratton & D''Antona 1989). Even if it is theoretically possible to obtain, under certain conditions, a high production of ^{7}Li by novae, there are no observational evidences of this type of contamination. On the one hand there are no indications of the presence of binary stars among the Li-rich giants due to the absence of radial velocities variations with a precision of 1 km s^{-1} (de Medeiros, Melo & Mayor 1996). On the other hand, they do not show the presence of a hot companion by means of UV (IUE) spectra (de la Reza & da Silva 1995).

3.2. The Planets/Brown Dwarf Engulfing Scenario

The idea by which giant stars, which had already been depleted by the first dredge-up process, can gain larger Li abundances by accreting planets has been proposed by Alexander (1967). This model has been discussed, introducing the possibility of engulfing also brown dwarf companions, by Gratton & D''Antona (1989) and Brown et al. (1989). This scenario has only recently been developed in detail by Siess & Livio (1999a, 1999b). In the following we will refer only to these last two publications. Depending on the mass of the accreting object, major perturbations of the giant star will occur. The main consequences are the following: a) formation and ejection of the CS producing this way FIR excesses. These authors maintain the same scenario of de la Reza et al. (1996, 1997) introducing only the engulfing planets/brown dwarf process at the beginning of the Li cycle. b) increase of the rotational velocity of the giant due to the transfer of angular momentum of the planet. For a typical $1 M_\odot$ giant with a radius of $20 R_\odot$ the resulting rotational velocity, due to the accretion of a $\sim 5 M_{Jup}$ object, will be of ~ 5.5 km s^{-1}. Larger bodies such as brown dwarfs will produce much faster rotating giants. c) Li, Be and B abundances will increase in the giant, but because of their different nuclear fragilities, the effect will be more sensitive for Li. d) the star metallicity will increase. e) Because of the deposition of a new mass, increasing this way the potential energy of the star, an expansion of the stellar envelope will be produced. An accretion of an even larger body can trigger nuclear burning at the base of the convective layer (HBB) in the AGB phase. In general, the engulfing process is more sensitive in the AGB because the stellar envelope is smaller than in the case of the RGB phase. Nevertheless, the accretion of nearby planets as is the case of 51 Peg systems, are expected to be more important in the RGB phase. In any case, no triggering of the CBP is obtained for the low-mass giants. f) Due to the increased rotation, a dynamo activity can be produced resulting in X-ray emission. In principle, at least qualitatively, due to different time scales of all these processes, they are not expected to appear all simultaneously. In a first approach, this scenario appears attractive because something similar (that is a non-simultaneity) is found for a collection of fast rotating giants when compared to their Li abundances, FIR excesses and X-ray emissions (Drake et al. 2000).

Nevertheless, some important major effects which can observationally be tested deserve more attention in this scenario. The most important one is the impossibility of this scenario to explain the presence of very high Li abundances of several RGB stars with abundances larger than the interstellar Li abundance (de la Reza & da Silva 1995; Pavlenko, Savanov & Yakovina 1999; Terra 1997). Also, very low rotation velocities (< 2 km s^{-1}) and very strong Li abundances ($\log \varepsilon(\text{Li}) = 3.70$) as is the case of the giant HD 19745 (de la Reza & da Silva 1995; Terra 1997; Terra et al. 2000) remain even less explainable. A definite test will be made soon (de la Reza, Cunha, & Smith in preparation) using the Hubble telescope. This will consist in measuring the boron UV lines. If the engulfing process is possible, we expect the boron (the least fragile among the primordial elements) lines to be larger due to the addition of the boron from the star and that of the planet. This will be especially the case for the limiting high rotating, very Li-rich giant HD 9746 ($v \sin i \sim 8$ km s^{-1}).

4. Non-⁻LTE Lithium Abundances Determinations

There are several physical reasons why Non-LTE (NLTE) is important in the line formation and Li abundance determinations in Li-strong giants. These are: a) the core of the line appears sometimes saturated and is formed in the very external layers of the photosphere b) the chromospheric radiation affects the ionization of the neutral lithium atom. c) the LTE Li abundance determinations often give different values for different Li lines.

The importance of the chromosphere comes from the fact that its UV radiation is critical for the neutral Li atom ionization. In fact, the radiation below 3500 Å producing the ionization from the first excited level 2S largely controls the ionization equilibrium of the neutral Li atom, due to a large photoionization cross-section corresponding to this bound-free transition.

Carlsson et al. (1994) discussed the NLTE Li atom line formation in cool stars generally. NLTE applications to specific cases of Li-strong K giants were made by de la Reza & da Silva (1995); Pavlenko, Savanov, & Yakovina (1999), and Abia, Pavlenko, & de Laverny (1999). In some cases quite large Li abundances have been obtained, much larger than the interstellar Li abundance of $\log \varepsilon(\text{Li}) \sim 3.1$ transforming these giants in important candidates for sources of Li in the Galaxy. The Li NLTE abundances presented by de la Reza & da Silva (1995) are, however, somewhat overestimated, because the UV opacity has not directly been taken into account producing this way an extra overionization. New calculations resulting in somewhat reduced abundance values remain, however, larger than the interstellar value and larger than LTE Li abundances. These new values are the result of a self-consisting iterative methodology taking into account the chromospheric effects (Terra 1997; Terra, de la Reza, & Batalha 2000). In a first step, the chromosphere is fixed by introducing appropriate parametric sources of the UV continuum opacities and by fitting, on the one hand, the observed UV (IUE) continuum and, on the other hand, the observed Hα line profile. This last calculation is made into the iteration scheme, in a separate NLTE calculations of the hydrogen lines. Hereafter, all possible Li lines (the resonance Li I line at 6708 Å, and the secondary lines at 6104 Å and 8123 Å) are calculated in NLTE, using the same Li abundance for all the lines as a main free parameter. The iteration is repeated until a convenient convergence is obtained.

The methodology followed by Pavlenko, Savanov & Yakovina (1999) is different in the sense that no direct chromospheric effects have been taken into account and NLTE calculations have been realized separately for each line. A satisfying solution was obtained when the Li abundances were similar for all the lines. A common case of study of these two methods is the case of the Li K giant HD 9746, where the resulting values are: $\log \varepsilon(\text{Li}) = 3.60 \pm 0.3$ in the case of Pavlenko et al. and $\log \varepsilon(\text{Li}) = 3.80$ in the case of Terra (1997). Considering very high derived NLTE Li abundances, one impressive case is that of the field K giant PDS 68, which appears to be the richest Li K giant known up to now with a value of $\log \varepsilon(\text{Li}) = 4.60$ (Terra 1997). This giant star presents an observed secondary Li I line at 6104 Å with an equivalent width of 280 mÅ!

5. Conclusions and Perspectives

Considering the different scenarios we believe that we can begin to make our choices. The engulfing planets/brown dwarfs scenario have a very strong limitation. At least, as far as it has been developed until now, this scenario is unable to explain the existence of low mass RGB stars with Li abundances larger than that of the interstellar medium. Naturally it is expected that the interstellar Li abundance is the one of the engulfed objects in this scenario. Only two solutions appear to solve this problem 1) to accept accreting bodies containing Li abundances larger by one or two orders of magnitude respect to the interstellar medium. This appears to be quite artificial. 2) the possibility that the engulfing process could trigger the CBP. This, however, appears not to be the case in the actual model. On the other hand, due to the intrinsic necessity of the giant to shallow massive bodies in the RGB stage in order to get a sensitive contamination, this results in an increasing of its rotation velocity. In this way, the existence of very Li-rich low-mass and low rotation giants remains even less explainable in this scenario. In any case a near future test, mentioned before, using the observed boron lines will definitely discard or not this scenario.

Under these circumstances we prefer the internal scenarios. Nevertheless, in this case, at least for the low-mass giants, the physical mechanism able to form rapidly a circumstellar shell remains to be found. Maybe a natural source of internal energy can be found in the dissipation of a fast rotating core. We must also not forget that a powerful convection can be the source of a CS as has been shown to be the case in Betelgeuse (Lim et al. 1998).

Recent observations are giving more clues to the understanding of the Li problem. This is the case of the discovery of very Li-strong giants in clusters. Hill & Pasquini (1999) discovered one very Li-strong K giant in the old open cluster Berkeley 21. This kind of discoveries is important to better establish the evolutionary age at which the Li cycle appears. Among globular clusters a very Li-strong RGB giant has been detected in M3 (Kraft et al. 1999) and moderately Li-strong ones in NGC 362 (Smith, Shetrone, & Keane 1999) and in M5 (Carney, Fry & Gonzalez 1998). For the case of M3, and using the ^{7}Li prompt enrichment - mass loss scenario, de la Reza et al. (2000) have shown that a large quantity of ^{7}Li can be produced enriching the intracluster matter of this cluster during the whole life of M3. This new Li is probably lost, by ram pressure, into the galactic disk when M3 crosses several times this disk contributing this way to its Li enrichment. They also suggest that for these low-metal giants, the CBP producing a fresh ^{7}Li enrichment, could be related to the nucleosynthesis of Na and Al explaining this way the observed star to star variations of these elements.

Is the prompt ^{7}Li - mass loss scenario also into action in the more advanced AGB stage? This remains to be explored. In any case, all the scenario ingredients are present in carbon giants such as a variety of strengths of Li lines including the most extreme Li lines known up to now, as is the case of WZ Cas, T Ara and IY Hya. Also, these giants present very clearly almost perfect spherical detached CS as is the case of TT Cyg (Olofsson et al. 2000). There are also several C giants presenting multiple CSs (Mauron & Huggins 1999). Another group that deserves special attention are the carbon J stars, which could be in a peculiar stage of evolution. These stars are characterized by containing moderately strong Li lines and by being enriched with ^{13}C (see Lorenz-Martins & Drake and also Abia & Isern in these proceedings). When J and cool type N carbon giants are placed in an IRAS diagram, they are concentrated in a region presenting more excesses in the 60 microns fluxes. If there is a Li cycle, it is much shorter than that of the K giants. This situation being compatible with their larger observed CS expansion velocities (≥ 10 km s^{-1}).

Up to now there is no a clear relation between ^{13}C and ^{7}Li enrichments. The first results for K giants (da Silva, de la Reza & Barbuy 1995; Drake 1998, see also Drake et al. in these proceedings) present no correlations indicating that probably enrichments in ^{13}C and ^{7}Li are acting in different time scales. Another group that deserves attention are the Post-AGB giants. Even if the statistics are poor, some of these stars, such as the Frosty Leo nebula present a relatively strong Li line.

Two important consequences, one related to the internal mixing and the other to episodic mass loss are the following: As more ^{7}Li is created more ^{3}He is destroyed. The cosmological consequences of this, in limiting the interval of abundances of ^{3}He as a ' "barionmeter˜" ", are discussed by Sackmann & Boothroyd (1999). The episodic mass loss for which a specific physical propulsion mechanism remains to be found (thermal pulses are proposed for this purpose in the AGB phase) introduces the notion that mass loss could be an essentially variable phenomenon and that this one begin already in the RGB phase. This does not prevent a less strong continuous mass loss from being also present.

One of the advantages of the prompt ^{7}Li enrichment - mass loss scenario consists in bringing a physical picture of the complete path, joining the element production source in the core of stars to the interstellar medium. By considering the Li cycles at the RGB phase and by establishing that similar cycles are also present in the AGB phase, we will dispose of a complete ensemble of observed ^{7}Li sources in order to have a more realistic model to study the evolution of Li in the Galaxy.

References

Abia, C., Isern, J. 1997, MNRAS, 289, L11

Abia, C., Pavlenko, Y., & de Laverny, P. 1999, A&A, 351, 273

Alexander, J.B. 1967, The Observatory, 87, 238

Brown, J.A., Sneden, C., Lambert, D.L., & Dutchover, E. 1989, ApJS, 71, 293

Cameron, A.G.W. 1955, ApJ, 121, 144

Cameron, A.G.W. & Fowler, W.A. 1971, ApJ, 164, 111

Carlsson, M., Rutten, R.J., Bruls, J.H.M.J., & Shchukina, N.G. 1994, A&A, 288, 860

Carney, B.W., Fry, A.M., & Gonzalez, G. 1998, AJ, 116, 2984

Castilho, B.V. et al. 1999, A&A, 345, 249

Charbonnel, C. 1994, A&A, 282, 811

da Silva, L., de la Reza, R., & Barbuy, B. 1995, ApJ, 448, L41

de la Reza, R. & da Silva, L. 1995, ApJ, 439, 917

de la Reza, R., Drake, N.A., & da Silva, L. 1996, ApJ, 456, L115

de la Reza, R., Drake, N.A., da Silva, L., Torres, C.A.O., & Martin, E. L. 1997, ApJ, 482, L77

de la Reza, R., da Silva, L., Drake, N.A, & Terra, M.A. 2000, ApJ, (Letters May)

de Medeiros, J.R. Melo, C.H. F., & Mayor, M. 1996, A&A, 309, 465

Drake, N.A. 1998, PhD Thesis, ⁻ Observatório Nacional, ⁻ Rio de Janeiro

Drake, N.A., de la Reza, R. da Silva, L., & Lambert, D.L. 2000, submitted to the AJ

Fekel, F.C. 1988, A Decade of UV Astronomy with the IUE Satellite, ESA, SP-281, Vol. 1, p. 331

Fekel, F.C., & Balachandran, S. 1993, ApJ, 403, 708

Fekel, F.C., Webb, R.A., White, R.J., & Zuckerman, B. 1996, ApJ, 462, L95

Gratton, R.G., & D`'Antona, F. 1989, A&A, 215, 66

Gregório-Hetem, J., Lepine, J.R.D., Quast, G.R., Torres, C.A.O., & de la Reza, R. 1992, AJ, 103, 549

Gregório-Hetem, J., Castilho, B.V., & Barbuy, B. 1993, A&A, 268, L25

Hill, V., & Pasquini, L. 1999, A&A, 348, L21
Kraft, R.P. et al. 1999, ApJ, 518, L53
Lim, J., Carilli, C.L., White, S.M., Beasley, J., & Marson, R.G. 1998, Nature, 392, 575
Mac Kellar, 1940, PASP, 52, 407
Mauron, N., & Huggins, P.J. 1999, A&A, 349, 203
Mazzitelli, I., D`'Antona, F., & Ventura, P. 1999, A&A, 348, 846
Olofsson, H. et al. 2000, A&A, 353, 583
Pavlenko, Ya. V., Savanov, I.S., & Yakovina, L.A. 1999, Astron. Rep., 43, 671
Sackmann, I-J. & Boothroyd, A.I. 1992, ApJ, 392, L71
Sackmann, I-J. & Boothroyd, A.I. 1999, ApJ, 510, 217
Siess, L. & Livio, M. 1999a, MNRAS, 304, 925
Siess, L. & Livio, M. 1999b, MNRAS, 308, 1133
Skinner, C.J. et al. 1995, ApJ, 444, 861
Smith, V.V., & Lambert, D.L. 1989, ApJ, 345, l75
Smith, V.V., Shetrone, M.D. & Keane, M.J. 1999, ApJ, 516, L73
Terra, M.A. 1997, PhD Thesis, Observatório Nacional, ⁻ Rio de Janeiro
Terra, M.A., de la Reza, R., & Batalha, C. 2000, in preparation
Wallerstein, G. & Sneden, C. 1982, ApJ, 255, 577

The properties of the PDS Li-rich giant stars [1]

C. A. O. Torres and G. R. Quast

Laboratório Nacional de Astrofísica/MCT, CP 21, 37504-360 Itajubá, MG, Brazil

R. de la Reza and L. da Silva

Observatório Nacional/MCT, Rua General José Cristino 77, 20921-030, Rio de Janeiro, Brazil

Abstract. In the Pico dos Dias Survey (PDS), devised to search for young stellar objects, we found also K giants presenting moderate to strong Lithium lines associated with IRAS sources. But there are some gK with weak or absent Li lines too. There is a dichotomy in the distribution of the IRAS spectral indices between both kinds of K stars. A model of an expanding shell with a mass-dependent velocity may explain this behavior, if the photospheric Li depletion takes only ∼2500 yrs.

1. Introduction

In spite of the infrared spectral criteria used in the the Pico dos Dias Survey (PDS), conceived to search for young stellar objects (Gregorio-Hetem et al., 1992), we found about 40 K giants in the direction of the selected IRAS sources, more than 20 of them presenting moderate to strong Lithium lines. A preliminary result was presented by de la Reza et al. (1997). To explain this result, de la Reza et al. (1996) developed a model where an internal Li enrichment during the red giant branch stage is followed by a prompt mass-loss event, responsible for the infrared excesses.

As the PDS was recently finished (Torres, 1998), we present now a more complete analysis of the K giants found.

2. The K giants of the PDS

To clarify the infrared properties of the K giants we restrict the sample by including giants in only two cases, taking into account the possibility of chance coincidences of the fairly abundant K stars with the IRAS sources within the positional error ellipses:
a) gK with Lithium equivalent widths (W_{Li}) larger than 0.07 Å and being the most probable optical counterpart of the IRAS source.

---

[1]Based on observations made at the Observatório do Pico dos Dias, operated by MCT/Laboratório Nacional de Astrofísica, Brazil

b) gK with very weak or absent Li line having no other obvious possible counterpart of the IRAS source.
Actually, most of the IRAS sources eliminated in this way show H_α and forbidden line emissions in the sky spectra and are thus probably associated with HII regions. We retained 17 and 11 stars of cases (a) and (b), respectively. They are presented in Tables 1 and 2, with some of their properties, where [12] means the IRAS flux in the 12 μm band, in Jy. The spectral indices are defined as $\alpha = \log\lambda F/\log\lambda$, α_1 being between 12 and 25 μm and α_2 between 25 and 60 μm. Two of the stars in Table 1 belong to an unpublished extension of the PDS, using the IRAS Faint Source Catalog.

Table 1. Properties of the PDS K giants with the Li I line $\lambda 6707$

PDS	IRAS	W_{Li}	V	[12]	α_1	α_2	SpT	A_V
003	03062-6538	-0.49	9.4V	0.30	0.31	-1.27	K0III	0.39
[a]	F04376-3238 [b]	-0.4:	10.37	0.26	-0.73	-1.65	K0III:	0
132	07227-1320	-0.25	12.59	2.47	-0.04	-1.69	K8III	1.46
258	07419-2514	-0.10	12.32	0.19	-0.02	0.73	K2III	2.36
135	07456-4722	-0.19	11.01	2.11	-0.58	-1.76	K2III	1.23
260	07577-2806	-0.50	14.62	0.60	1.13	-1.74	K0Ib/III	2.96
[c]	F08359-1644[b]	-0.45	14	0.10	1.04	-1.38	K0III	-
354	12236-6302	-0.37	12.30	0.95	1.01	1.37	K1III	0.94
355	12327-6523	-0.28	10.40	8.60	-1.24	-0.54	K1III	4.6
365	13313-5838	-0.46	13.15	1.88	0.59	-1.69	K1III	1.46
068	13539-4153	-0.60	12.81	0.48	-0.26	-1.28	K2II-III	1.74
410	16086-5255	-0.15	13.83	5.70	0.85	-1.15	K1IIIp	1.84
432	16514-4625	-0.51	11.48	2.60	0.14	0.35	K2III	1.23
485	17596-3952	-0.20	12.10	0.48	0.13	-1.71	K1III	1.26
524	18334-0631	-0.14	12.50	7.86	0.29	0.21	K0II	4.8
562	19083+0119	-0.16	9.4V	2.34	1.58	-1.10	K2III	0.28
100	19285+0517	-0.33	10.45	4.74	-0.56	-1.75	K0III	1.87

[a] CD -32 1919
[b] IRAS faint souce catalogue
[c] GSC 6015-2379

3. The biases on the sample

In any analysis we must carefully take into account the observational limits of the sample. For the observed stars we imposed a magnitude limit in the GSC of 14 (Torres, 1998). As these stars are red and the GSC magnitudes are, in general, blue ones, the visual limit is ~13. Although some stars in Tables 1 and 2 are somewhat fainter, this limit represents fairly well our sample. For an unreddened giant with $M_V \sim 0$ this means a limit in distance of ~4 kpc. The measured interstellar extinctions (A_V) show that this is essentially correct, assuming an average extinction of 1 mag/kpc. On the other hand, the limit on IRAS fluxes is ~0.25 Jy. We can establish a spectral index between the visual

Table 2.   Properties of the PDS K giants with weak or absent Li I line

PDS	IRAS	V	[12]	α_1	α_2	SpT	A_V
299	09553-5621	12.85	0.74	-0.67	1.31	K0III	2.13
071	14422-8021	12.67	0.45	-0.16	-1.56	K0III	1.98
419	16227-4839	11.86	2.94	-1.17	0.63	K3III	0.77
434	16552-3050	13.46	2.46	0.97	-1.10	G7III	1.25
462	17399-3100	13.58	3.43	-0.76	-0.59	K5III	4.05
466	17442-2441	13.78	3.30	-0.27	-0.75	K1III	2.76
533	18397-0400	11.43	7.54	-0.92	0.22	G7III	2.59
542	18454-0731	12.89	0.60	-1.17	0.36	K0III	1.22
552	18559+0140	12.05	1.40	-1.18	1.50	K0III	0.54
573	19210+1715	10.86	0.59	-0.01	0.73	K0III	3.11
583	19365+2557	13.62	0.34	-0.33	0.93	G7III	1.79

and the 12 μm band (β). The mean de-reddened β index is $\sim$-0.5. Thus, the IRAS flux limits imply a mean de-reddened visual limit of $\sim$12.

This means that both, V-magnitude and infrared flux, impose quite similar limits. The selected PDS sources and their gK counterparts are close to these limits as a result of their low galactic density (we found $\sim$ 1 star/kpc^3). Therefore, we detected mainly stars with shell infrared fluxes relatively strong compared to the photospheric ones.

4. Comments on some individual objects

PDS 258 – This star has the weakest Li line in Table 1 and may be considered an intermediate case between the two kinds of gK. In Figure 1, it is the open circle nearest to the upper left corner. The CO cloud WB 1046 may be another identification for this IRAS source (Wouterloot & Brand, 1989).

PDS 354 – This star, with the highest α_2 value, has a strong Hα emission in the background sky. This position on the diagram is typical for HII regions and the infrared fluxes of the star may be, at least, contaminated. It is near PDS 355.

PDS 355 – It is behind the Coalsack and, as the reddening comes from it and the star may be as near as 200 pc. The infrared fluxes may be contaminated.

PDS 434 – Hu et al. (1993) classified it as G9III and found a compact reflection nebula wich reinforces the association with the IRAS source. Its position in Figure 1, the asterisk nearest to the lower right corner, is peculiar. This star may remind us that, in evolved stars, mass loss not only occurs during the Li enrichment process.

5. Discussion

In Figure 1 we plot the infrared spectral indices of the stars of both tables. We can see that there is a dichotomy in their distribution with respect to the Li abundance. In fact, almost all stars in Table 2 are at the left of the diagram. We explain this distribution as being the result of Li depletion in the stellar photospheres after the shell expulsion. There are some difficulties to explain

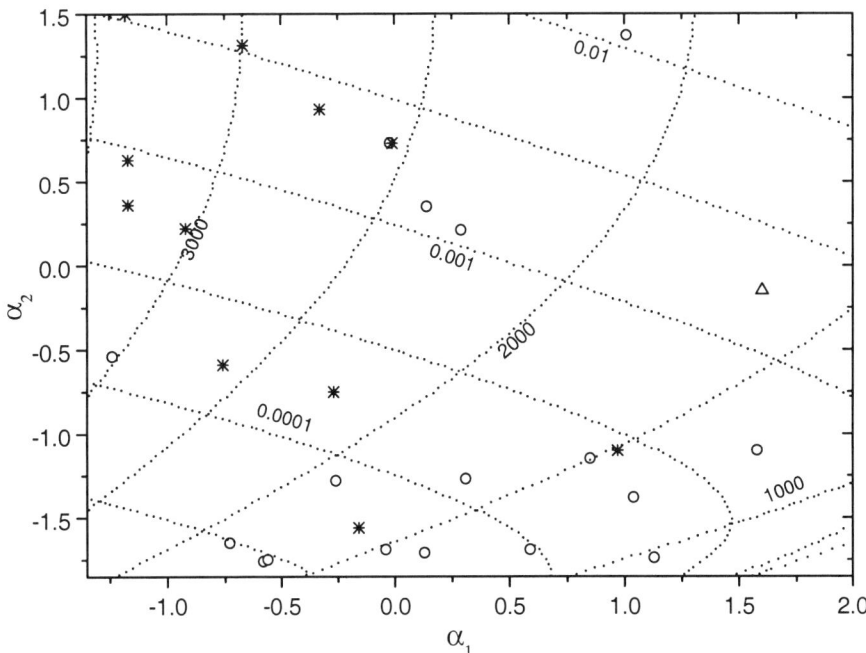

Figure 1. Diagram of IRAS spectral indices within the limits used in the PDS. Circles represent the stars with moderate to strong Li lines (Table 1) and asterisks those with weak to absent Li lines (Table 2). The triangle represents HDE233517, discovered by Fekel et al. (1996), that would belong to the PDS if it were extended to the northern hemisphere. Lines represent the expanding mass-dependent shell model. Isochrones are labeled with 1000 to 3000 yr and the evolutionary tracks for equal shell masses are labeled with 0.01 to 0.0001 $M_\odot$. This model was computed for a star with $R_* = 20 R_\odot$ and $T_{eff} = 4500$ K. As real stars may deviate from the above average parameters, individual positions relative to the model must be taken with caution. The dichotomy between the samples may be explained in this model by a Li depletion in the stellar photosphere in $\sim$2500 yr.

this by the model of de la Reza et al.(1996) as its isochrones are almost parallel to the α_1 axis. There are, of course, a lot of simplifying hypotheses in the model, one of them being the independence of the velocity of expansion of the shell on its mass. If we suppose that there may be such a dependence, we can find an empirical law that could explain the observational behavior. For example, it could be fairly well explained with a power law of the kind:

$$V = V_0 (M/M_0)^{1/4}$$

where V and M are the velocity of expansion and the mass of the shell. This is the model presented in Figure 1. According to this one the Li in K giants is depleted in $\sim$2500 yr if the normalization parameters are: $V_0 = 0.5$ km/s and $M_0 = 10^{-6} M_\odot$.

The fact that the Li depleted stars have, apparently, larger shell masses may be qualitatively explained by bias: first, the model forsees that the shell stays longer in the region of the graph where the Li depleted stars are and, second, the sample is biased towards more massive shells, because an evolved less massive shell has weaker infrared emission and would not have been selected in the PDS.

The Li-rich K giant stars are more concentrated in the lower part of the diagram. This represents the expected distribution in the diagram as it means only that low mass shells are more frequent than more massive ones. But it should be noted that the lack of stars older than 2000 yr, in the lower part of the diagram, may be the result of both the Li depletion and the weaker emission of the shell.

The semi-empirical expanding shell model presented here seems to describe quite well the behavior of the PDS K giants. Any model that attempts to explain the physical reasons of the shell ejection should reproduce the law (or a similar one) found by us.

The Li-rich giants are difficult to detected (in the PDS we found ~ 1 star/kpc^3). But considering the short time-scale of this phenomenon, it is probable that all gK will pass through this phase at least once. It may be one of the most important Li sources of the interstellar medium.

References

de la Reza, R., Drake, N. A., & da Silva, L. 1996, ApJ, 456, L115

de la Reza, R., Drake, N. A., da Silva, L., Torres, C. A. O., & Martin, E. L. 1997, ApJ, 482, L80

Fekel, F. C., Webb, R. A., White, R. J., & Zuckerman, B. 1996 ApJ462, L95

Gregorio-Hetem, J., Lépine, J. R. D., Quast, G. R., Torres, C. A. O., & de la Reza, R. 1992, AJ, 103, 549

Hu, J. Y., Slijkhuis, S., De Jong, T., Jiang, B. W. 1993, A&AS, 100, 413

Torres, C. A. O. 1998, Publicação Especial do Observatório Nacional, 10/99

Wouterloot, J. G. A., Brand, J. 1989, A&AS, 80, 149

Search for lithium-rich stars among G–K giants with IR–excess

G. Jasniewicz

GRAAL, UPRESA 5024, Université de Montpellier II, CC 72, 34095 Montpellier Cedex 05, France

M. Parthasarathy

Indian Institute of Astrophysics, Koramangala, Bangalore 560 034, India

P. de Laverny

Observatoire de la Côte d'Azur, Dpt. Fresnel, UMR 6528, B.P. 4229, 06304 Nice Cedex 4, France

F. Thévenin

Observatoire de la Côte d'Azur, Dpt. Cassini, UMR 6529, B.P. 4229, 06304 Nice Cedex 4, France

N. Mauron

GRAAL, UPRESA 5024, Université de Montpellier II, CC 72, 34095 Montpellier Cedex 05, France

M. Chadid

ESO, Casilla 19001, Santiago 19, Chile

Abstract. We report here new results about a search for Li-rich stars among 52 G–K giant stars which are known to have near IR excess. Eleven giants have been found to have $\log \epsilon(Li) \geq 1.0$. Five are new Li-rich stars. We suspect circumstellar shells around one of them, HD 219025. There is no clear correlation between Li-richness and rotation, or with binarity.

1. Introduction

This research is based on two observational facts. On the first hand there are a few G–K giant stars which are known to have IR excess, which is not the case of the majority of the giants. The origin of the dust responsible for the IR excess could be: sporadic mass-loss events from these stars (Zuckerman et al. 1995) during a possibly short-lived phase of evolution, a Vega-like disk heated by the star evolving into a giant (Judge et al. 1987; Plets et al. 1997), a hot spot produced by a nearby diffuse cloud (cirrus) locally heated by the star (Jura 1999).

On the other hand there are giant stars with an overabundance of lithium, which is not expected from the standard first dredge-up evolutionary models. From the histogram of Brown et al. (1989, Fig.13), concerning 644 normal field G-K giants, we deduce that only 11% of the stars have $\log \epsilon(Li) \geq 1.0$ in their sample. The theories and/or scenarios of Li enhancement involve: a self production (Sackmann and Boothroyd 1999; Charbonnel & Balachandran 2000) of Li without or with shell detachments, an accretion of planets or brown dwarfs by giant stars (Siess and Livio, 1999), mass and angular momentum tranfers in binary system as suggested by Barrado y Navascués et al (1998) for chromospherically active binary stars, etc...

The main question adressed in this work is: which correlation (if any) is there between IR-excess and Li abundance among late-type giants? This question has been adressed previously by other authors, especially by de la Reza and co-workers (de la Reza et al. 1996, 1997) who propose a scenario linking the high Li abundances of some K giant stars to the evolution of circumstellar shells. Besides, Fekel & Watson (1998) have selected 40 IR-excess giants from the list of Zuckerman et al. (1995) and shown that the percentage of giants with $\log \epsilon(Li) \geq 1.2$ is similar to that expected for normal field giants; however their conclusion is no more true if we adopt $\log \epsilon(Li) \geq 1.0$ for being Li-rich (see Sect. 3). A work similar to that of Fekel & Watson (1998) has also been simultaneously achieved in a totally independent way by us (Jasniewicz et al. 1999). In this communication new targets have been added to the list of Jasniewicz et al. (1999).

Another question adressed in this communication is the following: is there any evidence for a circumstellar shell around giant stars which are Li overabundant?

2. Targets and observations

The giant stars considered in this research come from various lists of stars with IRAS infrared excess: 29 stars come from Zuckerman et al. (1995), 14 stars from Oudmaijer et al. (1992), 5 stars from Stencel & Backman (1991), and 4 stars from Plets et al. (1997). These 52 objects were extracted from the 4 papers above with the following criteria: spectral type III and visual magnitude $m_V \leq 8.5$. The V-mag threshold was set in order to get high S/N ratio spectra with our instrumentation. Using the HIPPARCOS parallaxes available for 41 stars in our sample, we infer that the selected stars are not pre-main-sequence stars and that most of them are more luminous than the K giant clump (see the loci of some of our stars in the HR diagram by Charbonnel & Balachandran, this conference).

Observations were performed at ESO with the CAT-1.4m + CES spectrograph and a spectral resolution $\lambda/\delta\lambda=67\,000$, and at Haute Provence Observatory (France) with the 1.5m telescope and the Aurélie spectrograph with $\lambda/\delta\lambda=40\,000$. A few radial velocities were also measured with CORAVEL at OHP.

3. Analysis of spectroscopic data

Lithium abundances were determined by the spectral synthesis method for all the stars (see Lèbre et al. 1999 and Jasniewicz et al. 1999 for a general description). Updated MARCS model atmospheres computed with the code of Asplund et al. (1999) and of Bessel et al. (1998) were used. Considering uncertainties on model parameters, we estimate that the Li and Fe abundances are determined with an uncertainty better than ±0.2 dex.

In our sample of stars, we found 11 Li-rich giant stars ($\log \epsilon(Li) \geq 1.0$), say **21% of Li-rich stars** (see Table 1). Fekel and Watson (1998) have obtained a similar percentage of 20% of Li-rich giants ($\log \epsilon(Li) \geq 1.0$) in their sample of 40 stars from Zuckerman et al. (1995). Thus the rate of Li-rich stars in our modest-size sample of IR-excess giants is higher than the expected one for normal field giants, but we did not investigate the biases which could affect our sample. We also emphasize here the fact there are G-K giant stars with IR-excess without overabundance of Lithium and that, reciprocally, there are Li-rich G-K giant stars without IR-excesses: HD 157457, G8III Barium star; HD 112127, K2.5III possible Barium star; RS CVn stars, etc...

Table 1. **G AND K LI-RICH GIANT STARS**

HD	Sp. type	Teff	V_{rot}	[Fe/H]	$\log \epsilon(Li)$	Comments
HD 30834	K2III	4500	1.0	-0.5	2.4	no RV Var.
HD 80499	G8III	4980	15.0	-0.2	0.9	vis.binary
HD 89221	gG5	4980	1.0	-0.3	1.1	
HD 146850	K3III	4270	1.0	+0.4	2.0	No SB I ?
HD 152786	K3III	4270	1.0	+0.2	1.3	
HD 153751	G5III	5040	25.0	-0.3	1.2	RS CVn
HD 169689	G8III	4950	10.0	-0.2	1.0	SB I
HD 175492	G4III	5280	3.0	-0.2	1.3	SB I
HD 176884	K0III	4800	15.0	+0.2	1.2	no RV Var.
HD 190252	G8III	5200	1.0	-0.3	1.3	
HD 219025	K2III	4500	23.0	-0.1	3.0	RS CVn ?

All stars of Table 1 have far-IR excesses over the stellar continuum of a normal star. Five of them are new Li-rich stars:

HD 80499: IR excess at 25μ: 0.1 mag according to Oudmaijer et al. (1992)

HD 89221: IR excess at 25μ: 0.3 mag (Oudmaijer et al. 1992). Possibly a subgiant ($M_V = 3.7$)

HD 152786: IR excess at 60μ (Zuckerman et al. 1995; Plets et al. 1997)

HD 176884: High rotator with a high far-IR 60μ excess: 2.8 mag (Zuckerman et al. 1995; Plets et al. 1997; Oudmaijer et al. 1992)

HD 190252: IR excess at 25μ: 0.3 mag (Oudmaijer et al. 1992)

4. Search for correlations

- Is there any link between Li-overabundance and binarity ?

  Some Li rich giant stars are known to be members of binary systems:
  HD 169689 and HD 175492: spectroscopic binaries, the first one being of Algol type (P=385 days).
  HD 153751: high rotating RS CVn eclipsing binary, P=39.48d according to de Medeiros and Udry (1999).

  The case of HD 219025 is uncertain: it is a high rotating active star, possibly a RS CVn eclipsing binary, with a very high far-IR excess 60μ. This star is similar to the Li-rich star HD 233517. A link between these stars and the FK Comae-type stars (probably resulting from coalescence of binaries) is an open question.
  A possible link between high rotational velocity, Li abundance and engulfing of planets are discussed by Siess and Livio (1999).

  In return, some Li-rich giant stars are not known as binaries (at least short period binaries). Some of them are:
  HD 9746: Chromospherically active star with Prot=2,36days, presently not known as a binary.
  HD 30834 and HD 176884: no radial velocity variations.
  HD 146850: Li-rich according to Fekel and Watson (1998) and Jasniewicz et al. (1999). Detected RV variable by Fekel and Watson (1998); not in this work.

  Thus, the correlation between Li-richness and binarity is not clear.

- There is also no clear relation between Li abundance and $^{12}C/^{13}C$ ratio in Li-rich giants (da Silva et al. 1995). According to Jasniewicz et al. (1999), the $^{12}C/^{13}C$ ratio ranges from 6 to 28 similar to that found for normal K-giants.

- Fekel and Balachandran (1993) suggest a connection between rapid rotation and high Li abundance in giants. Besides, Li abundance does not depend on age for giant components of chromospherically active binary systems, but is closely related to stellar rotation (Barrado y Navascués et al. 1998).
  From the other hand, the 12 Li-rich giants selected by de Medeiros et al. (1996), except for one star, show normal rotational velocities with respect to typical Li-normal giants (de Medeiros et al. 1996).
  Using the data from this work (V_{rot} from CORAVEL) concerning IR-excess giants and data from a list of Li-rich K giant stars taken from literature, we find no clear correlation between Li abundance and rotation.

5. Search for circumstellar gas of IR-excess giants

According to the de la Reza et al. (1996, 1997) scenario, G–K giants might suffer episodic mass loss events in connection with Li production. This mass loss would produce IR emitting dust. If this is true, it is plausible that for some of these giants the expanding circumstellar gas can be detected through resonance absorption lines such as the NaI doublet. Typically, if the envelope has a mass loss rate of 10^{-8} $M_\odot$ yr^{-1}, an expansion velocity of 50 km s^{-1}, an inner radius of 10 R_* or $\sim$250 $R_\odot$ and a fraction of NaI of only 1 percent of total Na, one expects a NaI D1 equivalent width of $\sim$0.6 Å, which is well detectable. Absorption sodium lines may also be expected in alternative scenarios, as in the case of evaporating Vega-like icy discs, or circumbinary discs, or a local interstellar cloud.

Consequently, we acquired high-resolution spectra (R=40 000) of the NaI doublet for a group a Li-rich and Li-normal giants taken from our sample of IR-excess giants. This group comprises 7 Li-rich and 21 not Li-rich stars. We searched for both blue and red-shifted absorption components. We detected 13 stars with a clear blue or red-shifted component, and 5 with a distorted photospheric profile.

A difficult issue is how to distinguish the interstellar or circumstellar origin of any absorption component. In order to evaluate statistically the occurence of interstellar lines, we searched for hot stars which could be as close angularly as possible to each giant star, and for which interstellar lines are documented. Generally, 2 or 3 such hot stars were found within about 3 degrees. This angular separation represents 6 pc at a distance of 100 pc, and corresponds to an average size for diffuse interstellar clouds. Analysis of photometric and spectroscopic data on the hot stars allowed to locate them either beyond or in front of the cool giant, and to evaluate if common insterstellar features (with similar velocities) can be expected. Indeed, a careful examination of each case led us to conclude that the NaI absorption components are probably of interstellar origin in the large majority of giants. However, we noted that the high galactic latitude Li-rich star HD 219025 (b_{gal}=46 °; K2III) displays prominent blue-shifted NaI components with expansion velocity of 48km s^{-1} for which a circumstellar origin appears plausible, although this is not definitively established.

6. Conclusions

In order to study a possible link between Li overabundance and IR excess, a sample of 52 G-K giants *with IRAS infrared excess* has been considered. High resolution spectra of the LiI λ6707 line were obtained, revealing that 11 stars, i.e. $\sim$20%, are Li-rich with $\log \epsilon(Li) \geq 1.0$. Our sample is certainly not homogeneous and may suffer from various biases, but it can be noted that the above fraction is identical with that found by Fekel and Watson (1998) in a similar study. This fraction is also twice the fraction of Li-rich stars found from a sample of *normal* giants, according to the study of Brown et al.(1989).

With the goal of investigating the de la Reza scenario of a mass loss event connected with Li production, a search for NaI circumstellar absorption lines has also been carried out on a subsample of 28 giants. It seems that most of the observed NaI lines towards these objects can be due to intervening interstellar clouds, since they also appear with same velocities towards hot stars located within a few degrees. An interesting exception could be HD 219025 (K2III) which is very Li-rich ($\log \epsilon(Li) = 3.0$), and might possess circumstellar Na lines.

Acknowledgments: We thank very much Corinne Charbonnel for very fruitful discussions on the text at the Conference.

References

Asplund, M., Gustafsson, B., Kiselman, D., & Eriksson, K. 1999, A&A, 318, 521

Barrado y Navascues, D., de Castro, E., Fernandez-Figueroa, M.J., Cornide, M., & Garcia Lopez, R.J. 1998, A&A, 337, 739

Bessell, M.S., Castelli, F., & PLez, B. 1998, A&A, 333, 231

Brown, J.A., Sneden, C., Lambert, D.L., & Dutchover, E.Jr 1989, ApJS, 71, 293

Charbonnel, C., & Balachandran, S. 2000, this volume

da Silva, L., de la Reza, R., & Barbuy, B. 1995, ApJ, 448, 410

de la Reza, R., Drake, N.A., & da Silva, L., Torres, C.A.O., & Martin, E.L. 1997, ApJ, 482, 77

de la Reza, R., Drake, N.A., & da Silva, L. et al. 1996, ApJ 456, 1150

de Medeiros, J.R., & Udry, S. 1999, 346, 532

de Medeiros, J.R., Melo, C.H.F., & Mayor, M. 1996, 309, 465

Fekel, F.C., & Balachandran, S. 1993, ApJ, 403, 708

Fekel, F.C., & Watson, L.C. 1998, AJ, 116, 2466

Jasniewicz, G., Parthasarathy, M., de Laverny, P., Thévenin, F. 1999, A&A, 342, 831

Judge, P.G., Jordan, C., & Rowan-Robinson, M. 1987, MNRAS, 224, 93

Jura, M., 1999, ApJ, 515, 706

Lèbre, A., de Laverny, P., de Medeiros, J.R., Charbonnel, C., & da Silva ,L. 1999, A&A, 345, 936

Oudmaijer, R.D., van der Veen, W.E.C.J., Waters, L.B.F.M., Trams, N.R., Waelkens, C., & Engelsman E. 1992, A&AS, 96, 625

Plets, H., Waelkens, C., Oudmaijer, R.D., & Waters, L.B.F.M. 1997, A&A, 323, 513

Sackmann, I.J., & Boothroyd, A.I. 1999, ApJ, 510, 217

Siess, L., & Livio, M. MNRAS, 308, 1133

Stencel, R.E., Backman, D.E. 1991, ApJS, 75, 905

Zuckerman, B., Sungsoo, S.K., & Liu, T. 1995, ApJ, 446, 79

Be vs. Li Abundance in Li-rich Giants: an evidence of Li production in Red Giants

B. V. Castilho

Departamento de Astronomia, Instituto Astronômico e Geofísico, Universidade de São Paulo, CP 3386, 01060-970, Brazil

Abstract. We present Be abundances derived through spectral synthesis for two Li-rich giants and a Li-normal giant, plus α Cen A and B, based on observations of the Be II λ 3130.420 and 3131.066 Å lines obtained with the CASPEC spectrograph at the ESO 3.6m telescope[1].

The values derived for the two Li-rich giants (HD146850, HD787) agree with the one for the Li-normal giant (HD220321), and shows that Be was depleted in these stars (by > 90% i.e. by a factor of 10 or larger from the initial Pop I value ($logN(Be) = 1.4$)).

This result implies that the original Li in these stars was almost completely destroyed, and that Li is most probably produced in red giants.

1. Introduction

About 2% of the red giants show lithium abundances significantly larger than expected by dilution due to convection and, in some of them, the lithium abundance reaches values similar (and even larger) than the ISM value $logN(Li) = 3.3$.

Two main interpretations of the Li-rich giants are possible: i) the initial Li has been somehow preserved (perhaps by the inhibition of the classical mixing), ii) on the contrary, a mixing, deeper than the classical one, took place in some (or all) giants, leading to a dilution of the superficial lithium but sometimes overcompensating for it by an internal production of lithium, with transport to the surface (by the Cameron-Fowler mechanism).

Lithium and Beryllium are burned in stars in the deep (hot) layers of the stars, being preserved in the external layers. A shallow mixing of the external layers will deplete lithium, and a deeper one will deplete also Be. The Be abundance determination in Li-rich giants could give important information about the origin of the Li observed in these stars.

---

[1] Observations collected at the European Southern Observatory - ESO, Chile.

2. Data and Calculations

2.1. Observations

Observations were carried out using the 3.6m ESO telescope with the CASPEC Spectrograph equipped with the long camera centered at λ 3640 Å (order 156), covering the orders 131 to 190. The Be II λ 3130 Å lines are found at the order 181.

The slit was chosen to have 180x350 μm, corresponding to 2.4" on the sky, and to a resolving power R $\sim$ 32000 (see Castilho et al. 1999 for more details).

The final S/N ratio for the giants ranges from 40 to 20, due mainly to the low atmospheric transmission and low CCD sensitivity at these wavelengths. The spectrum of η Cen was inspected for telluric lines. We observed two Li-rich giants, one Li-poor giant and α Cen A and B as comparison stars.

2.2. Stellar Parameters and Models

Stellar parameters for the program stars were adopted from the literature, as indicated in Table 1. Model atmospheres employed have been interpolated in tables computed with the MARCS code by Gustafsson et al. (1975) and Edvardsson et al. (1993). The α Cen A and B synthetic spectra were also computed with models by Kurucz (1992) and we obtain a very good agreement with the above calculations.

Table 1. Adopted stellar parameters for program stars

Star	T_{eff}	log g	[Fe/H]	v_t	ref.
α Cen A	5800	4.40	0.10	1.0	Primas et al. (1997)
α Cen B	5350	4.50	0.10	1.0	Primas et al. (1997)
HD220321	4490	2.73	-0.40	1.3	McWilliam (1990)
HD146850	4000	1.50	-0.30	1.6	Castilho et al. (1995)
HD787	3890	1.74	0.03	1.5	McWilliam (1990)

2.3. Spectrum Synthesis

The spectrum synthesis calculations assume LTE, and the models described above.

Our list of atomic lines was built trying to adopt only the most reliable data, using the list of identified lines by Moore et al. (1966), and including the updated list and accurate data for Fe I by Nave et al. (1994). Oscillator strengths for atomic lines are adopted from laboratory determinations whenever available (see Castilho et al. 1999) otherwise they were obtained by fitting the solar spectrum.

For all molecular systems the line lists by R. Kurucz (CD ROM 18) were adopted, where we recomputed the 'molecular oscillator strengths', by recomputing their Honl-London factors, employing literature values for the Franck-Condon factors and the electronic oscillator strengths.

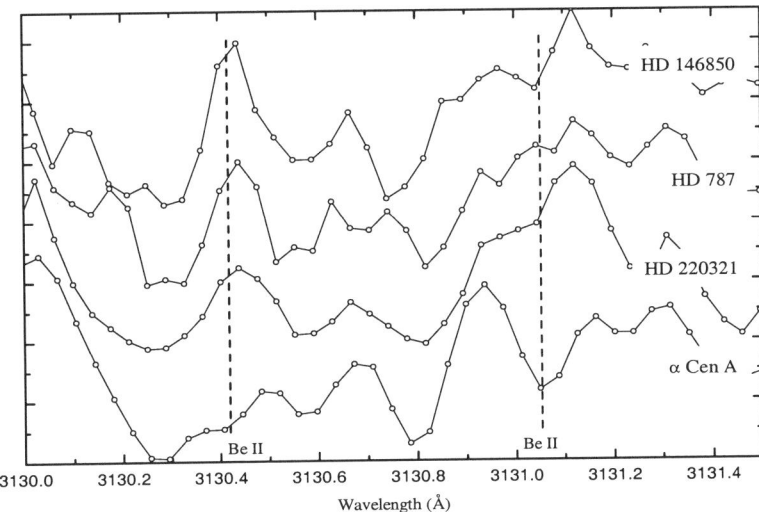

Figure 1. Comparison between α Cen A (bottom) where we can see the Be II lines and the other three giants where the lines are nearly absent.

3. Results

In Table 2 we report the derived Be abundances (Castilho et al. 1999) and literature values of Li abundances, including for reference the Be and Li abundances of the Sun according to Grevesse, Noels & Sauval (1996).

Table 2. Be abundances (present work) and Li abundances (literature)

Star	logN(Be)	logN(Li)	ref. for logN(Li)
α Cen A	1.20±0.1	1.37	King et al. (1997)
α Cen B	0.80±0.2	≤0.4	Chmielewski et al. (1992)
HD 220321	0.40±0.3	≤-0.2	Brown et al. (1989)
HD 146850	-0,50±0.4	1.6 (1.9nlte)	Castilho et al. (1995)
HD 787	0.00±0.4	2.2 (3.1nlte)	de la Reza & da Silva (1995)
Sun	1.15±0.1(6)	1.16±0.10	Grevesse et al. (1996)

The Be abundance estimated for the three red giant stars show that the Be abundance has been very depleted (by >90% i. e. by a factor of 10 or larger) from the initial Pop I value (logN(Be)= 1.4). By computing the Be abundance for HD 220321 and HD 787 using the atmospheric parameters used by Brown

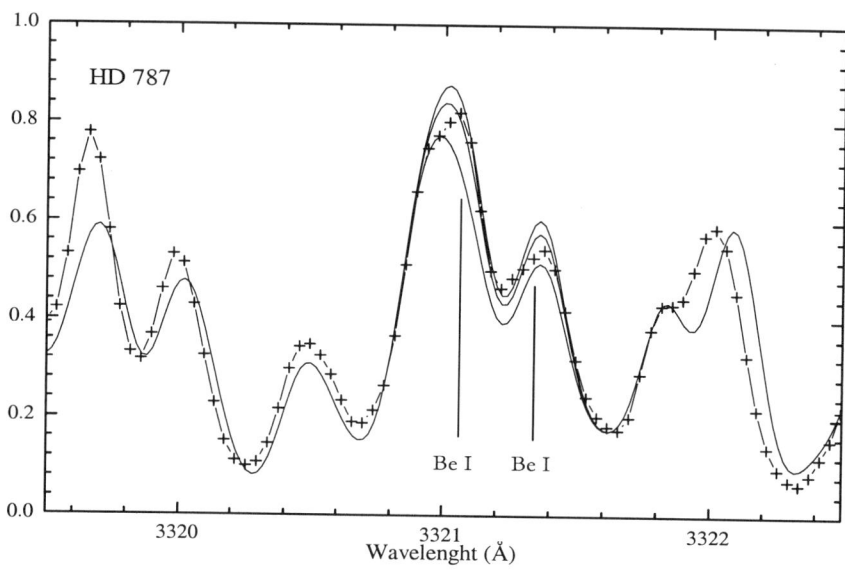

Figure 2. HD787 (+ +) spectrum and three synthetic spectra overimposed, with Be abundances logN(Be) = -1.0, 0.0(best fit), and 1.42.

et al. (1989) T_{eff} = 4510 K and 4220 K; log(g)=2.3 and 1.5 and [Fe/H]=-0.32 and 0.07, we find Be abundances of logN(Be)= 0.1 and -0.2 respectively.

In Fig. 1 we compare the observed spectrum of α Cen A, where we can see the Be II lines, to the other three giants where the Be II lines are essentially absent.

Our results for Be show good agreement with recent calculations by C. Charbonnel, carried out for giants of [Fe/H] = 0.0 and -0.5 and masses between 1.2 and 2.0 $M_\odot$, where mean values of Be/Be$_\odot$ ~ 0.1 and Li/Li$_\odot$ ~ 0.05; our depletions for Li are larger than predicted by the models, but an extra Li depletion is expected.

4. Be I lines

The determination of Be abundance through Be I lines would be a good confirmation of our results for Be II lines. There are three lines of Be I in the λ 3321 Å region: λ 3321.010 Å(χ_{ex} = 2.725 eV, log gf = -1.465), λ 3321.081 Å(χ_{ex} = 2.725 eV, log gf = -0.982) and λ 3321.340 Å(χ_{ex} = 2.725 eV, log gf = -0.773). All three lines are blended at R = 32000, and the strongest one at λ 3321.340 Å is blended to an Fe I line even at higher resolutions.

Although the Be I/Be II fraction in 4000 K giants is higher than in solar type stars, and we should expect strong lines, due to the severe depletion of Be in the program stars, we can not see the Be I lines even with R = 32000 and S/N $\approx$ 150.

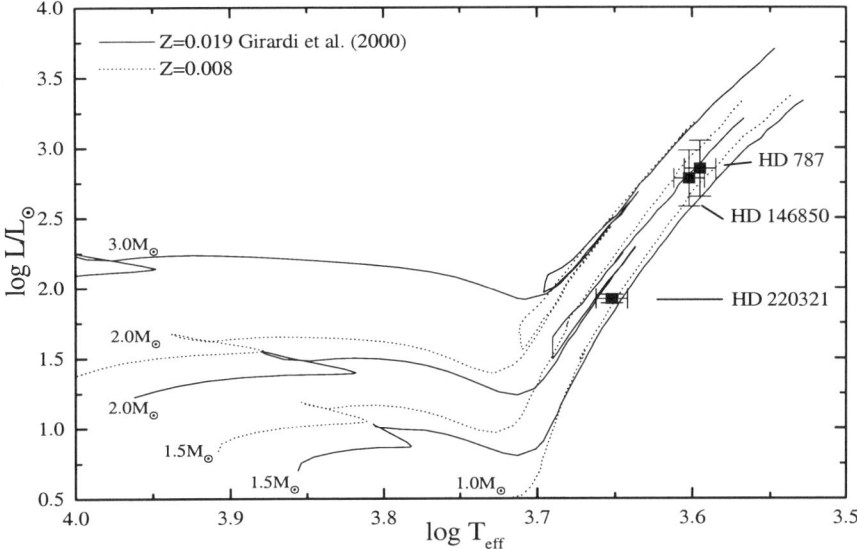

Figure 3. Position of the program stars in the HR diagram

The upper limits obtained for the three giants agree with the abundances determined from Be II lines. In Figure 2 we show the Be I region for HD 787 together with three synthetic spectra, with Be abundances logN(Be) = -1.0, 0.0(best fit), and 1.42.

5. Evolutionary Stage and Mass

In order to determine the position of our program stars in the H-R diagram and their masses we have determined the bolometric magnitude for them using the Hipparcos parallaxes to determine the distance, the interstellar reddening was estimated following the procedure used by Bond (1980), and the bolometric corrections were taken from Lejeune, Cuisinier & Buser (1998).

The luminosities for the giants where then calculated using the classical relation and the masses were determined overimposing the evolutionary tracks by Girardi et al. (2000) to the H-R diagram. In Table 3 we show the results for the three giants and in figure 3 we show the position of the stars in the H-R diagram. The error bars in figure 3 represent the error on the luminosities due parallaxes errors and a 100 K error in the temperatures.

Table 3. Luminosity and mass for the program stars

Star	π (m")	r (pc)	log L/L$_\odot$	M (M$_\odot$)
HD 787	5.33 ±0.87	187	2.85±0.2	1.9 ±1.0
HD 146850	3.77 ±0.85	265	2.78±0.2	1.3 ±1.0
HD 220321	20.14 ±0.72	50	1.92 ±0.03	1.1 ±0.5

6. Conclusion

The small Be abundance found in the Li-rich giants suggests a Be depletion, since the Be depletion found in the giants is larger than the one found in the region of the dip. Moreover, it is not likely that these two stars had by chance a low Be initial abundance, in spite of the fact that the Be abundance in Pop I stars shows some spread. Therefore, Be is not preserved, and its depletion should be due to a rather deep mixing, implying that the original Li in these stars must have been strongly depleted, as in the case of HD 220321. As a consequence, the high Li abundance found in the two Be-poor Li-rich giants studied here (and in all Li-rich giants) is probably due to a further Li production (cf. Sackmann & Boothroyd 1999).

References

Bond, H. E. 1980, ApJS, 44, 517

Brown, J. A., Sneden, C., Lambert, D. L., & Dutchover, E. 1989, ApJS, 71, 293

Castilho, B. V., Gregorio-Hetem, J., & Barbuy, B. 1995, A&A, 297, 503

Castilho, B. V., Spite, F., Barbuy B., Spite, M., & De Medeiros, J.R., Gregorio-Hetem, J. 1999, A&A, 345, 249

Chmielewski, Y., Friel, E., Cayrel de Strobel, G., & Bentolila, C. 1992, A&A, 263, 219

de la Reza, R., & da Silva, L. 1995, ApJ, 439, 917

Edvardsson, B., Andersen, J., Gustafsson, B. et al. 1993, A&A, 275, 101

Girardi, L., Bressan, A., Bertelli, G., & Chiosi, C. 2000, A&AS *in press*

Grevesse, N., Noels, A., & Sauval, J. 1996, in ASP Conf. Ser. 99, eds. S.S. Holt, G. Sonneborn, p. 117

Gustafsson, B., Bell, R. A., Eriksson, K., & Nordlund, A. 1975, A&A, 42, 407

King, J., Deliyannis, C. P., Hiltgen, D. D., Stephens, A., Cunha, K., & Boesgaard, A. M. 1997b, AJ, 113, 1871

Kurucz, R. L. 1992, in IAU Symp. 149, eds. B. Barbuy, A. Renzini, Kluwer Acad. Press, p. 225

Lejeune, T., Cuisinier, F., & Buser, R. 1998, A&AS, 130, 65

Moore, C. E., Minnaert, M. G., & Houtgast, J. 1966, The Solar Spectrum from 2935 to 8870 Å, Second Revision of Rowland's Preliminary Table of Solar Spectrum Wavelengths, N.B.S. Monog. 61 (Washington, D.C.: Government Printing Office)

Nave, G., Johansson, S., Learner, R. C. M., Thorne, A. P., & Brault, T. J. 1994, ApJS, 94, 221

Primas, F., Duncan, D. K., Pinsonneault, M. H., Deliyannis, C. P., & Thorburn, J. A. 1997, ApJ, 480, 784

Sackmann, I.-J., & Boothroyd, A. I. 1999, ApJ, 510, 217

The Interstellar Lithium Isotope Ratio toward Per OB2

David C. Knauth and Steven R. Federman

The University of Toledo, Department of Physics and Astronomy, Toledo, Ohio, USA

David L. Lambert

The University of Texas, Department of Astronomy, Austin, Texas, USA

Philippe Crane

Dartmouth University, Department of Physics and Astronomy, Hanover, New Hampshire, USA and NASA Headquarters, Washington, DC, USA

Abstract. We are conducting a survey on the ^{7}Li/^{6}Li ratio in interstellar space in order to seek limits on the variation in this ratio. The analysis is based on the technique we adopted in extracting the ^{11}B/^{10}B ratio. This technique uses a line of comparable strength, from a species likely to occupy the same volume of the interstellar cloud, as a template for separating velocity components within the line profile. For our study of the ^{7}Li/^{6}Li ratio, the K I line at 4044 Å serves as the velocity template. Our initial focus is on the variation in the ^{7}Li/^{6}Li ratio around the star-forming region IC 348. Our high-resolution observations of the Li I lines toward o and ζ Per show remarkably different isotope ratios: ^{7}Li/^{6}Li = 2–4 and 11, respectively, where the Solar System ratio is 12.3 and cosmic ray spallation yields a ratio of about 2. The significance of the very low ratio toward o Per is that it is essentially the value predicted for cosmic rays through spallation reactions. The direction to o Per passes closer to IC 348, a site of massive star formation, than does the line of sight to ζ Per. Furthermore, our analysis of OH column densities (Federman, Weber, & Lambert 1996) showed that the cosmic ray flux through o Per's diffuse clouds is higher than average, presumably reflecting the nearby presence of IC 348.

1. Introduction

The abundance of lithium is an important ingredient for studies of the chemical evolution of the Galaxy involving Big Bang Nucleosynthesis (BBN), Galactic cosmic ray (GCR) spallation reactions, and synthesis in Type II supernovae (SN II). Models of BBN show that no light element heavier than ^{7}Li is produced in significant amounts. BBN was the main source of ^{7}Li production in the early Galaxy. GCR spallation reactions involve high energy cosmic rays interacting with interstellar nuclei. These reactions account for the relative abundances of present day ^{6}Li, ^{9}Be, ^{10}B but only about 10% of the ^{7}Li and about half the ^{11}B (Fields, Olive, & Schramm 1994; Meneguzzi, Audouze, & Reeves 1971; Meneguzzi & Reeves 1975; Pagel 1997; Ramaty et al. 1996, 1997; Reeves et al. 1973; Reeves 1974). These calculations yield an isotopic ratio ^{7}Li/^{6}Li of 2.0 and ^{11}B/^{10}B of 2.5, compared to the Solar System values of 12.3 and 4.05, respectively (Anders & Ebihara 1982).

This suggests the need for an additional source of ^{7}Li and ^{11}B production in the present day Galaxy. The search for an additional source of ^{7}Li focused attention on the abundance of lithium in stars. Luminous intermediate-mass AGB stars are predicted and observed to synthesize ^{7}Li (Smith & Lambert 1989). High lithium abundances are occasionally seen in red giants (Brown et al. 1989), suggesting lithium production during some phase of stellar evolution. Neutrino induced spallation reactions in core collapse SN II have been shown to produce both ^{7}Li and ^{11}B (Woosley et al. 1990). A higher flux of low energy cosmic rays have been invoked (e.g., Ramaty et al. 1996) to account for a higher ^{11}B/^{10}B ratio, since ^{11}B is preferentially produced over ^{10}B at lower energies.

The interstellar medium (ISM) is another environment useful in studying light element synthesis. Neutral lithium has a low ionization potential. The dominant ion in the neutral ISM is Li II, which cannot be observed. This fact and the uncertainty in the amount of depletion onto interstellar grains make the total abundance difficult to determine with any precision. Interstellar lithium was detected toward ζ Per, ϵ Aur, and ζ Oph by Hobbs (1984) and vanden Bout et al. (1978) and toward σ Sco, β^1 Sco and 55 Cyg by Snell & vanden Bout (1981). These early detections of interstellar Li I showed absorption with equivalent widths on the order of a few mÅ. Published determinations of the ^{7}Li/^{6}Li ratio in the local ISM show a ratio that in some clouds may differ from that of the Solar System (12.3 ± 0.3, Anders & Ebihara 1982). Lemoine et al. (1993, 1995), and Meyer, Hawkins & Wright (1993) reported interstellar ^{7}Li/^{6}Li ratios of $12.5^{+4.3}_{-3.4}$ toward ρ Oph, $6.8^{+1.4}_{-1.7}$ toward ζ Oph, and $5.5^{+1.3}_{-1.1}$ toward ζ Per.

Generally there exists multiple interstellar clouds on any given line of sight. The velocity structure has to be well known before an accurate lithium isotope ratio can be obtained (Lemoine et al. 1993, 1995). Therefore, it is useful to find lines of sight with relatively simple velocity structure (i.e., one or two interstellar clouds on a line of sight). In addition, it is highly desirable to obtain data on another species, which has similar properties to lithium and resides in the same portion of the interstellar cloud, to use as a template of the velocity structure. Our method differs slightly from the technique of Lemoine et al. (1993, 1995). Lemoine et al. used the resonance line of K I (λ 7699) as their velocity template. This line of K I is much too strong to use as an effective velocity template. For this reason we obtained data on K I λ 4044 which is of comparable strength to the lithium lines and therefore a better template. Moreover, high signal to noise (SNR), high resolution spectra of the Li doublet lines at 6707 Å are required, since the fine structure splitting for ^{7}Li I is comparable to the isotope shift of ^{6}Li I ($\sim$ 0.160 Å), resulting in a blend of the ^{7}Li and ^{6}Li lines (see Figure 1). We have obtained high SNR, high resolution spectra of interstellar lithium toward ζ Per and o Per in the Perseus OB2 association as part of a study to improve our understanding of light element synthesis.

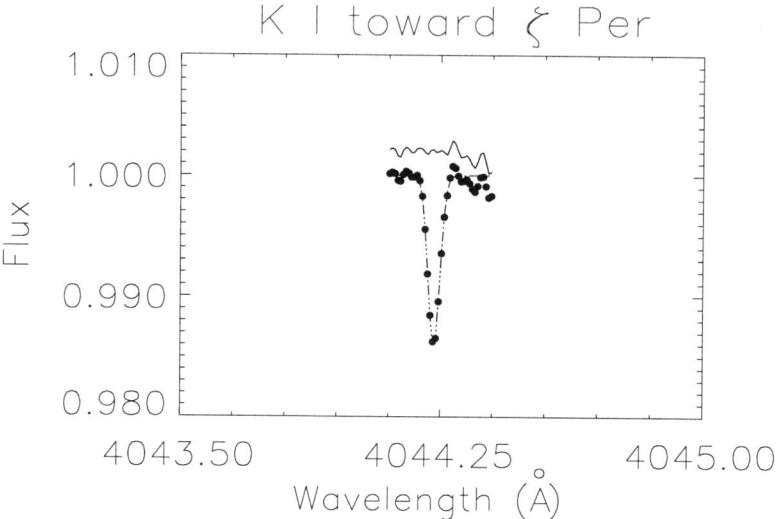

Figure 1. These graphs show interstellar K I and Li I toward ζ and o Per. The data points are the filled circles. The dot-dashed line is the fit to the line profiles using the K I line for the velocity structure. The solid line is the residuals of the fit. The vertical dashes in the o Per K I plot show the positions of the two velocity components. The vertical dashes in the ζ Per Li I plot show the relative positions of the ^{7}Li and ^{6}Li fine structure lines.

2. Observations

The observations were acquired with the 2.7 meter telescope and the "2dcoudé" spectrograph at McDonald Observatory (Tull et al. 1995). We observed ζ Per (B1 Ib; $V = 2.85$; $v\,sini = 59$ km s^{-1}) in 1996 November and 1997 January and o Per (B1 III; $V = 3.83$; $v\,sini = 85$ km s^{-1}) in 1998 January. High resolution spectra (R ~ 170,000) in two spectral regions were obtained, at 4044 Å for K I and 6707 Å for Li I. The spectra were reduced in a standard manner with NOAO SUN/IRAF (Revision 2.10.4). The stellar images were corrected for the bias and were flat fielded to remove any instrumental effects. The scattered light, which was negligible, was removed before aperture extraction. The extracted spectra were then placed on a wavelength scale with spectra from a Th-Ar hollow cathode which were taken periodically through the night. The final spectra were Doppler corrected and coadded yielding a final SNR of approximately 2500:1 for each stellar spectrum. Figure 1 displays the observations and profile fits for K I and Li I toward ζ and o Per. The data exhibit one interstellar velocity component toward ζ Per and two components toward o Per. The Li spectrum for gas toward o Per clearly reveals a non-solar isotope ratio: substantial absorption from ^{6}Li is evident after taking note of the 2:1 fine structure ratio for the ^{7}Li doublet and the two velocity components seen in λ 4044

3. Analysis

Comparison of observed and synthetic profiles was accomplished with a code (J. Zsargó, unpublished) in which the adjustable parameters of the synthetic profiles were changed until a minimum χ^2 was obtained. In determining the synthetic profile of the interstellar lines, we included the hyperfine structure splitting for both ^{6}Li and ^{7}Li (Sansonetti et al. 1995). The hyperfine structure splitting of K I λ 4044 was found to be negligible ($\sim$ 0.002 mÅ) and was not included in the analysis. The profiles of the unblended ^{7}Li I line and K I line were fitted first to obtain the b-value, column density (N) and lsr velocity of the line. The b-value and velocity were then fixed and used in the fit of the Li I profile, allowing for a relatively straightforward determination of the isotope ratio. This method was successfully used in our determination of interstellar ^{11}B/^{10}B (Federman et al. 1995; Lambert et al. 1998).

Table 1. ^{7}Li and ^{6}Li Results

Star	Isotope	b-value km s^{-1}	v_{lsr} km s^{-1}	N cm^{-2}	Ratio
ζ Per	^{7}Li	1.7	6.5	$(3.5 \pm 0.1) \times 10^9$	
	^{6}Li	1.7	6.5	$(3.3 \pm 0.9) \times 10^8$	10.3 ± 3.1
o Per	^{7}Li	2.1	6.5	$(2.6 \pm 0.1) \times 10^9$	
	^{6}Li	2.1	6.5	$(7.2 \pm 1.1) \times 10^8$	3.6 ± 0.3
o Per	^{7}Li	0.7	3.5	$(5.5 \pm 0.5) \times 10^8$	
	^{6}Li	0.7	3.5	$(3.2 \pm 0.5) \times 10^8$	1.8 ± 0.3

The fitting routine was performed twice, once using the relatively unblended ^{7}Li I line for the profile parameters and the second time using the K I line. The results of the two fits yielded agreement in column densities at the 1 σ level, leading to a heightened confidence in our method. The final results toward each star are the weighted average of the isotope ratio determined by the two methods. The isotope ratio and profile parameters are displayed in Table 1.

Another useful measure of lithium is the total abundance (N(Li)/N(H)). Derivation of the total interstellar abundance is no trivial task because knowledge of the Li II abundance and depletion mechanisms is necessary. Since the ionization potential of Li II is 75.6 eV, we need only consider Li I and Li II. An estimate for Li II abundance requires the electron density. Information on N(C II) and N(H) exists, and it is straightforward to calculate the electron density ($n_e = n$ (N(C II)/N(H))). N(H) is the total proton column density, represented by (N(H) = N(H I) + 2N(H$_2$)).

Since no precise N(C II) measurements exist for the direction toward o Per, the weighted mean interstellar ratio of N(C II)/N(H) = $(1.42 \pm 0.13) \times 10^{-4}$ (Sofia et al. 1997) was utilized, yielding a value of N(C II) = $(2.3 \pm 0.8) \times 10^{17}$ cm^{-2}. The value of N(C II) is $(1.84 \pm 0.32) \times 10^{17}$ cm^{-2} toward ζ Per (Cardelli et al. 1996). The values of N(H), $(16.1 \pm 5.6) \times 10^{20}$ cm^{-2} toward o Per and $(15.8 \pm 4.7) \times 10^{20}$ cm^{-2} toward ζ Per, were obtained from Bohlin et al. (1978) and an n of 800 and 700 cm^{-3} toward o and ζ Per, respectively, were used

(Federman et al. 1994). We then computed $N(\text{Li}) = N(\text{Li I})[G/(\alpha n_e)]$. A value of 41 was used for the photoionization rate to recombination rate coefficient, G/α (White 1986). The resulting total Li abundances are $(7.1 \pm 2.7) \times 10^{-10}$ toward o Per and $(11.1 \pm 2.0) \times 10^{-10}$ toward ζ Per. Since a Li isotope ratio of $\sim$ 2 toward o Per suggests newly processed Li in this direction, the comparable Li abundances are unexpected. The two abundances, however, are very similar to previous results for the ISM. Vanden Bout et al. (1978) derived an abundance $(8.8 \pm 3.4) \times 10^{-10}$ toward ζ Per and $(1.8 \pm 0.3) \times 10^{-9}$ toward ϵ Aur. Traub & Carleton (1973) derived an abundance of $(2.9 \pm 0.7) \times 10^{-10}$ toward ζ Oph. There remains the uncertain correction for depletion onto grains.

4. Discussion and Conclusions

The interstellar Li isotope ratio does vary from line of sight to line of sight and ranges from the GCR spallation reaction value of 2 to the Solar System value. The results for the two components toward o Per, 1.8 ± 0.3 and 3.6 ± 0.3, show that there exists a higher flux of high energy cosmic rays (CR) toward o Per than toward ζ Per, where the ratio is 10.3 ± 3.1. The isotope ratio toward ζ Per was previously determined to be $5.5^{+1.3}_{-1.1}$ (Meyer et al. 1993), and to within the errors, the two results agree. The total lithium abundances derived in this work compare well with other determinations for the ISM. Since a weighted interstellar average of C II was utilized in determining n_e toward o Per, there is additional uncertainty in the derived lithium abundance. This may contribute to the contradictory results involving newly synthesized Li and the Li elemental abundance.

The line of sight toward o Per lies closer than that of ζ Per to a site of star formation, IC 348, where CR's might arise. From OH and HD chemistry, Federman et al. (1996) found approximately an order of magnitude higher CR flux toward o Per. Their analysis was performed with O and D abundances precisely determined with HST. In essence, $\zeta_H \propto N(\text{OH})/N(\text{O})$, $N(\text{HD})/[\xi_D N(\text{H}_2)]$, was used to determine the CR ionization rate. This provides independent confirmation for a larger CR flux toward o Per.

Our initial results for o and ζ Per suggest other directions now being pursued. Further measurements of $^7\text{Li}/^6\text{Li}$, at ultra high resolution, are needed to fully resolve the velocity structure (≈ 1 km s^{-1}) on lines of sight to other stars. Measurements of $^{11}\text{B}/^{10}\text{B}$ toward the stars are useful in helping to disentangle the various synthesis routes and in this way further constrain models of light element nucleosynthesis.

References

Anders, E., & Ebihara, M. 1982, Geochim. Cosmoschim. Acta, 46, 2363

Bohlin, R. C., Savage, B. D., & Drake, J. F. 1978, ApJ, 224, 132

Brown, J. A., Sneden, C., Lambert, D. L., & Dutchover, E. Jr. 1989, ApJS, 71, 293

Cardelli, J. A., Meyer, D. M., Jura, M., & Savage, B. D. 1996, ApJ, 467, 334

Fields, B. D., Olive, K. A., & Schramm, D. N. 1994, ApJ, 435, 185

Federman, S. R., Strom C. J., Lambert, D. L., Cardelli, J. A., Smith, V. V., & Joseph, C. L. 1994, ApJ, 424, 772
Federman, S. R., Lambert, D. L., Cardelli, J. A., & Sheffer, Y. 1995, Nature, 381, 764
Federman, S. R., Weber, J., & Lambert, D. L. 1996, ApJ, 463, 181
Hobbs, L. M. 1984, ApJ, 286, 252
Lambert, D. L., Sheffer, Y., Federman, S. R., Cardelli, J. A., Sofia, U. J., & Knauth, D. C. 1998, ApJ, 494, 614
Lemoine, M., Ferlet, R., Vidal-Madjar, A., Emerich, C., & Bertin, P. 1993, A&A, 269, 469
Lemoine, M., Ferlet, R., & Vidal-Madjar, A. 1995, A&A, 298, 879
Meneguzzi, M., Audouze, J., & Reeves, H. 1971, A&A, 15, 337
Meneguzzi, M., & Reeves, H. 1975, A&A, 40, 99
Meyer, D. M., Hawkins, I., & Wright, E. L. 1993, ApJ, 409, L61
Pagel, B. E. 1997, Nucleosynthesis and Chemical Evolution of Galaxies, (United Kingdom: Cambridge University Press), 260
Ramaty, R., Kozlovsky, B., & Ligenfelter, R. E. 1996, ApJ, 456, 525
Ramaty, R., Kozlovsky, B., Ligenfelter, R. E., & Reeves, H. 1997, ApJ, 488, 730
Reeves, H., Audouze, J., Fowler, W. A., & Schramm, D. N. 1973, ApJ, 179, 909
Reeves, H. 1974, ARA&A, 12, 437
Sansonetti, C. J., Richou, B., Engleman, Jr., R., & Radziemski, L., J. 1995, Phys. Rev. A, 52, 2682
Smith, V. V., & Lambert, D. L. 1989, ApJ, 345, L75
Snell, R. L., & vanden Bout, P. A. 1981, ApJ, 250, 160
Sofia, U. J., Cardelli, J. A., Guerin, K. P., & Meyer, D. M. 1997, ApJ, 482, L105
Traub, W. A., & Carleton, N. P. 1973, ApJ, 184, L11
Tull, R. G., MacQueen, P. J., Sneden, C., & Lambert, D. L. 1995, PASP, 107, 251
Vanden Bout, P. A., Snell, R. L., Vogt, S. S., & Tull, R. G. ApJ, 1978, 221, 598
White, R. E. 1986, ApJ, 307, 777
Woosley, S. E., Hartmann, D. H., Hoffman, R. D., & Haxton, W. C. 1990, ApJ, 356, 272

The Light Elements and Their Evolution
IAU Symposium, Vol. 198, 2000
L. da Silva, M. Spite, J. R. de Medeiros, eds.

New Determination Method of Primordial Li Abundance

T. Kajino, T.-K. Suzuki, S. Kawanomoto, and H. Ando

National Astronomical Observatory
The Graduate University for Advanced Studies
Mitaka, Tokyo 181-8588, Japan
Department of Astronomy, University of Tokyo
Bunkyo-ku, Tokyo 113-0033, Japan

Abstract. We discuss the primordial nucleosynthesis in lepton asymmetric Universe models. In order to better estimate the universal baryon-mass density parameter Ω_b, we try to remove the uncertainty from the theoretical prediction of primordial ^{7}Li abundance. We propose a new method to determine the primordial ^{7}Li by the use of isotopic abundance ratio ^{7}Li/^{6}Li in the interstellar medium which exhibits the minimum effects of the stellar processes.

1. Introduction

Recent spectral and photometric observations of Type Ia supernovae at high redshifts (Riess et al. 1998; Perlmutter et al. 1999) have raised a possibility that the cosmic expansion is accelerated. For a flat cosmology these data have χ^2-minimum around $\Omega_0 \approx 0.3$ and $\Omega_\Lambda \approx 0.7$, allowing Hubble time ~15Gyr which is not inconsistent with the age of the Milky Way constrained from the observations of the oldest globular clusters.

Cosmological model for primordial nucleosynthesis provides independent method to determine Ω_0. The Big-Bang nucleosynthesis model (Copi et al. 1995) predicts $0.04 \leq \Omega_b h_{50}^2 \leq 0.08$. Combining this value with X-ray observations of rich clusters that indicate $0.3 h_{50}^{-3/2} \approx \Omega_b/\Omega_0$ (Bahcall et al. 1995; White et al. 1993), total Ω_0 turns out to be $\Omega_0 h_{50}^{1/2} \approx 0.1 \sim 0.3$, which is consistent with flat cosmology.

However, in the determination of Ω_b, a difficulty has been imposed by recent detections of a low deuterium abundance, $2.9 \times 10^{-5} \leq$ D/H $\leq 4.0 \times 10^{-5}$, in Lyman-α clouds along the line of sight to high red-shift quasars (Burles & Tytler 1998ab). Primordial abundance of ^{7}Li is constrained from the observed "Spite plateau", $0.91 \times 10^{-10} \leq$ ^{7}Li/H $\leq 1.91 \times 10^{-10}$ (Ryan et al. 2000a), and the ^{4}He abundance by mass, $0.226 \leq Y_p \leq 0.247$ (Olive et al. 1999), from the observations in the HII regions. In order to satisfy these abundance constraints by single Ω_b value, one has to assume an appreciable depletion in the observed abundance of ^{7}Li, which is still controversial both theoretically and observationally. We are now forced to critically study the uncertainty. An independent method to determine the primordial ^{7}Li is also desirable.

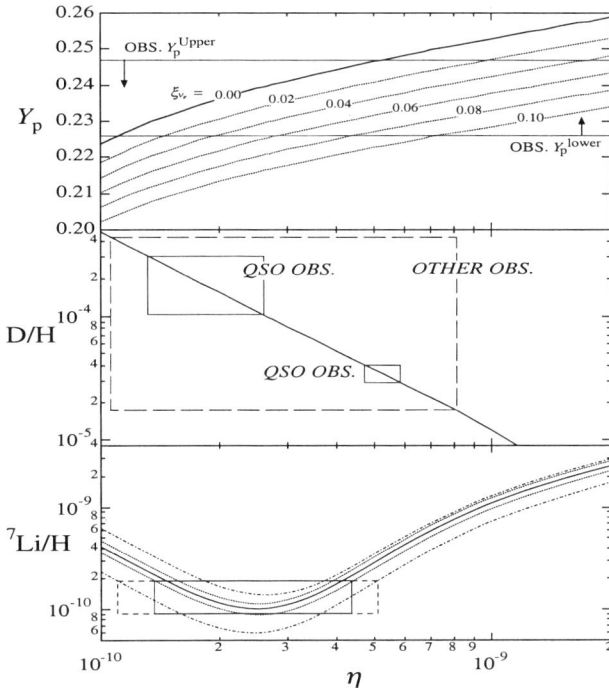

Figure 1. Y_p, D/H and ^{7}Li/H vs. η for various neutrino degeneracy parameters ξ_{ν_e}. Observed constraints on D/H are from Rugers & Hogan (1996) (upper) and Burles & Tytler (1998ab) (lower). Theoretical curves for ^{7}Li/H show the 2σ uncertainties based on two different error estimates of the reaction rates. See text for details.

2. Primordial Nucleosynthesis

2.1. ^{4}He vs. D and Neutrino Degeneracy

Shown in Fig. 1 is the comparison between the observed abundance constraints on ^{4}He, D/H, and ^{7}Li/H and the calculated curves in the homogeneous Big-Bang model as a function of η, where $\eta = n_B/n_\gamma$ and $\Omega_b h_{50}^2 = \eta\ 1.464 \times 10^8$. Solid curves display the theoretical prediction of primordial abundances in the standard particle model for neutrino, which preserves the lepton symmetry $L_\nu = 0$.

There is now a good collection of abundance information on the ^{4}He mass fraction, Y_p, in over 50 extragalactic HII regions, from which the upper limit on primordial abundance, $Y_p \leq 0.240$, and a systematic error, $\Delta Y_{sys} = 0.005$, were extracted. Unfortunately, for this upper limit one cannot find Ω_b to satisfy both abundance constraints on ^{4}He and D/H. (See the solid curves in Fig. 1.)

It has been recognized that ΔY_{sys} may even be larger (Izatov et al. 1994; Thuan 2000), making the upper limit as large as $Y_p \leq 0.247$. If this upper limit is adopted, the Universe model with $\eta \approx 5 \times 10^{-10}$ is marginally consistent with both abundance constraints. However, since even smaller value, $Y_p = 0.235 \pm 0.003$, in low-metallicity extragalactic HII regions has been reported by Peimbert & Peimbert (2000), this potential conflict is to be studied more carefully.

One possible solution is to introduce a lepton asymmetry. Theoretically, it is natural to assume that both baryon and lepton symmetries are simultaneously broken, $B \neq 0$ and $L_\nu \neq 0$, due to the CP violation in baryogenesis. $L_\nu \neq 0$ is fulfilled by neutrino degeneracy with non-zero ξ_{ν_e}, where $\xi_{\nu_e} = \mu_{\nu_e}/kT_\nu$ and μ_{ν_e} is the chemical potential of electron neutrino. Since neutrinos had energy density comparable to the densities due to photons and charged leptons in the early Universe, even a small degeneracy $0 < \xi_{\nu_e} \ll 1$ leads to an appreciable decrease in the neutron-to-proton number ratio, slightly faster acceleration of the Universal expansion, and a small increase of the weak-decoupling temperature. As a net result, ^{4}He abundance decreases with increasing ξ_{ν_e}, as shown in Fig. 1, while keeping D/H and ^{7}Li/H almost unchanged in logarithmic scale (Kajino & Orito 1998). Since the abundance constraint on primordial ^{4}He is more accurate than the other light elements, this helps determine the most likely ξ_{ν_e}. $\xi_{\nu_e} \sim 0.05$ can best fit the ^{4}He abundance as well as low deuterium abundance D/H$\sim 10^{-5}$, leaving inevitable requirement that the observed abundance level of Spite plateau, ^{7}Li/H$\sim 10^{-10}$, should be the result of depleted primordial abundance.

2.2. ^{7}Li vs. D

There are several input parameters in the primordial nucleosynthesis calculation. As the number of light neutrino families $N_\nu = 3$ and the neutron lifetime $\tau_n = 886.7 \pm 1.9$ s are known, the remaining major uncertainty arises from input nuclear reaction data. We did not take account of the effects of sterile neutrino which is a hypothetical particle for interpreting flavor mixing.

Laboratory cross section measurements ever done provide rather precise thermonuclear reaction rates for the production of D, T, ^{3}He, and ^{4}He. It however was claimed in literature (Smith et al. 1993) that the ^{7}Li abundance is strongly subject to large error bars associated with the measured cross sections for ^{4}He(^{3}H,γ)^{7}Li at $\eta \lesssim 2 \times 10^{-10}$ and ^{4}He(^{3}He,γ)^{7}Be at $3 \times 10^{-10} \lesssim \eta$. There are in fact several inconsistent data with one another, leading to large uncertainty in the primordial ^{7}Li, as displayed by long-dash-dotted curves in Fig. 1.

We studied these two reactions very carefully and concluded that the proper 2σ error bars could be $1/4 \sim 1/3$ of the previous ones (Kajino et al. 2000). This improvement owes mostly to, first, the new precise measurement (Brune et al. 1994) of the cross sections for ^{4}He(^{3}H,γ)^{7}Li and, second, the systematic theoretical studies of both reaction dynamics and quantum nuclear structures of ^{7}Li and ^{7}Be, whose validity is critically tested by electromagnetic form factors measured by high-energy electron scattering experiments.

When our recommended error estimate is applied to the determination of Ω_b in Fig. 1, we lose Ω_b value to explain both D/H and ^{7}Li/H simultaneously. If we allow for larger primordial ^{7}Li abundance in Population II halo stars because of possible lithium depletion for diffusion or rotation-induced mixing of matter (Deliyannis et al. 1998 ; Pinsonneault et al. 1992) or some systematic uncertainty in the model atmospheres (Kurucz 1995), we can recover the concordance. Taking depletion factor ≈ 2.5, $\Omega_b h_{50}^2 \approx 0.075$ best fits all abundance constraints in the homogeneous Big-Bang model. Note that larger $\Omega_b h_{50}^2 \approx 0.2$ is allowed in the inhomogeoeus Big-Bang model (Kajino & Orito 1998).

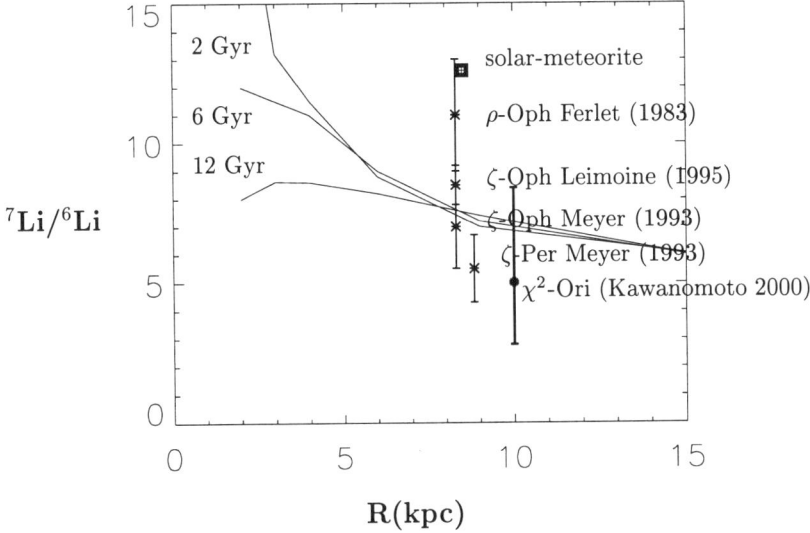

Figure 2. Calculated isotopic abundance ratios ^{7}Li/^{6}Li at various times as a function of the Galactocentric distance R, compared with observed data.

3. ^{7}Li/^{6}Li Ratio in the Interstellar Medium (ISM)

The lithium in the ISM is almost free from the complicated stellar processes. A diffuse cloud along the line of sight to ζOph was observed to show the lithium abundance depleted by 1.58dex from the meteoritic solar-system value 12.3. This is due to dust grain formation (Savage & Sembach 1996). The isotopic ratio is free from such condensation effects and represents the real ratio of chemical compositions in the gas phase. The D/H (Wannier 1980) and ^{3}He/H (Rood et al. 1995) abundance ratios in the ISMs have been observed over wide Galactocentric distance range 0≤R≤12kpc and used to constrain the primordial abundance of D/H (Dearborn et al. 1996), but the distribution of ^{7}Li/^{6}Li was poorly known.

3.1. Observation

Observations of isotopic abundance ratio, ^{7}Li/^{6}Li, have been performed by several groups (Ferlet & Dennefeld 1983, Lemoine et al. 1993, 1995, Meyer et al. 1993) only for the ISMs in our solar neighborhood. The observed ratio is less than 12.3 and larger than 2.1 which is a predicted GCR abundance ratio.

Using the Coude spectrograph of the 74-inch telescope at Okayama Astrophysical Observatory, Japan, we have succeeded for the first time in the determination of ^{7}Li/^{6}Li in the diffuse cloud along the line of sight to χ^2Ori, which is a member of OB association Gem-OB1, being located at R = 10kpc (Kawanomoto et al. 2000). The telescope performance was R=43,000 (with slit width of 100 μm), exp=50hours, and S/N=2,800.

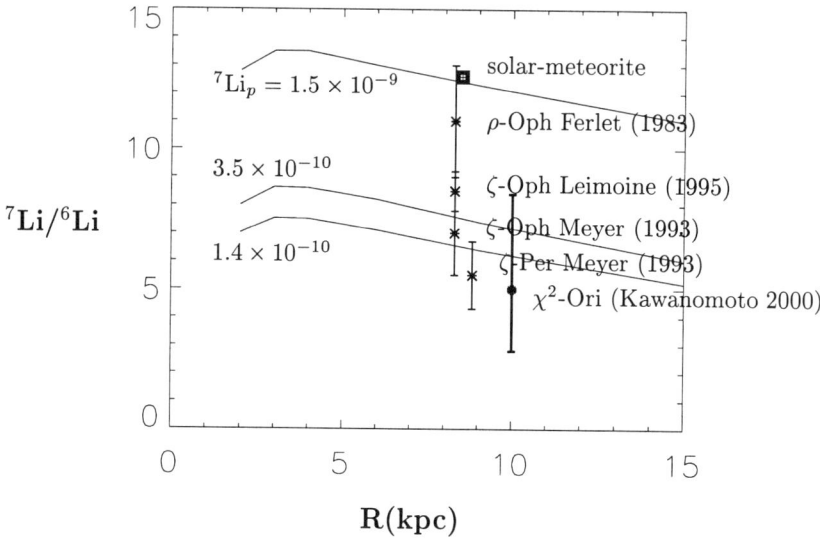

Figure 3. Calculated isotopic abundance ratios ^{7}Li/^{6}Li at the present time, $t_G = 12$Gyr, for various primordial abundances, $^7\text{Li}_p \equiv {}^7$Li/H(primordial), as a function of the Galactocentric distance R, compared with observed data.

We found a decreasing gradient of ^{7}Li/^{6}Li, as shown in Figs. 2 & 3. It is interpreted as a result of gradual extinction of the stellar production of ^{7}Li.

3.2. A New Method to Determine Primordial ^{7}Li

In order to study the sensitivity to the primordial lithium abundance, we have calculated Galactic chemical evolution (GCE) of lithium (Kawanomoto 2000). We adopted a hybrid model (Ryan et al. 2000b) of the inhomogeneous GCE model (Suzuki et al. 1999), which was constructed for the early evolution of metal-deficient stars, being smoothly connected with a simple one-zone GCE model for later evolution. Five different sources of lithium production are included in this model: Primordial nucleosynthesis, GCR interactions with ISM, ν-induced nucleosynthesis in Type II SNe, AGB star nucleosynthesis, and nova nucleosynthesis. We took an approximation that each ring having different Galactocentric distance evolves independently so that the observed present day star-formation-rate and the gas fraction are reproduced very well.

The calculated time variation of ^{7}Li/^{6}Li is shown as a function of R in Fig. 2. Remarkable decrease of ^{7}Li/^{6}Li in the inner region is caused by faster gas consumption for the star formation. It is discussed in literature that the meteoretic chemical compositions are peculier and different from those of ISM because they were possibly polluted by nearby AGB star. One might speculate another possibility that the solar-system might ever have moved outward over hundreds of turns of the Galactic disc, keeping high ^{7}Li/^{6}Li = 12.3 as it was in the original position when the solar system was isolated from viscous gas component at $t_G \approx 2 \sim 6$Gyr.

Figure 3 displays sensitivity of the ^{7}Li/^{6}Li ratio at the present time t_G=12Gyr to the primordial abundance of ^{7}Li. It is very sensitive to ^{7}Li$_p$. Except for old data point at ρ-Oph (Ferlet & Dennefeld 1983), which has the largest error bar among all data for the solar neighborhood, the observed ratios look more consistent with ^{7}Li$_p = (1.4 \sim 3.5) \times 10^{-10}$ than ^{7}Li$_p = 1.5 \times 10^{-9}$. More data with smaller error bars are highly desirable in order to convince the gradient of the ^{7}Li/^{6}Li ratio and to determine the primordial abundance of ^{7}Li in this method.

References

Bahcall, N.A., Lubin, L.M., & Dorman, V. 1995, ApJ 447, L81.
Brune, C.R., Kavanagh, R.W., & Rolfs, C. 1994, PR C50, 2205.
Burles, S., & Tytler, D. 1998a, ApJ 499, 699; 1998b, ApJ 507, 732.
Copi, C.J., Schramm, D.N., & Turner, M.S. 1995, ApJ 455, 95.
Dearborn, D.S.P. ,Steigman, G., & Tosi, M. 1996, ApJ 465, 887.
Deliyannis, P., et al. 1998, ApJ 498, L147.
Ferlet, R., & Dennefeld, M. 1983, ApJ 409, L61.
Izatov, Y.I., Thuan, T.X., & Lipovetsky, V.A. 1994, ApJ 435, 647.
Kajino, T., & Orito, M. 1998, Nucl. Phys. A629, 538.
Kajino, T., Orito, M., Sakai, K., & Deliyannis, P.C. 2000, in preparation.
Kawanomoto, S., Ando, H., Kajino, T., & Suzuki, T.-K. 2000, in preparation.
Kurucz, R.L. 1995, ApJ 452, 102.
Lemoine, M., et al. 1993, A&A 269, 469; 1995, A&A 298, 879.
Meyer, D.M., Hawkins,I., & Wright, E.L. 1993, ApJ 409, L61.
Olive, K., Steigman, G., & Walker, T. 1999, Phys. Rep., in press.
Peimbert, M., & Peimbert, A. 2000, astro-ph/0002120.
Perlmutter, S., et al. (Supernova Cosmology Project Team) 1999, ApJ 517, 565.
Pinsonneault, M.H., Deliyannis, C.P., & Demarque, P. 1992, ApJS 78, 179.
Rugers, M., & Hogan, C.J. 1996, ApJ 459, L1.
Riess, A., et al. (High-z Supernova Search Team) 1998, AJ 116, 1009.
Rood, R. et al. 1995, Light Element Abundances, (ed. P.Crane, Springer) 201.
Ryan, S., Beers, T., Olive, K., Fields, B., & Norris, J. 2000a, ApJ 530, L57.
Ryan, S.G., Kajino, T., Beers, T.C., Suzuki, T.-K., Romano, D., Matteucci, F., & Rosolankova, K. 2000b, ApJ, submitted.
Savage, B.D., & Sembach, K.R. 1996, ARA&A 34, 279.
Smith, M.S., Kawano, L.H., & Malaney, R.A. 1993, ApJS 85, 219.
Suzuki, T.-K., Yoshii, Y., & Kajino, T. 1999, ApJ 522, L125.
Thuan, T.X. 2000, in this volume.
Wannier, P.G. 1980, ARA&A 18, 366.
White, S.D.M., et al. 1993, Nature 366, 429.

Lithium in young open clusters

R. Pallavicini

Osservatorio Astronomico di Palermo, Italy

S. Randich

Osservatorio Astrofisico di Arcetri, Firenze, Italy

J. R. Stauffer

Smithsonian Astrophysical Observatory, Cambridge, MA, USA

S. C. Balachandran

Dept. of Astronomy, University of Maryland, College Park, MD, USA

Abstract. We present lithium abundances for ∼50 late-type members of the ∼ 30 − 50 Myr old open clusters IC 2602 and IC 2391 derived from high-resolution spectra obtained at ESO and CTIO. These data enlarge and extend to cooler temperatures previous Li surveys of these clusters by Stauffer et al. (1989) and Randich et al. (1997). We discuss the lithium vs. temperature distribution in the two clusters and we compare our results with those obtained for the Pleiades (∼120 Myr). The results are discussed in the general framework of lithium depletion in young clusters, from the pre-main sequence to the age of the Pleiades.

1. Introduction

Young open clusters with an age of ∼50 Myr, intermediate between pre-main sequence (PMS) stars (≤10 Myr) and older clusters like α Persei (∼ 70 Myr) and the Pleiades (∼120 Myr), are crucial for investigating the depletion of lithium during PMS and the early evolution on the MS (e.g. Jeffries 1999 and Pasquini 2000 for recent reviews). IC 2602 and IC 2391, with estimated ages of ∼ 30 − 50 Myr, cover precisely this age interval. We present in this paper Li abundances for ∼ 50 late-type stars in these two clusters, with membership confirmed on the basis of radial velocity measurements. These new data considerably enlarge and extend to later spectral types previous Li surveys by Stauffer et al. (1989) for IC 2391 and Randich et at. (1997) for IC 2602. Only a brief summary of this work is presented here; we refer to Randich et al. (2000) for a full account.

Stauffer et al. (1997) determined radial and rotational velocities, and Hα equivalenth widths (EWs) for candidate members of IC 2391. In the present study we have used the Stauffer et al. (1997) spectra to derive also lithium abundances; in addition, we have obtained new lithium data, as well as radial

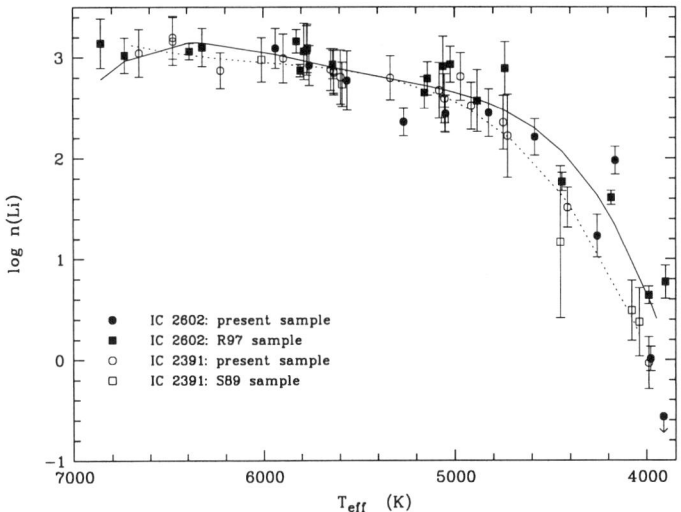

Figure 1. Li abundances vs. T_{eff} for IC 2602 (filled symbols) and IC 2391 (open symbols). Circles denote stars belonging to the present sample, while square represent stars in the samples of Randich et al. (1997; R97) and Stauffer et al. (1989; S89). Only stars warmer than 3800 K are plotted. The two curves are the regression curves of the log n(Li) vs. T_{eff} distributions of the two clusters (solid: IC 2602; dashed IC 2391).

and rotational velocities and Hα EWs, for a sample of IC 2602 stars. The latter sample extends to lower mass stars the previous study of IC 2602 by Randich et al. (1997). The new lithium data are combined with those of the previous surveys and compared with Li observations of the Pleiades. From a subsample of high S/N spectra, we have also determined for the first time the metallicity of the two clusters finding that it is nearly solar and similar to that of the Pleiades.

2. Observations and data analysis

The observations were obtained at the 4m Blanco telescope of the Cerro-Tololo Inter-American Observatory (CTIO) in January 1995 and at the 3.6m telescope of the European Southern Observatory (ESO) in April 1995.

We determined effective temperatures from the dereddened $(B-V)_\circ$ and $(V-I)_\circ$ colors adopting the scales of Bessel (1991), Soderblom et al. (1993) and Randich et al. (1997). Lithium abundances were determined using Soderblom et al. (1993) LTE curves of growth (COGs), after correcting the measured EWs of the LiI 6707.81 + Fe I 6707.44 blend for the contribution of the Fe I 6707.44 feature. NLTE corrections were applied using the code of Carlsson et al. (1994).

Average metallicities for the two clusters were determined from the EWs of 6 Fe I lines in the range $6700 \leq \lambda \leq 6735$ Å using the spectral synthesis

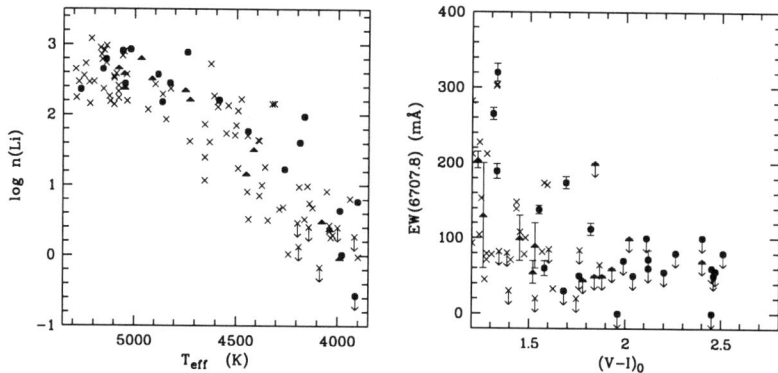

Figure 2. Comparison of the Li abundances of IC 2602 (filled circles) and IC 2391 (filled triangles) with the Pleiades (crosses). Only stars cooler than 5300 K are included in the figure. Log n(Li) vs. T_{eff} are plotted in the left-hand panel; EWs vs. $(V - I)_o$ are plotted in the right-hand panel.

code of Gratton & Sneden (1990) and a subsample of stars with the highest S/N spectra.

3. Results

Fig. 1 shows the derived Li abundances vs. T_{eff} for IC 2602 and IC 2391. The derived metallicities for the two clusters are [Fe/H] = -0.03 ± 0.01 and -0.0 ± 0.03, respectively, i.e. they are nearly solar and comparable to that of the Pleiades.

The comparison of the Li EWs and Li abundances for IC 2602 and IC 2391 stars with the Pleiades (Fig. 2) confirms the previous finding by Randich et al. (1997) that late G and early K stars in IC 2602 already present a star to star scatter in Li abundances similar to, albeit not as large as the one in the Pleiades. A scatter is also seen among late-K and M dwarfs. This indicates that the scatter is already present at ~50 Myr and must develop during the PMS. Unlike in the Pleiades, the scatter seen in IC 2602 shows no evident correlation with rotation. IC 2391 presents much less scatter than IC 2602, suggesting that the amount of scatter at any given age may depend on the individual cluster and possibly on different initial conditions.

Stars more massive than $\sim 1 M_\odot$ show no sign of depletion in both clusters, while cooler stars are all lithium depleted, with the amount of depletion increasing to cooler temperatures. The distribution of Li abundances vs. T_{eff}

is similar for the two IC clusters down to ~5000 K, but at lower temperatures the stars of IC 2391 tend to have less lithium than the stars of IC 2602 of similar temperature, i.e. the drop-off of lithium towards lower masses appears to start at an earlier color in IC 2391 than in IC 2602. This could suggest that IC 2391 is somewhat older than IC 2602, as recently proposed by Barrado et al. (1999) using the Li depletion boundary method, but this conclusion would hold only if lithium depletion is produced by standard mixing mechanisms (e.g. Pinsonneault 1997).

G and early K stars in IC 2602 are, on average, somewhat more lithium rich than their counterparts in the Pleiades. However the coolest stars in IC 2602 (and a fortiori in IC 2391) appear as depleted as the lowest lithium stars in the Pleiades. At the age of the Pleiades, stars in IC 2602 and IC 2391 are expected therefore to be more lithium depleted than stars of the same mass in the Pleiades. This suggests that lithium depletion might not be a single function of age for all clusters.

References

Barrado y Navascués, D., Stauffer, J.R., & Patten, B.M. 1999, ApJ, in press

Bessel, M.S. 1991, AJ 101, 662

Carlsson, M., Rutten, R.J., Bruls, J.H.M.J., & Shchukina, N.G. 1994, A&A 288, 860

Gratton, R.G., & Sneden, C. 1990, A&A 234, 366

Jeffries, R.J. 1999, in Stellar Clusters and Associations: Convection, Rotation, and Dynamos, eds. R. Pallavicini, G. Micela, & S. Sciortino, ASP Conf. Ser., in press

Pasquini, P. 2000, these Proceedings

Pinsonneault, M.H. 1997, ARA&A 35, 557

Randich, S., Aharpour, N., Pallavicini, R., Prosser, C.F., & Stauffer, J.R. 1997, A&A 323, 86

Randich, S., Pallavicini, R., Meola, G., Stauffer, J.R., & S.C. Balachandran 2000, A&A, submitted

Soderblom, D.R., Jones, B.F., Balachandran, S., Stauffer, J.R., Duncan, D.K., Fedele, S.B., & Hudon, J.D. 1993, AJ 106, 1059

Stauffer, J.R., Hartmann, L.W., Jones, B.F., & McNamara, B. 1989, ApJ 342, 285

Stauffer, J.R., Hartmann, L.W., Prosser, C.F., Randich, S., Balachandran, S., Patten, B.M., Simon, T., & Giampapa, M. 1997, ApJ 479, 776

The Light Elements and Their Evolution
IAU Symposium, Vol. 198, 2000
L. da Silva, M. Spite, J. R. de Medeiros, eds.

Li Abundance in Evolved Stars of NGC 6397

D.M. Allen[1], B.V. Castilho[1], L. Pasquini[2], B. Barbuy[1] and P. Molaro[3]

(1) Departamento de Astronomia, Instituto Astronômico e Geofísico, Universidade de São Paulo, Caixa Postal 3386, 01060-970, Brazil

(2) European Southern Observatory, Karl Schwarzschild Strasse 2, D-85748 Garching bei München, Germany

(3) Osservatorio Astronomico di Trieste, Via G.B. Tiepolo 11, I-34131 Trieste, Italy

Abstract. Five giants and 11 subgiants of the metal-poor globular cluster NGC 6397 are analysed. In this Poster we present the lithium abundances derived. The present Li abundances and those of turnoff stars by Pasquini & Molaro (1996) are complementary in terms of stellar evolution stage, and show the Li abundances decreasing off the main sequence along the red giant branch.

1. Introduction

The globular cluster NGC 6397 is among the most metal-poor clusters in the Galaxy with [Fe/H]≈-2.0 and it is the second nearest to us (2.2 kpc). We have gathered high-resolution (R ≈ 30000) spectra for five giants and eleven subgiants, in the spectral range $\lambda\lambda$ 4800-7500 Å. The observations of the giants were carried out with the CASPEC spectrograph at the ESO-3.6 m telescope and the subgiants with the EMMI spectrograph at the ESO-NTT. We present a detailed analysis of the 16 stars, providing their atmospheric parameters (T_{eff}, log g, [Fe/H]) and lithium abundances.

In the present work, we derive temperatures using Hα, and several photometric systems and calibrations, and discuss the inferred metallicities. Li abundances are very sensitive to effective temperatures. A decrease of Li abundance with evolutionary stage is clearly shown.

2. Stellar Parameters

For the 5 cool giants we gathered in the literature available colours, including (B-V), (V-K) and (J-K). The temperatures were derived by using colour-temperature relations by Lejeune, Cuisinier, & Buser (1998), McWilliam (1990), Buser & Kurucz (1992), Blackwell, Lynas-Gray, & Petford (1991), and Gratton, Carretta, & Castelli (1996).

For the subgiants we used the Hα profile, using both the ATLAS and MARCS models, as well as (b-y) Strömgren colours using calibrations by van-

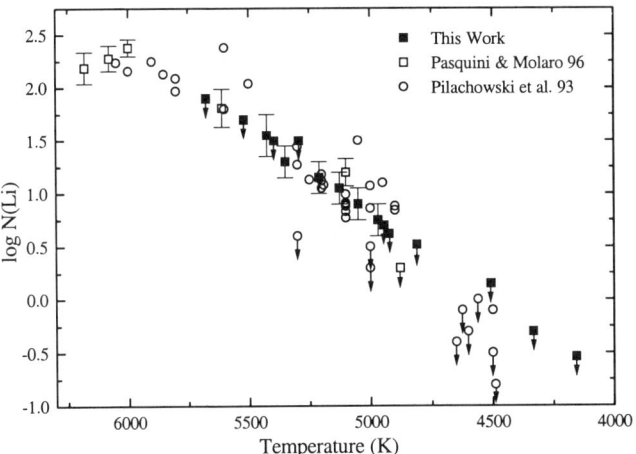

Figure 1. Li abundances vs. T$_{\text{eff}}$ for the NGC 6397 stars (filled squares, this work; open squares, Pasquini & Molaro 1996), and a subsample of the field subgiants from Pilachowski et al. (1993) (open circles). The arrows mean upper limit values.

denBerg & Bell (1985), Bergbusch & vandenBerg (1992), and Gratton et al. (1996).

For the giants there is reasonable agreement between different colours (except B-V) and calibrations. However there is strong discrepancy between different methods for the subgiants. For the present calculations we have adopted a mean of the derived temperatures. We obtain a mean metallicity of [Fe/H] = -2.0.

3. Li Abundances and Conclusions

The lithium abundances are derived by fitting synthetic spectra to the observed LiI λ 6707.8 Å line.

A decrease of the Li abundance with effective temperature is clearly seen in Figure 1 as expected by internal mixing and consequent Li destruction.

References

Blackwell, D. E., Lynas-Gray, A. E., & Petford, A. D. 1991, A&A, 245, 567
Bergbusch, P. A., & vandenBerg, D. 1992, ApJS, 81, 163
Buser, R., & Kurucz, R. 1992, A&A, 264, 557
Gratton, R. G., Carretta, E., & Castelli, F. 1996, A&A, 314, 191
Lejeune, T., Cuisinier, F., & Buser, R. 1998, A&AS, 130, 65
McWilliam, A. 1990, ApJS, 74, 1075
Pasquini, L., & Molaro, P. 1996, A&A, 307, 761
vandenBerg, D., & Bell, R. A. 1985, ApJS, 58, 711
Pilachowski, C. A., Sneden, C., & Booth, J. 1993, ApJ, 407, 699

Lithium depletion in a [Fe/H]= −3.4 star?

M. Spite, F. Spite and R. Cayrel

Observatoire de Paris-Meudon, CNRS UMR 8633, F-92195 Meudon Cedex, France

V. Hill

ESO, Garching bei München, D-85748, Germany

E. Depagne

Observatoire de Paris-Meudon, F-92195 Meudon Cedex, France

B. Nordström

Niels Bohr Institute, DK-2100 Copenhagen, Denmark

T.C. Beers

Michigan State University, Dept of Phys. and Astronomy, USA

Abstract. We present a determination of the lithium abundance from high quality spectra in an extremely metal poor star where the lithium line had not been detected.

1. Introduction

Very few halo stars are known to have a low lithium abundance in their atmospheres. Four are well established:

G66-30	[Fe/H] = −1.61	A(Li) < 1.5
G122-69	[Fe/H] = −2.52	A(Li) < 1.2
G139-8	[Fe/H] = −2.56	A(Li) < 1.5
G186-26	[Fe/H] = −2.80	A(Li) < 1.1

Moreover it has been claimed that, in CS 29527-15, an extremely metal poor star with [Fe/H]= −3.4, the lithium line would not be detectable: less than 10mÅ (Thorburn, 1994, Norris et al. 1997). As a consequence, CS29527-15 could be the most metal poor star with a very low lithium abundance in its atmosphere.

2. Observations

Since the cause of the lithium depletion in these stars is unknown, it is interesting to check the lithium abundance in CS 29527-15. We present here a new

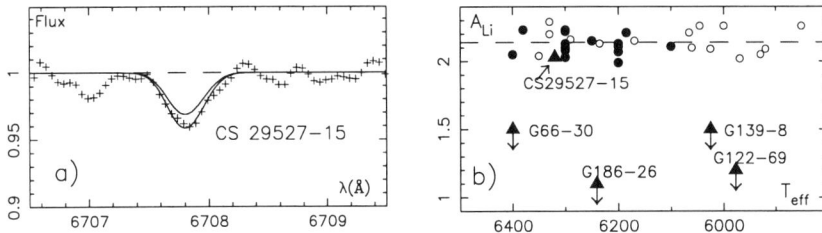

Figure 1. a) Comparison of the observed spectrum (crosses) with synthetic spectra computed with A(Li)=1.9 and 2.03.
b) A(Li) versus Teff. The filled circles represent known stars with [Fe/H]< −2.5, the open circles the stars with −2.5 <[Fe/H]< −1.6. The triangles represent the 5 stars with previously undetected lithium lines. The dashed line is the mean value of the lithium abundance in the halo stars (plateau). As is clearly seen, CS29527-15 has normal lithium abundance.

measurement of the lithium abundance in this star. Five spectra have been obtained with the EMMI spectrograph fed by the NTT, which represent a total exposure time of 10 hours. The resulting spectrum has a resolution of 30000 and the signal to noise ratio of ≈ 150.

3. Analysis and Discussion

The model has been interpolated in the grid of Edvardsson et al. (1993) computed with an updated version of the MARCS code with improved UV line blanketing. The temperature estimated from the profile of the wings of the hydrogen lines ($T_{\text{eff}} = 6250$K), is in rather good agreement with the value deduced from the $(B - V)_o$ color of the star ($T_{\text{eff}} = 6320$K following Norris et al. 1997). We adopted $T_{\text{eff}} = 6300$K.

In Fig.1a the synthesized profiles of the lithium line computed with different lithium abundances are compared to the observations. The best agreement is obtained for A(Li)= 2.03 ± 0.03. (This corresponds to an equivalent width of 16 ± 2mÅ). As is clearly seen in Fig.1b, CS29527-15 has normal lithium abundance.

Currently G 186-26 ([Fe/H]= −2.80) remains the most metal poor-star with a very low (A(Li) < 1.1) lithium abundance.

Different hypotheses have been considered to explain the lithium depletion: binarity, progeny of blue stragglers now evolving redward, (cf. Norris et al. 1997), but none of them can explain the low lithium abundance in every case.

References

Edvardsson B., Andersen J., Gustafsson B., Lambert D. L., Nissen P.E., Tomkin J., 1993, A&A 275, 101
Norris J.E., Ryan S.G., Beers T.C., 1997, ApJ 485, 370
Thorburn J., 1994, ApJ 421, 318

The Light Elements and Their Evolution
IAU Symposium, Vol. 198, 2000
L. da Silva, M. Spite, J. R. de Medeiros, eds.

Lithium in Metal Deficient K Giant Stars: The Absence of Dust Signature

Ramiro de la Reza[1], Licio da Silva[1], Natalia A. Drake[1,2] and Marco A. Terra[3]

[1] Observatório Nacional - Rio de Janeiro - 20921-400 Brazil

[2] Astronomical Institute - St. Petersburg University - St. Petersburg - 198904 Russia

[3] Observatório do Valongo - UFRJ - Rio de Janeiro - 20080-090 Brazil

Abstract. The ^{7}Li enrichment – mass loss scenario has been constructed using the strong connection found between solar metallicity Li-rich giants and IRAS sources. We show here that the Li-rich metal poor first-ascent red giants are not IRAS sources, although they are loosing Li enriched matter in form of gas.

1. The prompt ^{7}Li enrichment - mass loss scenario

By this scenario, suggested by de la Reza et al. (1996), all low mass giants ($M < 2.5 M_\odot$) suffer in the upper part of the red giant branch (RGB) a rapid ^{7}Li enrichment process of internal origin producing the formation of a Li-enriched circumstellar shell (CS) of gas and dust. This last one produces a signature in the IRAS colors. The CS detaches from the star transporting this way the new ^{7}Li into the interstellar medium. The fresh ^{7}Li remaining in the stars' photospheres is after depleted by convection. All these processes can be followed by means of closed loops (representing the gradually disappearance of the CS) in an IRAS diagram.

Smith (1998), collecting IRAS fluxes of metal deficient stars, concludes that mass loss in giants appears to be low due to the lack of observed IRAS fluxes. This is true for dust mass loss but not necessarily for the gas mass loss. Poor metal giants do not have sufficient IRAS fluxes because they do not have sufficient metals to produce enough dust. In fact, the dust-to-gas ratio, ψ, depends mainly on the abundances of metal dust producers and the drift velocity between gas and dust. If we consider the drift velocity as zero and the dust producers measured by the iron abundance, we then have $\psi \sim$ [Fe/H]. We can calculate the evolution of the CS in an IRAS diagram which is presented in the figure together with observed points corresponding to some Li-rich mild deficient (labeled by filled symbols) and Li-poor giants (open symbols; having fluxes limits). Recently spectra of the Li depleted giant HD 68298 have been obtained with FEROS at La Silla, Chile, under the Observatório Nacional-ESO agreement. From them, we determined for this star:
$T_{\text{eff}} = 4000$ K, $\log g = 0.75$, $\log \epsilon_{\text{Li}} = 0.2$ and [Fe/H]$= -0.4$

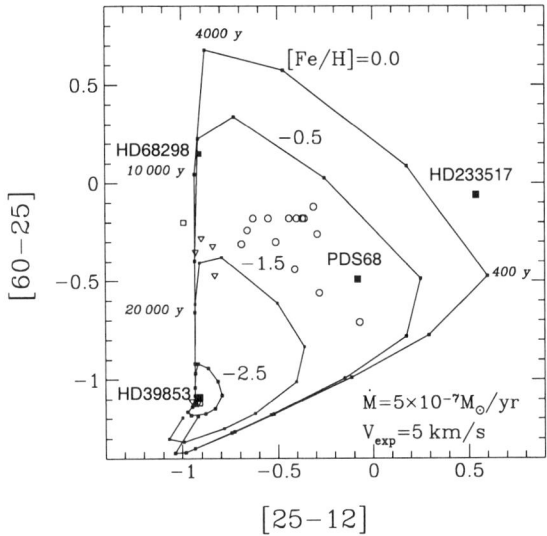

2. The scenario for deficient stars

The dependence of the dust optical depth, τ_{dust}, on the density of dust particles is given by $\tau_{dust} \sim \psi \cdot \dot{M}$, where $\dot{M}$ is the gas mass loss (de la Reza et al. 1996). If few IRAS fluxes are measured among the metal poor giants, this is because τ_{dust} has low values corresponding to the low values of ψ and not necessary to the low values of $\dot{M}$. The curves in the figure are calculated for $\dot{M} = 5 \times 10^{-7} M_\odot/\mathrm{yr}$. IRAS fluxes are observed only for mild metal deficient giants ([Fe/H]> -1.0). The small loops corresponding to very deficient giants explain why no IRAS fluxes are observed for these stars. Mass loss in very Li-rich and very metal poor giants must be put in evidence by other methods as detecting asymmetries in Na resonance lines.

As an example the very Li-rich RGB star discovered recently in the globular cluster M3 ([Fe/H]$= -1.5$) by Kraft et al. 1999 is very probably suffering a mass loss but is supposed not to present an important dust signature. If evolving giants are the main source of dust in M3, few dust must be present in this globular cluster. Penny et al. (1997) have found almost no dust in the central 14" region of M3.

Acknowledgments. R. de la R. and L. da S. thank CNPq for grants 301375/86-0 and 200580/97-3 respectively. N.A.D. and M.A.T. thank FAPERJ for the financial support under the contracts E-26/151.172/98 and E-26/150.571/98 respectively.

References

de la Reza, R., Drake, N.A., & da Silva, L. 1996, ApJ, 456, L115
Kraft, R.P., Peterson, R.C., Guhathakurta, P., Sneden, C., Fulbright, J.P., & Langer, G.E. 1999, ApJ, 518, L53
Penny, A.J., Evans, A., & Odenkirchen, M. 1997, A&A, 317, 694
Smith, G.H. 1998, PASP, 110, 1119

Lithium in Binary Systems with Evolved Components[1]

J. M. Costa[1], L. da Silva[2] and J. R. De Medeiros[1]

[1] *Universidade Federal do Rio Grande do Norte, Departamento de Física, 59072-970, Natal-RN, Brazil*

[2] *Observatório Nacional, 20921-030 ,Rio de Janeiro-RJ, Brazil*

Abstract. The analysis of 72 binaries with component of luminosity class III shows that the behavior of Li abundance in such systems follows the same pattern presented by their single counterparts. Binarity seems to affect the lithium dilution in systems presenting orbital period lower than about 100 days.

1. Introduction

Stellar lithium depletion depends on different physical parameters, for example, rotation, age, metallicity and mass. Rotation, in particular, is believed to play an important role in the mechanism controlling the dilution of lithium once stars evolve along the H-R diagram. Zahn (1994) has claimed that in tidally locked binary systems, namely binaries with short orbital period, lithium depletion would be inhibited due to tidal effects. Then we should expect some link between rotation and lithium abundance in such class of binary systems, because, at least in those with short orbital period, rotation is mostly controlled by tidal effects.

The present work brings lithium abundance for a large sample of spectroscopic binary sistems with evolved component, mostly of luminosity class III, along the spectral region F, G and K. On the basis of these data we study the behavior of the lithium content as a function of effective temperature, the link between rotation and lithium abundance and the possible role of tidal effects on lithium depletion in binary sistems with evolved component.

2. Observations and data analysis

The present sample is composed by 72 binary systems with evolved component. Observations of the lithium I resonance line were carried out using the CAT/CES at La Silla, ESO. The Li abundances were found by comparing the observed spectra with the synthetic ones calculated in LTE, for the range 6101.7-6104.3Å. Every know atomic lines in this range were considered and the gf values of the Li 6707.81Å line were taken from Andersen, Gustafsson & Lambert (1984).

---

[1] Based on observations made at the European Southern Observatory-ESO, La Silla, Chile

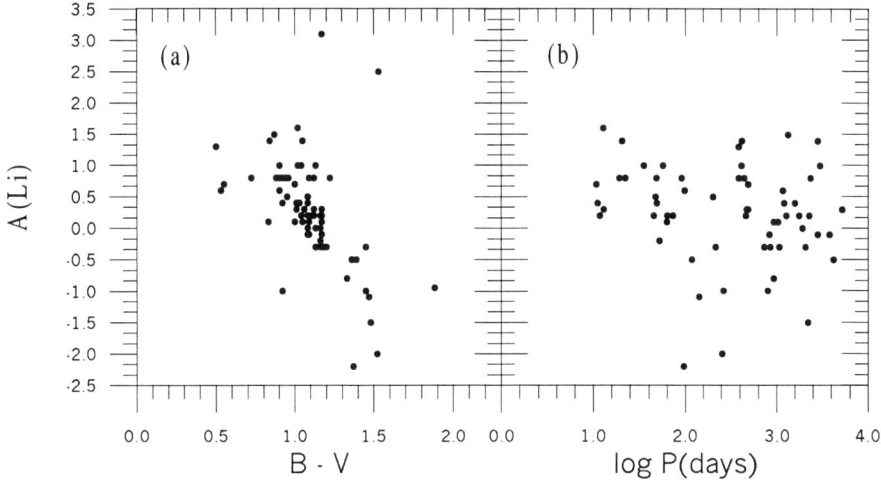

Figure 1. (a) Li abundance as a function of the color index ($B - V$) for binary systems with component of luminosity class III. (b) Li abundance versus orbital period for binary systems with component of luminosity class III.

Rotational velocities were taken from the Catalog of Rotational and Radial Velocities for Evolved Stars by De Medeiros and Mayor (1999).

3. Main Results

As shown in Figure 1a the behavior of the lithium abundance as a function of the effective temperature for binary systems with evolved component of luminosity class III, seems to follow the same feature presented by their single counterparts, namely a gradual decrease with decreasing temperatures.

Figure 1b shows that in binary systems with orbital period lower than 100 days, lithium abundance present a trend to be larger than about 0.0, in contrast with the binary systems presenting orbital period larger than 100 days. In addition, the spread in lithium abundance for binary systems with orbital period lower than about 100 days, shows also a trend to be smaller than the one for binary systems presenting orbital period larger than 100 days. Is this an indication that tidal effects are affecting the dilution of lithium?

References

Andersen, J., Gustafsson, B., & Lambert, D.L., 1984, A&A, 136, 65 De Medeiros J. R., Mayor M. 1999, A&A, 139, 433

Zahn J. P., 1994, A&A, 288, 829

The Light Elements and Their Evolution
IAU Symposium, Vol. 198, 2000
L. da Silva, M. Spite, J. R. de Medeiros, eds.

Lithium and rotation on the subgiant branch. A theoretical analysis of observations

J. D. do Nascimento Jr[1,4], C. Charbonnel[1], A. Lèbre,[2], P. De Laverny[3] and J. R. de Medeiros[4]

[1] *Laboratoire d'Astrophysique de Toulouse, 16 Avenue E. Belin, 31400 Toulouse, Fr*

[2] *GRAAL, Université Montpellier II, F-34095 Montpellier Cedex, Fr*

[3] *Observatoire de la Côte d'Azur, BP 4229, 06304 Nice CEDEX 4, Fr*

[4] *Departamento de Física, Universidade Federal do Rio Grande do Norte, 59072-970 Natal, R.N., Br*

Abstract. Lithium abundances, determined for 120 subgiant stars in Lèbre et al. (1999), are analyzed and compared with predictions for dilution. To this purpose, the evolutionary status of the sample as well as the individual masses have been determined. We look for the distributions of A(Li) and $V\sin i$ with mass when those stars evolve along the subgiant branch. Our results bring a new light on the Lithium and rotation discontinuities in evolved phases.

We investigate the physical processes that underline the lithium and rotational discontinuities along the subgiant branch based on new high resolution spectroscopic observations and precise rotational velocities. We have analyzed the Li and rotation observations for 120 Pop I F, G and K spectral types subgiant stars. We use rotational velocities given by De Medeiros & Mayor (1999) as well as the values derived by Lèbre et al. (1999) for $\log g$, A_{Li} and T_{eff} with their respective errors. We use the HIPPARCOS parallax measurements to locate precisely our objects in the HR diagram and determine the individual mass and evolutionary status. The tracks, mass and evolutionary stage determinations for the sample of stars are explained in do Nascimento et al. (2000a). The **lithium discontinuity** simply reflects the well-known dilution that occurs when the convective envelope starts to deepen after the turnoff and reaches the inner free-lithium layers. The dilution is a fast process, both in terms of age and effective temperature interval. We have compared the observed Li abundances with predictions of Li dilution caused by the deepening of the convective envelope on the subgiant branch. Stars with masses < 1.2 M$_\odot$ show a large range in abundance before the turnoff, indicating lithium depletion in the previous phases. Stars with masses between 1.2 and 1.5 M$_\odot$ show A_{Li} values in agreement with what is found in the open clusters. Many stars with masses higher than 1.5 M$_\odot$ (Fig. 1c,d) show lithium depletion up to two orders of magnitude before the start of the dilution at Log $T_{\text{eff}} \simeq 3.75$. Our observational result is in agreement with that one found by Balachandran (1990) for a few slightly evolved field stars originating from the hot side of the dip and showing

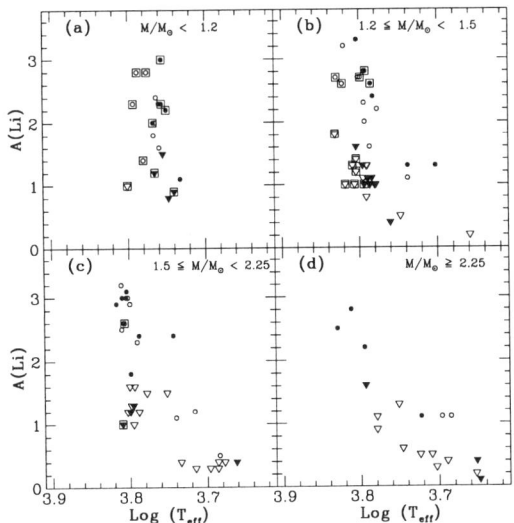

Figure 1. Lithium abundances as a function of $\log(T_{\text{eff}})$ for all our sample stars. Open and filled symbols represent single and binary stars respectively. The circles correspond to lithium detection while inverted triangles are for upper limits in the lithium abundance determination. Squares point out the main sequence stars

significant lithium depletion. This confirm the suggestion by Vauclair (1991) that some extra-lithium depletion occurs inside these stars when they are on the main sequence, even if its signature does not appear in the stellar surface at the age of the Hyades. Among some effects available one can quote an influence of rotation-induced mixing limited by a self-regulated hydrodynamical process in the main-sequence (see do Nascimento et al. 2000b and Vauclair 2000, on this meeting). For the rotation, we confirm that low mass stars leave the main sequence with a low rotational rate, while more massive stars are slowed only when reaching the subgiant branch. Our interpretation shows that lithium and rotation discontinuities seems to be independent.

References

Balachandran, S. 1990 ApJ 354, 310

De Medeiros J.R., Mayor M. 1999 A&AS 139, 443

do Nascimento, J.D.Jr, Charbonnel, C., Lèbre, A., de Laverny, P., de Medeiros, J.R., 1999, preprint (a)

do Nascimento, J.D.Jr., Vauclair, S. 2000, this conference (b)

Lèbre A., de Laverny P., De Medeiros J.R., Charbonnel C., da Silva L. 1999, A&A, 345, 936

Vauclair, S. 1991, in *IAU Symp. 145, Evolution of Stars: the Photospheric Abundance Connection (Michaud, G., Tutukov, A. eds.)* p. 327

Vauclair, S. 2000, this conference

Lithium abundances in Bright Giant Stars

Agnès Lèbre

GRAAL, Université Montpellier II, France

Patrick de Laverny

Dept Fresnel, Obs. Côte d'Azur, Nice, France

José Renan de Medeiros

Dfte-UFRN, Natal, Brazil

1. CORAVEL and High resolution spectroscopic observations

We present new high resolution spectroscopic data of the 6707.81 Å Li I line for 117 G and K Bright Giants (class of luminosity II). We derived Lithium abundances that we analysed along the stellar parameters: Teff, M_* and Vsini. With the CORAVEL spectrometers (at Observatoire de Haute Provence [OHP] and at European Southern Observatory [ESO]), De Medeiros & Mayor (2000) obtained radial velocities and Vsini with an uncertainty of about 0.3 kms^{-1} and 2.0 kms^{-1}, respectively. CORAVEL data also provide indication on the binary nature of our sample stars (32% are binary stars).
The λ 6707.81 Å Li I line was observed with the AURELIE spectrometer at OHP (R=$\lambda/\Delta\lambda$=45,000 and S/N>60) and with the CAT/CES at ESO-La Silla (R=$\lambda/\Delta\lambda$=60,000 to 80,000 and S/N>100).
We determine lithium abundances, A(Li), by the spectral synthesis method for all the stars of the sample (for a general description see Lèbre et al. 1999 and Jasniewicz et al. 1999). We estimate that the Li and iron abundances are determined with an uncertainty lower than 0.2 dex.

2. Lithium abundance, rotational velocity, Teff and stellar mass

For Bright Giants, the distribution of their Vsini values presents a sudden discontinuity located near the F8II spectral type (de Medeiros 1989, Ph.D Geneva Obs.). The derived A(Li) do not show any peculiar feature around this spectral type (as already found by Luck & Wepfer 1995, hereafter LW95), on the contrary to what has been observed for subgiant stars (Lèbre et al. 1999; do Nascimento et al. 2000). Fig. 1 presents the A(Li) we derived for our sample stars together with the data of LW95 (mainly devoted to F and G stars). We thus confirm the slow decline in A(Li) along Teff for F/G and K Bright Giants. Moreover, for G and K stars, we found that Li depletion is more important than in the theoretical predictions as the dilution factor is of order of several hundreds, while it is expected to range between 40–60 for 3 to 9 $M_\odot$ models (Iben 1965 & 1996).

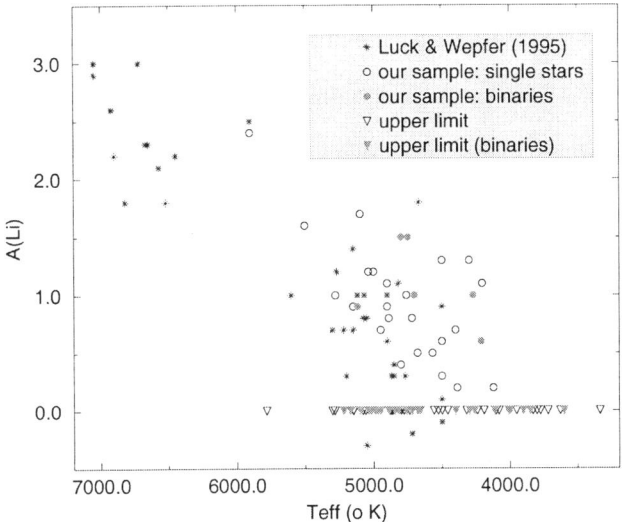

Figure 1. A(Li) along Teff for our sample stars together with the data of LW95. For our measurements upper limits are also indicated. Expected accuracy on Teff is ± 100 K for LW95 data and ± 200 K for ours. In both works, uncertainty on A(Li) is < 0.2 dex.

For our data we also infered the evolutionary status and the stellar masses, from HIPPARCOS data available at Simbad/CDS and from evolutionary tracks for 1.5 to 9 $M_\odot$ at solar metallicity (Schaller et al., 1992). We find no relation between mass and A(Li), and no difference between the Li behavior of single and binary Bright Giants, for a given Teff.

Up to date, only few giant stars (class of luminosity III) and Ia/Ib supergiants have been identified as "Super-Lithium Rich" stars (Brown et al. 1989). To explain these observed high values of A(Li) the most common scenario suggests that Li is produced in the inner layers of giant stars with rather well constrained mass and luminosity ranges (Sackmann & Boothroyd 1992 & 1999). Among our sample, 7 stars seem to be located in the required domain of mass and Mv, **but none of them present any sign of Li enrichment.**

references
Brown J., Sneden C., Lambert D.L., Dutchover E.Jr 1989, ApJS, 71, 293
Iben I. 1965, ApJ, 142, 1447 ; 1966, ApJ, 143, 483
Jasniewicz G. et al. 1999, A&A, 342, 831
Lèbre A. et al. 1999, A&A, 345, 936
Luck R.E., Wepfer G.G. 1995, AJ, 110, 2425
de Medeiros J.R., Mayor M. 2000, A&AS, 139, 433
do Nascimento J.D. et al. 2000, A&A (in press)
Sackmann I.J., Boothroyd A.I. 1992, ApJ, 392, L71 ; 1999, ApJ, 510, 217
Schaller G., Schaerer D., Meynet G., Maeder A. 1992, A&AS, 96, 269

Lithium in cool stars detected in EUV surveys

G. Tagliaferri and L. Pastori

Osservatorio Astonomico di Brera, Via Bianchi 46, 23807 Merate, I

G. Cutispoto

Osservatorio Astrofisico di Catania, V.le A. Doria 6, 95125 Catania, I

R. Pallavicini

Osservatorio Astronomico Palermo, P.za Parlamento 1 90134 Palermo I

Abstract. We selected a sample of active cool stars detected in the EUV band by the ROSAT WFC and performed spectroscopic and photometric observations. We inferred spectral type, luminosity class, distance, binary status, rotational velocity. Here we show the results of the Li abundances determination from the Li I 6707.8 Å spectral line.

1. Introduction

Large samples of stellar X-ray sources have been discovered serendipitously by the *Einstein*, EXOSAT and ROSAT Observatories. Optical follow-up studies have demonstrated that these samples are composed mainly of active stars like RS CVn and W UMa binaries, pre-main sequence and other very young stars, and BY Dra flare stars (Fleming et al. 1988; Tagliaferri et al. 1992, 1994; Favata et al. 1993; Pye et al. 1995; Jeffries 1995; Neuhäuser et al. 1997). The most interesting result of these surveys is that there seems to be an excess of young stars near ZAMS or even younger, with respect to what is predicted by the Galaxy models.

We defined a sample of active cool stars EUV–selected with the ROSAT WFC (Pye et al. 1995) and performed spectroscopic (Li I 6708Å, H_α and Ca II H&K lines) and photometric optical follow-up. Using these data and the information from the HIPPARCOS catalogs we derived accurate spectral type, rotational (vsini) and radial velocities and investigates the single or binary nature of these stars (see Cutispoto et al. 1999). Here we present the results from high–resolution spectra in the region of the Li I 6707.8 Å doublet.

2. Results

We compute Lithium abundances with n-LTE models and investigate the existence of correlations between the lithium abundances and other stellar parameters. We find that most of our single stars have very high Li abundances that

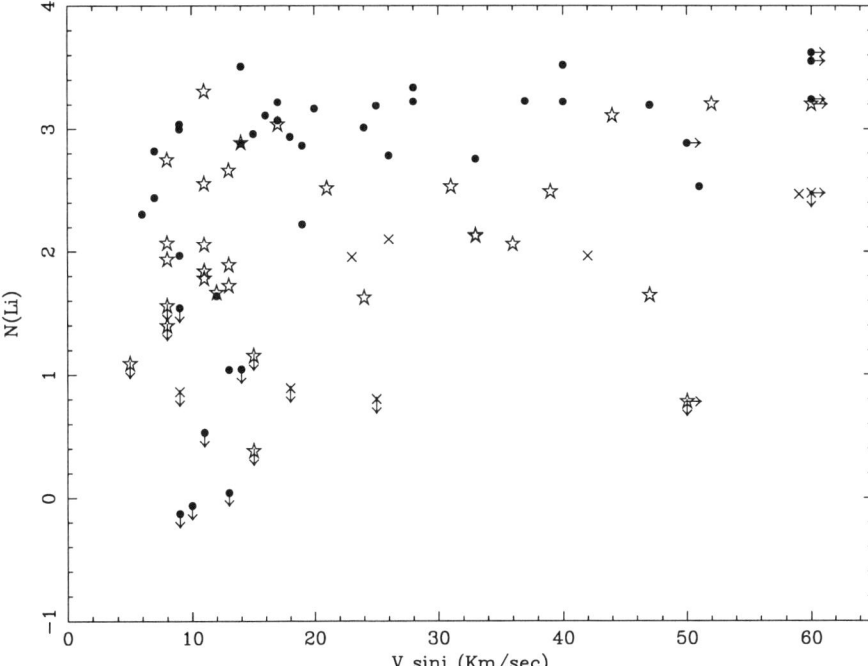

Figure 1. Li abundance vs vsini for our sample of stars. Dots are for single stars, stars are for binary systems and crosses are for binaries with an M star companion (the Li values are from the primary stars). Note that while for single stars with vsini smaller than $\sim 15 km\ s^{-1}$ the scatter in Li abundances is large, above this value they all have very high Li abundances. For binary stars this is not the case.

seems to correlate with EUV and X-ray emission and vsini (see Fig. 1). This is probably linked to the age of the stars, younger stars have higher Li abundances, higher rotation rate and so higher X-ray and EUV emission. For the binaries we see a larger scatter between these quantities. This can be explained with tidally locked effects, that imply higher rotation rate also for older stars.

References

Cutispoto, G., Pastori, L., Tagliaferri, G., et al. 1999, A&AS, 138, 87
Favata, F., Barbera, M., Micela, G., Sciortino, S. 1993, A&A, 277, 428
Fleming, T.A., Liebert, J., Gioia, I., Maccacaro, T. 1988, ApJ, 331, 958
Jeffries, R.D. 1995, MNRAS 273, 559
Neuhäuser, R., Torres, G., Sterzik, M.F., Randich, S. 1997, A&A, 325, 647
Pye, J.P., McGale, P.A., Allan, D.J., et al. 1995, MNRAS, 274, 1165
Tagliaferri, G., Cutispoto, G., Pallavicini, R., et al. 1994, A&A 285, 272

The Light Elements and Their Evolution
IAU Symposium, Vol. 198, 2000
L. da Silva, M. Spite, J. R. de Medeiros, eds.

Lithium abundance in late-type stars

L. Pompéia and B. Barbuy

Instituto Astronômico e Geofísico - USP. Av. Miguel Stefano, 4200, CEP 04301-904, São Paulo - Brazil

M. Grenon

Observatoire de Genève, CH-1290 Sauverny, Suisse

Abstract.

We have a list of nearby bulge-like turnoff stars with metallicities in the range -0.3 ≤ [Fe/H] ≤ +0.6, for which we have the absolute magnitude from Hipparcos, Geneva photometry (therefore temperature and metallicity), and radial velocity from Coravel (Grenon 1990, 1997). From Hipparcos data, the turnoff of these field stars indicate an age of 10-11 Gyr, which would be the age of the most metal-rich component of the bulge.

We obtained high resolution échelle spectra with FEROS, with the aim to carry out detailed analysis of these stars. In this paper we present the Li abundance for 40 of these metal-rich and old dwarf stars, as a function of their temperatures.

1. Observations

We present Li abundances for a sample of turnoff metal-rich stars, which seem to be a nearby component of the Galactic bulge. The characteristics of the sample were presented in Barbuy & Grenon (1990) and Grenon (1990, 1997). The sample is divided in two subsamples: a) stars more metal-rich than [Fe/H] = +0.3 (SMR or super metal rich stars), and b) stars in the range -0.8 ≤ [Fe/H] ≤ +0.3 (BGL or bulge-like stars). The stellar parameters $T_{\rm eff}$ and [Fe/H] adopted here were deduced from Geneva photometry.

The derivation of Li abundances is carried out for 40 stars of our sample, by computing synthetic spectra compared to observed spectra of the Li doublet at λ6707.776 Å and λ6707.927 Å. The gravity was assumed to be log g = 4.5 and the microturbulent velocity ξ = 1.0 kms^{-1}. The Li abundances are almost insensitive to variations in gravity and microturbulence velocity, and the errors in estimations are dominated by the temperature uncertainties. In the figures we show the Li abundances vs. temperature for SMR stars (Fig. 1a) and for BGL and Hyades stars (Fig. 1b).

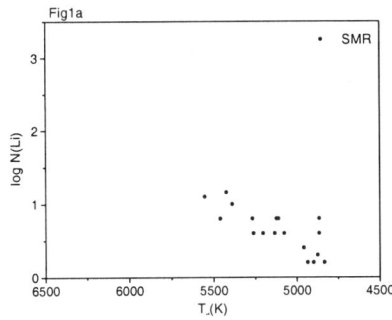
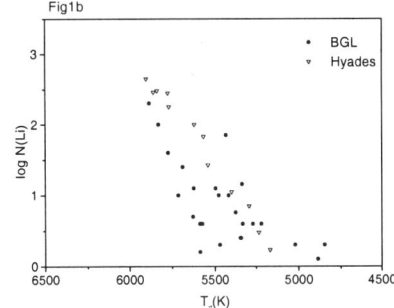

Figure 1. Lithium abundances for the SMR (Fig. 1a) and for BGL (circles) and Hyades (down triangles) stars (Fig. 1b).

2. Discussion

From Fig. 1a we note that all SMR stars are cooler than $T_{eff} < 5600K$, and have their Li abundance very depleted. Below this temperature the convective zone deepens to inner regions where lithium has been destroyed and dilution becomes important (Pilachowski et al. 1993). We found an increase from log N(Li) < 0.2 to 1.2 in the temperature range from 5000 to 5600 K.

For the less metal-rich star sample, BGL, Fig. 1b shows a large dispersion in Li abundances below 5600K. Stars hotter than this limit have a very steep increase in Li abundances with increasing temperatures. We can compare this behavior with a study of the Hyades by Cayrel et al. (1984), also shown in Fig. 1b. At similar temperatures, Hyades stars, which are younger and more metal-rich than BGL stars (700 Myr), have a higher Li abundance. The relation log N(Li) vs. T_{eff} is also steeper for the BGL stars, which agrees very well with the empirical results for the older solar-type clusters.

References

Barbuy B., Grenon M., 1990, in Bulges of Galaxies, 1st ESO / CTIO workshop, ESO Workshop and Conference Proceedings no 35, eds B. Jarvis, D. Terndrup, (ESO: Garching bei München), 83

Cayrel R., Cayrel de Strobel G., Campbell B., Däppen W. 1984, ApJ 283, 205

Grenon M. 1990, in Bulges of Galaxies, 1st ESO / CTIO workshop, ESO Workshop and Conference Proceedings no 35, eds B. Jarvis, D. Terndrup, (ESO: Garching bei München), 143

Grenon, M. 1997. In Highlights of Astronomy, ed Johannes Andersen, vol.11, 560

Pilachowski C. A., Sneden C., Booth J. 1993, ApJ 407, 699

Understanding the Li Production in AGB Stars: the J-type Stars

Carlos Abia

Dpt. Física Teórica y del Cosmos, Universidad de Granada, E-18071 Granada, Spain

Jordi Isern

Institut d'Estudis Espacials de Catalunya-C.S.I.C., c/ Gran Capitá 2-4, E-08034 Barcelona, Spain

Abstract. A full abundance analysis have been carried out in 12 galactic J-stars. The analysis shows: a) $^{12}C/^{13}C$ ratios are low, b) the abundance of s-process and rare-earth elements with respect to the stellar metallicity is nearly-solar, $[<heavy>/Fe] = 0.10 \pm 0.11$ c) Tc is not present in most of the stars although upper limits are placed in WZ Cas and WX Cyg, probably two SC-type stars, and d) all the stars are Li-rich, $\log \epsilon(Li) > 1$. The location of J-stars in an empirical H-R diagram when compared with that of the normal N-type carbon stars, suggests that J-stars may be in the early-AGB phase. The different mechanism proposed to be responsible of this Li production are briefly discussed.

1. Introduction

J-stars were defined by Gordon (1971) as carbon stars showing very strong isotopic CN & C_2 molecular bands implying usually very low $^{12}C/^{13}C$ ratios. There is observational evidence (e.g. Claussen et al. 1987) suggesting that most of galactic J-stars are low-mass objects, $M \leq 2\ M_\odot$. Among the observed galactic Li-rich AGB stars more than 60% are classified J-type (Boffin et al. 1993). However, current stellar models fail to produce low-mass, Li-rich AGB stars (see Sackmann, this volume). Indeed, the study of correlations between different chemical species might cast light on the Li production mechanism in these stars. Thus, we have performed an abundance analysis in 12 galactic J-type stars with known Hipparcos parallaxes, using high resolution and high S/N spectra. Details about the observations and analysis can be found in Abia & Isern (2000).

Figure 1 shows the location in the H-R diagram of the J-stars compared with normal (N) carbon stars and R-type stars. From this figure one can conclude that J-stars are in a transition phase between R- and N-type carbon stars. On the other hand, Figure 2 shows the Li vs. $^{12}C/^{13}C$ correlation found in our abundance analysis. Assuming that most of the J-stars studied here have $M \leq 2\ M_\odot$, is it possible to explain all these chemical properties within the standard models of stellar evolution?

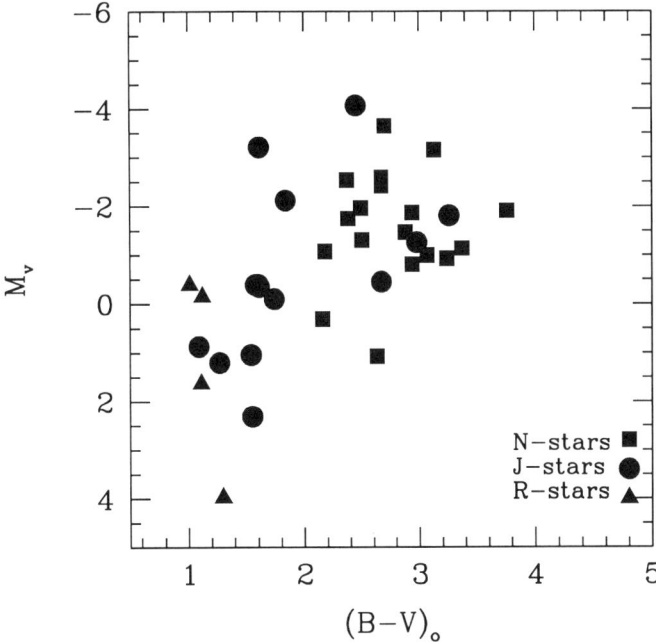

Figure 1. Observational H-R diagram for carbon stars. Data for N- and R-stars are taken from Alksnis et al. (1998).

1.1. AGB stars?

Current AGB models obtain C-rich (C/O> 1) and ^{13}C-rich envelopes in stars with M$\geq$ 4 M$_\odot$ through the operation of the successive He-shell flashes, 3th dredge-up episodes and hot bottom burning. These stars can also be Li-rich stars for a long period of time. However, they should be fairly luminous, M$_{bol}$ < -6 and present some s-nuclei enhancement. None of this is observed in the J-stars studied here. Furthermore, our objects are low-mass stars. Is there a non-standard mixing mechanism and cool-bottom processing during the early AGB phase in low-mass stars? (see Sackmann's paper in this volume).

1.2. Post He-core flash stars?

Simulations of the He-core flash in giants by Deupree & Wallace (1996) show that it is possible to mix enough core C-rich material into the envelope and transform the star into a carbon star. These authors claim that there is not s-nuclei production during the He-flash. However, no predictions are given concerning Li and ^{13}C production.

1.3. Binary system?

About 5−10% of J-star show silicate emission at 9.85 μm which can be explained through the existence of an O-rich circumstellar envelope accumulated in a disc around an unseen hypothetical companion (Lloyd-Evans 1991). However, how could Li survive during the mass-transfer and posterior mixing?

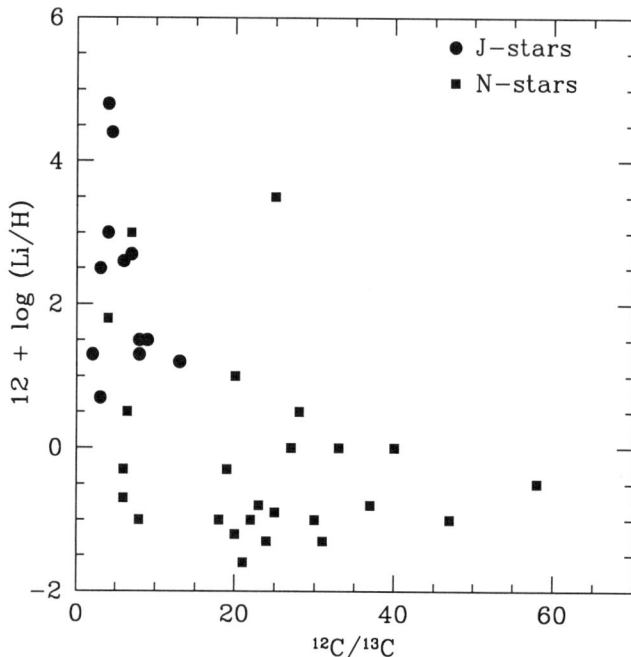

Figure 2. Li abundances vs. $^{12}C/^{13}C$ in the carbon stars sample.

2. Conclusions

J-stars are challenging objects for the theory of stellar evolution and in particular for the Li production in stars. No evolutionary scenario is able to explain all the observed chemical properties. Despite they are very frequently Li-rich objects, their contribution to the galactic Li is probably minimal due to their relatively low Li enhancements (see Abia & Isern 2000) and mass-loss rates observed, $\sim 10^{-7}$ $M_\odot$/yr. However, given the important uncertainties still existing in the derivation of the lithium abundance in carbon stars (Abia, Pavlenko & de Laverny 1999), this conclusion should be considered with caution.

References

Abia, C., Pavlenko, Ya., & de Laverny, P. 1999, A&A, 351, 273
Abia, C., & Isern, J. 2000, ApJ, (in press)
Alksnis, A., Balnlaus, A., Dzervitis, V. & Eglitis, I. 1998, A&A, 338, 273
Boffin, H.M.J., Abia, C., Isern, R., & Rebolo, R. 1993, A&AS, 102, 361
Claussen, M.J., Kleinmann, S.G., Joyce, R.R., & Jura, M. 1987, ApJS, 65, 385
Deupree, R.G., & Wallace, R.K. 1996, ApJ, 317, 214
Gordon, C.P. 1971, PASP, 83, 667
Lloyd-Evans, T. 1991, MNRAS, 249, 409

The Origin of the Lithium Rich Giants

Corinne Charbonnel
Laboratoire d'Astrophysique de l'Observatoire Midi-Pyrénées, Toulouse, France

Suchitra Balachandran
University of Maryland, USA

Abstract. We use Hipparcos parallaxes to determine the evolutionary status and the nature of the so-called Li-rich giants. These informations clearly support the hypothesis of internal nuclear lithium enrichment by an extra-mixing process at two distinct evolutionary phases.

1. The so-called Li-rich giants

About 1% of the G-K giants show unexpected strong lithium lines, some of these so-called Li-rich RGB even presenting abundances higher than the present interstellar medium value. Different propositions have been made to explain the Li-rich giant phenomenon. Some are related to external processes, like the contamination of the external layers of the giant by the debris of nova ejecta or by the engulfing of a planet. Other explanations refer to internal processes, like the preservation of the initial lithium content or fresh lithium production (see also De la Reza in this volume).

2. Clues on the nature of the Li-rich giants

When available, we use the Hipparcos data to determine the mass and evolutionary status of the Li-rich giants of the literature by comparing their position on the HRD with theoretical evolutionary tracks (Fig.1; see Charbonnel & Balachandran 2000). Among the twenty objects of the sample for which a precise determination of the location was possible, 8 appear to be Li-rich simply because they have not completed the standard first dredge-up dilution, and must thus be reclassified as Li-normal; these stars are the hexagons on the figure.

The remaining stars are confined in two special places in the HRD. First, the four stars with the highest Li abundance (black circles) are located at the place referred to as the bump in the luminosity function (shaded area). At this evolutionary point the outwardly moving hydrogen burning shell passes through the mean molecular weight discontinuity created by the first dredge-up. This permits access to a reservoir of ^{3}He rich material by an extra-mixing process which is actually known to significantly decrease the carbon isotopic ratio in low-mass RGB stars (e.g., Charbonnel et al. 1998, Charbonnel et al. in this volume).

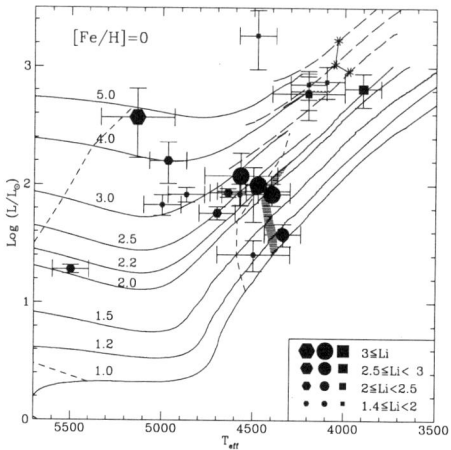

Figure 1. HR diagram for the Li-rich giants. The evolutionary tracks for [Fe/H]=0 are labelled by their mass. The short dashed lines and the asterisks delimit the first and second dredge-up respectively. The shaded region surrounds the location of the bump

We thus bring to the fore and clearly identify an episode of Li production at the RGB bump. Since no Li-rich RGB of low mass has been identified above the bump, we can conclude that the freshly synthetized Li is destroyed very quickly, as soon as the extra-mixing extends deep enough to change the surface carbon isotopic ratio.

Finally, five Li-rich stars are intermediate mass objects (black squares) which have ignited central He burning in a non-degenerate core before reaching the RGB bump. They are at present located on the early-AGB (dashed lines), close to the completion of the second dredge-up (asterisks). We thus identify an episode of Li production at this phase where the convective envelope is deepening but has not yet build a new molecular weight barrier between its base and the hydrogen burning shell; we expect the barrier to stop the extra-mixing so that this Li-rich episode is limited both in time and luminosity. We do not endorse any specific mechanism for the extra-mixing as none is evident from the observations.

Since both Li-rich episodes are very short, these stars should not contribute significantly to the Li enrichment of the ISM. We refer to Charbonnel & Balachandran (2000) for a more complete discussion.

References

Charbonnel C., Balachandran S., 2000, A&A in press, astro-ph/0005280
Charbonnel C., Brown J.A., Wallerstein G., 1998, A&A 332, 204
Charbonnel C., Deliyannis C.P., Pinsonneault M., 2000, this volume

The Light Elements and Their Evolution
IAU Symposium, Vol. 198, 2000
L. da Silva, M. Spite, J. R. de Medeiros, eds.

Detailed Analysis of Li-rich Giants

J. Gregorio-Hetem, B.V. Castilho and B. Barbuy

Universidade de São Paulo, CP 3386, São Paulo, SP, 01060-970, Brazil

F. Spite and M. Spite

Observatoire de Paris, DASGAL, UMR 8633 CNRS, F-92195 Meudon Cedex, France

Abstract. We have carried out a survey to detect Lithium-rich giants (LRGs). In Castilho et al. (1998) we reported the discovery of five red giants showing a strong Li I ($\lambda 670.78$ nm) line and six other ones showing a moderate Li I line. In this work we present the main results of a detailed analysis of 10 stars from our survey and 6 LRGs selected from the literature.

1. The Program Stars and Observational data

In our search for Lithium-Rich Giants (LRGs) we prepared a list of candidates with IRAS colours in a precise locus of the [12-25] vs. [25-60] diagram and obtained spectroscopy for more than a hundred of them (Castilho et al. 1998). In this Poster we present the detailed analysis of 10 stars, revealed in this survey to show a strong Li line, as well as 6 other previously known LRGs.

The observations obtained at ESO (Chile) provided high-resolution ($R \sim 58000$) spectra in regions from 535nm to 807nm. The medium-resolution spectra in the range $\lambda\lambda 650\text{-}680$nm were obtained at LNA (Brazil), and OHP (France), with $R \sim 20000$. Photometry in the UBVRI *Johnson-Cousins* system was obtained at the LNA.

2. Calculations and Results

Effective temperatures T_{eff} were estimated from the photometric calibrations by Bessell et al. (1998) and Lejeune et al. (1998) and checked against excitation equilibrium of FeI lines. Gravities ($\log g$) were obtained from ionization equilibrium of Fe I and Fe II lines. [Fe/H] and microturbulence velocities v_t were derived from FeI curves-of-growth.

The Li abundances were determined by fitting LTE synthetic spectra to the Li $\lambda 670.78$ nm and $\lambda 610.36$ nm (when available) lines. The non-LTE corrections by Carlsson et al. (1994) were then applied. Model atmospheres employed have been interpolated in the MARCS grids computed by Plez (1992) and Plez (1997, private communication).

In Castilho et al. (2000) we show that the strength of the Li lines is very dependent on $T_{\rm eff}$ and very little on gravity. A summary of the results obtained for our sample is reported in Table 1.

Table 1. Stellar parameters and Li abundance for the program stars.

Object	$T_{\rm eff}$	log g	[Fe/H]	v_t km.s^{-1}	logN(Li) ETL	NETL
HD 4893	4057	1.8	0.2	3.0	0.45	0.78
HD 44889	3775	0.4	-0.2	1.5	0.40	0.84
HD 65750	3600	0.6	-0.4	1.5	0.95	1.35
HD 90082	3686	0.0	-0.2	1.2	-0.10	0.18
HD 96195	3407	-0.5	0.2	1.5	0.10	0.38
GCSS 577	3300	0.0	0.0	2.0	0.25	0.70
HD 176588	3793	1.6	0.0	1.5	1.10	1.57
iras19012-0747	3810	1.5	0.0	1.5	2.50	2.55
iras19038-0026	3600	1.0	0.0	1.0	0.30	0.56
HD178168	4000	1.0	0.0	2.5	0.50	0.92
HD 787	3950	1.4	0.0	1.5	2.10	2.27
HD 19745	4750	2.9	0.1	1.2	3.85	3.65
HD 30238	3925	1.4	0.0	1.5	0.80	1.20
HD 31993	4350	2.4	0.1	3.0	1.65	1.84
HD 39853	3850	1.6	-0.3	1.5	2.80	2.95
HD 95799	4900	3.2	0.0	1.5	3.20	3.05

3. Conclusions

LRGs are characterized by IRAS colours indicating the presence of a dust envelope and by a high Li abundance, but the abundances of other elements are typical of normal red giants. This confirms the hypothesis that LRGs may correspond to an evolutionary stage of normal giants where Li and dust are produced.

References

Bessell, M. S., Castelli, F., Plez, B. 1998, A&A 333, 231

Carlsson, M., Rutten, R.J., Brus, J.H.M.J., Shchukina, N.G. 1994, A&A 288, 860

Castilho, B.V., Gregorio-Hetem, J., Spite, F., Spite, M., Barbuy, B. 1998, A&AS 127, 139

Castilho, B.V., Gregorio-Hetem, J., Barbuy, B., Spite, F., Spite, M. 2000, A&A, in preparation

Lejeune, T., Cuisinier, F., Buser, R. 1998, A&A 130, 65

Plez, B., Brett, J.M., Nordlund, Å. 1992, A&A 256, 551

The Light Elements and Their Evolution
IAU Symposium, Vol. 198, 2000
L. da Silva, M. Spite, J. R. de Medeiros, eds.

The Lithium Abundance and Mass Loss Rate in Galactic Super-Li-Rich Carbon and S Stars

D. A. Lubowich

Dept. of Physics, Hofstra U., Hempstead, NY & AIP, Melville, NY

Verne V. Smith

Dept. of Physics, Univ. Texas El Paso, El Paso, TX

B. E. Turner

National Radio Astronomy Observatory, Charlottesville, VA

R. Sahai

Jet Propulsion Laboratory, Pasadena, CA

Abstract. The super-Li-rich (SLR) stars are rare AGB stars containing enhanced Li/H ≈ 2000 times the ISM Li/H. In order to determine if mass loss from SLR stars is a significant source of ISM Li, we measured their Li/H and the mass loss rate. From the weak Li I 8126 Å line in 2/4 SLR C stars and 6/8 SLR S stars we obtained Li/H = $(0.2 - 5) \times 10^{-6}$ for SLR C stars and Li/H = $(0.2 - 5) \times 10^{-7}$ for the SLR S stars. We determined dM/dt from the expansion velocity and line profiles of the circumstellar CO. The $dM/dt = 1.3 \times 10^{-5} M_\odot/\text{yr}$, $2.6 \times 10^{-7} M_\odot/\text{yr}$, and $2.6 \times 10^{-8} M_\odot/\text{yr}$ for the C stars IY Hya, T Ara, and WZ Cas; and average $dM/dt = 5.5 \times 10^{-7} M_\odot/\text{yr}$ for the S stars. Thus mass loss from SLR stars is a significant source of Galactic Li.

1. Introduction

Super-Li-rich (SLR) stars are unique stars(4 C and 8 S) with strong lines of Li at 6707 Å (equivalent widths (EW) = (2-10) Å) and Li/H up to 3×10^{-6} or ≈ 2000 times the ISM Li/H (Denn, Luck, & Lambert 1991). These are AGB stars (2-6 $M_\odot$) where their envelopes are ejected by strong stellar winds (up to $10^{-4} M_\odot/\text{yr}$) giving birth to PN. In the Magellanic Clouds the SLR stars are the most luminous and massive stars (Smith et al., 1995).

Cameron & Fowler (1971) proposed that the enhanced Li in these stars is produced via $^3\text{He}(^4\text{He},\gamma)^7\text{Be}(e,\nu)^7\text{Li}$. The ^7Be is brought to the surface via convective diffusion from the hot-bottom envelope burning during the third dredge-up of the He-burning shell for all 3-6 $M_\odot$ AGB stars (Sackmann & Boothroyd 1992), resulting in a SLR phase lasting for 10^{4-5} years. However, the atmospheric Li will be transported back into the stellar interior and burned into He via $^7\text{Li}(p,\alpha)^4\text{He}$ at $T > 2 \times 10^6$ K unless there is mass loss or some other mechanism preventing the Li from being destroyed. Scalo (1976) proposed that mass loss from SLR stars is a major source of ISM lithium. The contribution of SLR stars to the ISM Li can be estimated from $(\text{Li/H}) \times (dM/dt) \times$ (lifetime of SLR phase) $\times$ (number of SLR stars).

2. Observations and results

The optical observations were done 22-24 July 1998 with the CTIO 4m telescope, the echelle grating, and T2KA CCD. The radio observations were done with the SEST 15m telescope (1991) and the NRAO 12m telecope (1991 & 1999). At CTIO we observed the C stars: IY Hya and T Ara and the S stars: RZ Sgr, T Sgr, VX Aql, CSS 703, CSS 861, and CSS 935 (CSS = Case Catalog of Galactic S stars, 2d edition). At SEST we observed T Ara, RZ Sgr, T Sgr, VX Aql, CSS 703, and CSS 935. At NRAO we observed IY Hya and WZ Cas. We will observe the WX Cyg (C*) and GS Per (S*) at NRAO in Jan. 2000; and CSS 583 and CSS 861 at SEST in April 2000.

The Li/H was determined from the weaker 8126 Å Li line and is (1.5-2) × larger than the Li/H determined from the strong variable 6707 Å line (Li/H in IY Hya varies by 30 ×; Boffin et al. 1993). Li/H = $(0.2 - 5) \times 10^{-6}$ for the SLR C stars and Li/H = $(0.2 - 5) \times 10^{-7}$ for the SLR S stars. It is not known if the variations in Li/H are due to the differences in mass, evolutionary phase, or lifetime of the SLR phase for the C and S stars.

The dM/dt is determined from the expansion velocity of the circumstellar CO and the IR luminosity. The dM/dt for the SLR C stars IY Hya (possible proto-PN), WZ Cas, and T Ara are: $dM/dt = 1.3 \times 10^{-5} M_\odot/\text{yr}$, $dM/dt = 2.6 \times 10^{-8} M_\odot/\text{yr}$, and $dM/dt = 2.6 \times 10^{-7} M_\odot/\text{yr}$ respectively (Knapp & Morris 1985; Olofsson et al. 1993). The dM/dt for the SLR S stars RZ Sgr, T Sgr, and VX Aql average $dM/dt = 5.5 \times 10^{-7} M_\odot/\text{yr}$ (Sahia & Liecht, 1995). CO was not detected in WX Cyg, CSS 703, or CSS 935.

Thus mass loss from the SLR stars are a major source of the ISM Li. The C stars yield more ISM Li because they have larger Li/H and dM/dt. However, if all AGB stars went through an SLR phase with the lowest Li/H and dM/dt, then the SLR stars would contribute a small portion of the ISM Li.

References

Boffin, H.M.J., Abia, C., Isern, J., & Rebelo, R., 1993, A&AS, 102, 361
Cameron, A. G. W. & Fowler. W. A. 1971, ApJ, 161, 11
Denn, G. R., Luck, R. E., & Lambert, D. L. 1991, ApJ, 377, 349
Knapp, G. R. & Morris, M. 1985, ApJ, 292, 640
Olofsson,H., Eriksson, K., Gustafsson, B., & Carlstrom, U. 1993, ApJS, 87, 267
Sackmann, I-J. & Boothroyd, A. I. 1992, ApJ, 392, L71
Sahai, R. & Liechti, S. 1995, A&A, 293, 198
Scalo. J. M. 1976, ApJ, 355, 18
Smith, V.V., Plez, B., Lambert, D. L., & Lubowich, D. A. 1995 ApJ, 441, 735

The Light Elements and Their Evolution
IA U Symposium, Vol. 198, 2000
L. da Silva, M. Spite, J. R. de Medeiros, eds.

Measurements of Li Abundance in a sample of T Tauri Stars

M. J. Sartori, J. Gregorio-Hetem, B. V. Castilho and J. R. D. Lépine

Universidade de São Paulo, CP 3386, São Paulo, SP, 01060-970, Brazil

Abstract. In this work we present the stellar parameters and the Li abundances, obtained by spectral synthesis, for a sample of 13 weak-line T Tauri stars.

1. Stellar Properties

In the frame of a study of kinematics and stellar properties of the pre-main sequence population of southern star forming regions, we are analyzing the evolutionary status of a sample T Tauri (TT) stars, on the basis of their stellar properties compared to their kinematics. In this work we studied 13 weak-line TT (WTT) stars of the Chamaeleon, Lupus, Ophiuchus and Upper Scorpius star-forming regions.

The absolute bolometric luminosities were estimated combining spectral types (S.T.), V and R with the distance of the associations, assuming intrinsic colours and bolometric corrections from Bessel, Castelli, & Plez (1998), and a S.T.-T_{eff} relation from de Jager & Nieuwenhuijzen (1987). The S.T. adopted were taken from literature (see Notes in Table 1) and from this work. For the star HT Lup the luminosity was calculated using the distance obtained from the HIPPARCOS parallax. The star WRA 488 is an isolated TT and does not have distance estimation. The ages and masses (see Table 1) were determined by comparison with D'Antona & Mazzitelli (1994) evolutionary tracks and isochrones which use CM convection and Alexander opacities.

2. Lithium Abundances

We obtained medium-resolution (R = 7500) Coudé spectra in the region of the Li I line at λ 6708 Å, using the 1.6m telescope at *Pico dos Dias* Observatory (MG, Brazil).

Li abundances were estimated comparing the observed spectra to LTE synthetic spectra computed with the code SPECTRUM (kindly made available by R. O. Gray) and Kurucz's (1992) stellar atmosphere models. The errors in logN(Li) are mainly due to the T_{eff} uncertainty, and are $\approx$ 0.3 dex. Our results are consistent with the Li depletion predicted by Mendes, D'Antona & Mazzitelli (1999) for 1.0, 0.8 and 0.6 $M_\odot$ stars.

Table 1. Stellar parameters and Li abundances of WTT stars

Object	S.T.	L ($L_\odot$)	Age (Myr)	Mass ($M_\odot$)	W_λ(Li) (mÅ)	logN(Li)	Notes
WRA 488	G6 V				199	3.0	a, i *
CHXR 3	K0 V	1.77	7.0	1.45	59	≤2.0	b, e
CHXR 8	G2 V	0.47	>100	0.9	134	3.0	b, i
CHXR 11	G8 V	2.52	7.0	1.6	135	2.3	b, i
CHX 18N	K2 V	0.70	15.0	1.1	534	3.5	b, i
Sz 41	K2 V	0.73	10.0	1.1	454	3.5	a, g
CoD-40 8434	K5 V	0.55	7.0	0.9	504	3.5	a, g
HT Lup	K2 V	3.95	0.7	1.1	472	3.5	a, f
Wa Oph/1	K2 IV	1.10	2.5	0.8	482	3.3	c, h
Wa Oph/3	K0 IV	1.91	2.5	1.1	534	3.4	c, h
V896 Sco	K7 V	0.90	1.5	0.6	520	3.1	a, g
CoD-24 12809	K3 V	1.37	3.0	1.0	529	3.6	a, i
Wa CrA/2	G5 IV	1.09	20.0	1.2	312	3.2	d, h

Notes: (V-R) and S.T. from: (a) Gregorio-Hetem et al. (1992), (b) Lawson, Feigelson, & Huenemoerder (1996), (c) Walter et al. 1994, (d) Walter et al. 1997, (e) Alcalá (1994), (f) Herbig (1977), (g) Torres (1998), (h) Walter (1986), (i) this work. (*) Isolated TT star with no distance determination.

References

Alcalá, J. M. 1994, PhD Thesis, Ruprecht-Karls-Univ., Heidelberg
Bessel, M. S., Castelli, F., & Plez, B. 1998, A&A, 333, 231
D'Antona, F., & Mazzitelli, I. 1994, ApJS, 90, 467
de Jager, C., & Nieuwenhuijzen, H. 1987, A&A, 177, 217
Gregorio-Hetem, J., Lépine, J. R. D., Quast, G. R., Torres, C. A. O., & de la Reza, R. 1992, AJ, 103, 549
Herbig, G. H. 1977, ApJ, 214, 747
Kurucz, R. L. 1992, in IAU Symp. 149, ed. B. Barbuy & A. Renzini (Kluwer Acad. Press), 255
Lawson, W. A., Feigelson E. D., & Huenemoerder, D. P. 1996, MNRAS, 280, 1071
Mendes, L. T. S., D'Antona, F., & Mazzitelli, I. 1999, A&A, 341, 174
Torres, C. A. O. 1998, PhD Thesis, Observatório Nacional, Brazil
Walter, F. M. 1986, ApJ, 306, 573
Walter, F. M., Vrba, F. J., Mathieu, R. D., Brown, A., & Myers, P. C. 1994, AJ, 107, 692
Walter, F. M., Vrba, F. J., Wolk, S. J., Mathieu, R. D., & Neuhäuser, R. 1997, AJ, 114, 1544

ABUNDANCE OF BERYLLIUM AND BORON

Luca Pasquini

Beryllium in the Sun: Re-Measurement and Implications

Suchitra C. Balachandran

Department of Astronomy, University of Maryland, College Park MD 20742, USA

Abstract. The solar beryllium abundance is important because it provides a constraint on the depth to which mixing has occurred below the surface convective zone. Unlike helioseismology which only maps the present-day Sun, the solar beryllium abundance provides an integrated picture of mixing over the entire history of the Sun. In this review I outline the logic involving the "missing UV opacity" that required that the solar beryllium abundance be re-determined. A brief summary of the empirical process of estimating the "missing UV opacity" is given along with a confirmation based on a recent re-calculation of the Fe I bound-free opacity. The addition of this opacity resulted in our finding that the solar beryllium abundance was meteoritic. The implications of this result in the context of mixing in solar-type stars is discussed.

1. Introduction

The fragile element lithium, which is destroyed at temperatures greater than 2.5 million K, is depleted by a factor of 140 in the solar photosphere relative to its meteoritic value. As the base of the solar convective zone is not hot enough to destroy lithium, this depletion requires surface material to be mixed to hotter temperatures below the convective zone. Such mixing is not predicted by the standard stellar models and the process which triggers the mixing is not understood. Non- standard stellar models (Pinsonneault et al. 1989; 1990; Charbonnel et al. 1994) have attempted to explain the depletion of lithium by linking it to a more ubiquitous process, the transport and dissipation of angular momentum. As a result of the contraction of the proto-stellar cloud, the young star is expected to be rotating rapidly when it arrives on the main sequence, and observations reveal that G and K stars in young clusters are often rapid rotators (see Stauffer 1991 for a review), in contrast to their older main sequence counterparts. It appears that at least a part of the angular momentum is lost while on the main sequence. The rotational models have suggested that the outer layers of the star are spun down initially by winds from the stellar surface, leaving the star in a state of differential rotation with radius. The transport of angular momentum is achieved via turbulence generated by shear forces between layers rotating at different rates . These models have predicted that a spread in lithium abundance will become apparent when stars with a range in initial angular momenta are spun down to the same final rotational velocity; a star with a larger initial angular momentum will undergo a greater amount of rotational spin-down, mixing and thus lithium depletion (Pinsonneault et al. 1990).

The principal interest in the solar photospheric abundance of beryllium (which is depleted at temperatures greater than 3.5 million K) is to understand whether this more robust element has also been depleted relative to its meteoritic value. A depletion of lithium and beryllium would imply that mixing below the surface convective zone was more than merely superficial. The rotational models, which use the present-day solar lithium abundance as a constraint, predict that beryllium would be depleted by a factor of two. And indeed, this is precisely the result obtained by two recent measurements of the photospheric solar beryllium (Primas et al. 1997 and King et al. 1997).

In this review I will explain our re-measurement of the solar beryllium abundance, provide some clues for the opacity sources in the UV and discuss the implications of our result.

2. The Opacity Problem and a Beryllium Re-Measurement

Our re-measurement of the solar beryllium abundance was motivated by our recognition that the "UV Opacity Problem" would impact upon all abundance measurements from UV lines. The "UV Opacity Problem" is the mismatch between observed and predicted solar UV fluxes first recognized by Holweger (1970) and Gustafsson et al. (1975); theoretical UV fluxes are larger than observed leading to the surmise that the UV opacity is underestimated. After the inclusion of several million additional UV lines, Kurucz (1992) suggested that the "Opacity Problem" was a result of the insufficient inclusion of line opacity and demonstrated, at low resolution, that a match to the solar flux was achieved. However Bell et al. (1994) re-examined the problem at high resolution and found that many strong lines in Kurucz's line list were not present in the solar spectrum. They showed that a removal of these spurious lines resulted in a reversal back to the original problem: the predicted flux was too large. They concluded that the "missing opacity" was either a veil of very weak lines or a continuous opacity.

The implications of ignoring the "missing opacity" are the following. With insufficient opacity, the predicted line is stronger than observed, requiring the abundance to be decreased to match the observed line profile. An excellent example of this is the determination of boron abundances in Boesgaard et al. (1998); boron is measured from B I lines near 2500 Å. Stars undepleted in lithium were found to have boron abundances down by 0.7 dex from the meteoritic value, a finding not explained by any mixing model. Recognizing this, Cunha et al. (2000) added Mg I bound-free opacity, the dominant opacity at this wavelength, to the analysis and found that stars undepleted in lithium were undepleted in boron as well. In reverse, if a solar oscillator strength is derived for a line, a spuriously large abundance will be calculated in, say, a metal-poor star in which contribution of the "missing opacity" may be negligibly small. The magnitude of the error will depend upon the magnitude of the "missing opacity". This logic formed the basis for our re-measurement of the solar beryllium abundance.

The impact of the "missing UV opacity" upon the solar beryllium abundance was recognized by Chmielewski et al. (1975). They determined that additional opacity had to be added to the synthetic fit to ensure that the spectrum of the center of the solar disk yielded the same abundance as that of the limb. They obtained a satisfactory but not complete match of the solar center and limb spectra and calculated log ϵ(Be) = 1.15 ± 0.2. Improvements in both spectral acquisition and opacity calculations since that study allow us to refine the opacity estimate.

The details of our re-measurement are given in Balachandran & Bell (1998) and only a brief summary is provided here. The solar spectrum used was the digital version of the Kurucz et al. (1984) solar atlas. Three solar model atmospheres were used: the Holweger-Muller empirical model (Holweger & Müller 1974), the OSMARCS (Edvardsson et al. 1993) and Kurucz (Kurucz 1993) theoretical models. With each model, the equivalent widths of the infrared OH lines from Grevesse et al. (1984) were analyzed to obtain the solar oxygen abundance. Fourteen clean OH lines were identified in the UV between 3100 and 3180 Å and each of these were forced to yield the same oxygen abundance as the IR-based value obtained from that particular model atmosphere. Additional opacity was required to fit all of the lines. The mean additional opacity was taken as an estimate of the "missing opacity". The value of the "missing opacity" was the same from each of the models. When this additional opacity was used in the analysis of the Be II lines at 3130 Å the beryllium abundance obtained was log ϵ(Be) = 1.40 ± 0.09, essentially equal to the meteoritic value. The additional opacity required was an augmentation of the hydrogenic opacities by a factor of 1.6. However this did not provide a good fit to limb darkening; an augmentation of the Fe I bound-free opacity (from Dragon & Mutschlecner 1980) of a factor of 30 provided both a better fit to the limb darkening and to the OH lines. The latter was chosen because of all the opacity sources which have some contribution at this wavelength, the Fe I bound-free opacity was the most uncertain.

We have recently been able to improve upon this empirical estimate of the "missing UV opacity" by including the revised Fe I bound-free opacity value of Bautista (1997) into our calculations. The details of this study are given in Bell, Balachandran & Bautista (2000) and, again, only a brief summary is provided here. Bautista's Fe I bound- frcc opacity was included in our re-calculation of the continuous UV flux using the OSMARCS model. Observed line-blocking was added from the digital version of the Kurucz et al. (1984) solar atlas and a comparison of this UV flux was made to the SOLSTICE solar flux data from Woods et al. (1996). It was found that between 3000 and 4000 Å a value of twice Bautista's (1997) opacity was required to obtain a good fit to the observed solar flux; this value is within the error limits set by Bautista for his calculations. The flux comparison was done at a very different resolution (2.5 Å) and over a much larger wavelength region (3000 - 4000 Å) than the OH-line fits (0.008 Åand 3100 - 3180 Å respectively) that were used to obtain the earlier empirical estimate for the "missing opacity". We will shortly incorporate Bautista's Fe I b-f opacity into our stellar synthesis code and recalculate the opacity estimate via the OH lines. However, the flux fit is essentially compatible with our finding that the solar beryllium abundance is meteoritic; twice Bautista's Fe I b-f opacity is roughly equivalent to twenty times the Dragon & Mutschlecner (1980) value.

3. Implications

Two important predictions form the backbone of the angular momentum transport and material mixing via turbulence models. First, the models predict that the present-day Sun should have a rapidly rotating core. Helioseismological data have been interpreted to show that angular momentum transport and dissipation has been far more efficient than predicted by the turbulence transport models; the present-day Sun down to $r=0.2$ $R_\odot$ is rotating slowly as a rigid body (Tomczyk, Schou, & Thompson 1995; Charbonneau et al. 1998). Alternatives to the transport of angular momentum via turbulence include magnetic fields (Charbonneau and MacGregor, 1993) and internal gravity waves (Kumar & Quataert 1997; Zahn et al. 1997). Modeling of both forms of transport are still in their infancy, but the internal gravity wave models suggest that solid body rotation would be achieved in 10 Myr.

The second important prediction of the rotation models is that because differential rotation is expected to persist through the radiative interior for a large fraction of the star's main sequence lifetime, slow, deep mixing occurs. When constrained by the solar lithium abundance, this results in a depletion in beryllium by a factor of two. The lack of beryllium depletion in the solar photosphere, *not just in the present-day Sun, but during its entire lifetime* provides a very strong constraint on these non-standard stellar models. Helioseismic models of the present-day Sun indicate the presence of only a thin shear layer (the tachocline) at the base of the convective zone. Both the internal gravity wave model (Brun, Turck-Chièze, & Zahn 1999) and the magnetic transport model (Barnes, Charbonneau & MacGregor 1999) are able to reproduce the combination of lithium depletion by a factor of 140 and no beryllium depletion by invoking turbulence at the solar tachocline. Barnes et al. (1999) particularly point out that hydrodynamic transport of angular momentum results in both lithium and beryllium depletion while hydromagnetic transport reproduces the helioseismological data and lithium depletion while producing no beryllium depletion.

It appears, therefore, that the two strong bases for the transport of angular momentum and mixing via shear turbulence are no longer supported.

Modest beryllium depletion has been cited in three Hyades G dwarfs (García López et al. 1995) with beryllium decreasing by 0.2 dex between 5600 and 5200 K. Given that the contribution of the Fe I b-f opacity increases with decreasing temperature, it is possible that this estimate may be due to the non-inclusion of sufficient Fe I bound-free opacity. Clearly additional calculations of the opacity contribution at different temperatures and a re-estimate of the beryllium abundance with its inclusion are crucial to determine if any depletion does occur in the G dwarfs.

The only dwarfs in which a real depletion of beryllium has been clearly measured are the F stars in the lithium 'dip' (Stephens et al. 1997; Deliyannis et al. 1998). It is clear that lithium is destroyed and not merely diffused out of sight in these stars (Balachandran 1995). Given the very shallow convective envelope and the relatively large lithium-preservation zones in these stars, the anomalous depletion of lithium requires considerable mixing below the surface convective zone. It is perhaps not surprising that such vigorous mixing is deep enough to result in beryllium depletion as well. The process by which the lithium-dip is produced is not understood, but the mixing mechanism may be completely disconnected from the process that causes lithium depletion in solar-type stars.

4. References

Balachandran, S. 1995, ApJ, 446, 203
Balachandran, S. C. & Bell, R. A. 1998, Nature, 392, 791
Barnes, G., Charbonneau, P., & MacGregor, K. B. 1999, ApJ, 511, 466
Bautista, M. 1997, A&AS, 122, 167
Bell, R. A., Balachandran, S. C. & Bautista, M. 2000, ApJL, submitted
Bell, R. A., Paltoglou, G., & Tripicco, M. J. 1994, MNRAS, 268, 771
Boesgaard, A. M., Deliyannis, C. P., Stephens, A., & Lambert, D. L. 1998, ApJ, 492, 727
Brun, A. S., Turck-Chièze, S., & Zahn, J.-P. 1999, ApJ, 525, 1032
Charbonneau, P. & MacGregor, K. B. 1993, ApJ, 417, 762
Charbonneau, P., Tomczyk, S., Schou, J., & Thompson, M. J. 1998, ApJ, 496, 1015
Charbonnel, C., Vauclair, S., Maeder, A., Meynet, G., & Schaller, G. 1994, A&A, 283, 155
Chmielewski, Y., Brault, J. W., & Müller, E. A. 1975, A&A, 42, 37
Cunha, K., Smith, V. V., Boesgaard, A. M., & Lambert, D. L. 2000, ApJ, 530, 939
Deliyannis, C. P., Boesgaard, A. M., Stephens, A., King, J. R., Vogt, S. S., & Keane, M. J. 1998, ApJ, 498, L147
Dragon, J. N., & Mutschlecner, J. P., 1980, ApJ, 239, 1045
Edvardsson, B. et al. 1993, A&A, 275, 101
García López, R. J., Rebolo, R., & Perez de Taoro, M. R. 1995, A&A, 302, 184
Grevesse, N., Sauval, A. J., & van Dischoeck, E. F. 1984, A&A, 141, 10
Gustafsson, B., Bell, R. A., Eriksson, K., & Nordlund, Å., 42, 407
Holweger, H. 1970, A&A, 4, 11
Holweger, H., & Müller, E. A. 1974, Solar Phys. 39, 19
King, J. R., Deliyannis, C. P., & Boesgaard, A., M., 1997, ApJ, 478, 778
Kumar, P., & Quataert, E. J. 1997, ApJ, 473, L143
Kumar, P., Talon, A., & Zahn, J.-P. 1999, ApJ, 520, 859
Kurucz, R. L. 1992, Rev. Mex. de Astromia y Astrofisica, 23, 181
Kurucz, R. L., 1993, CD Rom No. 13, SAO
Kurucz, R. L., Furenlid, I., Brault, J., & Testerman, L. 1984, Solar Flux Atlas from 296 to 1200 nanometers (Natl. Solar Observatory, Tucson)
Pinsonneault, M. H., Kawaler, S. D., Sofia, S., & Demarque, P. 1989, ApJ, 338, 424
Pinsonneault, M. H., Kawaler, S. D., & Demarque, P. 1990, ApJS, 74, 501
Primas, F., Duncan, D. K., Pinsonneault, M. H., Deliyannis, C. P., & Thorburn, J. A. 1997, ApJ, 480, 784
Stauffer, J. R. 1991, in Angular Momentum Evolution of Young Stars, eds. S. Catalano & J. R. Stauffer, (Dordrecht: Kluwer), p. 117
Stephens, A., Boesgaard, A. M., King, J. R., & Deliyannis, C. P. 1997, ApJ, 491, 339

Tomczyk, S., Schou, J., & Thompson, M. J. 1995, ApJ, 448, L57
Woods, T. N. et al. 1996, JGR, D6, 9541
Zahn, J.-P., Talon, S., & Matias, J. 1997, A&A, 322, 320

The Galactic Evolution of Beryllium

Ann Merchant Boesgaard

University of Hawaii, Institute for Astronomy, 2680 Woodlawn Drive, Honolulu, Hi 96822 U.S.A.

Abstract. The abundance of beryllium has been determined in unevolved stars over a range metal abundances in order to enhance our understanding of the chemical evolution of our Galaxy, cosmic-ray theory, and cosmology. Observations of 27 stars have been made with Keck I with HIRES at high spectral resolution (45,000) and high signal-to-noise ratios (60 - 110 typically). We find a remarkably linear relationship between log N(Be/H) and [Fe/H] with a slope of 0.96 (±0.04). Similarly, the relationship between log N(Be/H) and [O/H] is linear with a slope of 1.45 (±0.04). Beryllium increases at the same rate as Fe, but much faster than O. This provides constraints for and insights into models of Galactic chemical evolution. There is some evidence for an intrinsic spread in Be at a given [O/H] or [Fe/H]. There is no evidence of a plateau in Be at the lowest metallicities down to log N(Be/H) = -13.5.

1. Introduction

Unlike the heavier elements which are formed by nuclear fusion reactions in the interiors of stars at high–temperatures, Be is destroyed in stellar interiors. Instead of being formed by fusion of lower mass nuclei, it is widely accepted that Be is created by the breaking up of heavier nuclei in spallation reactions. There are interactions of high energy cosmic rays (H&He$_{CR}$) with abundant elements like C, N, O in the interstellar gas (CNO$_{ISM}$) as first proposed by Reeves, Fowler, & Hoyle (1970); whether these interactions take place in the immediate vicinity of supernovae or in the general interstellar gas is not completely clear. In the early Galaxy, there was little CNO$_{ISM}$ but plenty of CNO$_{CR}$ and this reaction was of greater *relative* importance then than now (Yoshii, Kajino & Ryan 1997). The evolution of the abundance of Be should reflect the production of CNO and cosmic rays and thus mirror the production of massive stars (which contribute the O atoms) and supernovae.

The determination of the amount of Be in the most metal-poor stars (expected to be the oldest stars), may reveal if any Be has been synthesized in the Big Bang. The standard model of Big Bang nucleosynthesis (BBN) produces only a little Be: (N(Be)/N(H) = 10^{-17}). But models that include inhomogeneities in the early universe could produce a different mixture of light elements than that in the standard models (see the review by Malaney & Mathews 1992).

Several papers have been published on Be abundances in halo stars including Rebolo et al. (1988), Ryan et al. (1991), Gilmore, Edvardsson, & Nissen (1991),

Ryan et al. (1992), Gilmore et al. (1992), Boesgaard & King (1993), Molaro et al. (1997), Boesgaard et al. (1999a). The relationship between log N(Be/H) and [Fe/H] appears to have a slope of 1, indicating that the environment for the spallation reactions is the vicinity of supernovae (SNII) with freshly minted CNO nuclei rather than the ambient interstellar medium.

We have observed stars with [Fe/H] values down to -3.0 to search for a Be plateau which might indicate a BBN component of Be. In the most metal-poor stars the Be II lines are expected to be very weak so spectra with high resolution and high signal-to-noise (S/N) were needed. We have observed Be in stars with an array of metallicities to investigate the Galactic evolution of Be. To this end we have also determined O abundances in these stars (Boesgaard et al. 1999b) and thus have a consistently determined data set of Be and O abundances in halo stars. The Be abundances could be determined to a precision of ± 0.1 dex.

2. Observations

The Be II resonance lines are at 3130.420 and 3131.065 Å in the ultraviolet (UV) spectral region accessible from the ground. For our Be observations we used the HIRES spectrometer (Vogt et al. 1994) on the Keck I telescope. To maximize the UV flux we have made observations as close as possible to the lowest air mass for each star. The resolution of the spectra is $\sim$45,000 (FWHM = 3 pixels) with an effective dispersion of 0.022 Å pix^{-1}. The signal-to-noise ratios were typically 60 - 110 per pixel.

We have spectra of 22 metal-poor stars. We also took exposures of the daytime sky (as a solar spectrum) and of three disk stars for comparison with the halo stars. For our faintest and most metal-poor star, BD -13 3442, our total integration time was 11 hours for a S/N of 130.

3. Data Analysis

Standard echelle reductions were carried out within the IRAF routines. In order to co-add the spectra from different times of night or from different nights/runs the cross-correlation techniques in IRAF were used to determine the spectral shifts. The continua were fit quite easily for the lowest metallicity stars since the blending features in this crowded spectral region are so weak. The spectra of these stars could be used to help identify the continuum high points in the stars with progressively stronger atomic and molecular features.

We invested much effort in the determination of the stellar parameters so they are as self-consistent as possible. The temperature indicators used are (b-y), (V-K), and (R-I). We applied corrections for interstellar reddening where needed. The values for [Fe/H] were taken from high spectral resolution determinations in the literature. These were 1) corrected to a solar value of log N(Fe/H) + 12.0 = 7.51 (Anders & Grevesse 1989), and 2) corrected for the temperature difference between the published values and our scale. One of the methods we employed to find gravities was to use published detailed analyses which used ionization balance to find log g.

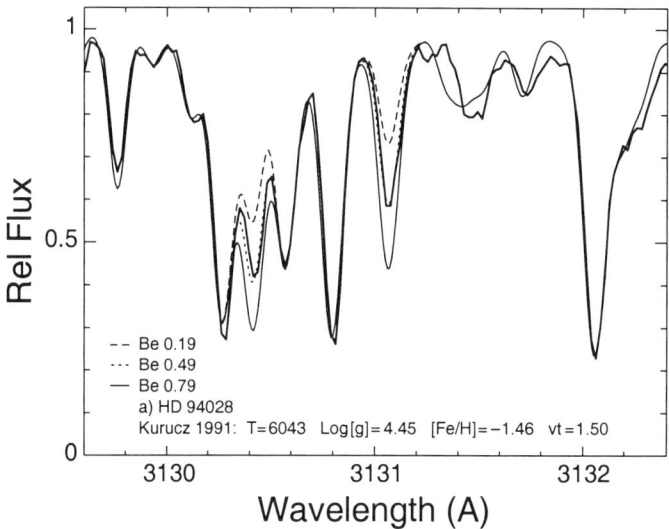

Figure 1. An example of the spectrum synthesis fit for HD 94028. They heavy solid line is the observed spectrum. The three syntheses are different from each other by factors of 2. Log N(Be/H) = −11.81: long dash; −11.51: dots; −11.21: light solid line. The best fit is −11.51.

4. Abundance Determinations

We have determined abundances from both spectrum syntheses and from the measured equivalent widths of Be II lines. In each method the results from the weaker, less blended line at 3131 Å are to be preferred. Kurucz (1993) model atmospheres have been used throughout.

For the equivalent width method we treated the λ3131.065 Be II line as a blend with 11 other atomic and molecular lines between 3131.015 and 3131.116 Å from the line list of King, Deliyannis & Boesgaard (1997) as weak possible blends. The spectrum synthesis calculations have been modified to include the effect of enhanced O. Our study of O abundances in these same stars (Boesgaard et al. 1999b) has enabled us to specify the O abundance. The agreement between the abundances determined by the equivalent width method and the spectrum synthesis method is very good. Figure 1 shows an example of the spectrum synthesis of HD 94028. The fit is excellent. This is significant because this star lies above the best fit in the Be vs Fe and Be vs O diagrams; it may indicate that there is an intrinsic spread in Be at a given Fe.

5. Results

The key results of this work are the trends of Be with Fe and with O. Figure 2 shows the relationship between log N(Be/H) +12.00 and [Fe/H]. The line

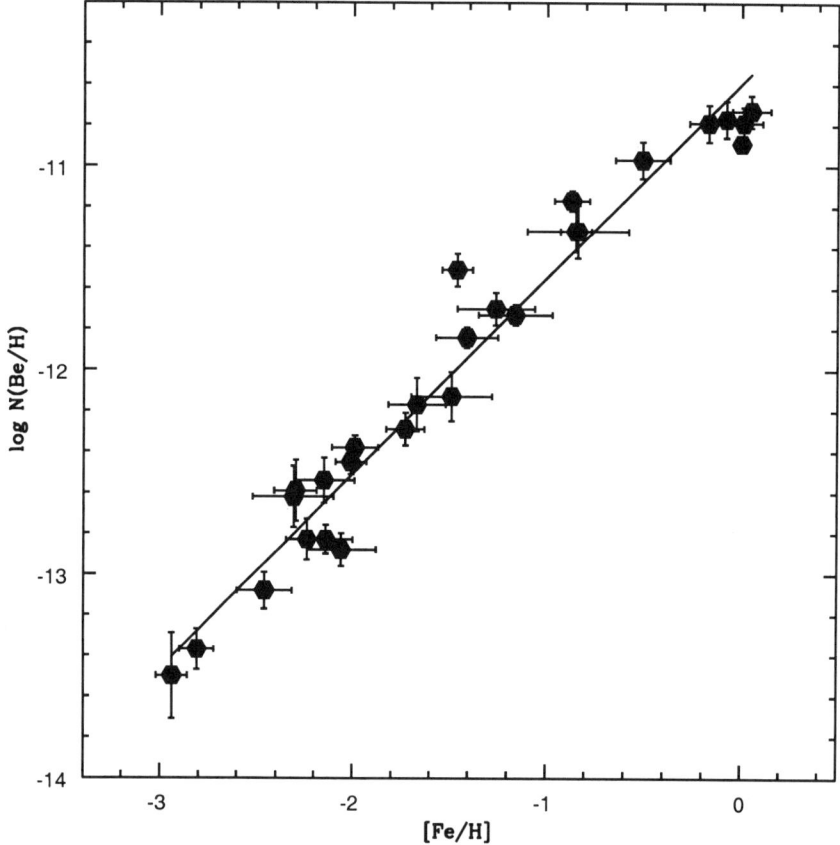

Figure 2. Beryllium abundances as log N(Be/H) vs [Fe/H] for the Keck observations. The line is the fit taking the error bars in both quantities into account.

through the points is a least squares fit which takes the errors in both coordinates into account. The equation for the line is:

log N(Be/H) = 0.96 (±0.04) [Fe/H] −10.59 (±0.03).

The slope is ∼1. As the Galaxy ages, Be and Fe increase together over three orders of magnitude in [Fe/H]. This is rather remarkable since they are formed by such different nucleosynthetic processes. Note that there is no indication of a plateau in the low metallicity, low Be corner of this plot. That is, there is no evidence for Big Bang-produced Be.

We know that there is some intrinsic spread in Be at the metallicities of solar-type stars (Boesgaard & King 1993, Boesgaard et al. 1999a) and we may be seeing some evidence of that here at lower metallicities, e.g. at [Fe/H] ∼−1.5 and ∼−2.2. One star stands out above the line at −1.46, −11.51; this is HD 94028 and, as mentioned above, the spectrum synthesis is excellent (see Figure 1). When this star is compared to HD 219617 at −1.49, −12.15, which has the

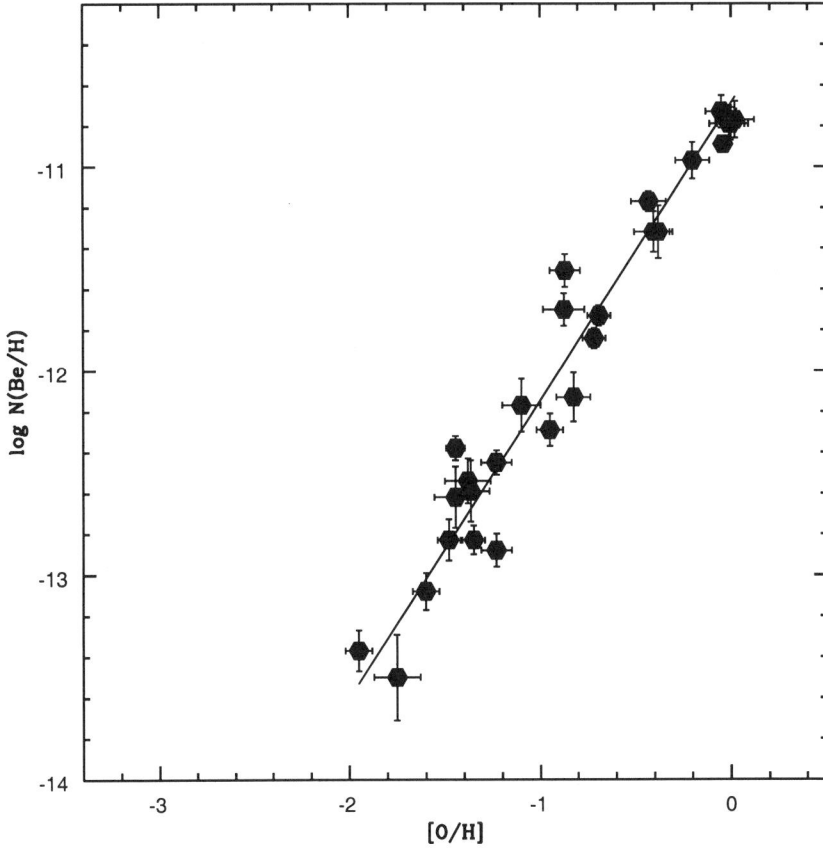

Figure 3. Beryllium abundances as log N(Be/H) vs [O/H] for the Keck observations. The line is the fit taking the error bars in both quantities into account.

same metallicity, temperature and log g, the difference in Be is large: a factor of 4.4, well beyond the errors.

Since Be is produced by spallation with CNO nuclei, it is important to examine the relationship between Be and O. We have determined O abundances from these same spectra and stellar parameters (Boesgaard et al. 1999b). Figure 3 shows the trend of Be with O; the line through the data points takes into account the errors in both quantities. Again, these are straight-line correlations, represented by

$$\log N(Be/H) = 1.46 \ (\pm 0.04) \ [O/H] \ -10.69 \ (\pm 0.04).$$

The relationship between Be and O is well-represented by a single straight line with a slope of 1.45. As O increases 100-fold, Be increases by 800 times.

We have made additional observations of Be in metal-poor stars with the 4-m at CTIO and with the Canada-France-Hawaii 3.6-m telescope (Deliyannis et al. 1995). The resolutions were 22,400 and 24,000 respectively, about half the

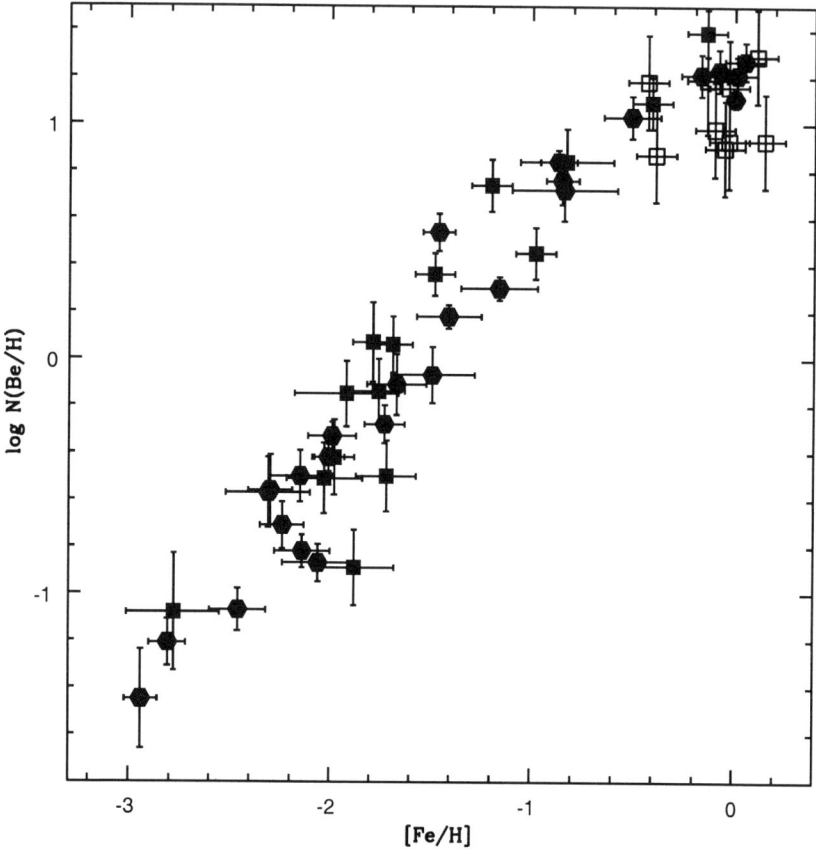

Figure 4. Beryllium abundances as log N(Be/H) vs [Fe/H] for the Keck observations are solid hexagons, for the CTIO and CFHT are solid squares, for the solar-type stars are open squares.

resolution of the Keck spectra. The signal-to-noise ratios in these spectra was 40-80. If we include this sample of stars with the Keck stars we add 22 more stars. In addition, we can add a few more solar type stars from our paper about correlated depletions of Li and Be (Deliyannis et al. 1998). The results from all these data are shown in Figure 4.

If Be is produced by cosmic ray protons breaking up CNO nuclei in the ambient interstellar medium, then, in the extreme case, the slope of the relationship between Be and O would be 2. This quadratic relationship results from the fact that the number of O atoms is proportional to the cumulative number of SNIIs (N) while the number of energetic proton cosmic rays is proportional to the instantaneous rate of SNIIs (dN). The abundance of the spallation products is therefore the integral of the product $N\,dN = kN^2$. Thus the spallation product (Be) vs O has a slope of two. However, chemical evolution effects, such as an outflow of mass from the halo, indicate that there would be a quadratic relation only at the very lowest metallicities (lower than the present sample), followed by a progressive shallowing of the slope to disk metallicities. From [Fe/H] between -1 and -2, the slope is close to 1.5. However, the slope would be 1 if Be is produced in the immediate vicinity of the SN IIs (the source of both the parent CNO atoms and the cosmic rays). The reaction CNO_{CR} + H&He$_{ISM}$ also produces a linear slope, the cosmic ray CNO being primary (Yoshii et al. 1997). Our slope for Be vs O is 1.5. This is consistent with the traditional mass outflow model where energetic protons and alpha particles impinge upon ambient CNO in the ISM, but less consistent with either of the linear slope scenarios just described.

We found no evidence of a Be plateau at abundances down to log N(Be/H) = -13.5. Therefore any primordial Be predicted by non-standard BBN must be below that value.

6. Summary

We obtained high-resolution and high S/N ratio spectra of 27 stars over a range of [Fe/H] of -3.0 to $+0.1$ with the Keck I telescope and HIRES in the blue and ultraviolet spectral region. We derived Be abundances in these stars from the Be II resonance lines by two methods which are in excellent agreement.

The special features of this work include 1) high spectral resolution, 2) high signal-to-noise ratios, 3) a self-consistent parameter set, 4) enhanced O/H in the spectrum synthesis. The results of the analysis show that Be and Fe increase together. The slope in log N(Be/H) vs [Fe/H] is 0.96 ± 0.04, so Be and Fe increase at the same rate. We find that Be increase faster than O. The slope between log N(Be/H) and [O/H] is 1.45 ± 0.04. As the abundant element, O, increases by 100 times, the trace element Be increases by 800 times.

Energetic cosmic rays interact (H&He$_{CR}$) with CNO nuclei to produce Be. Models with chemical evolution effects (e.g. mass outflow from the halo) predict a slope between log N(Be/H) and [O/H] of 1.5.

There is some evidence of an intrinsic spread in Be at a given [Fe/H] or [O/H]. This could result from different levels of production of Be in the interstellar medium in different sites. The stars formed in these different locales would have different Be contents, yet examples of each would be in the solar neighborhood now.

We found no evidence of even the beginning of a Be plateau at low values of [Fe/H]. This does not rule out an inhomogeneous early universe, however. The new large telescopes with sensitive UV spectrometers and detectors, should expand the search for this possible plateau. How much Be will we find at [Fe/H] = -4.0?

References

Anders, E., & Grevesse, N. 1989, Geochim.Cosmochim.Acta, 53, 197

Boesgaard, A. M. & King, J. R. 1993, AJ, 106, 2309

Boesgaard, A. M., Deliyannis, C. P., King, J. K., Ryan, S. G., Vogt, S. S. & Beers, T. C. 1999a, AJ, 117, 1549

Boesgaard, A. M., King, J. K., Deliyannis, C. P. & Vogt, S. S. 1999b, AJ, 117,492
Deliyannis, C. P., Boesgaard, A. M., King, J. R., & Duncan, D. 1995, *Ninth Cambridge Workshop on Cool Stars* eds. R. Pallavicini & A. Dupree, ASPCS, **109**, 679
Deliyannis, C. P., Boesgaard, A. M., Stephens, A., King, J. K., Vogt, S. S. & Keane, M. 1998, ApJ, 498, L147
King, J. R., Deliyannis, C. P., & Boesgaard, A. M. 1997, ApJ, 478, 778
Gilmore, G., Edvardsson, B., & Nissen, P. E. 1991, ApJ, 378, 17
Gilmore, G., Gustafsson, B., Edvardsson, B., & Nissen, P. E. 1992, Nature, 357, 379
Kurucz, R. L. 1993, private communication
Malaney, R. A. & Mathews, G. J. 1992, Phys. Rep., 229, 147
Molaro, P., Bonifacio, P., Castelli, F., and Pasquini, L. 1997, A&A, 319, 593
Rebolo, R., Molaro, P., and Beckman, J. E. 1988, A&A, 192, 192
Reeves, H., Fowler, W. A., Hoyle, F. 1970, Nature, 226, 727
Ryan, S. G., Norris, J. E. & Bessell, M. S. 1991, AJ, 102, 303
Ryan, S. G., Norris, J. E., Bessell, M. S., & Deliyannis, C. P. 1992, ApJ, 388, 184
Vogt, S. S. et al. 1994, Proc. SPIE, 2198, 362
Yoshii, Y., Kajino, T., & Ryan, S. G. 1997, ApJ, 485, 605

Galactic Evolution of Beryllium and Oxygen

Garik Israelian[1,2], Ramón J. García López[1,2] and Rafael Rebolo[1,3]

1.- *Instituto de Astrofísica de Canarias, E-38200 La Laguna, Tenerife, Spain*
2.- *Departamento de Astrofísica, Universidad de La Laguna, Av. Astrofísico Francisco Sánchez s/n, E-38071 La Laguna, Tenerife, Spain*
3.- *Consejo Superior de Investigaciones Científicas, Spain*

Abstract. We discuss the early evolution of beryllium and oxygen in our Galaxy by comparing abundances of these elements for halo and disk metal-poor stars. Both, O and Be rise as we go progressively to more metal-rich stars, showing a slope 0.41 ± 0.09 ([Be/O] vs [Fe/H]) for stars with [Fe/H]≤ -1. This relationship provides an observational constraint to the actually proposed Galactic Cosmic Ray theories.

1. Introduction

First attempts to measure beryllium abundances in metal-poor stars by Molaro & Beckman (1984) and Molaro, Beckman & Castelli (1984) demonstrated that stars in the early Galaxy formed with much lower Be abundances than in the present epoch. First detection of Be in metal-poor stars was achieved by Rebolo et al. (1988) and further studies by S. Ryan, G. Gilmore, A. Boesgaard, P. Molaro, R. J. García López and their respective collaborators revealed a clear linear correlation with iron.

Accelerated protons and α-particles in cosmic rays interact with ambient CNO in ISM and create Be. According to the standard Galactic Cosmic Ray (GCR) theory, these interactions in the general ISM should have given a quadratic relation between Be and O. Alternatively, spallation of cosmic ray CNO nuclei accelerated out of freshly processed material could account for the primary character of the observed early galactic evolution of Be. Another production site is the collective acceleration by SN shocks of ejecta-enriched matter in the interiors of superbubbles. In these two cases, the evolution of Be should reflect the production of CNO from massive stars. Oxygen is mostly produced by Type II SNe while iron is produced in both, Type II and in Type Ia SNe. The fact that Type Ia SNe have longer lifetime progenitors has been commonly used to argue that oxygen must be overabundant in very old stars. Observational evidence for high [O/Fe] ratios in many metal-poor stars has been reported over the last two decades.

Based on the study of [O I] lines at 6300 and 6363 Å in evolved stars (though the second line at 6363 Å is not visible in very metal-poor stars and the analysis is based *only on one line*), several authors have found that [O/Fe]= $0.3-0.4$ dex at [Fe/H]< -1 and is constant until [Fe/H]~ -3 (e.g. Barbuy 1988 and Kraft et al. 1992). In contrast with this result, oxygen abundances derived in unevolved stars using the O I IR triplet at 7774 Å (Abia & Rebolo 1989; Tomkin et al. 1992; King & Boesgaard 1995; and Cavallo, Pilachowski, & Rebolo 1997) point towards linearly increasing [O/Fe] values with decreasing [Fe/H] and reaching a ratio ~ 1 for stars with [Fe/H]~ -3. This may suggest a higher production of oxygen during the early Galaxy.

We discuss in this paper the comparison of these abundances with those derived from OH lines located in the near-UV part of the spectra of metal-poor stars, and their relation with beryllium abundances consistently derived from the same spectra.

2. Observations and Analyses

The observations were carried out in different runs using the UES ($R = \lambda/\Delta\lambda \sim$ 50000) of the 4.2-m WHT at the Observatorio del Roque de los Muchachos (La Palma), and the UCLES ($R \sim 60000$) of the 3.9-m AAT. The spectral region observed spanned typically from 3080 to 3300 Å, where several OH lines and the Be II doublet at 3131 Å are located. Details of the abundance analyses can be found in García López, Severino, & Gomez (1995) and Israelian, García López, & Rebolo (1998) for beryllium and oxygen, respectively. The stellar parameters play an important role in the abundance determinations from near-UV lines. Effective temperatures (T_{eff}) for our stars were estimated using the Alonso et al. (1996) calibrations versus $V - K$ and $b - y$ colors, which were derived by applying the infrared flux method (IRFM), and cover a wide range of spectral types and metal content. These temperatures were used to compute synthetic spectra around the Be II doublet and slightly modified, within the error bars provided by the calibrations, until obtaining a good reproduction of this region in the observed spectra. Metallicities were adopted from literature values obtained from high resolution spectra. Adopted gravities, derived using the accurate parallaxes measured by *Hipparcos* (ESA 1997), are larger by 0.28 dex in average than the values adopted by Israelian et al. (1998). This implies a mean small reduction of 0.09 dex in the oxygen abundances inferred from the OH lines with respect to that work, which does not affect significantly their original results.

Beryllium abundances reported here were obtained from the Be II resonance doublet located at 3130.421 and 3131.065 Å. The first line, which is also the strongest one, is severely blended with atomic and molecular lines of other species, and the abundance determination usually relies only on the other line, more isolated and weaker. The abundances used in this work are those presented in García López (1999).

3. Oxygen

Israelian et al. (1998) presented new oxygen abundances derived from near-UV OH lines (which form in the same layers of the atmosphere as [O I]) for 24 metal-poor stars. They have concluded that the [O/Fe] ratio of metal-poor stars increases from 0.6 to 1 between [Fe/H]=−1.5 and −3, with a slope of −0.31±0.11. Contrary to the previously accepted picture (see e.g. Bessell, Sutherland, &

Ruan 1991, who used older model atmospheres with a coarser treatment of the opacities in the UV), these new oxygen abundances derived from low-excitation OH lines, agreed well with those derived from high-excitation lines of the O I IR triplet at 7774 Å. The comparison with oxygen abundances derived using O I data from Tomkin et al. (1992) showed a mean difference of 0.00 ± 0.11 dex for the stars in common. Boesgaard et al. (1999a) made a similar analysis of several metal-poor stars using a different set of OH lines. They found a very good agreement with the results obtained by Israelian et al. (1998), and basically the same dependence of [O/Fe] versus metallicity. This is clearly seen in the upper panel of Figure 1.

The UV "missing opacity" problem discussed by Balachandran & Bell (1998), which could affect both oxygen and beryllium abundance determinations from these lines, has been studied recently by Allende Prieto & Lambert (2000). These authors have found a good agreement between T_{eff}s obtained from the IRFM and from the near-UV continuum for stars with $4000 \leq T_{eff} \leq 6000$ K when accurate *Hipparcos* gravities are used. This also agrees with our good reproduction of the near-UV spectral region using the IRFM temperatures. This result indicates that the model atmospheres used provide an adequate description of the near-UV continuum forming region. In any case, even if a not well understood opacity problem would exist as described by Balachandran & Bell, it would have a minor effect on the OH results since most of the stars in the samples of Israelian et al. and Boesgaard et al. are hotter than the Sun and very metal-poor. The corrections to oxygen abundances for individual stars would be lower than 0.15 dex, not changing significantly the [O/Fe] vs. [Fe/H] trend.

A new non-LTE analysis of the O I IR triplet for a sample of 38 metal-poor stars performed by Mishenina et al. (2000) gives consistent results with those of Abia & Rebolo (1989), Tomkin et al. (1992) and Kiselman (1993), and indicates that the mean value of the non-LTE correction in unevolved metal-poor stars is typically 0.1-0.2 dex. These authors confirmed the [O/Fe] vs [Fe/H] trend discussed by Israelian et al. (1998) and Boesgaard et al. (1999a) from the OH lines, without finding any trend of oxygen abundances with T_{eff} or $\log g$. It is also worthwhile to mention that the O I IR triplet is not affected by 3D effects, convection and small-scale inhomogeneities in the stellar atmosphere (Asplund et al. 1999). In addition, oxygen abundances derived form this triplet are not significantly affected by chromospheric activity either. The central panel of Fig. 1 shows a compilation of oxygen abundances derived using the IR triplet. The larger scatter observed as compared with the measurements based on OH lines can be associated with the different scales of stellar parameters (T_{eff}, gravities, and metallicities) adopted by the authors of each set of stars, and to the fact that some measurements have not been corrected for non-LTE effects. Very recently, Carretta, Gratton, & Sneden (2000) performed an independent analysis of 32 metal-poor stars hotter than 4600 K using the IR triplet, and provide LTE and non-LTE oxygen abundances which are significantly lower than those found in previous works. A preliminary attempt to understand the reasons for this discrepancy can be done by looking in detail into their most metal-poor star (BD +3°740, [Fe/H]= -2.66) where a surprisingly low oxygen abundance [O/Fe]= 0.38 is claimed. A recent study of stellar parameters based on the non-LTE analysis of iron lines (Thévenin & Idiart 1999), gives a lower effective temperature (by 140 K) and a higher gravity (by 0.3 dex) than the

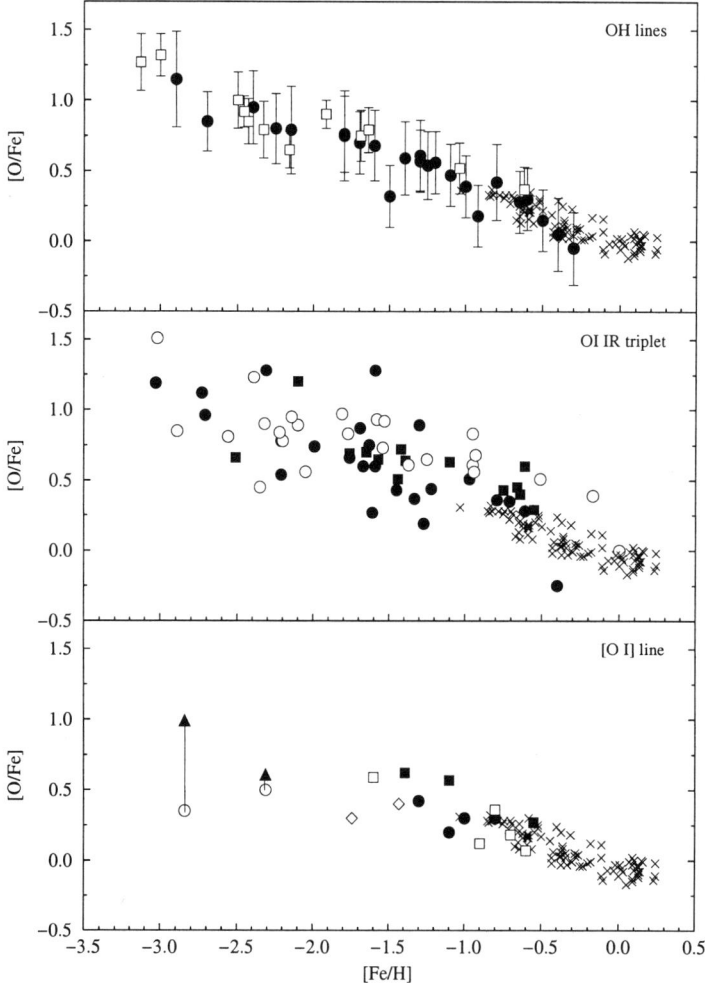

Figure 1. [O/Fe] vs. [Fe/H] for unevolved stars. Abundances from OH lines were derived by Israelian et al. (1998; filled circles) and Boesgaard et al. (1999a; open squares, corrected to the scale of stellar parameters adopted by Israelian et al.). Abundances from the IR triplet were derived in NLTE by Mishenina et al. (2000; filled circles), Cavallo et al. (1997; filled squares, corrected for NLTE effects by Mishenina et al.), and in LTE by Boesgaard et al. (1999; open circles). Finally, abundances from the [O I] line come from Spiesman & Wallerstein (1991; open diamonds), Spite & Spite (1991; open squares), Israelian et al. (1998; filled circles), Mishenina et al. (2000; filled squares), and Fulbright & Kraft (1999; open circles). Filled triangles indicate the change in abundances associated with the change in gravity according to the *Hipparcos* parallaxes for the two stars studied by Fulbright & Kraft. The abundances derived by Edvardsson et al. (1993; crosses) are shown in the three plots to indicate the trend in metal-rich stars.

values adopted by Carretta et al. for this star. Using these latter parameters we obtain an LTE oxygen abundance 0.4 dex higher than Carretta et al. (i.e. [O/Fe]$_{\rm LTE}$ = 1.05), and for a star with these parameters the non-LTE correction to the oxygen abundance is of the order of 0.05 dex (Mishenina et al. 2000), much lower than the 0.25 dex value used by Carretta et al. We therefore arrive at a value [O/Fe]$_{\rm NLTE}$ ~ 1.0, in good agreement with the OH determination by Boesgaard et al. (1999a). Corrections to the stellar parameters as inferred from the non-LTE analysis of Fe lines clearly have an impact on the oxygen abundances which we will address in a forthcoming paper.

Israelian et al. (1998) found four dwarfs in their sample for which oxygen abundances derived using [O I] were in good agreement with those derived from OH when *Hipparcos* gravities are used. Several oxygen measurements for unevolved stars based on the [O I] 6300 Å line are compiled in the lower panel of Fig. 1. This figure shows a similar trend than that observed for the abundances recently derived from forbidden lines by Carretta et al. (2000). The presence of a linear trend of [O/Fe] versus metallicity in Fig. 1 strongly depends on the only two measurements available at [Fe/H]≤ −2. These two measurements have been reported by Fulbright & Kraft (1999) for the subgiants BD +37°1458 and BD +23°3130, which were also considered by Israelian et al. (1998) and Boesgaard et al. (1999a; only BD +37°1458 in this case). The analysis carried out by Fulbright & Kraft is based on gravities derived from LTE iron ionization balance of these subgiants where it is well known that non-LTE effects are strong (Thévenin & Idiart 1999; see also Idiart & Thévenin, this conference). Allende Prieto et al. (1999) have shown that gravities derived using this technique in metal-poor stars do not agree with the gravities inferred from accurate *Hipparcos* parallaxes. They find that gravities are systematically underestimated when derived from ionization balances and that upward corrections of ~ 0.5 dex can be required at metallicities similar to those of our stars, in good agreement with Thévenin & Idiart. We remark here that any underestimation of gravities will also strongly underestimate the abundances inferred from the forbidden line. For the two stars under discussion our *Hipparcos* based gravities are 0.45 and 1.05 dex (for BD +37°1458 and BD +23°3130, respectively) higher than derived by Fulbright & Kraft, and would imply corrections in the oxygen abundances similar to those indicated in Fig. 1 (a detailed analysis would imply also the correction for the assumed metallicities). Our conclusion is that the uncertainties in the gravities of these subgiants allow the abundances inferred from the forbidden line to be consistent with those estimated from the OH lines or the triplet. Actually, consistency with the other oxygen indicators is achieved for the high gravities inferred from *Hipparcos* when consistent analyses are made, and this could be taken as an indication that the high gravities are indeed the correct ones.

4. Beryllium and oxygen

The dependence of log(Be/H) on [Fe/H] and on [O/H] (using the abundances derived from the OH UV lines) is essentially linear (García López 1999; Boesgaard et al. 1999b), but with different slopes: ~ 1.1 and ~ 1.5, respectively. No evidence of a primordial plateau of Be down to log(Be/H)=−13.5 is found.

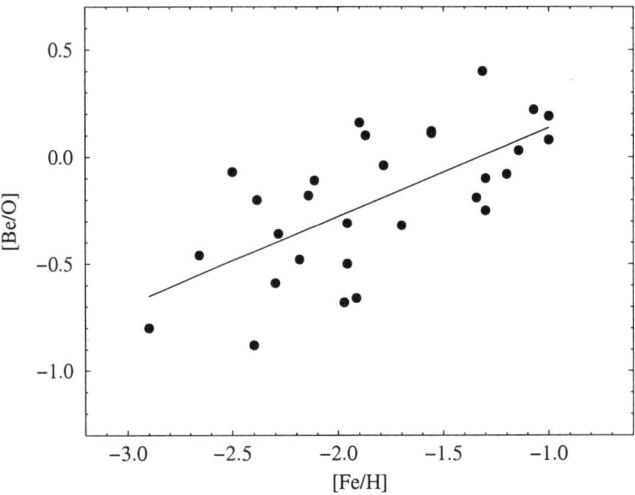

Figure 2. [Be/O] vs. [Fe/H] for unevolved metal-poor stars. Oxygen abundances used were derived from OH UV lines. The slope observed in the figure, 0.41 ± 0.09, provides an observational constraint to the GCR models.

Figure 2 shows the increase of the [Be/O] ratio with increasing metallicity and a slope of ∼ 0.4. This relation provides an observational constraint to the Galactic Cosmic Ray theories. Three types of GCR models exist at present which try to explain their observed evolution. These are 1) a pure primary GCR from superbubbles (Ramaty, this conference), 2) a hybrid model based on GCR and superbubble accelerated particles (Cassé, this conference), which could be accomplished by a pure superbubble model (Parizot & Drury, this conference), and 3) standard GCR (Olive, this conference). Apparently all these models can be adopted for both, variable and flat [O/Fe]. However, models presented by R. Ramaty and K. Olive show more consistency when variable [O/Fe] is adopted.

Chemical evolution models of the early Galaxy where stellar lifetimes are taken into account and assuming that Type Ia SN appear at a Galactic age of 30 million years can also explain the evolution of oxygen delineated in Fig. 1. (Chiappini et al. 1999.). The evolution of oxygen proposed in this paper also helps to understand the evolution of ^{6}Li versus [Fe/H] and the ^{6}Li/Be ratio at low metallicities in the framework of standard Galactic Cosmic Ray Nucleosynthesis (Fields & Olive 1999). In addition, Ramaty et al. (1999) have proposed that a delay between the effective deposition times into the ISM of Fe and O (only a fraction of which condensed in oxide grains) can explain a linear trend of [O/Fe].

It has been suggested (Vangioni-Flam & Cassé, this conference) to use magnesium as metallicity indicator instead of oxygen. However, given the existence of unevolved halo stars with negative [Mg/Fe] ratios (McWilliam 1997; Carney et al. 1997), this approach may not lead to better results. For example, the subdwarf BD+3°740 has [O/Fe]∼ 1 (see previous Section) while its [Mg/Fe]= −0.28 (Fuhrmann et al. 1995). Yield of Mg depends on the extent of mixing (Argast et al. 2000) and its primordial abundance can be changed due to the operation of the MgAl cycle.

References

Abia, C., & Rebolo, R. 1989, ApJ, 347, 186
Allende Prieto, C., García López, R. J., Lambert, D. L., & Gustafsson, B. 1999, ApJ, 527, 879
Allende Prieto, C., & Lambert, D.L. 2000, AJ, in press
Alonso, A., Arribas, S., & Martínez-Roger, C. 1996, A&AS, 313, 873
Argast, D., Samland, M., Gerhard, O. and Thielemann, F. 2000, A&A, in press
Asplund, M., Nordlund, Å, Trampedach, R., & Stein, R. 1999, A&A, 346, L17
Balachandran, S., & Bell, R. 1998, Nature, 392, 791
Barbuy, B. 1988, A&A, 191, 121
Bessell, M. S., Sutherland, R. S., & Ruan, K. 1991, ApJ, 383, L71
Boesgaard, A.M., Deliyannis, C. P., King, J.R., Ryan, S.G., Vogt, S.S. & Beers, T. 1999b, AJ, 117, 1549
Boesgaard, A.M., King, J.R., Deliyannis, C. P., & Vogt, S.S. 1999a, AJ, 117, 492
Carney, B., Wright, J., Sneden, C., Laird, J., Aguilar, L. & Latham, D. 1997, AJ, 114, 363
Carretta, E., Gratton, R. G., & Sneden, C. 2000, A&A, in press
Cavallo, R., Pilachowski, C., & Rebolo, R. 1997, PASP, 109, 226
Chiappini, C., Matteucci, F., Beers, T.C., & Nomoto, K. 1999, ApJ, 515, 226
Edvardsson, B., Andersen, J., Gustafsson, B., Lambert, D. L., Nissen, P. E., & Tomkin, J. 1993, A&A, 275, 101
ESA 1997, The Hipparcos and Tycho Catalogues, ESA SP-1200
Fields, B.D., & Olive, K.A. 1999, ApJ, 516, 797
Fuhrmann, K., Axer, M., & Gehren, T. 1995, A&A, 301, 492
Fulbright, J., & Kraft, R. 1999, AJ, 118, 527
García López, R.J. 1999, in LiBeB, Cosmic Rays, and Related X- and Gamma-Rays, Eds. R. Ramaty, E. Vangioni-Flam, M. Cassé, & K. Olive, ASP Conf. Series, 171, p. 77
García López, R.J., Severino, G., & Gomez, M.T. 1995, A&A, 297, 787
Israelian, G., García López, R.J., & Rebolo, R. 1998, ApJ, 507, 805
King, J.R. & Boesgaard, A.M. 1995, AJ, 109, 383
Kiselman, D. 1993, A&A, 275, 269
Kraft, R., Sneden, C., Langer, G., & Prosser, C. 1992, AJ, 104, 645
McWilliam, A. 1997, ARA&A, 35, 503
Mishenina, T., Korotin, S., Klochkova, V., & Panchuk, V. 2000, A&A, 353, 978
Molaro, P. & Beckman, J. 1984, A&A, 139, 394
Molaro, P., Beckman, J. & Castelli, F. 1984, ESA SP-219, 197

Ramaty, R., Vangioni-Flam, E., Cassé, M., & Olive, K. 1999, PASP, 111, 651
Rebolo, R., Molaro, P., Abia, C. & Beckman, J. 1988, A&A, 193, 193
Spiesman, W., & Wallerstein, G. 1991, AJ, 102, 1790
Thévenin, F., & Idiart, T. 1999, ApJ, 521, 753
Tomkin, J., Lemke, M., Lambert, D.L., & Sneden, C. 1992, AJ, 104, 1568

The Galactic Evolution of Boron

Francesca Primas

European Southern Observatory, Karl- Schwarzschildstr. 2, D-85748 Garching bei München

Abstract. Boron, together with lithium and beryllium, belongs to the group of the so-called *light elements*, the importance of which ranges from providing important tests to Big Bang nucleosynthesis scenarios to being useful probes of stellar interiors and useful tools to further constrain the chemical evolution of the Galaxy.

Since it became operative in the late eighties, the *Hubble Space Telescope* (HST) and its high- and medium-resolution spectrographs have played a key role in analyzing boron. Boron has now been observed in several stars and in the interstellar medium (ISM), providing important information in different fields of astrophysical research (nucleosynthesis, cosmic-ray spallation, stellar structure). In particular, determinations of boron in unevolved stars of different metallicity have allowed to study how boron evolves with iron.

After a general review of the current status of boron observations and of the major uncertainties affecting the measurements of its abundance, I will mainly concentrate on unevolved stars and discuss the "evolutionary" picture emerging from the most recent analyses and how its interpretation compares with theoretical expectations. A brief discussion on future prospects will conclude this contribution, showing how the field may evolve and improve.

1. Introduction

Abundances of the light elements Li, Be, and B play a critical role in understanding stellar mixing, Big-Bang nucleosynthesis (BBN), and galactic chemical evolution. The stellar structure interest (which will not be discussed here) stems from the fact that all three elements cannot survive in deep stellar interiors because they burn at progressively higher temperatures (~ 2.5, 3.5 and 5.0×10^6 K) at densities found in F and G stars near the base of the surface convection zone. Therefore circulation and destruction of the light elements can result in observable abundance changes, which become a powerful diagnostic for testing stellar structure models and internal mixing.

The "cosmological" interest, on the other hand, derives from the fact that the analysis of boron abundances as well as of the other two light nuclides in different types of stellar objects represents an additional test of the standard Big Bang nucleosynthesis theory, and can further constrain the galactic cosmic-ray (GCR) spallation scenario, responsible for Be and B formation. It is well known that standard Big Bang nucleosynthesis predicts primordial production of ^{7}Li, but not of ^{6}Li, ^{9}Be, ^{10}B, and ^{11}B, which have long been believed to arise from spallation reactions involving protons and/or α-particles colliding with the nuclei of the abundant elements carbon, nitrogen, and oxygen (CNO). The abundance from this source is expected to build up from near zero at the time of the formation of the Galaxy to the values observed in the interstellar medium today, at a rate which depends on the high energy particle flux and the abundance of the target nuclei CNO. From the analysis of light element abundances in stars of different metallicity it is possible to follow their galactic evolution and constrain which mechanisms are responsible for their production. This can be inferred, for instance, by investigating the slope of the trend Be,B versus Fe or by testing the presence of a knee in this trend and, if any, correlating it with plausible dynamical events in the formation of the halo according to the metallicity at which such change of slope is found. Until quite recently, however, there was very little information about the evolution of these elements throughout the history of the Galaxy. The prediction of a possible primordial production of Be and B in some inhomogeneous Big Bang models (cf Orito et al. 1997, and references therein) has helped in keeping alive the interest in how Be and B behave especially at the lowest metallicities alive. In other words, if a plateau were to be detected similarly to the one found in the case of lithium (the "Spite-plateau", cf Spite & Spite 1982).

As originally proposed by Reeves, Fowler, & Hoyle (1970), the rate of formation of ^{9}Be (as well as of ^{10}B) in GCR reactions is given by the product of the flux of high-energy protons times the cross sections for ^{9}Be formation by proton collision on the most abundant targets, ^{16}O and ^{12}C, times the abundance ratio of these targets with respect to hydrogen in space, i.e. :

$$dN_{Be}/dt \propto \chi(t) \times \int \sigma(E) \times \Phi(E,t)dE - loss\ terms \quad (1)$$

where $\chi(t)$ is the relative abundance of the target heavy elements, $\sigma(E)$ the reaction cross section (i.e. measured in a laboratory), which depend on the energy E of the cosmic rays, and $\Phi(E,t)$ the time-dependent cosmic-ray flux.

Based on the data available at that time, the approximate equality found between the product of the formation rate times the age of the Galaxy, on one hand, and the beryllium-to-hydrogen ratio measured in young stars ($\sim 10^{-11}$) was interpreted as a strong evidence for a major GCR contribution to some of the light elements. Indeed, the comparison of the ratios of the spallation cross sections of protons on O and C to the ratios of the stellar abundances of Li, Be and B shows that the GCR mechanism can satisfactorily account for the nuclei ^{6}Li, ^{9}Be, and ^{10}B. Cosmic-ray spallation in the general ISM has therefore been accepted for most of the past 25 years as the main site of production of these light isotopes, with the only exception of ^{7}Li (10% contribution) and ^{11}B (likely contribution from ν-spallation).

The build-up of Be and B in the Galaxy thus depends on the CR density, which presumably depends on the supernova rate and, in turn, on the star formation rate. The yields also depend on the rise of the (progenitor) CNO abundances and the decline of the gas mass fraction. Furthermore, the cumulative abundances are affected by the rates of infall of fresh (unprocessed) material and outflow, e.g. from supernova heating, and the time of formation of the halo and the disk. In such scenario, the light nuclide (boron, for instance) is expected to follow oxygen with a quadratic slope. In fact, since the primary process of spallation involves cosmic-ray protons and α-particles on interstellar heavy nuclei (mostly O), the increase of B is related to the supernova rate and the current O abundance. Being also O dependent on the supernova rate, the resulting dependence is quadratic.

2. Boron Observations

There are not many transitions available to measure boron abundances, and most of them fall in the ultraviolet part of the spectrum, thus requiring space-based observing facilities. This clearly explains why the Hubble Space Telescope had such a strong impact in the past decade on this specific field of research. The transitions most widely used for abundance measurements in stellar atmospheres are the resonance doublet of B I at 2497Å (typically used in cool stars, e.g. Duncan, Lambert, & Lemke 1993; Duncan et al. 1997; Primas et al. 1998, Primas et al. 1999) and the resonance line of B II at 1362Å (preferentially used in early-type stars, e.g. Venn, Lambert, & Lemke 1996). Moreover, Johansson et al. (1993) showed that the most favourable transition to attempt a determination of the isotopic ratio ^{11}B/^{10}B is the B I line at 2090Å, for which they predict an isotopic shift of 25mÅ (to be compared to the 8mÅ only in the case of the B I line at 2497Å). Rebull et al. (1999) have reported on their first attempt in such direction. Finally, at this symposium, Meléndez & Barbuy (2000) have shown that it may be possible to study boron in cool giants and supergiants via the infrared B I transitions at 1.166 and 1.624 μ.

2.1. Historical Background and Current Observational Status

Although the Hubble Space Telescope marked a new era of boron observations, there have been several analyses of boron before it was launched. One of the first studies of boron to appear in an official scientific journal is the work by Rowland & Tatnall (1895), "The Arc-spectra of the Elements.I. Boron and Beryllium" (in the first volume of the Astrophysical Journal). The first attempts of detecting boron in stellar atmospheres were of course devoted to the brightest and closest stars like the Sun (e.g. Nicholson & Perrakis 1928), and were then followed by more detailed analyses: in the Sun (e.g. Kohl, Parkinson, & Withbroe 1977), in Sirius (Praderie et al. 1977), and for the first time in A- and B-type stars (Boesgaard & Heacox 1978). The eighties were characterized by the International Ultraviolet Explorer and the first attempts to determine boron in metal-poor stars (Molaro 1987), whereas the Hubble Space Telescope clearly dominated the last decade of research. Thanks to this remarkable technological improvement, it finally became possible to start the first systematic analyses of boron. Duncan et al. (1992) took advantage of the first cycle of HST observations to show that the three metal-poor dwarf stars analyzed by them showed a surprising *linear* growth of boron abundances with increasing metallicity. A finding further confirmed by later analyses that included larger samples of data points (cf Duncan et al. 1997, Primas et al. 1999).

After 8 years of HST operations, the current status of boron observations includes, a part from the Sun and some other bright objects, a relevant number of early type galactic and extragalactic stars (cf Venn et al. 1996, and Cunha et al. 1997 respectively), galactic F- and G-type stars (cf Duncan et al. 1997; Primas et al 1999; García-López et al. 1999; Boesgaard et al. 1999), and different lines of sight in local diffuse interstellar clouds. Lambert et al. (1998), for instance, measured the boron isotopic ratio to be 3.4±0.7 dex in 3 different lines of sight (towards ζOph, κOri and δSco).

2.2. Difficulties Associated to Boron Measurements

Different factors affect the final accuracy of each boron measurement. In the first place, the quality of the observed data. All of the data available today for old galactic stars have been derived from the resonant doublet at 2500Å (mainly from the bluer of the 2 lines, the redder one being blended), hence they were collected with the Hubble Space Telescope. The choice of gratings is then limited to high or medium resolution, but very few spectra have been observed with the high-resolution echelle grating characterized by a nominal resolving power R$\sim$90,000 (with the Goddard High Resolution Spectrograph). Edvardsson et al. (1993) analyzed one of the well-known very metal-poor stars, HD 140283 ([Fe/H]=-2.60), with this instrumental set-up, and remarkably constrained the presence and strength of blending lines. But because this is a very time consuming choice, most of the boron data that will be shown and discussed here have been observed at R$\sim$30,000. Figure 1 shows the spectral region around 2500Å of an intermediate metallicity star (HD 94028, [Fe/H]$\sim$ -1.5). The continuous line represent the best fit to the data (here represented by photon statistics errorbars), whereas the dot-dashed lines were computed with ±0.20 dex change in boron abundance. The dotted line is the synthesis calculated without boron.

The second group of uncertainties is more related to the method of analysis. One of the main requirements for a proper analysis of the UV spectral region is the need of spectrum synthesis (cf Fig. 1). The crowding is severe and a high percentage of absorption features is blended. Such technique requires three main inputs: a reliable list of atomic and molecular lines, a grid of model atmospheres, and the determination of the main stellar parameters of the objects under investigation, i.e. effective temperature, gravity, and metallicity. Each of these items introduces different uncertainties, which must be carefully taken into account in the final computation of the uncertainties. On one side, one must include the uncertainties coming from laboratory measurements of the wavelengths and the oscillator strengths of each atomic and molecular transition, the physics behind the formation of the boron lines (and others) in the stellar atmosphere (Local Thermodynamic Equilibrium versus Non-Local Thermodynamic Equilibrium), and on the other hand one should keep in mind that current model atmospheres probably do not give a realistic description of a stellar atmosphere (1-dimensional versus 3-dimensional). Moreover, the effect and importance of possible blendings (like the Co I line at 2496.708Å very close to the bluer B I line at 2496.772Å) should also be quantified.

Despite this scenario may look quite discouraging, boron analyses carried out during the last few years have indeed introduced several improvements. Boron lines form under NLTE conditions (cf Kiselman & Carlsson 1994), hence corrections to the abundances usually determined under the simpler assumption of LTE must be (and have been) applied. The development of 3-D hydrody-

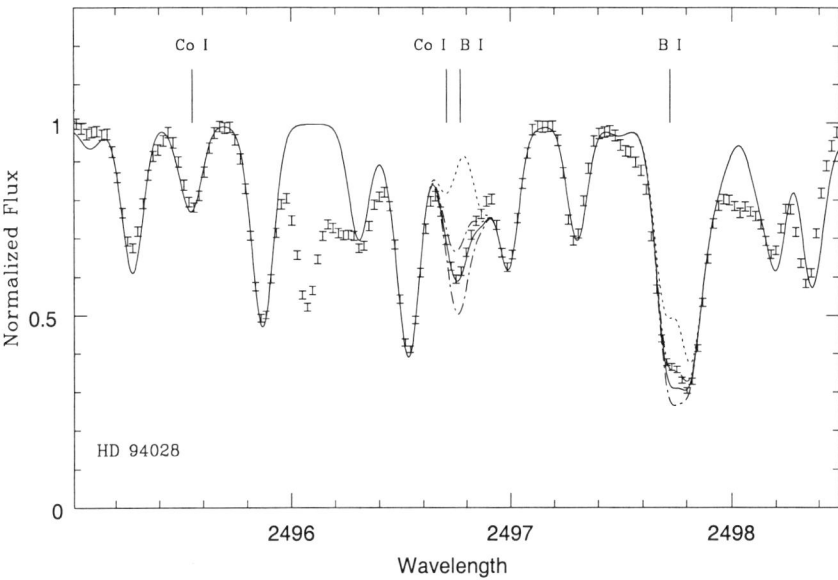

Figure 1. A portion of the spectrum of HD 94028 (cf Primas et al. 1999) around the B I resonance doublet. See text for more explanations.

namical models is in progress (cf Asplund et al. 1999; Asplund 2000): this will represent a remarkable step forward. As far as the problem of blendings is concerned, Peterson, Carney, & Smith (1999) suggested an improved solution for the treatment of the most important blend affecting B measurements: the neutral cobalt line at $\lambda 2496.708$. By taking advantage of the "unblended" (clean) Co I line at 2495.55Å, they were able to much better constain the cobalt abundance. Such an approach was readily implemented in the most recent analysis by Primas et al. (1999). Once all these different sources of uncertainty have been taken into account, typical error bars usually range between 0.10 and 0.25 dex.

Finally, it is important to notice that for the interpretation of the evolutionary trend of boron abundances, the size of the available data sample plays an important role too. Among the 3 light elements, boron abundances are indeed the least explored, mainly because its UV lines can be observed with space-based facilities only.

3. Observational Results

Because the increase of B and Be data during the past few years has been indeed remarkable (compared to 10 years ago, for instance), the study of Li, Be, and B evolutionary trends have become to be regarded as a powerful discriminant between different models of the chemical and dynamical evolution of the Galaxy.

The evolution of B has received considerable attention, since observations of its abundance in halo stars have shown that:

a) its relationship with respect to iron is a linear one, over 3 decades of metallicity;

b) it evolves with a constant ratio of B/Be~10–20, a clear indication of CR origin;

c) contrary to Li, but similarly to Be, it does not show any plateau value at low metallicity, i.e. no indication for a primordial origin;

d) differences in the boron (and beryllium) content between stars otherwise very similar have been detected.

Figures 2 and 3 represent some of these points: they show respectively how B_{NLTE} and the B_{NLTE}/Be ratio behave with metallicity. All the data plotted were derived from the resonant doublet at 2500Å, making use of the latest Kurucz model atmospheres (with the "approximate overshooting" option switched off) and his ATLAS and SYNTHE codes (Kurucz 1993). Details on this part of the analysis can be found in Primas et al. (1999). The χ^2-fit computed by Duncan et al. (1997) is superimposed just for comparison (it has a slope of 0.7). Most of the few points (filled circles) falling below the fit are slightly evolved stars for which the lower content of boron can be easily explained via stellar destruction and/or mixing (cf Primas et al. 1999). The only exception is the point at [Fe/H]=−1.80 for which the low boron content may be intrinsic (cf Primas et al. 1998). The two questionmarks have the purpose of reminding the reader that the detection of boron in these two very metal-deficient stars is under debate (cf García-López et al. 1998). The open triangle marked with "u.l." is an upper limit.

The first observable (point a) above) is difficult to interpret in the classical CR spallation scenario, in which B is a secondary element: a slope of 2 is in fact expected in the logarithmic plane B vs Fe, since supernovae are the primary source for both cosmic rays and O nuclei, and then in a separate process cosmic rays and O create B (and Be). Besides, adopting CR spectra similar to those observed in near-Earth space, this class of models underproduce the isotopic ratio $^{11}B/^{10}B$ observed in meteorites to be ~4 (cf Chaussidon & Robert 1994).

While the plateau showing near-uniformly lithium abundances in sufficiently warm halo stars points to a cosmological (non-galactic) origin for a substantial fraction of the observed lithium, the correlation of B (and Be) with metallicty in halo stars requires these elements to be made in the early Galaxy. Any primordial component must be lower than the abundances derived from current observations and would manifest itself as a plateau in the low-metallicity limit.

A first remark to the observed linearity was that the net rate of boron production does not seem to depend on the CNO abundances in the ISM. This led to the suggestion that the CR spallation most important for light elements production could be C and O nuclei colliding with ambient protons and α-particles (i.e., a "reverse" spallation reaction, which may decouple light element production from the metallicity of the ISM), probably in regions of massive star formation. Although able to reproduce the observed linearity, later calculations showed that it underproduces the LiBeB yields by at least one order of magnitude (e.g. Parizot & Drury 1999).

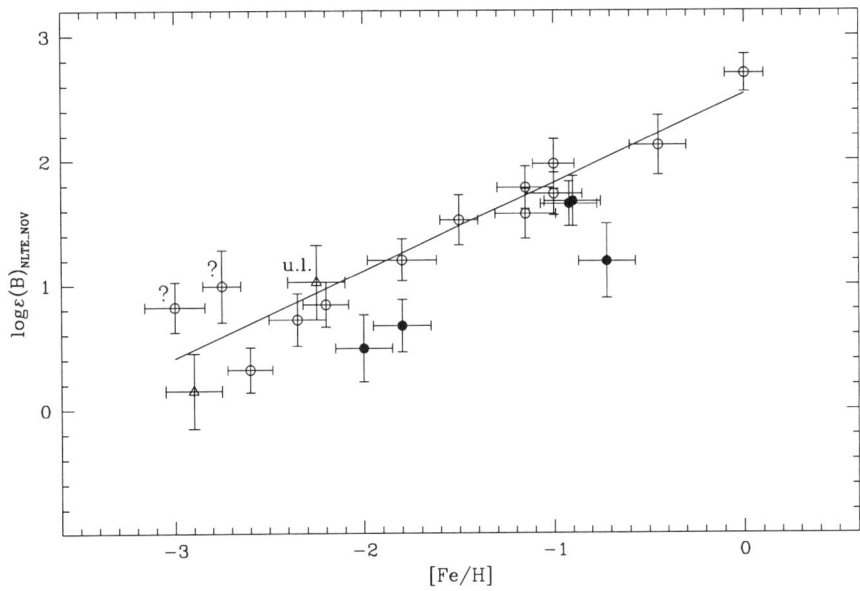

Figure 2. NLTE B abundances versus metallicity. The data points were taken from analyses of Primas et al. (1999, filled and open circles) and García-López et al. (1998, triangles). The data point at solar metallicity is the recent determination of photospheric solar boron by Cunha & Smith (1999).

It is probably safe to say that observations of B in low metallicity halo stars formed during the first 10^9 years of galactic evolution show that cosmic ray acceleration must have taken place in the early Galaxy. The B/Be ratio observed to be almost constant at all metallicities represents a strong evidence for CR spallation (cf Figure 3). The observed abundances of B relative to Fe, which in the early Galaxy is almost exclusively produced in Type II supernovae (SNe), strongly suggest that the cosmic ray acceleration is also related to such SNe with the particles being accelerated out of freshly nucleosynthesized matter before it mixes into the ambient, essentially non-metallic ISM. Unfortunately, the current data are not sufficient yet to further constrain the proposed scenarios. The latest developments, nicely reviewed at this meeting by several investigators (cf Beers, Cassé, Olive, Parizot, Ramaty, this volume) seem to explore two main scenarios: LiBeB production in superbubbles induced by the collective effect of SNe in OB associations and a two-component source of GCR in a SN-driven chemical evolution model of the Galactic halo in which both interstellar gas and fresh SN ejecta trapped in the shell undergo SN shock wave acceleration.

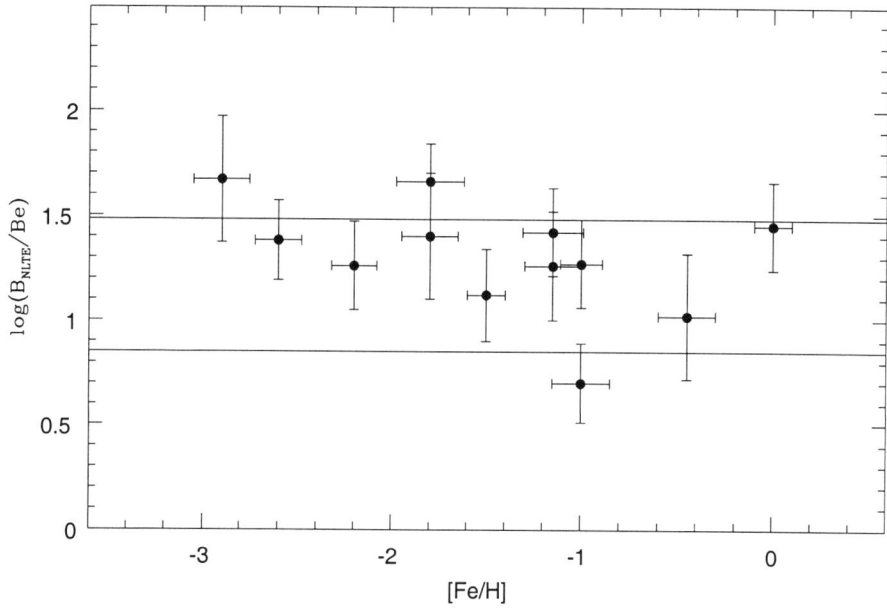

Figure 3. B_{NLTE}/Be versus metallicity for the same sample of stars displayed in Fig. 2, but restricting the sample to unevolved stars only.

However, despite all these attempts, the observed linearity still defies one simple theoretical explanation. The situation gets even more complicated for boron (compared to beryllium) because a significant contribution to ^{11}B from neutrino-spallation is very likely (and needed in order to match the observed meteoritic isotopic ratio). Thus, there are at least two main (probably complementary) mechanisms that need to be understood and disentangled.

4. Summary and Future Prospects

The abundance of B has now been measured in halo stars of metallicity as low as [Fe/H]=−3.0, showing that B/Fe is independent of metallicity: B follows Fe in direct proportion from the earliest times to the present, with little change of slope (if any) between halo and disk metallicities.

It is clear and well established that the cosmic ray spallation theory as originally proposed in the early 70s is unable to reproduce the observed trend. Such finding has triggered a re-analysis of this scenario in which B (and Be) are considered to be secondary elements, mostly formed by spallation reactions of galactic cosmic rays impinging on the interstellar medium. Unfortunately, although the trend of the B/Be ratio with metallicity suggests that cosmic ray spallation is indeed part of the game, the current data are not sufficient yet to discern among the several theoretical solutions that have been developed in the last few years. The differences in the assumptions are often too subtle to be thoroughly tested with the data sample currently available.

However, among the many aspects that still need to be explored in order to improve and narrow the constraints, there is plenty of room for progress. From the analytical point of view, as mentioned earlier, the availability of 3-D hydrodynamical model atmospheres in the near future and the constant improvements in the atomic physics of near-UV transitions will offer us a much more realistic description of the physics, dynamics and composition of stellar atmospheres.

A deeper knowledge of how [O/Fe] correlates with metallicity is also very important. The most direct way to interpret boron abundances is indeed to plot B (as well as Be) versus O, instead of Fe, because oxygen is directly involved in the same cosmic-ray spallation reactions responsible for the production of boron. Fields & Olive (1999) have shown that if the newly found [O/Fe] trend versus [Fe/H] is adopted (i.e. an increasing oxygen abundance as metallicity decreases, e.g., Israelian, García-López, & Rebolo 1999), then the observed linearity may still be reproduced within the context of the classical theory of cosmic-ray spallation. However, the determination of oxygen abundances is a very delicate matter and suffers from several uncertainties; therefore, until these high oxygen values in the most metal-deficient stars are further confirmed (ideally by using a larger selection of oxygen abundance indicators), any comparison between B and O can not be conclusive.

As far as observational efforts are concerned, there is no doubt that enlarging the sample of available data represents the next essential step, especially at the lowest metallicity end. The detection of boron in 2 out of the 3 most metal-poor objects ever observed with HST is still under debate (cf discussion of Fig. 2 in Section 3), hence few more data points around [Fe/H]~ -3.0 would be of invaluable help. Unfortunately, one of the biggest unknown here is the timescale on which such a progress will be realistically achieved for boron, whereas beryllium observations have indeed the big advantage of being observable from ground. The potentialities of the new Ultraviolet and Visible Echelle Spectrograph mounted on the ESO VLT Kueyen Telescope revealed during the period of first commissioning suggest that UVES (available to the scientific community starting April 1, 2000) will have a strong impact on this specific field of research.

References

Asplund, M., Nordlund, A., Trampedach, R., & Stein, R. 1999, A&A, 346L, 17

Asplund, M. 1999, this volume

Beers, T. C. 2000, this volume

Boesgaard, A. M., & Heacox, W. D. 1978, ApJ, 226, 888

Boesgaard, A. M., Deliyannis, C. P., King, J. R., Ryan, S. G., Vogt, S. S., Beers, T. C. 1999 AJ, 117, 1549

Cassé, M. 1999, this volume

Chaussidon, M., & Robert, F. 1994, Meteoritics, 29, 455

Cunha, K., Lambert, D. L., Lemke, M., Gies, D. R., & Roberts, L. C. 1997, ApJ, 478, 211

Cunha, K., & Smith, V. V. 1999, ApJ, 512, 1006

Duncan, D.K., Lambert, D. L., & Lemke, M. 1992, ApJ, 401, 584

Duncan, D. K., Primas, F., Rebull, L. M., Boesgaard, A. M., Delyiannis, C. P., Hobbs, L. M., King, J. R., & Ryan, S. G. 1997, ApJ, 488, 338

Edvardsson, B., Gustafsson, B., Johansson, S., G., Kiselman, D., Lambert, D. L., Nissen, P. E., & Gilmore, G. 1994, A&A, 290, 176

Fields, B. D., & Olive, K. A. 1999, ApJ, 516, 797

García-López, R., Lambert, D. L., Edvardsson, B., Gustafsson, B., Kiselman, D., & Rebolo, R. 1998, ApJ, 500, 241

Israelian, G., García-López, R. J., & Rebolo, R. 1998, ApJ, 507, 805

Johansson, S. G., Litzen, U., Kasten, J., & Kock, M. 1993, ApJ, 403L, 25

Kiselman, D., & Carlsson, M. 1996, A&A, 311, 680

Kohl, J. L., Parkinson, W. H., & Withbroe, G. L. 1977, ApJ, 212L, 101

Kurucz, R. L., 1993, CD-ROM # 1, 13, 18

Lambert, D. L., Sheffer, Y., Federman, S. R., Cardelli, J. A., Sofia, U. J, & Knauth, D. C. 1998, ApJ, 494, 614

Meléndez, J., & Barbuy, B. 2000, this volume

Molaro, P. 1987, A&A, 183, 241

Nicholson, S. B., & Perrakis, N. 1928, ApJ, 68, 327

Olive, K. 2000, this volume

Orito, M., Kajino, T., Boyd, R. N., & Mathews, G. J. 1997, ApJ, 488, 515

Parizot, E., & Drury, L. 1999, A&A, 346, 686

Parizot, E. 2000, this volume

Peterson, R. C., Carney, B. W., & Smith, H. 1999, ApJ, submitted

Praderie, F., Milliard, B., Pitois, M. L., & Boesgaard, A. M. 1977, ApJ, 214, 130

Primas, F., Duncan, D. K., & Thorburn, J. A. 1998, ApJ, 506L, 51

Primas, F., Duncan, D. K., Peterson, R. C., & Thorburn, J. A. 1999, A&A, 343, 545

Ramaty, R. 2000, this volume

Rebull, L. M., Duncan, D. K., Johansson, S. G., Thorburn, J. A., & Fields, B. D. 1998, ApJ, 507, 387

Reeves, H., Fowler, W. A., & Hoyle, F. 1970, nature, 226, 727

Rowland, H. A., & Tatnall, R. R. 1895, ApJ, 1, 14

Spite, F., & Spite, M. 1982, A&A, 115, 357

Venn, K. A., Lambert, D. L., & Lemke, M. 1996, A&A, 307, 849

The Abundance of Boron in Disk-Metallicity Stars

Katia Cunha[1,2]

[1] *Observatório Nacional - CNPq, Rio de Janeiro, Brazil*

[2] *University of Texas at El Paso, El Paso TX 79968*

Abstract. Although the behavior of boron versus metallicity has been probed in a fairly large sample of halo dwarfs with HST, it is only very recently that boron abundances have been derived systematically in solar metallicity dwarfs. This effort began with a re-analysis of the solar spectrum with modern atomic data and model atmospheres so that the Sun could be adopted as a standard for the calibration of a line list in the region of the B I transition at 2497 Å. The solar analysis indicates that boron is not depleted in the solar photosphere. From a subsequent study of a sample of 14 field F/G-dwarfs with roughly solar metallicities, it is found that the behavior of boron versus [Fe/H] follows the linear trend that is observed for the halo stars. The average B/Be obtained for solar metallicity stars is 27±5 compared to the solar ratio of 23. The determination of boron abundances in the young B-type and G-type stars of the Orion association reveals a behavior of boron and oxygen in Orion that is opposite of the positive correlation which is observed for the field stars: the boron and oxygen abundances are anticorrelated.

1. Introduction

The main sources of the light element boron in the Galaxy are spallation reactions occurring between energetic particles and interstellar nuclei. There are three categories of these processes. The first consists of accelerated protons and α-particles hitting CNO nuclei in the ISM to produce boron in the form of ^{10}B and ^{11}B. The second one is the so-called reverse process, where CNO nuclei, accelerated presumably by type II supernovae (SN II), hit interstellar H and He to produce also ^{10}B and ^{11}B. Another possibility is the production of boron in SN II from neutrino-induced nucleosynthesis (the ν-process). The neutrinos result from core collapse and produce primarily ^{11}B from interactions between ν's and ^{12}C in the C-rich shell of the progenitor star.

Observations of boron (together with beryllium) hold the possibility of discriminating between the various production mechanisms, and, in particular, the ν-process contribution. However, boron observations are still rather sparse. This is because boron is not observable from the ground, having strong transitions only in the ultraviolet, with B I observable in cooler stars (with roughly solar temperatures), and B II and B III in the hotter stars. The use of HST for UV spectroscopic analyses of boron in a variety of stars at different metallicities has provided the primary data set from which derived stellar boron abundances can be compared to model predictions for boron production and chemical evolution.

The first fairly large sample of stellar boron abundances was provided by Boesgaard & Heacox (1978) from Copernicus observations of B II at 1362 Å in 18 A and B field stars spanning a range in T_{eff} between 9000 and 25000K. Their analysis assumed the validity of LTE and an average of log $\epsilon(B)$=2.3 ± 0.2 was obtained for the boron abundances of the studied stars. The first detection of boron in the solar photosphere was reported by Kohl, Parkinson & Withbroe (1977). They analyzed the B I resonnance transition at 2497 Å in rocket spectra obtained at solar disk center. The derived boron abundance was log $\epsilon(B)$=2.6 ± 0.3.

After these two pioneer boron studies, it was only in the early 90's that routine access to boron spectroscopy in the ultraviolet became possible with the Hubble Space Telescope; data on the behavior of boron with [Fe/H] has since then increased considerably. In particular, the first extensive study of boron with metallicity was done by Duncan et al. (1997). From the analysis of B I in a sample of eight halo stars with metallicities ranging from [Fe/H]=-3.0 to -0.4 the surprising linear behavior of boron with metallicity was first established. In fact, this linear behavior had already been suggested from a previous study by Duncan, Lambert & Lemke (1992) that contained a small sample of 3 metal-poor dwarfs. Other studies expanded the boron data set, as well as probed boron abundances in dwarfs with the lowest metallicities (Primas et al. 1999 and García-López et al. 1998). All these observational results suggested that the production of boron in the Galaxy was directly related to the production of the heavier element Fe. The relation of boron and Fe for solar metallicity stars remained to be investigated. In the following section, we will discuss the re-analysis of the solar spectrum and the photospheric boron abundance in the Sun, the boron results for field dwarfs with roughly solar metallicity, and the boron abundances observed for a sample which constitutes a truly young population of the Galactic disk: B- and G- type stellar members of the Orion association.

2. Boron Abundances of Disk-Metallicity Stars

The spectra of solar-type solar-metallicity stars in the vicinity of the B I transition at 2497 Å are severely blended with large numbers of overlapping lines such that there are no regions free of line absorption. Moreover, knowledge of the atomic data in this ultraviolet region of the spectrum is still limited. The first attempts to synthesize the B I region in the spectra of dwarfs of roughly solar metallicity immediately revealed a significant mismatch between the adopted line list (in this case taken as the line list compiled by in Duncan et al. 1998) and the observed spectra. The line list, although it represented an updated compilation of the available atomic data, seemed to have a large number of missing lines that were present in the observed spectra and not matched by the synthetic one. A reasonable strategy was then to adopt the Sun as a standard and do a homogeneous and self-consistent analysis for all dwarfs with roughly solar metallicities that had been observed for boron with the Hubble Space Telescope.

2.1. The Photospheric Solar Boron Abundance

The first step towards a homogeneous boron abundance analysis of solar metallicity stars was to fine-tune the line list compiled by Duncan et al. (1998) in order to achieve the best possible fit to the disk-center spectrum of the Sun. The selection of the Sun as a standard ensures that a direct comparison can be done between disk stars and the Sun. The same solar spectrum that had been previously analyzed by Kohl et al. (1977) was also used by Cunha & Smith (1999) to re-derive the solar photospheric boron abundance. The adopted procedure to adjust the line list consisted of the following: when there was a line missing in the synthetic spectrum, a 'fake' Fe I line was added with an arbitrary excitation potential. The values of oscillator strengths for those lines which did not have accurate laboratory measurements were adjusted until the observed intensities could be matched by the synthesis, while the lines with accurate laboratory f-values were kept untouched.

The continous opacity in the spectral region of the B I transition (2500 Å) is dominated by the photoionization of Mg I. This is an important source of opacity that needs to be considered in boron abundance calculations, especially in metal-rich stars ([Fe/H]>-1). Synthetic spectra were calculated with the adoption of a more recent value for the photoionization cross-section of Mg I at 2500 Å that was taken from the Opacity Project; σ(Mg I)= 18 X 10^{-18} cm^2 (Butler, Mendozza & von Zeipen 1993). This value for the b-f cross-section is significantly lower than the published value from ~40 years ago that was adopted in the previous study of boron in the Sun: the experimental cross-section from Botticher (1958; σ(Mg I)= 45 X 10^{-18} cm^2). We note that the experimental value by Gingerich et al. (1971) with σ(Mg I)= 25 X 10^{-18} cm^2, as well as the theoretical value by Peach (1970; σ(Mg I)= 16 X 10^{-18} cm^2) became available in the early 1970's. In Cunha & Smith (1999) it is shown that the choice of the photoionization cross-section of Mg I has a measurable effect on the derived boron abundances in the Sun. They find that the adoption of lower values of σ(Mg I) in the calculations produces a better agreement between the solar observations and model intensities at the solar limb. For the higher values of σ(Mg I), the calculated model continuum intensities were already below the lowest possible definable continuum (the lowest possible continuum would be defined by the points of highest intensity in the observed spectrum). This inconsistency was also recognized in the calculations by Kohl et al. (1977), who argued that non-LTE effects in Mg I, that were not being considered, could be responsible for the effect.

This most recently derived photospheric boron abundance of log ϵ(B)= 2.70 with estimated statistical uncertainties of -0.12 and +0.21 (Cunha & Smith 1999) is in good agreement with the meteoritic abundance obtained by Zhai & Shaw (1994; log ϵ(B)=2.78), indicating that boron is not depleted in the Sun. This boron result is in line with the lack of Be depletion in the Sun as recently argued by Balachandran & Bell (1998). We note, however, that a modest Be depletion (of ~0.4 dex) would also be consistent with no boron depletion.

2.2. Boron in a Sample of Dwarfs with [Fe/H]>-1.0

A relatively large number of solar type dwarfs have been observed with the Goddard High Resolution Spectrograph (GHRS) in the spectral region that contains the B I transition at 2497 Å. Recently, Cunha et al. (2000a) selected 14 stars from the HST archive and analyzed the B I region to derive boron abundances from synthetic spectra calculated with the known sources of opacities. Their studied sample consisted of dwarfs which spanned a range in effective temperatures from 5650 to 6700K and metallicities [Fe/H] ranging between -0.75 and +0.15. Their boron analysis was done consistently relative to the Sun with the adoption of a line list adjusted in order to produce a good fit to the solar spectrum, as discussed above. Although the solar line list had the addition of fake Fe I lines (to properly fit the solar spectrum) it produced, in general, a very good fit of the B I region for most of the sample stars, even for those stars that were considerably hotter than the Sun. The good fits of the boron region obtained for stars spanning a large range in T_{eff} indicates that an accurate set of boron abundances can be derived.

One of the results of the Cunha et al.'s (2000a) study was that those sample stars with metallicities close to solar, showed boron abundances that were approximately solar as well. Concerning the boron-to-beryllium ratios for these stars, the average ratio obtained was 27±5, also in rough agreement with what is observed in the Sun (B/Be=23) and higher than the predictions of Galactic cosmic-ray models that find B/Be ~ 10-15.

In order to investigate the relation of boron and iron in disk stars, it was necessary to isolate, from the sample studied by Cunha et al. (2000a), those dwarfs that have undepleted Li (and Be) abundances. (Li is severely depleted before B starts to suffer significant depletion in the stellar interior). The behavior of boron and metallicity obtained then is shown in the top panel of Figure 1 where are gathered, from several studies in the literature, the boron abundances for the halo stars, as well as Cunha et al.'s results for the disk stars. The target stars shown represent nearly all the stars (with undepleted Li) that have been observed with the Hubble Space Telescope for boron. The results for the disk-metallicity stars (represented by filled circles) seem to indicate an extension of the behavior observed for the halo stars: a linear relation extends from the low metallicities in the halo to high metallicities around [Fe/H]=+0.15. A least-squares fit to all the abundance points in the figure (except the four points that represent the two stars with lowest metallicities and controversial boron detections) indicates a slope of ~0.9 with a correlation coefficient of 0.98. A change in slope at disk to halo metallicities does not seem to be obvious from this data set but has not been investigated in detail. However, it is important to note that the Fe abundances shown may suffer from departures from LTE and point out the recent non-LTE results by Thévenin & Idiart (1999) that indicate significant revisions to LTE Fe abundances derived for cool stars, especially with low metallicities. Adoption of their non-LTE Fe abundances would result in a different relation between B and Fe. (See discussion by Idiart & Thévenin in these procedings).

The relation of B with [Fe/H] can constrain the possible origins of boron; however, the production of boron is not directly related to the synthesis of Fe itself. Perhaps a more revealing comparison element is oxygen since a significant fraction of boron is produced primarily by cosmic-ray interactions with atoms in the ISM (spallation reactions between protons and α-particles with C, N, and O nuclei), thus the most direct metallicity indicator for the B abundance

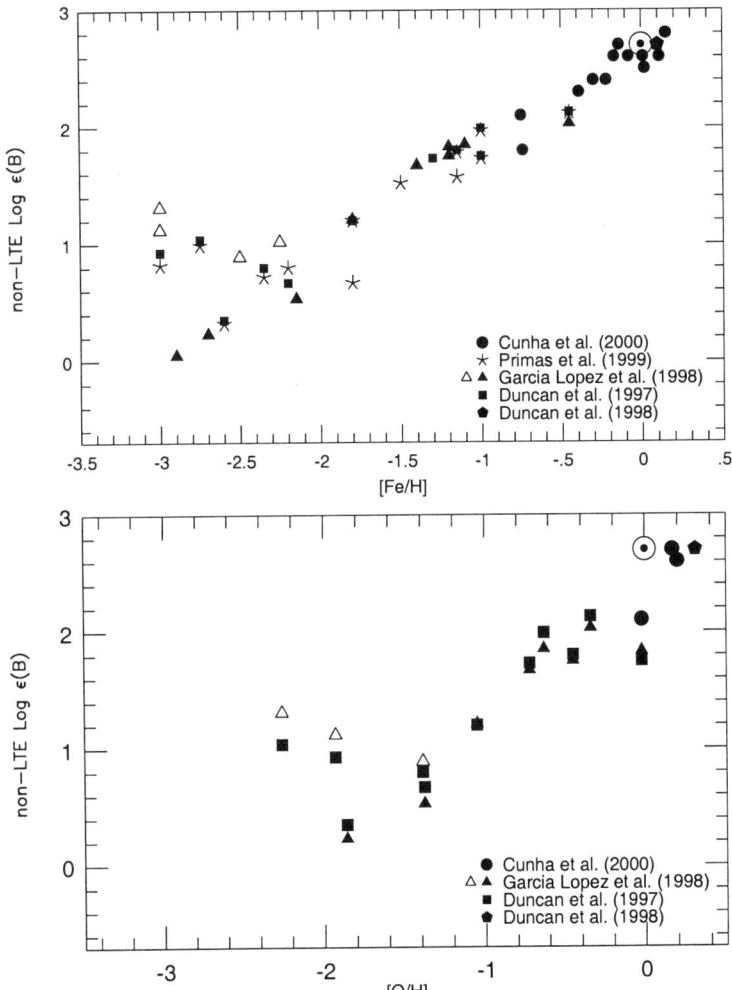

Figure 1. *Top panel*: The behavior of boron and iron in Galactic field halo and disk dwarfs with undepleted Li. The non-LTE boron and [Fe/H] abundances were gathered from the literature according to the references listed in the Figure. Because of the different values of [Fe/H] adopted by the different studies, a single star may not be represented by a unique [Fe/H]. *Bottom panel*: The behavior of boron and oxygen for a sample of halo and disk dwarfs. The adopted oxygen abundances represent a simple average of the oxygen abundances found in the literature and the non-LTE boron abundances are from the references listed in the Figure.

is the most abundant member of the CNO trio, which is oxygen. Abundances of oxygen, however, are more difficult to derive than iron. In particular, the oxygen abundances in metal poor stars are still a matter of debate and the flat behavior of [O/Fe] versus [Fe/H] for halo metallicities (below [Fe/H]$\sim$-1.0) is being questioned (see e.g. Israelian et al. 1998, Boesgaard et al. 1999, but also Fulbright & Kraft 1999).

Duncan et al. (1997) have collected the various oxygen abundance determinations from the literature for the metal poor stars in their sample and they find that the oxygen abundances for a given star show a considerable scatter. In the bottom panel of Figure 1 we plot the behavior of boron versus oxygen in field dwarfs and, unlike Duncan et al. (1997) who assumed a mean relation for [O/Fe] versus [Fe/H] in order to obtain oxygen abundances for their individual stars, here, we have taken the simple average of the different oxygen measurements for each star and plotted them versus their derived non-LTE boron abundances. The observed points in this figure suggest that the behavior of boron and oxygen for stars with oxygen abundances [O/H] larger than $\sim$ -2.0 is roughly linear with a slope of nearly 1.0. However, the reality of the suggested slope 1.0 linear relation (primary behavior) between boron and oxygen needs further investigation, due to the small number of oxygen abundance points and principally because the derived oxygen abundances from different sources are significantly discrepant. We note that Fields et al. (2000) identify the presence of a secondary component in the observed data for [O/H] $\gtrsim$ -1.4, when they assemble the different oxygen abundances determinations that adopt a consistent effective temperature scale. (The different stellar parameter scales were ignored in our discussion, where we have simply averaged the oxygen abundances obtained from different sources in the literature). In fact, Fields et al. conclude that this secondary component (a quadratic relation between boron and oxygen) dominates at high metallicities, being identifiable in two distinct and independant sets of adopted stellar parameter scales. For the low metallicity stars, the uncertainties in the derived oxygen abundances are larger. As pointed out by Fields et al. (2000) the behavior of boron and oxygen and the existence, or not, of a primary component (linear slope 1.0 relation) at low metallicities depends on the particular set of adopted stellar parameters used to calculate the oxygen abundances. Before any firmer conclusions can be reached about the presence (or not) of a primary plus a secondary component in the behavior of boron and oxygen, more (and more reliable) oxygen abundance points are needed.

At the metal poor end, not only are the oxygen abundances controversial, but also the boron abundances derived for the most metal poor dwarfs. From analysis of the same HST spectra, different studies do not agree on boron abundance values for low metallicity stars with $T_{eff} > 6000K$: García-López et al. (1998) argued that only boron abundance upper limits could be derived for such stars, while Duncan et al. (1997) and Primas et al. (1999) obtained boron abundance values for dwarfs with such effective temperatures and low metallicities. Depending on which set of boron abundances is taken (upper limits or detections) a distinct general behavior for boron versus oxygen could be inferred: if the boron abundance results from Duncan et al. (1997) and Primas et al. (1999) for the two stars with lowest [O/H] in Figure 1 are adopted, there would be some indication of a possible change of slope in boron versus oxygen at values of [O/H] $\lesssim$ -1.9. (A similar behavior would be seen for Fe in the top panel of this figure).

Moreover, there would exist a significant spread in the boron abundance at a given oxygen abundance. Note (in the bottom panel of Figure 1) that the two stars at [O/H] ∼ -1.9 (represented by filled squares) show a spread in boron of roughly 0.6 dex. However, if the boron abundaces for the two stars with lowest [O/H] in Figure 1 are in fact only upper limits, as argued by García-López et al. (1998), these could have boron abundances in agreement with the general trend which is observed for boron and oxygen in higher metallicities. But, of course, the 'real' trend also depends on the 'real' oxygen abundances for these stars. The final word on boron, and oxygen, is still to come with more observations of boron, especially in low metallicity stars, and reliable and self-consistent oxygen abundance determinations for the whole sample of stars observed for boron.

2.3. Boron Abundances in the Orion Association

An important step in trying to isolate the contributions of the different processes to the production of boron in the Galaxy comes from the study of a sample of stars from nearly the same birthplace and birthdate in the Galaxy but having different oxygen abundances. The Orion association is a perfect environment in this context as it contains member stars which have a spread in their oxygen abundances: the oxygen abundance spread amounts to a factor of ∼ 6 and is interpreted as the result from self-enrichment over the lifetime of the association ($\sim 10^7$ yr) by very massive SN II, whose dominant nucleosynthetic product is oxygen (Cunha & Lambert 1994 and Cunha, Smith & Lambert 1998).

In a previous study of boron in Orion stars based on HST GHRS spectra, Cunha et al. (1997) obtained spectra of B II at 1362 Å in a sample of four main-sequence B-stars in the Orion association. They selected four target stars, with two stars having low oxygen abundances and two having slightly higher oxygen abundances. The derived LTE abundances were rather low (when compared to the meteoritic value) and ranged between log ϵ(B)=1.6-2.0. However, the derivation of reliable B abundances (from B II) required a non-LTE analysis which indicated a large correction (up to ∼1.0 dex) to the derived LTE abundances. Concerning the behavior of boron and oxygen in this quartet of stars, the expectation would be a positive correlation, as observed for the general field stars. However, the particular situation in an OB association where there is evidence of self-enrichment has never been studied. The results obtained for the B-stars did not reveal a positive correlation of boron and oxygen, instead a slight negative trend was observed such that the boron abundances in the O-rich stars were some 0.4 lower than in the O-poor stars. But conclusions relied on the validity large non-LTE corrections applied to the LTE abundances. In this context, it is worth noting that the derived LTE abundances were similar to the interstellar abundances measured in the direction of the Orion association by Lambert et al. (1998: log ϵ(B)∼2.0), while the non-LTE corrections would increase the boron abundances in the Orion stars to the interval between logϵ(B)=2.5 and 2.9 (which encompasses the meteoritic value).

An important confirmation of the large non-LTE corrections derived for B II in Cunha et al. (1997) has now been obtained by Lambert et al. (2000) from calculations of boron abundances from B III at 2066 Å in one of the Orion targets that had been previously analyzed for B II. Their results indicate a high (close to meteoritic) B abundance, definitely not as low as the ISM, for the

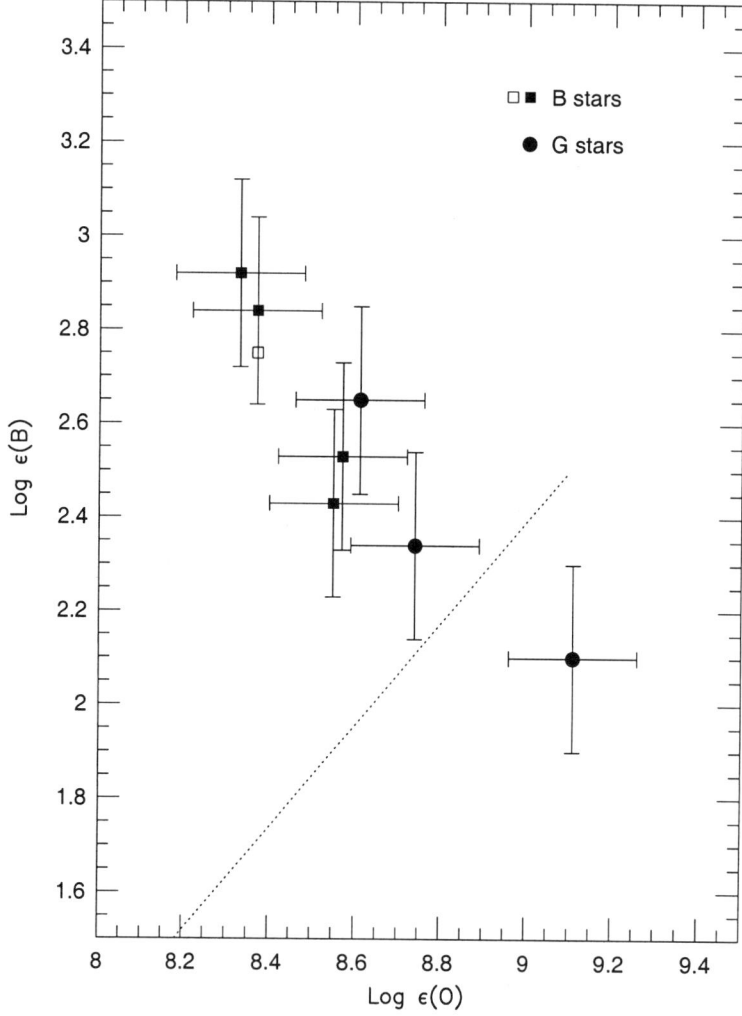

Figure 2. The anticorrelation of boron versus oxygen obtained for the young Orion association. The filled and open squares represent the four B-type stars analyzed for boron in Cunha et al. (1997) and Lambert et al. (2000), and for oxygen in Cunha & Lambert (1994); while the three G-type dwarfs analyzed in Cunha, Smith & Lambert (1999) and Cunha et al. (2000b) are represented by filled circles. Their oxygen abundances are from Cunha et al. (1998). The errorbars represent the estimated uncertainties in the derived boron and oxygen abundances. Also shown, for comparison, is the general trend observed for the field stars, with a least-squares fit (dotted line) calculated for the boron and oxygen abundance points in Figure 1. We have excluded from the fit the three stars that have boron upper limits in García-López et al. 1998) and adopted average boron values for each star.

studied stars. Unlike B II, B III is major ionization stage at the T_{eff}'s of the target stars with very small non-LTE corrections. These B III results bolster our confidence in the non-LTE abundances derived for the B-type stars.

As a further step to investigate the behavior of boron with oxygen in the young Orion members, Cunha et al. (1999) and Cunha et al. (2000b) analyzed three cooler lower mass stellar members of Orion (of spectral type G and masses ∼1-2M⊙) with oxygen abundance values similar to the oxygen-rich B-type stars and higher. These stars were observed with the GHRS and STIS on the HST and the observed sample was selected from Orion G-dwarfs with Li abundances that are undepleted (or nearly undepleted).

When the boron results obtained for the B-type stars in the Orion association are put together with the abundances obtained for the three G-type stars, a puzzling behavior seems to emerge. These boron results, plus the stars' respective oxygen abundances, are shown in Figure 2. The trend displayed is obvious: boron seems to decline with increasing oxygen. In principle, this observed anticorrelation of boron versus oxygen can set limits on boron production via ν-nucleosynthesis. Qualitatively, if the ν-process contribution were significant, one should expect that the boron abundance would increase with oxygen. But the observed trend for Orion is opposite. One possible explanation for the anticorrelation is to assume that the ν-process is negligible. However, this is not the only possibility. As discussed in Cunha et al. (2000b) the anticorrelation of boron and oxygen in Orion can be explained by a simple model in which two components of gas are poorly mixed: the SN II ejecta and the ambient medium which is B-enriched by spallation reactions. (According to superbubble (SB) models (Parizot 1998) boron production takes place in the SB supershells.) Therefore the ambient gas component could be significantly larger than the boron component ejected by the SN II. Such a scenario would still, qualitatively, accomodate the observed anticorrelation of boron and oxygen in the Orion association. A study of beryllium (together with boron) in Orion stellar members with undepleted Li would provide strong constraints on the ν-process contribution to boron production.

I would like to thank the organizers of the IAU Symposium 198 for the travel grant to attend the meeting.

References

Balachandran, S., & Bell, R. A. 1998, Nature, 392, 791

Boesgaard, A. M., & Heacox, W. D. 1978, ApJ, 226, 888

Boesgaard, A. M., Deliyannis, C. P., King, J. R., Ryan S. G., Vogt, S. S., & Beers T. C. 1999, AJ, 117, 492

Botticher, W. 1958 Z. Phys., 150, 336

Butler, K., Mendoza, C., & Zeipen, C. J. 1993, J. Phys. B, 26, 4409

Cunha, K., & Lambert, D. L. 1994, ApJ, 426, 170

Cunha, K., Lambert, D. L, Lemke, M., Gies, D. R., & Lewis, C. R. 1997, ApJ, 478, 211

Cunha, K., Smith, V. V., & Lambert, D. L. 1998, ApJ, 493, 195

Cunha, K., & Smith, V. V. 1999, ApJ, 512, 1006
Cunha, K., Smith, V. V., & Lambert, D. L. 1999, ApJ, 519, 844
Cunha, K., Smith, V. V., Boesgaard, A. M., & Lambert, D. L. 2000a, ApJ, Feb 20 issue
Cunha, K., Smith, V. V., Parizot, E., & Lambert, D. L. 2000b, ApJ submitted
Duncan, D. K., Lambert, D. L. & Lemke, M. 1992, ApJ, 401, 584
Duncan D. K., Primas, F., Rebull, L. M., Boesgaard, A. M., Deliyannis, C. P, Hobbs, L. M., King, J. R., & Ryan, S. G. 1997, ApJ, 488, 338
Duncan, D. K., Peterson, R. C., Thorburn, J. A., & Pinsonneault, M. H. 1998, ApJ, 499, 871
Fields, B. D., Olive, K. A., Vangioni-Flam, E., & Cassé, M. 2000, astro-ph/9911320
Fulbright, J. P., & Kraft, R. P. 1999, AJ, 118, 538
García-López, R. J., Lambert, D. L., Edvardsson, B., Gustafsson, B., Kiselman, D., & Rebolo, R. 1998, ApJ., 500, 241
Gingerich, O., Noyes, R. W., Kalkofen, W., & Cuny, Y. 1971, Sol. Phys., 18, 347
Israelian, G., García-López, R. J., & Rebolo, R. 1998, ApJ, 507, 805
Kohl, J. L., Parkinson, W. H., & Withbroe, G. L. 1977, ApJ, 212, L1 01
Lambert, D. L., Sheffer, Y., Federman, S. R., Cardelli, J. A., Sofia, U. J., & Knauth, D. C. 1998, ApJ, 494, 614
Lambert, D. L., Venn, K. A., Lemke, M., & Cunha, K. 2000, in preparation
Parizot, E. 1998, A&A, 331, 726
Peach, G. 1970, MmRAS, 73, 1
Primas, F., Duncan, D. K., Peterson, R. C., & Thorburn, J. A. 1999, A&A, 343, 545
Thévenin, F., & Idiart, T. P. 1999, ApJ, 521, 753
Zhai, M., & Shaw, D. M. 1994, Meteoritics, 29, 607

The Light Elements and Their Evolution
IAU Symposium, Vol. 198, 2000
L. da Silva, M. Spite, J. R. de Medeiros, eds.

The Light Elements Be and B as Stellar Chronometers in the Early Galaxy

Timothy C. Beers

Michigan State University, Dept. of Physics & Astronomy, E. Lansing, MI 48824 USA

Takeru K. Suzuki

University of Tokyo, Dept. of Astronomy, School of Science, University of Tokyo; Theoretical Astrophysics Division, National Astronomical Observatory, Mitaka, Tokyo, 181-8588 Japan

Yuzuru Yoshii

University of Tokyo, Institute of Astronomy, School of Science, University of Tokyo, Mitaka, Tokyo, 181-8588 Japan; Research Center for the Early Universe, School of Science, University of Tokyo Japan

Abstract. Recent detailed simulations of Galactic Chemical Evolution have shown that the heavy elements, in particular [Fe/H], are expected to exhibit a weak, or absent, correlation with stellar ages in the early Galaxy due to the lack of efficient mixing of interstellar material enriched by individual Type II supernovae. A promising alternative "chronometer" of stellar ages is suggested, based on the expectation that the light elements Be and B are formed primarily as spallation products of Galactic Cosmic Rays.

1. Introduction

It has become clear, from a number of lines of recent evidence, that the early evolution of the Galaxy is best thought of as a stochastic process. Within the first 0.5-1 Gyr following the start of the star formation process, chemical enrichment does not operate within a well-mixed uniform environment, as was assumed in the simple one-zone models that were commonly used in past treatments of this problem. Rather, the very first generations of stars are expected to have their abundances of heavy elements set by local conditions, which are likely to have been dominated by the yields from individual SNeII.

The seeds of this paradigm shift can be found in the observations, interpretations, and speculations of McWilliam et al. (1995), Audouze & Silk (1995), and Ryan, Norris, & Beers (1996). Models which attempt to incorporate these ideas into a predictive formalism have been put forward by Tsujimoto, Shigeyama, & Yoshii (1999; hereafter TSY), and Argast et al. (2000). Although they differ in the details of their implementation, and in a number of their assumptions, both of these models rely on the idea of enhanced star formation in the high-density shells of SN remnants, and the interaction of these shells of enriched material with a local ISM. The predictions which result are similar as well: (1) Both models are capable of reproducing the observed distributions of abundance (e.g., [Fe/H]) for stars in the tail of the halo metallicity distribution function (Laird et al. 1988; Ryan & Norris 1991; Beers 1999), and (2) Both models predict that the abundances of heavy elements, such as Fe, are not expected to show strong correlations with the ages of the first stars, at least up until an enrichment level on the order of [Fe/H]~ -2.0 is reached, i.e., at the time when mixing on a Galactic scale is possible (roughly 1 Gyr following the initiation of star formation).

Suzuki, Yoshii, & Kajino (1999; hereafter SYK, see also Suzuki, Yoshii, & Kajino, this volume) have extended the SN-induced chemical evolution model of TSY to include predictions of the evolution of the light element species ^{9}Be, ^{10}B, and ^{11}B, based on secondary processes involving spallative reactions with Galactic Cosmic Rays (hereafter GCRs). Recently, Suzuki, Yoshii, & Beers (2000) have considered the extension of this model to the prediction of ^{6}Li and ^{7}Li, and demonstrate that they naturally reproduce the recently detected slope in the abundance of Li in extremely metal-poor stars noted by Ryan, Norris, & Beers (1999; see also Ryan this volume). It is particularly encouraging that the same stochastic star-formation models which reproduce the observed trends of some (but not all) heavy elements, such as Eu, Fe, etc., also obtain predictions of the light element abundance distributions that match the available observations quite well, with a minimum of parameter tweaking.

In this contribution we summarize one of the more interesting predictions of the TSY/SYK class of models, that the abundances of the light elements Be and B (hereafter, BeB) might be useful as stellar chronometers in the early Galaxy (a time when the heavy element "age-metallicity" relationships are not operating due to the lack of global mixing). It appears possible that, with refinement of the modeling, and adequate testing, observations of BeB for metal-poor stars may provide a chronometer with "time resolution" on scales of tens of Myrs.

2. The Essence of the Model

In this section we would like to briefly explain our model of SN-induced star formation and chemical evolution. After formation of the very FIRST generation of (Pop. III) stars, with atmospheres containing gas of primordial abundance, the most massive of these stars exhaust their core H, and explode as SNeII. Following the explosion a shock is formed, because the velocity of the ejected material exceeds the local sound speed. Behind the shock the swept-up ambient material in the ISM accumulates to form a high-density shell. This shell cools in the later stages of the lifetime of a given SN remnant (SNR) and is a suitable site for the star formation process to occur. The SNR shells are expected to be distributed randomly throughout the early and rapidly evolving halo, and the shells do not easily merge with one another because of the large available volume. As a result, each SNR keeps its identity and the stars which form there reflect the abundances of material generated by their "parent" SN. TSY present this model, and describe the input assumptions, in more quantitative detail. Figure 1 provides a cartoon illustration of the processes which we discuss herein.

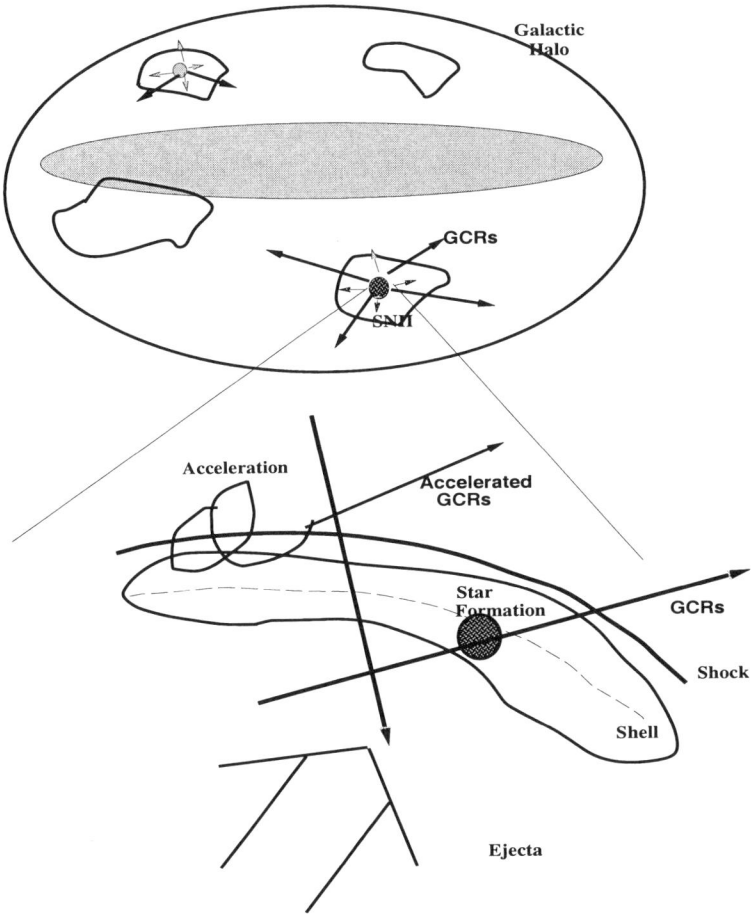

Figure 1. A simplified view of the early stages of chemical evolution in the Galactic halo. In the lower cutout we show star formation being triggered in SNR shells. See text for more detail.

One of the most important results of the TSY model is that stellar metallicity, especially [Fe/H], cannot be employed as an age indicator at these early epochs. Thus, to consider the expected elemental abundances of the metal-poor stars which form at a given time, a *distribution* of stellar abundances must be constructed, rather than adopting a global average abundance under the assumption that the gas of the ISM is well mixed. SYK constructed such a model, coupled with the model of SN-induced chemical evolution, which considers the evolution of the light elements.

SYK proposed that GCRs arise from the mixture of elements of individual SN ejecta and their swept-up ISM, with the acceleration being due to the shock formed in the SNR. GCRs originating from SNeII propagate faster than the material trapped in the clouds of gas making up the early halo. As a result, GCRs are expected to achieve uniformity throughout the halo faster than the general ISM, with its patchy structure. It follows that the abundances of BeB, which are mainly produced by spallation processes of CNO elements involving GCRs, are expected to exhibit a much tighter correlation with time than those of heavy elements, synthesized through stellar evolution and SN explosions.

We note that alternative models for the origin of spallative nucleosynthesis products have been developed which rely on the existence of *spatially correlated* SNeII in superbubbles of the early ISM (see Parizot & Drury 1999, and this volume). The superbubble model predicts a locally homogeneous production of both heavy and light elements, and the variety of stellar abundances which are observed are explained by the differing diffusion processes of metal-rich ([Fe/H] ~ -1) shells swept-up by the bubble and mixed with a metal-poor ([Fe/H] ~ -4) ISM. Tests of the "isolated" SN models vs. the superbubble models are expected to be conducted in the near future.

3. Abundance Predictions of the Model

Figure 2 shows the predicted behavior of the abundance of [Fe/H], log(Be/H), and log(B/H), as a function of time, over the first 0.6 Gyrs of the evolution of the early Galaxy, according to the model of SYK. At any given time (note that "zero time" is set by the onset of star formation, not the beginning of the Universe) the range of observed BeB is substantially less than that of Fe, owing to the global nature of light element production. For example, at time 0.2 Gyrs, the expected stellar [Fe/H] extends over a range of 50, while that of log (BeB/H) is on the order of 3-7.

During early epochs Fe is produced *only* by SNeII, and most of the Fe observed in stars formed in SNR shells originates from that contributed by the parent SN, because of uniformly low Fe abundance in the ISM at that time. Thus, the expected [Fe/H] of stars born at that time will exhibit a rather large range, reflecting differences in Fe yields associated with the different masses of the progenitor stars. On the other hand, according to the SYK model, most of the BeB is produced by spallation reactions of CNO nuclei involving globally transported GCRs. The observed abundances of BeB in metal-poor stars which formed at this time should reflect the global nature of their production, and the correlation between time and BeB abundance is expected to be much better than that found for heavier species.

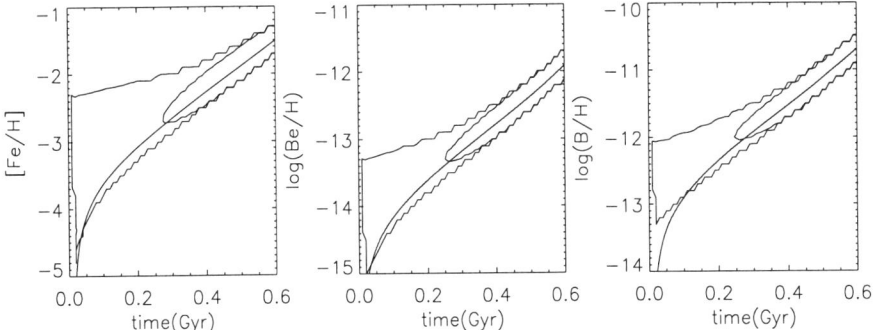

Figure 2. Predicted distribution of abundance for three elements, relative to H, for long-lived stars born at the indicated time *following* the initiation of star formation. The distributions have been convolved with Gaussians with $\sigma = 0.15$ dex to take into account expected observational errors. The two contours, from the inside to the outside, correspond to probability density 10^{-3} and 10^{-5} within the unit area $\Delta t = 10 (\text{Myr}) \times \Delta \log(\text{element}/\text{H}) = 0.1$. The solid lines show the predicted ISM gas abundances of each element.

In Table 1, we use the predictions from SYK, and the stellar abundance data from Boesgaard et al. (1999) for Be, to put forward "bold" estimates of stellar ages (since the onset of star formation). We note that these numbers are meant to be indicative, not definitive, predictions, as further tests of the model and its underlying assumptions still remain to be carried out. We have ordered the table according to estimated (Be) time since the onset of star formation in the early Galaxy.

It is interesting to consider the implications of this strong age-abundance relationship for individual stars which have been noted in the literature as having "peculiar" BeB (or ^{7}Li for that matter) abundances, at least as compared to otherwise similar stars of the same [Fe/H], T_{eff}, and log g. The set of "twins" G64-12 and G64-37 have been noted as one example of stars with very low metallicity, and apparently similar T_{eff} and log g, which never-the-less, exhibit rather different abundances of ^{7}Li. Could this difference be accounted for by a difference in AGE of these stars ? Answering this question is of great importance, and hopefully will be resolved in the near future.

4. Can we Test This Model ?

Yes, but it will take some hard work. Obviously, if there exists an independent method with which to verify the relative age determinations predicted by this model, that would be ideal. Fortunately, there have been numerous refinements in models of stellar atmospheres, and their interpretation, which may make

Table 1. Predictions of Stellar "Ages" Based on Be Abundance

Star	[Fe/H]	log(Be/H)	Be "age" (Gyr)
BD−13:3442	−3.02	−13.49	0.22 (−0.07,+0.03)
BD+03:740	−2.89	−13.33	0.26 (−0.05,+0.03)
HD 140283	−2.56	−13.08	0.32 (−0.05,+0.02)
BD+37:1458	−2.14	−13.07	0.32 (−0.05,+0.02)
HD 84937	−2.20	−12.94	0.35 (−0.05,+0.03)
BD+26:3578	−2.32	−12.79	0.39 (−0.05,+0.03)
BD+02:3375	−2.39	−12.80	0.39 (−0.05,+0.03)
BD−04:3208	−2.35	−12.69	0.41 (−0.05,+0.03)
HD 19445	−2.10	−12.55	0.45 (−0.04,+0.03)
HD 64090	−1.77	−12.49	0.46 (−0.04,+0.03)
BD+20:3603	−2.22	−12.47	0.46 (−0.04,+0.03)
BD+17:4708	−1.81	−12.40	0.48 (−0.04,+0.03)
HD 219617	−1.58	−12.15	0.54 (−0.04,+0.02)
HD 74000	−2.05	−12.10	0.55 (−0.04,+0.02)
HD 103095	−1.37	−12.04	0.56 (−0.04,+0.02)
HD 194598	−1.25	−11.88	0.59 (−0.03,+0.01)
BD+23:3912	−1.53	−11.92	0.59 (−0.03,+0.01)
HD 94028	−1.54	−11.55	> 0.60

this feasible (see Fuhrmann 2000). In order to apply the methods described by Fuhrmann, one requires high-resolution, high-S/N spectroscopy of individual stars. It is imperative that the present-generation 8m telescopes (VLT, SUBARU, GEMINI, HET) obtain this data, so that this, and other related questions, may be addressed with the best possible information.

Another feasible test would be to compare the abundances of BeB with [Fe/H], and other heavy elements, for a large sample of stars with [Fe/H] < −2.0. If the superbubble model is the correct interpretation, with an implied locally homogeneous production of the light elements, then one might expect to find correlations between the abundances of various heavy element species (including those other than Fe and O) and BeB. Simultaneous observations of light and heavy elements for stars of extremely low abundance are planned with all the major 8m telescopes, so it should not be too long before a sufficiently large sample to carry out this test is obtained.

One can also seek, as we have, confirmatory evidence in the predicted behavior of ^{7}Li vs. [Fe/H] (Suzuki et al. 2000).

5. Other Uses for This Model

If the model we have considered here can be shown to be correct, there are several new avenues of investigation which are immediately opened. For example, if one were able to "age rank" stars on the basis of their BeB abundances, one could refine alternative production mechanisms for the light element Li which are not driven by GCR spallation, including the SN ν−process and/or production via a giant-branch Cameron-Fowler mechanism (see Castilho et al. , this volume), in stellar flares, etc..

Furthermore, since BeB nuclei are more difficult to burn than Li nuclei, one could imagine a powerful test for the extent to which depletion of Li has operated in metal-poor dwarfs, with important implications for the Li constraint on Big Bang Nucleosynthesis (BBN). Realistic modeling of BeB evolution at early epochs may also help distinguish between predictions of standard BBN, non-standard BBN, and the accretion hypothesis (see Yoshii, Mathews, & Kajino 1995).

An age ranking of metal-poor stars based on their BeB abundances, in combination with measurements of their alpha, iron-peak, and neutron-capture elements, would open the door for an unraveling of the mass spectrum of the progenitors of first generation SNeIIs, and allow one to obtain direct constraints on their elemental yields as a function of mass, a key component to models of early nucleosynthesis.

Acknowledgments. TCB expresses gratitude to the IAU for support which enabled his attendance at this meeting, and acknowledges partial support from the National Science Foundation under grant AST 95-29454. TCB also wishes to express his congratulations to the LOC and SOC for a well-run, scientifically stimulating, and marvelously located meeting. YY acknowledges a Grant-in-Aid from the Center of Excellence (COE), 10CE2002, awarded by the Ministry of Education, Science, and Culture, Japan.

References

Argast D., Samlund, M., Gerhard, O.E., & Thielemann, F.-K. 2000, A&A, in press
Audouze J., & Silk, J. 1995, ApJ, 451, L49
Beers, T.C. 1999, in Third Stromlo Symposium: The Galactic Halo, eds. B. Gibson, T. Axelrod, & M. Putman, (ASP, San Francisco), 165, p. 206
Boesgaard, A.M., Deliyannis, C.P., King, J.R., Ryan, S.G., Vogt, S.S., & Beers, T.C. 1999, AJ, 117, 1549
Fuhrmann, K. 2000, in The First Stars, Proceedings of the Second MPA/ESO Workshop, eds. A. Weiss, T. Abel, & V. Hill (Springer, Heidelberg), in press
Laird, J.B., Carney, B.W., Rupen, M.P., & Latham, D.W. 1988, AJ, 96, 1908
McWilliam, A., Preston, W., Sneden, C., & Searle, L. 1995, AJ, 109, 2757
Ryan, S.G., Norris, J.E., & Beers, T.C. 1996, ApJ, 471, 254
Ryan, S.G., Norris, J.E., & Beers, T.C. 1999, ApJ, 523, 654
Ryan, S.G., & Norris, J.E. 1991, AJ, 101, 1865
Suzuki, T.K., Yoshii, Y., & Kajino, T. 1999, ApJ, 522, L125 (SYK)
Suzuki, T.K., Yoshii, Y., & Beers, T.C. 2000, ApJ, submitted
Tsujimoto, T., Shigeyama, T., & Yoshii, Y. 1999, ApJ, 519, L63 (TSY)
Yoshii, Y., Mathews, G.J., & Kajino, T. 1995, ApJ, 447, 184

A very reduced upper limit on the interstellar abundance of beryllium[1]

Guillaume Hébrard

Institut d'Astrophysique de Paris, CNRS, 98 bis boulevard Arago, F-75014 Paris, France

Martin Lemoine

DARC, Observatoire de Paris-Meudon, France

Roger Ferlet and Alfred Vidal-Madjar

Institut d'Astrophysique de Paris, France

Abstract. We present the results of observations of the $\lambda 3130$Å interstellar absorption doublet of ^{9}Be II in the direction of ζ Per. The data were obtained at the Canada-France-Hawaii 3.6m telescope using the Gecko spectrograph at a resolving power $\sim 110\,000$ and a signal-to-noise ratio ~ 2000. The ^{9}Be II lines are not detected and we obtain an upper limit on the equivalent width $W \leq 30\mu$Å. This upper limit is 7 times below the lowest upper limit ever reported hitherto. The derived interstellar abundance is $(^9$Be/H$) \leq 7 \times 10^{-13}$; it corresponds to an upper limit $\delta_{Be} \leq -1.5$ dex on the depletion factor of ^{9}Be.

1. Introduction

Beryllium is created in Big Bang nucleosynthesis with an extremely low primordial abundance. Subsequently, it is solely formed in spallation reactions of galactic cosmic rays (GCR) interacting with interstellar atoms, and is thoroughly destroyed through astration of interstellar gas. This simple scenario allows to account for the observed Pop I abundance of ^{9}Be, $(^9$Be/H$)_{PopI} \simeq 1.3 \times 10^{-11}$ (Boesgaard 1976), the solar abundance $(^9$Be/H$)_\odot \simeq 1.4 \times 10^{-11}$ (Chmielewski et al. 1975) and the meteoritic abundance $(^9$Be/H$)_{met} \simeq 2.6 \times 10^{-11}$ (Anders & Grevesse 1989). For this reason, ^{9}Be together with ^{6}Li, which shares a similar evolutionary picture, are used as tracers of cosmic ray spallation activity.

From ground observatories, beryllium can only be observed through the resonance ^{9}Be II doublet at 3130Å. It has never been detected in the ISM [see Hébrard et al. (1997) for a review of reported upper limits]. Beryllium depletion factor, largely unknown, is predicted to be ~ -0.2 dex in a correlation with first ionization potential, and ~ -1.5 dex in a correlation with condensation temperature (Boesgaard 1985).

---

[1]Based on observations collected at the Canada-France-Hawaii Telescope, Hawaii, USA.

We report here on our observations of the interstellar absorption of ^{9}Be II in the direction of ζ Per. Further details can be found in Hébrard et al. (1997).

2. Observation and result

Our observations were conducted in January 1994 and October 1995 at the Canada-France-Hawaii 3.6m telescope. We used the spectrograph Coudé f/4 Gecko at high resolving power $\lambda/\Delta\lambda \simeq 110\,000$. We reach a signal-to-noise ratio of ~ 2000 per pixel in the vicinity of the expected line, for a total integration time of 25h. The ^{9}Be II doublet was not detected on our spectra.

The absence of detection at such a high signal-to-noise ratio and resolution translates into a very reduced upper limit on the beryllium column density. Indeed, the limiting detectable equivalent width at 3 σ is $W_{lim} \simeq 30\,\mu\text{Å}$, implying the upper limit on the column density $N(^9\text{Be II}) \leq 1.0 \times 10^9$ cm^{-2}.

We assume now that at least 90% of the interstellar beryllium is present in the first ionization stage ^{9}Be II (Boesgaard 1985). This is supported by the ratios between ionization stages of others elements. For example, in this same line of sight, $N(\text{Mg I})/N(\text{Mg II}) \leq 10^{-2}$ and $N(\text{S III})/N(\text{S II}) \leq 10^{-3}$ (Snow 1977). Taking the hydrogen column density toward ζ Per $N(\text{H}) = 1.6 \times 10^{21}$ cm^{-2} (Savage et al. 1977), we thus deduce an upper limit of the interstellar abundance for ^{9}Be toward ζ Per: $(^9\text{Be/H})_{\zeta\,Per} \leq 7 \times 10^{-13}$.

3. Conclusion

Our interstellar abundance of ^{9}Be is at least 35 times less than the cosmic abundance, $(^9\text{Be/H})_{cosm} \simeq 2.6 \times 10^{-11}$. It corresponds to a depletion factor $\delta_{Be} \leq -1.5$ dex. This is a new and much more stringent upper limit compared to previous ones ($\delta_{Be} \leq -0.4$, Boesgaard 1985).

Our present upper limit favours the Field (1974) model of dust grain formation in stellar material. In effect, the predicted depletion for the condensation temperature of ^{9}Be is ~ -1.5 dex while the Snow (1975) model of dust grain formation by chemical trapping predicts $\delta_{Be} \simeq -0.2$ dex.

References

Anders, E., & Grevesse, N. 1989, Geochim. Cosmochim. Acta 53, 197
Boesgaard, A. M. 1985, PASP 97, 37
Boesgaard, A. M. 1976, ApJ 210, 466
Chmielewski, Y., Müller, E. A., & Brault, J. W. 1975, A&A 42, 37
Field, G. B. 1974, ApJ 187, 453
Hébrard, G., Lemoine, M., Ferlet, R., & Vidal-Madjar, A. 1997, A&A 324, 1145
Savage, B. D., Bohlin, R. C., Drake J. F., & Budich, W. 1977, ApJ 216, 291
Snow, T. P. Jr. 1975, ApJ 202, L87
Snow, T. P. Jr. 1977, ApJ 216, 724

The Show

STELLAR KNOWLEDGE TO AND FROM LIGHT ELEMENTS

Francesca Matteucci, with Roger Cayrel, on the right, and Renan de Medeiros and Monique Spite, on the left

The Light Elements and Their Evolution
IAU Symposium, Vol. 198, 2000
L. da Silva, M. Spite, J. R. de Medeiros, eds.

Effects of Photospheric Temperature Inhomogeneities on Lithium abundance Determinations (2D)

Roger Cayrel

Observatoire de Paris, DASGAL, 61, Observatoire de Paris F75014 Paris, France

Matthias Steffen

Astrophysikalisches Institut Potsdam, An der Sternwarte 16 D-14482 Potsdam, Germany

Abstract. Based on detailed 2D radiation hydrodynamics (RHD) simulations, we have investigated the effects of photospheric temperature inhomogeneities induced by convection on spectroscopic determinations of the lithium abundance. Computations have been performed both for the solar case and for a metal-poor dwarf. NLTE effects are taken into account, using a five-level atomic model for Li I. Comparisons are presented with traditional 1D models having the same effective temperature and gravity. The net result is that, while LTE results differ dramatically between 1D and 2D models, especially in the metal-poor case, this does not remain true when NLTE effects are included: 1D/2D differences in the inferred NLTE Li abundance are always well below 0.1 dex. The present computations still assume LTE in the continuum. New computations removing this assumption are planned for the near future.

1. Introduction

It would be an offense to the audience here to pretend to explain why it is important to determine accurately the abundance of ^{6}Li and ^{7}Li in the oldest stars. In this respect, we have nothing to add to the exposition by François Spite (this volume). We will right away mention the two major problems which may cast doubts on our real knowledge of the actual initial abundance of Li in the oldest stars, and consequently in the primordial matter. (i) While standard models of the internal structure of metal-poor dwarfs do not deplete ^{7}Li, more sophisticated models including rotationally induced mixing (Pinsonneault et al. 1992) have predicted that the measured abundance in the photosphere is 5 to 10 times less than the initial abundance representative of Big Bang material. (ii) On top of that, Kurucz (1995) claimed that the hot and cold convective structures produce large effects in metal-poor stellar photospheres, where the convection zone reaches the line formation layers. The claimed effect is an overionization of Li by a factor of 10, leading to an underestimation of the abundance of Li when derived from the resonance line of Li I (λ 670.8 nm) in the usual way.

If these two statements are correct, the true abundance of Li in primordial matter is 50 to 100 times higher than the value derived from 1D, LTE models of halo subdwarfs so far. The first factor of 5 to 10 has been discussed in a previous paper by Ryan (this symposium), and shown to be likely much smaller, of the order of 1 to 1.4. We shall not come back to this point, which we consider as very well treated.

Before this symposium, a single paper (Asplund et al. 1999) has dealt with the question of the other factor of 10 claimed by Kurucz (1995), whose arguments were based on a simplified two-column model. In contrast, the work by Asplund et al. relies on realistic 3D hydrodynamical models, similar to the simulations of the solar granulation (Stein & Nordlund 1998), but with parameters appropriate for two metal-poor stars: HD 140283 and HD 84937, both subgiants. The computation of the lithium resonance line was made under the assumption of LTE, and the correction to be applied to the Li abundance derived from standard 1D models was found to be large, of the order of -0.2 to -0.35 dex. Note that these corrections have the opposite sign as Kurucz's prediction! However, Kiselman (1997, 1998) had shown, in the solar case, that NLTE and LTE computations lead to significantly different values of equivalent widths of the Li I λ 670.8 nm line over hot and cold structures (see Fig. 3 of his 1997 paper, top panel).

For this reason, we decided to undertake NLTE radiation hydrodynamics computations for the case of a metal-poor star, and we report here on the results of this investigation. In the next section we recall former work related to simulations of the solar granulation, a useful benchmark for checking the theory, but not directly applicable to metal-poor stars. In section 3 we describe the assumptions underlying the construction of the 2D RHD models used for the spatially resolved computation of the lithium resonance line. Section 4 gives the description of the NLTE treatment of the Li atom, and section 5 summarizes our results and compares them to those presented by M. Asplund (this symposium). Finally, our conclusions are listed in section 6.

2. Former work at solar metallicity

While there is only one paper dealing with multidimensional atmospheres for metal-poor stars (cited above), there are several studies for the solar case, aimed at understanding the variation of the continuous radiation intensity (granulation), and the behavior of spectral lines across the solar granulation pattern. Several of these works use snapshots from 3D simulations by Stein & Nordlund (1998), such as Kiselman (1997,1998) and Uitenbroek (1998). Gadun & Pavlenko (1997) use their own 2D simulations.

Of particular interest for us are the papers dealing with the combined effects of multidimensional structures and NLTE (Kiselman and Uitenbroek). It is clear from Kiselman (1997) that the NLTE behavior of the Li I λ 670.8 nm resonance line is drastically different from its LTE behavior, in 2D as well as in 3D models. The result which is the most relevant for us is the difference of 30 per cent on the predicted mean equivalent width $\langle W \rangle$ of the line, leading to a similar change in the derived lithium abundance (NLTE/LTE abundance correction +0.15 dex). But another interesting difference is the reverse behavior of the equivalent width W as a function of surface continuum brightness I_c. While LTE computations

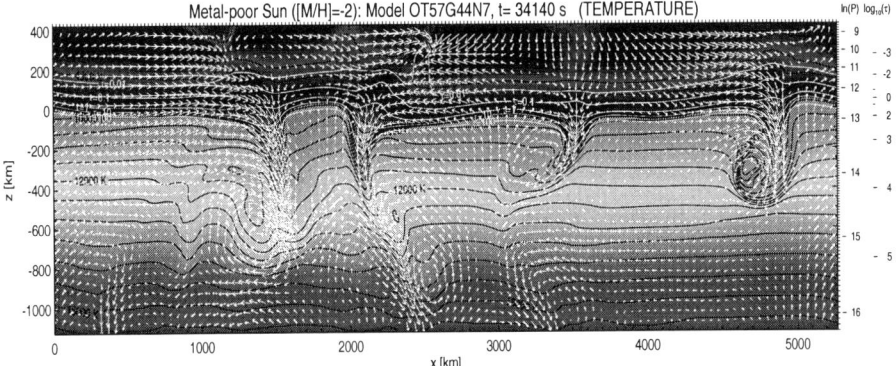

Figure 1. Snapshot from the metal-poor Sun simulation, showing the velocity field (arrows) and temperature structure (black contour lines). White lines are curves of iso-optical depth. The origin of the geometrical scale (left) corresponds to $\tau_{\mathrm{Ross}} = 1$; scales at right refer to average gas pressure P and average optical depth τ_{Ross}. This simulation was done on a 210x106 grid (tick marks at right and top), with a vertical extent of about 1500 km ($\approx 7 H_p$), and a horizontal period of 5250 km (upper and lower boundaries 'open', lateral boundaries periodic).

result in a strongly positive slope in the W versus I_c diagram (with a large scatter around the mean relation), NLTE computations show a slightly negative slope and a much tighter (anti-)correlation between I_c and W. This reflects the fact that the population of the Li I levels is much more controlled by the local temperature in LTE than in NLTE, where, for weak lines, the photoionization rates play the dominant role. So, even if the general conclusion of the above mentioned papers is that, in the solar case, the abundance determination of Li is not strongly affected by the combined effects of temperature fluctuations and NLTE, in comparison to what is obtained with classical 1D models having the same effective temperature and gravity, it appears unsafe to compute abundances from multidimensional stellar atmospheres based on the assumption of LTE line formation.

3. 2D radiation hydrodynamics models

Our LTE and NLTE computations have been performed on the basis of several snapshots from a 2D numerical simulation of convection in a stellar envelope having the same effective temperature and gravity as the Sun, but a 100 times smaller metallicity. Basically, the time dependent equations of hydrodynamics are solved for a compressible fluid, with an energy equation including 3 terms: turbulent and shock dissipation of kinetic energy, diffusive transport of heat, and radiative energy exchange. The main limitation of the code is the restriction of the flow to two spatial dimensions. Magnetic fields and rotation are ignored.

Apart from these simplifications, as much realistic physics as possible is included. The equation of state accounts for ionization of hydrogen and helium as well as H_2 molecule formation, opacities have been adapted from Kurucz's ATLAS code and include line absorption. For the computation of the radiative energy balance, we employ a multi-dimensional, non-local, frequency-dependent radiative transfer scheme, actually solving the transfer equation along 26880 independent rays of various inclinations, using an efficient modified Feautrier method (Feautrier 1964). At the bottom boundary, inflowing matter has a given specific entropy, which is adjusted to produce the prescribed effective temperature of the atmosphere. Energy dissipation on small scales is roughly modeled by introducing a subgrid scale eddy viscosity, depending on the grid resolution and local velocity gradients in the usual way. Details of the employed hydrodynamics code can be found in Ludwig et al. (1994) and Freytag et al. (1996). Fig. 1 shows a sample snapshot from our metal-poor Sun simulation. Note the complex velocity pattern and the occurrence of very strong temperature gradients. The relevant region for the formation of the Li I line is the $\tau = 0.1$ contour line.

4. NLTE computation of the Li I resonance line

The computation of the Li I spectrum is greatly simplified by the fact that all lines of Li I are weak, Li being a trace element. So the radiation field in the line is, to first approximation, the same as the continuous radiation field, a single iteration being sufficient for taking care of the small perturbation of the monochromatic radiation field brought about by the line.

We have, as a first step, approximated the Li I atomic configuration by a five level atom, exactly as done by Uitenbroek (1998). This leaves six permitted bound-bound transitions, and five photoionization rates needing the computation of the continuous radiation field at frequencies above the threshold, until the contribution of the product of photoionization cross-section and mean intensity of the radiation becomes negligible. Because in the UV the contribution of lines to the opacity is important, we have used Kurucz's Opacity Distribution Functions (ATLAS 9) for the relevant metallicity. This multiplies the computation time by 12, as each opacity bin is subdivided into 12 subintervals. So, for each snapshot, the transfer equations must be solved for about 12×120 wavelengths, along 26880 different rays (note that these extensive computations are done only after the actual hydrodynamical simulation for a few selected snapshots). After this, it is possible to compute all the coefficients in the equations of statistical equilibrium (see Mihalas 1970, p. 144). Once the departure coefficients b_i are evaluated, the line can be computed using the source function:

$$S = \frac{\kappa_c}{\kappa_l + \kappa_c} B_\nu + \frac{\kappa_l}{\kappa_l + \kappa_c} S_l$$

where:

$$S_l = S_{ij} = \frac{2h\nu_{ij}^3}{c^2} \frac{1}{(b_i/b_j)\exp(h\nu_{ij}/kT) - 1}$$

is the line source function. The departure coefficient for the lower level is b_i and the one for the upper level b_j. The other notations are standard. Subscript "c" stands for continuum, and "l" for "line". Note that in NLTE the expression of the partition function is modified and becomes:

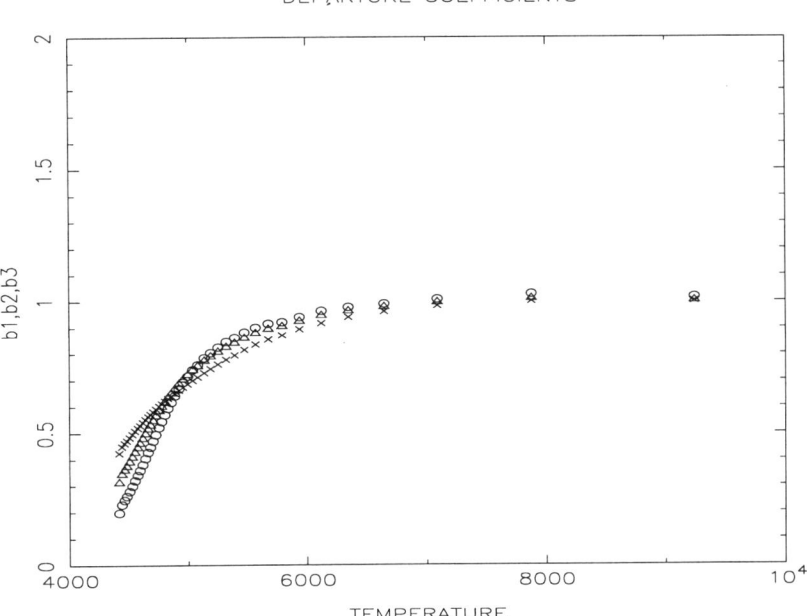

Figure 2. Variation of the first three departure coefficients with depth (actually temperature in this graph), based on an 1D ATLAS 9 model of the solar atmosphere (mixing-length parameter $\alpha = 0.5$). Circles refer to the ground level (b_1), triangles to the 2p level (b_2), and crosses to the 3s level (b_3). As expected, all three coefficients become very close to 1 in the deep layers, and reach values below 0.5 in surface layers, as found by Carlsson et al. (1994). The adopted logarithmic abundance of Li is 2.2, on the scale log(nH)= 12.0.

$$U = \sum_i b_i g_i \exp(-\chi_i/kT)$$

where the g_i and χ_i are the statistical weight and the excitation energy of level i, respectively.

5. Results and discussion

We have first tested our program on a Kurucz ATLAS 9 1D solar model to see whether the b_i had the expected behavior, already computed by Carlsson et al. (1994). Fig. 2 shows the depth-dependence of the first 3 departure coefficients, applying to levels 2s, 2p and 3s, respectively.

Next, we have computed the equivalent width of the Li I resonance line for the Kurucz 1D solar model and for two 2D solar snapshots, still for the logarithmic Li abundance 2.2. Fig. 3 shows the variation of the equivalent width

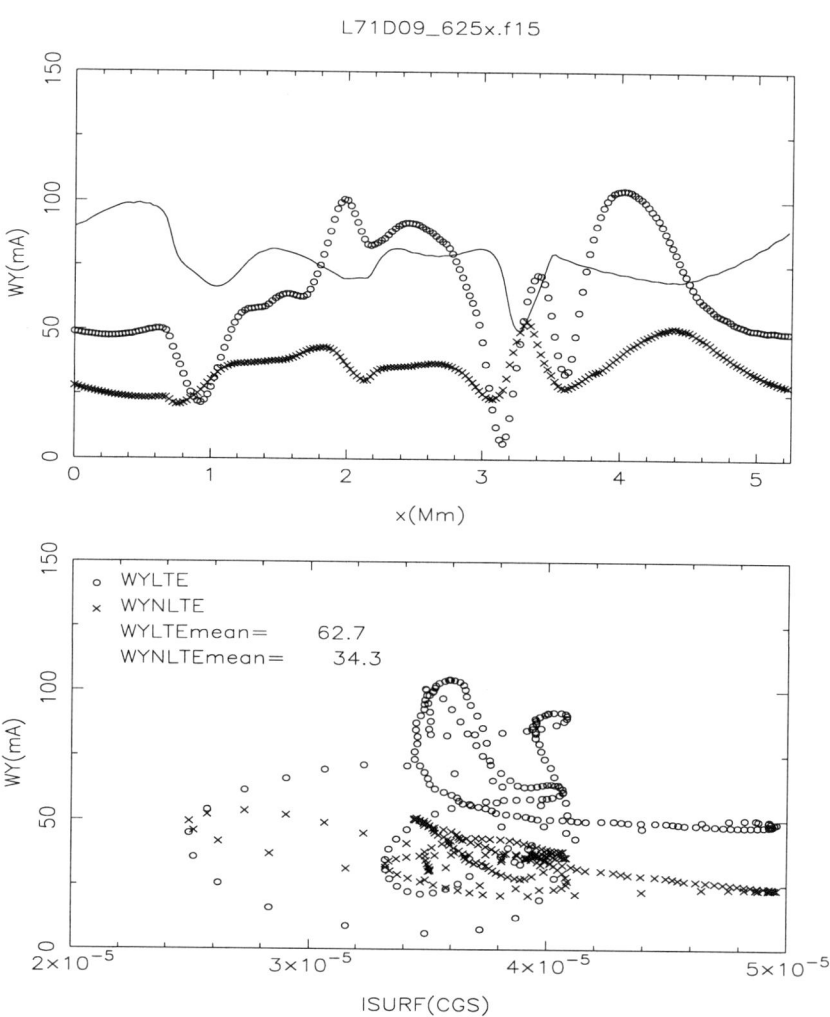

Figure 3. **Top:** LTE (circles) and NLTE (crosses) equivalent widths of the Li I resonance line as a function of horizontal position x, for a representative snapshot from a 2D simulation of *solar granulation*. The thin line gives the continuum intensity at λ 670.8 nm in arbitrary units. **Bottom:** Same equivalent widths as a function of continuum intensity ISURF. All data derived from vertical rays (μ=1).

Table 1. LTE and NLTE mean equivalent width $\langle W \rangle$ [mÅ] of the Li I λ 670.8 line for 2 different snapshots from a 2D hydrodynamical simulation of *solar surface convection*, obtained from the horizontally averaged *flux* spectrum. For comparison, the equivalent widths resulting for a 1D ATLAS9 reference model, computed with the same line formation code, are given in the first row. The results from Kieslman (1997) given in the last rows have been rescaled to the same Li abundance of 2.2; they refer to intensity ($\mu = 1$).

1D model or RHD snapshot	$\langle W \rangle$ (LTE)	$\langle W \rangle$ (NLTE)
1D Kurucz ($\alpha = 0.5$)	43	32
2D, L71D09-605	59	35
2D, L71D09-625	71	36
Kiselman 1D (OSMARCS)	44	42
Kiselman 3D	52	37

W (for the *intensity* normal to the surface) of the Li I 670.8 nm line over the simulated granulation pattern, both as a function of the horizontal position x (top) and as a function of the continuum intensity I_c (bottom) for one particular snapshot. Note the wide variation of W computed in LTE, compared to the much more limited excursion of W computed in NLTE. The mean equivalent widths for the *flux* spectrum integrated over the full length of the sample are given in Table 1 for the two solar snapshots and for the 1D reference model having the same effective temperature, gravity and (solar) metallicity. In each case, the line is computed in LTE as well as in NLTE. We note that, in NLTE, the results for the 2D snapshots do not differ significantly from the 1D case.

Finally, we have carried out a similar procedure for our metal-poor stellar example, again computing the equivalent width of the Li I resonance line for a Kurucz 1D model and for five snapshots from our 2D hydrodynamical simulation; as before, a Li abundance of 2.2 was adopted. Fig. 4 shows the variation of the equivalent width W of the Li I resonance line over the stellar granulation pattern for the typical snapshot displayed in Fig. 1. The difference between LTE and NLTE is even more pronounced than in the solar case, the NLTE correlation between W and I_c being much tighter and of opposite sign compared to LTE.

The mean equivalent widths derived from the horizontally averaged *flux* spectrum are listed in Table 2 for the 5 snapshots and for the Kurucz 1D reference model. In LTE, the 1D/2D difference is huge (granulation abundance correction ≈ -0.45 dex). But remarkably, the 2D NLTE line strengths show very little dispersion and do not indicate any significant offset with respect to the 1D case. An obvious conclusion from these results is that the 2D LTE computations are way off, strongly underestimating the Li abundance. In NLTE, the error introduced by representing the inhomogeneous stellar atmosphere by a flux-constant 1D Kurucz model appears to be almost negligible.

The mechanism behind the spatial variation of the line strength is clearly identified on Figs 1 and 4: hot granules produce at the same time a steeper temperature gradient and a lower temperature in the line formation region. In LTE, the latter leads to an overpopulation of the lower level of the transition

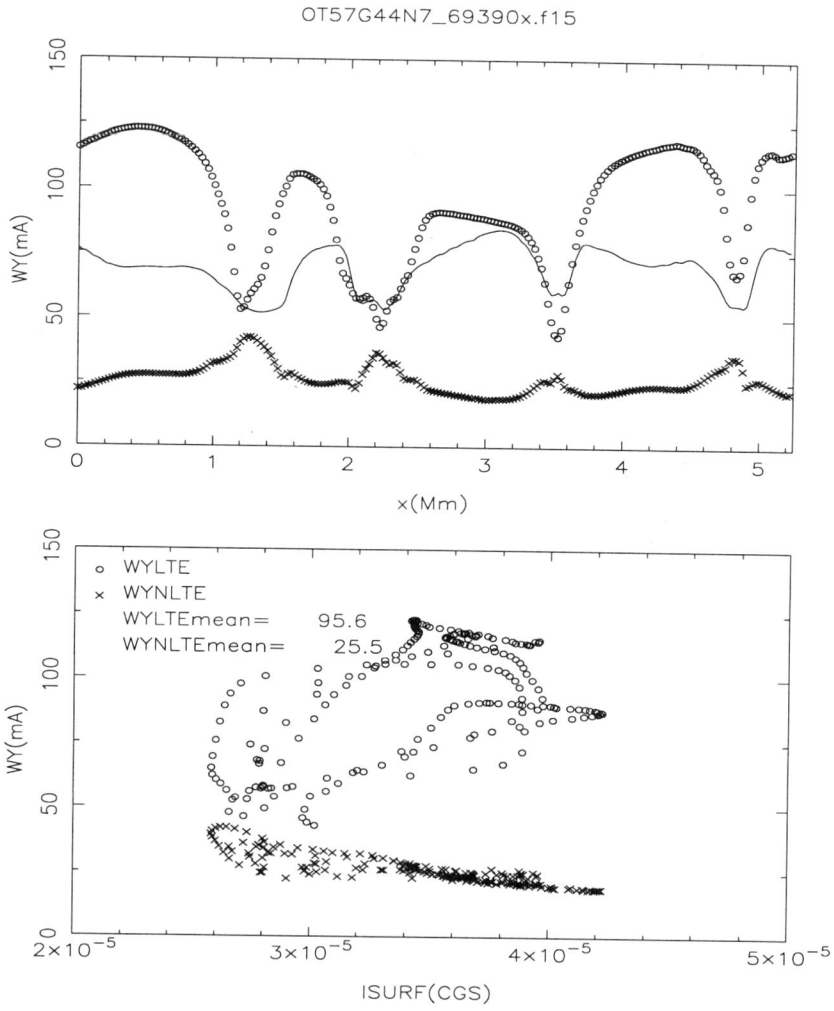

Figure 4. Same plots as Fig. 3, but for the snapshot from the *metal-poor star simulation* shown in Fig. 1

due to a shift of the ionization equilibrium towards neutral particles (Saha equation). Since both effects enhance the equivalent width of the line, the correlation between continuum intensity and LTE line strength is clearly positive (there is considerable dispersion around the mean $W(I_c)$ relation, however, due to the presence of inclined thermal inhomogeneities). This result is inverse of what is expected for a set of hot and cool *radiative equilibrium atmospheres* lined up side by side: here the line would weaken in the hot atmosphere, because the population of the lower level is the dominant factor. It is clear that the reasoning of Kurucz fails, essentially because the actual vertical stratification of hot and cold regions has little to do with the stratification of hot and cool radiative equilibrium atmospheres.

Table 2. Same as Table 1, but for 5 different snapshots from a 2D hydrodynamical simulation of surface convection in a *metal-poor star* ($[M/H] = -2$) with solar effective temperature and gravity. For comparison, the results for the corresponding 1D ATLAS9 model, computed with the same line formation code, are given in the first row.

1D model or RHD snapshot	$\langle W \rangle$ (LTE)	$\langle W \rangle$ (NLTE)
1D Kurucz ($\alpha = 0.5$)	38.4	26.1
T57G44N7-43108	116.0	27.7
T57G44N7-69390	115.4	27.1
T57G44N7-83752	112.7	27.3
T57G44N7-87007	110.9	27.7
T57G44N7-97644	108.2	26.5

A positive W-I_c correlation alone is not sufficient to explain the LTE result that the mean equivalent width is much larger in 2D than in 1D: as long as the fluctuations of the line strength with temperature remain in the linear domain, the mean equivalent width is not affected by the temperature fluctuations. But as the population of the ground level N_0 of Li I varies *exponentially* with T, a nonlinearity sets in: symmetrical temperature fluctuations produce asymmetrical population variations. Assume that Li is almost completely ionized and consider only the exponential factor of the Saha equation for simplicity. Then

$$N_0(T) = N_0(T_0)\exp(+\chi/kT) = N_0(T_0) * \left\{ 1 - qs + qs^2 + \frac{1}{2}q^2 s^2 + \ldots \right\}$$

where χ is the ionization potential and we have defined $q \equiv \chi/kT_0$ and $s \equiv (T - T_0)/T_0$, T_0 being the mean temperature. Taking the horizontal average assuming symmetrical temperature variations ($\langle s \rangle = 0$), we obtain $\langle N \rangle \approx N_0(T_0) * (1 + 0.5 q^2 \langle s^2 \rangle)$ ($\sqrt{\langle s^2 \rangle} \leq 0.04$; $q \approx 12$). So the mean of W is biased towards larger equivalent widths in an inhomogeneous atmosphere. This explains part of the 1D/2D difference found in our numerical computations. The main contribution, however, is attributed to the lower mean photospheric temperature in the 2D model as a result of adiabatic cooling due to overshooting: $T_0(2D) < T(1D)$.

In NLTE, the local temperature plays little role. Rather, the radiation field is the dominant factor. Hence, photoionization overionizes the lower level over a hot granule, and the equivalent width becomes smaller. As a result, NLTE equivalent widths are anti-correlated with the continuum surface brightness. The variation of W is much smaller than in LTE, because the angle-averaged radiation field depends only weakly on horizontal position at the height of line formation.

Finally, we would like to mention the work by M. Asplund, who has also presented his NLTE line formation calculations for Li I on this symposium. His results are based on the 3D snapshots that he had used previously for the LTE investigation of the subgiants HD 140283 and HD 84937. He reached, on this completely independent set of hydrodynamical models, and with a different NLTE code, the same conclusions as we did: NLTE line formation in multidimensional models is quite different from the LTE case, in the way that the NLTE multi-dimensional Li abundance is much closer to the abundance derived from classical flux-constant 1D models. A detailed comparison of our results shows that the remaining differences can be traced to the use of a Li I atom with 5 levels in our case, and with 20 levels in the case of Asplund et al. (this volume). Another difference is that the temperature inhomogeneities are somewhat enhanced in 2D models with respect of what occurs in 3D models, leading to correspondingly larger LTE granulation abundance corrections. But this is only a minor point. Certainly, the two groups agree that LTE Li abundance determinations relying on multidimensional hydrodynamical simulations of convection in metal-poor dwarfs are highly misleading.

6. Conclusions

1. The statement of Kurucz (1995) that abundances of lithium derived from standard 1D models of metal-poor stellar atmospheres is too small by a factor of 10 is not supported by actual multidimensional NLTE computations. Even the sign of the correction is doubtful, and the error is well below 0.1 dex, both according to our investigation and the one presented by M. Asplund on this symposium.

2. LTE abundance determinations based on inhomogeneous atmospheres are strongly discouraged. They produce large "granulation abundance corrections" due to non-linear effects in the direction opposite to Kurucz' prediction, but the actual NLTE line formation mechanism couples the population of the atomic levels more closely to the mean radiation field than to the local temperature.

3. The combination of multidimensional models with NLTE line formation for the Li I λ 670.8 nm resonance line leads to the same lithium abundance as that derived from NLTE analysis with flux-constant 1D models, abundance differences being less than 0.1 dex. However, this result must not be hastily generalized to other atoms with a different atomic structure.

4. An obvious future improvement is to extend the NLTE analysis to the continuum, which has be assumed here to be in LTE. We plan to do that in the near future. If it turns out that the H^- ion is affected by NLTE, this will raise a new question: should such effects be included already in the radiation hydrodynamics code, which determines the amplitude of the thermodynamical fluctuations?

References

Asplund, M., Nordlund, A, Trampedach, R., Stein, R.F. 1999, A&A, 346, L17
Carlsson, M., Rutten, R.J., Bruls, J.H.M.J., Shchukina, N.G. 1994, A&A, 288, 860
Freytag, B., Ludwig, H.-G., Steffen, M. 1996, A&A, 313, 497
Feautrier. P. 1964, C.R. Acad. Sci. Paris, 258, 3189
Gadun A.S., Pavlenko, Ya. V. 1997, A&A, 324, 281

Kiselman, D. 1997, ApJ, 489, L107
Kiselman, D. 1998, A&A, 333, 732
Kurucz, R.L. 1995, ApJ, 452, 102
Ludwig, H.-G., Jordan, S., Steffen, M. 1994, A&A, 284, 105
Mihalas, D. 1970 *Stellar Atmospheres* W. H. Freeman & Co.
Pinsonneault, M.H., Deliyannis, C. P., Demarque, P. 1992, ApJS, 78, 179
Stein, R.F., Nordlund, Å, 1998, ApJ, 499, 914
Uitenbroek, H. 1998, ApJ, 498, 427

The Light Elements and Their Evolution
IAU Symposium, Vol. 198, 2000
L. da Silva, M. Spite, J. R. de Medeiros, eds.

The light elements in the light of 3D hydrodynamical model atmospheres

Martin Asplund

Uppsala Astronomiska Observatorium, Box 515, SE-751 20 Uppsala, Sweden

Abstract. The influence of stellar granulation on the line formation of the light elements in metal-poor stars has been investigated by means of 3D hydrodynamical model atmospheres. Due to a lucky (?) coincidence, the effects due to the lower photospheric temperatures with 3D models compared with classical 1D model atmospheres are almost exactly balanced by a pronounced over-ionization of Li I, as concluded from detailed 3D NLTE calculations. Additionally, the effects of convective line asymmetries on ^{6}Li isotope determinations have been investigated and found to be of minor importance.

1. Introduction

Stellar abundances can function as archaeological remains to trace the cosmic, galactic and stellar evolution and therefore play a central role in astrophysics and cosmology. It is however important to realize that stellar abundances are in fact not *observed* but rather *deduced* from an observed spectrum using a multitude of more or less uncertain input data and simplifying assumptions. Atomic data (gf-values, opacities, equation-of-state etc), models of the stellar atmospheres (1D, hydrostatic and radiative equilibrium, mixing length theory etc), and radiative transfer assumptions (1D, LTE, micro- and macroturbulence etc) are merely a selection of all the necessary ingredients, which may introduce unpleasant systematic errors.

Recently it has become possible to compute realistic 3D, time-dependent, hydrodynamical surface convection simulations of solar-type and metal-poor stars (e.g. Stein & Nordlund 1998; Asplund et al. 1999) as an alternative to classical 1D hydrostatic model atmospheres, which should remove several of the uncertainties hampering abundance analyses. These 3D model atmospheres are highly successful in reproducing observational diagnostics of the Sun, such as helioseismology, detailed line shapes and asymmetries, and granulation properties (e.g. Stein & Nordlund 1998; Asplund et al. 2000b). Of particular interest is to apply such models to the line formation of light elements in metal-poor stars, which is discussed here.

2. 3D hydrodynamical model atmospheres of metal-poor stars

3D model atmospheres of stellar granulation have been computed with a time-dependent, compressible, radiative-hydrodynamics code which solves the equations of mass, momentum and energy conservation (Stein & Nordlund 1998; Asplund et al. 1999, 2000b). State-of-the-art equation-of-state, which accounts for ionization, excitation and dissociation, and opacities, including the effects of lines, have been used (Mihalas et al. 1988; Gustafsson et al. 1975; Kurucz 1993). In order to obtain a realistic atmospheric structure it is crucial to correctly describe the energy exchange between radiation and gas, which has been included through a simultaneous solution of the 3D LTE radiative transfer. Simulations appropriate for various solar-type stars and metal-poor stars have been performed with physical dimensions covering ≥ 10 granules at any time. It is noteworthy that the simulations contain no adjustable free parameters besides the stellar parameters, which have been estimated from the IR flux method ($T_{\rm eff}$), Hipparcos parallaxes ($\log g$) and spectroscopy ([Fe/H]). In particular there are no mixing length parameters involved.

The predictions from the solar simulations agree excellently with observations both in terms of detailed line shapes, asymmetries and shifts, granulation properties, and helioseismological constraints (e.g. Stein & Nordlund 1998; Asplund et al. 2000b). In particular the line profiles are perfectly matched without invoking any micro- and macroturbulence as a direct consequence of the Doppler shifts introduced by the self-consistent convective velocity fields. The only free parameter in 3D abundance analyses of the Sun is therefore the elemental abundance of the line in consideration, which should result in more secure abundance determinations. For stars, one additionally needs the rotational velocity, which however only affects the line *shapes* and not the line *strengths*.

The mean temperature stratification remains close to the radiative equilibrium expectations for solar-metallicity 3D model atmospheres but depart significantly for low-metallicity simulations. The temperature in the line-forming layers is determined mainly by a competition between adiabatic cooling and radiative heating due to absorption of photons in spectral lines. In metal-poor stars there are significantly fewer and weaker lines available which shifts the balance towards lower temperatures since the cooling remains essentially the same (Asplund et al. 1999). As a consequence the temperature in the optically thin layers are much lower than in classical 1D model atmospheres which enforce radiative equilibrium with a difference amounting to as much as 1000 K. Thus, the dependence of the temperature structure on [Fe/H] is opposite to that assumed with 1D models.

3. 3D LTE abundances of metal-poor stars

The drastically different temperature structures in 3D compared with 1D naturally have a large impact on the line formation and thus on the derived abundances (Asplund et al. 1999). Lines of neutral minority species, low-excitation transitions and strong lines are formed in the higher layers and thus feel the low temperatures there, making them stronger for a given abundance. In particular, the (LTE) Li abundances in metal-poor stars may have been over-estimated by

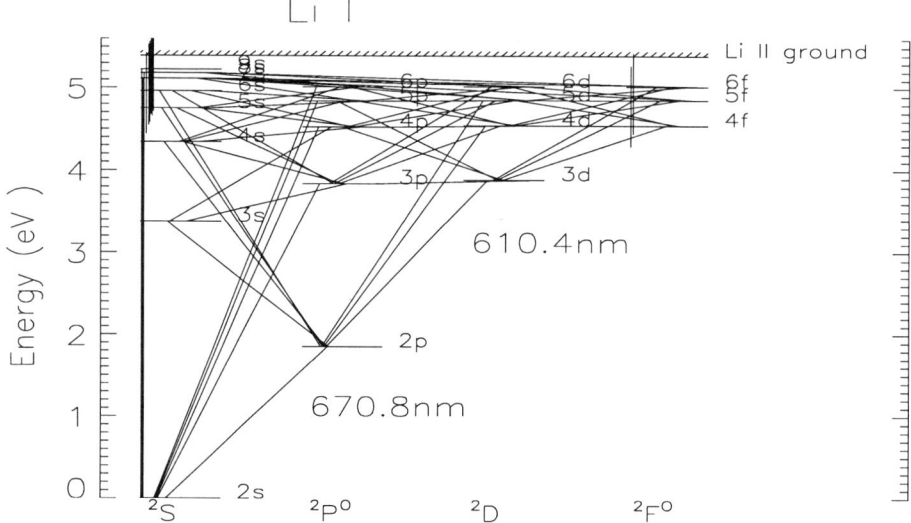

Figure 1. The Li model atom used for the 3D NLTE calculations

0.2-0.35 dex with 1D model atmospheres (but see Sect. 4). Likewise, abundances derived from B I and Fe I lines may be over-estimated by similar amount. On the other hand, Be II, O I and Fe II lines are formed in the deep atmospheric layers, which causes the 1D abundances to be underestimated by 0.05-0.1 dex.

The full impact of these LTE results in terms of galactic chemical evolution is impossible to assess here but a few remarks are in order. Although OH lines have not yet been investigated it is anticipated that the derived O abundances using 3D models will decrease relative to with 1D models, which would make the recently determined [O/Fe] vs [Fe/H] relation using OH lines (Israelian et al, these proceedings) flatter. Since the [Be/H] vs [Fe/H] behaviour should remain essentially unchanged, the [Be/H] vs [O/H] trend will become slightly shallower, making it more consistent with a primary origin. Furthermore, since B I and Be II react in opposite ways, the B/Be ratio may previously have been over-estimated.

It must be remembered, however, that these results are based on the assumption of LTE and thus caution must be exercised. One may worry that the steep temperature gradients may be prone to significant departures from LTE, in particular over-ionization, for species like Li I, B I and Fe I; majority species like O I and Fe II should be more immune to such NLTE effects. Indeed Fe I appears to be over-ionized, in which case it seems likely that also Li I would be. Furthermore, spatially resolved spectra of the Sun clearly demonstrates that the Li I line formation process is far from in LTE (Kiselman & Asplund 2000).

4. 3D NLTE Li I line formation

In order to investigate possible departures from LTE in the Li I line formation, detailed 3D NLTE calculations have been performed (Asplund & Carlsson 2000) from a few selected snapshots of the metal-poor convection simulations

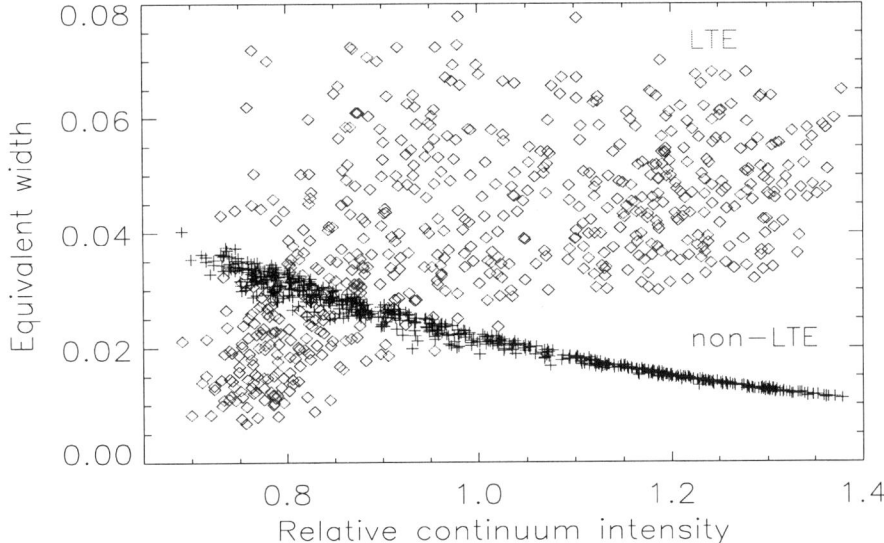

Figure 2. The variation of the Li I 670.8 nm line strength over the granulation pattern in the HD 140283 simulation in LTE (diamonds) and NLTE (crosses). Notice the remarkably small scatter in NLTE

of HD 140283 and HD 84937 (Asplund et al. 1999) with a 3D-version of the NLTE-code MULTI (Botnen 1997). For the purpose a 21-level Li atom with 70 bound-bound and 20 bound-free radiative transitions (Fig. 1), identical to the one adopted by Carlsson et al. (1994) for 1D NLTE calculations. Additionally similar calculations have been done for the Sun with very similar results to previous investigations (Kiselman 1997).

Similar to the solar case line formation process for the Li I 670.8 nm line in the 3D metal-poor model atmospheres is far from in LTE, as shown in Fig. 2. In LTE the line strength depend only on the local gas temperature and therefore the line is strong where the temperature is low, i.e. in general over up-flowing regions (high continuum intensities) due to the reversal of the temperature contrast in the optically thin layers (Stein & Nordlund 1998). As a consequence the LTE line strength tends to increase with the continuum intensity when considering spatially resolved spectra but with a large scatter due to the sensitivity to the local temperature. In contrast without the assumption of LTE the Li I line has the opposite behaviour and show essentially no scatter, reflecting that the line is completely governed by the non-local properties of the radiation field rather than local conditions. The main 3D NLTE effect is clearly over-ionization.

Also when considering spatially averaged profiles there are significant departures from LTE, which in turn translates to large abundance corrections compared with the 3D LTE results. Due to the over-ionization of Li I, the theoretical profile is much weaker than under the assumption of LTE, i.e. *the Li abundance in metal-poor stars is larger than estimated with 3D LTE calculations by about 0.3-0.4 dex*. Thus the effects of the lower photospheric temperatures in 3D models compared with 1D models for Li (Asplund et al. 1999) is almost completely cancelled by departures from LTE, leaving a net "3D effect" (3D

NLTE - 1D LTE) of ≤ 0.1 dex *for HD 140283 and HD 84937*. It must be emphasized however that the findings are still preliminary and that sofar only a couple of snapshots of two metal-poor simulations have been investigated. Considering the two large and partly compensating systematic errors (3D and NLTE) it is much too early to adopt the claimed trend in Li abundance with [Fe/H] from a 1D LTE analysis (Ryan, these proceedings), since the amount of overionization will depend on the stellar parameters. Clearly additional 3D NLTE Li calculations for a range of metallicities and $T_{\rm eff}$ are urgently needed.

The 3D NLTE calculations presented here have been independently confirmed by Cayrel & Steffen (these proceedings). They obtain very similar results in spite of their restrictions to 2D model atmospheres and their use of a much smaller Li atom (5 levels). Even if 2D model atmospheres tend to overestimate the temperature inhomogeneities (Asplund et al. 2000a) compared with 3D the overall conclusions are the same, namely that the Li line formation is determined mainly by the non-local properties of the continuum radiation field.

5. Lithium isotope ratios in metal-poor stars

Detections of the ^{6}Li isotope in metal-poor stars help constrain the primordial Li abundance from Big Bang nucleosynthesis, limit the allowed mixing during the stellar lifetime, and provide additional information about cosmic ray spallation of Li, Be and B (Nissen, these proceedings). The determinations of ^{6}Li abundances are based on the increased width and asymmetry of the Li I 670.8 nm doublet introduced by the isotope shift. Since classical 1D model atmospheres can not predict the inherent convective line asymmetries and must invoke ad-hoc macroturbulence line broadening, the conclusions suffer from unfortunate uncertainties. An attractive alternative is provided by 3D model atmospheres which are able to predict the convective line asymmetries, leaving any possible remaining line asymmetry to be attributed to ^{6}Li. Furthermore, no ad-hoc extra broadening parameters like micro- and macroturbulence must be specified (Asplund et al. 2000).

In order to verify the claimed detection of ^{6}Li in the metal-poor halo star HD 84937 (Smith et al. 1993), a similar analysis has been performed based on both 1D and 3D model atmospheres using the same observed spectrum. The 3D profiles were averaged over a sufficiently long time-sequence to produce statistically significant results. Due to computing-time considerations departures from LTE were not considered here. The stellar broadening (1D: macroturbulence+rotation, 3D: rotation) was determined from Ca I and Fe I lines of similar strength. Due to the large isotope shift in relation to the convective line asymmetries, almost identical results were obtained with 1D and 3D models: ^{6}Li/^{7}Li $= 0.04 \pm 0.02$, in agreement with previous determinations. This finding thus lends confidence to the reality of the detection of the isotope in HD 84937.

A very high S/N spectrum of the metal-poor halo star G271-162 obtained during the commissioning of UVES on VLT/UT2 has been analysed in a similar fashion with a suitable 3D model. The Li feature reveals a possible detection of ^{6}Li at the level of ^{6}Li/^{7}Li $= 0.02 \pm 0.01$ (Nissen et al. 2000). The smaller ^{6}Li abundance in G271-162 than in the almost identical HD 84937 may indicate an intrinsic scatter in ^{6}Li/^{7}Li in the ISM at a given metallicity.

6. Concluding remarks

It is clear from comparison with observations (e.g. helioseismology, granulation properties, line shapes and asymmetries) that 3D hydrodynamical model atmospheres are highly realistic without invoking any free parameters. Abundance analyses based on such 3D models should therefore in general be more reliable than corresponding investigations using classical 1D model atmospheres. In particular for metal-poor stars there are significant differences to 1D models which translate to a large impact on the emergent stellar spectra.

Even though the detailed 3D NLTE results for Li presented here apparently almost completely compensate the effects due to the use of hydrodynamical model atmospheres (Asplund et al. 1999) making them very similar to the 1D LTE predictions, it is important to realize that this is more a lucky coincidence than a general truth. It is by no means correct to say that 3D LTE abundances are erroneous in general, even if it may be true for Li (Cayrel & Steffen, these proceedings). Lines of other species, like ionized species, high-excitation transitions or very weak lines, are much more immune to departures from LTE, whether with 1D or 3D model atmospheres. It should be remembered that 1D analyses contain additional systematic errors by not taking into account properly the convection and will therefore *in general* produce less reliable results than 3D LTE investigations. Naturally, however, those results must also be verified by detailed 3D NLTE calculations whenever possible.

The full version of the present talk with additional figures can be found at: http://www.astro.uu.se/~martin/talks/Natal99

Acknowledgments. The author greatly appreciates fruitful collaboration with M. Carlsson, P.E. Nissen, Å. Nordlund, R.F. Stein, and R. Trampedach, without whom the above-mentioned projects would never have been possible.

References

Asplund, M., & Carlsson M. 2000, A&A, submitted
Asplund, M., Nordlund, Å., Trampedach, R., & Stein, R.F. 1999, A&A 346, L17
Asplund, M., Ludwig, H.-G., Nordlund, Å., & Stein, R.F. 2000a, A&A, in press
Asplund, M., Nordlund, Å., Trampedach, R., & et al. 2000b, A&A, in press
Botnen, A.V. 1997, Cand. Sci. thesis, University of Oslo
Carlsson, M., Rutten, R.J., Bruls, J.H.M.J., & et al. 1994, A&A, 288, 860
Gustafsson, B., Bell, R.A., Eriksson, K., & Nordlund, Å. 1975, ApJ, 42, 407
Kiselman, D. 1997, ApJ, 489, L107
Kiselman, D., & Asplund M. 2000, in: Cool stars, stellar systems and the Sun, 11th Cambridge workshop, in press
Kurucz, R.L. 1993, CD-ROM, private communication
Mihalas, D., Däppen, W., & Hummer, D.G. 1988, ApJ, 331, 815
Nissen, P.E., Asplund, M., Hill, V., & D'Odorico, S. 2000, A&A, in press
Smith, V.V., Lambert, D.L., & Nissen, P.E. 1993, ApJ, 408, 262
Stein, R.F., & Nordlund, Å. 1998, ApJ, 499, 914

The Light Elements and Their Evolution
IAU Symposium, Vol. 198, 2000
L. da Silva, M. Spite, J. R. de Medeiros, eds.

Formation of the optical spectra of the coolest M- and L-dwarfs and lithium abundances in their atmospheres

Yakiv V. Pavlenko

Main Astronomical Observatory of NAS, Golosiiv woods, 03680, Kyiv-127, Ukraine

Abstract. Theoretical aspects of modeling of spectra of late M- and L-dwarfs are discussed. We show, that the processes of formation of spectra of M- and L-dwarfs are basically different. Instead of the case of M-dwarfs, atoms of Ti and VO should be depleted into grains in the atmospheres of L-dwarfs. Overall shape of the L-dwarf spectra is governed by the K I + Na I resonance line wings of the huge strength. To fit lithium lines observed in spectra of the coolest dwarfs we used two additional suggestions: a) there are some *extra* depletions of molecular species absorbed in the optical spectra of L-dwarfs; b) there may be (a few?) additional ("dusty"?) opacity sources in their atmospheres. Problems of lithium line formation and the "natural" limitation of their use for the "lithium test" for the case of L-dwarfs are discussed.

1. Introduction

A few definitions of the "unconventional" spectral classification of low mass stars and substellar objects are used in this paper:

M-dwarfs are objects (stars + young brown dwarfs) with $4000 > T_{eff} > 2200$ K. Their spectra are governed by molecular bands of TiO and VO (in the optical part of the spectrum), H_2O (in the red).

L-dwarfs are objects (brown dwarfs + stars) with $1000 < T_{eff} < 2200$ K (cf. Martin et al. 1997). Optical spectra of L-dwarfs are governed by the K I + Na I lines and molecular bands of CrH + FeH (in the optical spectrum), H_2O (in the red).

T-dwarfs are super-giant, planet-like objects (big Jupiters?) with $T_{eff} < 1000$ K. Their spectra are formed by the "dusty opacities" and methane + H_2O + ...? absorption (Strauss et al. 1999, Burgasser et al. 1999)

For the time being most of the known brown dwarfs are actually recognized by the detection of the Li I resonance doublet in their spectra (Rebolo et al. 1996, Martín et al. 1997a, Kirkpatrick et al. 1999). Indeed, temperatures in the interiors of brown dwarfs are not high enough to burn lithium (Rebolo et al. 1992).

2. Procedure

The computations of synthetical spectra of M- and L-dwarfs are carried out by program WITA5, which is a modified version of the program WITA31 used by Pavlenko (1997). The modifications were aimed to incorporate "dusty effects" that affect the chemical equilibrium and radiative transfer processes in very cool atmospheres.

We have used the set of Tsuji's (1999) "dusty" (C-type) LTE model atmospheres. These models were computed for the case of segregation phase of dust and gas.

Chemical equilibrium was computed for the mix of $\approx$100 molecular species. To take into account the effect of the oversaturation, we reduced the abundances of those molecular species down to the equilibrium values (Pavlenko 1998).

In L-dwarf atmospheres the additional opacity (AdO) could appear due to molecular and/or dust absorption and/or scattering. We have modelled the additional opacity with a simple law of the form $a_\circ \, (\nu/\nu_\circ)^N$, with $N = 1 - 4$ (see Pavlenko, Zapatero Osorio & Rebolo 2000 for more details).

3. M-dwarf spectra

Lithium lines observed in spectra of the late M-dwarfs are well known tracers of their evolution. Completely convective M-dwarfs are very effective lithium destroyers, therefore cool pre-main sequence (PMS) stars are expected to preserve their initial lithium only during their first few million's years (Fig.3, see also Magazzu, Rebolo & Pavlenko 1992, Oppenheimer et al. 1997, Pavlenko (1997, 1997a), Pavlenko & Oppenheimer 1998).

Lithium lines in spectra of M-dwarfs are formed at the background of mighty TiO bands (Fig.1). Only cores of the saturated Li lines may be observed in the real spectra (Pavlenko et al. 1995). To estimate of the abundance of lithium one may use "pseudoequivalent widths" of lithium lines, i.e. W_λ measured in respect to the local pseudocontinuum formed by molecular bands around Li lines (see also Pavlenko 1997a).

Sure, the better way of the quantitatively determination log N(Li) is the use of synthetical spectra (Fig. 2).

4. L-dwarf spectra

Due to depletion of the Ti and VO into grains a structure of the optical spectra of L-dwarfs becomes more simple in comparison with M-dwarfs. The overall spectral energy distribution (SED) is governed by absorption of resonance doublets of K I and Na I which have pressure broadened wings extended up to thousands Å(Fig.3). Furthermore, Pavlenko et al. (2000) showed that L-dwarf optical spectra are affected by the additional ("dusty") absorption and/or scattering.

Our computations show that lithium lines are very sensitive to the additional absorption (AdO) that we need to incorporate in the spectral synthesis if we want to explain the observed broad spectral energy distribution(Fig.4). In Table 1 we give the predicted equivalent widths W_λ of the Li I resonance doublet at 670.8 nm for L-dwarf's model atmospheres (2000–1000 K) considered in this

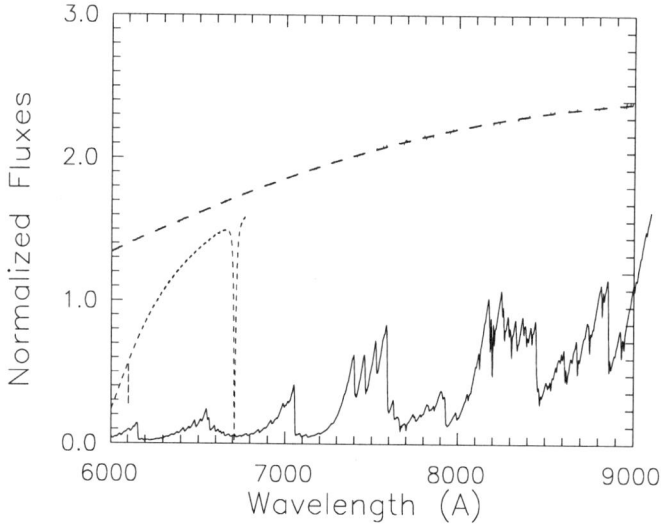

Figure 1. Comparison of the observed spectrum of the Pleiade's young brown dwarf Teide1 (Rebolo et al. 1996) with theoretical spectra computed using C-model atmosphere ($T_{\rm eff}$=2600 K, log g =5.0) of Tsuji (1999). The strongest atomic lines which may be observed in M- and L- spectra and the real (theoretical) continuum are shown.

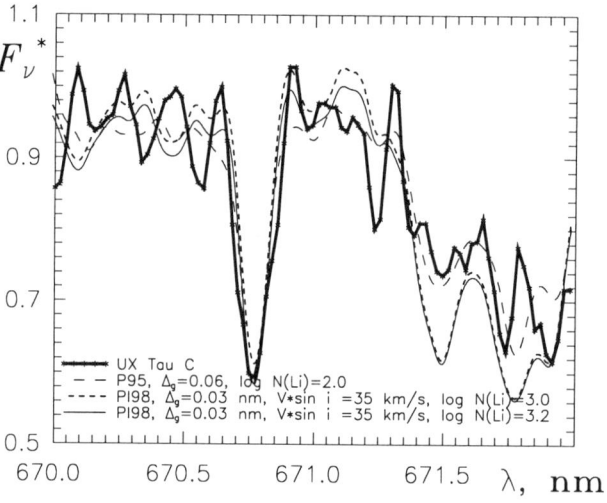

Figure 2. Fit to the observed Li I resonance doublet in UX Tau C spectrum (Magazzu et al. 1991) with the list of TiO lines of Plez (1998) and model atmosphere 3100/4.5 (Allard & Hauschildt 1995)

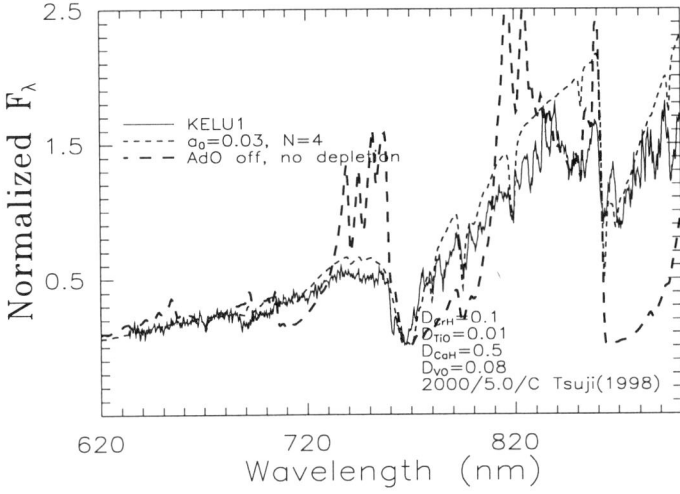

Figure 3. Fit of theoretical SED's computed for C-model atmosphere 2000/5.0 (Tsuji 1999) to Kelu1 spectrum. D factors showed in the Fig. are used to simulate the *extra* depletion of several species into grains.

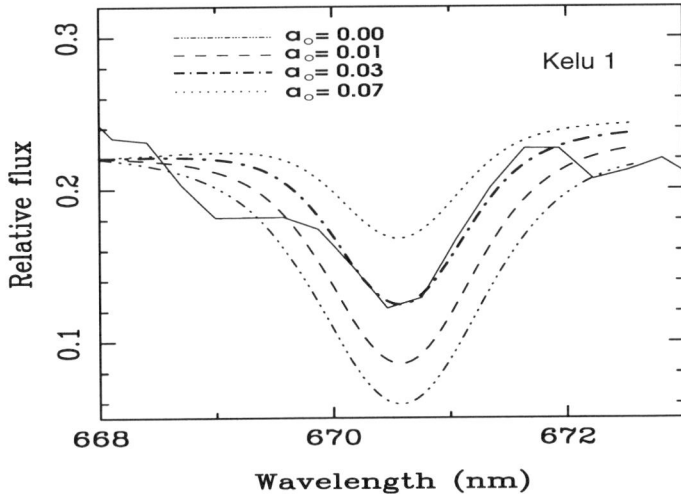

Figure 4. Fit to Li I λ 6708 nm resonance doublet in Kelu1 spectrum. Computations were carried out for log N (Li) = 3.0 and different parameters of the AdO

work. We found:

— In the AdO-free case (second column in the table), we would expect for the "cosmic" values of log N(Li) rather strong neutral Li resonance lines in the spectra of objects as cool as DenisP J0205–1159 and Gl 229B.

— The chemical equilibrium of Li-contained species still allow a sufficient number of Li atoms to produce a rather strong resonance feature.
— Our computations indicate that L-dwarfs with moderate dust opacities should show the Li I resonance doublet if they had preserved this element from nuclear burning, and consequently the lithium test can still be applied.
— Temporal variations of the dusty opacities may originate some kind of "meteorological" phenomena occurring in these cool atmospheres. Lithium lines (as well other lines) may be severely affected by the effect (see Pavlenko et al. 2000 for more details).

Table 1. Equivalent widths of the Li I resonance doublet at 670.8 nm computed for the C-type Tsuji's (1999) model atmospheres, cosmic Li abundance ($\log N(\text{Li}) = 3.2$) and gravity $\log g = 5.0$.

T_{eff}	$a_\circ$		
	0.00	0.01	0.10
(K)	W_λ(Å)		
1000	17	8	0.6
1200	30	12	0.7
1400	42	21	0.9
1600	40	24	1.6
2000	23	16	3.6

5. Conclusions

Finally, we have arrived to the following conclusions:

- Processes of formation of Li lines in L- and M-spectra differ significantly:
  – for M-dwarfs the main problem to be solved is blending of lithium lines by molecular lines,
  – in the case of L-dwarfs we deal with a menagerie of different processes: depletion of lithium atoms into molecules and grains, "dusty effects", meteorological phenomena, stratification effects, etc...

- We can fit the optical spectra of L-dwarfs in the frame of our simple model.

- Using our model we may perform a numerical analysis of the L-dwarf spectra (at least in the sense of the Li abundance determination).

- The basic algorithm of the "lithium test" may be used even for the assessment of the coolest L-dwarfs.

6. Acknowledgements

I thank IAU, LOC and SOC of the IAU Symposium N 198 for the financial support of my participation. I'm grateful to R. Rebolo and M.R. Zapatero Ozorio (IAC) for the fruitful collaboration and for providing the observational data in electronic form; to T.Tsuji, F. Allard and P.Hauschildt for providing model atmospheres in digital form. Partial financial support was provided by the Spanish DGES project no. PB95-1132-C02-01.

References

Allard, F., Hauschildt, P.H. 1995 ApJ, 445, 433

Burgasser, A. J., Kirkpatrick, J. D., Brown, M. E., et al. 1999, ApJ, 522, L65

Basri, G., Martín, E. L., 1999. AJ, 118, 2460.

Kirkpatrick, J. D., Reid, I. N., Liebert, J., et al. 1999, ApJ, 519, 802

Magazzu, A., Martín, E. L., Rebolo, R. Ya. 1991, A&A, 249, 149.

Magazzu, A. Rebolo, R., Pavlenko, Ya. 1992, ApJ,

Martín, E. L., Basri, G., Delfosse, X., Forveille, T. 1997, A&A, 327, L29

Oppenheimer, B., Basrii, G., Nakajima, T., Kulkarni, S.L. 1997, AJ, 113, 296.

Pavlenko, Ya.V. 1997, Ap&SS, 253, 43.

Pavlenko, Y. V. 1997a. Astron. Rept., 41, 537.

Pavlenko, Ya. V. 1998, Astron. Reports, 42, 787

Pavlenko, Ya.V., Zapapero Ozorio, M.R., Rebolo, R. 2000, A&A, in press (astro-ph 0001060)

Pavlenko, Ya.V., Oppenheimer, B. 1988, ASP Conf. Ser. 154, 1768.

Pavlenko Ya.V. 1998b, Atronom. Report, 42, 787.

Pavlenko et al. Pavlenko, Ya. V., Rebolo, R., Martín, E. L, García López, G. 1995, A&A, 303, 807.

Plez, B. 1998, A&A, 337, 495.

Rebolo R., Martín, E.L., Magazzu, A. 1992,

Rebolo R., Martín, E.L., Magazzu, A. 1992, ApJ, 389, L83.

Rebolo, R., Martín, E.L., Basri, G., G.W. Marcy, G.W. And Zapatero Osorio, M.R. 1996, ApJ, 469, L53.

Strauss, M. A., Fan, X., Gunn, J. E. et al. 1999, ApJ, 1999, 522, L61

Tsuji, T. 1999a, in Low-Mass Stars and Brown Dwarfs in Stellar Clusters and Associations, La Palma, CUP, in press

Constraints on Stellar Hydrodynamics from Abundance Anomalies of LiBeB and Metals

Georges Michaud[1], Jacques Richer[1] and Olivier Richard[1]

Département de physique, Université de Montréal, Montréal, Canada, H3C 3J7

Abstract. The availability of large atomic data bases has made it possible to calculate stellar evolution models taking into detailed account the atomic diffusion of all important contributors to opacity. The radiative accelerations and the opacity are continuously calculated during evolution taking the abundance changes of 28 species into account. This leads to the first self-consistent stellar evolution models for A and F stars. In A and F stars an iron-peak convection zone appears.

The calculated abundance anomalies are very similar to those observed in AmFm stars in open clusters except that they are larger by a factor of about 3. To reduce the calculated anomalies to the observed ones, an additional source of turbulence (or some other hydrodynamical process) must be introduced. The mixed zone must extend about 5 times deeper than the iron convection zone. Detailed comparisons to a few AmFm stars have been carried out.

The LiBeB abundances observed in clusters give additional information. The abundances of the 28 species offer considerable constraints on the models. Various potential turbulence models have been introduced in a stellar evolution code and results of evolutionary calculations for Li gap stars are discussed in the light of the constraints offered by the abundances of LiBeB and metals. The radiative accelerations of LiBeB have also been recalculated taking the effect of changing metal abundances into account. This modifies the expected Li gap in the absence of turbulence.

1. Introduction

Helioseismology results show that atomic diffusion is occurring in the sun and suggest that some weak turbulence is also needed in order to account for the small Li abundance (Proffitt & Michaud 1991; Chaboyer et al. 1995; Richard et al. 1996; Vauclair this conference). The presence of some turbulence Vauclair (1988) is also suggested by the progressive decrease of the Li abundance with the age of clusters as discussed elsewhere in this conference. Helioseismology results also show that, below the convection zone, the sun is rotating nearly as a solid body. This cannot be explained by any model yet suggested for the

---

[1]CEntre de Recherche en Calcul Appliqué (CERCA), 5160 boul. Décarie, bureau 400, Montréal, PQ, CANADA H3X 2H9

turbulent transport of angular momentum. Taken together these facts show that while turbulence is probably an important particle transport process, none of the suggested models passes the test of the constraints imposed by helioseismology. The origin of turbulence in stars remains unknown.

Our approach is first to calculate stellar evolution models with all the physics of particle transport that is known from first principles. In radiative zones this implies atomic diffusion including gravitational settling, thermal diffusion and radiative accelerations, in addition to the purely diffusive term. We call these models the *basic models*. In a second step, simple turbulence models are introduced in order to better reproduce the observed surface abundances. Other hydrodynamical processes, such as mass loss or magnetic fields, could also be important but they will not be discussed here.

In this paper, results of stellar evolution are first presented for the basic models. This leads to the development of iron convection zones in stars of 1.5 $M_\odot$ and larger and increases the depth of the convection zone in stars of 1.3 to 1.5 $M_\odot$. Turbulence is then introduced in order to reduce the calculated anomalies in the basic models of A stars to the observed abundance anomalies in AmFm stars. The LiBeB observations will be used to constrain the turbulence in F and G stars and a summary of the constraints abundance anomalies put on turbulence in main sequence stars will be presented.

2. Basic models and iron convection zones

The basic models include the physics generally found in standard stellar evolution models, with the addition of atomic diffusion, including gravitational settling, thermal diffusion, radiative accelerations and the diffusive term. The detailed treatment of diffusion transport is described in (Turcotte et al. 1998). The largest contribution comes from the drift part of the diffusion velocity whose two most important terms are gravity and the opposing radiative acceleration:

$$v_g \propto A g_{\rm rad} - \left(A - \frac{Z}{2} - \frac{1}{2} \right) g. \qquad (1)$$

When $g_{\rm rad} > g$, the involved element diffuses upward. A precise determination of radiative accelerations is essential; these are treated as described in Richer et al. (1998). For each atomic species (28), at each zone in the star (1000–1500), at each time step (500–1000 for an evolutionary sequence) integrals are carried out, at four points of the OPAL grid, over the spectral form (10^4 frequency points) of the OPAL data. The Rosseland mean opacities are simultaneously recalculated. The OPAL data contains more than one gigabyte of data most of which must be in core memory during the calculations.

Furthermore, for stellar evolution calculations, it is in practice required to use the Burgers (Burgers 1969) formulation of the calculations of diffusion velocities since the hydrogen and helium contributions to momentum exchange with metals are of the same order. For the transport velocities, one ends up solving 56 ($= 2 \times 28$) coupled, non-linear, highly stiff, differential equations.

On one SGI R-12000 processor, it takes us less than 30 minutes to calculate the evolution of a typical standard model. However it takes some 3–5 days on four such processors to calculate the evolution of one of our basic models. More

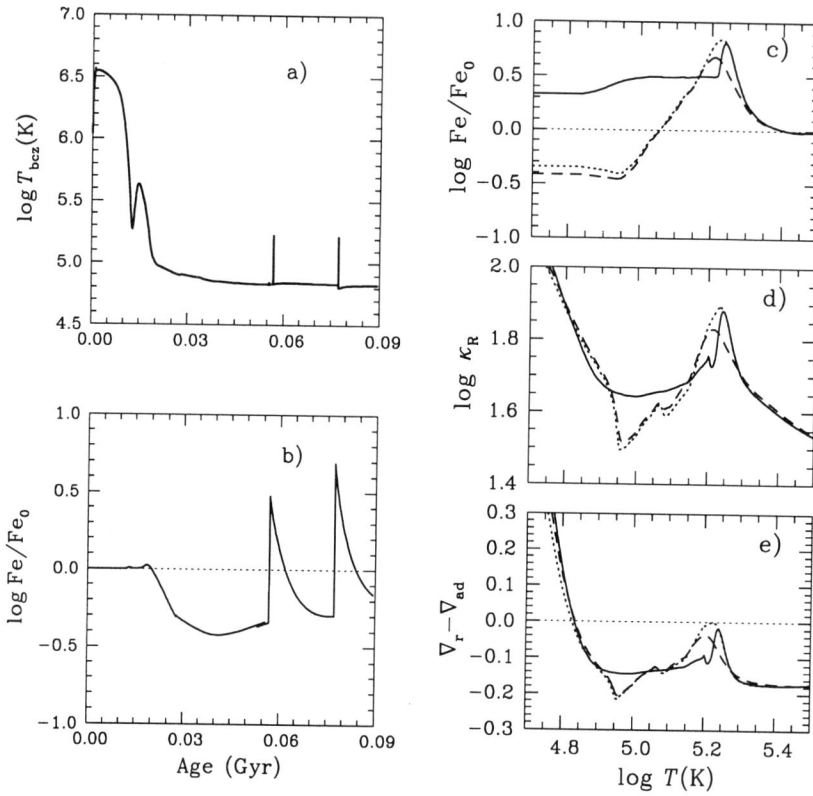

Figure 1. The figures **a** and **b** show respectively the evolution of the temperature at the bottom of the deepest surface convection ($T_{\rm bcz}$) zone and of the surface Fe abundance in a $1.5 M_\odot$ model. The figures **c**, **d**, and **e** show at three ages (44 Myr: dashed line, 56.9 Myr: dotted line, and 57.9 Myr: solid line) the Fe abundance, the Rosseland mean opacity, and the difference between the radiative and adiabatic gradients as a function of temperature.

than 96 % of the computing time (prior to parallelization) is spent on the atomic data.

Models have been computed from 1.0 to $4.0 M_\odot$ from the pre-main-sequence to the base of the giant branch. The solar model was used to calibrate the mixing length. More than 100 models have been computed up to now.

2.1. Iron convection zones

Probably the most striking difference between the basic models and standard stellar evolution models is the iron (or more properly iron-peak) convection zone they develop. Their natural appearance may be seen for instance in a $1.5 M_\odot$ model. On Figure 1a and 1b is shown the temperature at the bottom of the deepest surface convection ($T_{\rm bcz}$) zone as well as the surface Fe abundance in this model for the first 90 Myr of its evolution. At 57 and 76 Myr, spikes occur in the $T_{\rm bcz}$. It increases suddenly by slightly more than a factor of 2. A third spike was about to occur at 90 Myr. The structural changes that occurred in the model may be seen by looking at Figure 1c, 1d and 1e. Between 44 and 56.9 Myr, a progressive increase of the Fe abundance occurred at a temperature of about 170 000 K. This occurs, without any arbitrariness, as a consequence of the temperature dependence of $g_{\rm rad}$(Fe) which is larger than gravity for $T >$ 170 000 K and becomes smaller below this temperature so that Fe is pushed from the deeper interior and accumulates around 170 000 K. The Rosseland mean opacity follows the trend of the Fe abundance while the difference between the radiative and adiabatic temperature gradients reaches zero at that location at 56.9 Myr. An Fe convection zone then appears briefly. Complete mixing of the mass above the Fe convection zone was then imposed. This is partly justified by the results of Latour et al. (1981) who showed that between superficial hydrogen and helium convection zones the mixing is essentially complete. This will be seen below to lead to a simple model for the AmFm stars. The imposed mixing forced redistribution of the Fe accumulated at $T = 170\,000$ K up to the stellar atmosphere region thus reducing its abundance and the Rosseland opacity at 170 000 K; the Fe convection zone consequently disappears. The zone is now stable and Fe starts to accumulate again below the shallower convection zone. This process repeats itself until enough Fe has accumulated for the Fe convection zone to remain in spite of the mixing with the surface. As can be seen in Figure 1b, each mixing event causes the surface iron abundance to increase suddenly, then to relax exponentially back to a slowly increasing baseline level.

Alternatively, one could choose to impose no mixing above the Fe convection zone. It then remains and grows once it has formed. In either case one has an Fe convection zone; its existence is not dependent on the mixing between the convection zones. While the no mixing assumption might be more appropriate the first time the Fe convection zone appears, our assumption of complete mixing above the Fe convection zone appears more appropriate when it is fully developed. It is supported by the accuracy with which the observed abundances are reproduced in the AmFm stars.

3. The AmFm stars

It appears compatible with the statistics of stellar rotation that all non magnetic A and early F stars ($7000 < T_{\rm eff} < 10000$ K) be Am or Fm stars (Abt & Morrell 1995). The AmFm phenomenon is consequently not a fringe phenomenon. Its appearance must be the consequence of the normal stellar evolution of slowly rotating stars. The brightest star in the sky, Sirius A, is an Am star.

The basic model leads, with no arbitrary parameter, to abundance anomaly patterns that are very similar to those observed in AmFm stars. However, since the predicted anomalies are generally more important than the observed ones, a simple turbulent transport coefficient was introduced (Richer et al. 2000) to determine the properties of turbulent transport that can reduce these predictions to the observed levels:

$$D_T = \omega D({\rm He})_0 \left(\frac{\rho_0}{\rho}\right)^n \quad (2)$$

where $n = 2$, 3 or 4, and $D({\rm He})_0$ is the atomic diffusion coefficient of He at density ρ_0. The turbulent diffusion coefficient is then equal to ω times the atomic diffusion coefficient of He at the reference density in the model of interest. It decreases inward as ρ increases with an exponent that was varied between 2 and 4. It was found that the value of the exponent had little importance, so long as it was larger than, or equal to about 2. For AmFm stars, what matters is the mass which is mixed by turbulence. It is approximately given, in our models, by the mass above the point where $D_T = 2D(\text{He})$.

A detailed comparison of recent abundance determinations on Sirius to surface abundances calculated with various turbulence mixing is shown in Figure 18 of Richer et al. (2000). Five different groups of astronomers have, in eight papers, studied the surface chemical abundances of Sirius over the last seventeen years. When an element has been studied by different groups there appear variations in the determined abundance. Generally all groups agree as to the over- or underabundance of an element but the size of the determined anomaly varies often significantly. Of the 28 species we calculate, the abundances of 16 have been determined. Since our models depend on only one parameter, the mass of the mixed zone, reproducing the abundance anomalies of all measured species is a very severe test. The models used to compare with Sirius all have an age of 250 Myr, compatible with the T_{eff} of Sirius B and reasonable assumptions as to the original parameters of the binary system. The models differ by the size of the mixed zone which goes, for the models on Figure 18 of Richer et al. (2000), from $\Delta M/M_* \sim 3 \times 10^{-6}$ to $\Delta M/M_* \sim 1 \times 10^{-5}$. The model passes the test of the comparison very well. Of the 16 abundances, 12 are well reproduced by the model, 3 are not so well reproduced and one is a very uncertain observation. The model was also used to compare with AmFm stars in clusters. These have the advantage that their age is better known and that it is possible to get a handle on the original abundances through the abundances observed on late F or G stars. In Richer et al. (2000) comparisons were made to stars of the Hyades, the Pleiades, and Preasepe, with as much success as in Sirius except that the test is less rigorous since the abundances of fewer elements were observed.

We do not wish to claim that atomic diffusion coupled to turbulent transport are the only contributors to the AmFm phenomenon. However the accuracy of the observations will have to be improved before it is justified to look for the role of other processes. Such observations should now become possible in galactic clusters.

4. LiBeB abundances and turbulence

4.1. Radiative accelerations of lithium

LiBeB are not in the OPAL data since they do not contribute sufficiently to the Rosseland mean opacity. For the basic models their radiative accelerations were calculated in a consistent manner with the spectra of OPAL, including also the various corrections determined by Richer et al. (1997). The atomic data needed to calculate them is available and details are given in Richer et al. (in preparation). Figure 2 shows OPAL spectra of Fe and of the total opacity along with the Li spectrum. The frequency interval covers about one third of the total frequency range of the OPAL spectra. The most important Li line at that temperature is at $u \sim 6.7$, where the total opacity is dominated by a large number of Fe lines.

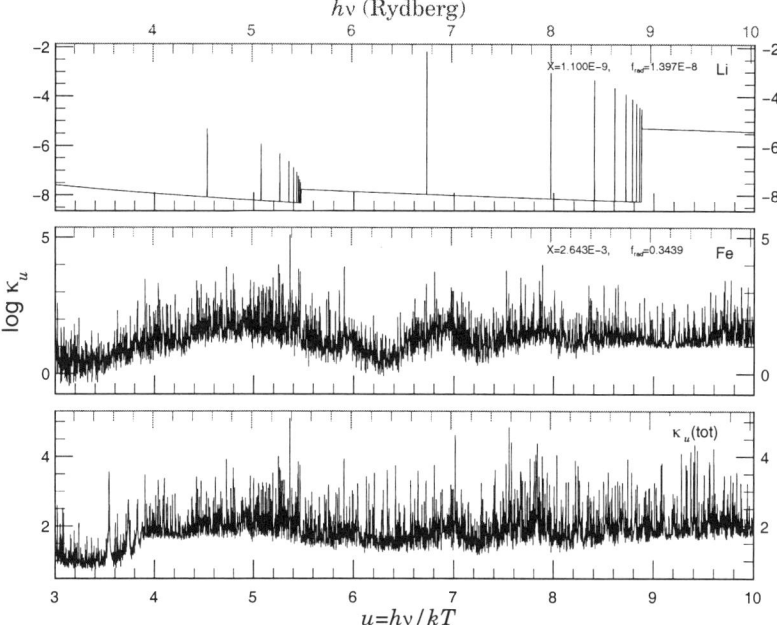

Figure 2. Iron OPAL spectra and the total OPAL opacity (in $cm^2 g^{-1}$) at a point in a stellar model where $\log T = 5.2$ and $\log \rho = -5.4$; the bulk of the radiative flux goes through this u range. Also shown in the top panel is the Li absorption spectrum calculated at the same resolution. The mass fraction of Li and Fe are indicated, as well as the fraction of the total radiative flux each one absorbs.

The effect of taking into detailed account the competition for photons by the 28 species is shown on Figure 3. The Li radiative acceleration is shown there at four time steps along with the Fe abundance. While at the beginning of the evolution, the radiative acceleration on Li is greater than gravity over a certain mass interval, it becomes rapidly smaller everywhere as a consequence of the increase of the Fe abundance which reduces the flux of photons available to Li. Furthermore the maximum of the radiative acceleration turns out to be in the Fe convection zone so that it would not influence transport there.

The use of models in which metals diffuse in a self-consistent manner strongly modifies the atomic diffusion of Li. The radiative accelerations of Li differ in important ways from those of Richer & Michaud (1993). Similar effects occur for Be and B whose radiative accelerations are shown in Richer et al. (in preparation).

4.2. Li and Be as constraints on turbulence

In Figure 4 is shown, at the age of the Hyades, the abundance of Li at the surface of stars between 5500 and 8600 K. Small open circles and triangles are respectively the observed abundances and upper limits in the Hyades (Boesgaard &

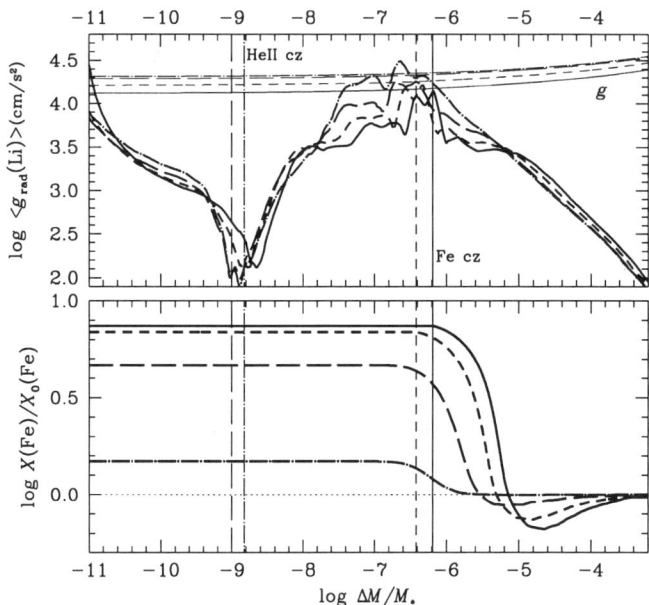

Figure 3. Lithium radiative acceleration (thick lines in the upper graph) and Fe abundance as a function of mass integrated from the surface, at different ages (10 Myr: dot-dashed lines, 70 Myr: long dashed lines, 232 Myr: short dashed lines, and 411 Myr: solid lines), in a $2\,M_\odot$ model with weak turbulence ($\omega = 50$, $n = 3$). Gravity is shown in the upper graph as thin lines for each age. The vertical thin lines in both graphs show the depth of the surface convective zone.

Tripicco 1986; Boesgaard & Budge 1988; Thorburn et al. 1993). The calculated models are identified on the figure.

The solid line links the models calculated with a mixing that reproduces approximately the surface abundances of AmFm stars. The same turbulent diffusion coefficient was used for all masses; no effort was made at optimizing the agreement for individual stars. The agreement for the five main-sequence AmFm stars of the Hyades (the five open circles between 8200 and 7200 K) is quite satisfactory. The reduction of the Li abundance is by a factor of 2–3 at the age of the Hyades. The reduction factor is the same down to 6700 K and it progressively approaches 1.0 at 6500 K. Overabundances of metals accompany the reduction of the Li abundance in these models (see Figs. 19 to 21 of Richer et al. 2000). While some of the stars of the Li gap (those where Li is underabundant by a factor of about 3) may have been affected by the same process that leads to the AmFm stars, this does not seem to be possible for all Li gap stars since the underabundances of Li remain small in these models. Furthermore, the accompanying overabundances of metals do not appear to be present in all Li gap stars.

The dashed line links models with no mixing beyond that occurring in convection zones. The only transport process is atomic diffusion. It is seen to

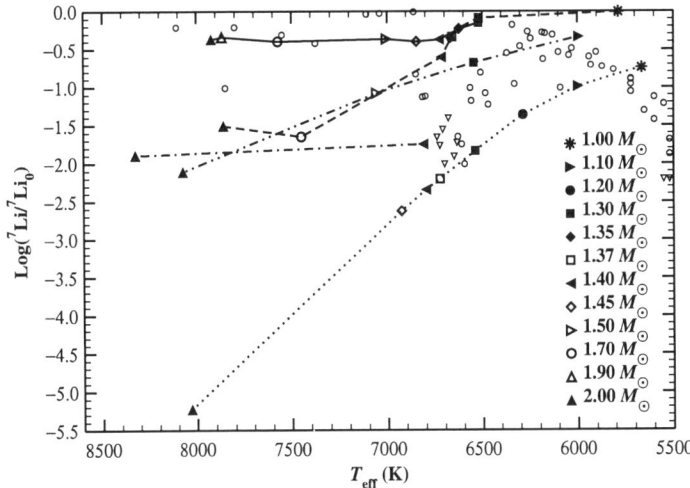

Figure 4. Surface lithium variation at the age of the Hyades. The small open circles and triangles represent respectively the measured values and the upper limits on lithium abundance. The solid line links models calculated with a mixing that reproduces approximately the surface abundances of the AmFm stars. The dashed line links models without turbulence where diffusion takes place below the Fe convection zone. The dotted line links models that were calculated with $n = 2.5$ and a large enough turbulence to destroy lithium by more than a factor of 100 at the bottom of the Li gap of the Hyades. The dashed-double dotted line links models calculated with the same $n = 2.5$ exponent as the models for the dotted line but with ω reduced by a factor 4. The dot-dashed line links models calculated with a $n = 0.72$ exponent that reproduces Li destruction in giant stars.

lead to much larger underabundance factors for Li than the model appropriate for the AmFm stars. The star at $T_{\rm eff} = 7900\,{\rm K}$ with a 1/10 Li underabundance may have less turbulence than other AmFm stars and so a Li abundance closer to the no-turbulence result. The larger underabundance factors occur at higher $T_{\rm eff}$ than the observed Li gap. It is also at a $T_{\rm eff}$ higher by 800 K than the bottom of the Li gap obtained by Richer & Michaud 1993. This shift is caused by Fe convection zones: they increase the depth of the total surface convection zone at a given $T_{\rm eff}$. Because of the increased depth of the convection zone, Li is underabundant only by a factor of 3 at $T_{\rm eff} = 6700\,{\rm K}$ (the $1.4M_\odot$ model). It is only in the more massive models that the mass in and above the Fe convection zone is small enough to allow underabundances by a factor of 30 at the age of the Hyades (in the $1.70 M_\odot$ star). That there should appear large underabundances of Li in this model is also related to the reduction of $g_{\rm rad}({\rm Li})$ caused by the increased Fe abundance as described above.

The dotted line links models that were calculated with $n = 2.5$ (see Eq. 2) and a large enough value of ω to destroy lithium by slightly more than a factor of 100 at the $T_{\rm eff}$ where the Li gap of the Hyades is deepest. It reproduces

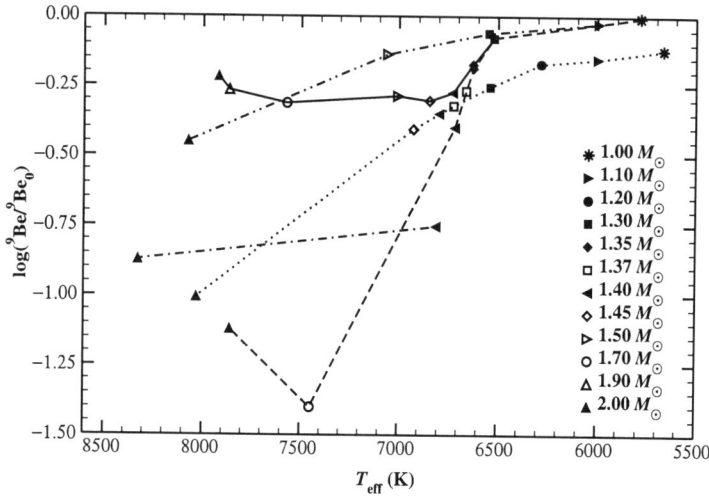

Figure 5. Surface beryllium variation at the age of the Hyades. The same convention as in Figure 4 is used for the lines.

approximately the shape of the cool side of the Li gap. As one considers lower $T_{\rm eff}$ stars, the distance between the bottom of the convection zone and the Li burning area is reduced favoring Li destruction but the mass of the surface Li reservoir increases more rapidly than the net destruction rate and thus the Li surface anomaly is reduced. However this model still destroys Li by far too large a factor in normal stars hotter than the Li gap ($T_{\rm eff} > 6700\,{\rm K}$). The dashed-double dotted line links models calculated with the same exponent as the models used to calculate the dotted line, but with a reduced factor ω. It allows to evaluate the range of acceptable ω values.

The dot-dashed line links models calculated with a $n = 0.72$ exponent that reproduces the Li destruction presumably occurring on the main sequence but observed in giant stars (Talon et al., in preparation). It is seen to lead to more destruction than observed in the hotter stars of the Hyades.

A similar figure is shown for Be (Fig. 5). No Be abundance have been determined in Hyades stars. The field Be abundances are discussed elsewhere in this conference by Deliyannis (see also Deliyannis et al. 1998). The solar Be abundance is discussed by Balachandran. Since the Be superficial abundance is usually much closer to the original abundance than Li, this implies a much smaller destruction for Be than for Li. In turn this implies a lower limit for n that is probably larger than $n = 0.72$. The extent to which the field star Li and Be abundances put limits on n will be further discussed in Richer et al. (in preparation).

5. Constraints on turbulence

The preceding discussions establish a number of constraints that turbulent transport must satisfy: 1) Abundance anomalies in AmFm stars (slowly rotating A and F stars) imply mixing of the outer $\sim 10^{-5}$ of the star's mass. 2) The existence of "normal" A and F stars (more rapidly rotating A and F stars) requires mixing of at least 3×10^{-4} to 10^{-3} of the star's mass if the anomalies are not to exceed 0.1 dex and of not more than 10^{-2} if Li is not to be completely destroyed. 3) The Li gap implies *partial mixing* of the outer 10^{-2} of the mass since that is the depth where Li is destroyed. This mixing is probably not complete however since turbulence has to lead to less Li destruction on the cool side of the Li gap. If the mixing were complete, no Li would remain. 4) While Be is less destroyed than Li, it is partly destroyed. The partial Li and Be destruction imply a density dependence of turbulence ($n \gtrsim 2$; see Eq. 2). The turbulent transport close to the region of Li burning is also constrained by observations of Li abundance in giant stars as described at this conference by Charbonnel. 6) The sun requires similar turbulence as required for Li and Be destruction in F stars. This turbulence is weak enough for the gravitational settling of He to lead to a 10% reduction of the He abundance in the solar convection zone.

These constraints can now be improved upon by more accurate abundance determinations in cluster stars. This is within reach of the new generation of large telescopes. Stellar evolution modeling also needs to be improved. The results presented here show that it is within reach, using available computers and atomic data bases.

References

Abt, H. A., & Morrell, N. I. 1995, ApJS, 99, 135
Boesgaard, A. M., & Budge, K. G. 1988, ApJ, 332, 410
Boesgaard, A. M., & Tripicco, M. J. 1986, ApJ, 302, L49
Burgers, J. M. 1969, *Flow equations for composite gases*, (New York: Academic Press)
Chaboyer, B., Demarque, P., & Pinsonneault, M. H. 1995, ApJ, 441, 865
Deliyannis, C. P., Boesgaard, A. M., Stephens, A., King, J. R., Vogt, S. S., & Keane, M. J. 1998, ApJ, 498, L147
Latour, J., Toomre, J., & Zahn, J.-P. 1981, ApJ, 248, 1081
Proffitt, C. R., & Michaud, G. 1991, ApJ, 380, 238
Richard, O., Vauclair, S., Charbonnel, C., & Dziembowski, W. A. 1996, A&A, 312, 1000
Richer, J., & Michaud, G. 1993, ApJ, 416, 312
Richer, J., Michaud, G., & Massacrier, G. 1997, A&A, 317, 968
Richer, J., Michaud, G., Rogers, F. J., Iglesias, C. A., Turcotte, S., & LeBlanc, F. 1998, ApJ, 492, 833
Richer, J., Michaud, G., & Turcotte, S., 2000, ApJ, in press
Thorburn, J. A., Hobbs, L. M., Deliyannis, C. P., & Pinsonneault, M. H. 1993, ApJ, 415, 150
Turcotte, S., Richer, J., Michaud, G., Iglesias, C. A., & Rogers, F. J. 1998, ApJ, 504, 539
Vauclair, S. 1988, ApJ, 335, 971

The Light Elements and Their Evolution
IAU Symposium, Vol. 198, 2000
L. da Silva, M. Spite, J. R. de Medeiros, eds.

Transport Phenomena and Light Element Abundances in the Sun and Solar Type Stars

Sylvie Vauclair

Laboratoire d'Astrophysique, 14 av. Ed. Belin, 31400-Toulouse, France

Abstract.
The observations of light elements in the Sun and Solar type stars give special clues for understanding the hydrodynamical processes at work in stellar interiors. In the Sun ^{7}Li is depleted by 140 while ^{3}He has not increased by more than $\cong$ 10% in 3 Gyrs. Meanwhile the inversion of helioseismic modes lead to a precision on the sound velocity of about .1%. The mixing processes below the solar convection zone are constrained by these observations. Lithium is depleted in most Pop I solar type stars. In halo stars however, the lithium abundance seems constant in the "spite plateau" with no observed dispersion, which is difficult to reconcile with the theory of diffusion processes. In the present paper, the various relevant observations will be discussed. It will be shown that the μ-gradients induced by element settling may help solving the "lithium paradox".

1. Introduction

Element diffusion and mixing processes in stellar interiors are now widely constrained, first by detailed observations of abundances, second by helio and asteroseismology. In most cases however, pure microscopic diffusion in stars would lead to abundance variations much larger than those observed : mild macroscopic motions in stellar radiative zones are definitely needed to account for the observations. This gives strong constraints on the kind of mixing processes allowed. Other constraints come from the consequences of the nuclear reactions occuring in stellar interiors : in some cases stellar mixing from the atmosphere down to the regions of nuclear processing is needed to explain the observed element abundances. This is the case, for example, to account for the depletion of lithium in the Sun and solar type stars.

Lithium observations in main-sequence population I field stars and galactic clusters show a large abundance dispersion which has been extensively studied in the literature (see reviews by Deliyannis 2000, Charbonnel 2000, Michaud 2000 and Pinsonneault 2000). The lithium abundance decreases for decreasing effective temperature below 5500K and the depletion increases with increasing age. This is generally attributed to the deepening of the convective zone, associated with some mild mixing process connecting the bottom of the convective zone with the nuclear destruction region.

Lithium is also depleted in F-type stars (the so-called "Boesgaard dip"). Several possible reasons have been invoked to explain this feature, most related to mixing and nuclear destruction. Element segregation has been proved negligible here as it would lead to unobserved variations of metal abundances (Turcotte et al 1998) and beryllium (Boesgaard 2000).

On the other hand, observations of lithium in main-sequence population II field stars show remarkably constant abundances, with a very small dispersion (e.g. Bonifacio and Molaro 1997) Why is lithium destroyed in Pop I stars while it does not seem destroyed in Pop II stars?

For the same effective temperatures, the convective zone is smaller in Pop II stars than in Pop I stars because of their smaller metallicity. Meanwhile they have a smaller rotation velocity on the average. This could explain why the lithium destruction induced by nuclear reactions is smaller in these stars than in Pop I stars. However the element segregation is more important for smaller densities and smaller rotation, so that this process should lead to a visible lithium depletion, which is not observed (Vauclair and Charbonnel 1995 and 1998). This represents the so-called "lithium paradox". Here we suggest that the influence of μ-gradients on the rotation-induced mixing may help solving this paradox.

2. Competition between rotation induced mixing and element diffusion

In rotating stars, the equipotentials of "effective gravity" (including the centrifugal acceleration) have ellipsoidal shapes while the energy transport still occurs in a spherically symetrical way. The resulting thermal imbalance must be compensated by macroscopic motions: the so-called "meridional circulation" (Von Zeipel 1924). The stellar regions outside the convective zones cannot be in complete radiative equilibrium. They are subject to entropy variations given by :

$$\rho T \left(\frac{\partial S}{\partial t} + \mathbf{u} \cdot \nabla S \right) = -\nabla \cdot \mathbf{F} + \rho \varepsilon_n$$
$$= \rho \varepsilon_\Omega \ (\neq 0) \quad (1)$$

where $\mathbf{F}$ represents the heat flux, ε_n the nuclear energy production and ε_Ω an energy generation rate which results from sources and sinks of energy along the equipotentials.

The vertical component of the meridional velocity u_r is computed as a function of ε_Ω in the stationary regime (from eq. 1):

$$u_r = \left(\frac{P}{C_p \rho T} \right) \frac{\varepsilon_\Omega}{g} \quad (2)$$

which, for a perfect gas, reduces to:

$$u_r = \frac{\varepsilon_\Omega}{g} \frac{\nabla_{ad}}{\nabla_{ad} - \nabla + \nabla_\mu} \quad (3)$$

where g represents the local gravity, ∇_{ad} and ∇ the usual adiabatic and real ratios $\left(\frac{d \ln T}{d \ln P} \right)$ and ∇_μ the mean molecular weight contribution $\left(\frac{d \ln \mu}{d \ln P} \right)$.

The expression of ε_Ω is computed by expanding the right-hand-side of eq. (1) on a level surface and writing that its mean value vanishes.

Mestel (1953, 1957 and 1965) pointed out that, in the presence of vertical μ-gradients, ε_Ω contains two kinds of terms : those related to the resulting horizontal variations of μ: the so-called "μ-induced currents" E_μ and those independent of μ, the so-called "Ω-induced currents" E_Ω. The expression of ε_Ω obtained in this case has been derived in detail by Maeder and Zahn (1998), who took into account several effects which were not included in the previous computations: more general equations of state instead of perfect gas law, presence of a thermal flux induced by horizontal turbulence, non-stationary cases.

Vauclair (1999) discussed more simple expressions, valid only for negligible differential rotation. In this case μ-currents are opposite to Ω-currents in most of the star and ε_Ω may be written :

$$\varepsilon_\Omega = \left(\frac{L}{M}\right)(E_\Omega + E_\mu) P_2(\cos\theta) \qquad (4)$$

with:

$$E_\Omega = \frac{8}{3}\left(\frac{\Omega^2 r^3}{GM}\right)\left(1 - \frac{\Omega^2}{2\pi G\bar\rho}\right) \qquad (5)$$

$$E_\mu = \frac{\rho_m}{\bar\rho}\left\{\frac{r}{3}\frac{d}{dr}\left[\left(H_T\frac{d\Lambda}{dr}\right) - (\chi_\mu + \chi_T + 1)\Lambda\right] - \frac{2H_T\Lambda}{r}\right\} \qquad (6)$$

Here $\bar\rho$ represents the density average on the level surface ($\simeq \rho$) while ρ_m is the mean density inside the sphere of radius r; H_T is the temperature scale height; Λ represents the horizontal μ fluctuations $\frac{\tilde\mu}{\mu}$; χ_μ and χ_T represent the derivatives:

$$\chi_\mu = \left(\frac{\partial \ln\chi}{\partial \ln\mu}\right)_{P,T} \quad ; \quad \chi_T = \left(\frac{\partial \ln\chi}{\partial \ln T}\right)_{P,\mu} \qquad (7)$$

Vertical μ-gradients may occur in stars due to two different processes : first the nuclear reactions which occur in the stellar cores, second the helium settling which occurs in the outer layers. The importance of the first process in reducing or even suppressing the meridional motions has been demonstrated several times in the literature (e.g. Huppert and Spiegel 1977). The second process on the other hand has not been extensively studied. We claim here that it may play a crucial role for understanding the lithium problem in Pop I and Pop II stars.

3. Application to Pop II stars

Computations of μ-currents induced by the helium settling in halo stars have been performed by Vauclair 1999 and Théado and Vauclair 2000 a and b. We found that, for slow rotation, μ-currents cancel Ω-currents for very small concentration gradients, corresponding to μ-gradients of order 10^{-15} cm^{-1}.

Let us summarize the situation of a slowly rotating star in which element settling leads to an increase of the μ-gradient below the outer convection zone. At the beginning, the star is homogeneous and meridional circulation can occur, leading to upward flows in the polar regions and downward flows in the equatorial parts (except in the very outer layers where the Gratton-Öpik term becomes important, which we do not discuss here). The μ-currents, opposite to the classical Ω-currents, are first negligible. The μ-gradients increasing with time because of helium settling, the order of magnitude of the μ-currents also increases until it reaches the value for which the circulation vanishes.

This does not occur all at once: as the μ-gradient decreases with depth below the convective zone, we expect that the meridional circulation freezes out step by step (see figure 1 of Théado and Vauclair 2000a). An equilibrium situation may be reached, in which the temperature and mean molecular weight gradients along the level surfaces are such that Ω-currents and μ-currents cancel each other.

Once it is reached, this equilibrium situation is quite robust. Suppose that some mechanism leads to a decrease of the μ-gradient: then $|E_\mu|$ becomes smaller than $|E_\Omega|$ and the circulation tends to be restablished in the $|E_\Omega|$ direction, thereby restoring the original μ gradient. Suppose now that the μ-gradient is increased. Then $|E_\mu|$ becomes larger than $|E_\Omega|$ and the circulation begins in the E_μ direction. Here again the original gradient is restored.

When the meridional circulation is frozen below the convective zone, helium settling could proceed further; however, due to the increase of the diffusion time scale with depth, this would modify the μ-gradient. We may thus expect that μ-currents would take place and restore the original equilibrium gradient, thereby strongly reducing the microscopic diffusion (Théado and Vauclair 2000b). This self-regulating process could be the reason for the low dispersion of the lithium abundance in the lithium plateau of halo stars.

4. Discussion : Pop I versus Pop II stars

There are many observations in stars which give evidences of mixing processes occuring below the outer convective zones as, for example, the lithium depletion observed in the Sun and in galactic clusters. The process we have described above should not apply in all these stars. The reason could be related to the rapid rotation of young stars on the ZAMS and to their subsequent rotational braking.

The abundance determinations in the solar photosphere show that lithium has been depleted by a factor of about 140 compared to the protosolar value while beryllium has not been depleted by more than a factor 2, and maybe much less, as discussed by Balachandran and Bell (1997). These values represent strong constraints on the mixing processes in the solar interior.

Observations of the ^{3}He/^{4}He ratio in the solar wind and in the lunar rocks (Geiss 1993, Geiss and Gloecker 1998) show that this ratio may not have increased by more than $\cong$ 10% since 3 Gyr in the Sun. While the occurence of some mild mixing below the solar convective zone is needed to explain the lithium depletion , the ^{3}He/^{4}He observations put a strict constraint on its efficiency. The only way to obtain such a result is to postulate a mild mixing, which would be efficient down to the lithium nuclear burning region but not too far below, to preserve the original ^{3}He abundance. The efficiency of this mixing should also decrease with time, as the ^{3}He peak itself builts up during the solar life.

It is interesting to compute the minimum enhancement of the ^{3}He/^{4}He ratio implied by the lithium observed depletion. Vauclair and Richard 1998 showed that it is possible to deplete lithium by a factor larger than 100 as observed and not increase ^{3}He/^{4}He by more than 5 percent since the solar origin. In this case beryllium is only depleted by about 10 percent.

Such a confined mixing zone is also needed from helioseismology : although the introduction of pure element settling in the solar models considerably improves the consistency with the seismic Sun, some discrepancies do remain, particularly below the convective zone where a "spike" appears in the sound velocity (Richard et al 1996, Turck-Chièze et al. 1998). It has been shown that this behavior may be due to the helium gradient which would be too strong in case of pure settling. Mild macroscopic motions below the convective zone slightly decrease this gradient and helps reducing the discrepancy (Richard et al 1996, Corbard et al 1998, Brun et al 1998). The helium profiles directly obtained from helioseismology (Basu 1998, Antia and Chitre 1998) show indeed a helium gradient smoother than the gradient obtained with pure settling.

The constraints implied by both the helioseismic inversions and abundance determinations in the Sun converge towards the existence of a small mild mixing region below the convective zone, which would extend down to a depth of the order of one scale height. The implied mixing region must be very mild, with diffusion coefficients of 10^3 - 10^4 only. It must also be completely deconnected from the solar core. No mixing can indeed be allowed down to the nuclear energy production region as it would lead to a sound velocity incompatible with helioseismology. In particular the mixing processes invoked by Morel and Schatzman 1996 to decrease the neutrino fluxes are excluded by helioseismology (Richard and Vauclair 1997).

Mixing processes localized at the boundary between convective and radiative regions include overshooting and regions of large differential rotation like the "tachocline" below the solar convective zone. Up to now, overshooting was generally treated in the models simply as a continuation of the convective zone on a fraction of a pressure scale height. Recent parametrisations use a diffusion coefficient which decreases exponentially with decreasing radius (Freytag et al 1996). The tachocline, which represents in the present Sun the small boundary between the region of large differential rotation (in the convective zone) and the region of solid rotation (in the radiative zone below) is also treated as a mixed layer with an exponentially decreasing diffusion coefficient (Brun et al 1998, Richard 1999). Results are encouraging, although more sophisticated numerical simulation including 2-D abundance variations would be needed to go further.

In any case, the self-regulating process that we have discussed for halo stars in section 3 would not apply below the convective zone in the Sun and solar type stars because of the differential rotation which takes place there. Such a differential rotation would not be expected in halo stars if we suppose that they always rotated slowly and thus did not suffer large transport of angular momentum. The different behavior for the lithium abundance in Pop I and Pop II stars could thus be directly related to their rotation history.

References

Antia, H.M., Chitre, S.M., 1998, *A&A* **339**, 239

Balachandran, S.C., Bell, R.A., 1997, *American Astronomical Society Meeting* **191**, 7408

Basu, S., 1998, *M.N.R.A.S.* **298**, 719

Boesgaard, A.M., 2000, to be published in *The 11th Cambridge Workshop on cool stars, stellar systems and the sun, Challenges for the New Millenium*

Bonifacio P., Molaro P., 1997, MNRAS, 285, 847

Brun, A.,S., Turck-Chieze, S., Morel, P., 1998, *ApJ* **506**, 113

Charbonnel, C., 2000, this meeting

Corbard,T., Berthomieu, G., Provost, P., Morel, P., 1998, *A&A* **330**, 1149

Deliyannis, C., 2000, this meeting

Freytag,B., Ludwig,H., Steffen M., 1996, *A&A* **313**, 497

Geiss, J.: 1993, *Origin and Evolution of the Elements*, ed. Prantzos, Vangioni-Flam & Cassé (Cambridge Univ. Press), **90**

Geiss, J., Gloecker, G., 1998, *Space Science Reviews* **84**, 239

Huppert, H.E., Spiegel, E.A., 1977, *ApJ* **213**, 157

Maeder A., Zahn J.-P., 1998, *A&A* **334**, 1000

Mestel L., 1953, *M.N.R.A.S.* **113**, 716

Mestel L., 1957, *ApJ* **126**, 550

Mestel L., 1965, Stellar Structure, in Stars and StellarSystems, vol 8, ed. G.P. Kuiper, B.M. Middlehurst, Univ. Chicago Press, 465

Michaud, G., 2000, this meeting

Pinsonneault, M., 2000, this meeting

Richard, O., 1999, *PhD thesis* , University of Toulouse

Richard, O., Vauclair, S., Charbonnel, C., Dziembowski, W.A., 1996, *A&A* **312**, 1000

Richard, O., Vauclair, S., 1997, *A&A* **322**, 671

Théado, S., Vauclair, S., 2000 a, this meeting

Theado, S., Vauclair, S., 2000 b, preprint

Turck-Chieze, S., Basu, S., Berthomieu, G., Bonnano, A., Brun, A.S., Christensen-Dalsgaard, J., Gabriel, M., Morel, P., Provost, J., Turcotte, S., The Golf Team, 1998, in *Structure and Dynamics of the Interior of the Sun and Sun-like Stars* ESA Publications Division, SP-418, 555

Turcotte, S., Richer, J., Michaud, G. Iglesias, C.A., Rogers, F.J., 1998, *ApJ* **504**, 539

Vauclair, S., 1999, *A&A* **351**, 973

Vauclair, S., Charbonnel, C., 1995, *A&A* **295**, 715

Vauclair, S., Charbonnel, C., 1998, *ApJ* **502**, 372

Vauclair, S., Richard, O.: 1998, in *Structure and Dynamics of the Interior of the Sun and Sun-like Stars* ESA Publications Division, SP-418, 427

Von Zeipel H., 1924, *M.N.R.A.S.* **84**, 665

AGB Stars Interferometric Signatures: Effects of possible Li-rich spots.

Patrick de Laverny

Observatoire de la Côte d'Azur, Département Fresnel UMR 6528, BP 4229, F-06304 Nice cedex 4, France

Bruno Lopez

Observatoire de la Côte d'Azur, Département Fresnel UMR 6528, France

Abstract. We propose to observe Asymptotic Giant Branch stars (AGB) with spectro-differential interferometric techniques in order to find new observational constraints to inner structure and evolutionary models of these stars. We examine the interferometric signatures created by possible heterogeneities on AGB stars surface due to local Li-enrichments or Li-rich spots and find that such heterogeneities, if they exist, could be detected with present or future interferometers.

In order to study interferometric signatures of an AGB star (i.e. visibilities and phases versus baseline), we first compute synthetic spectra of a typical carbon-rich star. For this purpose, we refer to Abia et al. (1999) where details of used model atmospheres, line lists and other input physics are given.

We consider a "normal" cool carbon star defined by T_{eff} = 3200 K, log g = 0, C/O = 1.1 and $^{12}C/^{13}C$ = 5. The star spectrum is then computed by combining the spectral contributions of: (i) an uniform disk representing the AGB surface with $R_{AGB} = 20 \times 10^{-3}$arcsec (i.e. $R_{AGB} = 400 R_\odot$ at 100pc) and very low Li abundance (A_{Li} = 0) as observed for most AGBs, and (ii) a not centered Li-rich spot with arbitrarily $R_{spot} = 20\% R_{AGB}$ and a very large local Li enrichment (A_{Li} = 4.5). Such a Li abundance for the spot corresponds to those presently found in the more Li-rich AGB known to date. The resulting spectrum is the combination of the "normal" AGB and the Li-rich spot spectra, taking into account their proper emitting surface. We find that the total stellar spectrum of the hypothetic AGB with a Li-rich spot matches the spectrum of a classic Li-poor AGB star. A star would thus be classified as Li-poor if observed with classical spectrometric technics, although a Li-rich spot could be present on its surface. Interferometric observations have then to be undertaken to detect such possible surface heterogeneities.

We have then calculated visibilities and phases versus the baseline in different spectral channels around the Li6708 transition. The moderate spectral resolution considered ($R < 10\,000$) corresponds to that offered by current/future interferometers as the *Grand Interféromètre à deux Télescopes* (Observatoire de la Côte d'Azur). In the following spectral channels, the star appearance changes indeed drastically:

1. In the core of the Li absorption line (6708Å), where the Li-rich spot does almost not contribute to the stellar flux, the total spectrum corresponds to the one emitted by an uniform disk with a dark spot on it. The interferometric fringes phase varies strongly with the baseline in the line core.

2. In the wing of the Li line, the Li-rich spot slightly contributes to the stellar flux. The star appears as an uniform disk with a grey spot on it.

3. Outside the Li line, the Li-rich spot does not affect anymore the stellar spectrum. The star looks like a completely uniform disk since it is spherically symetric and no structures are present at its surface. The interferometric fringes phase is then always equal to zero.

Therefore, from the core of the Li line to outside this transition, the photocenter of the studied star is strongly displaced due to changes of the Li-rich spot spectral contributions. This leads to strong variations of the AGB interferometric signatures, specially for the phase. A Li-rich spot, if it exists, could thus be easily detected by interferometric observations with a relative precision of $\sim 1\%$ for the visibility (such a precision measurement should be attained by the present/future interferometers). Furthermore, the variations of the interferometric fringes phase with respect to the baseline are much stronger (from $-\pi/2$ to $\pi/2$) and can therefore be measured much more easily. Phases differences or visibilities ratios between different spectral channels are indeed measured with the spectro-differential interferometric method. We can use this method here to get enough accurate relative measurements.

In summary, our simulations have shown that a very Li-rich spot on an AGB star with $R_{\rm spot} \sim 10-30\% R_{\rm AGB}$ or a larger spot ($R_{\rm spot} \sim 50\% R_{\rm AGB}$) with moderate Li-enrichment ($A_{Li} < 4$) does not strongly modify the stellar spectrum. Thus interferometric technics are requested to detect such configurations. We therefore propose that (i) AGB stars presently classified as Li-rich should have a large Li-enrichment over their whole surface, and (ii) AGB stars presently classified as Li-poor could have either no Li-enrichment at all **or** Li-rich spots could be present on their surface.

Other configurations and their interferometric signatures (several larger or smaller not centered Li-rich spots, line transitions of other elements interesting stellar evolution...) are under investigation (see also de Laverny & Lopez, 2000).

Acknowledgments. We thank IAU and the LOC for financial supports.

References

Abia C., Pavlenko Ya., de Laverny P., 1999, A&A 351, 273

de Laverny P., Lopez B., 2000, in "The changes in abundances in AGB stars", Mem. S.A.It, in the press.

Lithium abundances in main-sequence F stars and sub-giants

Jose Dias do Nascimento Jr, Sylvie Théado and Sylvie Vauclair

Laboratoire d'Astrophysique, 14 av. Ed. Belin, 31400 Toulouse, France

Abstract. The application to main-sequence stars of the rotation-induced mixing theory in the presence of μ-gradients leads to partial mixing in the lithium destruction region, not visible in the atmosphere. The induced lithium depletion becomes visible in the sub-giant phase as soon as the convective zone deepens enough. This may explain why the observed " lithium dilution " is smoother and the final dilution factor larger than obtained in standard models, while the lithium abundance variations are very small on the main sequence.

The observations of lithium in main-sequence stars on the hot side of the "Boesgaard dip" show a very small dispersion for normal stars while a light depletion (by a factor 3) is observed in Am stars (Burckhart and Coupry 2000). On the other hand, on the sub-giant branch, these stars present a lithium depletion larger than that predicted by the standard model (do Nascimento et al. 1999). These observations suggest that, while on the main sequence, the stars suffer in their internal layers a lithium destruction larger than the standard one : this extra-destruction, which must not appear at the surface in the main-sequence phase, is then dredged up during the subsequent evolution on the sub-giant branch (Vauclair 1991)

It has been suggested several times that the process responsible for this extra-depletion could be the result of rotation-induced mixing. Computations including such macroscopic motions as described by Zahn 1992 and Maeder & Zahn 1998 have recently been performed by Charbonnel and Talon 1999 and 2000. They show that the observations on the sub-giant branch can nicely be reproduced by such rotation-induced mixing. In their computationsi however, the effect of the microscopic diffusion of lithium was not introduced on the main-sequence, for the reason that in these stars the radiative acceleration may balance the lithium gravitational settling. For helium, on the contrary, the radiative acceleration is negligible : helium settling was then introduced but not taken into account while computing the meridional circulation velocity.

As shown by Mestel 1953, Maeder and Zahn 1998, Vauclair 1999, (see also Vauclair 2000 and Théado and Vauclair 2000), in the presence of vertical μ-gradients, the circulation velocity is the sum of two terms which leed to motions in the opposite direction, one which does not depend on μ (the so-called "Ω currents") and one which gathers the μ dependent terms (the "μ currents"). In case of helium gravitational settling, a "μ gradient" builts up which soon counteracts the standard meridional circulation and an equilibrium situation

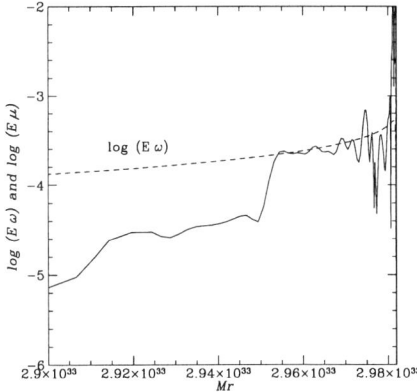 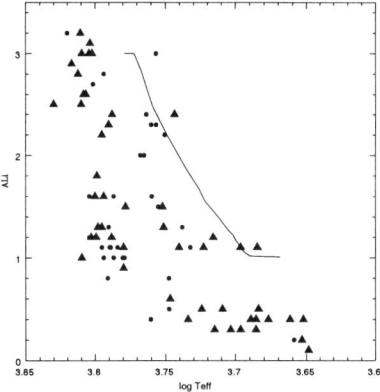

Figure 1. computations of the Ω-currents and μ-currents in a $1.5 M_\odot$ star with a rotation velocity of 40 km.s-1 and lithium evolution on the sub-giant branch obtained in this case, compared to the observations

may be reached, which could account for the fact that lithium is preserved on the main sequence, while extra-mixing occurs below the "frozen layer".

In the present paper, we have computed the evolution of a $1.5 M_\odot$ star taking into account the same effects as discussed in Théado and Vauclair 2000. We show that, when the opposite currents are taken into account, the layer just below the convection zone freezes out while mixing proceeds below. While evolving out of the main-sequence, dilution induced by the deepening of the convective zone leeds to a larger depletion than predicted by the standard model, reproducing the upper envelope of the observations. More computations are underway to extend these results to other masses and rotation parameters.

References

Burkhart, C., Coupry, M.F. 2000, preprint

Charbonnel, C., Talon, S. 1999, A&A **351**, 635

Charbonnel, C., Talon, S. 2000, this conference

Dias do Nascimento Jr., J.D., Charbonnel, C., Lèbre, A., de Laverny, P., de Medeiros, J.R., 1999, preprint

Maeder, A., & Zahn, J.-P. 1998, A&A, 334, 1000

Mestel L., 1953, *M.N.R.A.S.* **113**, 716

Théado and Vauclair 2000, this conference

Vauclair, S. 1991, in *IAU Symp. 145, Evolution of Stars : the Photospheric Abundance Connection (Michaud, G., Tutukov, A. eds.)* p. 327

Vauclair, S. 1999, A&A, 351, 973

Vauclair, S. 2000, this conference

Zahn, J.-P. 1992, A&A, 265, 115

The Light Elements and Their Evolution
IAU Symposium, Vol. 198, 2000
L. da Silva, M. Spite, J. R. de Medeiros, eds.

He abundance in Planetary Nebulae

R. Gruenwald and S. M. Viegas

IAG-USP; Av. Miguel Stefano, 4200; 04301-904 São Paulo, SP, Brazil

Abstract. Type I planetary nebulae (PNe), defined as those with high He (and N) abundances, generally present bipolar geometries and high stellar temperatures. The main goal of this paper is to check if the empirically derived He overabundance for Type I PNs is real, or if it is a consequence of geometrical effects due to the bipolarity and/or to the ionization stratification due to the high stellar temperature of these objects.

1. Introduction

The determination of the chemical abundance in PNe provides information on the products of stellar evolution, as well as on the enrichment of the interstellar medium, establishing constraints for stellar evolutionary models as well as for the study of the chemical evolution of galaxies. Chemical abundances in PNe are usually obtained using empirical methods (as for example in Peimbert & Torres-Peimbert 1987). Photoionization models also provide the gas chemical composition when applied to specific nebulae (for example, Harrington et al. 1982; Gruenwald, Viegas & Broguière 1997). Empirical methods, using a few bright emission-lines, are commonly used since they can easily be applied to a great number of objects. Following their abundance, PN are classified in types (Peimbert 1978; Peimbert & Torres-Peimbert 1983). In particular, Type I PNe are those with high He and N abundances, e.g. He/H $\geq$ 0.125 and $\log(N/O)$ > -0.30. In a study of morphological and physical properties of PNe, Corradi & Schwarz (1995) stated that bipolar nebulae have the hottest stars among PNe, and, except for two objects, all bipolar PNe for which chemical data are available are Type I. Therefore, objects with high He and N abundances have bipolar morphologies and high stellar temperatures. However, new evolutionary models fail to explain the high abundance of He and N in Type I PNe (van den Hoek & Groenewegen 1997). In this paper we analyze the effect of the geometry and of the stellar temperature on the empirical abundance determinations, in order to verify if the high abundances of He obtained for Type I PNe are real. For this, we apply our self-consistent 3D photoionization code (Gruenwald et al. 1997) to simulate planetary nebulae.

2. Method and results

For a wide range of nebular and stellar characteristics, we obtain the physical conditions in each point of the nebula, and calculate the resulting line intensity

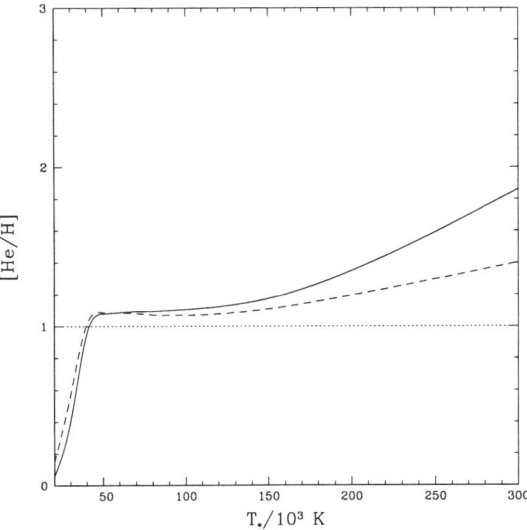

Figure 1. Fig. 1 - Ratio between the empirical and the adopted values for He/H versus the stellar temperature. The curves correspond to a nebula with uniform density ($n_H = 100$ cm^{-3}, $L_* = 3000 L_\odot$). The solid line represents the results obtained from line intensities corresponding to the whole nebula, while the dashed line corresponds to the results from a narrow slit crossing the nebula through its center.

ratios at different lines of sight to the nebula. From the calculated line intensity ratios we derive the abundances applying the empirical methods used in the literature. These abundances are then compared to those assumed in the models. The effect of the central stellar temperature can be seen in Figure 1. For low stellar temperatures ($T_* < 5\ 10^4$ K), the empirical He/H abundance ratio is lower than the real value, while for higher T_* ($> 10^5$ K) the opposite is true. This is due to the fact that the ionic distribution inside the nebula depends on T_*; for example, for low stellar temperatures the He^{++} zone is much smaller than the H$^+$ zone; for increasingly temperatures the He^{++} zone is relatively larger. A further consequence is that the obtained abundance depends also on the size and position of the slit used for the observation. In order to check for a geometrical effect, models for bipolar nebulae were also obtained, assuming a spherically symmetrical cloud with an equatorial torus of denser gas close to the ionizing star. Our results show that the abundance obtained from empirical methods can mimic an overabundance and/or the presence of abundance gradients of He in planetary nebulae with high stellar temperature. Geometrical effects strengthen this problem.

References

Corradi, R.L.M., & Schwarz, H.E. 1995, A&A, 293, 871

Gruenwald, R., Viegas, S.M., & Broguiere, D. 1997, ApJ, 480, 283

Harrington, J.P. et al. 1982, MNRAS, 199,517

Peimbert 1978, in IAU Symp. 76, Planetary Nebulae, ed. Y. Terzian (Dordrecht: Reidel), 215

Peimbert, M. & Torres-Peimbert, S. 1983, in IAU Symp. 76, Planetary Nebulae, ed. D.R. Flower (Dordrecht:Kluwer),233

Peimbert, M. & Torres-Peimbert, S. 1987, Rev.Mex.Astr.Astroph., 14, 540

van den Hoek, L.B. & Groenewegen, M.A.T. 1997, A&AS, 123, 305

The Light Elements and Their Evolution
IA U Symposium, Vol. 198, 2000
L. da Silva, M. Spite, J. R. de Medeiros, eds.

Non-LTE Effects in Berylium Abundances

T. P. Idiart
Universidade de São Paulo, IAG, Depto. de Astronomia
Av. Miguel Stefano 4200, São Paulo 01065-970, Brazil

F. Thévenin
Observatoire de la Côte d'Azur
B.P. 4229, 06304 Nice Cedex 4, France

Abstract. In this work we analyze the beryllium-iron chemical diagram from the point of view of non-LTE effects. Be abundances were re-calculated by considering non-LTE corrections in ionization equilibrium (logg) and Fe abundances ([Fe/H]). These corrections seem do not affect the linear relation between Be-Fe for metal-poor stars already found in the literature for LTE derived abundances.

1. Introduction

The analysis of the trends of abundances of light elements with respect to [Fe/H] for the oldest metal-poor stars is a direct way to provide some clues on their production mechanism and evolution.

In a recent work on non-LTE effects in iron abundances, Thévenin & Idiart (1999) (TI99) obtained that for metal-poor dwarf stars Fe abundances ([Fe/H]) are affected by significant non-LTE effects and, moreover, surface gravities (logg) derived by LTE analysis also need corrections. This logg corrections should be crucial for beryllium abundances determination, since Be II resonance lines normally used to estimate Be abundances are much sensitive to this stellar parameter.

In this work we examine the consequences of non-LTE corrections to logg and [Fe/H] for logN(Be/H) vs. [Fe/H] (or Be-Fe) diagram. In section 2 we present a short summary of our results obtained in TI99 for Fe and in section 3 the results for Be abundances.

2. Non-LTE Corrections for [Fe/H] and logg

TI99 performed statistical equilibrium calculations for Fe I and Fe II to estimate non-LTE effects in iron abundances. The main results are showed in figure 1 (see TI99 for details).

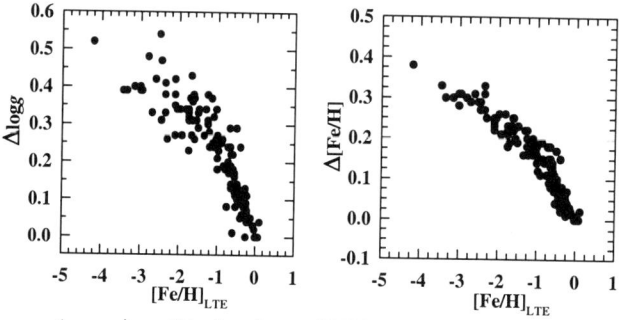

Figure 1. Amplitude of non-LTE logg and [Fe/H] corrections in function of LTE [Fe/H] for 136 subgiant to subdwarf stars.

3. Results for Be/H

We re-estimate N(Be/H) abundances for 21 stars also analised by Boesgaard et al.(1999) using T_{eff}, logg and [Fe/H] given by TI99. Be abundances were calculated assuming LTE conditions, since the non-LTE corrections are negligible for Be II lines considered here ($\lambda\lambda$ 3130 and 3131), as demonstrated by Garcia Lopez et al. (1995), for example. Figure 2 shows our results.

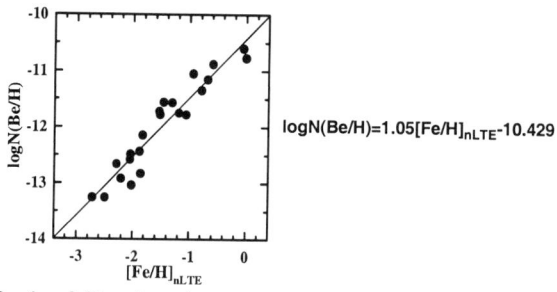

Figure 2. Derived Be abundances vs. non-LTE corrected [Fe/H].

We conclude that for the range of metal-poor objects $-3 <$ [Fe/H] < -1.5, non-LTE corrections for [Fe/H] compensate changes in Be abundances (as result of logg corrections) in the Be-Fe diagram, recovering the same linear behavior of LTE derived abundances (Boesgaard et al.1999). Similar results are found for Boron (see Primas 1999, this colloquium).

References

Thévenin,F. & Idiart,T. 1999, ApJ, 521, 753
Boesgaard,A.,Constantine,P.,King, J.,Ryan, S. Vogt,S., 1999, AJ, 117, 1549
Garcia Lopez,R., Severino,G. & Gomez,M., 1995, A&A, 297, 787
Primas F., 1999, this colloquium

The Light Elements and Their Evolution
IAU Symposium, Vol. 198, 2000
L. da Silva, M. Spite, J. R. de Medeiros, eds.

White Dwarf Probes of Interstellar Deuterium

Wayne Landsman

Raytheon ITSS, Code 681, NASA/GSFC, Greenbelt, MD 20771

Abstract.
We review the advantages of using hot white dwarfs (WDs) as probes of the deuterium abundance in the local interstellar medium. We then discuss advantages of the Space Telescope Imaging Spectrograph (STIS) for such observations, as compared with earlier observations with the Goddard High Resolution Spectrograph (GHRS). The GHRS Ly α profile of the white dwarf HZ 43 is probably modified by the hot "hydrogen wall" surrounding the Sun; but despite this complication, the sightline remains a promising one for an accurate determination of the deuterium abundance in the local interstellar medium.

1. Introduction

Studies of interstellar deuterium Lyα are limited to sightlines with a total column density N(H I) $\leq$ 18.5, in order to avoid saturation of the deuterium feature by the hydrogen absorption. The potential advantages of using nearby hot WDs to study deuterium in the local interstellar medium have long been known (e.g. Lemoine et al. 1996). Although the modeling of the WD Lyα profile is not unproblematic, it can be performed with higher confidence than is possible for chromospheric emission lines. The WD continuum can be used to probe the velocity component structure by allowing observations of metal lines with small thermal widths. Finally, EUV observations of a hot WD can provide an independent estimate of the total H I column density.

STIS has three advantages over GHRS for such WD observations. The spectral resolution achievable with STIS (R $\sim$ 170,000) is more than twice that of GHRS (R $\sim$ 80,000). The 2-dimensional echelle format of STIS allows 200 Å to be recorded in a single spectrum, compared to only 6 Å with GHRS. In particular, many important ISM lines, such as N I λ1200, Si II λ1190, λ1260, λ1304, and O I λ1302 are obtained simultaneously with Lyα. Finally, the two-dimensional format of STIS should allow a more reliable correction for scattered light. Although the GHRS and STIS profiles show excellent agreement for typical interstellar lines (Howk & Sembach 2000), the situation is less clear for a broad line profile such as Lyα. The discrepancy noted by Sahu et al. (1999) between the interstellar parameters derived from GHRS and STIS observations of G191B2B is likely due to a slight shift in the GHRS continuum level across the 3 Å of the profile (Sahu et al. 2000, in preparation). We currently have a STIS Cycle 9 program to study D/H toward the hot WDs GD 153 and HZ 43.

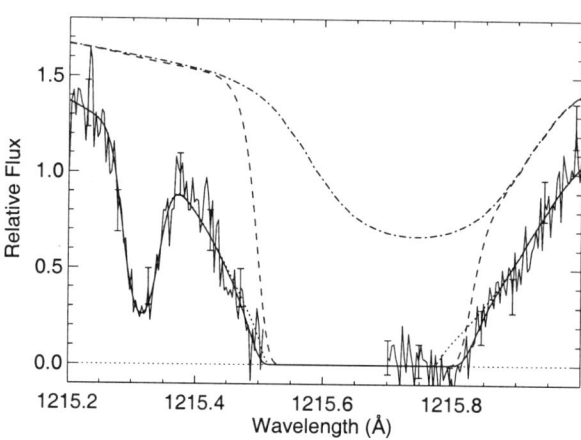

Figure 1. GHRS Lyα profile of HZ 43 and the model fit (solid line). Also shown are the non-LTE stellar profile (dot-dashed line), the hot hydrogen wall profile (dashed line), and the ISM profile (dotted line).

HZ 43 had been previously observed at modest S/N with GHRS by Landsman et al. (1995), who found that a good model fit was possible using a single ISM component, provided that one adopted a non-LTE Lyα stellar profile. However, as shown in Figure 1, a small discrepancy remains in fitting the wings of the saturated H I absorption, unless a hot ($\sim$ 20,000 K), low-density ($\sim 10^{15}$ cm^{-2}) component is included, such as would be expected from the hot hydrogen wall surrounding the Sun (e.g. Linsky & Wood 1996). Although there are large uncertainties in the derived density and temperature of the hydrogen wall toward HZ 43, the interstellar parameters are relatively well-determined, including both N(H I) = $(8.9 \pm 0.4) \times 10^{17}$ cm^{-2}), and D/H = $(1.6 \pm 0.25(2\sigma)) \times 10^{-5}$. The derived hydrogen column is in excellent agreement with EUVE results of Dupuis et al. (1995; N(H I) = $(8.7 \pm 0.6) \times 10^{17}$ cm^{-2}). The derived ISM temperature (T = 5700 ± 800 K) is more subject to systematic errors, but is in good agreement with the Orfeus results of Dupuis et al. (1998). The GHRS results on HZ 43 will be further discussed by Landsman et al. (2000, in preparation).

References

Dupuis, J. et al. 1995, ApJ, 455, 574
Dupuis, J. et al. 1998, ApJ, 500, L47
Howk, J.C., & Sembach, K.R. 2000, AJ, in press
Landsman, W., Sofia, U.J., & Bergeron, P. 1996, in Science with the Hubble Space Telescope - II, ed. P. Benvenuti, F.D. Macchetto & E.J. Schreier (STScI), 454
Lemoine, M. et al. 1996, A&A, 308, 601
Linsky, J.L. & Wood, B.E. 1996, ApJ, 463, 254
Sahu, M.S. et al. 1999, ApJ, 523, L159

IR Boron Lines in Stellar Spectra

J. Meléndez, B. V. Castilho and B. Barbuy

Instituto Astronômico e Geofísico, Universidade de São Paulo, Departamento de Astronomia, Caixa Postal 3386, 01060-970, Brazil

Abstract. We have computed synthetic spectra of the infrared B I transitions at 1.166 and 1.624 μm in order to examine the possibility of abundance determination by using these lines. We found that the IR boron lines are better observed in cool giants and supergiants. S/N > 150 and R ≈ 60000 are required in order to determine the boron abundances.

1. Introduction

The abundances of the primordial light elements Li, Be and B are important for the better understanding of big bang nucleosynthesis, chemical evolution of the Galaxy, and stellar evolution.

Boron stellar abundances are difficult to measure because the lines commonly used lie in the UV region (e.g. Duncan et al. 1998). On the other hand, the IR boron lines are very weak in the solar spectrum, and even using very high S/N data (S/N ≈ 1500) it is only marginally detected (Cunha & Smith 1999).

In this work we have investigated the behavior of the IR boron lines, in order to know if these lines can be observed in stars other than the sun.

2. Synthetic Spectra

For boron, we used transition probabilities from the recent critical data compilation of W.L. Wiese and J.F. Fuhr (1999, private com.). Atomic and molecular lines present in this region were taken from Meléndez & Barbuy (1999).

The synthetic spectrum is calculated assuming LTE. We employed the Kurucz model atmospheres, considering $4000 < T_{\rm eff} < 7000$ K, gravities $0.0 < \log g < 4.5$, and [M/H] = 0. The spectra have a FWHM resolution of ≈ 60000.

Our calculations showed that the IR B I lines are stronger in supergiants of $T_{\rm eff} \approx 4500$ K, with decreasing intensities for higher and lower temperatures. The B I lines at 1.166 (1.16600, 1.16625) μm are stronger than the lines at 1.624 μm. The line at 1.1660 μm is the strongest IR B I line, but it is blended with a carbon atomic line.

We have simulated observed spectra by adding noise to the synthetic spectra. In Fig. 1 the intensity variation of B lines is shown for different abundances. In this case, S/N > 150 is required to obtain the boron abundance. Higher S/N are needed in order to analyze other stars. For example, S/N > 300 is required for a K giant with $T_{\rm eff} = 4500$, log g = 1.5 and [M/H] = 0.

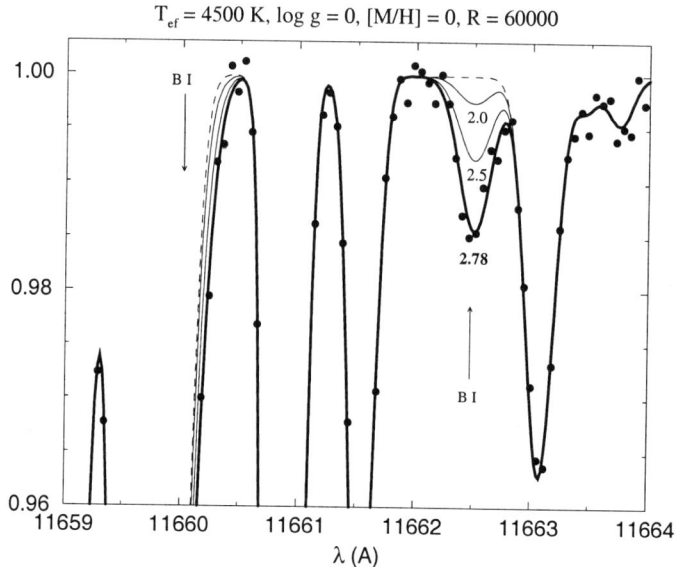

Figure 1. Synthetic spectra for no boron (- - -) and log N(B) = 2.0, 2.5 and 2.78. Points: log N(B) = 2.78 and S/N = 200.

3. Applications

Li, Be and B are fragile elements and are good tracers of the internal physics of stars. Gathering simultaneous data on these three elements can give us powerful tools for better understanding stellar evolution.

The B abundance determinations by Duncan et al. (1998) for 2 Hyades giants show a B depletion by a factor of 10, that agrees with the Be depletion found by Castilho et al. (1999) for 3 field Li-rich giants, indicating deep mixing.

4. Summary

We have investigated the possibility of detection of IR boron lines in stellar spectra. The strongest B I line in the IR is the 1.1660 μm-line, but it is blended. Alternatively, the 1.16625 μm-line could be used for abundance determinations. The most suitable conditions are for cool supergiants. S/N > 150 and R $\approx$ 60000 are required for this work. J. M. thanks Ph.D. fellowship 97/0109-8 (FAPESP).

References

Castilho, B. V., Spite, F., Barbuy B., Spite, M., De Medeiros, J.R., & Gregorio-Hetem, J. 1999, A&A, 345, 249

Cunha, K., & Smith, V. 1999, ApJ, 512, 1006

Duncan, D. K., Peterson, R. C., Thorburn, J. A. & Pinsonneault, M. H. 1998, ApJ, 499, 871

Meléndez, J., & Barbuy, B. 1999, ApJS, 124, 527

The Light Elements and Their Evolution
IAU Symposium, Vol. 198, 2000
L. da Silva, M. Spite, J. R. de Medeiros, eds.

Lithium in cool magnetic CP stars: Some new results of observations, using CAT (ESO) 2.6m (CrAO), (NOT) La Palma telescopes.

N. Polosukhina[1], D.Kurtz[2], M. Hack[3], P. North[4], I. Ilyin[5], J. Zverko[6]

[1] *Crimean Astrophysical Observatory Nauchny, Crimea, Ukraine*

[2] *University of Cape Town, Rondebosh 7700, South Africa*

[3] *Department of Astronomy, Trieste University, Italy*

[4] *Institut d'Astronomie de l'Universitè de Lausanne, CH 1290, Switzerland*

[5] *University of Oulu, P.O. Box 3000, 90401 Oulu, Finland*

[6] *Astronomical Institute of the Slovak Academy of Science, 05960 Tatranska Lomnica, The Slovak Republic*

1. Introduction

Lithium in cool magnetic CP stars in still poorly studied and estimations of the Li abundance in these stars are scarce. There is some evidence of variability of the LiI 6708 Å line, but this variability has not been studied systematicaly. Even the identification of the 6708 Å line with the LiI resonance doublet is still in doubt. This problem is important in the broader context of the Li abundance in various types of stars, as well as for deeper undersfanding of the magnetic star phonomenon itself. The reason for fhis is that the Li abundance in very sensitive to evolutionary status of the stars and their properties, such as the character and intensity of mixing processes.

2. Observations

The majority of the observations presented here were made by P. North with the European Southern Observatory (ESO) Coudé Auxiliary Telescope (CAT). The Coudé Echelle Spectrograph was used with resolving power R=3D 100000 and the S/N ratio for an individual spectrogram better than 100 per pixel in 1 σ level. The detector was the ESO CCD 34 with 2048 pixels along the dispersion. A Thorium-Argon lamp was used for the wavelength calibrations with accuracy better than 0.3 km/s. The wavelength range observed was 6675-6735 Å. The spectra were reduced by PN using the standard IRAF procedures. The observations made at the CrAO were a part of a long-term program, some results of which were published in Hack *et al.*(1997) and North *et al.*(1998). A Coudé spectrograph of the Shajn 2.6-m telescope was used; it is equipped with a CCD camera with a red-sensitive detector with a 600x400 pixel array. The linear dispersion is 2.5 Å/mm in the region of 6708 Å and R=3D65000.

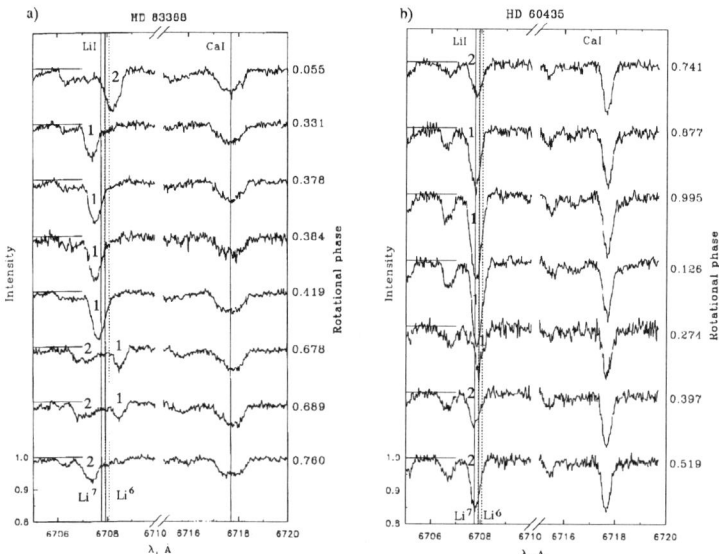

Figure 1. The spectra of ro Ap stars HD 83368 and HD 60435 in LiI 6708Å region in residual intensity scale. At the left side of each spectrum the position of the continium is shown by line. At the right side is indicated rotation phases. The doppler components of Li line 6708Å are marked 1,2.

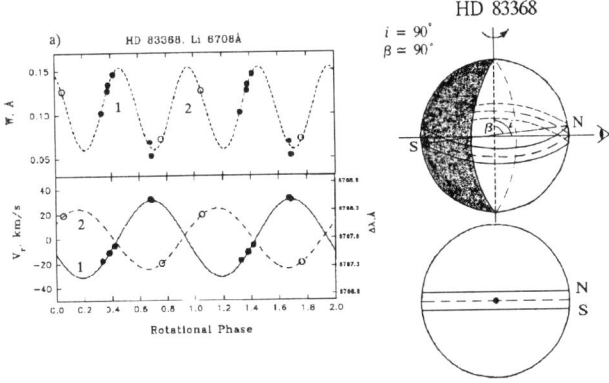

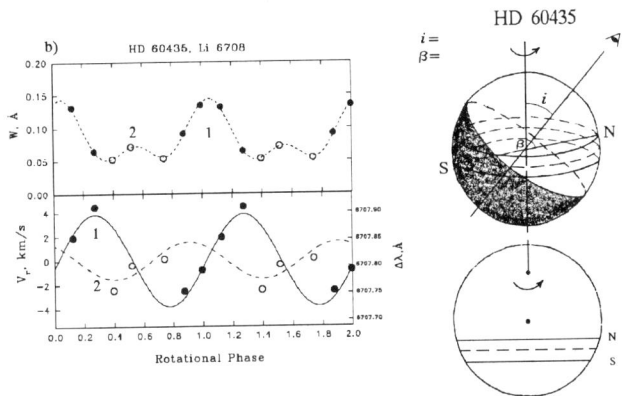

Figure 2. Variations of the equivalent widths EW(top) and of the radial velocities Vr of Li line 6708Å (bottom). The black dots indicate the spot 1, and circles - spot 2. At the right side is shown model oblique rotator for every star. N, S - poles of magnetic dipole are coincided with Li spots 1, 2.

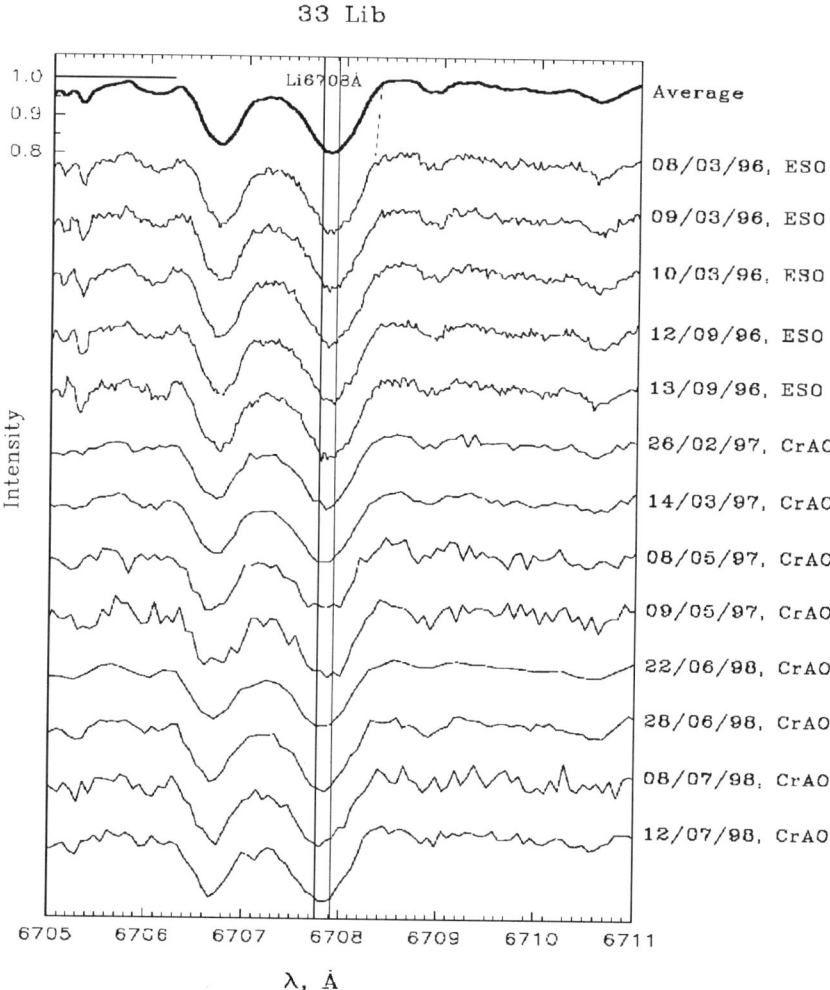

Figure 3. The spectra 33 Lib in LiI 6708 region in residnal intensity scale. Thick line is average spectrum. At the right side is indicated telescopes ESO(CAT) and CRAO(2.6m) are used for observations, and model oblique rotator for groupe 3 stars.

A typical S/N ratio is better than 100. The wavelength range was 6690-9730 Å. The spectra were reduced by DS using the package of S.Sergeev. The observations with the Nordic Optical Telescope (NOT) on La Palma, Spain, were carried out with the SOFIN echelle spectrograph (Tuominen, 1992) and two CCD cameras, yielding R=3D 80000 and 160000 in the 5500 - 8500 Å range and the S/N ratio 90-300 per pixel. A detailed description of the observations and the discussion of the line identification problems may be found in Hack et al. (1997), North et al. (1998), and Zverko et al. (1998).

3. Results and Conclusion.

The resonance LiI 6708 Å line was observed in spectra of twelve program Ap-CP-stars. The behaviour of the line 6708 Å permits us to distingnish four groupes of stars:

* groupe 1 – the line shows great variability of profile Li line and "W", "RV" with rotation phase (Fig 1a,b). This behaviour can be explained by the presence of two Li spots on the star's surface.=20 (HD 83386, HD 60435) (Fig 2a,b)

* group 2 – the line is variable, but the observations are too sparse to make a conclusion about the variability nature. (HD 188041, βCrB)

* group 3 – the line is strong, but nonvariable. (33 Lib, HD 134214, HD 166473) (Fig 3a)

* group 4 – the line was not detected (HD 42659, HD 80316, HD 118022 and HD 128898)

- The discovery of LiI-spots in HD 83368 and in 60435 is the first indication of spottiness in lithium distribution on surface of some cool magnetic=20 CP-stars.

- A good correlation between the positions of the spots, magnetic poles, and oscillation phenomena (HD 83368) indicates possible connections between the magnetic fields configuration and the local structure of star's atmosphere. (Polosukhina N. et al, 1999)

- We explained the behaviour of 6708 Å line using the model of the "spotted" oblique rotator, with different parameters for each star(Fig 2a,b). The anles "i" and "β" for different stars defermine the visibility of magnetic poles and spots, consequently, behaviour lithium line 6708 Å. HD 83368 - is unique star, the lithium are situated at the poles of the magnetic dipole. The magnetic dipole is placed hear equator's plane. The angles $i \simeq 90°$ and $\beta \simeq 90°$ too.=20 In the case of constancy of the line 6708 Å (group 3), observer sees spot near rotation pole $i \leq 30°$, $\beta \leq 30°$ (Fig 3b)

- The results of observations in Li I 6708 Å line and discovery Li spots on magnetic poles are good argee with prediction Babel theory, concern ambipolar diffusion of hydrogen in CP-stars with dipole structure of magnetic field (Babel J., Michaud G., 1991).

References

Hack M., Polosukhina N., Malanushenko V., Castelli F., 1997, A&A 319, 637
North P., Polosukhina N., Malanushenko V., Hack M., 1998, A&A 333, 644
Zverko J., Žižňovský J., North P., 1998, Contrib. Astron.Obs.Skalnaté Pleso 28, 109
Polosukhina N., North P., D.Kurtz D., Hack M., Zverko J., Ilyin I., Shakhovskoy D. 1999, A&A 351, 283-291
Babel J., Michaud G., 1991 Ap.J., 366, 560

Lithium Abundances in Solar-Type Stars[1]

L. da Silva

CNPq/Observatório Nacional, Brazil

G. F. Porto de Mello

Observatório do Valongo, Universidade Federal do Rio de Janeiro, Brazil

Abstract. We report Li abundances from the $\lambda 6707$ line for 19 nearby dwarf and subgiant solar-type stars. The unevolved stars in this sample present high (> 2.00) Li abundances. We found a few cases of subgiant stars which present high Li content. The Sun seems to be part of a population of nearly unevolved stars which have depleted their Li to a high degree: all other metal-normal, near ZAMS stars in our sample show higher than solar Li content. There seems to be no correlation of the degree of Li depletion with mass, atmospheric parameters or state of evolution: as an example we found a star (HR1532) almost identical to the Sun in its state of evolution and atmospheric parameters, but with over ten times the solar Li abundance. We propose that different histories of angular momentum distribution at star birth, and/or post-birth angular momentum evolution, may account for these differences.

1. Analysis and Results

Observations have been performed at the OPD[1], using the coudé spectrograph of the 1.60 m telescope was used to obtain S/N > 200, 0.20Å resolution spectra of 19 solar-neighborhood, solar-type stars. The observed spectral regions cover approximately 100Å each, centered at $\lambda 6050$, $\lambda 6150$ and $\lambda 6707$. The Sun was observed as a star by way of lunar spectra, with the same setup. Atmospheric parameters T_{eff}, log g, [Fe/H] and microturbulence velocities were obtained from the detailed analysis, differential with respect to the Sun, of the excitation & ionization equilibria of Fe (over 25 Fe I lines in average, 2 Fe II lines). We have also derived log g from absolute magnitudes based on HIPPARCOS parallaxes: excellent agreement was obtained between the two approaches. Luminosities were obtained by applying the bolometric corrections of Habets & Heintze (1981). Mean errors of the parameters thus obtained are: 70 K for the excitation T_{eff}, 0.30 dex and 0.06 dex for, respectively, the ionization and evolutionary log g ; 0.07 dex for [Fe/H] and 0.15 km/s for the microturbulent velocities. The lithium abundance was derived from the $\lambda 6707$ doublet by spectral synthesis of the ob-

---

[1]Based on observations collected at the Observatório do Pico dos Dias (OPD), Brazil, operated by the CNPq/Laboratório Nacional de Astrofísica .

served line profiles, using a program kindly made available by Monique Spite (Observatoire de Paris- Meudon). The error in the determination is estimated as 0.10 dex.

Table 1. Atmospheric parameters, Fe, Li abundances and luminosities for the program stars.

HR	T_{eff}	$\log g_{ion}$	$\log g_{evol}$	$\xi_{km/s}$	[Fe/H]	$\log L/L_\odot$	$\log N(Li)$
173	5270	3.75	3.84	1.35	-0.70	0.48	0.00
914	5020	3.66	3.60	0.93	-0.57	0.67	0.60
3138	5830	4.40	4.35	0.79	-0.27	0.08	0.30
8501	5750	4.27	4.32	1.30	-0.25	0.05	1.70
6998	5500	4.43	4.47	0.34	-0.16	-0.15	-0.50
1747	5960	4.21	4.25	1.67	-0.10	0.30	2.30
3862	6130	4.33	4.35	1.50	-0.08	0.24	2.53
77	5970	4.48	4.46	1.08	-0.07	0.07	2.30
SUN	5780	4.44	4.44	1.30	+0.00	0.00	1.15
3259	5380	4.38	4.45	1.07	+0.00	-0.19	0.30
2251	5950	4.36	4.32	1.47	+0.01	0.19	2.05
695	5830	3.87	4.01	1.37	+0.03	0.51	2.33
8635	5940	4.19	4.28	1.37	+0.04	0.25	2.60
8323	5900	4.48	4.42	1.08	+0.07	0.06	2.30
772	5710	4.00	4.05	1.53	+0.09	0.43	1.70
1532	5740	4.36	4.47	1.06	+0.09	-0.03	2.25
810	6130	4.42	4.43	1.74	+0.11	0.18	2.45
8700	5890	3.92	4.02	1.79	+0.19	0.53	2.55
1536	5990	4.30	4.12	1.61	+0.27	0.43	2.78
1856	6020	3.73	3.69	2.19	+0.27	0.99	1.20

2. Evolutionary State and Lithium Abundances

We have examined the distribution of Li abundances with state of evolution by plotting the stars in the theoretical HR diagrams of Schaerer et al. (1993 and references therein), roughly corresponding to metallicities of [Fe/H] = -0.37, +0.03 and +0.33. We show only the diagram corresponding to solar metallicity stars (figure 1). The metal-rich and metal-normal unevolved stars have preserved most of their Li, in contrast to the Sun, which is the only unevolved Li-poor star. The Sun may thus be considered part of a population of stars which have strongly depleted their Li (Pasquini et al. 1994). It has been suggested by King et al. (1997) that different histories of Li depletion for stars with very similar masses may be linked to the formation of a planetary system: as an example they cite the binary system 16 Cyg AB, both components being very solar-like (the B component being the planet harboring one), but differing in their Li content by a factor of ~ 5. We note that, in our sample, HR1532 is almost identical to the Sun in mass and atmospheric parameters but is more than ten times Li-richer than the Sun.

We have found a few cases of Li-rich subgiant stars. This may be understood by the "ressurgence" scenario (Dravins et al. 1993), in which the subgiant

Figure 1. Theoretical HR diagrams for the near solar metallicity stars. labeled with HR numbers and Li abundances.

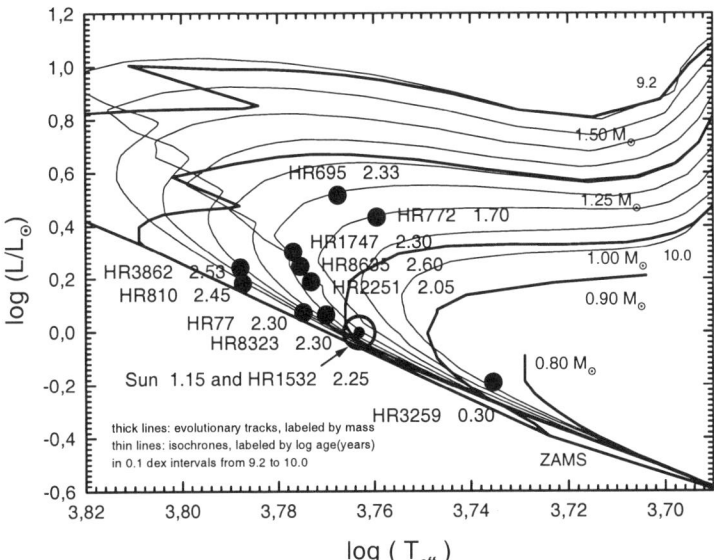

star dredges up to the surface Li that has been preserved below the convectively unstable surface layers, or else by the fact that these stars have maintained their Li abundance owing to low levels of depletion while on the main sequence (Randich et al. 1999). We have found that the metal-poor stars that are not yet subgiants are appreciably more Li-poor than their metal-richer peers. This may be explained by their being quite old stars: all of them lie in a narrow mass interval (0.80-0.90 $M_\odot$). They seem to have undergone different histories of Li depletion even among themselves, judging by their large Li abundance dispersion. The subgiant metal- rich and metal-normal stars have high to moderately high Li content, but the two metal-poor subgiants have much lower Li abundances. Whether they represent stars for which no "ressurgence" phenomenon was at work, or simply stars which depleted their Li while still close to the main-sequence, may not be decided with our current understanding.

References

Dravins D., Lindegren L., Nordlund A., Vandenberg D. A. 1993, ApJ, 403, 385
Habets G. M. J., Heintze J. R. W. 1981, A&AS, 46, 193
King J. R., Deliyannis C. P., Hiltgen D. D., Stephens A., Cunha K., Boersgaard A. M. 1997, AJ, 113, 1871
Pasquini L., Liu Q., Pallavicini R. 1994, A&A, 287, 191
Randich R., Gratton R., Pallavicini R., Pasquni L., Carretta E. 1999, A&A, 348, 487
Schaerer D., Meynet G., Maeder A., Schaller G. 1993, A&AS, 98, 253

On Meridional Circulation in Stars

Suzanne Talon, Georges Michaud and Alain Vincent

Département de Physique, Université de Montréal and CERCA, Canada

Abstract. Even though the existence of meridional currents in stars has been known for quite a long time (Eddington 1925, Vogt 1925), its exact structure as well as its influence on stellar evolution is still unclear. Some authors concentrated on finding the exact shape of meridional circulation in a rotating star, while others tried to model its effect on the chemical distribution in the interior. In all studies performed so far however, meridional circulation is considered in an asymptotic regime in which the advection of entropy by the meridional currents is supposed to balance exactly the source term of the non-zero radiative flux divergence. Other terms could however be added to that asymptotic regime which could turn out to dominate the transport of chemicals. We wish to present here preliminary results of 3D numerical simulation attempted to tackle this problem.

1. Calculation hypothesis

We are solving the full Navier-Stokes equations of hydrodynamics in the anelastic approximation, which consists of replacing the mass conservation equation by $\vec{\nabla} \cdot (\rho \vec{v}) = 0$, thus filtering out sound waves. The non-dimensionnal equations to solve are then expressing conservation of mass

$$\vec{\nabla} \cdot (\rho \vec{v}) = 0, \tag{1}$$

and momentum

$$\frac{\partial (\rho \vec{v})}{\partial t} + \vec{\nabla} \cdot \left(\rho \vec{v} \otimes \vec{v} + \mathcal{N}_1 P \bar{\bar{I}} - \frac{1}{\mathcal{R}e} \bar{\bar{\tau}} \right) = \frac{1}{\mathcal{F}r} \rho \vec{g}_{\text{eff}} - \frac{1}{\mathcal{R}o} \rho \vec{e}_z \times \vec{v} \tag{2}$$

and the evolution of entropy

$$\rho T \frac{\partial s}{\partial t} + \rho T \vec{\nabla} s \cdot \vec{v} = \left(\mathcal{N}_2 \bar{\bar{\tau}} \cdot \vec{\nabla} \right) \vec{v} + \vec{\nabla} \cdot \left(\frac{\chi}{\mathcal{R}e \mathcal{P}r} \vec{\nabla} T \right). \tag{3}$$

The full time-dependent set of equations is then solved using the finite volume method. This numerical problem is solved on a Cartesian grid; the star is simply put in a box. The solution is calculated for all points inside the star and fictitious points (outside the star) are used so that the boundary conditions are satisfied exactly *at the stellar surface*. The advantage of this method is that the discretized form of the equations remains quite simple. Furthermore, it allows to deal with changes in the shape of the star by simply redefining the position of the surface.

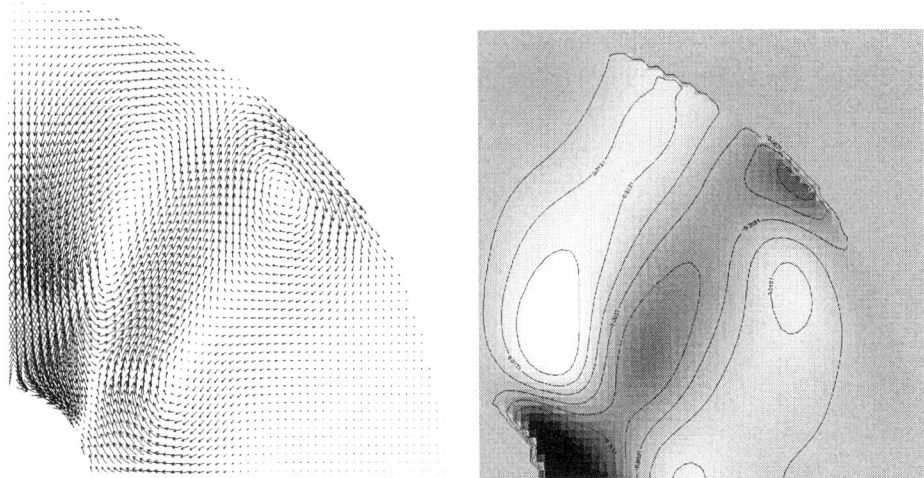

Figure 1. (left) Transitory meridionnal currents (right) Differential rotation in a meridian section

2. Preliminary results

Test simulations are currently performed at low spatial resolution in order to test the code. Results are shown for a $64 \times 64 \times 64$ simulation where only 1/8 of the star is simulated (symmetry about the equator is assumed and the flows going out from the $x = 0$ plane are injected back into the $y = 0$ plane).

The values used in this simulation are $\mathcal{R}e = 100$, $\mathcal{F}r = 5$, $\mathcal{P}r = 0.01$, $\mathcal{R}o = 1$, $\mathcal{A} = 0.03$, $\mathcal{N}_1 = 1$, $\mathcal{N}_2 = 10^{-6}$.

Figure 1 illustrates the meridionnal currents generated within the star due to thermal imbalance some time into the simulation (no assymptotic regime has been reached yet) as well as the amount of differential rotation at this point. The Coriolis force included in the calculation breaks the initial cell of meridionnal circulation in many smaller ones. Higher resolution simulations will permit to get closer to stellar values. However, simulating a real star will require a sub-grid model, which describes the smallest scales that are not resolved directly by the code. One then hopes to gain insight into the amount of mixing which can be generated in a star due to that circulation as well as to various hydrodynamical instabilities which can be triggered by the resulting differential rotation.

References

Eddington A.S., 1925, Observatory 48, 78
Vogt H., 1925, Astron. Nachr. 223, 229

The Light Elements and Their Evolution
IAU Symposium, Vol. 198, 2000
L. da Silva, M. Spite, J. R. de Medeiros, eds.

On the Link between Rotation and lithium in Giant stars

J. R. De Medeiros, J. D. Nascimento Jr, S. Sankarankutty, J. M. Costa, J. R. P. Da Silva and M. R. G. Maia

Universidade Federal do Rio Grande do Norte, Departamento de Física, 59072-970, Natal-RN, Brazil

1. Introduction

Late-type evolved stars are well known for the decrease with age of their rotation and lithium abundance. However, the root cause of this property as well as the relationship between rotation and lithium content in these stars are not yet completely established. In the present work, we study the link between rotation and lithium abundances in solar-type giant stars on the basis of a large sample of 380 stars of spectral type F, G and K.

2. Observational data

The data sample selected for the present investigation has as main characteristics the high precision of the rotational velocity and lithium abundance, as well as the large size of the sample.

Lithium abundances were taken from the following sources: 36 F and early-G stars from Wallerstein et al. (1994), 10 F stars from Balachandran (1990) and 334 late-G and K type giants from Brown et al. (1989). Rotational velocities were taken from De Medeiros and Mayor (1999).

3. Results

Figure 1 shows the behavior of lithium abundance as a function of rotational velocity for the single and binary stars of the present sample.

One observes that giant stars presenting the highest lithium content, typically stars earlier than the spectral type G0III, are also those with the greatest rotation rate. Stars located to the right of the drop in rotation, namely stars later than G0III, present as a rule, the lowest rotation rate and lithium content.

Such features indicate a trend leading to a correlation between rotation and lithium abundance along the giant branch. An additional important feature is the large spread in lithium abundance for low rotators. Stars with a $v \sin i$ lower than about 4.0 km s^{-1} show a wide range of lithium abundance values with $\log n(\text{Li})$ ranging from about -1.5 to the cosmic value, namely five orders of magnitude. Whereas, except for a few stars with $4.0 < v \sin i < 20.0$ km s^{-1}, most of the stars with enhanced rotation show an abundance of lithium around the cosmic value.

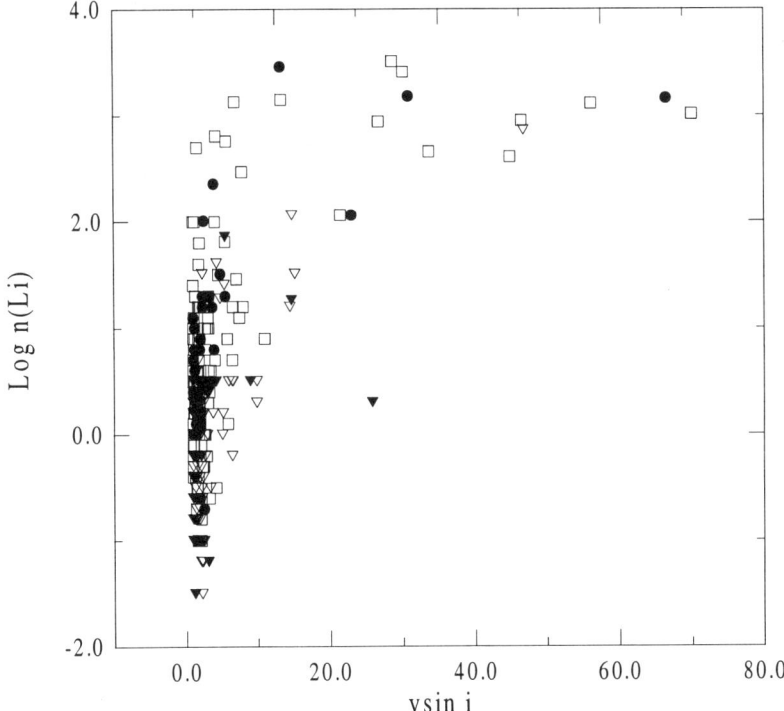

Figure 1. Lithium abundance as a function of rotational velocity (in units of km s^{-1}). Single stars are given as open squares while single stars with upper limit of Li abundances are represented by open triangles; binary systems are shown as filled circles whereas binary systems with upper limit of Li abundances are shown as filled triangles.

References

Balachandran S., 1990, ApJ **93**, 235

Brown J. A., Sneden C., Lambert D. L., Dutchover Jr. E., 1989, ApJS **71**, 293

De Medeiros J. R., Mayor M., 1999, A&AS, **139**, 433

Wallerstein G., Böhm-Vintense E., Vanture A. D., Gonzalez G., 1994, AJ **107**, 2211

Lithium-Rich K Giants with Infrared Excesses: Fundamental Parameters and CNO Abundances

Natalia A. Drake[1,2], Ramiro de la Reza[1] and Licio da Silva[1]

[1] *Observatório Nacional/MCT, R. José Cristino, 77 CEP: 20921-400, Rio de Janeiro, Brazil*

[2] *Astronomical Institute of St. Petersburg University, St. Petersburg, 198904 Russia*

Abstract. We present the results of an analysis of seven faint K giants with strong Li lines, recently discovered during the Picodos Dias urvey, from high-resolution and high signal-to-noise-ratio spectra. The fundamental parameters of these stars, such as effective temperature, surface gravity, metallicity and microturbulence, were determined. We also obtained LTE abundances of Li and CNO elements and measured the $^{12}C/^{13}C$ ratios by spectral synthesis using the lines of Li I ($\lambda 6104$ Å and $\lambda 6708$ Å), C_2, CN, and [O I]. The studied Li K stars are found to be red giant branch stars with masses around $1 M_\odot$.

1. Introduction

A small part of the red giant stars ($\sim 1\%$) have been found to have high lithium abundances and the great majority of these Li-rich K giants have IR excesses which can be caused by an existence of circumstellar shells. de la Reza, Drake, & da Silva (1996) and de la Reza et al. (1997) suggested that Li-rich giants are not peculiar stars and that all low mass ($\lesssim 2.5 M_\odot$) stars pass one or several times through a short Li-rich phase during the RGB stage. A Li enrichment mechanism acting in low mass red giants – cool bottom processing – was proposed by Sackmann & Boothroyd (1999). More discussion on this subject can be found in Sackmann & Boothroyd in these proceedings. In order to understand the evolution stage when Li enrichment and shell formation occur, we carried out a detailed spectral analysis of seven red giants with strong Li lines discovered during the Pico dos Dias Survey (PDS) (see Torres et al. in this Proceedings).

2. Observations

The high-dispersion spectrograms used in this analysis were obtained in Chile with the telescopes of 3.60 m at La Silla, ESO, (PDS 3=HD 19745, PDS 68, PDS 432) and of 4.0 m at Tololo, CTIO, (PDS 132, PDS 354, PDS 365, PDS 524). Standard NOAO's IRAF procedures were used for the data reduction.

3. Fundamental Stellar Parameters, Li Abundances and $^{12}C/^{13}C$

Using an iterative procedure we determined the effective temperatures and microturbulence velocities by obtaining iron abundances independent of excitation potentials and measured equivalent widths respectively. The surface gravities were derived from the ionization equilibrium. The photospheric models of Kurucz (1993) and Bell et al. (1976) were used. Because the faint giants studied here have no measured parallax, we adopt absolute magnitude values corresponding to their spectral and luminosity classes. Using obtained values of the temperature, superficial gravity and luminosity and photometric data we were able to estimate the stellar masses. To derive CNO and Li abundances and the $^{12}C/^{13}C$ isotopic ratios we realized a detailed LTE spectral synthesis analysis of each star in the spectral regions containing the lines of C_2, ^{12}CN, ^{13}CN, [O I] and Li I ($\lambda 6104$Å and $\lambda 6708$Å) using the last version of the MOOG program. The molecular line lists and the analysis details can be found in Drake (1998). The determined fundamental parameters, Li abundances and $^{12}C/^{13}C$ ratios are presented in the table. The values of $^{12}C/^{13}C$ are in the limits from 9 till 35 and are typical of the red giant branch stars. Comparing measured $^{12}C/^{13}C$ ratios and Li abundances we derived that the enrichment of ^{13}C relative to ^{12}C on the surfaces of Li-rich K giants is not correlated with that of ^{7}Li.

Table 1. Associated IRAS sources and determined stellar parameters

Star	IRAS	T_{eff} K	$\log g$	[Fe/H]	M $M_\odot$	$^{12}C/^{13}C$	$\log \epsilon$(Li) $\lambda 6708/\lambda 6104$
HD 19745	03062-6538	4700	2.3	0.08	1.2	30	3.9/3.9
PDS 68	13539-4153	4140	1.4	−0.40	1.2	20	3.3/3.9
PDS 132	07227-1320	3910	1.7	−0.01	2.0	12	0.8/≤1.5
PDS 354	12236-6302	4710	2.4	0.22	1.2	9	1.7/2.8
PDS 365	13313-5838	4540	2.2	−0.09	1.1	12	3.3/3.3
PDS 432	16514-4625	4580	2.0	0.28	0.7	35	2.9/3.4
PDS 524	18334-0631	4160	0.8	0.00	1.0	20	0.7/≤1.4

Acknowledgments. N.A.D. thanks FAPERJ for the financial support under the contract E-26/151.172/98 and R. de la R. and L. da S. thank CNPq for the grants 200580/97-3 and 301375/86-0 respectively.

References

Bell, R.A., Eriksson, K., Gustafsson, B., & Nordlund, Å. 1976, A&AS, 23, 37
de la Reza, R., Drake, N.A., da Silva, L. 1996, ApJ, 456, L115
de la Reza, R., Drake, N.A., da Silva, L., Torres, C.A.O., & Martin, E.L. 1997, ApJ, 482, L77
Drake, N.A. 1998, PhD Thesis, Observatório Nacional, Rio de Janeiro
Kurucz, R.L. 1993, ATLAS9, Smithsonian Astroph. Obs. CD-ROM no. 13
Sackmann, I.-J., & Boothroyd, A.I. 1999, ApJ, 510, 217

The Light Elements and Their Evolution
IAU Symposium, Vol. 198, 2000
L. da Silva, M. Spite, J. R. de Medeiros, eds.

Lithium as Probe of the Scenarios of the chemical enrichment of the Galaxy

P. François

ESO and DASGAL (Observatoire de Paris), Casilla 19001, Santiago 19, Chile

V. Hill

ESO and DASGAL-Observatoire de Paris, Karl Schwarzschild Strasse 2, Garching bei München, D-85748, Germany

M. Spite and F. Spite

DASGAL and UMR G8633 du CNRS, Observatoire de Paris, F-92195 Meudon CEDEX, France

Abstract. Jehin et al. (1999) find that, in a sample of moderately metal-poor stars, a group is rich in s elements, and they propose an enrichment by accretion of matter by the winds of AGB stars. We tried to check the implications for the lithium abundances.

1. Introduction

Recently Jehin et al. (1999) presented very careful determinations of abundances in a sample of 21 mildly metal-poor stars, ($-1.3 <$ [Fe/H]< -0.8), in particular, the abundances of the heavy elements. They show that the abundance of Europium, an element formed by the r process (rapid addition of neutrons) is uniformly correlated with the abundance of the α elements (like magnesium and titanium). This is expected, since both α elements and r elements are supposed to be produced and ejected by the same kind of massive supernovae (SN II).

On the contrary, the behaviour of the s elements (produced by slow addition of neutrons), relative to the α elements is not linear (cf. the behaviour of Yttrium in Fig. 1). In a first group of stars ($0.0<$ [Ti/Fe] < 0.23) there is a good correlation between [Y/Fe] and [Ti/Fe], but in the second group [Ti/Fe] is gathered in the narrow range $0.23 - 0.25$, whereas the spread in [Y/Fe] is large. This second group is (in average) Y-rich.

2. A scenario

Jehin et al. suggest a scenario based on two distinct phases of chemical enrichment: a first phase by the products of supernovae explosions of massive stars (SN II) a second phase by the stellar winds from intermediate mass AGB stars. Generally, the AGB produce both lithium and s process elements. In this case it can be expected that the second group (Y-rich) is also Li-rich. In some cases AGB stars produce s elements without lithium, and then their stellar wind will, on the contrary dilute Li and enhance Y in the observed stars.

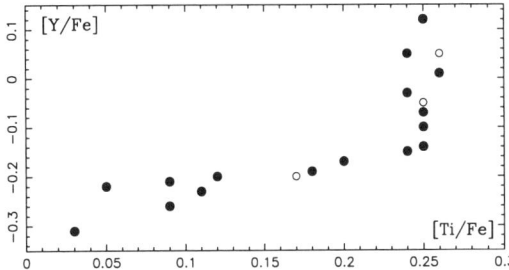

Figure 1. The first group ([Ti/Fe] < 0.23) shows a good correlation between [Y/Fe] and [Ti/Fe], the second group is Y-rich with a large Y spread. The open circles represent the stars where the lithium line could not been detected

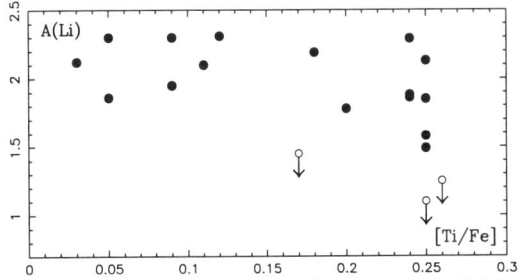

Figure 2. The first group has a rather constant Li abundance (typical of metal-poor stars), the second group ([Ti/Fe] > 0.23) has a larger spread in lithium and a lower mean abundance

We have measured the lithium abundance for 19 stars (out of the 21 stars). Most of the spectra have been obtained with FEROS at the ESO 1.5m telescope and three of them with the CFH telescope in Hawaii. The results are presented in Fig. 2 (similar to Fig. 1 but with the lithium abundance as ordinate). All these stars are dwarfs and their temperature is higher than 5700K.

The first group ([Ti/Fe] < 0.23) behaves like metal-poor stars : a rather constant Li abundance, but the second group ([Ti/Fe] > 0.23) has a larger spread and a smaller mean Li abundance suggesting an anticorrelation Li/Y by Li-dilution. But within this last group there is no clear correlation (or anticorrelation) star by star, between Y and Li. The temperature distributions in the two groups are rather similar, and thus the difference of lithium behaviour cannot be attributed to different temperature distributions. Our data do not bring a clear confirmation of the interpretation proposed by Jehin et al., the problem obviously deserves further investigation.

References

Jehin E., Magain P., Neuforge C., Noels A., Parmentier G., Thoul A. A. , 1999, A&A 341, 241

Peculiar J-type Carbon Stars and Li

S. Lorenz-Martins

Observatório do Valongo/UFRJ, Ladeira Pedro Antonio, 43, Rio de Janeiro, 20080-090 Brazil

N. A. Drake

Observatório Nacional/MCT, R. José Cristino, 77, Rio de Janeiro, 20921-400 Brazil

Astronomical Institute of St. Petersburg State University, St. Petersburg, 198904 Russia

Abstract. We determined the carbon and lithium abundances and carbon isotopic ratios for 4 peculiar carbon stars by means of spectral synthesis method. Li abundances were derived using Li resonance line at $\lambda 6707.8$ Å. For carbon abundance determinations we used the lines of the red system of CN molecule. Spectral region of $\lambda\lambda 7994 - 8030$ Å containing the lines of molecules ^{12}CN and ^{13}CN was used also for ^{12}C/^{13}C ratio measuring. The high-resolution spectra were obtained on the 1.52m telescope of ESO with the Fiber-fed Extended Range Optical Spectrograph (FEROS). These observations were obtained under the agreement between the CNPq-Observatório Nacional, Brazil, and ESO.

1. Introduction

The evolution along the AGB and third dredge-up transform an oxygen-rich giant in a carbon-rich star enhanced with s-process elements and ^{12}C. However, J-type carbon stars appear not to evolve in the same way as ordinary carbon stars, they have ^{12}C/^{13}C ratio about 4 and a lack of s-process elements. Furthermore, they present a very strong Li line at $\lambda 6707$ Å. These stars, like the carbon-rich stars, usually present the characteristic emission in the infrared at $11.3\,\mu$m due to SiC grains. However, this emission is not present in some J-type carbon stars (Little-Marenin 1986). These special stars, named *Peculiar J-type* carbon stars, present emission at $9.7\,\mu$m, which is due to silicate grains. This detection indicates that the circumstellar envelopes of peculiar J-type stars should be oxygen-rich, on the contrary to the photospheric chemical compositions characterized by C/O > 1.

2. Observations

The high-resolution spectra of our sample were obtained on the 1.52m telescope of ESO with the spectrograph FEROS in December 1998. The wavelength coverage in one exposure (object = sky) is $\lambda\lambda 3600 - 9200$ Å, with resolving power of $\lambda/\Delta\lambda = 48\,000$. The S/N ratio was about 80.

3. Results and Conclusions

Since their discovery, carbon stars with silicate emission have been suggested to be J-type carbon stars. However, $^{12}C/^{13}C$ ratios have been determined only for a very small number of peculiar J-type stars. In this work we determined for the first time the carbon isotopic ratios for 3 carbon stars with oxygen shells: C 749, C 1003 and C 1130. BM Gem, a well known J-type peculiar star was used for comparison. Synthetic spectra were calculated using the last version of the MOOG program (Sneden 1973). The molecular line lists for studied regions were taken from Drake (1998). Atmospheric models of Johnson (1982) calculated for superficial gravity of $\log g = 0.0$ were used for all stars under consideration. We determined also the radial velocities of the stars. In Table 1 we present the results obtained for our sample.

Table 1. Peculiar J-type C stars

Star	T_{eff}	[C]	[N]	[O]	C/O	$\log \epsilon(\text{Li})$	$^{12}C/^{13}C$	V_{rad} km/s
BM Gem	3000	0.36	−0.30	−0.01	1.002	1.2	10	82.8
C 749	2500	0.07	−0.70	−0.30	1.002	0.8	5	99.2
C 1003	3000	0.37	−0.30	0.00	1.003	1.0	8	38.0
C 1130	3000	0.37	−0.30	0.00	1.003	0.8	30	38.7

Our analysis showed that C 749 is a high velocity middle metal-poor star ([Fe/H]= −0.5), which seems indicate that it belongs to the thick disk stars. As concerned C 1130, we have found that it is not a J-type carbon star; its isotopic ratio is typical for ordinary carbon stars.

Although our sample is small, it seems that we have found an anti-correlation between Li and ^{13}C abundances. In addition, the C/O ratios are very close to unity, which seems indicate that these stars are very close to SC stars, in a transition phase between oxygen and carbon-rich stars. The idea of a transition scenario was proposed by several authors (e.g. Lorenz-Martins 1996). Meantime, it is necessary to take into account a larger sample.

Acknowledgments. S.L.M. and N.A.D. thank FAPERJ for financial support under the contracts E-26/150.726/99 and E-26/151.172/98 respectively.

References

Drake, N.A. 1998, PhD Thesis, Observatorio Nacional, Rio de Janeiro
Johnson, H. 1982, ApJ, 260, 254
Little-Marenin, I. 1986, ApJ, 307, L15
Lorenz-Martins, S. 1996, A&A, 314, 209
Sneden, C. 1973, *A current version of the LTE line analysis code MOOG*

The Behavior of the Rotational Velocity in Lithium-rich Evolved Stars

C. H. F. Melo[1], B. B. Soares[2], A. C. Miranda[2], J. R. P. Da Silva[2] and J. R. De Medeiros[2]

[1] *Geneva Observatory, Switzerland*

[2] *Universidade Federal do Rio Grande do Norte, Natal, Brazil*

Abstract. We analyse the behavior of the projected rotational velocity $v\sin i$ for a sample of 20 lithium-rich evolved stars. Most of these stars show normal rotational velocity with respect to the typical lithium-normal evolved stars of the same spectral type. Stars presenting enhanced rotation show also high activity level. No sign of binarity was found for these lithium-rich evolved stars.

1. Introduction

One of the most puzzling events in observational stellar astrophysics in recent years has been the discovery of the lithium-rich giant stars. Essentially such stars present an abnormally strong lithium feature for their spectral types, some of them possessing surface lithium abundance approaching the cosmic value of the interstellar medium and young main-sequence stars. The excess lithium content seen in the lithium-rich giants may be fresh lithium synthesized in the so-called beryllium transport mechanism or a preserved intrinsic lithium (e.g. Fekel & Balachandran 1993). In the present work we study the behavior of the rotational velocity for the lithium-rich giant stars, by using precise CORAVEL measurements.

2. Working Sample and Discussions

We have selected only giants with a lithium abundance larger than $\log n(\text{Li})=1.4$ because the standard models of convective dilution predict that lithium in red giants should not exceed that value. The entire sample of lithium-rich giants is presented in Table 1. The rotational velocity measurements were obtained from De Medeiros & Mayor (1999). For a more complete study of the behavior of the rotation in the lithium-rich giants, we have carried out a comparative study of the distribution of the rotational velocity values for such stars with the one for lithium-normal ($\log n(\text{Li}) < 1.4$) giant stars, as shows the Figure 1. To construct the distribution of rotational velocities for lithium-normal giants we have used a large sample of 333 G- and K-type giants with lithium abundances determined by Brown et al. (1989) for which we have now CORAVEL rotational velocity measurements. The main goal here was to control if these two data sets had been drawn from the same distribution function. For this analysis we have applied the Kolmogorov-Smirnov test. If we reject HD 9746, HD 31993 and HDE 233517 which are typically active stars, lithium-rich and lithium-normal giants show a

Figure 1. Cumulative distributions.

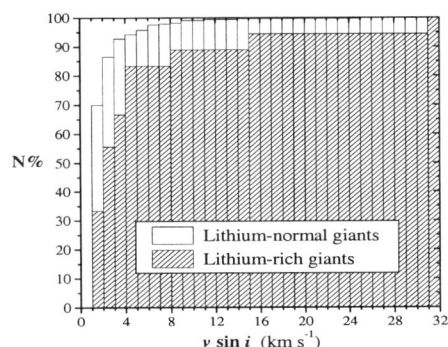

high probability to be drawn from the same parent population. The present analysis seems to indicate two important points: first, HD 9746, HD 31993 and HDE 233517 are abnormally high rotators in comparison with the other lithium-rich giants; and mainly, there is no significant difference between the rotational behavior of lithium-normal and non active lithium-rich giant stars. This study confirms the preliminary results obtained by De Medeiros et al. (1996).

Table 1. Lithium-rich giants.

Star	ST	$B-V$	$\log n(\text{Li})$	$v \sin i$ (km s^{-1})	Remark
HD 787	K5III	1.21	1.80	2.0	1; 3
HD 9746	K1III	1.21	2.70	8.7	1; 7
HD 19745	K0III	1.02	4.75	1.0	3
HD 30834	K2,5III	1.41	1.80	2.7	7
HD 31993	K2III	1.28	1.83	31.1	2
HD 39853	K5III	1.53	2.80	3.1	8
HD 40827	K1III-IV	1.10	1.60	1.9	1
HD 57669	K0III	1.22	1.50	4.5	1
HD 95799	G8III	1.32	3.05	2.1	2
HD 108471	G8III	0.93	2.00	4.1	1; 7
HD 112127	K2,5III	1.26	2.70	1.7	1; 7; 8
HD 121710	K3III	1.43	1.50	1.5	7
HD 148293	K2III	1.12	2.00	1.2	1; 7
HD 172365	F9Ib	0.79	2.20	3.3	9
HD 174104	G0Ib	0.73	3.50	4.8	9
HD 176588	K2III	1.78	1.57	2.1	2
HD 183492	K0III	1.05	2.00	1.0	1; 7
HDE 233517	K2III	1.32	3.30	15.0	6

(1.) Brown J. A. et al., 1989, ApJS 71, 293. (2.) Castilho B. V., Ph.D. Thesis, 1999, IAG/USP. (3.) De La Reza R. & Da Silva L., 1995, ApJ 439, 917. (6.) Fekel F. C., 1996, ApJ Letters 462, 195. (7.) Fekel F. C. & Balachandran S., 1993, ApJ 403, 708. (8.) Gratton R. G. & D'Antona F., 1989, A&A 215, 66. (9.) Luck R. E. & Sneden C., 1986, PASP 98, 310.

References

De Medeiros J. R., Mayor M., 1999, A&AS 139, 433.
De Medeiros J. R., Melo C. H. F. and Mayor M., 1996, A&A 309, 465.

Lithium in Post T Tauri Stars [1]

G. R. Quast and C. A. O. Torres

Laboratório Nacional de Astrofísica/MCT, CP 21, 37504-360 Itajubá, MG, Brazil

R. de la Reza and L. da Silva

Observatório Nacional/MCT, Rua General José Cristino 77, 20921-030, Rio de Janeiro, Brazil

Abstract. We analyze the Li depletion in the PTT-gap proposed by Martín (1997) using the young stars, with ages between 10 and 100 Myr, recently discovered by us during a survey around isolated T Tau stars.

1. Introduction

Martín (1997) proposed that the low-mass Post-T Tau stars (PTTS) should have Lithium equivalent widths (W_{Li}) between those of typical T Tau stars (TTS) and those of young main sequence stars of the Local Association – the PTT-gap. In X-ray surveys he found that ~15% of the X-ray stars are within this gap. We analyze the W_{Li} in the young associations found around the isolated TTS ER Eri [2], TW Hya and V4046 Sgr and also in two control areas where no TTS is known. The observational set-up is described in Torres et al. (2000).

2. The three New Nearby Associations

The TW Hya Association (TWA) – The first members of this group were found using as indicator IRAS sources (de la Reza et al., 1989; Gregorio-Hetem et al., 1992). Having now 21 stars, TWA is at a mean distance of 45 pc and has an age of ~10 Myr (Webb et al., 1999; Torres et al., 2000).
The Horologium Association (HorA) – This association, recently found by us using as indicator ROSAT sources (Torres et al., 2000), has 16 probable or possible members, is at a mean distance of ~60 pc and has an age of ~30 Myr.
The V4046 Sgr Association (VSA) – We found, using again ROSAT, 5 young stars around V4046 Sgr possibly associated with it. If these 6 stars form a new nearby - but somewhat loose - association (~100 pc), the age would be ~15 Myr.

---

[1] Based on observations made under the Observatório Nacional-ESO agreement for the joint operation of the 1.52 m ESO telescope and at the Observatório do Pico dos Dias, operated by MCT/Laboratório Nacional de Astrofísica, Brazil

[2] Actually, ER Eri was later identified as a RS CVn star

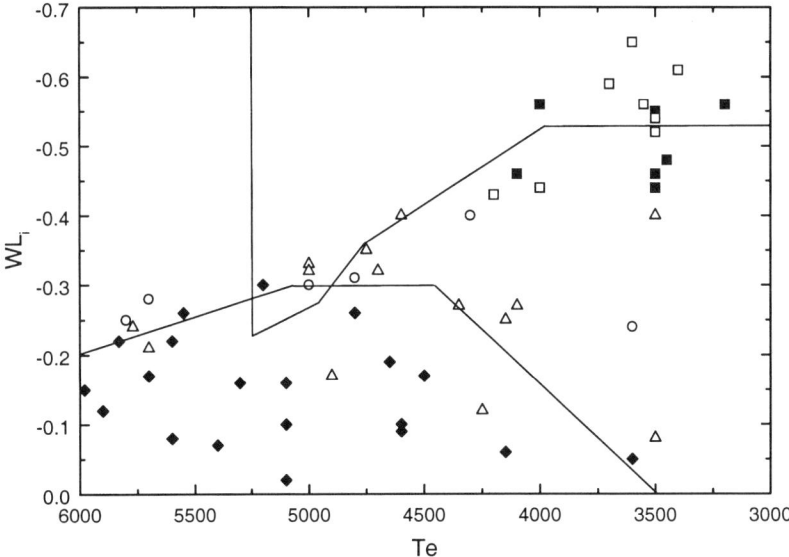

Figure 1. W_{Li} vs T_{eff} for the nearby associations: TWA (open squares, ours; filled from Webb et al. (1999)), VSA (circles) and HorA (triangles) and the observed young field stars (filled diamonds). The lines, defining the boundaries for classical TTS and representing the upper limit for Local Association stars, were taken from Martín (1997).

3. The distribution of the Li abundance

In Figure 1 we plot the three associations and the young field stars found during our surveys. We can see that the TWA is at the upper limit of the PTT-gap and the HorA is at the lower limit. This gap between the TWA and the HorA stars indicates the amount of Li depletion in ~20 Myr. As the Li depletion in convective evolutionary models for low mass stars is rather insensitive to opacity and atmospheric parameters, but depends critically on convection, these stars should restrict the values of convective parameters.

References

de la Reza, R., Torres, C. A. O., Quast, G., Castilho, B. V., & Vieira, G. L. 1989, ApJ, 343, L61

Gregorio-Hetem, J., Lépine, J. R. D., Quast, G. R., Torres, C. A. O., & de la Reza, R. 1992, AJ, 103, 549

Martín, E. L. 1997, A&A, 321, 492

Torres, C. A. O., da Silva, L., Quast, G. R., de la Reza, R., & Jilinski, E. 2000, AJ (submitted)

Webb, R. A., Zuckerman, B., Platais, I., Patience, J., White, R. J., Schwartz, M. J., & McCarthy, C. 1999, ApJ, 512, L63

Li in chromospherically active stars with large velocity components

H. J. Rocha-Pinto, B. V. Castilho and W. J. Maciel

Depto. de Astronomia, Instituto Astronômico e Geofísico, Universidade de São Paulo, CP 3386, 01060-970, São Paulo SP, Brazil

Abstract. We present lithium abundances for nine chromospherically young, kinematically old late-type stars. The data support the interpretation that these objects can be formed during the coalescence of a short-period binary.

1. Observational Sample and Lithium Ages

Some late-type dwarfs present high chromospheric activity and large velocity components. Their peculiar nature becomes more clearly demonstrated in an age-velocity diagram (Soderblom 1990). They do not follow the typical behavior of the other coeval objects. We use the acronym CYKOS to designate these chromospherically young, kinematically old stars. Their nature is the subject of this research.

Spectra were obtained with the 1.60-meter LNA telescope, mounted with a coudé spectrograph, with a resolution of $R \sim 25000$. The observations were made during August, 7-11, 1998, in the Li region ($\lambda\lambda 6627$-6759), for 15 stars, 9 of which are CYKOS. Lithium abundances were determined by spectral synthesis.

Figure 1 shows a lithium-depletion diagram, as a function of the stellar age and temperature. The curves indicate the expected lithium abundance for the objects with ages from 0 to 4 Gyr, according to Soderblom (1983). All objects shown are late-type chromospherically active dwarfs, with ages supposedly lower than 1.5 Gyr. Note that CYKOS generally show systematic lithium depletion, implying ages greater than 2 Gyr. The same does not occur for the normal stars, which present signs of high Li. Moreover, the anomalous depletion found in CYKOS cannot be assigned to the effects of the Boesgaard-Tripico dip.

2. CYKOS and Red Stragglers

The red stragglers (word proposed by Poveda, Allen, & Herrera 1996) are expected to result from the coalescence of two low-mass stars (each having > 0.5 $M_\odot$) originally in a short-period binary system. The angular momentum loss, due to the high magnetic activity, favours the synchronization of the orbital and rotational momenta, forming first close synchronized binaries, which eventually evolve to the state of contact binaries.

The approximation can lead to the coalescence of the pair, giving birth to a larger star. The coalescence can operate in a timescale of 2.5 Gyr (Stępień

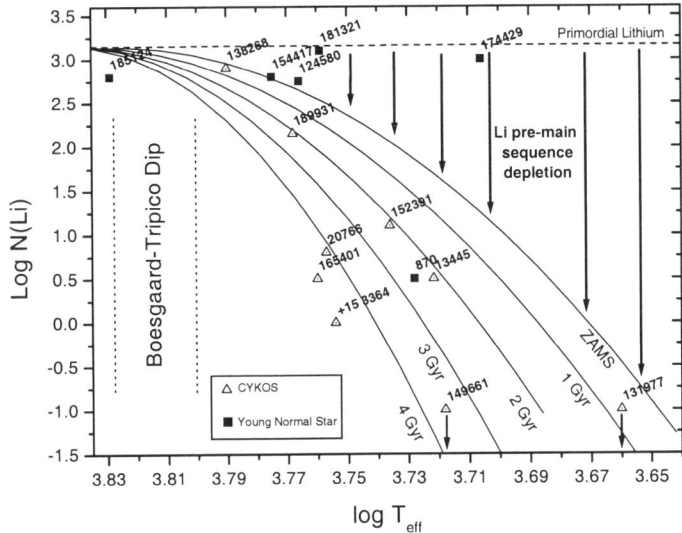

Figure 1. Lithium depletion diagram, adapted from Soderblom (1983). Lithium abundance for the stars in the sample are compared to depletion curves, which roughly define a lithium age for the stars. The labels of each point indicates its HD number, with exception of BD +15 3364

1995) for systems with initial orbital period of 2 days. The large rotation rate of the resulting star, connected to the stellar outer convective zone would produce a copious chromospheric activity, similar to that found in young stars. It would resemble a normal young star in most of its properties, even regarding its position on the ZAMS. However, it would inherit the same velocity components of the center-of-mass of the original binary system, and would have the light elements depleted due to the evolution before the coalescence. These are just the properties we observe in CYKOS.

References

Poveda, A., Allen, C., & Herrera, M. A. 1996, RMAA Ser.Conf., 5, 16
Soderblom, D. R. 1983, ApJS, 53, 1
Soderblom, D. R. 1990, AJ, 100, 204
Stępień, K. 1995, MNRAS, 274, 1019

The Ideal Stars for Exploration of Early-Epoch ^{7}Li Abundances

Silvia Rossi

Instituto Astronômico e Geofísico, Universidade de São Paulo, Av. Miguel Stefano 4200, 04301-904 São Paulo, SP Brazil

Timothy C. Beers

Department of Physics & Astronomy, Michigan State University, E. Lansing, MI 48824

Abstract.
We discuss estimates of stellar temperature, and the assemblage of a "critical set" of target low-metallicity stars, from the HK Survey of Beers and collaborators, that can be used in the near future to put additional strong constraints on models for lithium production in the early Galaxy, and for inference of its primordial value.

1. Introduction

Among the light elements produced in the Big Bang lithium plays a unique role, in the sense that its primordial value can be inferred from examination of the surface chemical composition of the most metal-poor stars in the Galaxy. Recent observations (e.g., Ryan, Norris & Beers 1999) have shown that a sample of 22 halo main-sequence stars with [Fe/H] ≤ -2.3 exhibits an extremely small intrinsic dispersion (essentially zero) in their Li abundances, *after* taking into account the presence of a significant slope in the A(Li) vs. [Fe/H] relation on the order of 0.12 ($\pm$ 0.02) dex per dex. The tiny intrinsic dispersion puts tight constraints on allowed surface-depletion mechanisms which may have been in operation, and these authors infer a primordial abundance of A(Li$_p$)= 2.09 dex, with systematic uncertainties up to 0.1 dex (see also Ryan this volume). Further exploration of this problem requires suitable stellar targets of extremely low metal abundance. We discuss the selection of such a sample.

2. Observational Requirements, the Calibration Sample, and Temperature Estimates

Stellar Li abundance is derived via a model-dependent abundance analysis, which relies in turn on accurate estimates of the physical parameters of the stars under consideration. Ryan et al. (1999) argue that errors in T$_{\rm eff}$ represent the largest contributions to the remaining uncertainties which must be addressed in order to improve estimates of the primordial lithium abundance. Although it has been suggested that stellar temperatures obtained from application of the Infrared Flux Method (IRFM; Alonso, Arribas, & Martinez-Roger 1996) should exhibit a minimum *systematic* error in estimates of stellar temperatures, individual *random* errors remain rather high, on the order of 100-150 K. Alternative approaches, such as the analysis of observed color indices which Ryan et al. employed, suffer from unknown (and potentially large) systematic errors, and leave open the possibility of errors due to reddening corrections in their determination. Thus, we have explored a alternative method for temperature estimation which makes use of an index, $HP2$, obtained from an "band-switched" pseudo-equivalent-width measurement of the Balmer line Hδ from 1-2 Å resolution spectroscopy (see Beers et al. 1999 for more details). This index is reddening free, and for the main-sequence turnoff stars which provide the best Li abundance estimates, it is insensitive to small differences in stellar surface gravity.

The sample of Beers et al. (1999) includes 104 dwarfs with [Fe/H] ≤ -2.0 and with medium-resolution spectra having S/N (at Hδ) $> 20/1$, and with available estimates of surface temperatures in the range $4500 < T_{\rm eff} < 7000$ K based on the IRFM. We have explored two sets of regression models: (a) a quadratic model, which does not require dependence on stellar metallicity, and (b) a piecewise-linear model, where a small dependence on [Fe/H] is required for those stars with $HP2 < 2$ Å (the cooler ones). Model (b) is superior for the higher temperatures of importance for Li abundance estimation. For the stars in our sample with $T_{\rm eff} \geq 6000$ K, the estimated scatter of the regression results indicates that temperatures on the IRFM scale can be inferred with a random error of 75–100 K, which represents a significant improvement. Beers et al. (2000) discuss these issues in more detail.

Application of Model (b) to the large sample of stars from HK Survey follow-up of Beers and collaborators identifies 80 stars with [Fe/H] ≤ -2.5 (including 10 stars with [Fe/H] < -3.0) and with estimated temperatures in the range considered by Ryan et al. (1999) ($6100 \leq T_{\rm eff} \leq 6300$ K). These stars should prove to be excellent targets for obtaining refined constraints on A(Li$_p$), once high-S/N, high-resolution analyses of their Li abundances are carried out with present-generation 8m telescopes such as VLT, SUBARU, and (soon) GEMINI.

Acknowledgments. TCB expresses gratitude to the IAU for support which enabled his attendance at this meeting, and acknowledges partial support from the National Science Foundation under grant AST 95-29454. TCB also wishes to express his congratulations to the LOC and SOC for a well-run, scientifically stimulating, and marvelously located meeting. SR acknowledges financial support from Fapesp.

References

Alonso, A., Arribas, S., & Martinez-Roger, C. 1996, A&A, 117, 227

Beers, T.C., Rossi, S., Norris, J.E., Ryan, S.G., & Schefler, T. 1999, AJ, 117, 981

Beers, T.C., Rossi, S., Nissen, P.E., Schuster, W., Beaver, M., & Kinemuchi, K. 2000, in preparation.

Ryan, S.G., Norris, J.E., & Beers, T.C. 1999, ApJ, 523, 654

The Light Elements and Their Evolution
IAU Symposium, Vol. 198, 2000
L. da Silva, M. Spite, J. R. de Medeiros, eds.

Meridional circulation, turbulence and lithium in sub-giants originating from the hot side of the dip

Suzanne Talon

Départ. de Physique, Université de Montréal, and CERCA, Canada

Corinne Charbonnel

Laboratoire d'Astrophysique de Toulouse, France

Abstract. We present the impact of meridional circulation and shear turbulence on the evolution of the lithium abundance at the surface of evolved stars originating from the hot side of the Li Dip. We show that our fully consistent treatment of the same hydrodynamical processes which can account for C and N anomalies in B type stars (Talon et al. 1997) and for the shape of the hot side of the Li dip in open clusters (Talon & Charbonnel 1998) also explains Li observations in stars with T_{eff} higher than 7000K on the main sequence as well as in their evolved counterparts (see also Charbonnel & Talon 1999).

1. The problem of Li in subgiants

We focus here on stars originating from the hot side of the Li dip (early-F and A types). Such stars are known to have Li abundances close to the cosmic value at the age of the Hyades. However, their subgiant counterparts show significant Li depletion much before dilution by the dredge-up takes place in standard models. These data reflect different degrees of depletion of Li inside the star while on the main sequence which do not show up at the stellar surface at the age of the Hyades.

2. The models vs the observations

We show how rotational mixing, based on wind driven meridional circulation and shear, modifies the Li abundance. We treat meridional circulation as a truly advective process (cf. Zahn 1992), whereas shear is treated via a turbulent viscosity (Talon & Zahn 1997). We also consider the settling of those elements for which radiative forces are known to be small (He, C, N, O, Ne and Mg). We built models with masses ranging from 1.5 to $2.2\,M_\odot$ with a solar metallicity. Initial velocities range from 50 to 150 km.s^{-1}. We consider two different cases. (1) Models conserving their global angular momentum (see also Charbonnel & Talon 1999). Such models do preserve most of their surface lithium at the age of the Hyades while more depletion occurs in their interior (compared to standard models), resulting in extra surface depletion on the sub-giant branch. (2) Stars

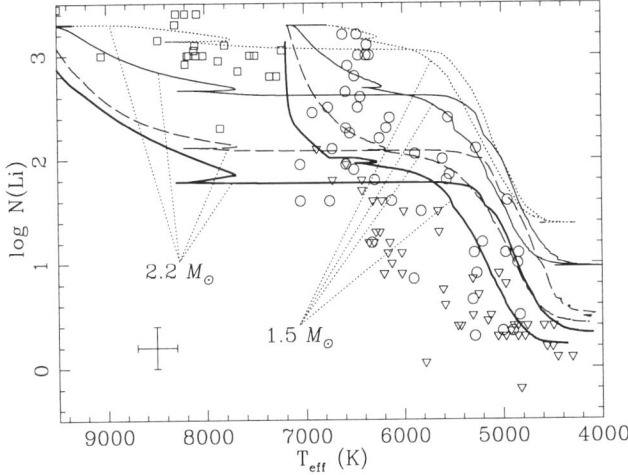

Figure 1. Evolution of surface lithium with effective temperature. The solid, dashed and dashed-dotted lines correspond respectively to the models with initial rotation velocities of 100, 50 and 150 km.sec^{-1}. Heavy lines are for models loosing a small amount of angular momentum on the main sequence. Dots and triangles : sub-giants observed by Lèbre et al. (1999) and Wallerstein et al. (1994). Squares : main sequence stars observed by Burkhart & Coupry (1998, 2000).

loosing a very small amount of angular momentum during their main sequence lifetime, resulting into a very small slowdown compatible with the measurements of surface velocities at the age of the Hyades. In such a case, more depletion occurs, leading to earlier depletion of lithium on the sub-giant branch. Figure 1 shows the comparison of our theoretical predictions with observations in sub-giants. Whereas the models which completely conserve their angular momentum predict fairly well the final amount of mixing, that mixing occurs somewhat too late. Including a small amount of braking (which agrees with the observations) takes care of that problem.

References

Burkhart C., Coupry M.F., 1998, A&A 338, 1073
Burkhart C., Coupry M.F., 2000, A&A, accepted
Charbonnel C., Talon S., 1999, A&A, 351, 635
Lèbre A., de Laverny P., De Medeiros J.R., Charbonnel C., da Silva L., 1999, A&A 345, 936
Talon S., Charbonnel C., 1998, A&A, 335, 959
Talon S., Zahn J.P., Maeder A., Meynet G., 1997, A&A 322, 209
Talon S., Zahn J.P., 1997, A&A 317, 749
Wallerstein G., Bohn-VItense E., Vanture A.D., Gonzalez G., 1994, AJ, 107, 221
Zahn J.-P., 1992, A&A 265, 115

On the Formation of Lithium Emission Lines in Nova Shells

Marcos Diaz

Instituto Astronômico e Geofísico - Universidade de São Paulo, 04301-904, São Paulo, SP, Brazil

Abstract. The role of nova outbursts in the galactic lithium production is still controversial. Photoionization model calculations of nova shells are presented in this contribution, aiming to identify the optimal post-outburst phase for detecting lithium emission lines from nova ejecta which may confirm its lithium enhancement. Predictions for the Li I λ6708 Å flux are made as a function of the envelope geometry, its physical parameters and properties of the central ionizing source.

1. Introduction

Over the last two decades theoretical models of thermonuclear runaways (TNR) in novae have predicted a physical scenario where significant amounts of ^{7}Li can be formed (Starrfield et al. 1978). According to recent hydrodynamic calculations, the convective time-scale in the core-envelope interface was found to be shorter than the life-time of ^{7}Be, allowing ^{7}Li enhancements of 10^2 to 10^3 with respect to solar values (Hernanz et al. 1996). Although lithium absorption lines have been detected in the secondary spectrum of a few black hole binaries, there is no observational confirmation of the nova ^{7}Li production mechanism in cataclysmic binaries. Depending on the conditions during outburst such a process may result in a significant contribution to the Galaxy ISM lithium abundance.

2. Nova Shell Models

The line forming region in the nova expanding shell evolves rapidly and therefore requires a time-resolved description. On the other hand, the intrinsic scatter in nova basic parameters has to be taken into account when searching for possible episodes of lithium line production. Our nova envelope models share many common fixed parameters which are described bellow. Average chemical abundances for non-ONeMg novae (Gehrz et al. 1998) were employed with a constant ^{7}Li enhancement of 1000 times the solar value. Typical maximum and minimum nova shell expansion velocities of 600 and 1000 km/s define the shell geometry as a function of time in the impulsive ejection approximation. The mass distribution inside the envelope is defined by a radial power density law with index α = -0.9 (Diaz et al. 1995). An extensive model grid containing more than 13000 points was used to study the complex line flux behavior over a wide range of shell masses ($10^{-6} \leq M_{shell} \leq 10^{-4} M_\odot$), time after maximum ($\leq 5000$

days), and central source temperatures ($80000 \leq T_{eff} \leq 350000 K$) and luminosities ($10^{35} \leq L_c \leq 10^{38.5} erg/s$). The ionizing spectrum is approximated by NLTE, high-gravity stellar model atmospheres (Rauch 1997). Finally, the effect of gas condensations in the nova envelope (Williams 1992) was investigated by randomly adding Gaussian globules to the background density law. For a few control points the line transfer was calculated under the Sobolev approximation yielding unimportant discrepancies in the line ratios when compared to the static case. All the calculations described in this contribution were made by the code RA3D which uses CLOUDY (Ferland et al. 1998) as a subroutine.

3. Observational Perspectives

The strongest emission line in the optical (Li I $\lambda 6707$Å) is mainly formed by collisional excitation from the ground state in the shell environment. The simulations indicate a well defined set of constraints for guiding deep spectroscopic observations of nova remnants. Our results suggest that the 6707 line emission is strongly suppressed in luminous and super-eddington remnants (i.e. in very fast novae) and increases for lower luminosity remnants ($L_c \lesssim 10^{36.7}$ erg/s). In the case of hot sources, which usually display high ionization spectra, such a emission is restricted to ~18 months after the visual maximum. Relatively low temperature sources ($T_{eff} \lesssim 130,000$ K) may produce significant 6707 line flux during an extended period of time after maximum ($\lesssim 3.0$ yr.). Such a timescale opens the possibility for spatially resolved diagnostic of nearby envelopes (Diaz 1998). Novae with low ejection efficiencies (i.e. having low nebular densities) present smaller emissivities while shells with mass above 10^{-5} $M_\odot$ produce stronger emission episodes along their evolution. Due to the low first ionization potential of lithium, most of the line originates in almost neutral gas. This explains the significant effect seen when the fractional mass in globules is increased to $\gtrsim 30\%$. These clumpy models suggest that the 6707/Hβ ratio may raise by one order of magnitude in a heterogeneous medium, depending on condensation properties. In general, the predicted 6707 relative flux is a small quantity reaching, in favorable situations, values around $10^{-3} \times$ Hβ. The maximum expected fluxes are close to 10^{-15} $erg.s^{-1}.cm^{-2}$ for an unabsorbed remnant at 1.0 kpc.

References

Diaz, M. 1998, in Science with Gemini, eds. Barbuy, Lapaset, Baptista, & Fernandes, (Santa Catarina:UFSC).

Diaz, M., Williams, R., Phillips, M., & Hamuy, M. 1995, MNRAS, 277, 959.

Ferland, G. 1998, Korista, K., Vernet, D., Fergunson, J., Kingdon, J., & Verner, E. 1998, PASP, 110, 761.

Gehrz, R., Truran, J., Williams, R., & Starrfield, S. 1998, PASP, 110, 3.

Hernanz, M., Jose, J., Coc, A., & Isern, J. 1996, ApJ, 465, L27.

Rauch, T. 1997, A&A, 320, 237.

Starrfield, S., Truran, J., Sparks, W., & Arnould, M. 1978, ApJ, 222, 600.

Williams, R.E. 1992, ApJ, 376, 721.

The Light Elements and Their Evolution
IAU Symposium, Vol. 198, 2000
L. da Silva, M. Spite, J. R. de Medeiros, eds.

Self-regulated hydrodynamical process in halo stars: a possible explanation of the lithium plateau

Sylvie Théado and Sylvie Vauclair

Laboratoire d'Astrophysique, 14 av. Ed. Belin, 31400 Toulouse, France

Abstract. It has been known for a long time (Mestel 1953) that the meridional circulation velocity in stars, in the presence of μ-gradients, is the sum of two terms, one due to the classical thermal imbalance (Ω-currents) and the other one due to the induced horizontal μ-gradients (μ-induced currents, or μ-currents in short). In the most general cases, μ-currents are opposite to Ω-currents. Vauclair (1999) has shown that such processes can, in specific cases, lead to a quasi-equilibrium stage in which both the circulation and the helium settling is frozen. Here we present computations of the circulation currents in halo star models, along the whole evolutionary sequences for four stellar masses with a metallicity of [Fe/H] = -2. We show that such a self-regulated process can account for the constancy of the lithium abundances and the small dispersion in the Spite plateau.

¿From spectroscopic observations, the lithium abundance in main sequence Pop II field stars with effective temperatures larger than 5500 K is remarkably constant, with a very low dispersion if any (Spite & Spite 1982; Spite et al. 1996; Bonifacio & Molaro 1997; Molaro 1999), while large lithium abundance dispersions do occur for Pop I stars. We claim that the reason for this behavior may be due to the self-regulating process in slowly rotating stars as described by Vauclair (1999).

In rotating stars, the equipotentials of "effective gravity" (including the centrifugal acceleration) have ellipsoidal shapes while the energy transport still occurs in a spherically symetrical way. The resulting thermal imbalance must be compensated by macroscopic motions: the so-called "meridional circulation"(Von Zeipel 1924; Mestel 1953; Maeder & Zahn 1998).

In the presence of vertical μ-gradients, the circulation velocity is the sum of two terms, one which does not depend on μ (the so-called "Ω currents") and one which gathers the μ dependent terms (the "μ currents").

In the present paper, we have computed the Ω-currents and the μ-currents along the evolutionary track of a .75 solar mass halo stars. All the parameters included in the computations are the same as for the solar models (Richard 1999). The horizontal μ gradients are derived using Zahn (1992) theory of anisotropic turbulence (see Vauclair (1999 and 2000) for details). The lithium variations with time are then computed within the same framework.

The μ-currents increase with time below the convective zone because of helium settling (it also increases in the core because of nuclear reactions). An equilibrium situation soon occurs below the convection zone, for which the two

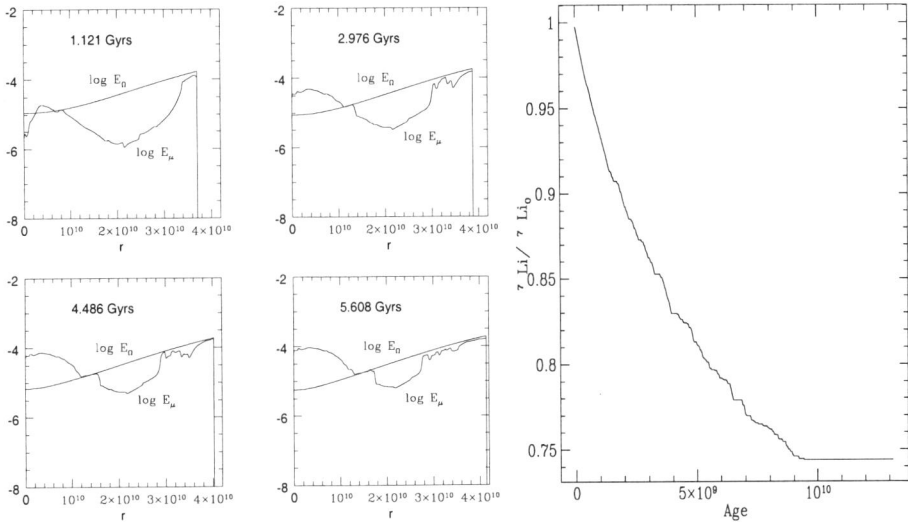

Figure 1. Computations of the Ω-currents and μ-currents and lithium abundance variation with time in a $0.75 M_\odot$ halo stars with [Fe/H]=-2.

currents become equal. Then, the circulation freezes out as well as the gravitational settling. Lithium decreases very slowly and remains constant when the whole star is "frozen". The depletion is not larger than 25%. This can explain the very small dispersion observed in the Spite plateau.

There are many observations in stars which give evidences of mixing processes occuring below the outer convection zones as, for example, the lithium depletion observed in the Sun and in galactic clusters. The process we have described here should not apply in all these stars. The reason could be related to the rapid rotation of young stars on the ZAMS and to their subsequent rotational braking and differential rotation, which is not supposed here to take place in halo stars.

References

Bonifacio, P., & Molaro, P. 1997, MNRAS, 285, 847
Maeder, A., & Zahn, J.-P. 1998, A&A, 334, 1000
Mestel, L. 1953, Mon. Not. R. Astron. Soc., 113, 716
Molaro, P. 1999, preprint
Richard, O. 1999, phD Thesis, University of Toulouse
Spite, M., & Spite F. 1982, A&A, 115, 357
Spite, F., Francois, P., Nissen, P.E., & Spite, M. 1996, ApJ, 408, 262
Vauclair, S. 1999, A&A, 351, 973
Vauclair, S. 2000, this conference
Von Zeipel, H. 1924, MNRAS, 84, 665
Zahn, J.-P. 1992, A&A, 265, 115

Audience view

EVOLUTION OF THE LIGHT ELEMENTS

Hubert Reeves presenting his conclusions.

The Light Elements and Their Evolution
IAU Symposium, Vol. 198, 2000
L. da Silva, M. Spite, J. R. de Medeiros, eds.

Evolution of D and ^{3}He in the Galaxy

Monica Tosi

Osservatorio Astronomico di Bologna, Via Ranzani 1, I-40127 Bologna, Italy

Abstract. The predictions of Galactic chemical evolution models for D and ^{3}He are described in connection with those on the other Galactic quantities for which observational constraints are available.

Models in agreement with the largest set of data predict deuterium depletions from the Big Bang to the present epoch smaller than a factor of 3 and do not allow for D/H primordial abundances larger than $\sim 4 \times 10^{-5}$. Models predicting higher D consumption do not reproduce other observed features of our Galaxy.

If both the primordial D and ^{3}He are low, models assuming that 90% of low-mass stars experience an extra-mixing during the red giant phase reproduce all the ^{3}He observed abundances. The same percentage allows to fit also the observed carbon isotopic ratios, thus supporting the self-consistency of the extra-mixing mechanism.

1. Introduction

In this review, I will try to describe what Galactic chemical evolution models tell us about the evolution and the primordial abundances of D and ^{3}He, and what, in turn, D and ^{3}He may tell us about stellar and Galactic evolution. In particular, I wish to emphasize that the light elements should not be treated separately, but should always be considered together with the other more diffuse elements, to better constrain their evolution.

The reason why Galactic chemical evolution models are required to derive the primordial D and ^{3}He abundances from the observed ones is that all the objects where the two elements are measurable are relatively young (the oldest being the sun with an age of 4.5 Gyr) and have therefore formed, with the only exception of high-redshift clouds, from an ISM whose chemical composition had been modified by the previous stellar generations. To infer the primordial abundances from these measurements, it is thus necessary to take into account the effects of the various cycles of gas astration and gas return, and the variations of the ISM chemical composition due to stellar nucleosynthesis and gas flows occurring up to the time when the observed objects have formed. This is accomplished by chemical evolution models.

D and ^{3}He are obviously related to each other, since all the D which enters a star is immediately burnt into ^{3}He (Reeves et al. 1973). However, the problems faced when studying their Galactic evolution are quite different, and I will thus treat them separately in this paper.

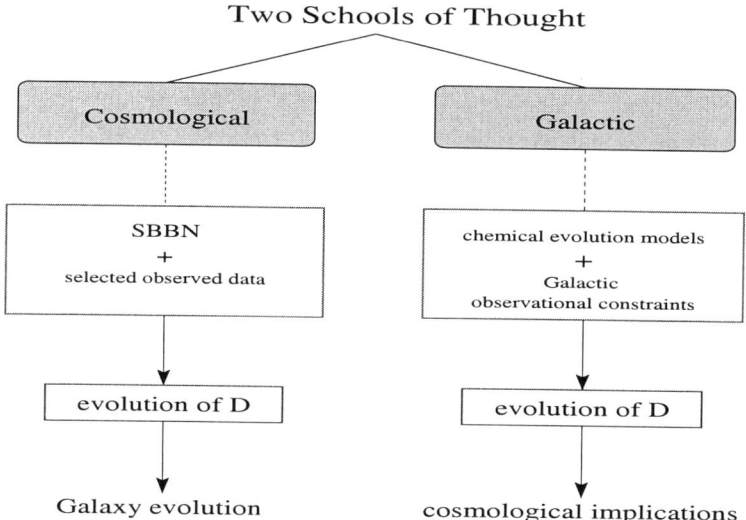

Figure 1. Sketch of the two main approaches to study the evolution of D.

2. D evolution

Since D is completely destroyed inside stars already in pre-main sequence phase, if we consider the Big Bang nucleosynthesis as the only source of D, the amount of D which can be present in any galactic region and at any place is that contained in gas which has never been through stars. In other words, the fraction of primordial D surviving at any epoch and in any region is equal to the fraction of virgin gas there. Hence, in principle, to infer the primordial abundance of D from its present one it would be sufficient to know the current fraction of gas which has not entered a star yet (Steigman & Tosi 1995). Unfortunately, we do not know the fraction of pristine gas even in the most local medium and we must therefore rely on Galactic chemical evolution models to derive the D evolution.

Historically, there are two schools of thought on how to proceed in studying the D evolution, as sketched in Fig.1. The first one, in chronological order, can be referred to as the *Cosmological School*. The approach of this school is to start from standard Big Bang nucleosynthesis (SBBN) prescriptions, select the observational constraints on D which can be considered reliable, and infer from these sets of data what the D evolution must have been in the Galaxy. They then build models of Galaxy evolution able to reproduce the inferred trend of D vs time, and the predictions of such models on the other Galactic quantities are a by-product.

The other school, which I will call *Galactic*, follows the opposite approach. We start from chemical evolution models of our Galaxy, select only those which are able to reproduce the largest set of observational constraints, and take the predictions on D only from these selected models. The consequences of these

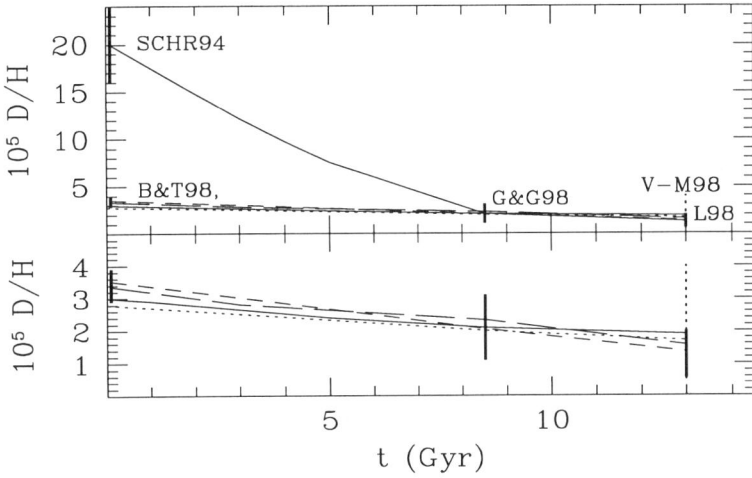

Figure 2. Top panel: Observational abundances of D as a function of the target formation epoch. The steep curve sketches the local evolution of D as proposed by the *Cosmological School*, the other curves show the predictions of chemical evolution models in agreement with the largest set of observed Galaxy properties. Bottom panel: blow-up of the lower part of the top panel. See text for details.

predictions on the primordial D abundance and on cosmology are a by-product. Were we living in the best of all possible worlds, the two approaches should provide the same results. Instead, their predictions are quite different from each other.

Fig.2 shows the D abundances derived from all the available observations and plotted as a function of the supposed epoch of formation of the observed objects. All the error bars are 2σ. The two vertical bars at t=0 represent the ranges of values derived by Songaila et al. (1994, hereinafter SCHR94) and Burles & Tytler (1998, B&T98) from high-redshift, low-metallicity, absorbers on the line of sight of distant QSO's. The bar at 8.5 Gyr represents the value of the Protosolar Cloud inferred by Geiss & Gloeckler (1998, G&G98) from solar system data. The solid bar at 13 Gyr shows the range of abundances derived by Linsky (1998, L98) for the local ISM, while the length of the dotted bar shows the possible cloud-to-cloud variations suggested by Vidal-Madjar, Ferlet & Lemoine (1998, V-M98).

Since the *Cosmological School* was founded when the primordial ^{4}He was definitely supposed to be low ($Y_P \simeq 0.23$ in mass fraction), and since SBBN predicts the primordial D to be anti-correlated with Y_P, members of this school obviously thought that the only reliable measures of D in almost primeval systems, like high-redshift, low-metallicity absorbers, were those leading to high D abundances. They thus thought that the natural evolution of deuterium with time is that connecting the SCHR94 value with the local ones and sketched by the solid line in Fig.2, i.e. a D destruction by one order of magnitude from the primordial to the present abundance.

To obtain such a high D destruction during the Galaxy evolution, one must invoke a high star formation rate (SFR), which is usually assumed to occur at the earliest epochs, because all the observational evidences are against high SFR at relatively recent times. These high SFRs (and their related metal enrichment), inevitably imply a large overproduction of the heavy elements with respect to the observed stellar abundances, unless compensated by mechanisms able to reduce the excess of metals, by diluting or removing them from the Galaxy. For this reason, models with high D destruction usually invoke infall of metal poor gas and galactic winds powered by supernovae explosions, sometimes coupled with variations in the initial mass function. There have been several attempts to find viable Galactic models with strong deuterium depletion, but no scenario consistent with all the Galactic data has been found. For instance, in their pioneering work, Vangioni-Flam & Audouze (1988) concluded that they excessively overproduced the metals, and Scully et al. (1997), in order to obtain the desired D without overproducing the metals, ended up with a present local SFR at least one order of magnitude lower than observed. Tosi et al. (1998) have tested all the possible combinations of the various parameters (SFR, infall, winds, etc.) and have always found significant inconsistencies in the models with high D destruction: metal overabundance with wrong galactocentric distribution, or metallicity distribution of the G-Dwarfs in the solar neighbourhood completely at odds with the observed one, or abundance ratios in halo or disk stars different from the observed ones (e.g. [O/Fe] vs [Fe/H]), or SFR inconsistent with the observed range. In no way have we been able to find a fairly self-consistent model with high D destruction.

The *Galactic School* works instead on chemical evolution models able to reproduce as well as possible the largest set of observed Galactic features. Thanks to the improvements both on the observational and on the theoretical sides, good chemical evolution models of the Milky Way nowadays can reproduce the average distribution of the following list of observed features (see e.g. Tosi 1996 and 2000, Boissier & Prantzos 1999 for references):

- current distribution with Galactocentric distance of the SFR,
- current distribution with Galactocentric distance of the gas and star densities,
- current distribution with Galactocentric distance of element abundances as derived from HII regions and from B-stars,
- distribution with Galactocentric distance of element abundances at slightly older epochs, as derived from PNe II,
- age-metallicity relation not only in the solar neighbourhood but also at other distances from the center,
- metallicity distribution of G-dwarfs in the solar neighbourhood,
- local Present-Day-Mass-Function (PDMF),
- relative abundance ratios (e.g. [O/Fe] vs [Fe/H]) in disk and halo stars.

When one compares with each other all the models in better agreement with these data (e.g. Tosi 1996), the striking result is that they all predict essentially the same deuterium evolution, in spite of the fact that they are computed by different people, with different assumptions on the input parameters and with different numerical codes. The bottom panel of Fig.2 shows an updated version of the comparison: the plotted models are from Galli et al. (1995a, short-dashed line), Dearborn, Steigman, & Tosi (1996, solid line), Chiappini & Matteucci

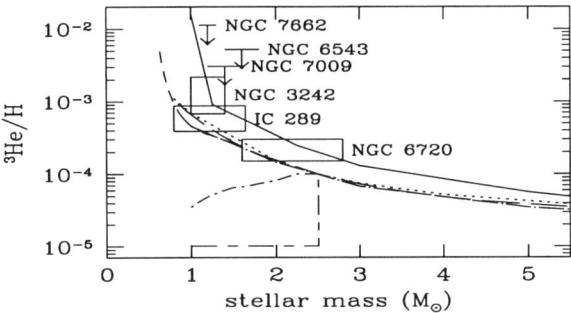

Figure 3. ³He yield as a function of the stellar initial mass. The curves show various stellar nucleosynthesis predictions, the boxes and arrows the PNe with high ³He (see Galli et al. 1997 for details).

(1996, long-dashed line) and Boissier & Prantzos (1999, dotted line). All the shown curves fit very well the average abundances derived for the local ISM, the pre-solar nebula and the high-redshift absorbers by B&T98. They all show a fairly moderate (a factor from 1.5 to 3, at most) D destruction during the Galaxy lifetime, and therefore suggest that the primordial D abundance should be low: $2 \leq (D/H)_P \times 10^5 \leq 4$.

This homogeneity of predictions is not a chance effect, but the consequence of the circumstance that all these models fit equally well the observational data on the present SFR, gas and mass densities, and chemical abundances, which necessarily implies that they predict similar fractions of pristine gas and, therefore, of surviving primordial deuterium.

Our current knowledge on the Galactic evolution of D can thus be summarized as follows: Models predicting high deuterium destruction cannot account for all the observed Galactic properties; models able to reproduce the largest set of Galactic properties all predict low deuterium destruction and, hence, low primordial D.

3. ³He evolution

³He has a more complex evolution than D, because it is produced not only during the Big Bang but also inside stars, during the main sequence phase. This early stellar production may be however largely compensated by further nuclear processing in subsequent phases. Standard stellar nucleosynthesis studies predict that, at the end of the star life, the ³He present in the initial stellar composition is significantly destroyed in massive stars, but preserved or even strongly enhanced in lower mass stars, and that the ³He net yield is a steeply decreasing function of the stellar initial mass, with a large net production in stars below 2–2.5 $M_\odot$ (see e.g the monothonic curves in Fig.3). This behaviour was known since the late sixties (e.g. Iben 1967), and already in 1976 Rood, Steigman, & Tinsley noticed that it leads to overproduce the solar abundance. Only in the mid-nineties, however, with the advent of more detailed combinations of ³He yields

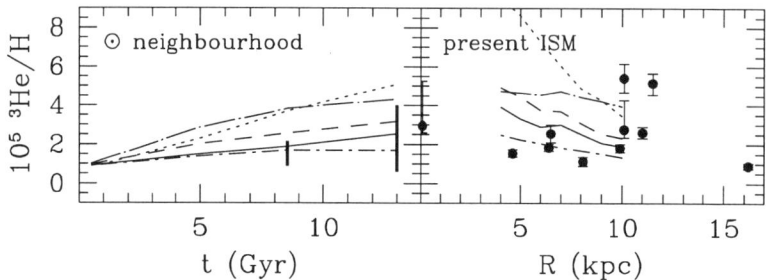

Figure 4. Comparison between the observed abundances of ^{3}He and the predictions of model Tosi-1 with CBP in 0, 90 or 100% of low-mass stars. Left hand panel: Evolution of ^{3}He/H in the solar neighbourhood. Right hand panel: Corresponding radial distribution at the present epoch. See text for symbols and references on the observational data.

with galactic chemical evolution models (Vangioni-Flam, Olive & Prantzos 1994, Galli et al. 1995a, Dearborn, Steigman & Tosi 1996) it became apparent that the results on ^{3}He of standard nucleosynthesis studies are definitely inconsistent both with the solar and with the ISM observed abundances. This inconsistency is found with any type of Galactic evolution models, including those in agreement with all the other observational constraints (e.g. Tosi 1996) and was emphasized by several groups at the Elba meeting on the Light Element Abundances in 1994 (Cassé & Vangioni-Flam 1995, Galli et al. 1995b, Tosi, Steigman & Dearborn 1995). In that occasion, Michel Cassé concluded with what has been the most popular refrain on ^{3}He ever since: ^{3}He delendum est, like the city of Carthago for the ancient Roman M.P. Cato Censor.

The most probable solution to the ^{3}He problem is less drastic than that applied to Carthago by the Romans and was proposed already in 1995 (Charbonnel 1995, Hogan 1995). It consists in the further ^{3}He processing into heavier elements favoured by an extra-mixing occurring in the red giant phase of low-mass stars (see both Charbonnel and Sackman, this volume). When low-mass stars are assumed to experience this extra-mixing and the so-called Cool Bottom Processing (CBP), Galactic evolution models do not overproduce ^{3}He anymore and fit well the observed solar and HII region abundances (Tosi 1996). The question is: in what fraction of low-mass stars CBP should occur to best fit all the data, taking into account that Bania, Rood et al. (this volume) measure in a few PNe a high ^{3}He perfectly consistent with the predictions of standard stellar nucleosynthesis (Fig.3) ? Galli et al. (1997) showed that the fraction should be larger than 80% to fit the ^{3}He abundances observed in the solar system (Geiss & Gloeckler 1998), in PNe and in HII regions (Rood et al. 1995 and this volume).

Fig.4 shows the predictions of the best of models Tosi-1 (see Tosi 1988, Dearborn et al. 1996) when 0% (dotted line in both panels), 90% (solid lines), or 100% (short-dash-dotted lines) of stars with $M \leq 2.5\ M_\odot$ are assumed to follow Sackman & Boothroyd's (1999) prescriptions for CBP ^{3}He depletion. For the remaining low-mass stars, as well as for all the intermediate and high-mass ones, the ^{3}He yield is taken from Dearborn et al. (1996). The dotted, solid and short-dash-dotted lines correspond to models assuming as initial abundances $(D/H)_p = 3 \times 10^{-5}$ and $(^3He/H)_p = 1 \times 10^{-5}$. The dashed lines show the predictions of the same model with 90% CBP, when only the initial D is changed to $(D/H)_p = 10 \times 10^{-5}$, while the long-dash-dotted lines correspond to $(D/H)_p = 20 \times 10^{-5}$.

The vertical bars in the left hand panel represent the ranges of ^{3}He abundances (at 2σ) derived by Geiss & Gloeckler (1998) and Gloeckler & Geiss (1998) for the Protosolar and the Local Interstellar Clouds, here assumed to be representative of the local ISM, 4.5 Gyr ago and now, respectively. The data points in the right hand panel show the ^{3}He abundances derived by Rood et al. (1995) from HII region radio observations. It is apparent that the models assuming 90% and 100% of low-mass stars with CBP fit quite well all the data when the initial D is sufficiently low. The CBP depletion is however insufficient to compensate the ^{3}He overproduction if the initial D/H, subsequently turned into ^{3}He, is higher than a few 10^{-5}, in which case, first the observed protosolar abundance, and then also the local ISM one, cannot be reproduced any more. This is a further argument in favour of the low primordial deuterium resulting from the previous section and from Tytler's (this volume) discussion of the observations at high redshift.

Hence, if $(D/H)_p \simeq 3 \times 10^{-5}$, the ^{3}He problem is solved if 90% of low-mass stars burn it during the extra-mixing occurring in their red giant phase. In fact, we can simultaneously reproduce the low ^{3}He abundances of the solar region and of HII regions at any Galactocentric distance, and the high abundance of NGC 3252 and the other PNe measured by Rood et al., which would consequently be associated to the remaining fraction (10%) of stars without deep mixing.

Since the deep mixing depletes not only ^{3}He, but also the $^{12}C/^{13}C$ ratio (see Charbonnel and Sackman, this volume), it is important to check the self-consistency of the solution by comparing the model predictions with the carbon isotopic ratio. Charbonnel and do Nascimento (1998) find indeed that more than 90% of 191 field and cluster red giants present carbon ratios significantly lower than the $^{12}C/^{13}C=25$ predicted by standard nucleosynthesis. What we also want to check are the predictions of chemical evolution models. This has been done by Palla et al. (2000, hereinafter PBSTG) with a two-folding approach: a) we have compared the available observational data on the carbon isotopic ratio with the corresponding predictions of chemical evolution models assuming the deep mixing in various percentages of low-mass stars; b) we have observed ^{12}C and ^{13}C in 28 PNe in mm-waves and compared the derived ratios with those predicted by stellar nucleosynthesis.

Fig.5 shows what model Tosi-1 predicts for the carbon ratio when the ^{12}C and ^{13}C adopted yields are from Boothroyd & Sackman (1999) for low-mass stars with CBP, from Marigo (2000) for low and intermediate-mass stars without CBP, and from Limongi, Chieffi & Straniero (2000) for massive stars. Equivalent results are described by PBSTG for stellar yields from other sources. The dotted line shows that without extra-mixing in low-mass stars the $^{12}C/^{13}C$ ratio is overpredicted with respect to both the abundances observed in the sun and in molecular clouds (assumed to be representative of the present disk abundances). Vice versa a good agreement is achieved if the fraction of stars with CBP is as high as possible (recall that one cannot assume 100% because of the few

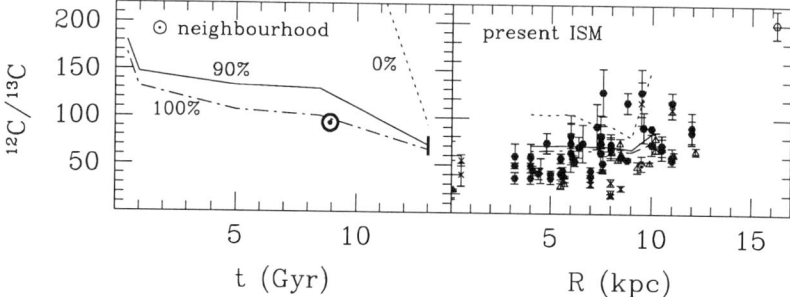

Figure 5. Comparison between observed carbon isotopic ratios and the predictions of model Tosi-1 with CBP in the indicated fraction of low-mass stars (PBSTG). The data refer to the sun and to molecular clouds in the Galactic disk (see PBSTG for references). Left hand panel: Evolution in the solar neighbourhood. Right hand panel: Corresponding radial distribution at the present epoch.

PNe with high ^{3}He). As discussed by PBSTG, the amount of predicted ^{12}C and ^{13}C strongly depends not only on the extra-mixing assumptions but also (mostly) on the assumptions for the nucleosynthesis in intermediate-mass stars, which are the major contributors to the ISM enrichment of the carbon isotopes. However, we can safely conclude that the observed carbon ratios are always better reproduced by models adopting high percentages of low-mass stars with CBP.

This is also supported by the comparison of the ^{12}C/^{13}C derived by PBSTG for the PNe where ^{13}C was actually measurable with the carbon ratio predicted for stars right before the ejection of the PN by various nucleosynthesis studies. The left panel of Fig.6 shows that most of the data points (triangles with associated error on the progenitor mass estimate) present carbon ratios lower than those expected from standard nucleosynthesis. The right hand panel shows that the measured carbon ratios are consistent with the predictions of the CBP models at the end of the red giant phase (unfortunately, no nucleosynthesis models are available yet up to the pre-PN phase, and one cannot perform a more appropriate comparison with the PNe observed ratios).

Hence, with deep mixing in ~90% of low-mass stars one can reproduce the abundances of ^{3}He observed in the sun, in the ISM and in PNe and the ^{12}C/^{13}C measured in the sun, in red giants, in the ISM and in PNe. We can then conclude that this mechanism appears to be a very promising process, which needs to be further investigated, both to individuate its possible causes and to check its effects of later stellar evolution phases.

Our current knowledge on the Galactic evolution of ^{3}He can be summarized as follows: All its available observational abundances can be explained if a) its primordial abundance is low, $(^3\text{He/H})_p \simeq 1 \times 10^{-5}$, b) the deuterium primordial abundance is also low, and c) deep mixing occurs in almost all low-mass stars.

Acknowledgments. Most of what has been described in this review results from the collaborations with C. Chiappini, D. Galli, F. Matteucci, F. Palla, L.

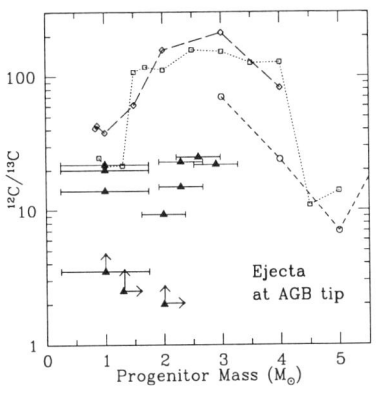
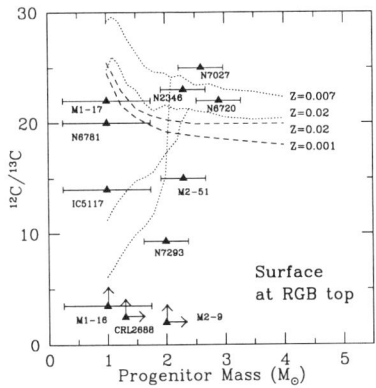

Figure 6. Comparison between the carbon isotopic ratio measured by PBSTG in PNe and the predictions from stellar nucleosynthesis. Left hand panel: composition just before the PN ejection predicted without deep mixing by Forestini & Charbonnel (1997, short-dashed curve), van den Hoek & Groenewegen (1997, dotted curve), and Marigo (2000, long-dashed curve). Right hand panel: composition predicted with and without deep mixing at the end of the red giant phase (the dotted lines refer to Boothroyd & Sackman 1999, the short-dashed ones to Charbonnel 1994).

Stanghellini and G. Steigman, and I thank them all for their help. Interesting conversations with C. Charbonnel and N. Prantzos have also been very useful. I thank the organizers for such an enjoyable and successful symposium. This work has been partially supported by the Italian COFIN98-MURST at Arcetri.

References

Boissier, S., & Prantzos, N. 1999, MNRAS, 307, 857

Boothroyd, A.I., & Sackman, I.-J. 1999, ApJ, 510, 232

Burles, S., & Tytler, D. 1998, in Primordial Nuclei and their Galactic evolution, N.Prantzos, M.Tosi, & R. von Steiger eds, Space Sci.Rev.84, 65, B&T98

Cassé, M., & Vangioni-Flam, E., 1995, in The Light Element Abundances, P.Crane ed. (Springer, De), p.44

Charbonnel, C. 1994, A&A, 282, 811

Charbonnel, C. 1995, ApJ, 453, L41

Charbonnel, C., & do Nascimento, J.D.Jr 1998, A&A, 336, 915

Chiappini, C., & Matteucci, F. 1996, in From Stars to Galaxies: The Impact of Stellar Physics on Galaxy Evolution, C. Leitherer, U. Fritze-von-Alvensleben, & J. Huchra eds, PASPConf.Ser., 98, 541

Dearborn, D.S.P., Steigman, G., & Tosi, M. 1996, ApJ465, 887 (DST96)

Forestini, M., & Charbonnel, C. 1997, A&AS, 123, 241

Galli, D., Palla, F., Ferrini, F., & Penco, U. 1995a ApJ, 443, 536
Galli, D., Palla, F., Ferrini, F., & Straniero, O. 1995b, in The Light Element Abundances, P.Crane ed. (Springer, De), p.224
Galli, D., Stanghellini, L., Tosi, M., & Palla F. 1997 ApJ, 477, 218
Geiss, J., & Gloeckler, G. 1998, in Primordial Nuclei and their Galactic evolution, N.Prantzos, M.Tosi, & R. von Steiger eds, Space Sci.Rev.84, 239
Hogan, C.J. 1995, ApJ, 441, L17
Iben, I. 1967, ApJ, 147, 650
Limongi, M., Chieffi, A., Straniero, O. 2000, in The chemical evolution of the Milky Way: stars versus clusters, F. Matteucci and F. Giovannelli eds (Kluwer, Holland) in press
Linsky, J.L. 1998, in Primordial Nuclei and their Galactic evolution, N.Prantzos, M.Tosi, & R. von Steiger eds, Space Sci.Rev.84, 285 L98
Marigo, P. 2000, in The chemical evolution of the Milky Way: stars versus clusters, F. Giovannelli and F. Matteucci eds (Kluwer, Holland) in press, astro-ph/9912341
Palla, F., Bachiller, R., Stanghellini, L., Tosi, M., & Galli, D. 2000, A&A, in press, astro-ph/9912086, PBSTG
Reeves, H., Audouze, J., Fowler, W.A., & Schramm, D.N. 1973, ApJ, 179, 979
Rood, R.T., Bania, T.M., Wilson, T.L., & Balser, D.S. 1995, in The light elements abundances, P.Crane ed. (Springer, De), p.201
Rood, R.T., Steigman, G., & Tinsley, B.M. 1976, ApJ, 207, L57
Sackman, I.-J., & Boothroyd, A.I. 1999, ApJ, 510, 232
Scully, S., Cassé, M., Olive, K.A., & Vangioni-Flam, E. 1997, ApJ, 476, 521
Songaila, A., Cowie, L.L., Hogan, C.J., Rugers, M. 1994, Nature, 368, 599, SCHR94
Steigman, G., & Tosi, M. 1995, ApJ, 453, 173
Tosi, M. 1988, A&A, 197, 33
Tosi, M. 1996, in From Stars to Galaxies: The Impact of Stellar Physics on Galaxy Evolution, C. Leitherer, U. Fritze-von-Alvensleben, & J. Huchra eds, PASPConf.Ser. 98, 299
Tosi, M. 2000, in The chemical evolution of the Milky Way: stars versus clusters, F. Giovannelli and F. Matteucci eds (Kluwer, Holland) in press, astro-ph/9912370
Tosi, M., Steigman, G., & Dearborn, D.S.P. 1995, in The Light Element Abundances, P.Crane ed. (Springer, De), p.228
Tosi, M., Steigman, G., Matteucci, F., & Chiappini, C. 1998, ApJ, 498, 226
van den Hoek, L.B., & Groenewegen, M.A.T. 1997, A&AS, 123. 305
Vangioni-Flam, E., & Audouze, J. 1988, A&A, 193, 81
Vangioni-Flam, E., Olive, K.A., & Prantzos, N. 1994, ApJ, 148, 3
Vidal-Madjar, A., Ferlet, R., Lemoine, M. 1998, in Primordial Nuclei and their Galactic evolution, N.Prantzos, M.Tosi, R. von Steiger eds, Space Sci. Rev. 84, 297, V-M98

The Light Elements and Their Evolution
IAU Symposium, Vol. 198, 2000
L. da Silva, M. Spite, J. R. de Medeiros, eds.

Implications of Early Cooling Flows and Galactic Winds for the Evolution of Deuterium

Amancio C. S. Friaça

Instituto Astronômico e Geofísico, USP, Brazil

Abstract. The deuterium abundances in high-redshift QSO absorption-line systems could be an important constraint in models of galaxy formation. Here we investigate the role of galactic winds and massive cooling flows present during the formation of galaxies on the evolution of deuterium abundance. Destruction factors are calculated and the time and spatial scales for the dispersal through galactic winds of the processed deuterium-depleted gas are presented and related to the D/H determinations for QSO absorption-line systems. The calculations are derived from a chemodynamical model within a scenario in which the absorbers are located inside the hot halo of a young galaxy.

1. Introduction

In order to derive the primordial D abundance from that observed in nearby astronomical objects, it is needed to account for the chemical evolution undergone by gas and stars since the big bang nucleosynthesis until now. The chemical evolution of deuterium is very simple: as soon as it is incorporated into stars, it is completely destroyed during star formation, being burned into ^{3}He during the pre-main-sequence evolution. Therefore, the deuterium observed in galaxies, the ISM, the Sun or the solar system is reduced by a destruction factor with respect to the primordial D. In order to recover the primordial D abundance, one needs to know which is the destruction factor for each astrophysical environment. For instance, the abundances observed in the solar system – $(D/H)_\odot = (2.1\pm0.5)\times 10^{-5}$ (Geiss & Gloecker 1998) – and in the local interstellar medium – $D/H)_{ISM} = (1.5\pm0.2)\times 10^{-5}$ (Linsky 2000) – are lower limits to the primordial deuterium abundance (both $(D/H)_\odot$ and $(D/H)_{ISM} < (D/H)_P$)

One would like to have systems as pristine as possible to get $(D/H)_P$. QSO absorption line systems would be good candidates for such systems, since they are at high redshifts (small cosmic age) and have low metallicities. Therefore, one would expect that the D abundance in these objects should be very nearly the primordial value. However, the so called "high-D, low-D dispute" on the D abundance in QSO absorption line systems has cast doubts on the identification of the D abundances in QSO absorption line systems with the primordial D abundance. The "high-D, low-D dispute" arises from the discrepancy between low deuterium abundances found by Tytler and collaborators – D/H= 3.0 – 4.7 × 10^{-5} (Olive, Steigman, & Walker 1999) – and claims of high abundances by other authors D/H= $(20 \pm 5) \times 10^{-5}$ (Webb et al. 1997). The low-D values are closer to those found in the sun and in the local ISM, and would imply modest destruction factors in the solar neighborhood if they are to be taken as representing the primordial deuterium. As a matter of fact, both model-independent arguments and extensive chemical modeling indicate D destruction factors of of 3 or less, thus favoring the low-D values (Tosi et al. 1998).

On the other hand, even though the QSO absorption line systems have very low metal abundances, their D content must still be considered only a lower limit to the primordial deuterium abundance, and it is possible that some QSO absorvers have already undergone significant chemical evolution, thus reducing the D abundance. Only detailed chemical evolution modeling would allow one to know how much the chemical evolution has affected the D abundances in QSO absorption line systems. Since the QSO absorption line systems were D has been detected have H I column densities typical of Lyman limit systems, we can assume that they are located inside galaxies ($r \approx R_e$), or in their galactic halos (Viegas & Friaça 1995). In this connection, this work addresses the effect on D abundance of massive cooling flows and galactic winds during the early evolution of galaxies. In addition, we can use the D depletion as a probe of the galaxy formation process.

Within the scenario in which the QSO absorption line systems are located inside halos of young elliptical galaxies, we use a chemodynamical model for evolution of galaxies (Friaça & Terlevich 1998) to obtain the D abundance evolution inside and around young galaxies. In particular, we calculate the destruction factors and the time and spatial scales for the dispersal of the processed deuterium-depleted gas.

2. The Chemodynamical Model

The chemodynamical model combines a multi-zone chemical evolution solver with 1-D hydrodynamics to follow the evolution of a galaxy since the stage of gaseous protogalaxy. The galaxy, assumed to be spherical, is subdivided into several spherical zones and the hydrodynamical evolution of its ISM is calculated. Then, taking into account the gas flow, the chemical evolution equations are solved for each zone, giving the evolution of the abundances of six chemical species (He, C, N, O, Mg, Fe). The model galaxy is the sum of three components: gas, stars and a dark halo (with masses M_g, M_*, M_h, respectively). The gas and the stars exchange mass through star formation and stellar mass losses (supernovae, planetary nebulae, and stellar winds). The stars formed are allowed to relax after one free-fall time to a King distribution $\rho_*(r) = \rho_{*0}[1+(r/r_c)^2]^{-3/2}$, where ρ_{*0} and r_* are the central stellar density and the stellar core radius, respectively. ρ_{*0} and r_c are related to the central stellar velocity dispersion σ_* by the virial condition $4\pi G \rho_{*0} r_c = 9\sigma_*^2$. The dark halo has no interplay with the gas and the stars, and it is given by a static mass density distribution $\rho_h(r) = \rho_{h0}[1 + (r/r_h)^2]^{-1}$, where ρ_{h0} is the halo central density and r_h is the halo core radius. Both the stellar distribution and the dark halo are truncated at a common tidal radius r_t. Details of the hydrodynamics and restoring of gas by evolved stars (stellar lifetimes, remnants, SN Ia and SN II rates, nucleosynthesis prescriptions) can be found in Friaça & Terlevich (1998). The models considered here are run until a present-day time $t_G = 13$ Gyr.

3. Results

Table 1. Galactic wind and deuterium destruction results

					10 kpc		100 kpc	
M_G	r_h	M_*	t_w	M_w	t_3	t_{100}	t_3	t_{100}
	(kpc)	($10^{11}\,M_\odot$)	(Gyr)	($10^{11}\,M_\odot$)	(Gyr)			
0.1	2.5	0.0615	0.305	0.0437	1.16	1.20	1.31	1.32
1	2.5	1.32	1.09	0.72	1.23	1.24	1.54	1.56
2	3.5	2.38	1.17	1.42	1.04	1.08	1.70	1.73
5	5	5.90	1.25	4.22	1.34	1.39	2.02	2.04
10	7	11.6	1.51	6.45	1.74	1.89	2.71	2.82

In order to investigate the evolution of deuterium during the early evolution of elliptical galaxies, we have built a sequence of galaxy models parameterized according to the total (initial) luminous mass inside the tidal radius, $M_G = M_g + M_*$ and r_h. The galaxy is initially purely gaseous ($M_g(t=0) \equiv M_G$). For all the models $M_h/M_G = 3$, and $r_t = 28 r_h$. In addition, the model galaxies follow a Faber-Jackson relation, $\sigma_* = 200(L_B/L_B^*)^{1/4}$ km s^{-1}. The star formation law is $\nu = \nu_0 (\rho/\rho_0)^{x_{SF}}$, with $x_{SF} = 1/2$, where $\nu_0 = 10$ Gyr^{-1} and ρ_0 is the initial gas density averaged inside r_h. For this star formation law, the time scale for star formation is proportional to the local dynamical time. The normalization $\nu_0 = 10$ Gyr^{-1} implies a $\sim 10^8$ yr timescale for star formation required by chemical evolution models in order to reproduce the suprasolar [Mg/Fe] ratio in giant ellipticals. We also assume a Salpeter IMF between 0.1 and 100 $M_\odot$ and a SN I binary parameter $A_{SNI} = 0.1$ (see Friaça & Terlevich 1998 for details).

Table 1 shows the results of the models. The first column identifies the model. The models are named by M_G in units of $10^{11}\,M_\odot$. Column (2) gives r_h. Column (3) shows the present-day (at $t_G = 13$ Gyr) stellar mass of the galaxy. Columns (4) and (5) exhibit properties of the galactic wind which appears in all models: t_w, the time of the onset of the galactic wind; and M_w, total gas mass ejected by the wind until t_G. The following columns give the depletion times i.e., the times when D is depleted by a factor of 3 (t_3) at $r = 10$ kpc and $r = 100$ kpc, and by a factor of 10 (t_{10}) at the same radii. As in Friaça & Terlevich (1998) we consider as fiducial model the model with $M_G = 2 \times 10^{11}\,M_\odot$. The fiducial model has $L_B = 2.4 \times 10^{10}\,L_\odot$, i.e., somewhat fainter than the break luminosity of the Schechter luminosity function ($L_B^* = 3.7 \times 10^{10}\,L_\odot$), and, therefore, is representative of the population of elliptical galaxies.

Figure 1 shows the chemical abundances of the deuterium and iron as a function of radius for the fiducial model at several epochs. Note that a little earlier ($t = 1$ Gyr) than the onset of the galactic wind ($t_w = 1.17$ Gyr), there is a significant D depletion inside the galaxy ($r \lesssim 10$ kpc) and at the same time the Fe abundance has become suprasolar in this region.

In Figure 2, we can see the evolution of D destruction factor ($X_D/X_{D,P}$) for the fiducial model at several radii, illustrating that the D-depletion time scale

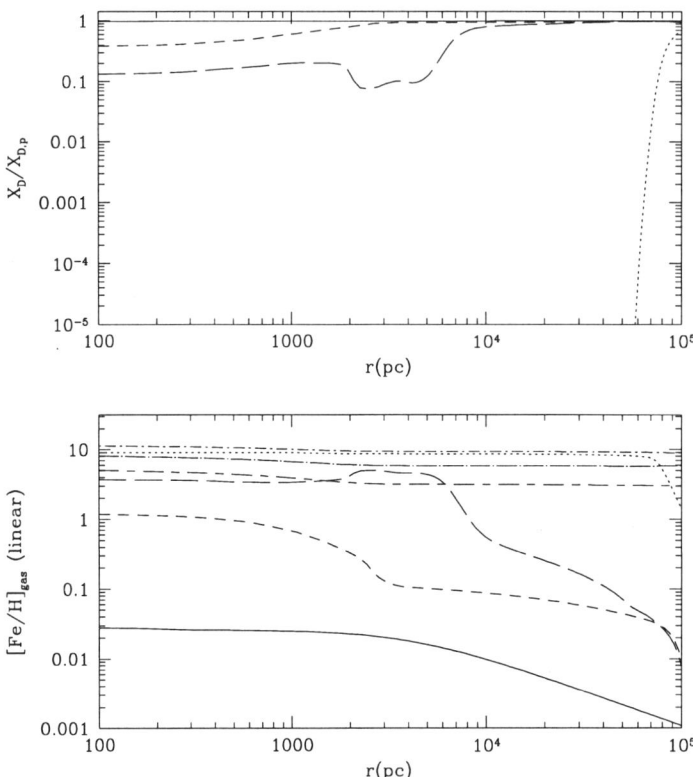

Figure 1. Gas abundance profiles for the fiducial model at several epochs: 0.03 Gyr (solid line), 0.35 Gyr (short-dashed), 1 Gyr (long-dashed), 1.6 Gyr (dotted), 1.9 Gyr (dot-short-dashed), 4.4 Gyr (dot-long-dashed), 13 Gyr (short-dashed-long-dashed).

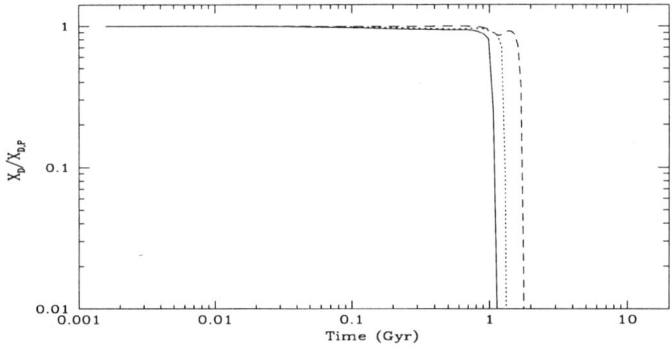

Figure 2. The evolution of D destruction factor for the fiducial model at several radii: 10 kpc (solid line), 30 kpc (dotted line), 100 kpc (dashed line).

for a given radius is very short. Also Table 1 allows one to assess how fast is the D-depletion process. For the fiducial model the time scale for a depletion by a factor 10 is ≈ 1 % larger than that by a factor 3. We have also calculated how much material in the galactic wind has been processed by stellar evolution. 1.55×10^{11} $M_\odot$ of gas has been expelled from the wind, of which 8.87×10^{10} $M_\odot$ is primordial gas. The deuterium in the total mass removed by the wind is depleted with respect to the primordial value by an amount $X_D = 0.57 X_{D,P}$.

4. Conclusions

During the pre-galactic wind stage, there is only a modest destruction of D. For the fiducial model, $X_D = 0.73 X_{D,P}$ at $r = 10$ kpc and $t = 1$ Gyr. Note that time scale for metal enrichment is very short (for the fiducial model, [Fe/H]=-0.29 at $r = 10$ kpc and $t = 1$ Gyr), and, therefore, high values of D-depletion are forbidden by the typical subsolar metallicities derived for the QSO absorption line systems in which deuterium has been detected.

Significant D destruction occurs only after the the galactic wind has been established, typically 0.5-1 Gyr later. In addition, the total amount of gas ejected by the galactic wind has a $\sim 1/2 - 1/2$ primordial-star processed mixture.

Acknowledgments. ACSF acknowledges support from the Brazilian agencies CNPq, FAPESP, and FINEP (through the program PRONEX).

References

Friaça, A. C. S, & Terlevich, R. J. 1998, MNRAS, 289, 399

Geiss, J. & Gloecker, G. 1998, Space Sci.Rev., 84, 239

Linsky, J. L. 2000, these proceedings

Olive, K. A., Steigman, G., & Walker, T.P. 1999, astro-ph/9905320

Tosi, M., Steigman, G., Matteucci, F., Chiappini, C., 1998, ApJ, 498, 226

Viegas, S.M., & Friaça, A. C. S, 1995, MNRAS, 272, L35

Webb, J. K., Carswell, R. F., Lanzetta, K. M., Ferlet, R., Lemoine, M., Vidal-Madjar, A. & Bowen, D. V. 1997, Nature, 388, 250

The Light Elements and Their Evolution
IAU Symposium, Vol. 198, 2000
L. da Silva, M. Spite, J. R. de Medeiros, eds.

The Evolution of ^{3}He, ^{4}He and D in the Galaxy

Cristina Chiappini

Depto. Astronomia - Observatório Nacional, R. Gal. José Cristino 77 - CEP 20921-400 - Rio de Janeiro - RJ - Brazil

Francesca Matteucci

Dipartimento di Astronomia, Università di Trieste, Via G. B. Tiepolo 11 - 34100 Trieste, Italy

Abstract. In this work we present the predictions of a modified version of the "two-infall model" (Chiappini et al. 1997 - CMG) for the evolution of ^{3}He, ^{4}He and D in the solar vicinity, as well as their distributions along the Galactic disk. In particular, we show that when allowing for extra-mixing process in low mass stars (M< 2.5 $M_\odot$), as predicted by Charbonnel and do Nascimento (1998), a long standing problem in chemical evolution is solved, namely: the overproduction of ^{3}He by the chemical evolution models as compared to the observed values in the sun and in the interstellar medium. Moreover, we show that chemical evolution models can constrain the primordial value of the deuterium abundance and that a value of $(D/H)_p < 3 \times 10^{-5}$ is suggested by the present model. Finally, adopting the primordial ^{4}He abundance suggested by Viegas et al. (1999), we obtain a value for $\Delta Y/\Delta Z \simeq 2$ and a better agreement with the solar ^{4}He abundance.

1. Introduction

As discussed by Tosi (this volume), chemical evolution models are useful both to derive the primordial abundances of D, ^{3}He and ^{4}He and to give informations on stellar nucleosynthesis. In this work we show the predictions of the two-infall model (CMG) for the chemical evolution of the above elements in the solar vicinity and for their distribution along the galactic disk. We adopt a new version of the two-infall model which includes the contribution by novae enrichment and the new proposed mechanism of extra-mixing in low mass stars (Charbonnel, Sackman, this meeting). The model was calibrated to the solar galactocentric distance of 8 kpc (we were still adopting 10 kpc in CMG to better compare our predictions with the ones of Matteucci and François 1989). This model assumes two main infall episodes for the formation of the halo (and part of the thick disk) and thin disk, respectively. The timescale for the formation of the thin disk is much longer than that of the halo, implying that the infalling gas forming the thin disk comes not only from the halo but mainly from the intergalactic medium. The timescale for the formation of the thin disk is assumed to be a function of the galactocentric distance, leading to an inside-out picture for the Galaxy disk buildup. The two-infall model differs from other models in the literature mainly in two aspects: i) it considers an almost independent evolution between the halo and thin disk components (see also Pagel & Tautvaisiene 1995) and ii) it assumes a threshold in the star formation process (Kennicutt 1989). The last point has important consequences for the predicted abundance gradients (Chiappini, Matteucci & Romano 2000 - CMR).

2. Results

2.1. The solar Vicinity

Our present model differs from those of CMG in i) the adopted yields for the low and intermediate mass range stars which are now taken from van den Hoek & Groenewegen (1997) instead of Renzini & Voli (1981); ii) the fact that now we are including the explosive nucleosynthesis from nova outbursts (see Romano et al. 1999) and iii) the adopted solar galactocentric distance of 8 kpc. Moreover, in the present model we adopt a primordial helium-4 abundance of 0.241 (by mass) instead of 0.23 as recently suggested by Viegas et al. (1999).

Our model is in good agreement with what is called the minimum set of observational constraints, among which the most important is the G-dwarf metallicity distribution (see Tosi 2000 for a recent review; see Figure 1 and Tables 1 and 2).

Table 1. Observed and predicted quantities at $R_{g,\odot}$ and $t = t_{now}$

	CMR	CMG	Observ.
Metal-poor/total stars (%)	4 %	6-13 %	2 - 10 %
SNIa (century^{-1})	0.4	0.3	0.3 ± 0.2
SNII (century^{-1})	1.16	0.78	1.2 ± 0.8
$\Psi(R_{g,\odot},t_{now})$ (M$_\odot$ pc^{-2} Gyr^{-1})	2.6	2.6	2-10
$\sigma_g(R_{g,\odot},t_{now})$ (M$_\odot$ pc^{-2})	7.0	7.0	6.6 ± 2.5
σ_g / σ_T $(R_{g,\odot},t_{now})$	0.13	0.14	0.05-0.20
$\sigma_{inf}(R_{g,\odot},t_{now})$(M$_\odot$ pc^{-2} Gyr^{-1})	1.0	1.0	1.0
0.7	0.18-3.0		
Nova Outbursts (yr^{-1})	21	-	20-30
$X_2(P)/X_2(now)$	1.55	1.5	< 3

The primordial abundances by mass of D and ^{3}He were taken to be 4.4 × 10^{-5} and 2.0 × 10^{-5} respectively. While the D primordial value is an upper limit (as can be seen in Figure 2a) the ^{3}He is a lower limit (see Figure 2b). As can be seen in Figure 2a, the observations of the local interstellar medium (ISM) and the solar system represent tight constraints to the deuterium primordial abundance. In fact, models that can reproduce the bulk of the observational data predict only a modest D destruction (in our case a factor $\simeq$ 1.6; see Tosi et al. 1998).

Table 2. Solar Abundances by Mass (* at 4.5 Gyrs ago)

Element	*CMG	*CMR	Anders & Grevesse (1989)
H	0.73	0.71	0.70
D	4.6 (-5)	3.3 (-5)	4.8 (-5)
^{3}He	10.0 (-5)	2.1 (-5)	2.9 (-5)
^{4}He	2.5 (-1)	2.71 (-1)	2.75 (-1)
^{12}C	1.8 (-3)	3.5 (-3)	3.0 (-3)
^{16}O	7.3 (-3)	7.2 (-3)	9.6 (-3)
^{14}N	1.4 (-3)	1.6 (-3)	1.1 (-3)
^{13}C	4.8 (-5)	4.8 (-5)	3.7 (-5)
^{20}Ne	0.9 (-3)	1.0 (-3)	1.6 (-3)
^{24}Mg	2.5 (-4)	2.5 (-4)	5.1 (-4)
Si	7.0 (-4)	7.0 (-4)	7.1 (-4)
S	3.1 (-4)	3.0 (-4)	4.2 (-4)
Ca	3.9 (-5)	3.9 (-5)	6.2 (-5)
Fe	1.43 (-3)	1.34 (-3)	1.27 (-3)
Cu	8.20 (-7)	7.8 (-7)	8.4 (-7)
Zn	2.4 (-6)	2.3 (-6)	2.1 (-6)
Z	1.4 (-2)	1.6 (-2)	1.9 (-2)

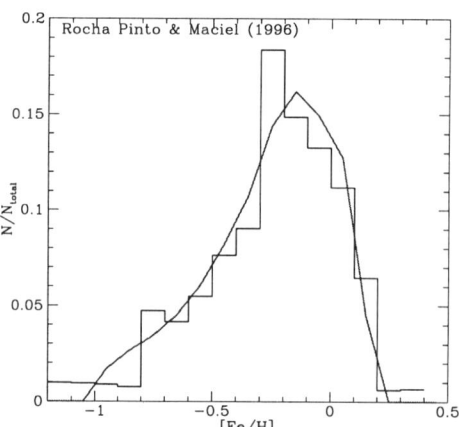

Figure 1. G-dwarf metallicity distribution at the solar vicinity. Data from Rocha-Pinto & Maciel 1996. The curve shows our best model prediction

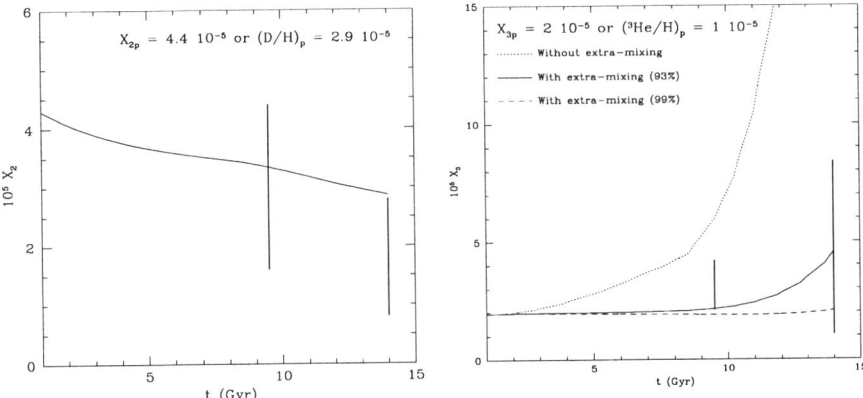

Figure 2. a) D and b) ^{3}He evolution (by mass) as predicted by the present model. The bars at 9.5 Gyr (4.5 Gyrs ago) and 14 Gyr (age of the Galaxy) represent the value of solar (Geiss & Gloeckler 1998 - 2σ) and ISM (Linsky 1998 - 2σ) measured abundances respectively. In the present model we are assuming a halo formation timescale of 1 Gyr.

With respect to our old predictions for the solar vicinity (CMG), we now have better agreement between the predicted solar abundances of ^{3}He, ^{12}C and ^{4}He (Table 2). The improvement in the predicted ^{12}C is due to the fact that we adopt the new nucleosynthetic yields from van den Hoek & Groenewegen (1997) for low and intermediate mass stars. Moreover, the adoption of a higher Y primordial abundance (where Y is the ^{4}He abundance by mass) leads to a better agreement with the solar value. The improvement in the predicted ^{3}He is due to the fact that we are taking into account the recently proposed extra-mixing mechanism in low mass stars ($M < 2.5 M_\odot$, eg. Charbonnel & do Nascimento 1998). In Figure 2b we show models with different assumptions with respect to the extra-mixing mechanism. The model without extra-mixing clearly overproduces ^{3}He with respect to both solar and ISM observed abundances (even with the lower limit primordial abundance of ^{3}He adopted in the present model). The two other curves show our predictions for models which assume that 93% (solid line) and 99% (dashed-line) of stars with M< 2.5 M$_\odot$ completely destroy their ^{3}He, respectively. A model assuming that, when a star suffers extra-mixing (93%) a fraction of $\simeq 1/100$ of its ^{3}He is preserved, gives essentially the same result as the solid line model shown in Figure 2b. The predicted Y vs Z and Y vs O/H are shown in Figures 3a and 3b respectively.

2.2. The Galactic Disk

In this section we show our results for the predicted abundance distributions of D (Figure 4a), ^{3}He (Figure 4b) and ^{4}He (Figure 5). The predicted gradient for D is positive and steep. This is due to the faster evolution of the inner disk regions as compared with the outer parts (which are still in the process of formation thus having an almost primordial composition). In fact, of the various elements, D is probably the most sensitive to radial variations in the timescale

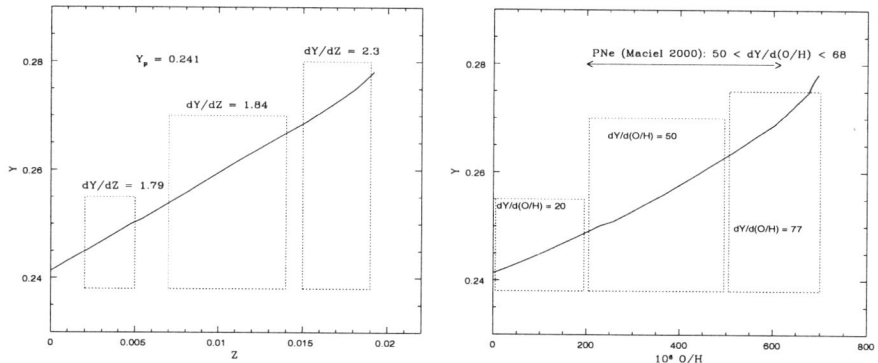

Figure 3. Helium by mass, Y, as a function of a) global metallicity, Z and b) O/H. The observed value obtained from planetary nebulae in the 200-600 10^6 O/H range is also shown.

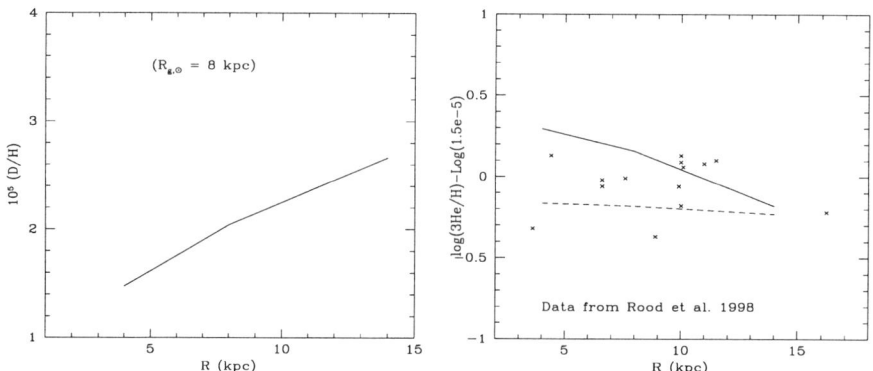

Figure 4. a) Predicted D gradient; b) Predicted ^{3}He gradient. The curves are labelled as in figure 2b

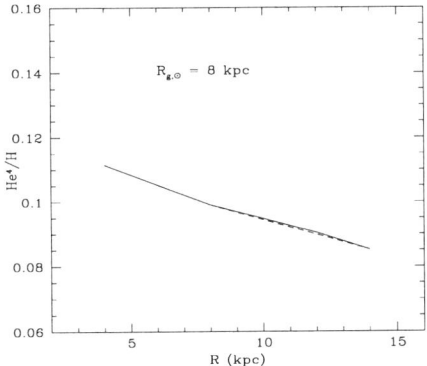

Figure 5. Predicted ^{4}He/H abundance gradient

of disk formation. A hope for the future is to have some D abundance measurements in regions outside the solar vicinity. This would certainly represent a very important constraint to the disk-formation mechanism!

In Figure 4b the predicted ^{3}He abundance gradient is shown. It can be seen that the assumption that 99% of the low mass stars suffer extra-mixing leads to a flat gradient (dashed line). The solid line (93% of low mass stars destroy their ^{3}He) is in marginal agreement with the data, but it is still acceptable. Again, in this case we see that this gradient is sensitive to the adopted timescales of disk formation. In fact, in the inner regions (older in the inside-out scenario) the contribution of low mass stars for the ^{3}He enrichment of the ISM has been more important than in the outer regions, and a negative gradient is predicted. More data on ^{3}He abundance at different galactocentric distances are welcome and would be very important to better constrain our models and the low-mass stellar nucleosynthesis as well.

Finally, in Figure 5 the ^{4}He gradient is shown. The predicted gradient is $\simeq -0.003$ dex/kpc over the 4-14kpc galactocentric range. This value is in agreement with the results presented by Maciel (this meeting) based on disk planetary nebulae.

Acknowledgments. C. C. thanks C. Charbonnel, M. Tosi, W. Maciel and B. Pagel for fruitful discussions during the meeting. The authors thank D. Romano who introduced in our code the new yields of van den Hoek & Groenewegen (1997). C. C. acknowledges financial support by CNPq and FAPERJ (Brazil).

References

Anders, E. & Grevesse, N. 1989, Geochim. Cosmochim. Acta 53, 197
Charbonnel, C., do Nascimento, J.D. Jr. 1998, A&A 336, 915
Charbonnel, C. 2000 (this volume)
Chiappini, C., Matteucci, F., Gratton, R. 1997, ApJ 477, 765
Chiappini, C., Matteucci, F., Romano, D. 2000 (in preparation)

Geiss, J. & Gloeckler, G. 1998, in Primordial Nuclei and their Galactic evolution, N. Prantzos, M. Tosi & R. von Steiger eds., Space Sci. Rev. 84, 239

Kennicutt, R. C. Jr. 1989, ApJ 344, 685

Linsky, J. L. 1998, in Primordial Nuclei and their Galactic evolution, N. Prantzos, M. Tosi & R. von Steiger eds., Space Sci. Rev. 84, 239

Maciel, W. J. 2000 (this volume)

Matteucci, F. & François, P. 1989, MNRAS 239, 885

Pagel, B.E.J. & Tautvaisiene, G. 1995, MNRAS 276, 505

Renzini, A. & Voli, M. 1981, A&A 94, 175

Rocha-Pinto, H.J. & Maciel, W. J. 1996, MNRAS 279, 447

Romano, D., Matteucci, F., Molaro, P., Bonifacio, P. 1999, A&A 352, 117

Sackman, J. 2000 (this volume)

Tosi, M. 2000 (this volume)

Tosi, M. 2000, in The chemical evolution of the Milky Way: stars versus clusters, F. Giovanelli and F. Matteucci eds., Kluwer (in press)

Tosi, M., Steigman, G., Matteucci, F., Chiappini, C. 1998, ApJ 498, 226

van den Hoek, L. B. and Groenewegen, M. A. T. 1997, A&AS 123, 305

Viegas, M. S., Gruenwald, R. & Steigman, G. 1999 (astroph/9909213)

The Light Elements and Their Evolution
IAU Symposium, Vol. 198, 2000
L. da Silva, M. Spite, J. R. de Medeiros, eds.

The Evolution of ^{4}He and LiBeB

Keith A. Olive
Theoretical Physics Institute, School of Physics and Astronomy, University of Minnesota, Minneapolis MN 55455, USA

Abstract. Our understanding of the evolution of ^{4}He and ^{7}Li depends critically on the available data for these two elements at low metallicity. In particular, the degree to which there is a slope in an abundance vs metallicity regression can help determine the evolution of He, C, N, and O in dwarf galaxies in the case of ^{4}He, and cosmic-ray induced nucleosynthesis of LiBeB in our own galaxy in the case of ^{7}Li. Recent data and their implications will be discussed.

1. Introduction

There is a relatively large set of data available on the big bang nucleosynthesis (BBN) element isotopes of ^{4}He and ^{7}Li at low metallicity. It is common, particularly in the case of ^{4}He, to perform a linear regression on the data with respect to some metallicity tracer such as O/H or Fe/H. The intercept of such a regression can be directly related to the primordial abundance of that isotope, and the slope of the regression offers important clues to the nature of its chemical evolution. While one can not necessarily justify a *linear* relation from first principles, generally due to the quality of the data at low metallicity, such an approximation is acceptable. In fact, using the currently available ^{4}He and ^{7}Li data[1], it is easy to show that a linear regression is significantly better than a weighted mean, yet more complicated fits using additional parameters generally do not yield a statistically significant improvement in the fit.

Our inferences of the primordial abundances and evolution of the light elements are clearly tied to the quality of the data and our understanding of the systematic uncertainties in the derived abundances. Evolution is one the effects which is responsible for systematic uncertainties. In the case of ^{4}He, the environment of the HII system is not pristine and includes non-primordial ^{4}He. In addition, the true elemental abundances of ^{4}He may be clouded due to effects such as underlying stellar absorption, collisional excitation, or flourecence. In the case of ^{7}Li, abundances are contaminated by the non-primordial contribution of ^{7}Li from galactic cosmic-ray nucleosynthesis (GCRN) and uncertainties concerning the degree of stellar depletion of ^{7}Li in pop II, halo stars.

---

[1]Using, for example, the data of Izatov and Thuan (1998) on ^{4}He and Ryan, Norris, & Beers (1999) on ^{7}Li.

2. ⁴He

The ⁴He abundance has been determined from observations of HeII → HeI recombination lines in a large sample of extragalactic HII regions (Pagel et al. 1992; Skillman & Kennicutt 1993; Skillman et al. 1994; Izotov, Thuan, & Lipovetsky 1994,1997; Izotov & Thuan 1998). Since ⁴He is produced in stars along with heavier elements such as Oxygen, it is expected that the primordial abundance of ⁴He can be determined from the intercept of the correlation between Y and O/H, namely $Y_p = Y(\text{O/H} \to 0)$. A detailed analysis of the combined data (Olive, Skillman, & Steigman 1997; Fields & Olive 1998) found an intercept corresponding to a primordial abundance

$$Y_p = 0.238 \pm 0.002 \pm 0.005 \qquad (1)$$

The first uncertainty is purely statistical and the second uncertainty is an estimate of the systematic uncertainty in the primordial abundance determination. The helium abundance used in this analysis was determined using electron densities n obtained from SII data. Izotov, Thuan, & Lipovetsky (1994,1997) and Izotov & Thuan (1998) proposed a method based on several He emission lines to "self-consistently" determine the electron density. This method yields a higher primordial value

$$Y_p = 0.244 \pm 0.002 \pm 0.005 \qquad (2)$$

Our interpretation of the evolution of ⁴He depends heavily on the slope of the ⁴He abundance with respect to a tracer element such as O/H and/or N/H. While models of chemical evolution tend to give relatively low slopes ($\Delta Y/\Delta(\text{O/H}) \sim 20 - 60$), the He data based on SII densities gives a much larger slope ($\Delta Y/\Delta(\text{O/H}) \sim 110 \pm 25$), whereas the self-consistent method gives ($\Delta Y/\Delta(\text{O/H}) \sim 47 \pm 26$). The model calculations (Fields & Olive 1998, and references therein) depend crucially on the assumed yields of N in the AGB phase and on assumptions concerning hot-bottom burning. Many of the models attempting to reproduce the higher He slopes also rely on significant amounts of outflow in these dwarf galaxies.

As can be ascertained from the brief discussion above, the method of analysis has a huge impact on both the determination of the primordial ⁴He abundance and the slope of the He vs O/H regression. Therefore, rather than discuss specific chemical evolution models in detail here, I will discuss some of the key sources of the uncertainties in the He abundance determinations and prospects for improvement.

The He abundance is always quoted relative to H, e.g., He line strengths are measured relative to $H\beta$. The H data must first be corrected for underlying absorption and reddening. Beginning with an observed line flux $F(\lambda)$, and an equivalent width $W(\lambda)$, we can parameterize the correction for underlying stellar absorption as

$$X_A(\lambda) = F(\lambda)\left(\frac{W(\lambda) + a}{W(\lambda)}\right) \qquad (3)$$

The parameter a is expected to be relatively insensitive to wavelength. A reddening correction is applied to determine the intrinsic line intensity $I(\lambda)$ relative

The Evolution of ^{4}He and LiBeB

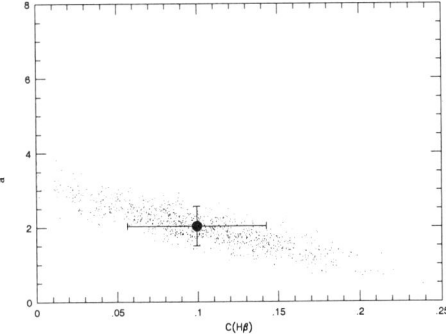

Figure 1. A Monte Carlo determination of the underlying absorption a (in Å), and reddening parameter $C(H\beta)$, based on synthetic data.

to $H\beta$

$$X_R(\lambda) = \frac{I(\lambda)}{I(H\beta)} = \frac{X_A(\lambda)}{X_A(H\beta)} 10^{f(\lambda)C(H\beta)} \quad (4)$$

where $f(\lambda)$ represents an assumed universal reddening law and $C(H\beta)$ is the correction factor to be determined. By comparing $X_R(\lambda)$ to theoretical values, $X_T(\lambda)$, we determine the parameters a and $C(H\beta)$ self consistently (Olive & Skillman, 2000), and run a Monte Carlo over the input data to test the robustness of the solution and to determine the systematic uncertainty associated with these corrections.

In Figure 1 (from Olive & Skillman 2000), I show the result of such a Monte-Carlo based on synthetic data with an assumed correction of 2 Å for underlying absorption and a value for $C(H\beta) = 0.1$. The synthetic data were assumed to have an intrinsic 2% uncertainty. While the mean value of the Monte-Carlo results very accurately reproduces the input parameters, the spread in the values for a and $C(H\beta)$ are generally a factor of 2 larger than one would have derived from the direct solution due to the covarience in a and $C(H\beta)$.

The uncertainties found for $H\beta$ must next be propagated into the analysis for ^{4}He, for which we follow an analogous procedure to that described above (Olive & Skillman 2000). We again start with a set of observed quantities: line intensities $I(\lambda)$ which include the reddening correction previously determined along with its associated uncertainty which includes the uncertainties in $C(H\beta)$; the equivalent width $W(\lambda)$; and temperature t. The Helium line intensities are scaled to $H\beta$ and the singly ionized helium abundance is given by

$$y^+(\lambda) = \frac{I(\lambda)}{I(H\beta)} \frac{E(H\beta)}{E(\lambda)} \left(\frac{W(\lambda) + a'}{W(\lambda)}\right) \frac{1}{(1+\gamma)} \frac{1}{f(\tau)} \quad (5)$$

where $E(\lambda)/E(H\beta)$ is the theoretical emissivity scaled to $H\beta$. The expression (5) also contains a correction factor for underlying stellar absorption, parameterized now by a', a density dependent collisional correction factor, $(1 + \gamma)^{-1}$,

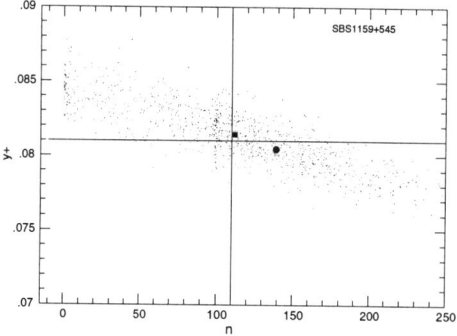

Figure 2. A Monte Carlo determination of the helium abundance and electron density (in cm^{-3}). Solutions for a' and τ are not shown here.

and a flourecence correction which depends on the optical depth τ. Thus y^+ implicitly depends on 3 unknowns, the electron density, n, a', and τ.

One can use 3-6 lines to determine the weighted average helium abundance, $\bar{y}$. From $\bar{y}$, we can calculate the χ^2 deviation from the average, and minimize χ^2, to determine n, a', and τ. Uncertainties in the output parameters are also determined.

This procedure differs somewhat from that proposed by ITL, in that the χ^2 above is based on a straight weighted average, where as ITL minimize the difference of a ratio of He abundances (to one wavelength, say $\lambda 4471$) to the theoretical ratio. When the reference line is particularly sensitive to a systematic effect such as underlying stellar absorption, this uncertainty propagates to all lines this way. In our case, the individual uncertainties in the line strengths are kept separate.

Finally, as in the case for the hydrogen lines, we have performed a Monte-Carlo simulation of the data to test the robustness of the solution for n, a', and τ (Olive & Skillman 2000). In Figure 2, I show the result of a single case based on the data of Izatov and Thuan (1998) for SBS1159+545. Here, the helium abundance and density solutions are displayed. The vertical and horizontal lines show the position of the IT solution. The circle shows the position of the our solution to the minimization, and the square shows the position of the mean of the Monte-Carlo distribution. The spread shown here is significantly greater than the uncertainty quoted by IT.

3. ^{7}Li

The population II abundance of ^{7}Li has been determined by observations of over 100 hot, halo stars, and is found to have a very nearly uniform abundance (Spite & Spite 1982). For stars with a surface temperature $T > 5500$ K and a metallicity less than about 1/20th solar, the abundances show little or no dispersion beyond that which is consistent with the errors of individual measurements. The Li data from Bonifacio & Molaro (1997) indicate a mean ^{7}Li abundance of

$$\text{Li/H} = (1.6 \pm 0.1) \times 10^{-10} \tag{6}$$

The small error is statistical and is due to the large number of stars in which ^{7}Li has been observed.

There is, however, an important source of systematic error due to the possibility that Li has been depleted in these stars, though the lack of dispersion in the Li data limits the amount of depletion. In fact, as discussed by Sean Ryan (these proceedings, and Ryan, Norris, & Beers, 1999, hereafter RNB) a small observed slope in Li vs Fe and the tiny dispersion about that correlation indicates that depletion is negligible in these stars. Furthermore, the slope may indicate a lower abundance of Li than that in (6).

For reference, the weighted mean of the ^{7}Li abundance in the RNB sample is [Li] = 2.12 ([Li] = log ^{7}Li/H + 12). It is common to test for the presence of a slope in the Li data by fitting a regression of the form [Li] = $\alpha + \beta$ [Fe/H]. The RNB data indicate a rather large slope, $\beta = 0.07 - 0.16$ and hence a downward shift in the "primordial" lithium abundance Δ[Li] = $-0.20 - -0.09$. Models of galactic evolution which predict a small slope for [Li] vs. [Fe/H], can produce a value for β in the range $0.04 - 0.07$ (Ryan et al. 2000). Of course, if we would like to extract the primordial ^{7}Li abundance, we must examine the linear (rather than log) regressions. For Li/H = $a' + b'$Fe/Fe$_\odot$, we find $a' = 1 - 1.2 \times 10^{-10}$ and $b' = 40 - 120 \times 10^{-10}$. A similar result is found fitting Li vs O. Overall, when the regression based on the data and other systematic effects are taken into account a best value for Li/H was found to be (Ryan et al. 2000)

$$\text{Li/H} = 1.23 \times 10^{-10} \tag{7}$$

with a plausible range between $0.9 - 1.9 \times 10^{-10}$.

Figure 3 shows the different Li components for a model with $(^7\text{Li/H})_p = 1.23 \times 10^{-10}$. The linear slope produced by the model is $b' = 65 \times 10^{-10}$, and is independent of the input primordial value (unlike the log slope given above). The model (discussed in detail in Fields & Olive, 1999a,b and Fields et al. 2000) includes in addition to primordial ^{7}Li, lithium produced in galactic cosmic ray nucleosynthesis, (primarily $\alpha + \alpha$ fusion) in addition to ^{7}Li produced by the ν-process during type II supernovae. As one can see, these processes are not sufficient to reproduce the population I abundance of ^{7}Li, and additional production sources are needed (see e.g. Matteucchi, these proceedings).

4. Concordance

Bearing in mind the degree of uncertainty in the derived primordial abundances, one can test the concordance of ^{4}He and ^{7}Li with the prediction of BBN. This is best summarized in a comparison of likelihood functions as a function of the one free parameter of BBN, namely the baryon-to-photon ratio η. By combining the theoretical predictions (and its uncertainties) with the observationally determined abundances discussed above, we can produce individual likelihood functions (Fields et al. 1996) which are shown in Figure 4. A range of primordial

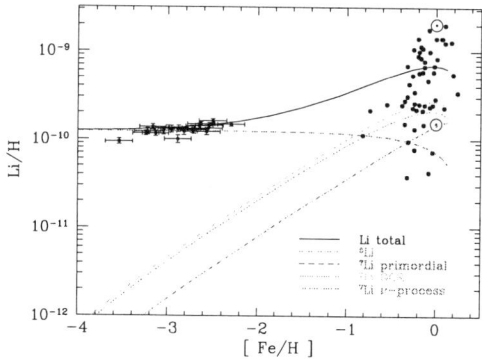

Figure 3. Contributions to the total predicted lithium abundance from the adopted GCE model of Fields & Olive (1999a,b), compared with low metallicity stars (RNB) and a sample of high metallicity stars. The solid curve is the sum of all components.

^{7}Li values are chosen based on the analysis in Ryan et al. (2000). The double peaked nature of the ^{7}Li likelihood functions is due to the presence of a minimum in the predicted lithium abundance in the expected range for η. For a given observed value of ^{7}Li, there are two likely values of η. As the lithium abundance is lowered, one tends toward the minimum of the BBN prediction, and the two peaks merge. Also shown are both values of the primordial ^{4}He abundances discussed above. As one can see, at this level there is clearly concordance between ^{4}He, ^{7}Li and BBN.

5. LiBeB

The production of ^{7}Li by galactic cosmic-ray nucleosynthesis shown in Figure 3, is accompanied by the production of the heavier intermediate elements Be and B (Reeves, Fowler, & Hoyle 1970; Meneguzzi, Audouze, & Reeves 1971). Standard GCRN is dominated by interactions originating from accelerated protons and α's on CNO in the ISM, and predicts that BeB should be "secondary" versus the spallation targets, giving Be $\propto$ O^2 (Vangioni-Flam, Cassé, Audouze, & Oberto 1990). However, this simple model was challenged by the observations of BeB abundances in Pop II stars, and particularly the BeB trends versus metallicity. Measurements showed that both Be and B vary roughly *linearly* with Fe, a so-called "primary" scaling. If O and Fe are co-produced (i.e., if O/Fe is constant) then the data clearly contradicts the canonical theory, i.e. BeB production via standard GCR's.

These observations led to the creation of many new models of cosmic-ray nucleosynthesis (Cassé, M., Lehoucq, R., & Vangioni-Flam, E. 1995; Vangioni-Flam et al. 1996; Ramaty et al. 1997, 1999) which include a dominant primary component of BeB. Such models were discussed here by Cassé, Parizot and Ramaty.

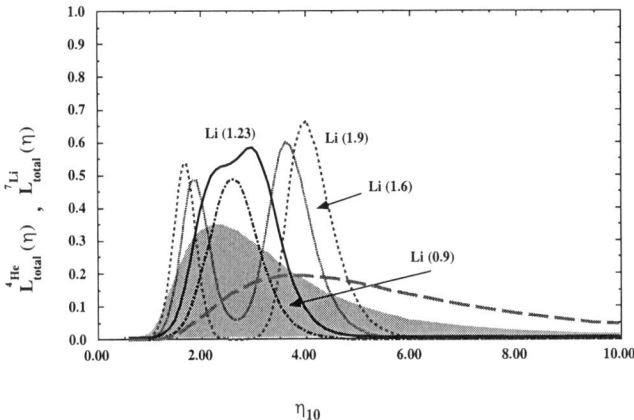

Figure 4. Likelihood distributions for four values of primordial ^{7}Li/H ($10^{10} \times\ ^7$Li = 1.9 (*dashed*), 1.6 (*dotted*), 1.23 (*solid*), and 0.9 (*dash-dotted*)), and for ^{4}He (*shaded*) for which we adopt $Y_p = 0.238 \pm 0.002 \pm 0.005$ (Eq. (1)). Also shown by the long dashed curve is the likelihood function based on the ^{4}He abundance from Eq. (2).

As was discussed here by Deliyannis and Israelian, there is growing evidence that the O/Fe ratio is *not* constant at low metallicity (Israelian, García-López, & Rebolo 1998; Boesgaard et al. 1999a), but rather increases towards low metallicity. This trend offers another solution to resolve discrepancy between the observed BeB abundances as a function of metallicity and the predicted secondary trend of GCR spallation (Fields & Olive 1998). As noted above, standard GCR nucleosynthesis predicts Be $\propto$ O^2, while observations show Be $\sim$ Fe, roughly; these two trends can be consistent if O/Fe is not constant in Pop II. A combination of standard GCR nucleosynthesis, and ν-process production of ^{11}B is consistent with current data.

The determination of abundances from raw stellar spectra requires stellar atmosphere models. The atmospheric models require key input parameters, notably the effective temperature T_{eff} and surface gravity g, and assumptions regarding the applicability of local thermodynamic equilibrium (LTE). Unfortunately, there is no standard set of stellar parameters for the halo stars of interest. In practice, different groups derive abundances via different procedures, which give similar results but retain systematic differences. The systematic differences in the data can in fact obscure the BeB-OFe trends one seeks. Thus, to derive meaningful BeB fits, one must systematically and consistently present abundances derived under the same assumptions and parameters for stellar atmospheres.

It is not possible to overly stress the importance of reliable stellar data. The Balmer-line data appear to be self-consistent, and are probably the most reliable. However, because other scales such as those based on the IRFM are commonly used, I would like to point out that there are significant differences in the reported data. To illustrate the point consider for example the case of the star BD 3° 740. From Axer et al. (1994), whose data is based on the Balmer line method we find this star to have $(T_{\text{eff}}, \ln g, [\text{Fe/H}]) = (6264, 3.72, -2.36)$. The beryllium and oxygen abundances for this star was reported by Boesgaard et al. (1999b) and (1999a). When adjusted for these stellar parameters, we find $[\text{Be/H}] = -13.36$, and $[\text{O/H}] = -1.74$. In contrast, the stellar parameters from Alonso et al. (1996a) based on the IRFM (IRFM1) are (6110,3.73,-2.01) with corresponding Be and O abundances of -13.44 and -2.05. Garcia-Lopez et al. 1998 use a calibrated IRFM (IRFM2) based on Alonso et al. (1996b) and take (6295,4.00, -3.00). For these choices, we have $[\text{Be/H}] = -13.24$ and $[\text{O/H}] = -1.90$. Notice the extremely large range in assumed metallicities and the difference in the two so-called IRFM temperatures. While this may not be a typical example of the difference in stellar parameters, it is differences such as this (and this star is not unique) that accounts for the difference in our results and the implications we must draw from them. Uncertainties in [Fe/H] in particular, make modeling extremely difficult. This is especially true of one attempts to model the correlations of BeBO with respect to Fe/H.

Below, we will present results based for the available BeBOFe data based on three methods of analysis. We will refer to these as the Balmer line data and the IRFM1,2 data. Complete results of this analysis can be found in Fields et al. (2000). There are a total of 36 stars with low metallicity OH data. Of these, roughly 2/3 have available data using one of the systematic methods described (Balmer, IRFM1, IRFM2). In each case, one finds a significant slope for [O/Fe] vs [Fe/H] ranging from -0.32 to -0.51.

Of key importance to the modeling of the BeB evolution is the determination of a primary or secondary source for the BeB isotopes. Primary vs secondary is typically ascertained by fitting the BeB data versus a tracer element. Historically, Fe/H was used even though the actual production of LiBeB is independent of [Fe/H]. This is justifiable so long as [O/Fe] is constant. As one can see in the tables below, the data seem to indicate that Be is mostly primary with respect to Fe, and secondary with respect to O/H. (IRFM2 should be considered suspect as the derived parameters were obtained outside the limits of validity of the calibration.) This is what one would expect if [O/Fe] is not constant as the OH data now indicate. B on the other hand shows primary evolution with respect to [O/H] and an even flatter evolution with respect to Fe/H. This too is expected if the ν-process plays a significant role in the production of ^{11}B.

Table 1. Slopes for Be versus Fe and O.

method	number	tracer	slope	number	tracer	slope
Balmer	22	Fe	1.39 ± 0.16	19	O	1.78 ± 0.19
IRFM1	22	Fe	1.23 ± 0.14	21	O	1.83 ± 0.19
IRFM2	18	Fe	1.18 ± 0.11	18	O	1.36 ± 0.09

The models for primary and secondary production of BeB are all physical. What is unclear however, is which is dominant over the history of the Galaxy and at what epoch. If both mechanisms are operative, it is reasonably certain that primary mechanisms should dominate in the early Galaxy and that secondary mechanisms should dominate later. The cross-over or break point can be determined from the data (in principle) by fitting to both linear and quadratic

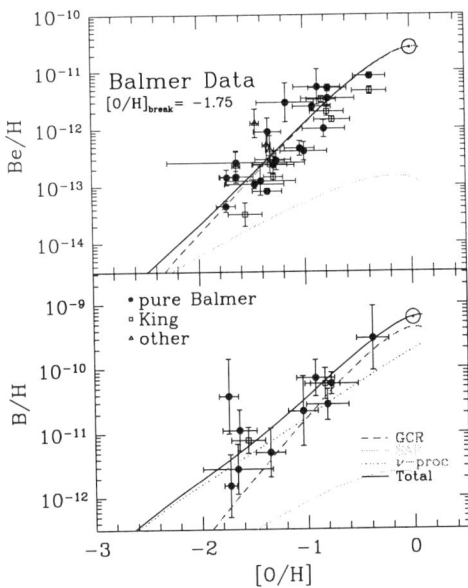

Figure 5. Be vs O (*top panel*) and B vs O (*bottom panel*). Data shown are the Balmer points, which are found to have a break point as indicated. Models are adjusted to have the break point and O/Fe slope of these data.

components,

$$\frac{A}{H} = \left(\frac{A}{H}\right)_\odot \left[\alpha_1 \frac{O/H}{(O/H)_\odot} + \alpha_2 \left(\frac{O/H}{(O/H)_\odot}\right)^2\right] \quad (8)$$

for $A \in$ BeB. The resulting coefficients and break points for Be are found in Table 3. As one can see, for Balmer and IRFM1, the break point occurs at low [O/H], indicating that most of the evolution in the observed data has been secondary. To fully resolve this issue, a larger and systematic data set is required.

Finally, in Figure 5 (from Fields *et al.* 2000), the evolution of BeB with respect to O/H is shown in a simple closed box model of chemical evolution. In addition to standard GCR nucleosynthesis, a primary component based on

Table 2. Slopes for B versus Fe and O.

method	number	tracer	slope	number	tracer	slope
Balmer	11	Fe	0.78 ± 0.22	10	O	1.23 ± 0.32
IRFM1	9	Fe	0.73 ± 0.19	9	O	0.98 ± 0.28
IRFM2	11	Fe	0.72 ± 0.14	11	O	1.02 ± 0.16

Table 3. Break points for Be versus O.

number	method	α_1	α_2	$[O/H]_{eq}$
19	Balmer	0.042 ± 0.003	2.30 ± 0.70	-1.75
21	IRFM1	0.034 ± 0.034	2.14 ± 0.53	-1.79
18	IRFM2	0.111 ± 0.031	2.57 ± 0.76	-1.37

the superbubble accelerated particle spectrum of Bykov (1999) is included along with the neutrino-process for ^{11}B (Fields et al. 2000). The secondary GCR cosmic-ray flux is normalized by the solar abundance of Be and is consisitent with the present cosmic-ray flux scaled by the star formation rate. The break point (from Table 3) determined the relative scaling of the primary component to the secondary one, and the neutrino process is scaled to the solar ^{11}B/^{10}B ratio.

Acknowledgments. I would like to thank my collaborators T. Beers, M. Cassé, B. Fields, J. Norris, S. Ryan, E. Skillman, and E. Vangioni-Flam whose work has been summarized here. This work was supported in part by DoE grant DE-FG02-94ER-40823 at the University of Minnesota.

References

Alonso, A., Arribas, S., & Martinez-Roger, C. 1996a, A & AS, 117, 227
Alonso, A., Arribas, S., & Martinez-Roger, C. 1996b, A & A, 313, 873
Axer, M., Fuhrmann, K., & Gehren, T. 1994, A & A, 291, 895
Boesgaard, A.M., King, J.R., Deliyannis, C.P., & Vogt, S.S. 1999a, AJ, 117, 492
Boesgaard, A.M., Deliyannis, C.P., King, J.R., Ryan, S.G., Vogt, S.S., & Beers, T.C. 1999b, AJ, 117, 1549
Bonifacio, P. & Molaro, P. 1997, MNRAS, 285, 847
Bykov, A.M. 1999, in "LiBeB, cosmic rays and related X-and Gamma- Rays", eds. Ramaty et al. , ASP, vol. 171, p. 146
Cassé, M., Lehoucq, R., & Vangioni-Flam, E. 1995, Nature, 373, 38
Fields, B. D., Kainulainen, K., Olive, K. A., & Thomas, D. 1996, New Astron., 1, 77
Fields, B.D. & Olive, K.A. 1998, ApJ, 506, 177
Fields, B.D. & Olive, K.A. 1999, ApJ, 516, 797
Fields, B.D. & Olive, K.A. 1999, New Astron., 4, 255
Fields, B.D., Olive, K.A., Vangioni-Flam, E. & Cassé, M. 2000, astro-ph/9911320.
García-López, R.J., et al. , Ap.J. 500 (1998)
Israelian, G., García-López, R.J., & Rebolo, R. 1998, ApJ, 507, 805
Izotov, Y.I. & Thuan, T.X. 1998, ApJ, 500, 188.
Izotov, Y.I., Thuan, T.X., & Lipovetsky, V.A. 1994 ApJ 435, 647
Izotov, Y.I., Thuan, T.X., & Lipovetsky, V.A. 1997, ApJS, 108, 1
Meneguzzi, M. , Audouze, J., & Reeves, H. 1971, A&A, 15, 337

Olive, K.A., Steigman, G. & Skillman, E.D. 1997, ApJ, 483, 788

Olive, K.A. & Skillman, E. 2000, in preparation.

Pagel, B.E.J., Simonson, E.A., Terlevich, R.J. & Edmunds, M. 1992, MNRAS, 255, 325

Ramaty, R., Kozlovsky, B., Lingenfelter, R.E., & Reeves, H. 1997, ApJ, 488, 730

Ramaty, R., Scully, S., Lingenfelter, R. and Kozlovsky, B. 1999 (astro-ph/9909021)

Reeves, H., Fowler, W.A., & Hoyle, F. 1970, Nature, 226, 727

Ryan, S.G., Beers, T.C., Olive, K.A., Fields, B.D., & Norris, J.E. 2000, ApJL, in press (astro-ph/9905211)

Ryan, S.G., Norris, J.E., & Beers, T.C. 1999, ApJ, 523, 654

Skillman, E., & Kennicutt 1993, ApJ, 411, 655

Skillman, E., Terlevich, R.J., Kennicutt, R.C., Garnett, D.R., & Terlevich, E. 1994, ApJ, 431, 172

Spite, F. & Spite, M. 1982, A&A, 115, 357

Vangioni-Flam, E., Cassé, M., Audouze, J., & Oberto, Y. 1990, ApJ, 364, 568

Vangioni-Flam, E., Cassé, M., Fields, B.D., & Olive, K.A. 1996, ApJ, 468, 199

Stellar and GCR Production of Lithium in the Milky Way

F. Matteucci

Department of Astronomy, University of Trieste, Via G.B. Tiepolo, 11, I-34131 Trieste, Italy

D. Romano

SISSA/ISAS, Via Beirut, 2-4, I-34014 Trieste, Italy

Abstract. Lithium production from several stellar sources (C-stars, massive AGB stars, Type II supernovae and novae) as well as from galactic cosmic rays (GCR) is included in a succesfull model for the chemical evolution of the Milky Way in order to predict the evolution of the ^{7}Li abundance (A(Li) = log(^{7}Li/H) + 12) as a function of [Fe/H]. From comparison with the oservational data we infer the following conclusions: 1) Li production from novae seems to be necessary to explain the steep rise of the Li abundance for metallicities larger than [Fe/H] = − 1.0 dex, 2) Li production from SNe II should be less than assumed before in order to reproduce the long plateau observed for A(Li) for low metallicities, 3) Li production from C-stars is negligible relative to Li production from massive AGB stars which is instead a necessary Li- source, 4) Li production from GCR should contribute by no more than 20% to the solar Li abundance.

1. Introduction

The observed A(Li) vs. [Fe/H] in dwarf stars shows that the Li abundance in Pop I and T Tauri stars is at least a factor of ten higher than in Pop II stars.

This fact can be interpreted in different ways:

- either the primordial Li abundance is high (that observed in Pop I stars) implying that this element is progressively destroyed inside stars and that non-standard Big Bang nucleosynthesis is required,

- or the primordial Li abundance is low and coincides with that measured in Pop II stars which show a remarkable uniformity of Li abundance (the so-called Spite plateau). In such a case, beyond the fact that the estimated primordial Li abundance is in agreement with the predictions of the standard Big Bang nucleosynthesis, an efficient production of Li during the Galaxy lifetime is required to explain the Li abundance in Pop I stars.

In the following we will adopt this point of view.

A recent data compilation by Romano et al. (1999) concerning lithium abundances in disk and halo dwarfs with $T_{eff} \geq 5700$ K has shown:

- a well defined plateau extending perhaps up to [Fe/H] ≥ − 0.5 dex,

- a general trend of an increasing Li abundance with increasing [Fe/H] (> − 1.5 dex) reaching quite high Li abundances in the T Tauri stars and in the local interstellar medium (ISM).

These observational data (Figure 1) together with the fact that the ^{7}Li/^{6}Li ratio in the solar system is ∼ 12 as opposed to the value of this ratio expected from the GCR production (∼ 2), suggest that a substantial fraction of the ^{7}Li in the Galaxy should have a stellar origin (see Reeves 1993). The data also show a large spread in the Li abundance of disk stars whereas the spread observed in the plateau is much smaller. The smaller spread in Pop II stars can be probably explained as an effect of metallicity. D'Antona and Mazzitelli (1984) suggested that main-sequence Li depletion is inhibited in low metallicity stars.

On the other hand, main-sequence Li destruction occurs in high metallicity stars because of deeper surface convection due to the higher opacity of Pop I stars. Therefore, the spread in Pop I stars is mainly due to age effects, the older stars being the most Li depleted. Because of this, models of galactic chemical evolution predicting the Li abundance in the ISM should fit the upper envelope of the A(Li) vs. [Fe/H] data.

In this paper, we adopt a succesfull model of chemical evolution of the Milky Way reproducing the majority of observational constraints (Chiappini et al. 1997, 1999 where a detailed description can be found), and we use it to compute the evolution of the Li abundance in the interstellar medium (ISM) by taking into account all the possible stellar Li-producers as well as the Li produced by spallation of GCR. The results are then compared with the observational data in order to put constraints on the origin of lithium.

2. Stellar Li-sources

2.1. Observations of Li-rich stars

Observations of massive AGB and M supergiants in the Magellanic Clouds indicate the presence of a large Li abundance together with enhanced s-process elements, from which one infers that they are in their thermal pulsing (TP) phase, but they are not C-stars (Smith and Lambert 1989, 1990; Plez et al. 1993).

On the other hand, Galactic C-stars with luminosities from $M_{bol} \simeq$ − 6 to − 3.5 show in some cases very high Li abundances (Abia et al. 1991; 1993) but the total amount of mass that they can loose is negligible to have an effect on Galactic Li production.

Therefore, there is a clear observational indication that low and intermediate mass stars ($1 \leq M/M_\odot \leq 8$) can be Li-producers, especially the more massive ones (see D'Antona and Matteucci 1991).

Concerning the other possible Li producers such as Type II SNe and novae, there are no observational indications about Li production. As a consequence, the nucleosynthesis prescriptions for these objects are only theoretical.

2.2. Nucleosynthesis prescriptions

C-stars($2 - 5\ M_\odot$)

Li is produced during thermal pulses and then destroyed: survival may last a few thermal pulses but the actual mass of produced and ejected Li is quite uncertain. We assume the same prescriptions adopted by D'Antona and Matteucci (1991) (A(Li) =3.85 dex per star).

High mass AGB stars

Sackmann and Boothroyd (1992) showed that stars of masses $4 - 6\ M_\odot$, in the luminosity range $-6 \leq M_{bol} \leq -7$, can reach A(Li)=4.5 dex in the envelope as a consequence of hot bottom burning. We assumed that each star produces either A(Li)=3.5 or 4.15 dex, in agreement with Matteucci et al. (1995) and in agreement with the observations.

Type II SNe

ν-induced nucleosynthesis can create a substantial amount of Li. The flux of neutrinos following the core collapse to form a neutron star becomes so great that it can induce substantial transmutation. μ and τ neutrinos excite heavy elements and He to particle unbound levels. The evaporation of a neutron or proton and the back reaction of these nucleons on other species alters the outcome of traditional nucleosynthesis calculations. ^{7}Li is partly produced in the H-shell but mostly in the He- and Si-shells (Woosley et al. 1990).

Novae.

Very recent nucleosynthesis calculations (José and Hernanz 1998) suggest that the explosive formation of ^{7}Li in novae can produce an average Li mass per nova of $< M_{Li} > = 1.8 - 7.5 \times 10^{-7}\ M_\odot$. The important aspect of novae is that they start restoring Li into the ISM only after a timescale of > 1 Gyr, due to the slow evolution leading to the nova outburst (white dwarfs in binary systems). Details on the nova evolution relative to the Galactic enrichment in lithium can be found in Romano et al. (1999).

Table 1.   Input parameters and results.

	C-stars	M-AGB	SNeII	novae	A(Li)$_{SS}$	A(Li)$_{ISM}$
A	3.85	4.15	WW95	no	2.91	2.90
B	3.85	4.15	WW95	yes	3.12	3.24
C	no	3.50	WW95/2	yes	2.95	3.12
C + GCRs	no	3.50	WW95/2	yes	3.21	3.39

3. Model Results

We have computed the evolution in time of the Li abundance by changing the nucleosynthesis prescriptions. Table 1 summarizes the input parameters of the different models and some results. In particular the first column of Table 1 indicates the model, the second the prescriptions adopted for C-stars with the assumed amount of Li produced by each star expressed in terms of A(Li), the same is for the massive AGB stars in column 3. Column 4 indicates the sources of the adopted prescriptions for type II SNe: in a couple of models the produced

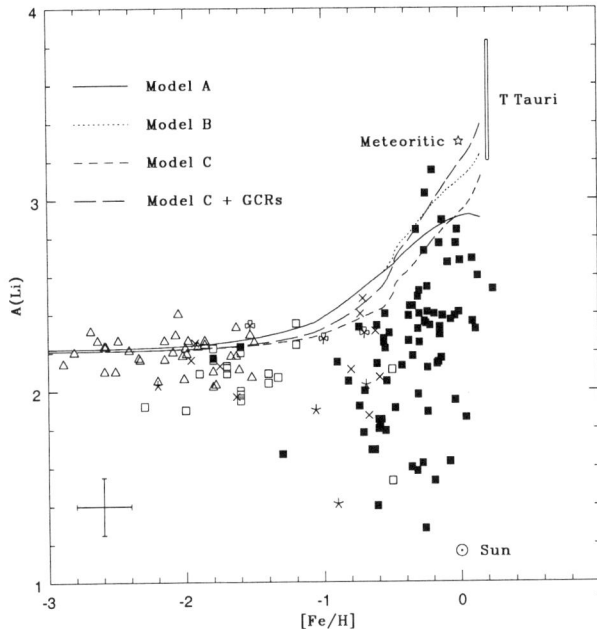

Figure 1. Observed and predicted A(Li) versus [Fe/H] in the solar neighbourhood. Data are from the compilation of Romano et al. 1999. The abundances of Li in the T Tauri are indicated as well well as the Li abundance in the sun. Models are labelled A, B, C, C+GCR according to the text.

lithium has been reduced to one half. Column 5 indicates if we have adopted the novae prescriptions or not (the Li mass produced by novae is the one mentioned in the previous paragraph). Finally in column 6 and 7 we report the predicted Li abundance for the ISM at the time of the formation of the solar system and for the local ISM, respectively. Concerning the production of ^{7}Li from GCR we assumed the results of Lemoine et al. (1998) and we computed only one model including all the Li prescriptions (stellar plus GCR).

In Figure 1 we show a comparison between the predictions of the models of Table 1 and the observational data. As mentioned before, a good model of chemical evolution should reproduce the upper envelope of these data ignoring the observed spread which indicates the progressive Li destruction inside stars and does not reflect the Li abundance of the ISM. An inspection of Figure 1 indicates that the models including a late Li production from novae are necessary to explain the steep increase of the Li abundance for [Fe/H] > − 1.0 dex (Models B, C, C+GCR). Among these models the one which also includes a late production of Li by GCR seems to be the best although the amount of Li contributed by GCR to the solar system Li abundance should not exceed 10 − 20% of the total. In particular the model presented here (Model C+GCR) contributes for more than 20% of the solar lithium. Therefore a model intermediate between Model B (no GCR production) and Model C+GCR would produce the best fit to the data. Another consideration arising from the inspection of Figure 1 is that the neutrino spallation process acting in SNe II seems to overproduce Li at low metallicities and that reducing this quantity by at least one half produces a better agreement with the data.

4. Conclusions

In this paper we have presented some predictions for the evolution of the abundance of ^{7}Li in the ISM of the Milky Way. We have assumed that the Li abundance observed in the local ISM as well as in young stars as T Tauri is the result of the stellar lithium production. We have envisaged several possible Li producers and suggested that novae or alternatively a stellar Li source acting at late times in the Galactic evolutionary history (t $>$ 1 Gyr) are necessary to explain the observational data. Other sources like massive AGB stars are also necessary to produce the solar system and present time observed Li abundance. Lithium production from GCR is not fundamental but it can help if is acting at late times and does not produce more than 10-20% of the total solar lithium as indicated by the meteoritic ^{7}Li/^{6}Li ratio as compared with what is expected from GCR Li production.

References

Abia, C., Boffin, H.M.J., Isern, J., & Rebolo, R. 1991, A&A, 245, L1
Abia, C., Isern, J., & Canal, R. 1993, A&A, 275, 96
Chiappini, C., Matteucci, F., & Gratton, R. 1997, ApJ, 477, 765
Chiappini, C., Matteucci, F., Beers, T.C., & Nomoto, K. 1999, ApJ, 515, 226
D'Antona, F., & Matteucci, F. 1991, A&A, 247, L37
D'Antona, F., & Mazzitelli, I. 1984, A&A, 138, 431
José, J., & Hernanz, M. 1998, ApJ, 494, 680
Lemoine, M., Vangioni-Flam, E., & Cassé, M. 1998, ApJ, 499, 735
Matteucci, F., D'Antona, F., & Timmes, F.X. 1995, A&A, 303, 460
Plez, B., Smith, V.V., & Lambert, D.L. 1993, ApJ, 418, 812
Reeves, H. 1993, A&A, 269, 166
Romano, D., Matteucci, F., Molaro, P., & Bonifacio, P. 1999, A&A, 352, 117
Sackmann, I.J., & Boothroyd, A.I. 1992, ApJ, 392, L71
Smith, V.V., & Lambert, D.L. 1989, ApJ, 345, L75
Smith, V.V., & Lambert, D.L. 1990, ApJ, 361, L69
Woosley, S.E., Hartmann, D.H., Hoffman, R.D., & Haxton, W.C. 1990, ApJ, 356, 272

Light Element Evolution at the Solar Neighborhood

Andreu Alibés, Javier Labay and Ramon Canal

Departament d'Astronomia i Meteorologia, Universitat de Barcelona, C/Martí i Franquès 1, 08028 Barcelona, Spain

Abstract. We present the Light Element Evolution resulting from our new Chemical Evolution model. The LiBeB evolution is correctly fitted by taking into account several sources: Big Bang, Galactic Cosmic Ray Nucleosynthesis, the ν-process, novae and AGB and C-stars.

1. Introduction

Our aim in this work has been to study the synthesis and evolution of the abundances of the light elements in the Solar Neighborhood from the birth of the Galaxy to present. Light elements have sources different from those of heavier elements. We have included GCR nucleosynthesis in a complete Galactic Chemical Evolution Model that takes into account 76 stable isotopes from hydrogen to zinc. Any successful LiBeB evolution model should also be compatible with other observational constraints like the age-metallicity relation, the G-dwarf distribution or the evolution of other elements. Some of the standard outputs of our chemical evolution model, which is in good agreement with those observational data, are the helium, carbon, nitrogen and oxygen evolution in the ISM. Those elements are the targets of the GCR made of the newly synthesized material in a type II supernova, accelerated by the shock wave.

2. Sources considered

We have considered several stellar and interstellar frameworks for the production of the light elements. The main source of ^{6}Li, ^{9}Be and ^{10}B is the *Galactic Cosmic Ray Nucleosynthesis*, due to spallation reactions between Cosmic Ray particles and ISM material. Following Ramaty et al. (1997), and using their LIBEB code, we have calculated the production rate by this mechanism, resulting from the ISM abundances calculated by our model being hit by GCR composed of material newly synthesized and ejected by a SNII. The source energy spectrum of the GCR is: $q(E) \propto \frac{p^{-2.2}}{\beta} e^{-\frac{E}{E_0}}$, with $E_0 = 10$ GeV/n.

Neutrino induced nucleosynthesis. Light elements are produced when the strong neutrino flux produced in a core collapse supernova goes through the material surrounding the core. Its yields, which depend on that neutrino flux, are quite uncertain. This mechanism produces ^{7}Li and ^{11}B and we have fixed the contribution of Woosley and Weaver (1995) yields by means of $(^{11}B/^{10}B)_\odot$.

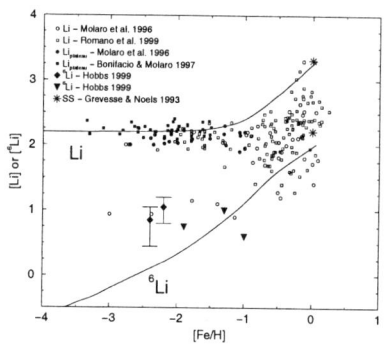
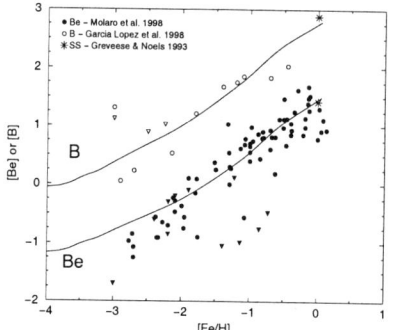

Novae are supposed to produce some ^{7}Li in each outburst. We calculate the outburst rate following the equation: $D \int_{m(t)+0.5}^{9.5} \frac{\phi(m_B)}{m_B} \int_{\mu_m}^{\mu_M} f(\mu) \psi(t - \tau_{m_B(1-\mu)} - t_{cool})$ where D imposes the actual outburst rate and $t_{cool} = 1$ Gyr is the cooling time for a white dwarf until it can produce the first nova outburst. The nova system we have considered is made of a white dwarf coming from an initial 1-8 $M_\odot$ primary star and a 0.5-1.5 $M_\odot$ secondary star. Each outburst ejects on average $1.03 \cdot 10^{-10}$ $M_\odot$ of ^{7}Li (José & Hernanz 1998).

AGB and *C-stars* are also ^{7}Li producers. We have included them as in Abia et al. 1993, using a present production rate of $P_7 = 10^{-8}$ $M_\odot \text{pc}^{-2} \text{Gyr}^{-1}$, in order to fit the lithium evolution. That rate is just 5 times larger than the one statistically determined by those authors, but lower than the one needed in their evolution model.

The results are shown in the figures.

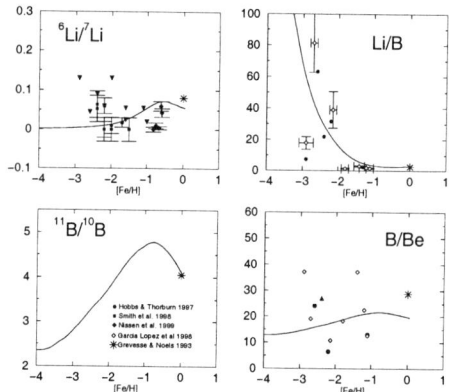

References

Abia, C., Isern, J., & Canal, R. 1993, A&A, 275, 96
Chiappini, C., Matteucci, F., & Gratton, R. 1997, ApJ, 477, 765
José, J., & Hernanz, M. 1998, ApJ, 494, 680
Matteucci, F., D'Antona, F., & Timmes, F.X. 1995, A&A, 303, 460
Ramaty, R., Kozlovsky, B., Lingenfelter, R.E., & Reeves, H. 1997, ApJ, 488, 730
Romano, D., Matteucci, F., Molaro, P., & Bonifacio, P. 1999, A&A, 352, 117
Woosley, S.E., & Weaver, T.A. 1995, ApJS, 101, 181

Evolution of ^{6}LiBeB in Inhomogeneous Early Galaxy

Takeru Ken Suzuki

Department of Astronomy, School of Science, University of Tokyo; Theoretical Astrophysics Division, National Astronomical Observatory, Mitaka, Tokyo, 181-8588 Japan; stakeru@th.nao.ac.jp

Yuzuru Yoshii

Institute of Astronomy, School of Science, University of Tokyo, Mitaka, Tokyo, 181-8588 Japan; Research Center for the Early Universe, School of Science, University of Tokyo

Toshitaka Kajino

Theoretical Astrophysics Division, National Astronomical Observatory, Mitaka, Tokyo, 181-8588 Japan

Abstract. Results of evolution of light elements (^{6}LiBeB) based on supernova (SN)-induced chemical evolution model are presented. We point out an important property of light elements as a cosmic clock in metal-poor stars.

Recent observations of old population II stars (Mcwilliam et al. 1995, Ryan et al. 1996) reveals that the their chemical compositions seem to reflect nearby SN events in the regions in which the stars were formed (Audouze, & Silk 1995). As a result, metallicity cannot be used as a reliable age indicator for the most metal-poor stars. Tsujimoto, Shigeyama, & Yoshii (1999) presented a SN-induced chemical evolution model based on these views. In their model all the stars at early epochs are born from SNR shells, as a result of their relatively high density. Their model accounts very well for the large observed scatter in the abundances of (some) heavy elements in PopII stars. Suzuki, Yoshii, & Kajino (1999) extended this model to investigate Be and B, which are mainly produced by spallation reactions involving Galactic cosmic rays (GCRs). The great advantage of this model is that it can treat the evolution of elements in an inhomogeneous, evolving, Galactic Halo in a self-consistent manner, without the need to set model parameters in an ad-hoc fashion.

Fig.1 shows our results for the SN-induced chemical evolution model of the expected frequency distributions of ^{6}LiBeB (vertical axis) and Fe (transverse axis) in the long-lived stars ($m < M_\odot$) with the available observations superposed. An important prediction for ^{6}LiBeB, according to this model, is that the abundances of these light elements in metal-deficient stars are expected to be better correlated with the time of formation of the stars than the abundance of heavy elements (a detailed discussion given by Suzuki et al. 2000; see also Beers et al. this volume). This is because these elements are synthesized by reactions involving GCRs which propagate rapidaly throughout the entire Halo.

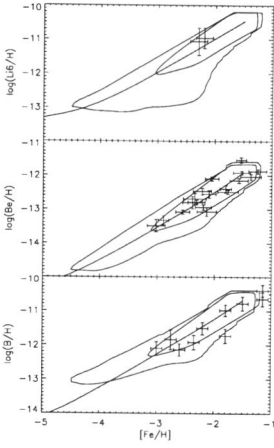

Figure 1. Predicted frequency distributions of long-lived stars in the [Fe/H]-log(L/H) planes, convolved with Gaussians having $\sigma = 0.15$ dex for Be, B, and Fe and $\sigma = 0.3$ dex for ^{6}Li, for comparison with the observations. The two contour lines, from the inside to the outside, correspond to those of constant probability density 10^{-3}, and 10^{-5} in unit area of Δ[Fe/H]=0.1×Δlog(L/H)=0.1. The solid line shows the [Fe/H]-log(L.E./H) relation in the gas. The crosses represent the data, with observational errors, taken from Smith et al. (1998; ^{6}Li), Boesgaard et al. (1999; Be), and Primas et al. (1999; B) and Duncan et al. (1997; B).

References

Audouze J., & Silk, J. 1995, ApJ, 451, L49

Boesgaard, A. M., Deliyannis, C. P., King, J. R., Ryan, S. G., Vogt, S. S., & Beers, T. C. 1999, AJ, 117, 1549

McWilliam, A., Preston, W., Sneden, C., & Searle, L. 1995, AJ, 109, 2757

Duncan, D. K., Primas, F., Rebull, L. M., Boesgaard, A. M., Deliyannis, C. P., Hobbs, L. M., King, J. R., & Ryan, S. G. 1997, ApJ, 488, 338

Primas, F., Duncan, D. K., Peterson, R. C., & Thorburn, J. A. 1999, A&A, 343, 545

Ryan, S. G., Norris, J. E., & Beers, T. C. 1996, ApJ, 471, 254

Smith, V. V., Lambert, D. L., & Nissen, P. E. 1998, ApJ, 506, 405

Suzuki, T. K., Yoshii, Y., & Kajino, T. 1999, ApJ, 522, L125

Suzuki, T. K. et al. 2000, in preparation

Tsujimoto, T., Shigeyama, T., & Yoshii, Y. 1999, ApJ, 519, L63

The Light Elements and Their Evolution
IAU Symposium, Vol. 198, 2000
L. da Silva, M. Spite, J. R. de Medeiros, eds.

One Zone Numerical Model for the Galactic Evolution of Lithium

Marco Terra[1] and Lilia I. Arany-Prado

Observatório do Valongo/Universidade Federal do Rio de Janeiro

Abstract. We analyze the evolution of ^{7}Li/H using the basic model for the galactic chemical evolution (GCE) described by Brown (1992) and a numerical procedure which allows to estimate the contribution of the red giants to the enrichment of lithium in the interstellar medium (ISM).

1. Motivation and Models

Since the discovery of the Li rich giants, stellar nucleosynthesis has been regarded as an important source for the galactic Li in conjunction with cosmic rays spallation and the Big Bang primordial nucleosynthesis. Also low mass stars represented by K giants became a promising source of Li (de la Reza et al. 1996; Gregorio-Hettem et al. 1993). Our goal is to evaluate the contribution of the red giants to the enrichment of Li in the ISM. Our numerical procedure considers the one zone GCE model, assumes the sudden mass loss approximation; the IMF and the turnoff mass of Scalo (1986); two different star formation rates (SFRs) (figure 1a): the one of Mathews & Schramm (M&S, see Brown 1992) and the maximum exponentially decreasing (ExpD, Miller & Scalo 1979). The mass of the remnants is taken from Iben & Renzini (1983). The galactic disk is supposed to be 12 Gyr old. We use the following mass ranges: 1-2.5$M_\odot$ for red giants; 2.5-7$M_\odot$ for AGB stars; and 7-62$M_\odot$ for supernovae (SNe).

2. Results and Comments

We establish the return mass fraction to the ISM, R(t) (Arany-Prado & Maciel, 1998), for each SFR (figure 1b). The numerical procedure allows to analyze each site separately. We assume as the maximum values for the lithium abundance in the stellar ejecta: 10^{-8} for AGB stars (Smith & Lambert 1990) and 10^{-10} for SNe (Brown et al. 1990). If we consider: a) the maximum Li abundance ejected by K giants as 2×10^{-8} (Terra et al. 2000); b) the observational remark of Brown et al. (1989) that near 2% of red giants are Li rich; c) the evolutionary argument from de la Reza et al. (1996), we conclude that the mean value for the lithium abundance ejected at any short time range by K giants is up to 4×10^{-9}. Taking into account the contribution to the ISM due to stellar sites plus cosmic ray spallation, the mean Li abundance in the red giants ejecta has to be about 3×10^{-9} in order to total Li fit the present abservational constraints. This is in good agreement with the arguments above. Furthermore, K giants become the most important site for Li production at about the fourth Gyr of the galactic

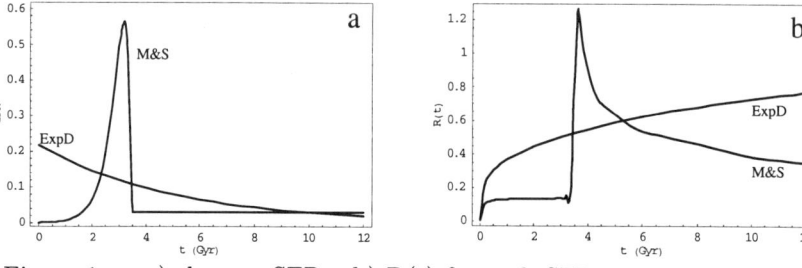

Figure 1. a) the two SFRs. b) R(t) for each SFR.

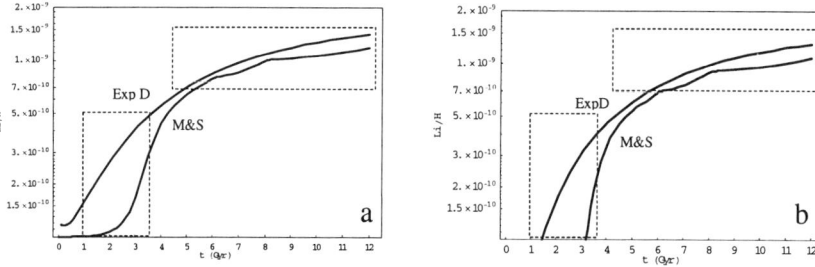

Figure 2. The evolution of ^{7}Li/H with time in the ISM considering two SFRs and the following: a) with Big Bang nucleosynthesis; b) only stellar nucleosynthesis. Boxes represent the observational constraints given by Brown (1992).

life. We show the results of our numerical models considering the primordial nucleosysnthesis (figure 2a) and also the arguments of Burbidge & Hoyle (1998), for whom only stellar nucleosynthesis is responsible for the elements generation (figure 2b).

Terra, M. would like to thank FAPERJ for financial support: project 150.571/98

References

Arany-Prado, L.I., Maciel, W., 1998, Rev.Mex.Ast.Astrof., 34, 21
Brown, L.E., 1992, ApJ, 389, 251
Brown, J.A., Sneden, C., Lambert, D.L., Dutchover, E.J., 1989, ApJS, 71, 293
Brown, L.E., Dearborn, D.S., Schramm, D.N., Larsen, J.T., Kurokawa, S., 1990, ApJ, 371, 648
Burbidge, G., Hoyle, F., 1998, ApJ, 509, L1
de la Reza, R., Drake, N., da Silva, L., 1996, ApJ, 456, L115
Gregorio-Hetem, J., Castilho, B.V., Barbuy, B., 1993, A&A, 268, L25
Iben, I., Renzini, A., 1983, ARA&A, 21, 271
Miller, G.E., Scalo, J.M. 1979, ApJS, 41, 513
Scalo, J M , 1986, Fund. Cosmic Phys., 11, 1
Smith, V., Lambert, D., 1990, ApJ, 361, L69
Terra, M., de la Reza, R., Batalha, C., 2000, in preparation

CONCLUSIONS

Bernard Pagel presenting his Conclusions

The Light Elements and Their Evolution
IAU Symposium, Vol. 198, 2000
L. da Silva, M. Spite, J. R. de Medeiros, eds.

Conclusions I

B. E. J. Pagel

Astronomy Centre, CPES, University of Sussex, Brighton BN1 9QJ, UK

Abstract. Some conclusions are drawn from the Conference as to Big Bang Nucleosynthesis, cosmic rays, AGB star evolution and various kinds of mixing processes in stars.

1. Introduction

The last symposium on Light Elements that I remember was held on Elba in 1994 (Crane 1995). Thinking of that meeting reminds me of how much we have lost through the untimely death of Dave Schramm, whose breadth and enthusiasm would have added so much to this meeting had he been spared.

The problems we face are to find the abundances of light elements (D, ^{3,4}He, ^{6,7}Li, ^{9}Be, ^{10,11}B) as functions of place and time and thereby find out about Big Bang nucleosynthesis (BBNS), Galactic cosmic rays and production and destruction by stars in the course of Galactic chemical evolution (GCE). In this quest we face a number of difficulties, notably abundance uncertainties due to errors in parameters, inhomogeneities and departures from LTE in stellar atmospheres, collisional effects, fluorescence and underlying absorption in H II regions affecting ^{4}He, and unexplained scatter, notably for D.

2. Elements relevant to BBNS

2.1. Deuterium and ^{3}He

Starting with deuterium, there is now a handful of convincing measurements at high red-shifts in Lyman-limit systems where there is no reason to doubt that the abundance is primordial. Garry Steigman gave comparable weight to the 'high D' (D/H $> 10^{-4}$, say) and 'low D' values, but in my view all the 'high D' estimates in the literature suffer from such technical flaws that we can dismiss them. The convincing work is that by Dave Tytler and his students using the Keck telescopes.

How low is 'low'? Here Sergei Levshakov has made an important contribution by introducing the idea of mesoturbulence, related to the still more sophisticated analysis of local clouds presented by Jeff Linsky. The upshot is that primordial D/H is between 3 and about 5×10^{-5}, corresponding to η_{10} between 4 and 6.

This range of η is also in agreement with measurements of ^{3}He (which seems to be pretty much neutral to GCE effects) in Galactic H II regions reported by Tom Bania and Bob Rood.

Coming to more local objects – the Solar System and the interstellar medium (ISM) – one expects a downward trend with time in the sense $D_{hi\ z} > D_{SS} > D_{local\ ISM}$ and a positive gradient with galactocentric distance. Some evidence for the latter was provided by the Galactic centre observations reported by Don Lubowich, which can be compared to published observations of the D I hyperfine structure line towards the anticentre by Chengalur, Braun & Burton (1997).

In the Solar System, George Gloeckler gave $D/H = 1.9 \times 10^{-5}$ at high heliocentric latitudes measured from the *Ulysses* satellite, while in the local ISM after allowing for emission-line profiles, hydrogen walls and other factors, Jeff Linsky gives 1.6×10^{-5}, all of which seems quite reasonable. Models of GCE allow the corresponding astration factor between 2 and 3, as reported by Monica Tosi and Cristina Chiappini, although this concordance may be taken as supporting those models as much as supporting the primordial D/H value.

All this is very nice, but there could be a fly in the ointment: Is D/H constant? Alfred Vidal-Madjar made an eloquent case for variations, but Meena Sahu has argued that, for stars within 100 pc or so, this case does not really hold up. Further afield, there are the obstinate cases of γ^2 Vel and δ Ori that have been around since the time of *Copernicus* and, as noted by George Sonneborn, these have not been explained away so far. Perhaps Sergei Levshakov could get to work on these cases. In any case, there will be more data from the *Fuse* mission in the near future.

2.2. ^{4}He

Pagel et al. (1992) derived a primordial helium abundance $Y_P = 0.228$, which is certainly too low; as transpired afterwards, we underestimated effects of underlying absorption lines in I Zw 18 and there could be other factors. Izotov & Thuan have presented the results of a splendid job based on a uniform set of observations with high signal:noise deriving $Y_P = 0.244$ and a plausible value of dY/dZ, consistent with results reported by Walter Maciel for planetary nebulae and with stellar data (Pagel & Portinari 1998). The error bars quoted by Izotov & Thuan are overly optimistic; we estimated a systematic error of up to 0.005, and even that may be optimistic.

There are various effects, e.g. ionization correction factors (as discussed by Sueli Viegas), collisional excitation of hydrogen and maybe some real inhomogeneity. Thus it would be useful to have observations with a higher spectral resolution, as Thuan plans to do, helping to check up on underlying absorption (as Keith Olive pointed out) and measuring lines from higher energy levels that are less sensitive to collisional excitation.

Manuel Peimbert, who – with Silvia – started this whole business back in 1974, presented some exremely nice work on NGC 346 in the SMC avoiding some of the problems, notably that of underlying absorption. They find a lower helium abundance $Y_{ngc\ 346} = 0.241 \pm 0.002$ extrapolating to $Y_P = 0.236$.

What are we to make of this? I think it best not to rely on one single object, however good, as there must be intrinsic scatter at some level. We should therefore hold our horses before deciding that there is a discrepancy with 'low' deuterium and standard BBNS with $N_\nu = 3$.

2.3. ^{7}Li

On ^{7}Li we have had no less than 25 papers and 18 posters. It is not possible to summarise all of them, and I apologise for ignoring many significant contributions. The classical Spite plateau, according to Sean Ryan's very careful discussion, survives previous attempts to superimpose a cosmic scatter, but on the other hand now seems to have a slight slope due to cosmic-ray production, as is to be expected from the presence of ^{6}Li that Poul Nissen reported on. The CR contribution to ^{7}Li at [Fe/H] = -2 is at least 5 per cent, possibly more.

Possible destruction factors in the atmospheres of Population II dwarfs and subgiants now seem to be very severely limited by the small scatter and presence of ^{6}Li (0.15 dex according to Marc Pinsonneault and 0.1 dex according to Sylvie Vauclair), although the relevant factors are not completely understood. Concerning the abundance determinations themselves, we have a reassurance at least in the particular case of the subgiant HD 140283 from Martin Asplund who showed that effects of atmospheric inhomogeneity are more or less cancelled out by non-LTE effects, and from a similar study described by Roger Cayrel. Is this like using Satan to cast out Beelzebub? In any case this is very impressive fundamental work dispensing with the usual fudge factors associated with model atmospheres.

Taka Kajino described new nuclear physics results that reassure us that the theoretical uncertainties in primordial abundances have not been underestimated. Thus with Ryan's estimate $1.96 \leq 12 + \log(\text{Li/H})_\text{P} \leq 2.38$ and the upper range of 'low' deuterium we have concordance in the range

$$4 \leq \eta_{10} \leq 5; \quad 0.014 \leq \Omega_\text{B} h^2 \leq 0.019, \tag{1}$$

in good agreement with what has been deduced on other grounds from the Lyman-α forest. The corresponding range of Y_P is between 0.242 and 0.247, and we heard from Hannu Kurki-Suonio that the simple homogeneous Big Bang is still the best model.

3. Lithium and Galactic chemical evolution

It has been clear for a long time that ^{7}Li in Population I needs stellar sources as well as Galactic cosmic rays (GCR) since, as Hubert Reeves has mentioned, the solar-system ratio ^{7}Li/^{6}Li = 12 is so high.

Fig 1 shows some GCE models invoking AGB stars, supernovae etc. by Matteucci et al. (1995) together with a very simple-minded model that just assumes ^{7}Li–^{7}Li $_\text{P} \propto$ Fe normalized to the Solar System. These models seem to have some quite desirable properties in relation to Sean Ryan's conclusions in that the Li abundance begins to rise noticeably at [Fe/H] = -2.5 and very noticeably between -2 and -1, whereas the more recent models shown in Fig 2 are a bit too flat in this region. On the other hand, there seems to be a steeper rise above [Fe/H] = -1, which suggests the influence of a class of stars having a still longer evolutionary lifetime than the SNIa reponsible for the bulk of the iron, and two of the models by Romano (1999) shown in Fig 2 do this by appealing to novae, but I think AGB stars of sufficiently low mass would do equally well, and this is relevant to several papers that we heard at this meeting.

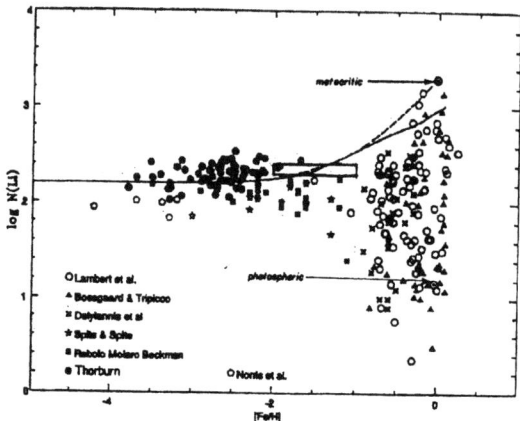

Figure 1. Stellar lithium abundances as a function of metallicity. The full-drawn curve shows the prediction of the GCE model by Matteucci, D'Antona & Timmes (1995), assuming contributions from carbon stars, massive AGB stars and SNII, while the broken-line curve gives the sum of a primordial component and an additional component proportional to iron and normalized to meteoritic abundance. The rectangle shows the range of undepleted lithium abundances at metallicities $-2 < $ [Fe/H] < -1 reported by Ryan at this conference. Adapted from Pagel (1997).

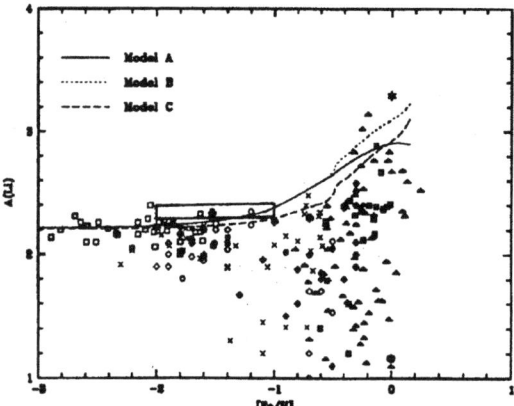

Figure 2. Stellar lithium abundances as a function of metallicity. The curves represent numerical models by Romano (1999): A with ^{7}Li production from carbon stars, massive AGB stars and SNII; B the same with addition of thermonuclear runaway in nova outbursts; C like B but without any contribution from carbon stars and a lower one from massive AGB stars and SNII. The rectangle shows the range of undepleted lithium abundances at metallicities $-2 < $ [Fe/H] < -1 reported by Ryan at this conference. Adapted from Romano (1999).

I may mention in particular the stimulating and eloquent presentation by Juliana Sackmann. There are basically two modes of ^{7}Li production by the Cameron–Fowler mechanism: hot-bottom burning in massive stars on the AGB; and cool-bottom burning in low-mass and preferably low-metallicity stars on the RGB. Things basically happen at the red-giant clump where the H-burning shell encounters H, ^{3}He rich material from the previous maximum penetration of the surface convection zone, μ-barriers are reduced and fresh ^{3}He becomes available to make ^{7}Be. This kind of source was also discussed by Corinne Charbonnel, as both a sink and a source of ^{7}Li (the example of a sink in the cooler component of Capella was discussed by George Wallerstein (1966) many years ago).

At this meeting, a wealth of observational evidence was presented, notably the PDS survey here in Brazil described by Ramiro de la Reza, Carlos Torres and Bruno Castilho. It seems that, at a critical stage, your K giant can emit a dust cloud detectable by IRAS, which expands, cools and disappears, so the star does a loop in the HR diagram possibly supplying fresh ^{7}Li to the interstellar medium in the process. However, as Gerard Jasniewicz pointed out, other mechanisms such as engulfing planets etc. are not ruled out.

Yet other aspects of lithium depletion/non-depletion have been discussed. Rafael Rebolo showed some beautiful results on brown dwarfs, where lithium supplies a useful diagnostic and alternative age estimator for clusters and he even suggests that Population II brown dwarfs could potentially supply an alternative estimate for primordial lithium. In any case, Yakiv Pavlenko now achieves impressive results with synthetic spectra.

There is considerable new data on lithium depletion in Galactic clusters. The theory, described by Constantine Deliyannis, envisages some sort of slow mixing generated by rotation accompanied by diffusion, and Georges Michaud gave a very detailed diffusion model for Am and Fm stars. Coming back to lithium, one may single out Vanessa Hill's study of NGC 2473 and 47 Tuc, two clusters of similar metallicity but different ages and with turnoffs on either side of the Boesgaard lithium gap. The depletions are similar in the two cases, supporting Luca Pasquini's conclusion that most of the depletion on the main sequence occurs in a relatively short time like 10^8 years.

Finally, we have the lithium isotope anomalies in the ISM, described by Dave Knauth who confirms previous suspicions of ratios as low as 2 towards o Per, which suggests a large contribution from cosmic rays. Is there any relation with the suspected D/H anomalies? Conversely, we heard from Francesca Primas about the halo star HD 160617 which has plateau-like lithium but anomalously low beryllium and boron, suggesting an unusually low exposure to cosmic rays. This star also has high nitrogen, but it is not clear whether that has any connection.

4. Elements produced by Galactic cosmic rays

Cosmic rays (GCR) are thought to be the main source of ^{6}Li, Be and B, although ^{11}B can also have a contribution from the neutrino process in supernovae. In the classical model of Reeves, Fowler & Hoyle (1970), the main process is the hitting of stationary interstellar CNO nuclei by relativistic protons and α-particles in the GCR, with a minor contribution from the inverse process, and this more-

or-less accounts for the abundances of these species in the Solar System apart from underpredicting the $^{11}B/^{10}B$ ratio. The latter might have been explained by postulating a low-energy GCR component, but the whole idea is challenged by the 'primary' behaviour of Be and B relative to iron at low metallicities.

If the O/Fe ratio steadily rises towards low metallicity as claimed by Israelian, Garcia-Lopez & Rebolo (1998) and by Boesgaard et al. (1999), then the trend relative to oxygen has a slope of about 1.5, intermediate between primary and secondary. Fields & Olive have managed to fit a basically secondary GCR model to this trend, but we heard from Etienne Parizot and Reuven Ramaty that the energetic difficulties associated with the secondary model still remain.

This problem bears on the origin of cosmic rays as well as on the enrichment of stars and the ISM in the early Galaxy. Etienne Parizot, Michel Cassé and Reuven Ramaty have discussed various forms of primary models which in general involve formation of stars in a superbubble wherein ejecta from one or more supernovae mix with the ambient ISM leading to energetic particles with CNO nuclei present, but models differ in the precise details of the mechanism and location of GCR acceleration.

I do not go into details on that, but just note that, in these situations, metallicity (however defined) is not a good clock, but rather a measure of the environment — how massive the supernova was and how far away from the low-mass stars that we observe now — so the 1.5 power of O-abundance (if indeed that is what it is, of which I am not yet convinced; cf. Fulbright & Kraft 1999) can be a measure of the relative abilities of the two elements to escape from the SN environment, and one expects a certain amount of scatter as indeed has been found for the r-process (Tsujimoto, Shigeyama & Yoshii 1999). Also, oxygen can vary relative to metals like magnesium, as a result of gas-dust separation.

Other related issues are the roles of a low-energy component and of the ν-process. ^{6}Li is very considerably enhanced relative to Be in low-metallicity stars compared to the Solar System (the exact amount depending on what view is taken of its depletion) and it seems that $\alpha - \alpha$ fusion is not enough to account for this, so a low-energy component may still be needed. Whether B/Be varies at all with metallicity is still an open question, and — as Francesca Primas informed us — isotopic data for boron are awaited.

Beryllium and boron, while more robust than lithium, nevertheless are destroyed at temperatures of 5 million K or so and therefore together with lithium provide important constraints on the depth of mixing, e.g. in the Sun, where we heard from Suchitra Balachandran that beryllium is quite undepleted, implying that mixing is confined to a relatively narrow layer below the outer convection zone. Another issue raised by this work is the UV opacity affecting OH lines, which were used by Balachandran & Bell to calibrate the opacity using theoretical f-values, whereas Israelian and Boesgaard et al. changed the f-values to fit solar data with Kurucz models. However, the changes are small, typically 0.1 dex or so, and so do not have a major bearing on the conflict between these authors and Fulbright & Kraft on the O/Fe ratio. I think Israelian had a good point on nLTE effects in gravity determination from ionization equilibrium; on the other hand I would not either trust HIPPARCOS parallaxes when they are so small. So in my opinion that issue remains open.

Boron is another element affected by UV opacity and there was an impressive treatment by Katia Cunha, both for the Sun, where again the photosphere is brought into agreement with meteorites, and for hot stars where nLTE effects are being brought under control. Again, boron seems to track iron, but not oxygen, in the Orion association — another little piece of data probably telling us something, but I don't know what.

So we have plenty of data, a little more understanding maybe, plenty of controversies and plenty to be done. It just remains to thank the organizers and our Brazilian hosts for the opportunity to enjoy this very lively meeting in such a splendid environment.

References

Boesgaard, A.M., King, J.R., Deliyannis, C.P. & Vogt, S.S. 1999, AJ, 117, 492

Chengalur, J.N., Braun, R. & Burton, W.B. 1997, A & A Lett., 318, L35

Crane, P. (ed.) 1995, ESO/EIPC Workshop: *The Light Element Abundances*, Springer, Berlin

Fulbright, J.P. & Kraft, R.P. 1999, AJ, 118, 527

Israelian, G., Garcia-Lopez, R.J. & Rebolo, R. 1998, ApJ, 507, 805

Matteucci, F., D'Antona, F. & Timmes, F.X. 1995, A&A, 303, 460

Pagel, B.E.J. 1997, *Nucleosynthesis and Chemical Evolution of Galaxies*, Cambridge University Press

Pagel, B.E.J. & Portinari, L. 1998, MNRAS, 298, 747

Pagel, B.E.J., Simonson, E.A., Terlevich, R.J. & Edmunds, M.G. 1992, MNRAS, 255, 325

Reeves, H., Fowler, W.A. & Hoyle, F. 1970, Nature, 226, 727

Romano, D. 1999, in R. Ramaty, E. Vangioni-Flam, M. Cassé & K. Olive (eds.), *LiBeB, Cosmic Rays, and Related X- and Gamma-Rays*, ASP Conf. Series, Vol. 171, p. 55

Tsujimoto, T., Shigeyama, T. & Yoshii, Y. 1999, ApJ, 519, L63

Wallerstein, G. 1966, ApJ, 143, 823

The Light Elements and Their Evolution
IAU Symposium, Vol. 198, 2000
L. da Silva, M. Spite, J. R. de Medeiros, eds.

Concluding Remarks II

Hubert Reeves

1. Introduction

The symposium has shown that our subject is well and very alive. It is progressing rapidly, thanks to a large amount of new observational data, obtained in particular by a young generation of competent astronomers. It is encouraging, incidentally, to note the large fraction of women in this generation. When I started in this field, some forty years ago, the feminine contribution was much smaller. I plan to review the state of the situation for each of the three nucleosynthesis processes responsible for the light elements: Big Bang Nucleosynthesis (BBN), galactic cosmic ray spallation of interstellar nuclei (GCR) and stellar nucleosynthesis. I will point out their successes, their riddles and possible avenues by which these riddles could be solved. I plan first to review the observations.

2. Observatio Results

2.1. Deuterium

We have three sets of historical data (fig 1), pertaining respectively to the present ISM (interstellar medium), the protosolar nebula (4.5 billion years ago), and the early universe (some 14 billion years ago). Linsky (1999), Vidal-Madjar, Ferlet and Lemoine (1997), Vidal-Madjar (1999) and Sahu (1999) have presented and discussed the D/H measurements in ISM, in front of bright stars. The estimated value of D/H is 1.5×10^{-5}, (+/- 0,10). Inhomogeneities of D/H in the ISM are reported. The FUSE satellite, recently put into orbit, is expected to clear up this delicate point.

An argument based on the measurement of the helium isotopic ratio in the solar system can be used to give an estimate of the D/H ratio in the galactic gas at the birth of the sun, 4,5 billion years ago (Geiss and Reeves 1972), on the assumption that the solar wind 3He is the sum of the protosolar D burned in 3He plus the protosolar 3He. In 1969 Johannes Geiss and his colleagues in Bern measured the 3He/4He ratio in the solar wind from an aluminum sheet brought back from the moon (Geiss and Gloeckler 1997). Since the solar convective zone is hot enough to burn D+p -¿ 3He, it was clear that the solar wind contains the sum of the primordial D and 3He of the protosolar nebula. But was it possible that, during the past history of this zonal material, some 3He had been burned in 4He? The presence of 9Be in the solar photosphere at essentially the meteoritic (i.e. protosolar value) argued against this possibility. In the surface convective zone of the sun, the destruction of the 9Be occurs at lower temperature than the 3He burning reactions. Hence any 3He depletion would have been accompanied by an even stronger 9Be depletion. This does not seem to

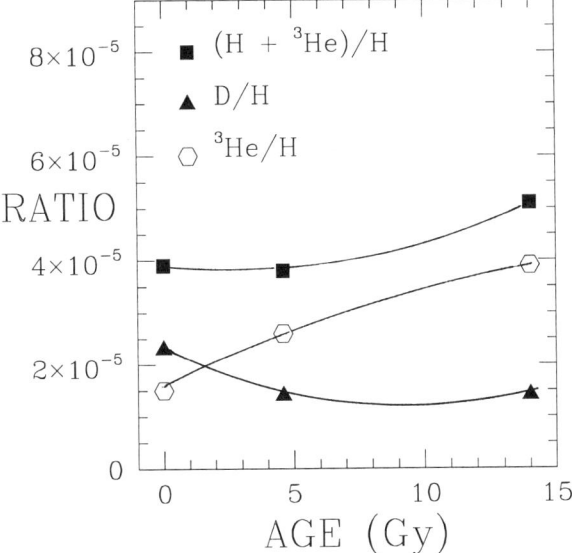

Figure 1. Deuteron and ^{3}He abundances as a function of time from BBN (14 Gy ago) till today (adapted from Geiss and Gloeckler 1997). The value of ^{3}He at 14 Gy is not an observation but results from BBN calculations. The apparent variation of this nuclide at later time is well within the uncertainties.

be the case. The protosolar 3He/4He needes to pursue this argument is obtained from the Jupiter value (discussed later). Subtracting it from the solar wind value one obtains a protosolar D/H = 1.9 (+or-0,5)x10-5 (Gloeckler 99) . This is quite in agreement with the Jupiter D/H value which have been reported by Donahue and Nieman (97) at (2.6 +/-0.7)x 10-5. Other jovian values (Gautier and Owen 1989), (Bjoraker et al 1986), (Carlson et al 1983), (Encrenaz et al 1996,) , range from 2 to 5x10-5 . Because of possible chemical fractionations of hydrogen in the jovian atmosphrer , one cannot readily obtain the protosolar value from these measurements. Deuterium measurements by line absorpions on remote intergalactic clouds in front of quasars have led to a period of confusion . A high value D/H of a few times 10-4 was first obtained by Songaila et al (1997) (discussed by Hogan 1997) A lower value of (2,5 (+/-0,2) x10-5) was later reported by Tytler (1997) .An analysis of the Tytler data by Levshakov et al (1997), (1999) give a D/H value 4,5x10-5. There appears to be a consensus for such low values against the early high values. The possibility of spatial variations of deuterium in these absorbing clouds is discussed by Vidal-Madjar et al (1997) .Again the situation should cleared up with the recently launched FUSE satellite.

2.2. Helium-3

A measurement of 3He/4He= 2.48 (+/-0,65) X 10-4 in the ISM near the solar system has been obtained by (Geiss and Gloeckler (1997) . Rood et al (1997) have measured 3He/4He values in ionized hydrogen nebulae (HII regions) with different oxygen abundances, situated at various galactocentric distances . These abundances are in approximate agreement with the Jupiter value : 3He/4He = 1,03 (+or-0,06)x 10 -4 (Donahue and Nieman 1997) and with the Geiss and Gloeckler (1997) ISM value . With a dispersion of a factor three , there appears to be no systematic trends with galactocentric distance and with the abundance of elements other than hydrogen and helium (called " metallicity" in the astronomer's jargon) . Geiss and Gloeckler (1997) mention the small increase between the birth of the solar system and the present gas .

2.3. Helium-4

Helium-4 is generated in the Big Bang and also by Main Sequence stars. The observed fractional mass abundances range from 24the gradual effect of stellar energy generation in galaxies. The search for the primordial yield is conducted in objects which have been least contaminated by stellar synthesis, such as the extra galactic HII regions (blue compact galaxies) , as is evidenced by their comparatively low abundance of heavier isotopes (Peimbert1999). One strategy consists of a plot of He versus O or C or N , which is tentatively extrapolated to zero value for these stellar-generated atoms. Another estimate is obtained by averaging on the objects with the lowest metal abundances. (Kunth and Sargent 1983) (Pagel 1989, 1992), Pagel et al 1992) , (Campbell 1992), (Fuller et al 1992) (Thuan and Izotov 1997) (Thuan 1999) (Viegas 1999) . A primordial value of the helium mass fraction of 0,230-0.245 is obtained. There is much debate regarding the uncertainties attached to this number.

2.4. Lithium -7

New measurements of its abundances in stars have been reported by (Pasquini 1999), (Hill 1999). (Kajino 1999),(Randich 1999), (de la Reza1999), (Malik1999), (Thevenin1999) , (Torres1999). One of the best arguments in favor of a Big Bang contribution to the abundance of 7Li comes from the fact that, for old stars, 7LI/H 1,5x10-10, is independent of the metallicity , for Fe/H ratio varying from 10-3.5 to almost 10-1 solar (Spite and Spite 1982ab); (Spite et al 1997) (Hobbs and Duncan 1987), (Rebolo et al 1988, 1992). The "flatness" of the plateau is discussed by (Cayrel 1997, 1999) and (Spite 1997)(Molaro 1999).(Ryan 1999) To obtain the Big Bang yield from these observations one has to estimate the lithium depletion from surface processes in these old stars. This turns out to be a difficult problem with remaining unsolved features (Michaud and Charbonneau 1991) (Vauclair 1999) (Michaud 1999). Nevertheless the flatness of the plateau (with a dispersion of less than a factor of two), suggest a rather small amount of surface stellar depletion, less than a factor of two, at best comparable to the dispersion.

2.5. Beryllium-9

This atom has been observed in a large number of young (Pop 1) stars (Boesgaard and Heacox1978),(Boesgaard and Ryan 1999) (Israelian1999). The mean value is in good agreement with the solar value Be/H=1,4 x10-11 measured by Chmielsvski et al 1975, Balachandran (1999) has shown than this value does not differ by more than 20carbonaceous chondrite) value (Grevesse and Noels 1992).

In old (PopII) stars beryllium shows a systematic decrease with decreasing stellar metallicity (Gilmore et al 1991) (Gilmore et al 1992, Gilmore 1992) ; (Ryan et al1991) (Ryan 1992, 1999) Rebolo et al 1992) (Ryan 1999) . The slope of the Be/H vs Fe/H indicates a linear growth of Be with Fe during the galactic life. Unlike Li however , Be shows no sign of a plateau at very small metallicity.

2.6. Boron

With an abundance ratio of B/Be =20 to 40, the B/H ratio behaves very much like the Be/H ratio (Boesgaard and Heacox(1978))(Primas et al 1999) (Primas 1999) . It has recently shown (Cunha and Smith 1999) (Cunha 1999) that the solar value is in agreement with the meteoritic value B/H 7 x 10-10, (Anders and Grevesse 1989) ,thereby solving a longstanding reported discrepancy between the two values. . With decreasing metallicity , B/H decreases down to 10-12 with no sign of a plateau. . As in the case of Be , the slope of the B/H vs Fe/H is close to one showing again a linear growth of B with Fe along the galactic life. The boron isotopic ratio in the solar system is 11B/10B = 4,05 0,05.(Shima and Honda 1962) with variations of a few percent Chaussidon and Robert 1998 .Measurements of 11B/10B 3.6 have been made on two stars (Nissen 1999) . In the interstellar medium the situation appears to be confused. (Rebull et al 1998)(Knauth 1999)

3. Big Bang Nucleosynthesis

The aim of the theory of Big Bang Nucleosynthesis is to account for the abundances of the nuclides D, 3He,4He and 7Li in the very early universe , prior to the birth of galaxies and stars. (Schramm 1992) (Steigman 1999) (Kurki-Suonio1999). Over the past three decades, BBN has progressively reached a high degree of credibility. By providing initial galactic abundances of these nuclides it has become a major tool for the study of stellar and galactic evolution. It also serves as a test-bench for new cosmological ideas. Big Bang nucleosynthesis has one free parameter : the baryonic number h (the ratio of the number of nucleons to the number of photons) or equivalently Wb the ratio of the baryonic density to the critical density. The main success of BBN comes from the fact that with h = (5 +/-1) x10-10 corresponding to Wb = 0.04 +/-0.15BBN one can satisfactorily account for the primordial abundances of the three nuclides (D, 4He and 7Li) while being compatible with 3He (for which we have , however no primordial abundance measurement ; the value of 3He/H at BBN in fig 1 has been computed with the value of h = (5 +/-1) x10-10) Let us call "cosmological window " the range of baryonic densities allowed by astronomical observations. The density Wlum of luminous matter (stars and hot gases) is evaluated at 0.003 . This is clearly a lower limit to the universal nucleon density. On the other hand, the total matter density Wmat, estimated by its gravitational manifestations (in Einstein theory, all forms of matter and energy exert gravitational effects), is close to 0.3; an upper limit to Wb. Thus the BBN calculated Wb manages to account for the abundances of the primordial nuclides while being compatible with the cosmological window (0.003¡Wb¡0.3) This coherence between the nuclear physics and astronomical domains is another strong argument in favor of the Big Bang theory. An evaluation of Wb can also be obtained from the analysis of the absorption lines in the spectra of remote quasars. These lines are ascribed to diffuse clouds of matter in intergalactic space. The computations, although somewhat model-dependant, yield indeed values of Wb compatible with the BBN value. One last point before we close the subject of the Big Bang. Recent experiments (Normile 1998), have given a mass of approximately 0.1 eV for one of the three varieties of neutrinos ,(probably the tau neutrino) . The others varieties are most likely even less massive. This insures that all three neutrinos are relativistic at BBN (kT = one Mev) , and thus justifies the standard model for which the choice of three relativistic neutrinos. This was previously an assumption.

4. BBN as a tool for stellar and galactic evolution

4.1. Deuterium

During galactic evolution,deuterium is systematically destroyed by astration, thus we expect a gradual depletion of its BBN abundance as a function of time , as more galactic gas is incorporated in stars and later ejected with D free gas . This depletion is plausibly tempered by the continuous infall of extragalactic gas on the galactic disk , with presumably pristine D (Lubovich 1999) . Several authors (Tosi 1997), (Prantzos et al 1992), (Prantzos 1996) , (Matteuci 1999), (Olive 1999), (Chiappini 1999), (Vangioni-Flam, Olive and Prantzos 1994) (Olive 1999) have presented consistent galactic evolution models incorporating relevant astronomical observations such as the abundance ratios of heavier elements, the galactic mass fractions of stars and gaseous nebulae, as a function of age and radial distance from the center of the galaxy .These evolution models include up-to-date nuclear and atomic data for the various elements and isotopes of interest. The initial input of these models are 1) the rate of star formation 2) the Initial Mass Function and 3) the rate of infall of extragalactic matter. In this respect the observations of (Beers 1999) on the population of very metal poor stars are of great interest. The general conclusion is that models which accounts satisfactorily for the ensemble of observations generate only a modest amount of D depletion - at best by a factor of three- , in good agreement with the data (fig1).

4.2. Helium-3

This nucleus is produced in significant quantity both by BBN and by stellar synthesis, as an intermediate step in hydrogen burning . In stars, 3He is produced during the Main Sequence phase by the (D+p) reaction. In the hot centers , 3He is further transformed in 4He by the 3He+3He or 3He+4He reactions . In the outer stellar layers however the temperature is high enough to generate 3He but not to destroy it. Stellar models show there an accumulation of 3He reaching values of 3He/4He 10-2 As the star moves toward the red giant branch, a fraction of these nuclei is burned into 4He while another fraction is convected to the surface and ejected by stellar winds. Other processes in novae or other advanced stages of stellar evolution may further generate significant amounts of 3He. The final fate of an 3He is governed by stellar internal processes, from the Main Sequence to the Giant Branch and the Planetary Nebula phases, through the effects of various dredge-up and mixings. The conventional stellar models predict an important yield of 3He for stars smaller than 2 solar masses. Observations of Planetary Nebulae (PN) (Rood et al 1997) have confirmed this predictions with some detections at the level of 3He/4He 10-3. The problem lies with the theoretically expected stellar contribution of these PN during the life of the galaxy. According to conventional models , values of 3He/4He 10-3 should be observed in the interstellar medium , quite in disagreement with the measurements. Charbonnel et al (1997),Charbonnel (1999) have discussed the relation between this problem and the problems raised by the observations of 12C/13C ratio and of N and O abundances in PN . The discrepancy could be solved by the operation of a new mixing phase in the Red Giant Branch , which would account for the destruction of 3He and also for the unexpectedly low 12C/13C observed in some PN . The large scatter in the observed abundances of 3He makes it difficult to estimate the relative contributions of BBN and stellar evolution to its abundance. The BBN value, shown in fig 1, has been computed in the frame of BBN using the baryonic number obtained from D, 4He and 7Li. Helium-4 After the BBN production of an helium fractional mass of 0,24, the main effect of all the Main Sequence stars has been to increase its galactic value to 0,30 implying a transformation of 64He corresponding to an energy release of 0,4 MeV per nucleon.

4.3. Lithium-7

The BBN contribution to 7Li dominates the galactic gas abundance in the first billions of years. The flatness of the curve however has far reaching implications for the hydrodynamics of stellar surface layers. The effects of atomic (microscopic) diffusion below the convective zone should results in lithium abundances varying with the mass and surface temperature of the star in which they are observed (contrary to observations) unless these effects are neutralized by others phenomena. These as been a problem for many years. Several groups (Delyannis 99) (Pinsonneault et al 1992, Pinsonneault 1999, Vauclair 99) have studied the possible influence of various physical phenomena such as mixing via differential rotations , gravity waves and stellar winds . In all cases the flatness of the curve imposes important constraints on the efficiency of these processes. The flatness of the lithium abundance curve as a function of metallicity argues against heavy stellar astration and hence against important D depletion for metallicity less than one tenth solar. When the galactic mass fraction of heavy elements became larger than one part in a thousand (one tenth the solar value) a new stellar source (Sackman 1999) managed to increase the abundance of lithium by an extra factor of ten. The observations of lithium abundances in the surface of evolved stars of the Asymptotic Giant Branch (AGB stars) at a value appreciably larger than PopI value of 2×10^{-9} suggest that these stars have contributed, by mass ejection, to this enrichment of galactic lithium . Other candidates such as novae and supernovae have also been suggested. Convincing quantitative models of the galactic enrichment are still lacking. The GCR production of 7Li , evaluated through the abundance of 9Be, never dominates the abundance curve.
. Lithium has been observed in a large number of stars in various states of evolution. The study of the abundance variations is a very important tool for the elucidation of the physics of stars and galactic evolution.

5. GCR origin of LI Be B

We consider first the case of 9Be. It is a typical case of a nucleus for which we know of no low energy production mechanism. We have to resort to spallation reactions in interstellar space. The rate of formation of 9Be by GCR is given by the product of the flux of high energy protons (16 cm-2sec-1) times the cross-sections for 9Be formation by proton collision on the most abundant targets 16O and 12C (5mb) times the abundance ratio of these targets to hydrogen (10-3) in space. As mentioned by Parizot 1999 in his review talk, the approximate equality between the product of the formation rate times the age of the galaxy, on the one hand, and the beryllium to hydrogen ratio in recent stars (10-11) is the best evidence for a major GCR contribution to 9Be and by the same token to some of the light elements of the group Li Be B (Reeves et al 1970). To pursue this idea , the important parameters are the excitation functions for spallation reactions induced by protons and alphas. In principle all the nuclei with A¿11 in interstellar space are target candidates. In practice, only 16O , 12C and marginally 14N are abundant enough to contribute appreciably. Alpha + alpha reactions may have been important in the early days of the galaxy (Montmerle 1977)(Steigman and Walker 1992). Thanks mostly to the pioneering work at the Orsay laboratory in France, the important excitation functions are now known with sufficient accuracy ((reviewed in Reeves 1974 and Read and Viola 1985) . The number of thermalized nuclides added to the interstellar gas per unit time is the sum of two contributions. First : spallation of interstellar heavy nuclei by fast protons and alphas. In this case , the recoil energy of the spallation products is small (a few MeV per nucleon) and they suffer no further destruction or loss. Second: spallation of fast heavy nuclei by interstellar H and He . The spallation products are partially destroyed during their deceleration by electron collisions. To compare theory and observations , it is convenient to study separately two phases of the galactic life. First : the "recent" era (the last ten billion years or so, corresponding more or less to PopI stars) and second: the early days of the galaxy (PopII stars).

6. The recent galactic era

Comparing the ratios of the spallation cross-sections of protons on O and C to the ratios of the stellar abundances of Li Be B confirms the view that some of the light nuclei are generated by GCR. The analysis, shows that the GCR mechanism satisfactorily accounts for 6Li, 9Be , 10B , gives a major contribution to the abundance of 11B and a minor (10abundance of 7Li (Reeves et al 1970)(Meneguzzi et al 1971)(Reeves et al 1973) Another source of 7Li is clearly required. A smaller discrepancy exists also for the boron isotopic ratio which will be discussed separately. The early galactic era. The study of the GCR generated elements in PopII stars may yield important clues to the physical conditions accompanying the formation of galaxies. (Ryan et al 1991,1992) (Rebolo et al 1988; 1992) ; (Spite et al 1991); (Duncan 1997) ; (Gilmore et al 1991), (Smith et al 1992) .(Primas1999) (Cunha 1999) . The growth of the Be/H and B/ H as a function of the metallicity (Fe/H) of the galaxy appears to be linear . The implications of this behaviour are still under discussions.This result suggests that the ratio of fast (CNO/ HHe) was much larger in the early GCR than today (Cass 1999) (Ramaty 1999) (Parizot 1999). In the astrophysical context, this situation could occur in the "superbubble scenario". Regions of massive star formation (OB associations, starbursts), containing numerous supernovae ejecta enriched in heavy elements, are assumed to be seat of intense shock wave acceleration processes.

7. The isotopic composition of lithium

Four cases of stellar 6Li idenfifications in old Pop II stars are reported, all at the levels of 5the observational lower limit. Considering the fact that 6Li is far more fragile than 7Li, these detections are often presented as a strong argument for the case of no large depletion of 7Li in those stars. One point however remains worrysome. Given that the theoretically possible range 6Li /7Li extends almost to 100all four values are at the lower observational limit is somewhat suspicious. Systematic uncertainties may be larger than expected. Finding even one other Pop II star with an higher value of 6Li /7Li would ease the situation. If confirmed, the detection of 6Li in metal poor stars would provide the first evidence for the contribution of a+a reactions to the abundance of Li . It would also show that the GCR contributed a few percent to the Pop II lithium abundance.

Inhomogeneities in the 6Li /7Li ratios in the present interstellar medium ranging from 0.08 (the protosolar value) to almost 0.5 have been reported by Lemoine et al (1992), Federman (1999), (Knauth 1999), in regions of star formation (Ophiucus and Persei) . The effect of strong irradiation of the interstellar medium by fast particles- producing local increase of 6Li- have been suggested, supported by observations of locally increased ionization rates (by a factor of ten over the mean interstellar value) (Federman 1999). However to produce a noticeable increase of 6Li would require a time- integrated flux comparable to the galactic time integrated flux ($10^{10}y$). With (only) a factor ten increase in the ionization rate , the duration of this extra flux would have to last one GY (10^9y), far longer than the duration of a star forming process (10^7y). Local variations of 7Li could come from recent release of stellar-made 7Li by evolved stars. In this case the low protosolar 6/7 ratio (0.08) (implying a comparatively strong local contamination in 7Li) would be difficult to reconcile with the fact that the protosolar 7Li/H is in good agreement with the observed values found in young stars (implying no strong solar inhomogeneities). In summary the interpretation of these local lithium isotopic inhomogeneities poses a serious challenge to theorist!

8. The isotopic composition of boron

One long standing problem is the difference between the protosolar boron isotopic ratio (11B/10B= 4.0 (+or- 0.1) and the value 2.5 generated by the Galactic Cosmic Rays, with mean energy around one GeV . Given the shape of the spallation cross-sections leading to these isotopes at low energy it was proposed that the contribution of lower energy particles, unobservable from the inner solar system, could account for the difference (Reeves and Meyer 1978)(Ramaty et al 1999). Another solution exist : the neutrino disintegration of 12C which would generate far more 11B than 10 B. The process would take place in exploding stars (supernovae) in which intense fluxes of neutrinos emitted from collapsing cores (Woosley 1990) are later reabsorbed by the outer layers , thereby expelling them in the form of remnants. More isotopic boron ratio measurements in stars and in ISM are needed to decide if a neutrino contribution is mandatorily requires. Notice that , contrary to the case of the lithium isotiopic ratio, we do not expect differential depletion of B on the Main-Sequence (except at the very cold end).

9. Stellar nucleosynthesis

Stellar contributions are required for 7Li and perhaps also for 11B. The existence of lithium rich stars (more than the present cosmic abundanes 7Li/H ¿ 3X10-9) such as WZ Cassiopea and others, give is a strong motivation to the idea that stellar winds from such stars could have enriched the ISM with respect to the BBN value. The observations from (Castilho 1999) of giant stars with abundant lithium together with depleted Be and B confirms the occurence of lithum production mechanisms in advanced stages of stellar evolution. The formation mechanisms in stars of the Red Giant Branch, Asymptotic Giant Branch , novae ,and supernovae avec been discussed by (Sackman 1999) and (Mateucci 1999) A report of infrared observations, from 25 to 60 microns, of lithium rich giants has been presented by de la Reza (1999). This study give access to shell ejections , the very mechanism by which the lithium atoms are most likely expelled and spread in the ISM.

References

Asplund, M., 1999 IAU Symposium 198 Natal , Brazil nov 1999
Anders, E., and N. Grevesse, 1989 Geochim. Cosmochim.Acta , 53, 197
Balachandran, S.C. ,Bell,R.A., 1998 Nature 392 791 .
Beers, T.C., 1999 IAU Symposium 198 Natal , Brazil nov 1999
Bjoraker , G. et al 1986 Icarus, 66 579.
Boesgaard, A.M.,and W.D. Heacox, 1978,Astrophys. J. 226 888
Boesgaard, A.M., and Ryan S.G., 1999 IAU Symposium 198 Natal , Bra
Campbell, A., Astrophys. J. 1992 401 157
Carlson, B.et al 1993 J.Geophys.Rev. 98, 5251.
Cass , M., 1999 IAU Symposium 198 Natal , Brazil nov 1999
Castilho, B.V. 1999 IAU Symposium 198 Natal , Brazil nov 1999
Cayrel, R 1998, ISSI Workshop on Primordial Nuclei and Their Galactic Evolution, Space Science Reviews 84, N. Prantzos, M, Tosi and R.von Steiger editors (Kluwer Academic Press).
Cayrel, R. 1999 IAU Symposium 198 Natal , Brazil nov 1999
Charbonnel,C et al 1997 ISSI Workshop on Primordial Nuclei and Their Galactic Evolution. Berne Switzerland 6-10 May 1999, Space Science Review

Kluwer Academic Press May 1997 N. Prantzos, M, Tosi and R.von Steiger editors .

Charbonnel, C., 1999 IAU Symposium 198 Natal , Brazil nov 1999

Chaussudon, M. et Robert, F 1998 Earth Planet. Sci. Lett. 164, 577

Chiappini, C., 1999 IAU Symposium 198 Natal , Brazil nov 1999

Chmielevski, Y., Muller, F.H., and Brault, J.W. 1975 Astron. Astrophys. 42, 37

Cunha, K. and Smith, V, ApJ.1999 512 1006

Cunha , K., 1999 IAU Symposium 198 Natal , Brazil nov 1999

de la Reza , R., 1999 IAU Symposium 198 Natal , Brazil nov 1999

Delyannis, C., 1999 IAU Symposium 198 Natal , Brazil nov 1999

Donahue, T.M. and Nieman, H.B. 1998, ISSI Workshop on Primordial Nuclei and Their Galactic Evolution, Space Science Reviews 84, N. Prantzos, M, Tosi and R.von Steiger editors (Kluwer Academic Press).

Duncan, D 1998, ISSI Workshop on Primordial Nuclei and Their Galactic Evolution, Space Science Reviews 84, N. Prantzos, M, Tosi and R.von Steiger editors (Kluwer Academic Press).

Encrenaz, Th . et al., 1996 , A -A 315, L397

Federman 1999 IAU Symposium 198 Natal , Brazil nov 1999

Fuller, G.M., R.N. Boyd, and J.D. Kalen, 1991, ApJ. 371, L11 et al 1992

Gautier ,D., and T. Owen, 1989, in Origin and Evolution of atmospheres, S.K. Atreya et al eds.(University of Arizona Press). p. 487

Geiss, J., P. Eberhardt, F. Buhler, J. Meister, and P.Signer, 1970, J. Geophys. Res. 75, 5972. a Geiss, J., and H. Reeves,1972, Astron. Astrophys. 18, 126,

Geiss, J . and Gloecker, G 1998, ISSI Workshop on Primordial Nuclei and Their Galactic Evolution, Space Science Reviews 84, N. Prantzos, M, Tosi and R.von Steiger editors (Kluwer Academic Press).

Gilmore, G., B. Edvardsson,and P.E. Nissen, 1991, Astrophys. J. 378, 17.

Gilmore, G., B. Gustafsson, B. Edvardsson, and P. Nissen, 1992, Nature 357, 379.

Gilmore, G., Symposium in " On the Origin and Evolution of the elements", Paris, June 1992, Prantzos, N., E. Flam, and M.Cass,. (Cambridge University Press).

Gloeckler, G. 1999 IAU Symposium 198 Natal , Brazil nov 1999

Grevesse, N and Noels,A 1992 in " On the Origin and Evolution of the elements", Paris, June 1992, Prantzos, N., E. Flam, and M.Cass,. (Cambridge University Press).

Hill, V., 1999. IAU Symposium 198 Natal , Brazil nov 1999

Hobbs, L.M., and D.K.Duncan,1987, Astrophys. J. 317, 796. Hogan , C.J. 1998, ISSI Workshop on Primordial Nuclei and Their Galactic Evolution, Space Science Reviews 84, N. Prantzos, M, Tosi and R.von Steiger editors (Kluwer Academic Press).

Israelian G., 1999 IAU Symposium 198 Natal , Bra

Kajino ,T., Knauth,D 1999 IAU Symposium 198 Natal , Brazil nov 1999
Knauth,D 1999 IAU Symposium 198 Natal , Brazil nov 1999
Kunth, D., and W. Sargent, 1983, Astrophys. J. 273, 81
Kurki-Suonio, 1999 IAU Symposium 198 Natal , Bra
Lemoine, M., R. Ferlet, A. Vidal-Madjar, C. Emerich, P. Bertin, 1992, to appear in Astron. Astrophys. also in " On the Origin and Evolution of the elements", Paris, June 1992, Prantzos, N., E. Flam, and M.Cass, (Cambridge University Press).
Levshakov , S.A., et al 1998, ISSI Workshop on Primordial Nuclei and Their Galactic Evolution, Space Science Reviews 84, N. Prantzos, M, Tosi and R.von Steiger editors (Kluwer Academic Press).
Lcvshakov. S., 1999 IAU Symposium 198 Natal , Brazil nov 1999
Linsky, L.J. 1998, ISSI Workshop on Primordial Nuclei and Their Galactic Evolution, Space Science Reviews 84, N. Prantzos, M, Tosi and R.von Steiger editors (Kluwer Academic Press).
Linsky, J. 1999 IAU Symposium 198 Natal , Brazil nov 1999
Lubovich, D., 1999 IAU Symposium 198 Natal , Brazil nov 1999
Mallik,S.V., 1999 IAU Symposium 198 Natal , Bra
Matteucci, F 1999 IAU Symposium 198 Natal , Brazil nov 1999
Meneguzzi, M., J. Audouze, and H. Reeves, 1971, Astron. Astrophys., 15, 337,
Meneguzzi, M., and H . Reeves, Astron. Astrophys., 1975, 40, 91
Michaud, G., and P . Charbonneau, 1991, Space Science Reviews, 57, 1
Michaud, G.1999 IAU Symposium 198 Natal , Brazil nov 1999
Molaro., P. 1999 IAU Symposium 198 Natal , Brazil nov 1999
Montmerle , T., 197,7 Astrophys. J., 217, 872 .
Nakano, J., and K. Sato, Contributed paper of Annual Meeting of Phys. Soc. Japan ,1990, Oct.
Nissen,P.E. 1999 IAU Symposium 198 Natal , Brazil nov 1999
Normile, D., 1998 Science. 280 1690.
Olive, K., 1999 IAU Symposium 198 Natal , Brazil nov 1999
Pagel , B. E. J., and E. A. Simonson, 1989, Rev. Mex. Astr. Astrophys. ,18, 153
Pagel , B. E. J., and E. A. Simonson, R. J. Terlevich , and M. G. Edmunds, 1992, Mon.Not. R. Astr. Soc., 255, 325.
Pagel, B. E. J ,1992, in " On the Origin and Evolution of the elements", Paris, June 1992, Prantzos, N., E. Flam, and M. Cass, (Cambridge University Press).
Parizot, E 1999 IAU Symposium 198 Natal , Brazil nov 1999
Pasquini, I, 1999 IAU Symposium 198 Natal , Brazil nov 1999
Peimbert, M., 1999 IAU Symposium 198 Natal , Brazil nov 1999
Pinsonneault, M. H., C.P. Delyannis, and P. Demarque, 1992, Astrophys. J., Supp 78, 179.
Pinsonneault, M.H. 1999 IAU Symposium 198 Natal , Brazil nov 1999

Prantzos, N., 1998, ISSI Workshop on Primordial Nuclei and Their Galactic Evolution, Space Science Reviews 84, N. Prantzos, M, Tosi and R.von Steiger editors (Kluwer Academic Press).

Prantzos, N., M. Cass, and E. Viangoni-Flamm, Symposium in " On the Origin and Evolution of the elements", Paris, June 1992, Prantzos, N., E. Flam, and M.Cass,. (Cambridge University Press).

Primas,F.,Duncan,D.K.,Peterson,R.C.and Thoburn, J.A.1999 A& A 343, 545

Primas,F.1999 IAU Symposium 198 Natal , Brazil nov 1999

Ramaty, R., Koslovsky, B., Lingenfelter,R.E and Reeves, H 1997, ApJ ,488, 730.

Ramaty,R., 1999 IAU Symposium 198 Natal , Brazil nov 1999

Read, S. M., and V. E. Jr Viola, 1984, . At. Data . Nucl. Tables, 31,359.

Rebolo , R., P. Molaro, and J. E. Beckman, 1988, Astron. Astrophys. ,192, 192.

Rebolo, R., 1990,Proceedings of the Cambridge Conference on Elements in the cosmos

Rebolo, R., R. J. Garcia-Lopez, E. L. Martin, J. E. Beckman, C. D. McKeith, J.K. Webb, and Y.V. Pavlenko, 1992, in " On the Origin and Evolution of the elements", Paris, June 1992, Prantzos, N., E. Flam, and M.Cass, (Cambridge University Press).

Reeves, H., W. A. Fowler, and F. Hoyle, 1970, Nature (London), 226, 727

Reeves, H., J. Audouze, W.A. Fowler, and D.N. Schramm, 1973, Astrophys. J. , 177, 909

Reeves, H., 1974, Ann. Rev. Astr. Ap, 12, 437.

Reeves, H., and J.P. Meyer, 1978, Astrophys. J., 226, 613.

Rood, R. T., T. M. Bania, and T. L. Wilson, 1992, Nature (London), 355, feb 13

Rood, R.T. et al 1998, ISSI Workshop on Primordial Nuclei and Their Galactic Evolution, Space Science Reviews 84, N. Prantzos, M, Tosi and R.von Steiger editors (Kluwer Academic Press).

Rood, R.T. and Bania, T.M. 1999 IAU Symposium 198 Natal , Brazil nov 1999

Ryan, S. G., J. E. Norris, and M. S. Bessel, 1991, Astron. J., 102, 303.

Ryan, S. G., in " On the Origin and Evolution of the elements", Paris, June 1992, Prantzos, N., E. Flam, and M.Cass, (Cambridge University Press).

Ryan,S.G. 1999 IAU Symposium 198 Natal , Brazil nov 1999

Sackman, J., 1999 IAU Symposium 198 Natal , Brazil nov 1999

Sahu, K., 1999 IAU Symposium 198 Natal , Brazil nov 1999

Schramm, D. N., 1992, in " On the Origin and Evolution of the elements", Paris, June 1992, Prantzos, N., E. Flam, and M.Cass, (Cambridge University Press).

Smith , V. V., D. L. Lambert, and P. E. Nissen, 1992, "The 7/6 ratio in two metal poor stars", Astron.Astrophys.J. 408, 262 Astrophys. J. , sept 1992

Songaila , A. et al, 1997, Nature, 385, 137.

Spite, M., and F. Spite, 1982a, Nature (London), 297, 483.

Spite, F., and M. Spite,1982b, Astron. Astrophys., 115, 337.

Spite et al 1998, ISSI Workshop on Primordial Nuclei and Their Galactic Evolution, Space Science Reviews 84, N. Prantzos, M, Tosi and R.von Steiger editors (Kluwer Academic Press).

Spite, F 1999 IAU Symposium 198 Natal , Brazil nov 1999

Steigman, G., and T. P. Walker, 1992, Astrophys. J., Lett., 385, L13.

Steigman, G., and M.Tosi, 1992, Astrophys. J. ,401, 150.

Steigman, G. 1999 IAU Symposium 198 Natal , Brazil nov 1999

Thevenin,F 1999 IAU Symposium 198 Natal , Brazil nov 1999

Thuan, T,X and Isotov, Y, I 1998, ISSI Workshop on Primordial Nuclei and Their Galactic Evolution, Space Science Reviews 84, N. Prantzos, M, Tosi and R.von Steiger editors (Kluwer Academic Press).

Thuan, T.X., 1999 IAU Symposium 198 Natal , Brazil nov 1999

Torres, C.A.O.,1999 IAU Symposium 198 Natal , Bra

Tosi, M., 1998, ISSI Workshop on Primordial Nuclei and Their Galactic Evolution, Space Science Reviews 84, N. Prantzos, M, Tosi and R.von Steiger editors (Kluwer Academic Press).

Tosi , M., 1999 IAU Symposium 198 Natal , Brazil nov 1999

Tytler, D. 1998, ISSI Workshop on Primordial Nuclei and Their Galactic Evolution, Space Science Reviews 84, N. Prantzos, M, Tosi and R.von Steiger editors (Kluwer Academic Press).

Tytler, D 1999 IAU Symposium 198 Natal , Brazil nov 1999

Vangioni-Flam, E., Olive, K. and Prantzos, N. 1994, ApJ, 148, 3. Vauclair, S., and Charbonnel, C., 1998 ApJ 502, 372,

Vauclair, S., 1999 IAU Symposium 198 Natal , Brazil nov 1999

Viegas, S;, 1999 IAU Symposium 198 Natal , Brazil nov 1999

Vidal-Madjar, A Ferlet, R and Lemoine , M 1998, ISSI Workshop on Primordial Nuclei and Their Galactic Evolution, Space Science Reviews 84, N. Prantzos, M, Tosi and R.von Steiger editors (Kluwer Academic Press).

Vidal-Madjar., A., 1999 IAU Symposium 198 Natal , Brazil nov 1999

Walker, T.P., G.J. Mathews, and V.E. Viola, 1985, Astrophys. J. , 299, 745

Woosley, S.E., D.H. Hartman, R.D. Hoffman, and W.C. Haxton, 1990, Astrophys. J., 356, 272.

INTERNATIONAL ASTRONOMICAL UNION (IAU) SYMPOSIA VOLUMES

Published by

Astronomical Society of the Pacific

INTERNATIONAL ASTRONOMICAL UNION (IAU) SYMPOSIA
Published by the Astronomical Society of the Pacific

PUBLISHED: 1999

Vol. No. 190   NEW VIEWS OF THE MAGELLANIC CLOUDS
eds. You-Hua Chu, Nicholas B. Suntzeff, James E. Hesser, and David A. Bohlender
ISBN: 1-58381-021-8

Vol. No. 191   ASYMPTOTIC GIANT BRANCH STARS
eds. T. Le Bertre, A. Lèbre, and C. Waelkens
ISBN: 1-886733-90-2

Vol. No. 192   THE STELLAR CONTENT OF LOCAL GROUP GALAXIES
eds. Patricia Whitelock and Russell Cannon
ISBN: 1-886733-82-1

Vol. No. 193   WOLF-RAYET PHENOMENA IN MASSIVE STARS AND STARBURST GALAXIES
eds. Karel A. van der Hucht, Gloria Koenigsberger, and Philippe R. J. Eenens
ISBN: 1-58381-004-8

Vol. No. 194   ACTIVITY IN GALAXIES AND RELATED PHENOMENA
eds. Yervant Terzian, Daniel Weedman, and Edward Khachikian
ISBN: 1-58381-008-0

PUBLISHED: 2000 (*) Asterisk indicates editors are still working on the volume as of September 2000

Vol. No. 195   HIGHLY ENERGETIC PHYSICAL PROCESSES AND MECHANISMS FOR EMISSION FROM ASTROPHYSICAL PLASMAS
eds. P. C. H. Martens, S. Tsuruta, and M. A. Weber
ISBN: 1-58381-038-2

Vol. No. 196*   PRESERVING THE ASTRONOMICAL SKY
eds. R. J. Cohen and W. T. Sullivan

Vol. No. 197   ASTROCHEMISTRY: FROM MOLECULAR CLOUDS TO PLANETARY SYSTEMS
eds. Y. C. Minh and E. F. van Dishoeck
ISBN: 1-58381-034-X

Vol. No. 198   THE LIGHT ELEMENTS AND THEIR EVOLUTION
eds. L. da Silva, M. Spite, and R. de Medeiros
ISBN: 1-58381-048-X

Vol. No. 199*   THE UNIVERSE AT LOW RADIO FREQUENCIES
eds. A. Pramesh Rao

****

Vol. XXIVA   TRANSACTIONS OF THE INTERNATIONAL ASTRONOMICAL UNION
REPORTS ON ASTRONOMY 1996-1999
ed. Johannes Andersen
ISBN: 1-58381-035-8

INTERNATIONAL ASTRONOMICAL UNION (IAU) SYMPOSIA
Published by the Astronomical Society of the Pacific

Complete lists of proceedings of past IAU Meetings are maintained at the
IAU Web site at the URL: http://www.iau.org/publicat.html

Volumes 32 - 189 in the IAU Symposia Series may be ordered from
Kluwer Academic Publishers
P. O. Box 117
NL 3300 AA Dordrecht
The Netherlands

All other book orders or inquiries concerning volumes listed should be directed to the:

Astronomical Society of the Pacific Conference Series
390 Ashton Avenue
San Francisco CA 94112-1722 USA

Phone: 415-337-2126
Fax: 415-337-5205
E-mail: catalog@aspsky.org
Web Site: http://www.aspsky.org